EDITION
**11**

# College Physics

## Volume 1

### Raymond A. Serway
*Emeritus, James Madison University*

### Chris Vuille
*Embry-Riddle Aeronautical University*

WITH CONTRIBUTIONS FROM
### John Hughes
*Embry-Riddle Aeronautical University*

 Cengage

Australia • Brazil • Canada • Mexico • Singapore • United Kingdom • United States

*College Physics,* **Eleventh Edition**
**Volume 1**
**Raymond A. Serway and Chris Vuille**

Product Director: Dawn Giovanniello

Product Manager: Rebecca Berardy Schwartz

Content Developers: Ed Dodd,
Michael Jacobs, Ph.D.

Product Assistant: Caitlin N. Ghegan

Marketing Manager: Tom Ziolkowski

Senior Content Project Manager: Tanya Nigh

Digital Content Specialist: Justin Karr

Senior Art Director: Cate Barr

Manufacturing Planner: Doug Bertke

Production Service and Compositor:
Cenveo® Publisher Services

Intellectual Property Project Manager:
Nick Barrows

Intellectual Property Analyst:
Christine Myaskovsky

Photo and Text Researcher:
Lumina Datamatics, Ltd.

Text Designer: Dare Porter

Cover Designer: Liz Harasymczuk

Cover Image: Jakataka/DigitalVision Vectors/
Getty Images

For product information and technology assistance, contact us at
**Cengage Customer & Sales Support, 1-800-354-9706**
**or support.cengage.com.**

For permission to use material from this text or product, submit all
requests online at **www.copyright.com.**

Library of Congress Control Number: 2016952167

ISBN: 978-1-305-96551-5

**Cengage**
200 Pier 4 Boulevard
Boston, MA 02210
USA

Cengage is a leading provider of customized learning solutions
with employees residing in nearly 40 different countries and sales in more
than 125 countries around the world. Find your local representative at:
**www.cengage.com.**

To learn more about Cengage platforms and services, register or access
your online learning solution, or purchase materials for your course,
visit **www.cengage.com.**

Printed in the United States of America
Print Number: 06      Print Year: 2022

We dedicate this book to our wives, children, grandchildren, relatives, and friends who have provided so much love, support, and understanding through the years, and to the students for whom this book was written.

# Contents Overview

# Contents

vi

# About the Authors

**Raymond A. Serway** received his doctorate at Illinois Institute of Technology and is Professor Emeritus at James Madison University. In 2011, he was awarded an honorary doctorate degree from his alma mater, Utica College. He received the 1990 Madison Scholar Award at James Madison University, where he taught for 17 years. Dr. Serway began his teaching career at Clarkson University, where he conducted research and taught from 1967 to 1980. He was the recipient of the Distinguished Teaching Award at Clarkson University in 1977 and the Alumni Achievement Award from Utica College in 1985. As Guest Scientist at the IBM Research Laboratory in Zurich, Switzerland, he worked with K. Alex Müller, 1987 Nobel Prize recipient. Dr. Serway was also a visiting scientist at Argonne National Laboratory, where he collaborated with his mentor and friend, the late Sam Marshall. Early in his career, he was employed as a research scientist at the Rome Air Development Center from 1961 to 1963 and at the IIT Research Institute from 1963 to 1967. Dr. Serway is also the coauthor of *Physics for Scientists and Engineers*, ninth edition; *Principles of Physics: A Calculus-Based Text*, fifth edition; *Essentials of College Physics*, *Modern Physics*, third edition; and the high school textbook *Physics*, published by Holt, Rinehart and Winston. In addition, Dr. Serway has published more than 40 research papers in the field of condensed matter physics and has given more than 60 presentations at professional meetings. Dr. Serway and his wife Elizabeth enjoy traveling, playing golf, fishing, gardening, singing in the church choir, and especially spending quality time with their four children, nine grandchildren, and a great grandson.

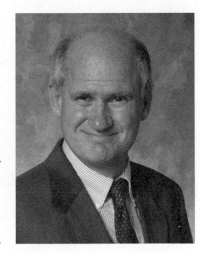

**Chris Vuille** is an associate professor of physics at Embry-Riddle Aeronautical University (ERAU), Daytona Beach, Florida, the world's premier institution for aviation higher education. He received his doctorate in physics at the University of Florida in 1989. While he has taught courses at all levels, including postgraduate, his primary interest and responsibility has been the teaching of introductory physics courses. He has received a number of awards for teaching excellence, including the Senior Class Appreciation Award (three times). He conducts research in general relativity, astrophysics, cosmology, and quantum theory, and was a participant in the JOVE program, a special three-year NASA grant program during which he studied neutron stars. His work has appeared in a number of scientific journals and in *Analog Science Fiction/Science Fact* magazine. In addition to this textbook, he is the coauthor of *Essentials of College Physics*. Dr. Vuille enjoys playing tennis, swimming, yoga, playing classical piano, and writing science fiction; he is a former chess champion of St. Petersburg and Atlanta and the inventor of x-chess. His wife, Dianne Kowing, is Chief of Optometry at a local VA clinic. He has a daughter, Kira, and two sons, Christopher and James, all of whom love science.

# Preface

*College Physics* is written for a one-year course in introductory physics usually taken by students majoring in biology, the health professions, or other disciplines, including environmental, earth, and social sciences, and technical fields such as architecture. The mathematical techniques used in this book include algebra, geometry, and trigonometry, but not calculus. Drawing on positive feedback from users of the tenth edition, analytics gathered from both professors and students, as well as reviewers' suggestions, we have refined the text to better meet the needs of students and teachers.

This textbook, which covers the standard topics in classical physics and twentieth-century physics, is divided into six parts. Part 1 (Topics 1–9) deals with Newtonian mechanics and the physics of fluids; Part 2 (Topics 10–12) is concerned with heat and thermodynamics; Part 3 (Topics 13 and 14) covers wave motion and sound; Part 4 (Topics 15–21) develops the concepts of electricity and magnetism; Part 5 (Topics 22–25) treats the properties of light and the field of geometric and wave optics; and Part 6 (Topics 26–30) provides an introduction to special relativity, quantum physics, atomic physics, and nuclear physics.

## Objectives

The main objectives of this introductory textbook are twofold: to provide the student with a clear and logical presentation of the basic concepts and principles of physics and to strengthen their understanding of them through a broad range of interesting, real-world applications. To meet those objectives, we have emphasized sound physical arguments and problem-solving methodology. At the same time we have attempted to motivate the student through practical examples that demonstrate the role of physics in other disciplines.

## Changes to the Eleventh Edition

The text has been carefully edited to improve clarity of presentation and precision of language. We hope that the result is a book both accurate and enjoyable to read. Although the overall content and organization of the textbook are similar to the tenth edition, numerous changes and improvements have been made in preparing the eleventh edition. Some of the new features are based on our experiences and on current trends in science education. Other changes have been incorporated in response to comments and suggestions offered by users of the tenth edition. The features listed here represent the major changes made for the eleventh edition.

### WebAssign for *College Physics*, Eleventh Edition

WebAssign offers an extensive online program for physics to encourage the practice that's so critical for concept mastery. The meticulously crafted pedagogy and exercises in our proven texts become even more effective in WebAssign. WebAssign includes a highly customizable, interactive eBook. For *College Physics*, WebAssign includes:

- **Read It** links under each question quickly jump to the corresponding section of a complete **eBook**.

- **Watch It** links provide step-by-step instruction with short, engaging videos that are ideal for visual learners.

- **Master It Tutorials** show students how to solve a similar problem in multiple steps by providing direction along with derivation so students understand the concepts and reasoning behind the problem solving.

- **Active Examples** guide students through the process needed to master a concept.

- **Training Tutorials** offer students another training tool to assist them in understanding how to apply certain key concepts presented in a given chapter.

- **Pre-Lecture Explorations** combine **Active Figures** with conceptual and analytical questions that guide students to a deeper understanding and help promote a robust physical intuition.

- **Interactive Video Vignette** questions encourage students to address their alternate conceptions outside of the classroom. They include online video analysis and interactive individual tutorials to address learning difficulties identified by PER (Physics Education Research).

- **Cengage eBook** WebAssign has an eBook that lets you tailor the textbook to fit your course and connect with your students. You can remove and rearrange chapters in the table of contents and tailor assigned readings to match your syllabus exactly. Powerful editing tools let you change as much as you'd like—or leave it just like it is. You can highlight key passages or add sticky notes to pages to comment on a concept in the reading, and then share any of these individual notes and highlights with your students, or keep them personal. You can also edit narrative content in the textbook by adding a text box or striking out text. With a handy link tool, you can drop in an icon at any point in the eBook that lets you link to your own lecture notes, audio summaries, video lectures, or other files on a personal Web site or anywhere on the Web. A simple YouTube widget lets you easily find and embed videos from YouTube directly into eBook pages. The eBook helps students go beyond just reading the textbook. Students can also highlight the text and add their own notes or bookmarks.

  Use the **Textbook Edition Upgrade Tool** to automatically update all of your assignments from the previous edition to corresponding questions in this textbook.

  Please visit **webassign.net** to view an interactive demonstration of WebAssign.

## Organization by Topics

Our preparatory research for this edition showed that successful students don't just *read* physics, they *engage with* physics. While revising *College Physics*, we realized that students were using the textbook as a resource while working on their online homework, rather than as a narrative source. As we continued creating a variety of media, and other material to support our activity-based pedagogy, it became clear we were designing assessments around specific *topics*, guided by the fundamental learning objectives of those topics. Consequently, we switched from "chapters" to "topics" to emphasize the textbook's new place as part of an active, fully-integrated online experience.

## Vector Rearrangement

The topic of vectors has been moved to Topic 1 with other preliminary material. This rearrangement allows students to get comfortable with vectors and how they are used in physics well before they're needed for solving problems.

## Revision of Topic 4 (Newton's Law of Motion)

A revision to the discussion of Newton's laws of motion will ease students' entry into this difficult topic and increase their success. Here, the common contact forces are introduced early, including the normal force, the kinetic friction force, tension forces, and the static friction force. After finishing these new sections, students will already know how to calculate these forces in the most common contexts. Then, when encountering applications, they will suddenly find that many difficult, two-dimensional problems will reduce to one dimension, because the second dimension simply gives the normal and friction forces that they already understand.

### The System Approach Extended to Rotating Systems

The most difficult problems in first-year physics are those involving both the second law of motion and the second law of motion for rotation. Following an insight by one of the authors (Vuille) while teaching an introductory class, it turns out that these problems, involving up to four equations and four unknowns, can often be easily solved with one equation and one unknown! Vuille has put this technique in Topic 8 (Rotational Equilibrium and Dynamics). Not found in any other first-year textbook, this technique greatly reduces the learning curve in that topic by turning the hardest problem type into one of the easiest.

### New Conceptual Questions

One hundred and twenty-five of the conceptual questions in the text (25% of the total amount) are new to this edition; they have been developed to be more systematic and clicker-friendly.

### New End-of-Topic Problems

Hundreds of new problems have been developed for this edition, taking into account statistics on problem usage by past users.

## Textbook Features

Most instructors would agree that the textbook assigned in a course should be the student's primary guide for understanding and learning the subject matter. Further, the textbook should be easily accessible and written in a style that facilitates instruction and learning. With that in mind, we have included the following pedagogical features to enhance the textbook's usefulness to both students and instructors.

**Examples**  Each example constitutes a complete learning experience, with a strategy statement, a side-by-side solution and commentary, conceptual training, and an exercise. Every effort has been made to ensure the collection of examples, as a whole, is comprehensive in covering all the physical concepts, physics problem types, and required mathematical techniques. The examples are in a two-column format for a pedagogic purpose: students can study the example, then cover up the right column and attempt to solve the problem using the cues in the left column. Once successful in that exercise, the student can cover up both solution columns and attempt to solve the problem using only the strategy statement, and finally just the problem statement. The Question at the end of the example usually requires a conceptual response or determination, but they also include estimates requiring knowledge of the relationships between concepts. The answers for the Questions can be found at the back of the book. On the next page is an in-text worked example, with an explanation of each of the example's main parts.

**Artwork**  Every piece of artwork in the eleventh edition is in a modern style that helps express the physics principles at work in a clear and precise fashion. Every piece of art is also drawn to make certain that the physical situations presented correspond exactly to the text discussion at hand.

*Guidance labels* are included with many figures in the text; these point out important features of the figure and guide students through figures without having to go back and forth from the figure legend to the figure itself. This format also helps those students who are visual learners. An example of this kind of figure appears at the bottom of this page.

**Conceptual Questions**  At the end of each topic are approximately fifteen conceptual questions. The Applying Physics examples presented in the text serve as models for students when conceptual questions are assigned and show how the concepts can be applied to understanding the physical world. The conceptual questions provide the student with a means of self-testing the concepts presented in the topic. Some conceptual questions are appropriate for initiating classroom discussions. Answers to

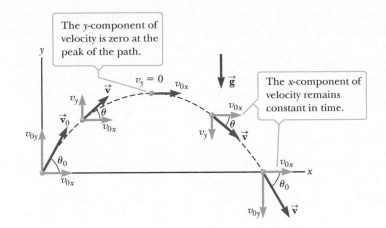

The *y*-component of velocity is zero at the peak of the path.

$v_y = 0$

The *x*-component of velocity remains constant in time.

**Figure 3.5**
The parabolic trajectory of a particle that leaves the origin with a velocity of $\vec{v}_0$. Note that $\vec{v}$ changes with time. However, the *x*-component of the velocity, $v_x$, remains constant in time, equal to its initial velocity, v0x. Also, $v_y = 0$ at the peak of the trajectory, but the acceleration is always equal to the free-fall acceleration and acts vertically downward.

The **Goal** describes the physical concepts being explored within the worked example.

The **Problem** statement presents the problem itself.

The **Strategy** section helps students analyze the problem and create a framework for working out the solution.

The **Solution** section uses a two-column format that gives the explanation for each step of the solution in the left-hand column, while giving each accompanying mathematical step in the right-hand column. This layout facilitates matching the idea with its execution and helps students learn how to organize their work. Another benefit: students can easily use this format as a training tool, covering up the solution on the right and solving the problem using the comments on the left as a guide.

**EXAMPLE 13.7   MEASURING THE VALUE OF *g***

**GOAL**  Determine *g* from pendulum motion.

**PROBLEM**  Using a small pendulum of length 0.171 m, a geophysicist counts 72.0 complete swings in a time of 60.0 s. What is the value of *g* in this location?

**STRATEGY**  First calculate the period of the pendulum by dividing the total time by the number of complete swings. Solve Equation 13.15 for *g* and substitute values.

**SOLUTION**

Calculate the period by dividing the total elapsed time by the number of complete oscillations:

$$T = \frac{\text{time}}{\text{\# of oscillations}} = \frac{60.0 \text{ s}}{72.0} = 0.833 \text{ s}$$

Solve Equation 13.15 for *g* and substitute values:

$$T = 2\pi \sqrt{\frac{L}{g}} \rightarrow T^2 = 4\pi^2 \frac{L}{g}$$

$$g = \frac{4\pi^2 L}{T^2} = \frac{(39.5)(0.171 \text{ m})}{(0.833 \text{ s})^2} = \boxed{9.73 \text{ m/s}^2}$$

**Remarks** follow each Solution and highlight some of the underlying concepts and methodology used in arriving at a correct solution. In addition, the remarks are often used to put the problem into a larger, real-world context.

**REMARKS**  Measuring such a vibration is a good way of determining the local value of the acceleration of gravity.

**QUESTION 13.7**  True or False: A simple pendulum of length 0.50 m has a larger frequency of vibration than a simple pendulum of length 1.0 m.

**EXERCISE 13.7**  What would be the period of the 0.171-m pendulum on the Moon, where the acceleration of gravity is 1.62 m/s²?

**ANSWER**  2.04 s

**Question** Each worked example features a conceptual question that promotes student understanding of the underlying concepts contained in the example.

**Exercise/Answer** Every Question is followed immediately by an exercise with an answer. These exercises allow students to reinforce their understanding by working a similar or related problem, with the answers giving them instant feedback. At the option of the instructor, the exercises can also be assigned as homework. Students who work through these exercises on a regular basis will find the end-of-topic problems less intimidating.

odd-numbered conceptual questions are included in the Answers section at the end of the book. Answers to even-numbered questions are in the *Instructor's Solutions Manual*.

**Problems**  All questions and problems for this revision were carefully reviewed to improve their variety, interest, and pedagogical value while maintaining their clarity and quality. An extensive set of problems is included at the end of each topic (in all, more than 2 100 problems are provided in the eleventh edition). Answers to odd-numbered problems are given at the end of the book. For the convenience of both the student and instructor, about two-thirds of the problems are keyed to specific sections of the topic. The remaining problems, labeled "Additional Problems," are not keyed to specific sections. The three levels of problems are graded according to their difficulty. Straightforward problems are numbered in **black**, intermediate level problems are numbered in blue, and the most challenging problems are numbered in **red**.

There are six other types of problems we think instructors and students will find interesting as they work through the text; these are indicated in the problems set by the following icons:

- **T Master It Tutorials** available in WebAssign help students solve problems by having them work through a stepped-out solution.
- **V Watch It** video solutions available in WebAssign explain fundamental problem-solving strategies to help students step through selected problems.
- **BIO Biomedical problems** deal with applications to the life sciences and medicine.
- **S Symbolic problems** require the student to obtain an answer in terms of symbols. In general, some guidance is built into the problem statement. The goal is to better train the student to deal with mathematics at a level appropriate to this course. Most students at this level are uncomfortable with symbolic equations, which is unfortunate because symbolic equations are the most efficient vehicle for presenting relationships between physics concepts. Once students understand the physical concepts, their ability to solve problems is greatly enhanced. As soon as the numbers are substituted into an equation, however, all the concepts and their relationships to one another are lost, melded together in the student's calculator. Symbolic problems train the student to postpone substitution of values, facilitating their ability to think conceptually using the equations. An example of a symbolic problem is provided here:

14. **S** An object of mass $m$ is dropped from the roof of a building of height $h$. While the object is falling, a wind blowing parallel to the face of the building exerts a constant horizontal force $F$ on the object. (a) How long does it take the object to strike the ground? Express the time $t$ in terms of $g$ and $h$. (b) Find an expression in terms of $m$ and $F$ for the acceleration $a_x$ of the object in the horizontal direction (taken as the positive $x$-direction). (c) How far is the object displaced horizontally before hitting the ground? Answer in terms of $m$, $g$, $F$, and $h$. (d) Find the magnitude of the object's acceleration while it is falling, using the variables $F$, $m$, and $g$.

- **QC Quantitative/conceptual problems** encourage the student to think conceptually about a given physics problem rather than rely solely on computational skills. Research in physics education suggests that standard physics problems requiring calculations may not be entirely adequate in training students to think conceptually. Students learn to substitute numbers for symbols in the equations without fully understanding what they are doing or what the symbols mean. Quantitative/conceptual problems combat this tendency by asking for answers that require something other than a number or a calculation. An example of a quantitative/conceptual problem is provided here:

5. **QC** Starting from rest, a 5.00-kg block slides 2.50 m down a rough 30.0° incline. The coefficient of kinetic friction between the block and the incline is $\mu_k = 0.436$. Determine (a) the work done by the force of gravity, (b) the work done by the friction force between block and incline, and (c) the work done by the normal force. (d) Qualitatively, how would the answers change if a shorter ramp at a steeper angle were used to span the same vertical height?

- ■ GP **Guided problems** help students break problems into steps. A physics problem typically asks for one physical quantity in a given context. Often, however, several concepts must be used and a number of calculations are required to get that final answer. Many students are not accustomed to this level of complexity and often don't know where to start. A guided problem breaks a problem into smaller steps, enabling students to grasp all the concepts and strategies required to arrive at a correct solution. Unlike standard physics problems, guidance is often built into the problem statement. For example, the problem might say "Find the speed using conservation of energy" rather than asking only for the speed. In any given topic, there are usually two or three problem types that are particularly suited to this problem form. The problem must have a certain level of complexity, with a similar problem-solving strategy involved each time it appears. Guided problems are reminiscent of how a student might interact with a professor in an office visit. These problems help train students to break down complex problems into a series of simpler problems, an essential problem-solving skill. An example of a guided problem is provided here:

62. GP Two blocks of masses $m_1$ and $m_2$ ($m_1 > m_2$) are placed on a frictionless table in contact with each other. A horizontal force of magnitude $F$ is applied to the block of mass $m_1$ in

**Figure P4.62**

Figure P4.62. (a) If $P$ is the magnitude of the contact force between the blocks, draw the free-body diagrams for each block. (b) What is the net force on the system consisting of both blocks? (c) What is the net force acting on $m_1$? (d) What is the net force acting on $m_2$? (e) Write the $x$-component of Newton's second law for each block. (f) Solve the resulting system of two equations and two unknowns, expressing the acceleration $a$ and contact force $P$ in terms of the masses and force. (g) How would the answers change if the force had been applied to $m_2$ instead? (*Hint:* use symmetry; don't calculate!) Is the contact force larger, smaller, or the same in this case? Why?

**Quick Quizzes** All the Quick Quizzes (see example below) are cast in an objective format, including multiple-choice, true–false, matching, and ranking questions. Quick Quizzes provide students with opportunities to test their understanding of the physical concepts presented. The questions require students to make decisions on the basis of sound reasoning, and some have been written to help students overcome common misconceptions. Answers to all Quick Quiz questions are found at the end of the textbook, and answers with detailed explanations are provided in the *Instructor's Solutions Manual*. Many instructors choose to use Quick Quiz questions in a "peer instruction" teaching style.

Quick Quiz

**4.4** A small sports car collides head-on with a massive truck. The greater impact force (in magnitude) acts on (a) the car, (b) the truck, (c) neither, the force is the same on both. Which vehicle undergoes the greater magnitude acceleration? (d) the car, (e) the truck, (f) the accelerations are the same.

**Problem-Solving Strategies** A general problem-solving strategy to be followed by the student is outlined at the end of Topic 1. This strategy provides students with a structured process for solving problems. In most topics, more specific strategies and suggestions (see example below) are included for solving the types of problems featured in both the worked examples and the end-of-topic problems.

> **PROBLEM-SOLVING STRATEGY**
>
> **Newton's Second Law**
> *Problems involving Newton's second law can be very complex. The following protocol breaks the solution process down into smaller, intermediate goals:*
>
> 1. **Read** the problem carefully at least once.
> 2. **Draw** a picture of the system, identify the object of primary interest, and indicate forces with arrows.
> 3. **Label** each force in the picture in a way that will bring to mind what physical quantity the label stands for (e.g., *T* for tension).
> 4. **Draw** a free-body diagram of the object of interest, based on the labeled picture. If additional objects are involved, draw separate free-body diagrams for them. Choose convenient coordinates for each object.
> 5. **Apply Newton's second law.** The *x*- and *y*-components of Newton's second law should be taken from the vector equation and written individually. This usually results in two equations and two unknowns.
> 6. **Solve** for the desired unknown quantity, and substitute the numbers.

This feature helps students identify the essential steps in solving problems and increases their skills as problem solvers.

**Biomedical Applications** For biology and pre-med students, BIO icons point the way to various practical and interesting applications of physical principles to biology and medicine. A list of these applications can be found on page xxi.

**MCAT Test Preparation Guide** Located on pages xxiii and xxiv, this guide outlines the six content categories related to physics on the new MCAT exam that began being administered in 2015. Students can use the guide to prepare for the MCAT exam, class tests, or homework assignments.

**Applying Physics** The Applying Physics features provide students with an additional means of reviewing concepts presented in that section. Some Applying Physics examples demonstrate the connection between the concepts presented in that topic and other scientific disciplines. These examples also serve as models for students when they are assigned the task of responding to the Conceptual Questions presented at the end of each topic. For examples of Applying Physics boxes, see Applying Physics 9.5 (Home Plumbing) on page 292 and Applying Physics 13.1 (Bungee Jumping) on page 433.

**Tip 4.3** Newton's Second Law Is a *Vector* Equation

In applying Newton's second law, add all of the forces on the object as vectors and then find the resultant vector acceleration by dividing by *m*. Don't find the individual magnitudes of the forces and add them like scalars.

Newton's third law ▶

BIO **APPLICATION**
Diet Versus Exercise in Weight-loss Programs

**Tips** Placed in the margins of the text, Tips address common student misconceptions and situations in which students often follow unproductive paths (see example at right). More than 95 Tips are provided in this edition to help students avoid common mistakes and misunderstandings.

**Marginal Notes** Comments and notes appearing in the margin (see example at the right) can be used to locate important statements, equations, and concepts in the text.

**Applications** Although physics is relevant to so much in our modern lives, it may not be obvious to students in an introductory course. Application margin notes (see example to the right) make the relevance of physics to everyday life more obvious by pointing out specific applications in the text. Some of these applications pertain to the life sciences and are marked with a BIO icon. A list of the Applications appears on page xxi.

**Style** To facilitate rapid comprehension, we have attempted to write the book in a style that is clear, logical, relaxed, and engaging. The somewhat informal and relaxed writing style is designed to connect better with students and enhance their reading enjoyment. New terms are carefully defined, and we have tried to avoid the use of jargon.

**Introductions** All topics begin with a brief preview that includes a discussion of the topic's objectives and content.

**Units** The international system of units (SI) is used throughout the text. The U.S. customary system of units is used only to a limited extent in the topics on mechanics and thermodynamics.

**Pedagogical Use of Color** Readers should consult the pedagogical color chart (inside the front cover) for a listing of the color-coded symbols used in the text diagrams. This system is followed consistently throughout the text.

**Important Statements and Equations** Most important statements and definitions are set in **boldface** type or are highlighted with a background screen for added emphasis and ease of review. Similarly, important equations are highlighted with a tan background to facilitate location.

**Illustrations and Tables** The readability and effectiveness of the text material, worked examples, and end-of-topic conceptual questions and problems are enhanced by the large number of figures, diagrams, photographs, and tables. Full color adds clarity to the artwork and makes illustrations as realistic as possible. Three-dimensional effects are rendered with the use of shaded and lightened areas where appropriate. Vectors are color coded, and curves in graphs are drawn in color. Color photographs have been carefully selected, and their accompanying captions have been written to serve as an added instructional tool. A complete description of the pedagogical use of color appears on the inside front cover.

**Summary** The end-of-topic Summary is organized by individual section heading for ease of reference. Most topic summaries also feature key figures from the topic.

**Significant Figures** Significant figures in both worked examples and end-of-topic problems have been handled with care. Most numerical examples and problems are worked out to either two or three significant figures, depending on the accuracy of the data provided. Intermediate results presented in the examples are rounded to the proper number of significant figures, and only those digits are carried forward.

**Appendices and Endpapers** Several appendices are provided at the end of the textbook. Most of the appendix material (Appendix A) represents a review of mathematical concepts and techniques used in the text, including scientific notation, algebra, geometry, and trigonometry. Reference to these appendices is made as needed throughout the text. Most of the mathematical review sections include worked examples and exercises with answers. In addition to the mathematical review, some appendices contain useful tables that supplement textual information. For easy reference, the front endpapers contain a chart explaining the use of color throughout the book and a list of frequently used conversion factors.

## Teaching Options

This book contains more than enough material for a one-year course in introductory physics, which serves two purposes. First, it gives the instructor more flexibility in choosing topics for a specific course. Second, the book becomes more useful as a resource for students. On average, it should be possible to cover about one topic each week for a class that meets three hours per week. Those sections, examples, and end-of-topic problems dealing with applications of physics to life sciences are identified with the BIO icon. We offer the following suggestions for shorter courses for those instructors who choose to move at a slower pace through the year.

*Option A:* If you choose to place more emphasis on contemporary topics in physics, you could omit all or parts of Topic 8 (Rotational Equilibrium and Rotational Dynamics), Topic 21 (Alternating-Current Circuits and Electromagnetic Waves), and Topic 25 (Optical Instruments).

*Option B:* If you choose to place more emphasis on classical physics, you could omit all or parts of Part 6 of the textbook, which deals with special relativity and other topics in twentieth-century physics.

### CengageBrain.com

To register or access your online learning solution or purchase materials for your course, visit **www.cengagebrain.com**.

### Lecture Presentation Resources

**Cengage Learning Testing Powered by Cognero** is a flexible, online system that allows you to author, edit, and manage test bank content from multiple Cengage Learning solutions, create multiple test versions in an instant, and deliver tests from your LMS, your classroom, or wherever you want.

### Instructor Resource Website for Serway/Vuille
*College Physics, Eleventh Edition*

The Instructor Resource Website contains a variety of resources to aid you in preparing and presenting text material in a manner that meets your personal preferences and course needs. The posted *Instructor's Solutions Manual* presents complete worked solutions for all end-of-chapter problems and even-numbered conceptual questions, answers for all even-numbered problems, and full answers with explanations for the Quick Quizzes. Robust PowerPoint lecture outlines that have been designed for an active classroom are available, with reading check questions and Think-Pair-Share questions as well as the traditional section-by-section outline. Images from the textbook can be used to customize your own presentations. Available online via **www.cengage.com/login**.

### Student Resources

To register or access your online learning solution or purchase materials for your course, visit **www.cengagebrain.com**.

*Physics Laboratory Manual,* **Fourth Edition** by David Loyd (Angelo State University). Ideal for use with any introductory physics text, Loyd's *Physics Laboratory Manual* is suitable for either calculus- or algebra/trigonometry-based physics courses. Designed to help students demonstrate a physical principle and teach techniques of careful measurement, Loyd's *Physics Laboratory Manual* also emphasizes conceptual understanding and includes a thorough discussion of physical theory to help students see the connection between the lab and the lecture. Many labs give students hands-on experience with statistical analysis, and now five computer-assisted data entry labs are included in the printed manual. The fourth edition maintains the minimum equipment requirements to allow for maximum flexibility and to make the most of preexisting lab equipment. For instructors interested in using some of Loyd's experiments, a customized lab manual is another option available through the Cengage Learning Custom Solutions program. Now, you can select specific experiments from Loyd's *Physics Laboratory Manual*, include your own original lab experiments, and create one affordable bound book. Contact your Cengage Learning representative for more information on our Custom Solutions program. Available with InfoTrac® Student Collections http://gocengage.com/infotrac.

*Physics Laboratory Experiments,* **Eighth Edition** by Jerry D. Wilson (Lander College) and Cecilia A. Hernández (American River College). This market-leading manual for the first-year physics laboratory course offers a wide range of class-tested experiments designed specifically for use in small to midsize lab programs. A series of integrated experiments emphasizes the use of computerized instrumentation and includes a set of "computer-assisted experiments" to allow students and instructors to gain experience with modern equipment. It also lets instructors determine the appropriate balance of traditional versus computer-based experiments for their courses. By analyzing data through two different methods, students gain a greater understanding of the concepts behind the experiments. The Eighth Edition is updated with four new economical labs to accommodate shrinking department budgets and thirty new Pre-Lab Demonstrations, designed to capture students' interest prior to the lab and requiring only widely available materials and items.

## Acknowledgments

In preparing the eleventh edition of this textbook, we have been guided by the expertise of many people who have reviewed one or more parts of the manuscript or provided suggestions. Prior to our work on this revision, we conducted a survey of professors who teach the course; their collective feedback helped shape this revision, and we thank them:

Brian Bucklein, *Missouri Western State University*
Brian L. Cannon, *Loyola University Chicago*
Kapila Clara Castoldi, *Oakland University*
Daniel Costantino, *The Pennsylvania State University*
John D. Cunningham, S.J., *Loyola University Chicago*
Jing Gao, *Kean University*
Awad Gerges, *The University of North Carolina at Charlotte*
Lipika Ghosh, *Virginia State University*
Bernard Hall, *Kean University*
Marc L. Herbert, *Hofstra University*
Dehui Hu, *Rochester Institute of Technology*
Shyang Huang, *Missouri State University*
Salomon Itza, *University of the Ozarks*
Cecil Joseph, *University of Massachusetts Lowell*
Bjorg Larson, *Drew University*
Gen Long, *St. John's University*
Xihong Peng, *Arizona State University*
Chandan Samantaray, *Virginia State University*
Steven Summers, *Arkansas State University—Newport*

We also wish to acknowledge the following reviewers of recent editions, and express our sincere appreciation for their helpful suggestions, criticism, and encouragement.

Gary B. Adams, *Arizona State University*; Ricardo Alarcon, *Arizona State University*; Natalie Batalha, *San Jose State University*; Gary Blanpied, *University of South Carolina*; Thomas K. Bolland, *The Ohio State University*; Kevin R. Carter, *School of Science and Engineering Magnet*; Kapila Calara Castoldi, *Oakland University*; David Cinabro, *Wayne State University*; Andrew Cornelius, *University of Nevada–Las Vegas*; Yesim Darici, *Florida International University*; N. John DiNardo, *Drexel University*; Steve Ellis, *University of Kentucky*; Hasan Fakhruddin, *Ball State University/The Indiana Academy*; Emily Flynn; Lewis Ford, *Texas A & M University*; Gardner Friedlander, *University School of Milwaukee*; Dolores Gende, *Parish Episcopal School*; Mark Giroux, *East Tennessee State University*; James R. Goff, *Pima Community College*; Yadin Y. Goldschmidt, *University of Pittsburgh*; Torgny Gustafsson, *Rutgers University*; Steve Hagen, *University of Florida*; Raymond Hall, *California State University–Fresno*; Patrick Hamill, *San Jose State University*; Joel Handley; Grant W. Hart, *Brigham Young University*; James E. Heath, *Austin Community College*; Grady Hendricks, *Blinn College*; Rhett Herman, *Radford University*; Aleksey Holloway, *University of Nebraska at Omaha*; Joey Huston, *Michigan State University*; Mark James, *Northern Arizona University*; Randall Jones, *Loyola College Maryland*; Teruki Kamon, *Texas A & M University*; Joseph

Keane, *St. Thomas Aquinas College*; Dorina Kosztin, *University of Missouri–Columbia*; Martha Lietz, *Niles West High School*; Edwin Lo; Rafael Lopez-Mobilia, *University of Texas at San Antonio*; Mark Lucas, *Ohio University*; Mark E. Mattson, *James Madison University*; Sylvio May, *North Dakota State University*; John A. Milsom, *University of Arizona*; Monty Mola, *Humboldt State University*; Charles W. Myles, *Texas Tech University*; Ed Oberhofer, *Lake Sumter Community College*; Chris Pearson, *University of Michigan–Flint*; Alexey A. Petrov, *Wayne State University*; J. Patrick Polley, *Beloit College*; Scott Pratt, *Michigan State University;* M. Anthony Reynolds, *Embry-Riddle Aeronautical University*; Dubravka Rupnik, *Louisiana State University*; Scott Saltman, *Phillips Exeter Academy*; Surajit Sen, *State University of New York at Buffalo*; Bartlett M. Sheinberg, *Houston Community College*, Marllin L. Simon, *Auburn University*; Matthew Sirocky; Gay Stewart, *University of Arkansas*; George Strobel, *University of Georgia*; Eugene Surdutovich, *Oakland University*; Marshall Thomsen, *Eastern Michigan University*; James Wanliss, *Presbyterian College*; Michael Willis, *Glen Burnie High School*; David P. Young, *Louisiana State University*

*College Physics*, eleventh edition, was carefully checked for accuracy by Grant W. Hart, *Brigham Young University*; Eugene Surdutovich, *Oakland University;* and Extanto Technology. Although responsibility for any remaining errors rests with us, we thank them for their dedication and vigilance.

Gerd Kortemeyer and Randall Jones contributed several end-of-topic problems, especially those of interest to the life sciences. Edward F. Redish of the University of Maryland graciously allowed us to list some of his problems from the Activity Based Physics Project. Andy Sheikh of Colorado Mesa University regularly sends in suggestions for improvements, clarifications, or corrections.

Special thanks and recognition go to the professional staff at Cengage Learning—in particular, Rebecca Berardy Schwartz, Ed Dodd, Susan Pashos, Michael Jacobs, Tanya Nigh, Janet del Mundo, Nicole Hurst, Maria Kilmek, Darlene Amidon-Brent, Cate Barr, and Caitlin Ghegan—for their fine work during the development, production, and promotion of this textbook. We recognize the skilled production service provided by Eve Malakoff-Klein and the staff at Cenveo® Publisher Services, and the dedicated permission research efforts of Ranjith Rajaram and Kanchana Vijayarangan at Lumina Datamatics.

Finally, we are deeply indebted to our wives and children for their love, support, and long-term sacrifices.

**Raymond A. Serway**
St. Petersburg, Florida

**Chris Vuille**
Daytona Beach, Florida

# Engaging Applications

Although physics is relevant to so much in our lives, it may not be obvious to students in an introductory course. In this eleventh edition of *College Physics,* we continue a design feature begun in the seventh edition. This feature makes the relevance of physics to everyday life more obvious by pointing out specific applications in the form of a marginal note. Some of these applications pertain to the life sciences and are marked with the BIO icon. The list below is not intended to be a complete listing of all the applications of the principles of physics found in this textbook. Many other applications are to be found within the text and especially in the worked examples, conceptual questions, and end-of-topic problems.

# Welcome to Your MCAT Test Preparation Guide

The MCAT Test Preparation Guide makes your copy of *College Physics*, eleventh edition, the most comprehensive MCAT study tool and classroom resource in introductory physics. The MCAT was revised in 2015 (see **www.aamc.org/students/applying/mcat/mcat2015** for more details); the test section that now includes problems related to physics is *Chemical and Physical Foundations of Biological Systems*. Of the ~65 test questions in this section, approximately 25% relate to introductory physics topics from the six content categories shown below:

**Content Category 4A:** Translational motion, forces, work, energy, and equilibrium in living systems

### Review Plan

**Motion**

■ **Topic 1, Sections 1.1, 1.3, 1.5, and 1.9–1.10**
Quick Quizzes 1.1–1.2
Examples 1.1–1.2, and 1.11–1.13
Topic problems 1–6, 15–27, and 54–71

■ **Topic 2, Sections 2.1–2.2**
Quick Quizzes 2.1–2.5
Examples 2.1–2.3
Topic problems 1–25

■ **Topic 3, Sections 3.1–3.2**
Quick Quizzes 3.1–3.5
Examples 3.1–3.6
Topic problems 1–19, 47, 50, 53, and 56

**Force and Equilibrium**

■ **Topic 4, Sections 4.1–4.4 and 4.6**
Quick Quizzes 4.1–4.9
Examples 4.1–4.12
Topic problems 1–31, 38, 40, 49, and 53

■ **Topic 8, Sections 8.1–8.3**
Quick Quiz 8.1
Examples 8.1–8.11
Topic problems 1–36, 85, 91, and 92

**Work**

■ **Topic 5, Sections 5.1 and 5.2**
Quick Quiz 5.1–5.2
Examples 5.1–5.3
Topic problems 1–18 and 27

■ **Topic 12, Section 12.1**
Quick Quiz 12.1
Examples 12.1–12.2
Topic problems 1–10

**Energy**

■ **Topic 5, Sections 5.2–5.7**
Quick Quizzes 5.2–5.7
Examples 5.3–5.14
Topic problems 9–58, 67, 73, 74, and 78

**Periodic Motion**

■ **Topic 13, Sections 13.7–13.9**
Examples 13.8–13.10
Topic problems 41–60

**Content Category 4B:** Importance of fluids for the circulation of blood, gas movement, and gas exchange

### Review Plan

**Fluids**

■ **Topic 9, Sections 9.1–9.3 and 9.5–9.9**
Quick Quizzes 9.1–9.2 and 9.5–9.7
Examples 9.1–9.16
Topic problems 1–64, 79, 80, 81, 83, and 84

**Gas phase**

■ **Topic 9, Section 9.5**
Quick Quizzes 9.3–9.4
Topic problems 8, 10, 14–15, and 83

■ **Topic 10, Sections 10.2, 10.4, and 10.5**
Quick Quiz 10.6
Examples 10.1–10.2 and 10.6–10.10
Topic problems 1–10 and 29–50

**Content Category 4C:** Electrochemistry and electrical circuits and their elements.

### Review Plan

**Electrostatics**

■ **Topic 15, Sections 15.1–15.4**
Quick Quizzes 15.1 and 15.3–15.5
Examples 15.1–15.5
Topic problems 1–39

■ **Topic 16, Sections 16.1–16.3**
Quick Quizzes 16.1–16.7
Examples 16.1–16.5
Topic problems 1–24

**Circuit elements**

■ **Topic 16, Sections 16.5–16.8**
Quick Quizzes 16.8–16.11
Examples 16.6–16.12
Topic problems 29–57

- **Topic 17, Sections 17.1 and 17.3–17.5**
  Quick Quizzes 17.1 and 17.3–17.6
  Examples 17.1 and 17.3–17.4
  Topic problems 1–32 and 34

- **Topic 18, Sections 18.1–18.3 and 18.8**
  Quick Quizzes 18.1–18.8
  Examples 18.1–18.3
  Topic problems 1–17

### Magnetism

- **Topic 19, Sections 19.1 and 19.3–19.4**
  Quick Quizzes 19.1–19.3
  Examples 19.1–19.4
  Topic problems 1–21

**Content Category 4D:** How light and sound
interact with matter

**Review Plan**

### Sound

- **Topic 13, Sections 13.7 and 13.8**
  Examples 13.8–13.9
  Topic problems 41–48

- **Topic 14, Sections 14.1–14.4, 14.6,
  14.9–14.10, and 14.12–14.13**
  Quick Quizzes 14.1–14.3 and 14.5–14.6
  Examples 14.1–14.2, 14.4–14.5,
  and 14.9–14.10
  Topic problems 1–36 and 54–60

### Light, electromagnetic radiation

- **Topic 21, Sections 21.10–21.12**
  Quick Quizzes 21.7 and 21.8
  Examples 21.8 and 21.9
  Topic problems 49–63 and 76

- **Topic 22, Sections 22.1**
  Topic problems 1–6

- **Topic 24, Sections 24.1, 24.4,
  and 24.6–24.9**
  Quick Quizzes 24.1–24.6
  Examples 24.1–24.4 and 24.6–24.8
  Topic problems 1–61

- **Topic 27, Section 27.3–27.4**
  Example 27.2
  Topic problems 15–23

### Geometrical optics

- **Topic 22, Sections 22.2–22.4 and 22.7**
  Quick Quizzes 22.2–22.4
  Examples 22.1–22.6
  Topic problems 7–44 and 52

- **Topic 23, Sections 23.1–23.3
  and 23.5–23.6**
  Quick Quizzes 23.1–23.6
  Examples 23.1–23.10
  Topic problems 1–46

- **Topic 25, Sections 25.1–25.6**
  Quick Quizzes 25.1–25.2
  Examples 25.1–25.8
  Topic problems 1–40, 60, and 62–65

**Content Category 4E:** Atoms, nuclear decay,
electronic structure, and atomic chemical
behavior

**Review Plan**

### Atomic nucleus

- **Topic 29, Sections 29.1–29.5 and 29.7**
  Quick Quizzes 29.1–29.3
  Examples 29.1–29.5
  Topic problems 1–35, 44–50, and 57

### Electronic structure

- **Topic 19, Section 19.10**

- **Topic 27, Sections 27.2 and 27.8**
  Examples 27.1 and 27.5
  Topic problems 9–14 and 35–40

- **Topic 28, Sections 28.2–28.3, 28.5,
  and 28.7**
  Quick Quizzes 28.1 and 28.3
  Examples 28.1 and 28.2
  Topic problems 1–30 and 37–41

**Content Category 5E:** Principles of chemical
thermodynamics and kinetics

**Review Plan**

### Energy changes in chemical reactions

- **Topic 10, Sections 10.1 and 10.3**
  Quick Quizzes 10.1–10.5
  Examples 10.3–10.5
  Topic problems 11–28

- **Topic 11, Sections 11.1–11.5**
  Quick Quizzes 11.1–11.5
  Examples 11.1–11.11
  Topic problems 1–50

- **Topic 12, Sections 12.1–12.2 and 12.4–12.6**
  Quick Quizzes 12.1 and 12.4–12.5
  Examples 12.1–12.3, 12.10–12.12, and
  12.14–12.16
  Topic problems 1–61, 73–74

# Units, Trigonometry, and Vectors

THE GOAL OF PHYSICS IS TO PROVIDE an understanding of the physical world by developing theories based on experiments. A physical theory, usually expressed mathematically, describes how a given physical system works. The theory makes certain predictions about the physical system which can then be checked by observations and experiments. If the predictions turn out to correspond closely to what is actually observed, then the theory stands, although it remains provisional. No theory to date has given a complete description of all physical phenomena, even within a given subdiscipline of physics. Every theory is a work in progress.

The basic laws of physics involve such physical quantities as force, velocity, volume, and acceleration, all of which can be described in terms of more fundamental quantities. In mechanics, it is conventional to use the quantities of **length** (L), **mass** (M), and **time** (T); all other physical quantities can be constructed from these three.

## 1.1 Standards of Length, Mass, and Time

To communicate the result of a measurement of a certain physical quantity, a *unit* for the quantity must be defined. If our fundamental unit of length is defined to be 1.0 meter, for example, and someone familiar with our system of measurement reports that a wall is 2.0 meters high, we know that the height of the wall is twice the fundamental unit of length. Likewise, if our fundamental unit of mass is defined as 1.0 kilogram and we are told that a person has a mass of 75 kilograms, then that person has a mass 75 times as great as the fundamental unit of mass.

In 1960 an international committee agreed on a standard system of units for the fundamental quantities of science, called **SI** (Système International). Its units of length, mass, and time are the meter, kilogram, and second, respectively.

### 1.1.1 Length

In 1799 the legal standard of length in France became the meter, defined as one ten-millionth of the distance from the equator to the North Pole. Until 1960, the official length of the meter was the distance between two lines on a specific bar of platinum–iridium alloy stored under controlled conditions. This standard was abandoned for several reasons, the principal one being that measurements of the separation between the lines were not precise enough. In 1960 the meter was defined as 1 650 763.73 wavelengths of orange-red light emitted from a krypton-86 lamp. In October 1983 this definition was abandoned also, and **the meter was redefined as the distance traveled by light in vacuum during a time interval of 1/299 792 458 second.** This latest definition establishes the speed of light at 299 792 458 meters per second.

### 1.1.2 Mass

**The SI unit of mass, the kilogram, is defined as the mass of a specific platinum–iridium alloy cylinder kept at the International Bureau of Weights and Measures at Sèvres, France** (similar to that shown in Fig. 1.1a). As we'll see in Topic 4, mass is a

> **Tip 1.1** No Commas in Numbers with Many Digits
>
> In science, numbers with more than three digits are written in groups of three digits separated by spaces rather than commas, so that 10 000 is the same as the common American notation 10,000. Similarly, $\pi = 3.14159265$ is written as 3.141 592 65.

◄ Definition of the meter

◄ Definition of the kilogram

**Figure 1.1** (a) International Prototype of the Kilogram, an accurate copy of the International Standard Kilogram kept at Sèvres, France, is housed under a double bell jar in a vault at the National Institute of Standards and Technology. (b) A cesium fountain atomic clock. The clock will neither gain nor lose a second in 20 million years.

quantity used to measure the resistance to a change in the motion of an object. It's more difficult to cause a change in the motion of an object with a large mass than an object with a small mass.

### 1.1.3 Time

Before 1960, the time standard was defined in terms of the average length of a solar day in the year 1900. (A solar day is the time between successive appearances of the Sun at the highest point it reaches in the sky each day.) The basic unit of time, the second, was defined to be $(1/60)(1/60)(1/24) = 1/86\ 400$ of the average solar day. In 1967 the second was redefined to take advantage of the high precision attainable with an atomic clock, which uses the characteristic frequency of the light emitted from the cesium-133 atom as its "reference clock." **The second is now defined as 9 192 631 700 times the period of oscillation of radiation from the cesium atom.** The newest type of cesium atomic clock is shown in Figure 1.1b.

### 1.1.4 Approximate Values for Length, Mass, and Time Intervals

Approximate values of some lengths, masses, and time intervals are presented in Tables 1.1, 1.2, and 1.3, respectively. Note the wide ranges of values. Study these tables to get a feel for a kilogram of mass (this book has a mass of about 2 kilograms), a time interval of $10^{10}$ seconds (one century is about $3 \times 10^9$ seconds), or 2 meters of length (the approximate height of a forward on a basketball team). Appendix A reviews the notation for powers of 10, such as the expression of the number 50 000 in the form $5 \times 10^4$.

Systems of units commonly used in physics are the Système International, in which the units of length, mass, and time are the meter (m), kilogram (kg), and second (s); the cgs, or Gaussian, system, in which the units of length, mass, and time are the centimeter (cm), gram (g), and second; and the U.S. customary system, in which the units of length, mass, and time are the foot (ft), slug, and second. SI units are almost universally accepted in science and industry and will be used throughout the book. Limited use will be made of Gaussian and U.S. customary units.

**Table 1.1** Approximate Values of Some Measured Lengths

|  | Length (m) |
| --- | --- |
| Observable Universe | $1 \times 10^{26}$ |
| Earth to Andromeda | $2 \times 10^{22}$ |
| Earth to Proxima Centauri | $4 \times 10^{16}$ |
| One light-year | $9 \times 10^{15}$ |
| Earth to Sun | $2 \times 10^{11}$ |
| Earth to Moon | $4 \times 10^{8}$ |
| Radius of Earth | $6 \times 10^{6}$ |
| World's tallest building | $8 \times 10^{2}$ |
| Football field | $9 \times 10^{1}$ |
| Housefly | $5 \times 10^{-3}$ |
| Typical organism cell | $1 \times 10^{-5}$ |
| Hydrogen atom | $1 \times 10^{-10}$ |
| Atomic nucleus | $1 \times 10^{-14}$ |
| Proton diameter | $1 \times 10^{-15}$ |

**Table 1.2** Approximate Values of Some Masses

|  | Mass (kg) |
| --- | --- |
| Observable Universe | $1 \times 10^{52}$ |
| Milky Way galaxy | $7 \times 10^{41}$ |
| Sun | $2 \times 10^{30}$ |
| Earth | $6 \times 10^{24}$ |
| Moon | $7 \times 10^{22}$ |
| Shark | $1 \times 10^{2}$ |
| Human | $7 \times 10^{1}$ |
| Frog | $1 \times 10^{-1}$ |
| Mosquito | $1 \times 10^{-5}$ |
| Bacterium | $1 \times 10^{-15}$ |
| Hydrogen atom | $2 \times 10^{-27}$ |
| Electron | $9 \times 10^{-31}$ |

**Table 1.3** Approximate Values of Some Time Intervals

|  | Time Interval (s) |
| --- | --- |
| Age of Universe | $5 \times 10^{17}$ |
| Age of Earth | $1 \times 10^{17}$ |
| Age of college student | $6 \times 10^{8}$ |
| One year | $3 \times 10^{7}$ |
| One day | $9 \times 10^{4}$ |
| Heartbeat | $8 \times 10^{-1}$ |
| Audible sound wave period[a] | $1 \times 10^{-3}$ |
| Typical radio wave period[a] | $1 \times 10^{-6}$ |
| Visible light wave period[a] | $2 \times 10^{-15}$ |
| Nuclear collision | $1 \times 10^{-22}$ |

[a]A *period* is defined as the time required for one complete vibration.

Some of the most frequently used "metric" (SI and cgs) prefixes representing powers of 10 and their abbreviations are listed in Table 1.4. For example, $10^{-3}$ m is equivalent to 1 millimeter (mm), and $10^3$ m is 1 kilometer (km). Likewise, 1 kg is equal to $10^3$ g, and 1 megavolt (MV) is $10^6$ volts (V). It's a good idea to memorize the more common prefixes early on: femto- to centi-, and kilo- to giga- are used routinely by most physicists.

**Table 1.4** Some Prefixes for Powers of Ten Used with "Metric" (SI and cgs) Units

| Power | Prefix | Abbreviation |
|---|---|---|
| $10^{-18}$ | atto- | a |
| $10^{-15}$ | femto- | f |
| $10^{-12}$ | pico- | p |
| $10^{-9}$ | nano- | n |
| $10^{-6}$ | micro- | $\mu$ |
| $10^{-3}$ | milli- | m |
| $10^{-2}$ | centi- | c |
| $10^{-1}$ | deci- | d |
| $10^1$ | deka- | da |
| $10^3$ | kilo- | k |
| $10^6$ | mega- | M |
| $10^9$ | giga- | G |
| $10^{12}$ | tera- | T |
| $10^{15}$ | peta- | P |
| $10^{18}$ | exa- | E |

# 1.2 The Building Blocks of Matter

A 1-kg ($\approx$ 2-lb) cube of solid gold has a length of about 3.73 cm ($\approx$ 1.5 in.) on a side. If the cube is cut in half, the two resulting pieces retain their chemical identity. But what happens if the pieces of the cube are cut again and again, indefinitely? The Greek philosophers Leucippus and Democritus couldn't accept the idea that such cutting could go on forever. They speculated that the process ultimately would end when it produced a particle that could no longer be cut. In Greek, *atomos* means "not sliceable." From this term comes our English word *atom*, once believed to be the smallest particle of matter but since found to be a composite of more elementary particles.

The atom can be naively visualized as a miniature solar system, with a dense, positively charged nucleus occupying the position of the Sun and negatively charged electrons orbiting like planets. This model of the atom, first developed by the great Danish physicist Niels Bohr nearly a century ago, led to the understanding of certain properties of the simpler atoms such as hydrogen but failed to explain many fine details of atomic structure.

Notice the size of a hydrogen atom, listed in Table 1.1, and the size of a proton—the nucleus of a hydrogen atom—one hundred thousand times smaller. If the proton were the size of a ping-pong ball, the electron would be a tiny speck about the size of a bacterium, orbiting the proton a kilometer away! Other atoms are similarly constructed. So there is a surprising amount of empty space in ordinary matter.

After the discovery of the nucleus in the early 1900s, questions arose concerning its structure. Although the structure of the nucleus remains an area of active research even today, by the early 1930s scientists determined that two basic entities—protons and neutrons—occupy the nucleus. The *proton* is nature's most common carrier of positive charge, equal in magnitude but opposite in sign to the charge on the electron. The number of protons in a nucleus determines what the element is. For instance, a nucleus containing only one proton is the nucleus of an atom of hydrogen, regardless of how many neutrons may be present. Extra neutrons correspond to different isotopes of hydrogen—deuterium and tritium—which react chemically in exactly the same way as hydrogen, but are more massive. An atom having two protons in its nucleus, similarly, is always helium, although again, differing numbers of neutrons are possible.

The existence of *neutrons* was verified conclusively in 1932. A neutron has no charge and has a mass about equal to that of a proton. Except for hydrogen, all atomic nuclei contain neutrons, which, together with the protons, interact through the strong nuclear force. That force opposes the strongly repulsive electrical force of the protons, which otherwise would cause the nucleus to disintegrate.

The division doesn't stop here; strong evidence collected over many years indicates that protons, neutrons, and a zoo of other exotic particles are composed of six particles called **quarks** (rhymes with "sharks" though some rhyme it with "forks"). These particles have been given the names *up, down, strange, charm, bottom,* and *top.* The up, charm, and top quarks each carry a charge equal to $+\frac{2}{3}$ that of the proton, whereas the down, strange, and bottom quarks each carry a charge equal to $-\frac{1}{3}$ the proton charge. The proton consists of two up quarks and one down quark (see Fig. 1.2), giving the correct charge for the proton, +1. The neutron is composed of two down quarks and one up quark and has a net charge of zero.

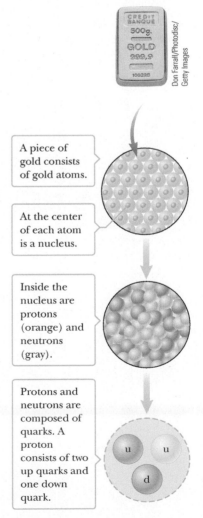

A piece of gold consists of gold atoms.

At the center of each atom is a nucleus.

Inside the nucleus are protons (orange) and neutrons (gray).

Protons and neutrons are composed of quarks. A proton consists of two up quarks and one down quark.

**Figure 1.2** Levels of organization in matter.

The up and down quarks are sufficient to describe all normal matter, so the existence of the other four quarks, indirectly observed in high-energy experiments, is something of a mystery. Despite strong indirect evidence, no isolated quark has ever been observed. Consequently, the possible existence of yet more fundamental particles remains purely speculative.

# 1.3 Dimensional Analysis

In physics the word *dimension* denotes the physical nature of a quantity. The distance between two points, for example, can be measured in feet, meters, or furlongs, which are different ways of expressing the dimension of *length*.

The symbols used in this section to specify the dimensions of length, mass, and time are L, M, and T, respectively. Brackets [ ] will often be used to denote the dimensions of a physical quantity. In this notation, for example, the dimensions of velocity $v$ are written $[v] = L/T$, and the dimensions of area $A$ are $[A] = L^2$. The dimensions of area, volume, velocity, and acceleration are listed in Table 1.5, along with their units in the three common systems. The dimensions of other quantities, such as force and energy, will be described later as they are introduced.

In physics it's often necessary to deal with mathematical expressions that relate different physical quantities. One way to analyze such expressions, called **dimensional analysis,** makes use of the fact that **dimensions can be treated as algebraic quantities.** Adding masses to lengths, for example, makes no sense, so it follows that quantities can be added or subtracted only if they have the same dimensions. If the terms on the opposite sides of an equation have the same dimensions, then that equation may be correct, although correctness can't be guaranteed on the basis of dimensions alone. Nonetheless, dimensional analysis has value as a partial check of an equation and can also be used to develop insight into the relationships between physical quantities.

The procedure can be illustrated by developing some relationships between acceleration, velocity, time, and distance. Distance $x$ has the dimension of length: $[x] = L$. Time $t$ has dimension $[t] = T$. Velocity $v$ has the dimensions length over time: $[v] = L/T$, and acceleration the dimensions length divided by time squared: $[a] = L/T^2$. Notice that velocity and acceleration have similar dimensions, except for an extra dimension of time in the denominator of acceleration. It follows that

$$[v] = \frac{L}{T} = \frac{L}{T^2}T = [a][t]$$

From this it might be guessed that velocity equals acceleration multiplied by time, $v = at$, and that is true for the special case of motion with constant acceleration starting at rest. Noticing that velocity has dimensions of length divided by time and distance has dimensions of length, it's reasonable to guess that

$$[x] = L = L\frac{T}{T} = \frac{L}{T}T = [v][t] = [a][t]^2$$

Here it appears that $x = at^2$ might correctly relate the distance traveled to acceleration and time; however, that equation is not even correct in the case of constant acceleration starting from rest. The correct expression in that case is $x = \frac{1}{2}at^2$.

**Table 1.5** Dimensions and Some Units of Area, Volume, Velocity, and Acceleration

| System | Area ($L^2$) | Volume ($L^3$) | Velocity (L/T) | Acceleration ($L/T^2$) |
|---|---|---|---|---|
| SI | $m^2$ | $m^3$ | m/s | $m/s^2$ |
| cgs | $cm^2$ | $cm^3$ | cm/s | $cm/s^2$ |
| U.S. customary | $ft^2$ | $ft^3$ | ft/s | $ft/s^2$ |

These examples serve to show the inherent limitations in using dimensional analysis to discover relationships between physical quantities. Nonetheless, such simple procedures can still be of value in developing a preliminary mathematical model for a given physical system. Further, because it's easy to make errors when solving problems, dimensional analysis can be used to check the consistency of the results. When the dimensions in an equation are not consistent, it indicates an error has been made in a prior step.

---

### EXAMPLE 1.1    ANALYSIS OF AN EQUATION

**GOAL**  Check an equation using dimensional analysis.

**PROBLEM**  Show that the expression $v = v_0 + at$ is dimensionally correct, where $v$ and $v_0$ represent velocities, $a$ is acceleration, and $t$ is a time interval.

**STRATEGY**  Analyze each term, finding its dimensions, and then check to see if all the terms agree with each other.

**SOLUTION**

Find dimensions for $v$ and $v_0$.

$$[v] = [v_0] = \frac{L}{T}$$

Find the dimensions of $at$.

$$[at] = [a][t] = \frac{L}{T^2}(T) = \frac{L}{T}$$

**REMARKS**  All the terms agree, so the equation is dimensionally correct.

**QUESTION 1.1**  True or False: An equation that is dimensionally correct is always physically correct, up to a constant of proportionality.

**EXERCISE 1.1**  Determine whether the equation $x = vt^2$ is dimensionally correct. If not, provide a correct expression, up to an overall constant of proportionality.

**ANSWER**  Incorrect. The expression $x = vt$ is dimensionally correct.

---

### EXAMPLE 1.2    FIND AN EQUATION

**GOAL**  Derive an equation by using dimensional analysis.

**PROBLEM**  Find a relationship between an acceleration of constant magnitude $a$, speed $v$, and distance $r$ from the origin for a particle traveling in a circle.

**STRATEGY**  Start with the term having the most dimensionality, $a$. Find its dimensions, and then rewrite those dimensions in terms of the dimensions of $v$ and $r$. The dimensions of time will have to be eliminated with $v$, because that's the only quantity (other than $a$, itself) in which the dimension of time appears.

**SOLUTION**

Write down the dimensions of $a$:

$$[a] = \frac{L}{T^2}$$

Solve the dimensions of speed for T:

$$[v] = \frac{L}{T} \quad \rightarrow \quad T = \frac{L}{[v]}$$

Substitute the expression for T into the equation for $[a]$:

$$[a] = \frac{L}{T^2} = \frac{L}{(L/[v])^2} = \frac{[v]^2}{L}$$

Substitute L = $[r]$, and guess at the equation.

$$[a] = \frac{[v]^2}{[r]} \quad \rightarrow \quad a = \frac{v^2}{r}$$

**REMARKS**  This is the correct equation for the magnitude of the centripetal acceleration—acceleration towards the center of motion—to be discussed in Topic 7. In this case it isn't necessary to introduce a numerical factor. Such a factor is often displayed explicitly as a constant $k$ in front of the right-hand side; for example, $a = kv^2/r$. As it turns out, $k = 1$ gives the correct expression. A good technique sometimes introduced in calculus-based textbooks involves using unknown powers of the dimensions. This problem would then be set up as $[a] = [v]^b[r]^c$. Writing out the dimensions and equating powers of each dimension on both sides of the equation would result in $b = 2$ and $c = -1$.

*(Continued)*

**QUESTION 1.2** True or False: Replacing $v$ by $r/t$ in the final answer also gives a dimensionally correct equation.

**EXERCISE 1.2** In physics, energy $E$ carries dimensions of mass times length squared divided by time squared. Use dimensional analysis to derive a relationship for energy in terms of mass $m$ and speed $v$, up to a constant of proportionality. Set the speed equal to $c$, the speed of light, and the constant of proportionality equal to 1 to get the most famous equation in physics. (Note, however, that the first relationship is associated with energy of motion and the second with energy of mass. See Topic 26.)

**ANSWER**  $E = kmv^2$  $\rightarrow$  $E = mc^2$ when $k = 1$ and $v = c$.

---

# 1.4 Uncertainty in Measurement and Significant Figures

Physics is a science in which mathematical laws are tested by experiment. No physical quantity can be determined with complete accuracy because our senses are physically limited, even when extended with microscopes, cyclotrons, and other instruments. Consequently, it's important to develop methods of determining the accuracy of measurements.

All measurements have uncertainties associated with them, whether or not they are explicitly stated. The accuracy of a measurement depends on the sensitivity of the apparatus, the skill of the person carrying out the measurement, and the number of times the measurement is repeated. Once the measurements, along with their uncertainties, are known, it's often the case that calculations must be carried out using those measurements. Suppose two such measurements are multiplied. When a calculator is used to obtain this product, there may be eight digits in the calculator window, but often only two or three of those numbers have any significance. The rest have no value because they imply greater accuracy than was actually achieved in the original measurements. In experimental work, determining how many numbers to retain requires the application of statistics and the mathematical propagation of uncertainties. In a textbook it isn't practical to apply those sophisticated tools in the numerous calculations, so instead a simple method, called *significant figures,* is used to indicate the approximate number of digits that should be retained at the end of a calculation. Although that method is not mathematically rigorous, it's easy to apply and works fairly well.

Suppose in a laboratory experiment we measure the area of a rectangular plate with a meter stick. Let's assume the accuracy to which we can measure a particular dimension of the plate is ±0.1 cm. If the length of the plate is measured to be 16.3 cm, we can only claim it lies somewhere between 16.2 cm and 16.4 cm. In this case, we say the measured value has three significant figures. Likewise, if the plate's width is measured to be 4.5 cm, the actual value lies between 4.4 cm and 4.6 cm. This measured value has only two significant figures. We could write the measured values as 16.3 ± 0.1 cm and 4.5 ± 0.1 cm. In general, **a significant figure is a reliably known digit** (other than a zero used to locate a decimal point). Note that in each case, the final number has some uncertainty associated with it and is therefore not 100% reliable. Despite the uncertainty, that number is retained and considered significant because it does convey some information.

Suppose we would like to find the area of the plate by multiplying the two measured values together. The final value can range between (16.3 − 0.1 cm)(4.5 − 0.1 cm) = (16.2 cm)(4.4 cm) = 71.28 cm² and (16.3 + 0.1 cm)(4.5 + 0.1 cm) = (16.4 cm)(4.6 cm) = 75.44 cm². Claiming to know anything about the hundredths place, or even the tenths place, doesn't make any sense, because it's clear we can't even be certain of the units place, whether it's the 1 in 71, the 5 in 75, or somewhere in between. The tenths and the hundredths places are clearly not significant. We have *some* information about the units place, so that number is significant. Multiplying the numbers at the middle of the uncertainty ranges gives (16.3 cm)

(4.5 cm) = 73.35 cm$^2$, which is also in the middle of the area's uncertainty range. Because the hundredths and tenths are not significant, we drop them and take the answer to be 73 cm$^2$, with an uncertainty of $\pm 2$ cm$^2$. Note that the answer has two significant figures, the same number of figures as the least accurately known quantity being multiplied, the 4.5-cm width.

Calculations as carried out in the preceding paragraph can indicate the proper number of significant figures, but those calculations are time-consuming. Instead, two rules of thumb can be applied. The first, concerning multiplication and division, is as follows: **In multiplying (dividing) two or more quantities, the number of significant figures in the final product (quotient) is the same as the number of significant figures in the *least accurate* of the factors being combined, where *least accurate* means *having the lowest number of significant figures.***

To get the final number of significant figures, it's usually necessary to do some rounding. If the last digit dropped is less than 5, simply drop the digit. If the last digit dropped is greater than or equal to 5, raise the last retained digit by one.[1]

Zeros may or may not be significant figures. Zeros used to position the decimal point in such numbers as 0.03 and 0.007 5 are not considered significant figures. Hence, 0.03 has one significant figure, and 0.007 5 has two.

When zeros are placed after other digits in a whole number, there is a possibility of misinterpretation. For example, suppose the mass of an object is given as 1 500 g. This value is ambiguous, because we don't know whether the last two zeros are being used to locate the decimal point or whether they represent significant figures in the measurement.

Using scientific notation to indicate the number of significant figures removes this ambiguity. In this case, we express the mass as $1.5 \times 10^3$ g if there are two significant figures in the measured value, $1.50 \times 10^3$ g if there are three significant figures, and $1.500 \times 10^3$ g if there are four. Likewise, 0.000 15 is expressed in scientific notation as $1.5 \times 10^{-4}$ if it has two significant figures or as $1.50 \times 10^{-4}$ if it has three significant figures. The three zeros between the decimal point and the digit 1 in the number 0.000 15 are not counted as significant figures because they only locate the decimal point. Similarly, trailing zeros are not considered significant. However, any zeros written after a decimal point, or between a nonzero number and before a decimal point, are considered significant. For example, 3.00, 30.0, and 300. have three significant figures, whereas 300 has only one. In this book, **most of the numerical examples and end-of-topic problems will yield answers having two or three significant figures.**

For addition and subtraction, it's best to focus on the number of decimal places in the quantities involved rather than on the number of significant figures. **When numbers are added (subtracted), the number of decimal places in the result should equal the smallest number of decimal places of any term in the sum (difference).** For example, if we wish to compute 123 (zero decimal places) + 5.35 (two decimal places), the answer is 128 (zero decimal places) and not 128.35. If we compute the sum 1.000 1 (four decimal places) + 0.000 3 (four decimal places) = 1.000 4, the result has the correct number of decimal places, namely four. Observe that the rules for multiplying significant figures don't work here because the answer has five significant figures even though one of the terms in the sum, 0.000 3, has only one significant figure. Likewise, if we perform the subtraction 1.002 − 0.998 = 0.004, the result has three decimal places because each term in the subtraction has three decimal places.

To show why this rule should hold, we return to the first example in which we added 123 and 5.35, and rewrite these numbers as 123.*xxx* and 5.35*x*. Digits written with an *x* are completely unknown and can be any digit from 0 to 9. Now we

**Tip 1.2** Using Calculators

Calculators are designed by engineers to yield as many digits as the memory of the calculator chip permits, so be sure to round the final answer to the correct number of significant figures.

---

[1]Some prefer to round to the nearest even digit when the last dropped digit is 5, which has the advantage of rounding 5 up half the time and down half the time. For example, 1.55 would round to 1.6, but 1.45 would round to 1.4. Because the final significant figure is only one representative of a range of values given by the uncertainty, this very slight refinement will not be used in this text.

line up 123.*xxx* and 5.35*x* relative to the decimal point and perform the addition, using the rule that an unknown digit added to a known or unknown digit yields an unknown:

$$123.xxx$$
$$+\quad 5.35x$$
$$\overline{\phantom{+}128.xxx}$$

The answer of 128.*xxx* means that we are justified only in keeping the number 128 because everything after the decimal point in the sum is actually unknown. The example shows that the controlling uncertainty is introduced into an addition or subtraction by the term with the smallest number of decimal places.

---

## EXAMPLE 1.3   CARPET CALCULATIONS

**GOAL**  Apply the rules for significant figures.

**PROBLEM**  Several carpet installers make measurements for carpet installation in the different rooms of a restaurant, reporting their measurements with inconsistent accuracy, as compiled in Table 1.6. Compute the areas for **(a)** the banquet hall, **(b)** the meeting room, and **(c)** the dining room, taking into account significant figures. **(d)** What total area of carpet is required for these rooms?

**Table 1.6** Dimensions of Rooms in Example 1.3

|  | Length (m) | Width (m) |
|---|---|---|
| Banquet hall | 14.71 | 7.46 |
| Meeting room | 4.822 | 5.1 |
| Dining room | 13.8 | 9 |

**STRATEGY**  For the multiplication problems in parts **(a)**–**(c)**, count the significant figures in each number. The smaller result is the number of significant figures in the answer. Part **(d)** requires a sum, where the area with the least accurately known decimal place determines the overall number of significant figures in the answer.

..................................................................................................................................

**SOLUTION**

**(a)** Compute the area of the banquet hall.

Count significant figures:

$$14.71 \text{ m} \;\rightarrow\; 4 \text{ significant figures}$$
$$7.46 \text{ m} \;\rightarrow\; 3 \text{ significant figures}$$

To find the area, multiply the numbers keeping only three digits:

$$14.71 \text{ m} \times 7.46 \text{ m} = 109.74 \text{ m}^2 \;\rightarrow\; \boxed{1.10 \times 10^2 \text{ m}^2}$$

**(b)** Compute the area of the meeting room.

Count significant figures:

$$4.822 \text{ m} \;\rightarrow\; 4 \text{ significant figures}$$
$$5.1 \text{ m} \;\rightarrow\; 2 \text{ significant figures}$$

To find the area, multiply the numbers keeping only two digits:

$$4.822 \text{ m} \times 5.1 \text{ m} = 24.59 \text{ m}^2 \;\rightarrow\; \boxed{25 \text{ m}^2}$$

**(c)** Compute the area of the dining room.

Count significant figures:

$$13.8 \text{ m} \;\rightarrow\; 3 \text{ significant figures}$$
$$9 \text{ m} \;\rightarrow\; 1 \text{ significant figure}$$

To find the area, multiply the numbers keeping only one digit:

$$13.8 \text{ m} \times 9 \text{ m} = 124.2 \text{ m}^2 \;\rightarrow\; \boxed{100 \text{ m}^2}$$

**(d)** Calculate the total area of carpet required, with the proper number of significant figures.

Sum all three answers without regard to significant figures:

$$1.10 \times 10^2 \text{ m}^2 + 25 \text{ m}^2 + 100 \text{ m}^2 = 235 \text{ m}^2$$

The least accurate number is 100 m², with one significant figure in the hundred's decimal place:

$$235 \text{ m}^2 \;\rightarrow\; \boxed{2 \times 10^2 \text{ m}^2}$$

**REMARKS** Notice that the final answer in part **(d)** has only one significant figure, in the hundred's place, resulting in an answer that had to be rounded down by a sizable fraction of its total value. That's the consequence of having insufficient information. The value of 9 m, without any further information, represents a true value that could be anywhere in the interval [8.5 m, 9.5 m), all of which round to 9 when only one digit is retained.

**QUESTION 1.3** How would the final answer change if the width of the dining room were given as 9.0 m?

**EXERCISE 1.3** A ranch has two fenced rectangular areas. Area A has a length of 750 m and width 125 m, and area B has length 400 m and width 150 m. Find **(a)** area A, **(b)** area B, and **(c)** the total area, with attention to the rules of significant figures. Assume trailing zeros are not significant.

**ANSWERS** **(a)** $9.4 \times 10^4 \, \text{m}^2$   **(b)** $6 \times 10^4 \, \text{m}^2$   **(c)** $1.5 \times 10^5 \, \text{m}^2$

In performing any calculation, especially one involving a number of steps, there will always be slight discrepancies introduced by both the rounding process and the algebraic order in which steps are carried out. For example, consider $2.35 \times 5.89/1.57$. This computation can be performed in three different orders. First, we have $2.35 \times 5.89 = 13.842$, which rounds to 13.8, followed by $13.8/1.57 = 8.789\,8$, rounding to 8.79. Second, $5.89/1.57 = 3.751\,6$, which rounds to 3.75, resulting in $2.35 \times 3.75 = 8.812\,5$, rounding to 8.81. Finally, $2.35/1.57 = 1.496\,8$ rounds to 1.50, and $1.50 \times 5.89 = 8.835$ rounds to 8.84. So three different algebraic orders, following the rules of rounding, lead to answers of 8.79, 8.81, and 8.84, respectively. Such minor discrepancies are to be expected, because the last significant digit is only one representative from a range of possible values, depending on experimental uncertainty. To avoid such discrepancies, some carry one or more extra digits during the calculation, although it isn't conceptually consistent to do so because those extra digits are not significant. As a practical matter, in the worked examples in this text, intermediate reported results will be rounded to the proper number of significant figures, and only those digits will be carried forward. In the problem sets, however, given data will usually be assumed accurate to two or three digits, even when there are trailing zeros. **In solving the problems, the student should be aware that slight differences in rounding practices can result in answers varying from the text in the last significant digit, which is normal and not cause for concern.** The method of significant figures has its limitations in determining accuracy, but it's easy to apply. In experimental work, however, statistics and the mathematical propagation of uncertainty must be used to determine the accuracy of an experimental result.

# 1.5 Unit Conversions for Physical Quantities

Sometimes it's necessary to convert units from one system to another (see Fig. 1.3). Conversion factors between the SI and U.S. customary systems for units of length are as follows:

$$1 \, \text{mi} = 1\,609 \, \text{m} = 1.609 \, \text{km} \qquad 1 \, \text{ft} = 0.304\,8 \, \text{m} = 30.48 \, \text{cm}$$

$$1 \, \text{m} = 39.37 \, \text{in.} = 3.281 \, \text{ft} \qquad 1 \, \text{in.} = 0.025\,4 \, \text{m} = 2.54 \, \text{cm}$$

A more extensive list of conversion factors can be found on the front endsheets of this book. In all the given conversion equations, the "1" on the left is assumed to have the same number of significant figures as the quantity given on the right of the equation.

Units can be treated as algebraic quantities that can "cancel" each other. We can make a fraction with the conversion that will cancel the units we don't want, and

**Figure 1.3** The speed limit is given in both kilometers per hour and miles per hour on this road sign. How accurate is the conversion?

multiply that fraction by the quantity in question. For example, suppose we want to convert 15.0 in. to centimeters. Because 1 in. = 2.54 cm, we find that

$$15.0 \text{ in.} = 15.0 \text{ in.} \times \left( \frac{2.54 \text{ cm}}{1.00 \text{ in.}} \right) = 38.1 \text{ cm}$$

The next two examples show how to deal with problems involving more than one conversion and with powers.

---

**EXAMPLE 1.4**    **PULL OVER, BUDDY!**

**GOAL**  Convert units using several conversion factors.

**PROBLEM**  If a car is traveling at a speed of 28.0 m/s, is the driver exceeding the speed limit of 55.0 mi/h?

**STRATEGY**  Meters must be converted to miles and seconds to hours, using the conversion factors listed on the front endsheets of the book. Here, three factors will be used.

· · · · · · · · · · · · · · · · · · · · · · · · · · · · · · · · · · · · · · · · · · · · · · · · · · · · · · · · · · · · · · · · · · · · · · · · · · · · · · · · · · · · · · · ·

**SOLUTION**

Convert meters to miles:

$$28.0 \text{ m/s} = \left( 28.0 \, \frac{\text{m}}{\text{s}} \right)\left( \frac{1.00 \text{ mi}}{1\,609 \text{ m}} \right) = 1.74 \times 10^{-2} \text{ mi/s}$$

Convert seconds to hours:

$$1.74 \times 10^{-2} \text{ mi/s} = \left( 1.74 \times 10^{-2} \, \frac{\text{mi}}{\text{s}} \right)\left( 60.0 \, \frac{\text{s}}{\text{min}} \right)\left( 60.0 \, \frac{\text{min}}{\text{h}} \right)$$

$$= 62.6 \text{ mi/h}$$

· · · · · · · · · · · · · · · · · · · · · · · · · · · · · · · · · · · · · · · · · · · · · · · · · · · · · · · · · · · · · · · · · · · · · · · · · · · · · · · · · · · · · · · ·

**REMARKS**  The driver should slow down because he's exceeding the speed limit.

**QUESTION 1.4**  Repeat the conversion, using the relationship 1.00 m/s = 2.24 mi/h. Why is the answer slightly different?

**EXERCISE 1.4**  Convert 152 mi/h to m/s.

**ANSWER**  67.9 m/s

---

**EXAMPLE 1.5**    **PRESS THE PEDAL TO THE METAL**

**GOAL**  Convert a quantity featuring powers of a unit.

**PROBLEM**  The traffic light turns green, and the driver of a high-performance car slams the accelerator to the floor. The accelerometer registers 22.0 m/s². Convert this reading to km/min².

**STRATEGY**  Here we need one factor to convert meters to kilometers and another two factors to convert seconds squared to minutes squared.

· · · · · · · · · · · · · · · · · · · · · · · · · · · · · · · · · · · · · · · · · · · · · · · · · · · · · · · · · · · · · · · · · · · · · · · · · · · · · · · · · · · · · · · ·

**SOLUTION**

Multiply by the three factors:

$$\frac{22.0 \text{ m}}{1.00 \text{ s}^2}\left( \frac{1.00 \text{ km}}{1.00 \times 10^3 \text{ m}} \right)\left( \frac{60.0 \text{ s}}{1.00 \text{ min}} \right)^2 = 79.2 \, \frac{\text{km}}{\text{min}^2}$$

· · · · · · · · · · · · · · · · · · · · · · · · · · · · · · · · · · · · · · · · · · · · · · · · · · · · · · · · · · · · · · · · · · · · · · · · · · · · · · · · · · · · · · · ·

**REMARKS**  Notice that in each conversion factor the numerator equals the denominator when units are taken into account. A common error in dealing with squares is to square the units inside the parentheses while forgetting to square the numbers!

**QUESTION 1.5**  What time conversion factor or factors would be used to further convert the answer to km/h²?

**EXERCISE 1.5**  Convert $4.50 \times 10^3$ kg/m³ to g/cm³.

**ANSWER**  4.50 g/cm³

# 1.6 Estimates and Order-of-Magnitude Calculations

Getting an exact answer to a calculation may often be difficult or impossible, either for mathematical reasons or because limited information is available. In these cases, estimates can yield useful approximate answers that can determine whether a more precise calculation is necessary. Estimates also serve as a partial check if the exact calculations are actually carried out. If a large answer is expected but a small exact answer is obtained, there's an error somewhere.

For many problems, knowing the approximate value of a quantity—within a factor of 10 or so—is sufficient. This approximate value is called an **order-of-magnitude** estimate and requires finding the power of 10 that is closest to the actual value of the quantity. For example, 75 kg $\sim 10^2$ kg, where the symbol $\sim$ means "is on the order of" or "is approximately." Increasing a quantity by three orders of magnitude means that its value increases by a factor of $10^3 = 1\ 000$.

Occasionally the process of making such estimates results in fairly crude answers, but answers ten times or more too large or small are still useful. For example, suppose you're interested in how many people have contracted a certain disease. Any estimates under ten thousand are small compared with Earth's total population, but a million or more would be alarming. So even relatively imprecise information can provide valuable guidance.

In developing these estimates, you can take considerable liberties with the numbers. For example, $\pi \sim 1$, $27 \sim 10$, and $65 \sim 100$. To get a less crude estimate, it's permissible to use slightly more accurate numbers (e.g., $\pi \sim 3$, $27 \sim 30$, $65 \sim 70$). Better accuracy can also be obtained by systematically underestimating as many numbers as you overestimate. Some quantities may be completely unknown, but it's standard to make reasonable guesses, as the examples show.

---

### EXAMPLE 1.6    BRAIN CELLS ESTIMATE

**GOAL** Develop a simple estimate.

**PROBLEM** Estimate the number of cells in the human brain.

**STRATEGY** Estimate the volume of a human brain and divide by the estimated volume of one cell. The brain is located in the upper portion of the head, with a volume that could be approximated by a cube $\ell = 20$ cm on a side. Brain cells, consisting of about 10% neurons and 90% glia, vary greatly in size, with dimensions ranging from a few microns to a meter or so. As a guess, take $d = 10$ microns as a typical dimension and consider a cell to be a cube with each side having that length.

........................................................................................................................

**SOLUTION**

Estimate the volume of a human brain:

$$V_{\text{brain}} = \ell^3 \approx (0.2 \text{ m})^3 = 8 \times 10^{-3} \text{ m}^3 \approx 1 \times 10^{-2} \text{ m}^3$$

Estimate the volume of a cell:

$$V_{\text{cell}} = d^3 \approx (10 \times 10^{-6} \text{ m})^3 = 1 \times 10^{-15} \text{ m}^3$$

Divide the volume of a brain by the volume of a cell:

$$\text{number of cells} = \frac{V_{\text{brain}}}{V_{\text{cell}}} = \frac{0.01 \text{ m}^3}{1 \times 10^{-15} \text{ m}^3} = \boxed{1 \times 10^{13} \text{ cells}}$$

........................................................................................................................

**REMARKS** Notice how little attention was paid to obtaining precise values. Some general information about a problem is required if the estimate is to be within an order of magnitude of the actual value. Here, knowledge of the approximate dimensions of brain cells and the human brain were essential to developing the estimate.

**QUESTION 1.6** Would $10^{12}$ cells also be a reasonable estimate? What about $10^9$ cells? Explain.

**EXERCISE 1.6** Estimate the total number of cells in the human body.

**ANSWER** $10^{14}$ (Answers may vary.)

---

**EXAMPLE 1.7    STACK ONE-DOLLAR BILLS TO THE MOON**

**GOAL**  Estimate the number of stacked objects required to reach a given height.

**PROBLEM**  How many one-dollar bills, stacked one on top of the other, would reach the Moon?

**STRATEGY**  The distance to the Moon is about 400 000 km. Guess at the number of dollar bills in a millimeter, and multiply the distance by this number, after converting to consistent units.

· · · · · · · · · · · · · · · · · · · · · · · · · · · · · · · · · · · · · · · · · · · · · · · · · · · · · · · · · · · · · · · · · · · · · · · · · · · · · · · · · · · · · · · · · · · · · · · · · · · · · · · · · · · · · · · · · · · · · · · · · · · · · · · ·

**SOLUTION**

We estimate that ten stacked bills form a layer of 1 mm. Convert mm to km:

$$\frac{10 \text{ bills}}{1 \text{ mm}}\left(\frac{10^3 \text{ mm}}{1 \text{ m}}\right)\left(\frac{10^3 \text{ m}}{1 \text{ km}}\right) = \frac{10^7 \text{ bills}}{1 \text{ km}}$$

Multiply this value by the approximate lunar distance:

$$\text{\# of dollar bills} \sim (4 \times 10^5 \text{ km})\left(\frac{10^7 \text{ bills}}{1 \text{ km}}\right) = \boxed{4 \times 10^{12} \text{ bills}}$$

· · · · · · · · · · · · · · · · · · · · · · · · · · · · · · · · · · · · · · · · · · · · · · · · · · · · · · · · · · · · · · · · · · · · · · · · · · · · · · · · · · · · · · · · · · · · · · · · · · · · · · · · · · · · · · · · · · · · · · · · · · · · · · · ·

**REMARKS**  That's within an order of magnitude of the U.S. national debt!

**QUESTION 1.7**  Based on the answer, about how many stacked pennies would reach the Moon?

**EXERCISE 1.7**  How many pieces of cardboard, typically found at the back of a bound pad of paper, would you have to stack up to match the height of the Washington Monument, about 170 m tall?

**ANSWER**  $\sim 10^5$ (Answers may vary.)

---

**EXAMPLE 1.8    NUMBER OF GALAXIES IN THE UNIVERSE**

**GOAL**  Estimate a volume and a number density, and combine.

**PROBLEM**  Given that astronomers can see about 10 billion light-years into space and that there are 14 galaxies in our local group, 2 million light-years from the next local group, estimate the number of galaxies in the observable universe. (*Note:* One light-year is the distance traveled by light in one year, about $9.5 \times 10^{15}$ m.) (See Fig. 1.4.)

**STRATEGY**  From the known information, we can estimate the number of galaxies per unit volume. The local group of 14 galaxies is contained in a sphere a million light-years in radius, with the Andromeda group in a similar sphere, so there are about 10 galaxies within a volume of radius 1 million light-years. Multiply that number density by the volume of the observable universe.

**Figure 1.4** In this deep-space photograph, there are few stars—just galaxies without end.

NASA, ESA, S. Beckwith (STScI) and the HUDF Team

· · · · · · · · · · · · · · · · · · · · · · · · · · · · · · · · · · · · · · · · · · · · · · · · · · · · · · · · · · · · · · · · · · · · · · · · · · · · · · · · · · · · · · · · · · · · · · · · · · · · · · · · · · · · · · · · · · · · · · · · · · · · · · · ·

**SOLUTION**

Compute the approximate volume $V_{lg}$ of the local group of galaxies:

$$V_{lg} = \tfrac{4}{3}\pi r^3 \sim (10^6 \text{ ly})^3 = 10^{18} \text{ ly}^3$$

Estimate the density of galaxies:

$$\text{density of galaxies} = \frac{\text{\# of galaxies}}{V_{lg}}$$

$$\sim \frac{10 \text{ galaxies}}{10^{18} \text{ ly}^3} = 10^{-17} \frac{\text{galaxies}}{\text{ly}^3}$$

Compute the approximate volume of the observable universe:

$$V_u = \tfrac{4}{3}\pi r^3 \sim (10^{10} \text{ ly})^3 = 10^{30} \text{ ly}^3$$

Multiply the density of galaxies by $V_u$:

$$\text{\# of galaxies} \sim (\text{density of galaxies})V_u$$

$$= \left(10^{-17} \frac{\text{galaxies}}{\text{ly}^3}\right)(10^{30} \text{ ly}^3)$$

$$= \boxed{10^{13} \text{ galaxies}}$$

**REMARKS** Notice the approximate nature of the computation, which uses $4\pi/3 \sim 1$ on two occasions and $14 \sim 10$ for the number of galaxies in the local group. This is completely justified: Using the actual numbers would be pointless, because the other assumptions in the problem—the size of the observable universe and the idea that the local galaxy density is representative of the density everywhere—are also very rough approximations. Further, there was nothing in the problem that required using volumes of spheres rather than volumes of cubes. Despite all these arbitrary choices, the answer still gives useful information, because it rules out a lot of reasonable possible answers. Before doing the calculation, a guess of a billion galaxies might have seemed plausible.

**QUESTION 1.8** About one in ten galaxies in the local group are not dwarf galaxies. Estimate the number of galaxies in the Universe that are not dwarfs.

**EXERCISE 1.8** **(a)** Given that the nearest star is about 4 light-years away, develop an estimate of the density of stars per cubic light-year in our galaxy. **(b)** Estimate the number of stars in the Milky Way galaxy, given that it's roughly a disk 100 000 light-years across and a thousand light-years thick.

**ANSWER** **(a)** 0.02 stars/ly$^3$    **(b)** $2 \times 10^{11}$ stars (Estimates will vary. The actual answer is probably about twice that number.)

# 1.7 Coordinate Systems

Many aspects of physics deal with locations in space, which require the definition of a coordinate system. A point on a line can be located with one coordinate, a point in a plane with two coordinates, and a point in space with three.

A coordinate system used to specify locations in space consists of the following:

- A fixed reference point $O$, called the *origin*
- A set of specified axes, or directions, with an appropriate scale and labels on the axes
- Instructions on labeling a point in space relative to the origin and axes

One convenient and commonly used coordinate system is the **Cartesian coordinate system,** sometimes called the **rectangular coordinate system.** Such a system in two dimensions is illustrated in Figure 1.5. An arbitrary point in this system is labeled with the coordinates $(x, y)$. For example, the point $P$ in the figure has coordinates $(5, 3)$. If we start at the origin $O$, we can reach $P$ by moving 5 meters horizontally to the right and then 3 meters vertically upward. In the same way, the point $Q$ has coordinates $(-3, 4)$, which corresponds to going 3 meters horizontally to the left of the origin and 4 meters vertically upward from there.

Positive $x$ is usually selected as right of the origin and positive $y$ upward from the origin, but in two dimensions this choice is largely a matter of taste. (In three dimensions, however, there are "right-handed" and "left-handed" coordinates, which lead to minus sign differences in certain operations. These will be addressed as needed.)

Sometimes it's more convenient to locate a point in space by its **plane polar coordinates** $(r, \theta)$, as in Figure 1.6. In this coordinate system, an origin $O$ and a reference line are selected as shown. A point is then specified by the distance $r$ from the origin to the point and by the angle $\theta$ between the reference line and a line drawn from the origin to the point. The standard reference line is usually selected to be the positive $x$-axis of a Cartesian coordinate system. The angle $\theta$ is considered positive when measured counterclockwise from the reference line and negative when measured clockwise from the reference line. For example, if a point is specified by the polar coordinates 3 m and $60°$, we locate this point by moving out 3 m from the origin at an angle of $60°$ above (counterclockwise from) the reference line. A point specified by polar coordinates 3 m and $-60°$ is located 3 m out from the origin and $60°$ below (clockwise from) the reference line.

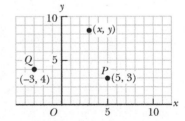

**Figure 1.5** Designation of points in a two-dimensional Cartesian coordinate system. Every point is labeled with coordinates $(x, y)$.

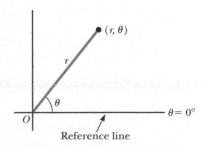

**Figure 1.6** The plane polar coordinates of a point are represented by the distance $r$ and the angle $\theta$, where $\theta$ is measured counterclockwise from the positive $x$-axis.

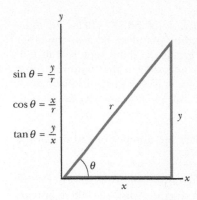

$\sin \theta = \dfrac{y}{r}$

$\cos \theta = \dfrac{x}{r}$

$\tan \theta = \dfrac{y}{x}$

**Figure 1.7** Certain trigonometric functions of a right triangle.

# 1.8 Trigonometry Review

Consider the right triangle shown in Figure 1.7, where side $y$ is opposite the angle $\theta$, side $x$ is adjacent to the angle $\theta$, and side $r$ is the hypotenuse of the triangle. The basic trigonometric functions defined by such a triangle are the ratios of the lengths of the sides of the triangle. These relationships are called the sine (sin), cosine (cos), and tangent (tan) functions. In terms of $\theta$, the basic trigonometric functions are as follows:[2]

$$\sin \theta = \frac{\text{side opposite } \theta}{\text{hypotenuse}} = \frac{y}{r}$$

$$\cos \theta = \frac{\text{side adjacent to } \theta}{\text{hypotenuse}} = \frac{x}{r} \qquad \text{[1.1]}$$

$$\tan \theta = \frac{\text{side opposite } \theta}{\text{side adjacent to } \theta} = \frac{y}{x}$$

For example, if the angle $\theta$ is equal to 30°, then the ratio of $y$ to $r$ is always 0.50; that is, $\sin 30° = 0.50$. Note that the sine, cosine, and tangent functions are quantities without units because each represents the ratio of two lengths.

Another important relationship, called the **Pythagorean theorem,** exists between the lengths of the sides of a right triangle:

$$r^2 = x^2 + y^2 \qquad \text{[1.2]}$$

Finally, it will often be necessary to find the values of inverse relationships. For example, suppose you know that the sine of an angle is 0.866, but you need to know the value of the angle itself. The inverse sine function may be expressed as $\sin^{-1}(0.866)$, which is a shorthand way of asking the question "What angle has a sine of 0.866?" Punching a couple of buttons on your calculator reveals that this angle is 60.0°. Try it for yourself and show that $\tan^{-1}(0.400) = 21.8°$. Be sure that your calculator is set for degrees and not radians. In addition, the inverse tangent function can return only values between $-90°$ and $+90°$, so when an angle is in the second or third quadrant, it's necessary to add 180° to the answer in the calculator window.

**Tip 1.3** Degrees vs. Radians

When calculating trigonometric functions, make sure your calculator setting—degrees or radians— is consistent with the angular measure you're using in a given problem.

The definitions of the trigonometric functions and the inverse trigonometric functions, as well as the Pythagorean theorem, can be applied to *any* right triangle, regardless of whether its sides correspond to $x$- and $y$-coordinates.

These results from trigonometry are useful in converting from rectangular coordinates to polar coordinates, or vice versa, as the next example shows.

---

**EXAMPLE 1.9** | **CARTESIAN AND POLAR COORDINATES**

**GOAL** Understand how to convert from plane rectangular coordinates to plane polar coordinates and vice versa.

**PROBLEM** (a) The Cartesian coordinates of a point in the $xy$-plane are $(x, y) = (-3.50 \text{ m}, -2.50 \text{ m})$, as shown in Figure 1.8. Find the polar coordinates of this point. (b) Convert $(r, \theta) = (5.00 \text{ m}, 37.0°)$ to rectangular coordinates.

**Figure 1.8** (Example 1.9) Converting from Cartesian coordinates to polar coordinates.

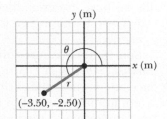

**STRATEGY** Apply the trigonometric functions and their inverses, together with the Pythagorean theorem.

---

[2]Many people use the mnemonic *SOHCAHTOA* to remember the basic trigonometric formulas: *Sine = Opposite/ Hypotenuse, Cosine = Adjacent/Hypotenuse,* and *Tangent = Opposite/Adjacent.* (Thanks go to Professor Don Chodrow for pointing this out.)

## SOLUTION

**(a)** Cartesian to polar conversion

Take the square root of both sides of Equation 1.2 to find the radial coordinate:

$$r = \sqrt{x^2 + y^2} = \sqrt{(-3.50\text{ m})^2 + (-2.50\text{ m})^2} = \boxed{4.30\text{ m}}$$

Use Equation 1.1 for the tangent function to find the angle with the inverse tangent, adding 180° because the angle is actually in the third quadrant:

$$\tan\theta = \frac{y}{x} = \frac{-2.50\text{ m}}{-3.50\text{ m}} = 0.714$$

$$\theta = \tan^{-1}(0.714) = 35.5° + 180° = \boxed{216°}$$

**(b)** Polar to Cartesian conversion

Use the trigonometric definitions, Equation 1.1.

$$x = r\cos\theta = (5.00\text{ m})\cos 37.0° = \boxed{3.99\text{ m}}$$

$$y = r\sin\theta = (5.00\text{ m})\sin 37.0° = \boxed{3.01\text{ m}}$$

**REMARKS** When we take up vectors in two dimensions in Topic 3, we will routinely use a similar process to find the direction and magnitude of a given vector from its components, or, conversely, to find the components from the vector's magnitude and direction.

**QUESTION 1.9** Starting with the answers to part **(b)**, work backwards to recover the given radius and angle. Why are there slight differences from the original quantities?

**EXERCISE 1.9** **(a)** Find the polar coordinates corresponding to $(x, y) = (-3.25\text{ m}, 1.50\text{ m})$. **(b)** Find the Cartesian coordinates corresponding to $(r, \theta) = (4.00\text{ m}, 53.0°)$.

**ANSWERS** **(a)** $(r, \theta) = (3.58\text{ m}, 155°)$   **(b)** $(x, y) = (2.41\text{ m}, 3.19\text{ m})$

---

## EXAMPLE 1.10    HOW HIGH IS THE BUILDING?

**GOAL** Apply basic results of trigonometry.

**PROBLEM** A person measures the height of a building by walking out a distance of 46.0 m from its base and shining a flashlight beam towards the top. When the beam is elevated at an angle of 39.0° with respect to the horizontal, as shown in Figure 1.9, the beam just strikes the top of the building. **(a)** If the flashlight is held at a height of 2.00 m, find the height of the building. **(b)** Calculate the length of the light beam.

**STRATEGY** Refer to the right triangle shown in the figure. We know the angle, 39.0°, and the length of the side adjacent to it. Because the height of the building is the side opposite the angle, we can use the tangent function. With the adjacent and opposite sides known, we can then find the hypotenuse with the Pythagorean theorem.

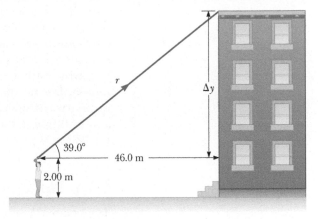

**Figure 1.9** (Example 1.10)

## SOLUTION

**(a)** Find the height of the building.

Use the tangent of the given angle:

$$\tan 39.0° = \frac{\Delta y}{46.0\text{ m}}$$

Solve for the height:

$$\Delta y = (\tan 39.0°)(46.0\text{ m}) = (0.810)(46.0\text{ m}) = 37.3\text{ m}$$

Add 2.00 m to $\Delta y$ to obtain the height:

height = $\boxed{39.3\text{ m}}$

**(b)** Calculate the length of the light beam.

Use the Pythagorean theorem:

$$r = \sqrt{x^2 + y^2} = \sqrt{(37.3\text{ m})^2 + (46.0\text{ m})^2} = \boxed{59.2\text{ m}}$$

(*Continued*)

**REMARKS** In the next section, right-triangle trigonometry is often used when working with vectors.

**QUESTION 1.10** Could the distance traveled by the light beam be found without using the Pythagorean theorem? How?

**EXERCISE 1.10** While standing atop a building 50.0 m tall, you spot a friend standing on a street corner. Using a protractor and dangling a plumb bob, you find that the angle between the horizontal and the direction to the spot on the sidewalk where your friend is standing is 25.0°. Your eyes are located 1.75 m above the top of the building. How far away from the foot of the building is your friend?

**ANSWER** 111 m

# 1.9 Vectors

Jon Feingersh/Stone/Getty Images

**Figure 1.10** A vector such as velocity has a magnitude, shown on the race car's speedometer, and a direction, straight out through the race car's front windshield. The mass of the car is a scalar quantity, as is the volume of gasoline in its fuel tank.

Physical quantities studied in this text fall into two main categories. One type, known as a *scalar quantity*, can be completely described by a single number (with appropriate units) giving its magnitude or size. Some common scalars are mass, temperature, volume, and speed. For example, a car's speed can be completely described by noting the number on its speedometer. A basketball's mass can be specified by measuring a single number with a scale.

The other type of quantity, known as a *vector quantity*, has both a magnitude and a direction. Velocity is a common vector quantity with a magnitude specifying how fast an object is moving and a direction specifying the direction of travel. For example, a car's velocity could be specified by noting it was traveling 60 miles per hour in the direction due north. A car traveling 60 miles per hour in an eastward direction would have a different velocity vector but the same scalar speed. Figure 1.10 illustrates each type of quantity.

In this book, symbols for scalar quantities are shown in italics (e.g., $m$ for mass and $T$ for temperature), and symbols for vector quantities are usually shown in bold with an arrow above the letter (e.g., $\vec{\mathbf{v}}$ for velocity). A vector's magnitude is a scalar quantity indicating its length and is shown in italics. For example, the scalar $v$ indicates the magnitude of vector $\vec{\mathbf{v}}$. In general, **a vector quantity is characterized by having both a magnitude and a direction.** By contrast, **a scalar quantity has magnitude, but no direction.** Scalar quantities can be manipulated with the rules of ordinary arithmetic. Vectors can also be added and subtracted from each other, and multiplied, but there are a number of important differences, as will be seen in the following sections.

## 1.9.1 Equality of Two Vectors

Two vectors $\vec{\mathbf{A}}$ and $\vec{\mathbf{B}}$ are equal if they have the same magnitude and the same direction. They need not be located at the same point in space. The four vectors in Figure 1.11 are all equal to each other. Moving a vector from one point in space to another doesn't change its magnitude or its direction.

## 1.9.2 Adding Vectors

**Tip 1.4** Vector Addition vs. Scalar Addition

$\vec{\mathbf{A}} + \vec{\mathbf{B}} = \vec{\mathbf{C}}$ differs significantly from $A + B = C$. The first is a vector sum, which must be handled graphically or with components, whereas the second is a simple arithmetic sum of numbers.

When two or more vectors are added, they must all have the same units. For example, it doesn't make sense to add a velocity vector, carrying units of meters per second, to a displacement vector, carrying units of meters. Scalars obey the same rule: It would be similarly meaningless to add temperatures to volumes or masses to time intervals.

Vectors can be added geometrically or algebraically. (The latter is discussed at the end of the next section.) To add vector $\vec{\mathbf{B}}$ to vector $\vec{\mathbf{A}}$ geometrically, first draw $\vec{\mathbf{A}}$ on a piece of graph paper to some scale, such as 1 cm = 1 m, so that its

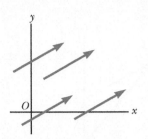

**Figure 1.11** These four vectors are equal because they have equal lengths and point in the same direction.

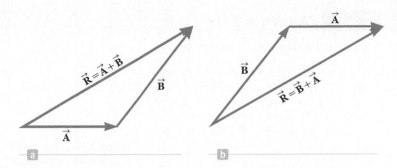

**Figure 1.12** (a) When vector $\vec{B}$ is added to vector $\vec{A}$, the vector sum $\vec{R}$ is the vector that runs from the tail of $\vec{A}$ to the tip of $\vec{B}$. (b) Here the resultant runs from the tail of $\vec{B}$ to the tip of $\vec{A}$. These constructions prove that $\vec{A} + \vec{B} = \vec{B} + \vec{A}$.

direction is specified relative to a coordinate system. Then draw vector $\vec{B}$ to the same scale with the tail of $\vec{B}$ starting at the tip of $\vec{A}$, as in Figure 1.12a. Vector $\vec{B}$ must be drawn along the direction that makes the proper angle relative to vector $\vec{A}$. The **resultant vector** $\vec{R} = \vec{A} + \vec{B}$ is the vector drawn from the tail of $\vec{A}$ to the tip of $\vec{B}$. This procedure is known as the **triangle method of addition.**

When two vectors are added, their sum is independent of the order of the addition: $\vec{A} + \vec{B} = \vec{B} + \vec{A}$. This relationship can be seen from the geometric construction in Figure 1.12b, and is called the **commutative law of addition.**

This same general approach can also be used to add more than two vectors, as is done in Figure 1.13 for four vectors. The resultant vector sum $\vec{R} = \vec{A} + \vec{B} + \vec{C} + \vec{D}$ is the vector drawn from the tail of the first vector to the tip of the last. Again, the order in which the vectors are added is unimportant.

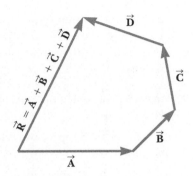

**Figure 1.13** A geometric construction for summing four vectors. The resultant vector $\vec{R}$ is the vector that completes the polygon.

### 1.9.3 Negative of a Vector

The negative of the vector $\vec{A}$ is defined as the vector that gives zero when added to $\vec{A}$. This means that $\vec{A}$ and $-\vec{A}$ have the same magnitude but opposite directions.

### 1.9.4 Subtracting Vectors

Vector subtraction makes use of the definition of the negative of a vector. We define the operation $\vec{A} - \vec{B}$ as the vector $-\vec{B}$ added to the vector $\vec{A}$:

$$\vec{A} - \vec{B} = \vec{A} + (-\vec{B}) \qquad [1.3]$$

Vector subtraction is really a special case of vector addition. The geometric construction for subtracting two vectors is shown in Figure 1.14.

### 1.9.5 Multiplying or Dividing a Vector by a Scalar

Multiplying or dividing a vector by a scalar gives a vector. For example, if vector $\vec{A}$ is multiplied by the scalar number 3, the result, written $3\vec{A}$, is a vector with a magnitude three times that of $\vec{A}$ and pointing in the same direction. If we multiply vector $\vec{A}$ by the scalar $-3$, the result is $-3\vec{A}$, a vector with a magnitude three times that of $\vec{A}$ and pointing in the opposite direction (because of the negative sign).

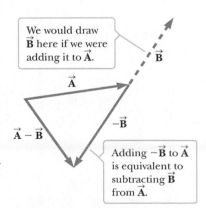

We would draw $\vec{B}$ here if we were adding it to $\vec{A}$.

Adding $-\vec{B}$ to $\vec{A}$ is equivalent to subtracting $\vec{B}$ from $\vec{A}$.

**Figure 1.14** This construction shows how to subtract vector $\vec{B}$ from vector $\vec{A}$. The vector $-\vec{B}$ has the same magnitude as the vector $\vec{B}$ but points in the opposite direction.

*Quick Quiz*

**1.1** The magnitudes of two vectors $\vec{A}$ and $\vec{B}$ are 12 units and 8 units, respectively. What are the largest and smallest possible values for the magnitude of the resultant vector $\vec{R} = \vec{A} + \vec{B}$ ? (a) 14.4 and 4 (b) 12 and 8 (c) 20 and 4 (d) none of these.

**EXAMPLE 1.11**  **TAKING A TRIP**

**GOAL**  Find the sum of two vectors by using a graph.

**PROBLEM**  A car travels 20.0 km due north and then 35.0 km in a direction 60.0° west of north, as in Figure 1.15. Using a graph, find the magnitude and direction of a single vector that gives the net effect of the car's trip. This vector is called the car's *resultant displacement*.

**STRATEGY**  Draw a graph and represent the displacement vectors as arrows. Graphically locate the vector resulting from the sum of the two displacement vectors. Measure its length and angle with respect to the vertical.

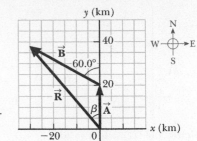

**Figure 1.15**  (Example 1.11) A graphical method for finding the resultant displacement vector $\vec{R} = \vec{A} + \vec{B}$.

**SOLUTION**

Let $\vec{A}$ represent the first displacement vector, 20.0 km north, and $\vec{B}$ the second displacement vector, extending west of north. Carefully graph the two vectors, drawing a resultant vector $\vec{R}$ with its base touching the base of $\vec{A}$ and extending to the tip of $\vec{B}$. Measure the length of this vector, which turns out to be about 48 km. The angle $\beta$, measured with a protractor, is about 39° west of north.

**REMARKS**  Notice that ordinary arithmetic doesn't work here: the correct answer of 48 km is not equal to 20.0 km + 35.0 km = 55.0 km!

**QUESTION 1.11**  Suppose two vectors are added. Under what conditions would the sum of the magnitudes of the vectors equal the magnitude of the resultant vector?

**EXERCISE 1.11**  Graphically determine the magnitude and direction of the displacement if a man walks 30.0 km 45° north of east and then walks due east 20.0 km.

**ANSWER**  46 km, 27° north of east

---

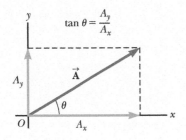

**Figure 1.16** Any vector $\vec{A}$ lying in the xy-plane can be represented by its rectangular components $A_x$ and $A_y$.

**Tip 1.5** x- and y-components

Equation 1.4 for the x- and y-components of a vector associates cosine with the x-component and sine with the y-component, as in Figure 1.17a. This association is due *solely* to the fact that we chose to measure the angle $\theta$ with respect to the positive x-axis. If the angle were measured with respect to the y-axis, as in Figure 1.17b, the components would be given by $A_x = A \sin \theta$ and $A_y = A \cos \theta$.

# 1.10 Components of a Vector

One method of adding vectors makes use of the projections of a vector along the axes of a rectangular coordinate system. These projections are called **components.** Any vector can be completely described by its components.

Consider a vector $\vec{A}$ in a rectangular coordinate system, as shown in Figure 1.16. $\vec{A}$ can be expressed as the sum of two vectors: $\vec{A}_x$, parallel to the x-axis, and $\vec{A}_y$, parallel to the y-axis. Mathematically,

$$\vec{A} = \vec{A}_x + \vec{A}_y$$

where $\vec{A}_x$ and $\vec{A}_y$ are the component vectors of $\vec{A}$. The projection of $\vec{A}$ along the x-axis, $A_x$, is called the x-component of $\vec{A}$, and the projection of $\vec{A}$ along the y-axis, $A_y$, is called the y-component of $\vec{A}$. These components can be either positive or negative numbers with units. From the definitions of sine and cosine, we see that $\cos \theta = A_x/A$ and $\sin \theta = A_y/A$, so the components of $\vec{A}$ are

$$A_x = A \cos \theta \qquad [1.4a]$$

$$A_y = A \sin \theta \qquad [1.4b]$$

These components form two sides of a right triangle having a hypotenuse with magnitude A. It follows that $\vec{A}$'s magnitude and direction are related to its components through the Pythagorean theorem and the definition of the tangent:

$$A = \sqrt{A_x^2 + A_y^2} \qquad [1.5]$$

$$\tan \theta = \frac{A_y}{A_x} \qquad [1.6]$$

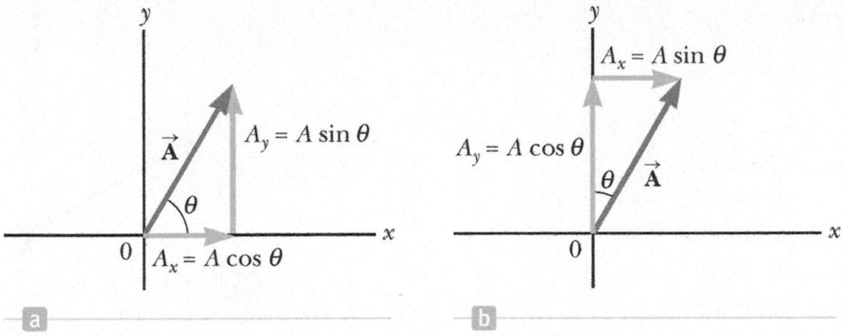

**Figure 1.17** The angle $\theta$ need not always be defined from the positive $x$-axis.

a

$A_y = A \sin \theta$
$\vec{A}$
$\theta$
$A_x = A \cos \theta$

b

$A_x = A \sin \theta$
$A_y = A \cos \theta$
$\theta$
$\vec{A}$

To solve for the angle $\theta$, which is measured counterclockwise from the positive $x$-axis by convention, the inverse tangent can be taken of both sides of Equation 1.6:

$$\theta = \tan^{-1}\left(\frac{A_y}{A_x}\right)$$

This formula gives the right answer for $\theta$ only half the time! The inverse tangent function returns values only from $-90°$ to $+90°$, so the answer in your calculator window will only be correct if the vector happens to lie in the first or fourth quadrant. If it lies in the second or third quadrant, adding $180°$ to the number in the calculator window will always give the right answer. The angle in Equations 1.4 and 1.6 must be measured from the positive $x$-axis. Other choices of reference line are possible, but certain adjustments must then be made. (See Tip 1.5 and Fig. 1.17.)

If a coordinate system other than the one shown in Figure 1.16 is chosen, the components of the vector must be modified accordingly. In many applications it's more convenient to express the components of a vector in a coordinate system having axes that are not horizontal and vertical but are still perpendicular to each other. Suppose a vector $\vec{B}$ makes an angle $\theta'$ with the $x'$-axis defined in Figure 1.18. The rectangular components of $\vec{B}$ along the axes of the figure are given by $B_{x'} = B \cos \theta'$ and $B_{y'} = B \sin \theta'$, as in Equations 1.4. The magnitude and direction of $\vec{B}$ are then obtained from expressions equivalent to Equations 1.5 and 1.6.

**Tip 1.6** Inverse Tangents on Calculators: Right Half the Time

The inverse tangent function on calculators returns an angle between $-90°$ and $+90°$. If the vector lies in the second or third quadrant, the angle, as measured from the positive $x$-axis, will be the angle returned by your calculator plus $180°$.

### Quick Quiz

**1.2** Figure 1.19 shows two vectors lying in the $xy$-plane. Determine the signs of the $x$- and $y$-components of $\vec{A}$, $\vec{B}$, and $\vec{A} + \vec{B}$.

**1.3** Which vector has an angle with respect to the positive $x$-axis that is in the range of the inverse tangent function?

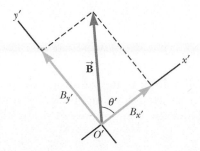

**Figure 1.18** The components of vector $\vec{B}$ in a tilted coordinate system.

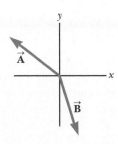

**Figure 1.19** (Quick Quizzes 1.2 and 1.3)

## EXAMPLE 1.12   HELP IS ON THE WAY!

**GOAL** Find vector components, given a magnitude and direction, and vice versa.

**PROBLEM** (a) Find the horizontal and vertical components of the $d = 1.00 \times 10^2$ m displacement of a superhero who flies from the top of a tall building along the path shown in Figure 1.20a. (b) Suppose instead the superhero leaps in the other direction along a displacement vector $\vec{B}$ to the top of a flagpole where the displacement components are given by $B_x = -25.0$ m and $B_y = 10.0$ m. Find the magnitude and direction of the displacement vector.

**STRATEGY** (a) The triangle formed by the displacement and its components is shown in Figure 1.20b. Simple trigonometry gives the components relative to the standard $xy$-coordinate system: $A_x = A \cos \theta$ and $A_y = A \sin \theta$ (Eqs. 1.4). Note that $\theta = -30.0°$, negative because it's measured clockwise from the positive $x$-axis. (b) Apply Equations 1.5 and 1.6 to find the magnitude and direction of the vector.

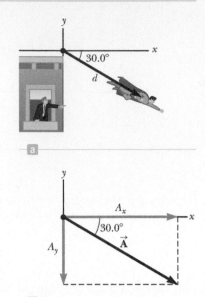

**Figure 1.20** (Example 1.12)

........................................................................................................

**SOLUTION**

**(a)** Find the vector components of $\vec{A}$ from its magnitude and direction.

Use Equations 1.4 to find the components of the displacement vector $\vec{A}$:

$$A_x = A \cos \theta = (1.00 \times 10^2 \text{ m}) \cos (-30.0°) = \boxed{+86.6 \text{ m}}$$

$$A_y = A \sin \theta = (1.00 \times 10^2 \text{ m}) \sin (-30.0°) = \boxed{-50.0 \text{ m}}$$

**(b)** Find the magnitude and direction of the displacement vector $\vec{B}$ from its components.

Compute the magnitude of $\vec{B}$ from the Pythagorean theorem:

$$B = \sqrt{B_x^2 + B_y^2} = \sqrt{(-25.0 \text{ m})^2 + (10.0 \text{ m})^2} = \boxed{26.9 \text{ m}}$$

Calculate the direction of $\vec{B}$ using the inverse tangent, remembering to add 180° to the answer in your calculator window, because the vector lies in the second quadrant:

$$\theta = \tan^{-1}\left(\frac{B_y}{B_x}\right) = \tan^{-1}\left(\frac{10.0}{-25.0}\right) = -21.8°$$

$$\theta = \boxed{158°}$$

........................................................................................................

**REMARKS** In part (a), note that $\cos (-\theta) = \cos \theta$; however, $\sin (-\theta) = -\sin \theta$. The negative sign of $A_y$ reflects the fact that displacement in the $y$-direction is *downward*.

**QUESTION 1.12** What other functions, if any, can be used to find the angle in part (b)?

**EXERCISE 1.12** (a) Suppose the superhero had flown 150 m at a 120° angle with respect to the positive $x$-axis. Find the components of the displacement vector. (b) Suppose instead the superhero had leaped with a displacement having an $x$-component of 32.5 m and a $y$-component of 24.3 m. Find the magnitude and direction of the displacement vector.

**ANSWERS** (a) $A_x = -75$ m, $A_y = 130$ m   (b) 40.6 m, 36.8°

---

### 1.10.1 Adding Vectors Algebraically

The graphical method of adding vectors is valuable in understanding how vectors can be manipulated, but most of the time vectors are added algebraically in terms of their components. Suppose $\vec{R} = \vec{A} + \vec{B}$. Then the components of the resultant vector $\vec{R}$ are given by

$$R_x = A_x + B_x \qquad \text{[1.7a]}$$
$$R_y = A_y + B_y \qquad \text{[1.7b]}$$

So $x$-components are added only to $x$-components, and $y$-components only to $y$-components. The magnitude and direction of $\vec{R}$ can subsequently be found with Equations 1.5 and 1.6.

Subtracting two vectors works the same way because it's a matter of adding the negative of one vector to another vector. You should make a rough sketch when adding or subtracting vectors, in order to get an approximate geometric solution as a check.

## EXAMPLE 1.13    TAKE A HIKE

**GOAL** Add vectors algebraically and find the resultant vector.

**PROBLEM** A hiker begins a trip by first walking 25.0 km 45.0° south of east from her base camp. On the second day she walks 40.0 km in a direction 60.0° north of east, at which point she discovers a forest ranger's tower. **(a)** Determine the components of the hiker's displacements in the first and second days. **(b)** Determine the components of the hiker's total displacement for the trip. **(c)** Find the magnitude and direction of the displacement from base camp.

**STRATEGY** This problem is just an application of vector addition using components, Equations 1.7. We denote the displacement vectors on the first and second days by $\vec{A}$ and $\vec{B}$, respectively. Using the camp as the origin of the coordinates, we get the vectors shown in Figure 1.21a. After finding $x$- and $y$-components for each vector, we add them "componentwise." Finally, we determine the magnitude and direction of the resultant vector $\vec{R}$, using the Pythagorean theorem and the inverse tangent function.

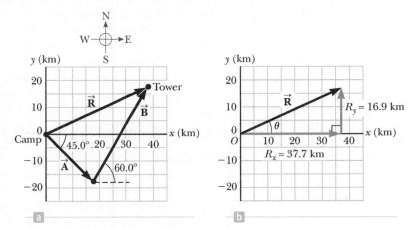

**Figure 1.21** (Example 1.13) (a) Hiker's path and the resultant vector. (b) Components of the hiker's total displacement from camp.

## SOLUTION

**(a)** Find the components of $\vec{A}$.

Use Equations 1.4 to find the components of $\vec{A}$:

$$A_x = A\cos(-45.0°) = (25.0 \text{ km})(0.707) = \boxed{17.7 \text{ km}}$$

$$A_y = A\sin(-45.0°) = -(25.0 \text{ km})(0.707) = \boxed{-17.7 \text{ km}}$$

Find the components of $\vec{B}$:

$$B_x = B\cos 60.0° = (40.0 \text{ km})(0.500) = \boxed{20.0 \text{ km}}$$

$$B_y = B\sin 60.0° = (40.0 \text{ km})(0.866) = \boxed{34.6 \text{ km}}$$

**(b)** Find the components of the resultant vector, $\vec{R} = \vec{A} + \vec{B}$.

To find $R_x$, add the $x$-components of $\vec{A}$ and $\vec{B}$:

$$R_x = A_x + B_x = 17.7 \text{ km} + 20.0 \text{ km} = \boxed{37.7 \text{ km}}$$

To find $R_y$, add the $y$-components of $\vec{A}$ and $\vec{B}$:

$$R_y = A_y + B_y = -17.7 \text{ km} + 34.6 \text{ km} = \boxed{16.9 \text{ km}}$$

**(c)** Find the magnitude and direction of $\vec{R}$.

Use the Pythagorean theorem to get the magnitude:

$$R = \sqrt{R_x^2 + R_y^2} = \sqrt{(37.7 \text{ km})^2 + (16.9 \text{ km})^2} = \boxed{41.3 \text{ km}}$$

Calculate the direction of $\vec{R}$ using the inverse tangent function:

$$\theta = \tan^{-1}\left(\frac{16.9 \text{ km}}{37.7 \text{ km}}\right) = \boxed{24.1°}$$

**REMARKS** Figure 1.21b shows a sketch of the components of $\vec{R}$ and their directions in space. The magnitude and direction of the resultant can also be determined from such a sketch.

**QUESTION 1.13**  A second hiker follows the same path the first day, but then walks 15.0 km east on the second day before turning and reaching the ranger's tower. Is the second hiker's resultant displacement vector the same as the first hiker's, or different?

**EXERCISE 1.13**  A cruise ship leaving port travels 50.0 km 45.0° north of west and then 70.0 km at a heading 30.0° north of east. Find **(a)** the components of the ship's displacement vector and **(b)** the displacement vector's magnitude and direction.

**ANSWERS**   **(a)** $R_x = 25.3$ km, $R_y = 70.4$ km    **(b)** 74.8 km, 70.2° north of east

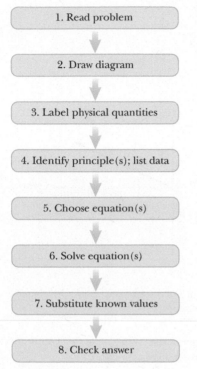

**Figure 1.22** A guide to problem solving.

# 1.11 Problem-Solving Strategy

Most courses in general physics require the student to learn the skills used in solving problems, and examinations usually include problems that test such skills. This brief section presents some useful suggestions to help increase your success in solving problems. An organized approach to problem solving will also enhance your understanding of physical concepts and reduce exam stress. Throughout the book, there will be a number of sections labeled "Problem-Solving Strategy," many of them just a specializing of the list given below (and illustrated in Fig. 1.22).

## 1.11.1 General Problem-Solving Strategy

### Problem
1. **Read** the problem carefully at least twice. Be sure you understand the nature of the problem before proceeding further.
2. **Draw** a diagram while rereading the problem.
3. **Label** all physical quantities in the diagram, using letters that remind you what the quantity is (e.g., $m$ for mass). Choose a coordinate system and label it.

### Strategy
4. **Identify** physical principles, the knowns and unknowns, and list them. Put circles around the unknowns. There must be as many equations as there are unknowns.
5. **Equations**, the relationships between the labeled physical quantities, should be written down next. Naturally, the selected equations should be consistent with the physical principles identified in the previous step.

### Solution
6. **Solve** the set of equations for the unknown quantities in terms of the known. Do this algebraically, without substituting values until the next step, except where terms are zero.
7. **Substitute** the known values, together with their units. Obtain a numerical value with units for each unknown.

### Check Answer
8. **Check** your answer. Do the units match? Is the answer reasonable? Does the plus or minus sign make sense? Is your answer consistent with an order of magnitude estimate?

This same procedure, with minor variations, should be followed throughout the course. The first three steps are extremely important, because they get you mentally oriented. Identifying the proper concepts and physical principles assists you in choosing the correct equations. The equations themselves are essential, because when you understand them, you also understand the relationships between the physical quantities. This understanding comes through a lot of daily practice.

Equations are the tools of physics: To solve problems, you have to have them at hand, like a plumber and his wrenches. Know the equations, and understand what they mean and how to use them. Just as you can't have a conversation without knowing the local language, you can't solve physics problems without knowing and understanding the equations. This understanding grows as you study and apply the concepts and the equations relating them.

Carrying through the algebra for as long as possible (substituting numbers only at the end) is also important, because it helps you think in terms of the physical quantities involved, not merely the numbers that represent them. Many beginning physics students are eager to substitute, but once numbers are substituted it's harder to understand relationships and easier to make mistakes.

The physical layout and organization of your work will make the final product more understandable and easier to follow. Although physics is a challenging discipline, your chances of success are excellent if you maintain a positive attitude and keep trying.

> **Tip 1.7** Get Used to Symbolic Algebra
>
> Whenever possible, solve problems symbolically and then substitute known values. This process helps prevent errors and clarifies the relationships between physical quantities.

---

**EXAMPLE 1.14** | A SHOPPING TRIP

**GOAL** Illustrate the Problem-Solving Strategy.

**PROBLEM** A shopper leaves home and drives to a store located 7.00 km away in a direction 30.0° north of east. Leaving the store, the shopper drives 5.00 km in a direction 50.0° west of north to a restaurant. Find the distance and direction from the shopper's home to the restaurant.

**STRATEGY** We've finished reading the problem (step 1), and have drawn a diagram (step 2) in Figure 1.23 and labeled it (step 3). From the diagram, we recognize that the vector $\vec{R}$ locating the restaurant's position is the sum of vectors $\vec{A}$ and $\vec{B}$ and identify (step 4) the principles involved: the magnitude and direction of a vector. Vectors $\vec{A}$ and $\vec{B}$ are known and vector $\vec{R}$ is unknown.

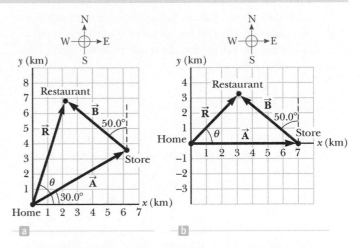

**Figure 1.23** (a) (Example 1.14) (b) (Exercise 1.14)

................................................................

**SOLUTION**

Express $\vec{R}$ as a vector sum (step 5):

$$\vec{R} = \vec{A} + \vec{B}$$

Solve symbolically for the components of $\vec{R}$ (step 6):

$$R_x = A_x + B_x \text{ and } R_y = A_y + B_y$$

Substitute the numbers, with units (step 7). The direction of $\vec{B}$ measured counterclockwise from the positive $x$-axis is $50.0° + 90.0° = 140.0°$:

$$R_x = (7.00 \text{ km}) \cos(30.0°) + (5.00 \text{ km}) \cos(140.0°) = 2.23 \text{ km}$$

$$R_y = (7.00 \text{ km}) \sin(30.0°) + (5.00 \text{ km}) \sin(140.0°) = 6.71 \text{ km}$$

Solve for the distance using the Pythagorean theorem:

$$R = \sqrt{R_x^2 + R_y^2} = \sqrt{(2.23 \text{ km})^2 + (6.71 \text{ km})^2} = \boxed{7.07 \text{ km}}$$

Solve for the direction:

$$\theta = \tan^{-1}\left(\frac{R_y}{R_x}\right) = \tan^{-1}\left(\frac{6.71 \text{ km}}{2.23 \text{ km}}\right) = \boxed{71.6°}$$

................................................................

**REMARKS** In checking (step 8), note that the units match and the answer seems reasonable. A sketch drawn at least approximately to scale can help determine whether or not the distance and angle are reasonable.

**QUESTION 1.14** What are the answers if both distances in the problem statement are doubled but the directions are not changed?

**EXERCISE 1.14** Find the distance and direction from the shopper's home to the restaurant if the store is located 7.00 km due east from home and the restaurant remains 5.00 km from the store in a direction 50.0° west of north.

**ANSWER** $R = 4.51$ km, $\theta = 45.4°$

---

## 1.1 Standards of Length, Mass, and Time

The physical quantities in the study of mechanics can be expressed in terms of three fundamental quantities: length, mass, and time, which have the SI units meters (m), kilograms (kg), and seconds (s), respectively.

## 1.2 The Building Blocks of Matter

Matter is made of atoms, which in turn are made up of a relatively small nucleus of protons and neutrons within a cloud of electrons. Protons and neutrons are composed of still smaller particles, called quarks.

## 1.3 Dimensional Analysis

Dimensional analysis can be used to check equations and to assist in deriving them. When the dimensions on both sides of the equation agree, the equation is often correct up to a numerical factor. When the dimensions don't agree, the equation must be wrong.

## 1.4 Uncertainty in Measurement and Significant Figures

No physical quantity can be determined with complete accuracy. The concept of significant figures affords a basic method of handling these uncertainties. A significant figure is a reliably known digit, other than a zero used to locate the decimal point. The two rules of significant figures are as follows:

1. When multiplying or dividing using two or more quantities, the result should have the same number of significant figures as the quantity having the fewest significant figures.

2. When quantities are added or subtracted, the number of decimal places in the result should be the same as in the quantity with the fewest decimal places.

Use of scientific notation can avoid ambiguity in significant figures. In rounding, if the last digit dropped is less than 5, simply drop the digit; otherwise, raise the last retained digit by one.

## 1.5 Unit Conversions for Physical Quantities

Units in physics equations must always be consistent. In solving a physics problem, it's best to start with consistent units, using the table of conversion factors on the front endsheets as necessary.

Converting units is a matter of multiplying the given quantity by a fraction, with one unit in the numerator and its equivalent in the other units in the denominator, arranged so the unwanted units in the given quantity are canceled out in favor of the desired units.

## 1.6 Estimates and Order-of-Magnitude Calculations

Sometimes it's useful to find an approximate answer to a question, either because the math is difficult or because information is incomplete. A quick estimate can also be used to check a more detailed calculation. In an order-of-magnitude calculation, each value is replaced by the closest power of ten, which sometimes must be guessed or estimated when the value is unknown. The computation is then carried out. For quick estimates involving known values, each value can first be rounded to one significant figure.

## 1.7 Coordinate Systems

The Cartesian coordinate system consists of two perpendicular axes, usually called the x-axis and y-axis, with each axis labeled with all numbers from negative infinity to positive infinity. Points are located by specifying the x- and y-values. Polar coordinates consist of a radial coordinate r, which is the distance from the origin, and an angular coordinate θ, which is the angular displacement from the positive x-axis (see Fig. 1.24).

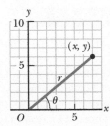

**Figure 1.24** A point in the plane can be described with Cartesian coordinates $(x, y)$ or with polar coordinates $(r, \theta)$.

## 1.8 Trigonometry Review

The three most basic trigonometric functions of a right triangle are the sine, cosine, and tangent, defined as follows:

$$\sin \theta = \frac{\text{side opposite } \theta}{\text{hypotenuse}} = \frac{y}{r}$$

$$\cos \theta = \frac{\text{side adjacent to } \theta}{\text{hypotenuse}} = \frac{x}{r} \qquad [1.1]$$

$$\tan \theta = \frac{\text{side opposite } \theta}{\text{side adjacent to } \theta} = \frac{y}{x}$$

The **Pythagorean theorem** is an important relationship between the lengths of the sides of a right triangle:

$$r^2 = x^2 + y^2 \qquad [1.2]$$

where $r$ is the hypotenuse of the triangle and $x$ and $y$ are the other two sides (see Fig. 1.25).

## 1.9 Vectors

A scalar quantity has a magnitude but no direction; a vector quantity has both magnitude and direction. Two vectors $\vec{A}$ and $\vec{B}$ can be added geometrically with the **triangle method.** The two vectors are drawn to scale on graph paper, with the tail of the second

| Figure 1.25 |

vector located at the tip of the first. The **resultant** vector is the vector drawn from the tail of the first vector to the tip of the second.

The negative of a vector $\vec{A}$ is a vector with the same magnitude as $\vec{A}$, but pointing in the opposite direction. A vector can be multiplied by a scalar, changing its magnitude, and its direction if the scalar is negative.

## 1.10 Components of a Vector

A vector $\vec{A}$ can be split into two components, one pointing in the $x$-direction and the other in the $y$-direction (see Fig. 1.26). These components form two sides of a right triangle having a hypotenuse with magnitude $A$ and are given by

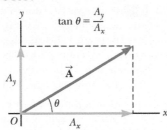

**Figure 1.26** A vector can be written in terms of components in the $x$- and $y$-directions.

$$A_x = A \cos \theta \qquad [1.4a]$$
$$A_y = A \sin \theta \qquad [1.4b]$$

The magnitude and direction of $\vec{A}$ are related to its components through the Pythagorean theorem and the definition of the tangent:

$$A = \sqrt{A_x^2 + A_y^2} \qquad [1.5]$$
$$\tan \theta = \frac{A_y}{A_x} \qquad [1.6]$$

In Equation 1.6, $\theta = \tan^{-1}(A_y/A_x)$ gives the correct vector angle only for vectors with $-90° < \theta < 90°$. If the vector has a negative $x$-component, $180°$ must be added to the answer in the calculator window.

If $\vec{R} = \vec{A} + \vec{B}$, then the components of the resultant vector $\vec{R}$ are

$$R_x = A_x + B_x \qquad [1.7a]$$
$$R_y = A_y + B_y \qquad [1.7b]$$

## CONCEPTUAL QUESTIONS

1. Estimate the order of magnitude of the length, in meters, of each of the following: (a) a mouse, (b) a pool cue, (c) a basketball court, (d) an elephant, (e) a city block.

2. What types of natural phenomena could serve as time standards?

3. Find the order of magnitude of your age in seconds.

4. An object with a mass of 1 kg weighs approximately 2 lb. Use this information to estimate the mass of the following objects: (a) a baseball, (b) your physics textbook, (c) a pickup truck.

5. BIO (a) Estimate the number of times your heart beats in a month. (b) Estimate the number of human heartbeats in an average lifetime.

6. Estimate the number of atoms in 1 cm³ of a solid. (Note that the diameter of an atom is about $10^{-10}$ m.)

7. BIO Lacking modern timepieces, early experimenters sometimes measured time intervals with their pulse. Why was this a poor method of measuring time?

8. For an angle $\theta$ measured from the positive $x$-axis, the values of $\sin \theta$ and $\cos \theta$ are always (choose one): (a) greater than $+1$ (b) less than $-1$ (c) greater than $-1$ and less than 1 (d) greater than or equal to $-1$ and less than or equal to 1 (e) less than or equal to $-1$ or greater than or equal to 1.

9. The left side of an equation has dimensions of length and the right side has dimensions of length squared. Can the equation be correct (choose one)? (a) Yes, because both sides involve the dimension of length. (b) No, because the equation is dimensionally inconsistent.

10. List some advantages of the metric system of units over most other systems of units.

11. BIO Estimate the time duration of each of the following in the suggested units in parentheses: (a) a heartbeat (seconds), (b) a football game (hours), (c) a summer (months), (d) a movie (hours), (e) the blink of an eye (seconds).

12. Suppose two quantities, $A$ and $B$, have different dimensions. Determine which of the following arithmetic operations *could* be physically meaningful. (a) $A + B$ (b) $B - A$ (c) $A - B$ (d) $A/B$ (e) $AB$

13. Answer each question yes or no. Must two quantities have the same dimensions (a) if you are adding them? (b) If you are multiplying them? (c) If you are subtracting them? (d) If you are dividing them? (e) If you are equating them?

14. Two different measuring devices are used by students to measure the length of a metal rod. Students using the first device report its length as 0.5 m, while those using the second report 0.502 m. Can both answers be correct (choose one)? (a) Yes, because their values are the same when both are rounded to the same number of significant figures. (b) No, because they report different values.

15. If $\vec{B}$ is added to $\vec{A}$, under what conditions does the resultant vector have a magnitude equal to $A + B$? Under what conditions is the resultant vector equal to zero?

16. Under what circumstances would a vector have components that are equal in magnitude?

## PROBLEMS

*See the Preface for an explanation of the icons used in this problems set.*

### 1.3 Dimensional Analysis

1. The period of a simple pendulum, defined as the time necessary for one complete oscillation, is measured in time units and is given by

$$T = 2\pi\sqrt{\frac{\ell}{g}}$$

where $\ell$ is the length of the pendulum and $g$ is the acceleration due to gravity, in units of length divided by time squared. Show that this equation is dimensionally consistent. (You might want to check the formula using your keys at the end of a string and a stopwatch.)

2. (a) Suppose the displacement of an object is related to time according to the expression $x = Bt^2$. What are the dimensions of $B$? (b) A displacement is related to time as $x = A \sin (2\pi ft)$, where $A$ and $f$ are constants. Find the dimensions of $A$. *Hint:* A trigonometric function appearing in an equation must be dimensionless.

3. **S** A shape that covers an area $A$ and has a uniform height $h$ has a volume $V = Ah$. (a) Show that $V = Ah$ is dimensionally correct. (b) Show that the volumes of a cylinder and of a rectangular box can be written in the form $V = Ah$, identifying $A$ in each case. (Note that $A$, sometimes called the "footprint" of the object, can have any shape and that the height can, in general, be replaced by the average thickness of the object.)

4. **V** Each of the following equations was given by a student during an examination: (a) $\frac{1}{2}mv^2 = \frac{1}{2}mv_0^2 + \sqrt{mgh}$ (b) $v = v_0 + at^2$ (c) $ma = v^2$. Do a dimensional analysis of each equation and explain why the equation can't be correct.

5. Newton's law of universal gravitation is represented by

$$F = G\frac{Mm}{r^2}$$

where $F$ is the gravitational force, $M$ and $m$ are masses, and $r$ is a length. Force has the SI units kg · m/s$^2$. What are the SI units of the proportionality constant $G$?

6. **Q|C** Kinetic energy $KE$ (Topic 5) has dimensions kg · m$^2$/s$^2$. It can be written in terms of the momentum $p$ (Topic 6) and mass $m$ as

$$KE = \frac{p^2}{2m}$$

(a) Determine the proper units for momentum using dimensional analysis. (b) Refer to Problem 5. Given the units of force, write a simple equation relating a constant force $F$ exerted on an object, an interval of time $t$ during which the force is applied, and the resulting momentum of the object, $p$.

### 1.4 Uncertainty in Measurement and Significant Figures

7. A rectangular airstrip measures 32.30 m by 210 m, with the width measured more accurately than the length. Find the area, taking into account significant figures.

8. Use the rules for significant figures to find the answer to the addition problem 21.4 + 15 + 17.17 + 4.003.

9. **V** A carpet is to be installed in a room of length 9.72 m and width 5.3 m. Find the area of the room retaining the proper number of significant figures.

10. **Q|C** Use your calculator to determine $(\sqrt{8})^3$ to three significant figures in two ways: (a) Find $\sqrt{8}$ to four significant figures; then cube this number and round to three significant figures. (b) Find $\sqrt{8}$ to three significant figures; then cube this number and round to three significant figures. (c) Which answer is more accurate? Explain.

11. How many significant figures are there in (a) 78.9 ± 0.2, (b) $3.788 \times 10^9$, (c) $2.46 \times 10^{26}$, (d) 0.003 2

12. The speed of light is now defined to be $2.997\,924\,58 \times 10^8$ m/s. Express the speed of light to (a) three significant figures, (b) five significant figures, and (c) seven significant figures.

13. A rectangle has a length of (2.0 ± 0.2) m and a width of (1.5 ± 0.1) m. Calculate (a) the area and (b) the perimeter of the rectangle, and give the uncertainty in each value.

14. The radius of a circle is measured to be (10.5 ± 0.2) m. Calculate (a) the area and (b) the circumference of the circle, and give the uncertainty in each value.

15. The edges of a shoebox are measured to be 11.4 cm, 17.8 cm, and 29 cm. Determine the volume of the box retaining the proper number of significant figures in your answer.

16. Carry out the following arithmetic operations: (a) the sum of the measured values 756, 37.2, 0.83, and 2.5; (b) the product $0.003\,2 \times 356.3$; (c) the product $5.620 \times \pi$.

### 1.5 Unit Conversions for Physical Quantities

17. The Roman cubitus is an ancient unit of measure equivalent to about 0.445 m. Convert the 2.00-m height of a basketball forward to cubiti.

18. A house is advertised as having 1 420 square feet under roof. What is the area of this house in square meters?

19. A fathom is a unit of length, usually reserved for measuring the depth of water. A fathom is approximately 6 ft in length. Take the distance from Earth to the Moon to be 250 000 miles, and use the given approximation to find the distance in fathoms.

20. A small turtle moves at a speed of 186 furlongs per fortnight. Find the speed of the turtle in centimeters per second. Note that 1 furlong = 220 yards and 1 fortnight = 14 days.

21. A firkin is an old British unit of volume equal to 9 gallons. How many cubic meters are there in 6.00 firkins?

22. Find the height or length of these natural wonders in kilometers, meters, and centimeters: (a) The longest cave system in the world is the Mammoth Cave system in Central Kentucky, with a mapped length of 348 miles. (b) In the United States, the waterfall with the greatest single drop is Ribbon Falls in California, which drops 1 612 ft. (c) At 20 320 feet, Mount McKinley in Alaska is America's highest mountain. (d) The deepest canyon in the United States is King's Canyon in California, with a depth of 8 200 ft.

23. A car is traveling at a speed of 38.0 m/s on an interstate highway where the speed limit is 75.0 mi/h. Is the driver exceeding the speed limit? Justify your answer.

24. A certain car has a fuel efficiency of 25.0 miles per gallon (mi/gal). Express this efficiency in kilometers per liter (km/L).

25. The diameter of a sphere is measured to be 5.36 in. Find (a) the radius of the sphere in centimeters, (b) the surface area of the sphere in square centimeters, and (c) the volume of the sphere in cubic centimeters.

26. **V** **BIO** Suppose your hair grows at the rate of 1/32 inch per day. Find the rate at which it grows in nanometers per second. Because the distance between atoms in a molecule is on the order of 0.1 nm, your answer suggests how rapidly atoms are assembled in this protein synthesis.

27. The speed of light is about $3.00 \times 10^8$ m/s. Convert this figure to miles per hour.

28. **T** A house is 50.0 ft long and 26 ft wide and has 8.0-ft-high ceilings. What is the volume of the interior of the house in cubic meters and in cubic centimeters?

29. The amount of water in reservoirs is often measured in acre-ft. One acre-ft is a volume that covers an area of one acre to a depth of one foot. An acre is 43 560 ft². Find the volume in SI units of a reservoir containing 25.0 acre-ft of water.

30. The base of a pyramid covers an area of 13.0 acres (1 acre = 43 560 ft²) and has a height of 481 ft (Fig. P1.30). If the volume of a pyramid is given by the expression $V = bh/3$, where $b$ is the area of the base and $h$ is the height, find the volume of this pyramid in cubic meters.

**Figure P1.30**

31. A quart container of ice cream is to be made in the form of a cube. What should be the length of a side, in centimeters? (Use the conversion 1 gallon = 3.786 liter.)

## 1.6 Estimates and Order-of-Magnitude Calculations

*Note:* In developing answers to the problems in this section, you should state your important assumptions, including the numerical values assigned to parameters used in the solution.

32. Estimate the number of steps you would have to take to walk a distance equal to the circumference of the Earth.

33. **V** **BIO** Estimate the number of breaths taken by a human being during an average lifetime.

34. **BIO** Estimate the number of people in the world who are suffering from the common cold on any given day. (Answers may vary. Remember that a person suffers from a cold for about a week.)

35. The habitable part of Earth's surface has been estimated to cover 60 trillion square meters. Estimate the percent of this area occupied by humans if Earth's current population stood packed together as people do in a crowded elevator.

36. **BIO** Treat a cell in a human as a sphere of radius 1.0 $\mu$m. (a) Determine the volume of a cell. (b) Estimate the volume of your body. (c) Estimate the number of cells in your body.

37. An automobile tire is rated to last for 50 000 miles. Estimate the number of revolutions the tire will make in its lifetime.

38. **BIO** A study from the National Institutes of Health states that the human body contains trillions of microorganisms that make up 1% to 3% of the body's mass. Use this information to estimate the average mass of the body's approximately 100 trillion microorganisms.

## 1.7 Coordinate Systems

39. **T** A point is located in a polar coordinate system by the coordinates $r = 2.5$ m and $\theta = 35°$. Find the $x$- and $y$-coordinates of this point, assuming that the two coordinate systems have the same origin.

40. A certain corner of a room is selected as the origin of a rectangular coordinate system. If a fly is crawling on an adjacent wall at a point having coordinates (2.0, 1.0), where the units are meters, what is the distance of the fly from the corner of the room?

41. Express the location of the fly in Problem 40 in polar coordinates.

42. **V** Two points in a rectangular coordinate system have the coordinates (5.0, 3.0) and (−3.0, 4.0), where the units are centimeters. Determine the distance between these points.

43. Two points are given in polar coordinates by $(r, \theta) = (2.00$ m, $50.0°)$ and $(r, \theta) = (5.00$ m, $-50.0°)$, respectively. What is the distance between them?

44. **S** Given points $(r_1, \theta_1)$ and $(r_2, \theta_2)$ in polar coordinates, obtain a general formula for the distance between them. Simplify it as much as possible using the identity $\cos^2 \theta + \sin^2 \theta = 1$. *Hint:* Write the expressions for the two points in Cartesian coordinates and substitute into the usual distance formula.

## 1.8 Trigonometry Review

45. **T** For the triangle shown in Figure P1.45, what are (a) the length of the unknown side, (b) the tangent of $\theta$, and (c) the sine of $\phi$?

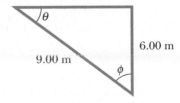

**Figure P1.45**

46. A ladder 9.00 m long leans against the side of a building. If the ladder is inclined at an angle of 75.0° to the horizontal, what is the horizontal distance from the bottom of the ladder to the building?

**47.** A high fountain of water is located at the center of a circular pool as shown in Figure P1.47. Not wishing to get his feet wet, a student walks around the pool and measures its circumference to be 15.0 m. Next, the student stands at the edge of the pool and uses a protractor to gauge the angle of elevation at the bottom of the fountain to be 55.0°. How high is the fountain?

**Figure P1.47**

**48.** **V** A right triangle has a hypotenuse of length 3.00 m, and one of its angles is 30.0°. What are the lengths of (a) the side opposite the 30.0° angle and (b) the side adjacent to the 30.0° angle?

**49.** In Figure P1.49, find (a) the side opposite $\theta$, (b) the side adjacent to $\phi$, (c) $\cos \theta$, (d) $\sin \phi$, and (e) $\tan \phi$.

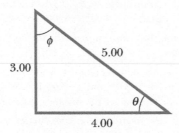

**Figure P1.49**

**50.** In a certain right triangle, the two sides that are perpendicular to each other are 5.00 m and 7.00 m long. What is the length of the third side of the triangle?

**51.** In Problem 50, what is the tangent of the angle for which 5.00 m is the opposite side?

**52.** **GP** **S** A woman measures the angle of elevation of a mountaintop as 12.0°. After walking 1.00 km closer to the mountain on level ground, she finds the angle to be 14.0°. Find the mountain's height, neglecting the height of the woman's eyes above the ground. *Hint:* Distances from the mountain ($x$ and $x - 1$ km) and the mountain's height are unknown. Draw two triangles, one for each of the woman's locations, and equate expressions for the mountain's height. Use that expression to find the first distance $x$ from the mountain and substitute to find the mountain's height.

**53.** A surveyor measures the distance across a straight river by the following method: starting directly across from a tree on the opposite bank, he walks $x = 1.00 \times 10^2$ m along the riverbank to establish a baseline. Then he sights across to the tree. The angle from his baseline to the tree is $\theta = 35.0°$ (Fig. P1.53). How wide is the river?

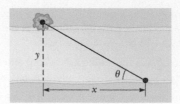

**Figure P1.53**

## 1.9 Vectors

**54.** Vector $\vec{A}$ has a magnitude of 8.00 units and makes an angle of 45.0° with the positive $x$-axis. Vector $\vec{B}$ also has a magnitude of 8.00 units and is directed along the negative $x$-axis. Using graphical methods, find (a) the vector sum $\vec{A} + \vec{B}$ and (b) the vector difference $\vec{A} - \vec{B}$.

**55.** Vector $\vec{A}$ has a magnitude of 29 units and points in the positive $y$-direction. When vector $\vec{B}$ is added to $\vec{A}$, the resultant vector $\vec{A} + \vec{B}$ points in the negative $y$-direction with a magnitude of 14 units. Find the magnitude and direction of $\vec{B}$.

**56.** **Q|C** An airplane flies $2.00 \times 10^2$ km due west from city A to city B and then $3.00 \times 10^2$ km in the direction of 30.0° north of west from city B to city C. (a) In straight-line distance, how far is city C from city A? (b) Relative to city A, in what direction is city C? (c) Why is the answer only approximately correct?

**57.** Vector $\vec{A}$ is 3.00 units in length and points along the positive $x$-axis. Vector $\vec{B}$ is 4.00 units in length and points along the negative $y$-axis. Use graphical methods to find the magnitude and direction of the vectors (a) $\vec{A} + \vec{B}$ and (b) $\vec{A} - \vec{B}$.

**58.** A force $\vec{F}_1$ of magnitude 6.00 units acts on an object at the origin in a direction $\theta = 30.0°$ above the positive $x$-axis (Fig. P1.58). A second force $\vec{F}_2$ of magnitude 5.00 units acts on the object in the direction of the positive $y$-axis. Find graphically the magnitude and direction of the resultant force $\vec{F}_1 + \vec{F}_2$.

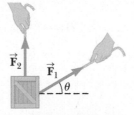

**Figure P1.58**

**59.** **V** A roller coaster moves $2.00 \times 10^2$ ft horizontally and then rises 135 ft at an angle of 30.0° above the horizontal. Next, it travels 135 ft at an angle of 40.0° below the horizontal. Use graphical techniques to find the roller coaster's displacement from its starting point to the end of this movement.

## 1.10 Components of a Vector

**60.** Calculate (a) the $x$-component and (b) the $y$-component of the vector with magnitude 24.0 m and direction 56.0°.

**61.** A vector $\vec{A}$ has components $A_x = -5.00$ m and $A_y = 9.00$ m. Find (a) the magnitude and (b) the direction of the vector.

**62.** A person walks 25.0° north of east for 3.10 km. How far due north and how far due east would she have to walk to arrive at the same location?

**63.** [V] The magnitude of vector $\vec{A}$ is 35.0 units and points in the direction 325° counterclockwise from the positive $x$-axis. Calculate the $x$- and $y$-components of this vector.

**64.** A figure skater glides along a circular path of radius 5.00 m. If she coasts around one half of the circle, find (a) her distance from the starting location and (b) the length of the path she skated.

**65.** A girl delivering newspapers covers her route by traveling 3.00 blocks west, 4.00 blocks north, and then 6.00 blocks east. (a) What is her final position relative to her starting location? (b) What is the length of the path she walked?

**66.** A quarterback takes the ball from the line of scrimmage, runs backwards for 10.0 yards, and then runs sideways parallel to the line of scrimmage for 15.0 yards. At this point, he throws a 50.0-yard forward pass straight downfield, perpendicular to the line of scrimmage. How far is the football from its original location?

**67.** A vector has an $x$-component of $-25.0$ units and a $y$-component of 40.0 units. Find the magnitude and direction of the vector.

**68.** A map suggests that Atlanta is 730. miles in a direction 5.00° north of east from Dallas. The same map shows that Chicago is 560. miles in a direction 21.0° west of north from Atlanta. Figure P1.68 shows the location of these three cities. Modeling the Earth as flat, use this information to find the displacement from Dallas to Chicago.

**Figure P1.68**

**69.** [T] The eye of a hurricane passes over Grand Bahama Island in a direction 60.0° north of west with a speed of 41.0 km/h. Three hours later the course of the hurricane suddenly shifts due north, and its speed slows to 25.0 km/h. How far from Grand Bahama is the hurricane 4.50 h after it passes over the island?

**70.** The helicopter view in Figure P1.70 shows two people pulling on a stubborn mule. Find (a) the single force that is equivalent to the two forces shown and (b) the force a third person would have to exert on the mule to make the net

force equal to zero. The forces are measured in units of newtons (N).

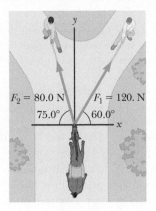

**Figure P1.70**

**71.** A commuter airplane starts from an airport and takes the route shown in Figure P1.71. The plane first flies to city $A$, located 175 km away in a direction 30.0° north of east. Next, it flies for 150. km 20.0° west of north, to city $B$. Finally, the plane flies 190. km due west, to city $C$. Find the location of city $C$ relative to the location of the starting point.

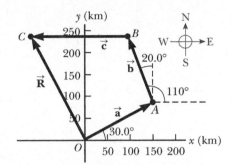

**Figure P1.71**

## Additional Problems

**72.** (a) Find a conversion factor to convert from miles per hour to kilometers per hour. (b) For a while, federal law mandated that the maximum highway speed would be 55 mi/h. Use the conversion factor from part (a) to find the speed in kilometers per hour. (c) The maximum highway speed has been raised to 65 mi/h in some places. In kilometers per hour, how much of an increase is this over the 55-mi/h limit?

**73.** The displacement of an object moving under uniform acceleration is some function of time and the acceleration. Suppose we write this displacement as $s = ka^m t^n$, where $k$ is a dimensionless constant. Show by dimensional analysis that this expression is satisfied if $m = 1$ and $n = 2$. Can the analysis give the value of $k$?

**74.** [V] Assume it takes 7.00 minutes to fill a 30.0-gal gasoline tank. (a) Calculate the rate at which the tank is filled in gallons per second. (b) Calculate the rate at which the tank is filled in cubic meters per second. (c) Determine the time interval, in hours, required to fill a 1.00-m³ volume at the same rate. (1 U.S. gal = 231 in.³)

75. **T** One gallon of paint (volume = $3.79 \times 10^{-3}$ m$^3$) covers an area of 25.0 m$^2$. What is the thickness of the fresh paint on the wall?

76. A sphere of radius $r$ has surface area $A = 4\pi r^2$ and volume $V = (4/3)\pi r^3$. If the radius of sphere 2 is double the radius of sphere 1, what is the ratio of (a) the areas, $A_2/A_1$ and (b) the volumes, $V_2/V_1$?

77. **T** Assume there are 100 million passenger cars in the United States and that the average fuel consumption is 20 mi/gal of gasoline. If the average distance traveled by each car is 10 000 mi/yr, how much gasoline would be saved per year if average fuel consumption could be increased to 25 mi/gal?

78. **V** In 2015, the U.S. national debt was about $18 trillion. (a) If payments were made at the rate of $1 000 per second, how many years would it take to pay off the debt, assuming that no interest were charged? (b) A dollar bill is about 15.5 cm long. If 18 trillion dollar bills were laid end to end around the Earth's equator, how many times would they encircle the planet? Take the radius of the Earth at the equator to be 6 378 km. (*Note:* Before doing any of these calculations, try to guess at the answers. You may be very surprised.)

79. (a) How many Earths could fit inside the Sun? (b) How many of Earth's Moons could fit inside the Earth?

80. **BIO** An average person sneezes about three times per day. Estimate the worldwide number of sneezes happening in a time interval approximately equal to one sneeze.

81. The nearest neutron star (a collapsed star made primarily of neutrons) is about $3.00 \times 10^{18}$ m away from Earth. Given that the Milky Way galaxy (Fig. P1.81) is roughly a disk of diameter $\sim 10^{21}$ m and thickness $\sim 10^{19}$ m, estimate the number of neutron stars in the Milky Way to the nearest order of magnitude.

Richard Payne/NASA

**Figure P1.81**

# Motion in One Dimension

**LIFE IS MOTION.** Our muscles coordinate motion microscopically to enable us to walk and jog. Our hearts pump tirelessly for decades, moving blood through our bodies. Cell wall mechanisms move select atoms and molecules in and out of cells. From the prehistoric chase of antelopes across the savanna to the pursuit of satellites in space, mastery of motion has been critical to our survival and success as a species.

The study of motion and of physical concepts such as force and mass is called **dynamics.** The part of dynamics that describes motion without regard to its causes is called **kinematics.** In this topic the focus is on kinematics in one dimension: motion along a straight line. This kind of motion—and, indeed, *any* motion—involves the concepts of displacement, velocity, and acceleration. Here, we use these concepts to study the motion of objects undergoing constant acceleration. In Topic 3 we will repeat this discussion for objects moving in two dimensions.

The first recorded evidence of the study of mechanics can be traced to the people of ancient Sumeria and Egypt, who were interested primarily in understanding the motions of heavenly bodies. The most systematic and detailed early studies of the heavens were conducted by the Greeks from about 300 BC to AD 300. Ancient scientists and laypeople regarded the Earth as the center of the Universe. This **geocentric model** was accepted by such notables as Aristotle (384–322 BC) and Claudius Ptolemy (AD 100 – c.170). Largely because of the authority of Aristotle, the geocentric model became the accepted theory of the Universe until the seventeenth century.

About 250 BC, the Greek philosopher Aristarchus worked out the details of a model of the solar system based on a spherical Earth that rotated on its axis and revolved around the Sun. He proposed that the sky appeared to turn westward because the Earth was turning eastward. This model wasn't given much consideration because it was believed that a turning Earth would generate powerful winds as it moved through the air. We now know that the Earth carries the air and everything else with it as it rotates.

The Polish astronomer Nicolaus Copernicus (1473–1543) is credited with initiating the revolution that finally replaced the geocentric model. In his system, called the **heliocentric model,** Earth and the other planets revolve in circular orbits around the Sun.

This early knowledge formed the foundation for the work of Galileo Galilei (1564–1642), who stands out as the dominant facilitator of the entrance of physics into the modern era. In 1609 he became one of the first to make astronomical observations with a telescope. He observed mountains on the Moon, the larger satellites of Jupiter, spots on the Sun, and the phases of Venus. Galileo's observations convinced him of the correctness of the Copernican theory. His quantitative study of motion formed the foundation of Newton's revolutionary work in the next century.

## 2.1 Displacement, Velocity, and Acceleration

### 2.1.1 Displacement

Motion involves the displacement of an object from one place in space and time to another. Describing motion requires some convenient coordinate system and a specified origin. A **frame of reference** is a choice of coordinate axes that defines

NASA/USGS

NASA/USGS

**Figure 2.1** (a) How large is the canyon? Without a frame of reference, it's hard to tell. (b) The canyon is Valles Marineris on Mars, and with a frame of reference provided by a superposed outline of the United States, its size is easily grasped.

the starting point for measuring any quantity–an essential first step in solving virtually any problem in mechanics (Fig. 2.1). In Figure 2.2a, for example, a car moves along the $x$-axis. The coordinates of the car at any time describe its position in space and, more importantly, its *displacement* at some given time of interest.

Definition of displacement ▶

The **displacement** $\Delta x$ of an object is defined as its *change in position* and is given by

$$\Delta x \equiv x_f - x_i \tag{2.1}$$

where $x_i$ is the coordinate of the object's initial position and $x_f$ is the coordinate of the object's final position. (The indices $i$ and $f$ stand for initial and final, respectively.)

**SI unit: meter (m)**

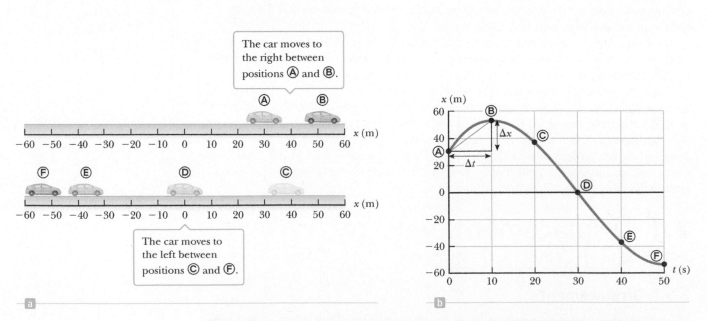

**Figure 2.2** (a) A car moves back and forth along a straight line taken to be the $x$-axis. Because we are interested only in the car's translational motion, we can model it as a particle. (b) Graph of position vs. time for the motion of the "particle."

We use the Greek letter delta, $\Delta$, to denote a change in any physical quantity. From the definition of displacement, we see that $\Delta x$ (read "delta ex") is positive if $x_f$ is greater than $x_i$ and negative if $x_f$ is less than $x_i$. For example, if the car moves from point Ⓐ to point Ⓑ so that the initial position is $x_i = 30$ m and the final position is $x_f = 52$ m, the displacement is $\Delta x = x_f - x_i = 52$ m $- 30$ m $= +22$ m. However, if the car moves from point Ⓒ to point Ⓕ, then the initial position is $x_i = 38$ m, the final position is $x_f = -53$ m, and the displacement is $\Delta x = x_f - x_i = -53$ m $- 38$ m $= -91$ m. A positive answer indicates a displacement in the positive x-direction, whereas a negative answer indicates a displacement in the negative x-direction. Figure 2.2b displays the graph of the car's position as a function of time.

Because displacement has both a magnitude (size) and a direction, it's a vector quantity, as are velocity and acceleration. Vector quantities will usually be denoted in boldface type with an arrow over the top of the letter. For example, $\vec{v}$ represents velocity and $\vec{a}$ denotes acceleration, both vector quantities. In this topic, however, it won't be necessary to use that notation because in one-dimensional motion an object can only move in one of two directions, and these directions are easily specified by plus and minus signs.

**Tip 2.1** Displacement and Distance

*Displacement* is a change in position. *Distance* is the magnitude of the displacement.

**Tip 2.2** Vectors Have Both a Magnitude and a Direction

Scalars have size. Vectors, too, have size, but they also indicate a direction.

## 2.1.2 Velocity

In everyday usage the terms *speed* and *velocity* are interchangeable. In physics, however, there's a clear distinction between them: speed is a scalar quantity, having only magnitude, whereas velocity is a vector, having both magnitude and direction.

Why must velocity be a vector? If you want to get to a town 70 km away in an hour's time, it's not enough to drive at a speed of 70 km/h; you must travel in the correct direction as well. That's obvious, but it shows that velocity gives considerably more information than speed, as will be made more precise in the formal definitions.

The **average speed** of an object over a given time interval is the length of the path it travels divided by the total elapsed time:

$$\text{Average speed} \equiv \frac{\text{path length}}{\text{elapsed time}}$$

**SI unit: meter per second (m/s)**

◀ Definition of average speed

This equation might be written with symbols as $v = d/t$, where $v$ represents the average speed (not average velocity), $d$ represents the path length, and $t$ represents the elapsed time during the motion. The path length is often called the "total distance," but that can be misleading because distance has a different, precise mathematical meaning based on differences in the coordinates between the initial and final points. Distance (neglecting any curvature of the surface) is given by the Pythagorean theorem, $\Delta s = \sqrt{(x_f - x_i)^2 + (y_f - y_i)^2}$, which depends only on the endpoints, $(x_i, y_i)$ and $(x_f, y_f)$, and not on what happens in between. The same equation gives the magnitude of a displacement. The straight-line distance from Atlanta, Georgia, to St. Petersburg, Florida, for example, is about 500 miles. If someone drives a car that distance in 10 h, the car's average speed is 500 mi/10 h = 50 mi/h, even if the car's speed varies greatly during the trip. If the driver takes scenic detours off the direct route along the way, however, or doubles back for a while, the path length increases while the distance between the two cities remains the same. A side trip to Jacksonville, Florida, for example, might add 100 miles to the path length, so the car's average speed would then be 600 mi/10 h = 60 mi/h. The magnitude of the average velocity, however, would remain 50 mi/h.

**Tip 2.3** Path Length vs. Distance

*Distance* is the length of a straight line joining two points. *Path length* is the length of an actual path traversed between two points, including any retracing of steps or deviations from a straight line.

| EXAMPLE 2.1 | THE TORTOISE AND THE HARE |

**GOAL** Apply the concept of average speed.

**PROBLEM** A turtle and a rabbit engage in a footrace over a distance of 4.00 km. The rabbit runs 0.500 km and then stops for a 90.0-min nap. Upon awakening, he remembers the race and runs twice as fast. Finishing the course in a total time of 1.75 h, the rabbit wins the race. **(a)** Calculate the average speed of the rabbit. **(b)** What was his average speed before he stopped for a nap? Assume no detours or doubling back.

**STRATEGY** Finding the overall average speed in part **(a)** is just a matter of dividing the path length by the elapsed time. Part **(b)** requires two equations and two unknowns, the latter turning out to be the two different average speeds: $v_1$ before the nap and $v_2$ after the nap. One equation is given in the statement of the problem ($v_2 = 2v_1$), whereas the other comes from the fact the rabbit ran for only 15 minutes because he napped for 90 minutes.

**SOLUTION**

**(a)** Find the rabbit's overall average speed.

Apply the equation for average speed:

$$\text{Average speed} \equiv \frac{\text{path length}}{\text{elapsed time}} = \frac{4.00 \text{ km}}{1.75 \text{ h}}$$

$$= 2.29 \text{ km/h}$$

**(b)** Find the rabbit's average speed before his nap.

Sum the running times, and set the sum equal to 0.25 h:   $t_1 + t_2 = 0.25 \text{ h}$

Substitute $t_1 = d_1/v_1$ and $t_2 = d_2/v_2$:   (1) $\dfrac{d_1}{v_1} + \dfrac{d_2}{v_2} = 0.25 \text{ h}$

Substitute $v_2 = 2v_1$ and the values of $d_1$ and $d_2$ into Equation (1):   (2) $\dfrac{0.500 \text{ km}}{v_1} + \dfrac{3.50 \text{ km}}{2v_1} = 0.25 \text{ h}$

Solve Equation (2) for $v_1$:   $v_1 = 9.0 \text{ km/h}$

**REMARKS** As seen in this example, average speed can be calculated regardless of any variation in speed over the given time interval.

**QUESTION 2.1** Does a doubling of an object's average speed always double the magnitude of its displacement in a given amount of time? Explain.

**EXERCISE 2.1** Estimate the average speed of the *Apollo* spacecraft in meters per second, given that the craft took five days to reach the Moon from Earth. (The Moon is $3.8 \times 10^8$ m from Earth.)

**ANSWER** ~ 900 m/s

Unlike average speed, **average velocity** is a vector quantity, having both a magnitude and a direction. Consider again the car of Figure 2.2, moving along the road (the *x*-axis). Let the car's position be $x_i$ at some time $t_i$ and $x_f$ at a later time $t_f$. In the time interval $\Delta t = t_f - t_i$, the displacement of the car is $\Delta x = x_f - x_i$.

Definition of average ▶ velocity

The average velocity $\bar{v}$ during a time interval $\Delta t$ is the displacement $\Delta x$ divided by $\Delta t$:

$$\bar{v} \equiv \frac{\Delta x}{\Delta t} = \frac{x_f - x_i}{t_f - t_i} \qquad [2.2]$$

**SI unit: meter per second (m/s)**

Unlike the average speed, which is always positive, the average velocity of an object in one dimension can be either positive or negative, depending on the sign of the displacement. (The time interval $\Delta t$ is always positive.) In Figure 2.2a, for example, the average velocity of the car is positive in the upper illustration—a positive sign indicating motion to the right along the *x*-axis. Similarly, a negative average velocity for the car in the lower illustration of the figure indicates that it moves to the left along the *x*-axis.

As an example, we can use the data in Table 2.1 to find the average velocity in the time interval from point Ⓐ to point Ⓑ (assume two digits are significant):

$$\overline{v} = \frac{\Delta x}{\Delta t} = \frac{52\text{ m} - 30\text{ m}}{10\text{ s} - 0\text{ s}} = 2.2\text{ m/s}$$

Aside from meters per second, other common units for average velocity are feet per second (ft/s) in the U.S. customary system and centimeters per second (cm/s) in the cgs system.

To further illustrate the distinction between speed and velocity, suppose we're watching a drag race from a stationary blimp. In one run, we see a car follow the straight-line path from Ⓟ to Ⓠ shown in Figure 2.3 during the time interval Δt, and in a second run a car follows the curved path during the same interval. From the definition in Equation 2.2, the two cars had the same average velocity because they had the same displacement $\Delta x = x_f - x_i$ during the same time interval Δt. The car taking the curved route, however, traveled a greater path length and had the higher average speed.

**Table 2.1** Position of the Car at Various Times

| Position | t (s) | x (m) |
|---|---|---|
| Ⓐ | 0 | 30 |
| Ⓑ | 10 | 52 |
| Ⓒ | 20 | 38 |
| Ⓓ | 30 | 0 |
| Ⓔ | 40 | −37 |
| Ⓕ | 50 | −53 |

**Figure 2.3** A drag race viewed from a stationary blimp. One car follows the rust-colored straight-line path from Ⓟ to Ⓠ, and a second car follows the blue curved path.

## Quick Quiz

**2.1** Figure 2.4 shows the unusual path of a confused football player. After receiving a kickoff at his own goal, he runs downfield to within inches of a touchdown, then reverses direction and races back until he's tackled at the exact location where he first caught the ball. During this run, which took 25 s, what is (a) the path length he travels, (b) his displacement, (c) his average velocity in the x-direction, and (d) his average speed?

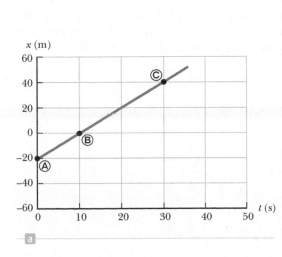

**Figure 2.4** (Quick Quiz 2.1) The path followed by a confused football player.

**Graphical Interpretation of Velocity** If a car moves along the x-axis from Ⓐ to Ⓑ to Ⓒ, and so forth, we can plot the positions of these points as a function of the time elapsed since the start of the motion. The result is a **position versus time graph** like those of Figure 2.5. In Figure 2.5a, the graph is a straight line because the car is moving at constant velocity. The same displacement Δx occurs in each time interval Δt. In this case, the average velocity is always the same and is equal to Δx/Δt. Figure 2.5b is

**Tip 2.4** Slopes of Graphs

The word *slope* is often used in reference to the graphs of physical data. Regardless of the type of data, the *slope* is given by

Slope = change in vertical axis / change in horizontal axis

Slope carries units.

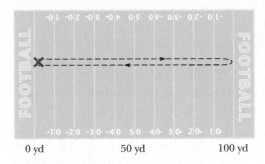

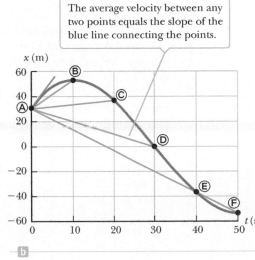

**Figure 2.5** (a) Position vs. time graph for the motion of a car moving along the x-axis at constant velocity. (b) Position vs. time graph for the motion of a car with changing velocity, using the data in Table 2.1.

a graph of the data in Table 2.1. Here, the position versus time graph is not a straight line because the velocity of the car is changing. Between any two points, however, we can draw a straight line just as in Figure 2.5a, and the slope of that line is the average velocity $\Delta x/\Delta t$ in that time interval. In general, **the average velocity of an object during the time interval $\Delta t$ is equal to the slope of the straight line joining the initial and final points on a graph of the object's position versus time.**

From the data in Table 2.1 and the graph in Figure 2.5b, we see that the car first moves in the positive $x$-direction as it travels from Ⓐ to Ⓑ, reaches a position of 52 m at time $t = 10$ s, then reverses direction and heads backwards. In the first 10 s of its motion, as the car travels from Ⓐ to Ⓑ, its average velocity is 2.2 m/s, as previously calculated. In the first 40 seconds, as the car goes from Ⓐ to Ⓔ, its displacement is $\Delta x = -37$ m $-$ (30 m) $= -67$ m. So the average velocity in this interval, which equals the slope of the blue line in Figure 2.5b from Ⓐ to Ⓔ, is $\bar{v} = \Delta x/\Delta t$ $=(-67$ m$)/(40$ s$) = -1.7$ m/s. In general, there will be a different average velocity between any distinct pair of points.

**Instantaneous Velocity**  Average velocity doesn't take into account the details of what happens during an interval of time. On a car trip, for example, you may speed up or slow down a number of times in response to the traffic and the condition of the road, and on rare occasions even pull over to chat with a police officer about your speed. What is most important to the police (and to your own safety) is the speed of your car and the direction it was going at a particular instant in time, which together determine the car's **instantaneous velocity.**

So in driving a car between two points, the average velocity must be computed over an interval of time, but the magnitude of instantaneous velocity can be read on the car's speedometer.

Definition of instantaneous ▶
velocity

The instantaneous velocity $v$ is the limit of the average velocity as the time interval $\Delta t$ becomes infinitesimally small:

$$v \equiv \lim_{\Delta t \to 0} \frac{\Delta x}{\Delta t}$$

[2.3]

SI unit: meter per second (m/s)

The notation $\lim_{\Delta t \to 0}$ means that the ratio $\Delta x/\Delta t$ is repeatedly evaluated for smaller and smaller time intervals $\Delta t$. As $\Delta t$ gets extremely close to zero, the ratio $\Delta x/\Delta t$ gets closer and closer to a fixed number, which is defined as the instantaneous velocity.

To better understand the formal definition, consider data obtained on our vehicle via radar (Table 2.2). At $t = 1.00$ s, the car is at $x = 5.00$ m, and at $t = 3.00$ s, it's at $x = 52.5$ m. The average velocity computed for this interval $\Delta x/\Delta t = (52.5$ m $-$ 5.00 m$)/(3.00$ s $-$ 1.00 s$) = 23.8$ m/s. This result could be used as an estimate for the velocity at $t = 1.00$ s, but it wouldn't be very accurate because the speed changes considerably in the 2-second time interval. Using the rest of the data, we can construct Table 2.3. As the time interval gets smaller, the average velocity more closely

**Tip 2.5** Average Velocity vs. Average Speed

Average velocity is *not* the same as average speed. If you run from $x = 0$ m to $x = 25$ m and back to your starting point in a time interval of 5 s, the average velocity is zero, whereas the average speed is 10 m/s.

**Table 2.2** Positions of a Car at Specific Instants of Time

| $t$ (s) | $x$ (m) |
|---------|---------|
| 1.00 | 5.00 |
| 1.01 | 5.47 |
| 1.10 | 9.67 |
| 1.20 | 14.3 |
| 1.50 | 26.3 |
| 2.00 | 34.7 |
| 3.00 | 52.5 |

**Table 2.3** Calculated Values of the Time Intervals, Displacements, and Average Velocities of the Car of Table 2.2

| Time Interval (s) | $\Delta t$ (s) | $\Delta x$ (m) | $\bar{v}$ (m/s) |
|-------------------|----------------|----------------|-----------------|
| 1.00 to 3.00 | 2.00 | 47.5 | 23.8 |
| 1.00 to 2.00 | 1.00 | 29.7 | 29.7 |
| 1.00 to 1.50 | 0.50 | 21.3 | 42.6 |
| 1.00 to 1.20 | 0.20 | 9.30 | 46.5 |
| 1.00 to 1.10 | 0.10 | 4.67 | 46.7 |
| 1.00 to 1.01 | 0.01 | 0.470 | 47.0 |

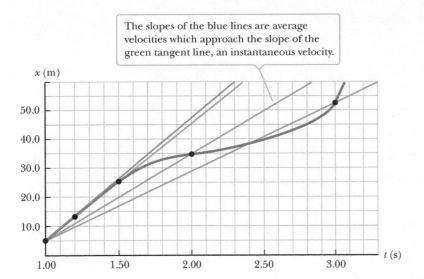

The slopes of the blue lines are average velocities which approach the slope of the green tangent line, an instantaneous velocity.

**Figure 2.6** Graph representing the motion of the car from the data in Table 2.2.

approaches the instantaneous velocity. Using the final interval of only 0.010 0 s, we find that the average velocity is $\bar{v} = \Delta x/\Delta t = 0.470\ \text{m}/0.010\ 0\ \text{s} = 47.0\ \text{m/s}$. Because 0.010 0 s is a very short time interval, the actual instantaneous velocity is probably very close to this latter average velocity, given the limits on the car's ability to accelerate. Finally, using the conversion factor on the front endsheets of the book, we see that this is 105 mi/h, a likely violation of the speed limit.

As can be seen in Figure 2.6, the chords formed by the blue lines gradually approach a tangent line as the time interval becomes smaller. **The slope of the line tangent to the position versus time curve at "a given time" is defined to be the instantaneous velocity at that time.**

**The instantaneous speed of an object, which is a scalar quantity, is defined as the magnitude of the instantaneous velocity.** Like average speed, instantaneous speed (which we will usually call, simply, "speed") has no direction associated with it and hence carries no algebraic sign. For example, if one object has an instantaneous velocity of $+15$ m/s along a given line and another object has an instantaneous velocity of $-15$ m/s along the same line, both have an instantaneous speed of 15 m/s.

◀ Definition of instantaneous speed

## EXAMPLE 2.2    SLOWLY MOVING TRAIN

**GOAL** Obtain average and instantaneous velocities from a graph.

**PROBLEM** A train moves slowly along a straight portion of track according to the graph of position vs. time in Figure 2.7a. Find **(a)** the average velocity for the total trip, **(b)** the average velocity during the first 4.00 s of motion, **(c)** the average velocity during the next 4.00 s of motion, **(d)** the instantaneous velocity at $t = 2.00$ s, and **(e)** the instantaneous velocity at $t = 9.00$ s.

**STRATEGY** The average velocities can be obtained by substituting the data into the definition. The instantaneous velocity at $t = 2.00$ s is the same as the average velocity at that point because the position vs. time graph is a straight line, indicating constant velocity. Finding the instantaneous velocity when $t = 9.00$ s requires sketching a line tangent to the curve at that point and finding its slope.

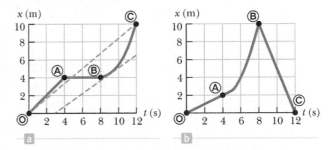

**Figure 2.7** (a) (Example 2.2) (b) (Exercise 2.2)

. . . . . . . . . . . . . . . . . . . . . . . . . . . . . . . . . . . . . . . . . . . . . . . . . . . . . . . . . . . . . . . . . . . . . . . . . . . . . . . . . . . . . . . . . . . . . . .

## SOLUTION

**(a)** Find the average velocity from Ⓞ to Ⓒ.

Calculate the slope of the dashed blue line:

$$\bar{v} = \frac{\Delta x}{\Delta t} = \frac{10.0\ \text{m}}{12.0\ \text{s}} = \boxed{+0.833\ \text{m/s}}$$

*(Continued)*

**(b)** Find the average velocity during the first 4 seconds of the train's motion.

Again, find the slope:

$$\bar{v} = \frac{\Delta x}{\Delta t} = \frac{4.00 \text{ m}}{4.00 \text{ s}} = +1.00 \text{ m/s}$$

**(c)** Find the average velocity during the next 4 seconds.

Here, there is no change in position as the train moves from Ⓐ to Ⓑ, so the displacement $\Delta x$ is zero:

$$\bar{v} = \frac{\Delta x}{\Delta t} = \frac{0 \text{ m}}{4.00 \text{ s}} = 0 \text{ m/s}$$

**(d)** Find the instantaneous velocity at $t = 2.00$ s.

This is the same as the average velocity found in **(b)**, because the graph is a straight line:

$$v = 1.00 \text{ m/s}$$

**(e)** Find the instantaneous velocity at $t = 9.00$ s.

The tangent line appears to intercept the $x$-axis at (3.0 s, 0 m) and graze the curve at (9.0 s, 4.5 m). The instantaneous velocity at $t = 9.00$ s equals the slope of the tangent line through these points:

$$v = \frac{\Delta x}{\Delta t} = \frac{4.5 \text{ m} - 0 \text{ m}}{9.0 \text{ s} - 3.0 \text{ s}} = 0.75 \text{ m/s}$$

. . . . . . . . . . . . . . . . . . . . . . . . . . . . . . . . . . . . . . . . . . . . . . . . . . . . . . . . . . . . . . . . . . . . . . . . . . . . . . . . . . . . . . . . . . . . . . . . . . . . . . . . . . . . . .

**REMARKS** From the origin to Ⓐ, the train moves at constant speed in the positive $x$-direction for the first 4.00 s, because the position vs. time curve is rising steadily toward positive values. From Ⓐ to Ⓑ, the train stops at $x = 4.00$ m for 4.00 s. From Ⓑ to Ⓒ, the train travels at increasing speed in the positive $x$-direction.

**QUESTION 2.2** Would a vertical line in a graph of position vs. time make sense? Explain.

**EXERCISE 2.2** Figure 2.7b graphs another run of the train. Find **(a)** the average velocity from Ⓞ to Ⓒ; **(b)** the average velocity from Ⓞ to Ⓐ and the instantaneous velocity at any given point between Ⓞ and Ⓐ; **(c)** the approximate instantaneous velocity at $t = 6.0$ s; and **(d)** the average velocity on the open interval from Ⓑ to Ⓒ and instantaneous velocity at $t = 9.0$ s.

**ANSWERS** **(a)** 0 m/s  **(b)** both are +0.5 m/s  **(c)** 2 m/s  **(d)** both are −2.5 m/s

---

**Figure 2.8** A car moving to the right accelerates from a velocity of $v_i$ to a velocity of $v_f$ in the time interval $\Delta t = t_f - t_i$.

### 2.1.3 Acceleration

Going from place to place in your car, you rarely travel long distances at constant velocity. The velocity of the car increases when you step harder on the gas pedal and decreases when you apply the brakes. The velocity also changes when you round a curve, altering your direction of motion. The changing of an object's velocity with time is called **acceleration.**

**Average Acceleration** A car moves along a straight highway as in Figure 2.8. At time $t_i$ it has a velocity of $v_i$, and at time $t_f$ its velocity is $v_f$, with $\Delta v = v_f - v_i$ and $\Delta t = t_f - t_i$.

Definition of average ▶
acceleration

The average acceleration $\bar{a}$ during the time interval $\Delta t$ is the change in velocity $\Delta v$ divided by $\Delta t$:

$$\bar{a} \equiv \frac{\Delta v}{\Delta t} = \frac{v_f - v_i}{t_f - t_i} \qquad [2.4]$$

**SI unit: meter per second per second (m/s²)**

For example, suppose the car shown in Figure 2.8 accelerates from an initial velocity of $v_i = +10$ m/s to a final velocity of $v_f = +20$ m/s in a time interval of 2 s.

(Both velocities are toward the right, selected as the positive direction.) These values can be inserted into Equation 2.4 to find the average acceleration:

$$\overline{a} = \frac{\Delta v}{\Delta t} = \frac{20 \text{ m/s} - 10 \text{ m/s}}{2 \text{ s}} = +5 \text{ m/s}^2$$

Acceleration is a vector quantity having dimensions of length divided by the time squared. Common units of acceleration are meters per second per second ((m/s)/s, which is usually written m/s$^2$) and feet per second per second (ft/s$^2$). An average acceleration of $+5$ m/s$^2$ means that, on average, the car increases its velocity by 5 m/s every second in the positive $x$-direction.

For the case of motion in a straight line, the direction of the velocity of an object and the direction of its acceleration are related as follows: **When the object's velocity and acceleration are in the same direction, the speed of the object increases with time. When the object's velocity and acceleration are in opposite directions, the speed of the object decreases with time.**

To clarify this point, suppose the velocity of a car changes from $-10$ m/s to $-20$ m/s in a time interval of 2 s. The minus signs indicate that the velocities of the car are in the negative $x$-direction; they do *not* mean that the car is slowing down! The average acceleration of the car in this time interval is

$$\overline{a} = \frac{\Delta v}{\Delta t} = \frac{-20 \text{ m/s} - (-10 \text{ m/s})}{2 \text{ s}} = -5 \text{ m/s}^2$$

The minus sign indicates that the acceleration vector is also in the negative $x$-direction. Because the velocity and acceleration vectors are in the same direction, the speed of the car must increase as the car moves to the left. Positive and negative accelerations specify directions relative to chosen axes, not "speeding up" or "slowing down." The terms *speeding up* or *slowing down* refer to an increase and a decrease in speed, respectively.

> **Tip 2.6** Negative Acceleration
>
> Negative acceleration doesn't necessarily mean an object is slowing down. If the acceleration is negative and the velocity is also negative, the object is speeding up!

> **Tip 2.7** Deceleration
>
> The word *deceleration* means a reduction in speed, a slowing down. Some confuse it with a negative acceleration, which can speed something up. (See Tip 2.6.)

### Quick Quiz

**2.2 True or False?** (a) A car must always have an acceleration in the same direction as its velocity. (b) It's possible for a slowing car to have a positive acceleration. (c) An object with constant nonzero acceleration can never stop and remain at rest.

An object with nonzero acceleration can have a velocity of zero, but only instantaneously. When a ball is tossed straight up, its velocity is zero when it reaches its maximum height. Gravity still accelerates the ball at that point, however; otherwise, it wouldn't fall down.

**Instantaneous Acceleration** The value of the average acceleration often differs in different time intervals, so it's useful to define the **instantaneous acceleration,** which is analogous to the instantaneous velocity discussed in Section 2.1.2.

The instantaneous acceleration $a$ is the limit of the average acceleration as the time interval $\Delta t$ goes to zero:

$$a \equiv \lim_{\Delta t \to 0} \frac{\Delta v}{\Delta t} \qquad [2.5]$$

**SI unit: meter per second per second (m/s$^2$)**

◄ Definition of instantaneous acceleration

Here again, the notation $\lim\limits_{\Delta t \to 0}$ means that the ratio $\Delta v/\Delta t$ is evaluated for smaller and smaller values of $\Delta t$. The closer $\Delta t$ gets to zero, the closer the ratio gets to a fixed number, which is the instantaneous acceleration.

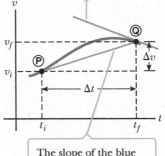

The car moves with different velocities at points Ⓟ and Ⓠ.

**a**

The slope of the green line is the instantaneous acceleration of the car at point Ⓠ (Eq. 2.5).

The slope of the blue line connecting Ⓟ and Ⓠ is the average acceleration of the car during the time interval $\Delta t = t_f - t_i$ (Eq. 2.4).

**b**

**Figure 2.9** (a) A car, modeled as a particle, moving along the x-axis from Ⓟ to Ⓠ, has velocity $v_{xi}$ at $t = t_i$ and velocity $v_{xf}$ at $t = t_f$. (b) Velocity vs. time graph for an object moving in a straight line.

Figure 2.9, a **velocity versus time graph,** plots the velocity of an object against time. The graph could represent, for example, the motion of a car along a busy street. The average acceleration of the car between times $t_i$ and $t_f$ can be found by determining the slope of the line joining points Ⓟ and Ⓠ. If we imagine that point Ⓠ is brought closer and closer to point Ⓟ, the line comes closer and closer to becoming tangent at Ⓟ. The **instantaneous acceleration of an object at a given time equals the slope of the tangent to the velocity versus time graph at that time.** From now on, we will use the term *acceleration* to mean "instantaneous acceleration."

In the special case where the velocity versus time graph of an object's motion is a straight line, the instantaneous acceleration of the object at any point is equal to its average acceleration. That also means the tangent line to the graph overlaps the graph itself. In that case, the object's acceleration is said to be *uniform,* which means that it has a constant value. Constant acceleration problems are important in kinematics and are studied extensively in this and the next topic.

### Quick Quiz

**2.3** Parts (a), (b), and (c) of Figure 2.10 represent three graphs of the velocities of different objects moving in straight-line paths as functions of time. The possible accelerations of each object as functions of time are shown in parts (d), (e), and (f). Match each velocity vs. time graph with the acceleration vs. time graph that best describes the motion.

**Figure 2.10** (Quick Quiz 2.3) Match each velocity vs. time graph to its corresponding acceleration vs. time graph.

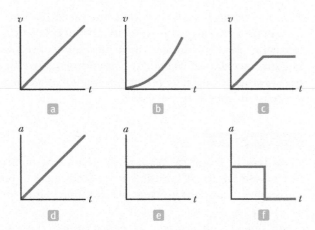

---

**EXAMPLE 2.3** **CATCHING A FLY BALL**

**GOAL** Apply the definition of instantaneous acceleration.

**PROBLEM** A baseball player moves in a straight-line path in order to catch a fly ball hit to the outfield. His velocity as a function of time is shown in Figure 2.11a. Find his instantaneous acceleration at points Ⓐ, Ⓑ, and Ⓒ.

**STRATEGY** At each point, the velocity vs. time graph is a straight line segment, so the instantaneous acceleration will be the slope of that segment. Select two points on each segment and use them to calculate the slope.

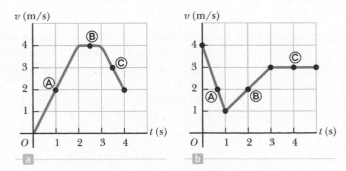

**Figure 2.11** (a) (Example 2.3) (b) (Exercise 2.3)

---

**SOLUTION**

Acceleration at Ⓐ.

The acceleration at Ⓐ equals the slope of the line connecting the points (0 s, 0 m/s) and (2.0 s, 4.0 m/s):

$$a = \frac{\Delta v}{\Delta t} = \frac{4.0 \text{ m/s} - 0}{2.0 \text{ s} - 0} = \boxed{+2.0 \text{ m/s}^2}$$

Acceleration at Ⓑ.

$\Delta v = 0$, because the segment is horizontal:

$$a = \frac{\Delta v}{\Delta t} = \frac{4.0 \text{ m/s} - 4.0 \text{ m/s}}{3.0 \text{ s} - 2.0 \text{ s}} = \boxed{0 \text{ m/s}^2}$$

Acceleration at Ⓒ.

The acceleration at Ⓒ equals the slope of the line connecting the points (3.0 s, 4.0 m/s) and (4.0 s, 2.0 m/s):

$$a = \frac{\Delta v}{\Delta t} = \frac{2.0 \text{ m/s} - 4.0 \text{ m/s}}{4.0 \text{ s} - 3.0 \text{ s}} = \boxed{-2.0 \text{ m/s}^2}$$

**REMARKS** For the first 2.0 s, the ballplayer moves in the positive $x$-direction (the velocity is positive) and steadily accelerates (the curve is steadily rising) to a maximum speed of 4.0 m/s. He moves for 1.0 s at a steady speed of 4.0 m/s and then slows down in the last second (the $v$ vs. $t$ curve is falling), still moving in the positive $x$-direction ($v$ is always positive).

**QUESTION 2.3** Can the tangent line to a velocity vs. time graph ever be vertical? Explain.

**EXERCISE 2.3** Repeat the problem, using Figure 2.11b.

**ANSWER** The accelerations at Ⓐ, Ⓑ, and Ⓒ are $-3.0 \text{ m/s}^2$, $1.0 \text{ m/s}^2$, and $0 \text{ m/s}^2$, respectively.

# 2.2 Motion Diagrams

Velocity and acceleration are sometimes confused with each other, but they're very different concepts, as can be illustrated with the help of motion diagrams. A **motion diagram** is a representation of a moving object at successive time intervals, with velocity and acceleration vectors sketched at each position, red for velocity vectors and violet for acceleration vectors, as in Figure 2.12. The time intervals between adjacent positions in the motion diagram are assumed equal.

A motion diagram is analogous to images resulting from a stroboscopic photograph of a moving object. Each image is made as the strobe light flashes. Figure 2.12 represents three sets of strobe photographs of cars moving along a straight roadway from left to right. The time intervals between flashes of the stroboscope are equal in each diagram.

In Figure 2.12a, the images of the car are equally spaced: The car moves the same distance in each time interval. This means that the car moves with *constant positive velocity* and has *zero acceleration*. The red arrows are all the same length (constant velocity) and there are no violet arrows (zero acceleration).

In Figure 2.12b, the images of the car become farther apart as time progresses and the velocity vector increases with time, because the car's displacement between adjacent positions increases as time progresses. The car is moving with a *positive velocity* and a constant *positive acceleration*. The red arrows are successively longer in each image, and the violet arrows point to the right.

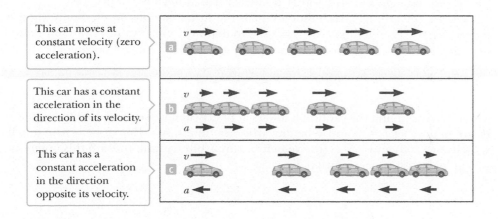

This car moves at constant velocity (zero acceleration).

This car has a constant acceleration in the direction of its velocity.

This car has a constant acceleration in the direction opposite its velocity.

**Figure 2.12** Motion diagrams of a car moving along a straight roadway in a single direction. The velocity at each instant is indicated by a red arrow, and the constant acceleration is indicated by a purple arrow.

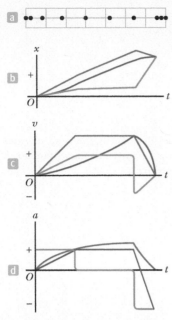

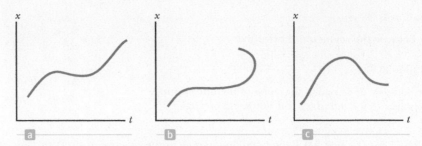

**Figure 2.13** (Quick Quiz 2.4) Which position vs. time curve is impossible?

In Figure 2.12c, the car slows as it moves to the right because its displacement between adjacent positions decreases with time. In this case, the car moves initially to the right with a constant negative acceleration. The velocity vector decreases in time (the red arrows get shorter) and eventually reaches zero, as would happen when the brakes are applied. Note that the acceleration and velocity vectors are *not* in the same direction. The car is moving with a *positive velocity*, but with a *negative acceleration*.

Try constructing your own diagrams for various problems involving kinematics.

**Figure 2.14** (Quick Quiz 2.5) Choose the correct graphs.

### Quick Quiz

**2.4** The three graphs in Figure 2.13 represent the position vs. time for objects moving along the *x*-axis. Which, if any, of these graphs is not physically possible?

**2.5** Figure 2.14a is a diagram of a multiflash image of an air puck moving to the right on a horizontal surface. The images sketched are separated by equal time intervals, and the first and last images show the puck at rest. (a) In Figure 2.14b, which color graph best shows the puck's position as a function of time? (b) In Figure 2.14c, which color graph best shows the puck's velocity as a function of time? (c) In Figure 2.14d, which color graph best shows the puck's acceleration as a function of time?

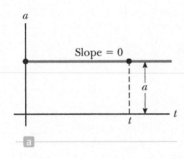

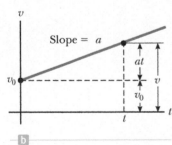

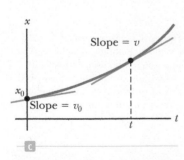

**Figure 2.15** A particle moving along the *x*-axis with constant acceleration *a*. (a) the acceleration vs. time graph, (b) the velocity vs. time graph, and (c) the position vs. time graph.

## 2.3 One-Dimensional Motion with Constant Acceleration

Many applications of mechanics involve objects moving with *constant acceleration*. This type of motion is important because it applies to numerous objects in nature, such as an object in free fall near Earth's surface (assuming air resistance can be neglected). A graph of acceleration versus time for motion with constant acceleration is shown in Figure 2.15a. **When an object moves with constant acceleration, the instantaneous acceleration at any point in a time interval is equal to the value of the average acceleration over the entire time interval.** Consequently, the velocity increases or decreases at the same rate throughout the motion, and a plot of *v* versus *t* gives a straight line with either positive, zero, or negative slope.

Because the average acceleration equals the instantaneous acceleration when *a* is constant, we can eliminate the bar used to denote average values from our defining equation for acceleration, writing $\bar{a} = a$, so that Equation 2.4 becomes

$$a = \frac{v_f - v_i}{t_f - t_i}$$

The observer timing the motion is always at liberty to choose the initial time, so for convenience, let $t_i = 0$ and $t_f$ be any arbitrary time *t*. Also, let $v_i = v_0$ (the initial velocity at $t = 0$) and $v_f = v$ (the velocity at any arbitrary time *t*). With this notation, we can express the acceleration as

$$a = \frac{v - v_0}{t}$$

or

$$v = v_0 + at \quad \text{(for constant } a\text{)} \tag{2.6}$$

Equation 2.6 states that the acceleration $a$ steadily changes the initial velocity $v_0$ by an amount $at$. For example, if a car starts with a velocity of $+2.0$ m/s to the right and accelerates to the right with $a = +6.0$ m/s$^2$, it will have a velocity of $+14$ m/s after 2.0 s have elapsed:

$$v = v_0 + at = + 2.0 \text{ m/s} + (6.0 \text{ m/s}^2)(2.0 \text{ s}) = +14 \text{ m/s}$$

The graphical interpretation of $v$ is shown in Figure 2.15b. The velocity varies linearly with time according to Equation 2.6, as it should for constant acceleration.

Because the velocity is increasing or decreasing *uniformly* with time, we can express the average velocity in any time interval as the arithmetic average of the initial velocity $v_0$ and the final velocity $v$:

$$\overline{v} = \frac{v_0 + v}{2} \quad \text{(for constant } a\text{)} \tag{2.7}$$

Remember that this expression is valid only when the acceleration is constant, in which case the velocity increases uniformly.

We can now use this result along with the defining equation for average velocity, Equation 2.2, to obtain an expression for the displacement of an object as a function of time. Again, we choose $t_i = 0$ and $t_f = t$, and for convenience, we write $\Delta x = x_f - x_i = x - x_0$. This results in

$$\Delta x = \overline{v}t = \left(\frac{v_0 + v}{2}\right)t$$

$$\Delta x = \tfrac{1}{2}(v_0 + v)t \quad \text{(for constant } a\text{)} \tag{2.8}$$

We can obtain another useful expression for displacement by substituting the equation for $v$ (Eq. 2.6) into Equation 2.8:

$$\Delta x = \tfrac{1}{2}(v_0 + v_0 + at)t$$

$$\Delta x = v_0 t + \tfrac{1}{2}at^2 \quad \text{(for constant } a\text{)} \tag{2.9}$$

This equation can also be written in terms of the position $x$, since $\Delta x = x - x_0$. Figure 2.15c shows a plot of $x$ versus $t$ for Equation 2.9, which is related to the graph of velocity versus time: The area under the curve in Figure 2.15b is equal to $v_0 t + \tfrac{1}{2}at^2$, which is equal to the displacement $\Delta x$. In fact, **the area under the graph of $v$ versus $t$ for any object is equal to the displacement $\Delta x$ of the object.**

Finally, we can obtain an expression that doesn't contain time by solving Equation 2.6 for $t$ and substituting into Equation 2.8, resulting in

$$\Delta x = \tfrac{1}{2}(v + v_0)\left(\frac{v - v_0}{a}\right) = \frac{v^2 - v_0^2}{2a}$$

$$v^2 = v_0^2 + 2a\,\Delta x \quad \text{(for constant } a\text{)} \tag{2.10}$$

Equations 2.6 and 2.9 together can solve any problem in one-dimensional motion with constant acceleration, but Equations 2.7, 2.8, and, especially, 2.10 are sometimes convenient. The three most useful equations — Equations 2.6, 2.9, and 2.10 — are listed in Table 2.4.

The best way to gain confidence in the use of these equations is to work a number of problems. There is usually more than one way to solve a given problem, depending on which equations are selected and what quantities are given. The difference lies mainly in the algebra.

**Table 2.4** Equations for Motion in a Straight Line Under Constant Acceleration

| Equation | Information Given by Equation |
|---|---|
| $v = v_0 + at$ | Velocity as a function of time |
| $\Delta x = v_0 t + \frac{1}{2}at^2$ | Displacement as a function of time |
| $v^2 = v_0^2 + 2a\Delta x$ | Velocity as a function of displacement |

Note: Motion is along the $x$-axis. At $t = 0$, the velocity of the particle is $v_0$.

---

## PROBLEM-SOLVING STRATEGY

### Motion in One Dimension at Constant Acceleration

*The following procedure is recommended for solving problems involving accelerated motion.*

1. **Read** the problem.
2. **Draw** a diagram, choosing a coordinate system, labeling initial and final points, and indicating directions of velocities and accelerations with arrows.
3. **Label** all quantities, circling the unknowns. Convert units as needed.
4. **Equations** from Table 2.4 should be selected next. All kinematics problems in this topic can be solved with the first two equations, and the third is often convenient.
5. **Solve** for the unknowns. Doing so often involves solving two equations for two unknowns.
6. **Check** your answer, using common sense and estimates.

> **Tip 2.8** Pigs Don't Fly
>
> After solving a problem, you should think about your answer and decide whether it seems reasonable. If it isn't, look for your mistake!

Most of these problems reduce to writing the kinematic equations from Table 2.4 and then substituting the correct values into the constants $a$, $v_0$, and $x_0$ from the given information. Doing this produces two equations—one linear and one quadratic—for two unknown quantities.

---

## EXAMPLE 2.4 | THE DAYTONA 500

**GOAL** Apply the basic kinematic equations.

**PROBLEM** A race car starting from rest accelerates at a constant rate of $5.00 \text{ m/s}^2$. **(a)** What is the velocity of the car after it has traveled $1.00 \times 10^2$ ft? **(b)** How much time has elapsed? **(c)** Calculate the average velocity two different ways.

**STRATEGY** We've read the problem, drawn the diagram in Figure 2.16, and chosen a coordinate system (steps 1 and 2). We'd like to find the velocity $v$ after a certain known displacement $\Delta x$. The acceleration $a$ is also known, as is the initial velocity $v_0$ (step 3, labeling, is complete), so the third equation in Table 2.4 looks most useful for solving part **(a)**. Given the velocity, the first equation in Table 2.4 can then be used to find the time in part **(b)**. Part **(c)** requires substitution into Equations 2.2 and 2.7, respectively.

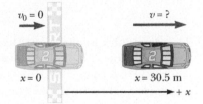

**Figure 2.16** (Example 2.4)

···················································································

**SOLUTION**

**(a)** Convert units of $\Delta x$ to SI, using the information in the inside front cover.

$$1.00 \times 10^2 \text{ ft} = (1.00 \times 10^2 \text{ ft})\left(\frac{1 \text{ m}}{3.28 \text{ ft}}\right) = 30.5 \text{ m}$$

Write the kinematics equation for $v^2$ (step 4):

$$v^2 = v_0^2 + 2a\,\Delta x$$

Solve for $v$, taking the positive square root because the car moves to the right (step 5):

$$v = \sqrt{v_0^2 + 2a\,\Delta x}$$

Substitute $v_0 = 0$, $a = 5.00 \text{ m/s}^2$, and $\Delta x = 30.5$ m:

$$v = \sqrt{v_0^2 + 2a\,\Delta x} = \sqrt{(0)^2 + 2(5.00 \text{ m/s}^2)(30.5 \text{ m})}$$
$$= \boxed{17.5 \text{ m/s}}$$

**(b)** How much time has elapsed?

Apply the first equation of Table 2.4:

$$v = at + v_0$$

Substitute values and solve for time $t$:

$$17.5 \text{ m/s} = (5.00 \text{ m/s}^2)t$$

$$t = \frac{17.5 \text{ m/s}}{5.00 \text{ m/s}^2} = \boxed{3.50 \text{ s}}$$

**(c)** Calculate the average velocity in two different ways.

Apply the definition of average velocity, Equation 2.2:

$$\bar{v} = \frac{x_f - x_i}{t_f - t_i} = \frac{30.5 \text{ m}}{3.50 \text{ s}} = \boxed{8.71 \text{ m/s}}$$

Apply the definition of average velocity in Equation 2.7:

$$\bar{v} = \frac{v_0 + v}{2} = \frac{0 + 17.5 \text{ m/s}}{2} = \boxed{8.75 \text{ m/s}}$$

**REMARKS** The answers are easy to check. An alternate technique is to use $\Delta x = v_0 t + \frac{1}{2}at^2$ to find $t$ and then use the equation $v = v_0 + at$ to find $v$. Notice that the two different equations for calculating the average velocity, due to rounding, give slightly different answers.

**QUESTION 2.4** What is the final speed if the displacement is increased by a factor of 4?

**EXERCISE 2.4** Suppose the driver in this example now slams on the brakes, stopping the car in 4.00 s. Find **(a)** the acceleration, **(b)** the distance the car travels while braking, assuming the acceleration is constant, and **(c)** the average velocity.

**ANSWERS** **(a)** $-4.38 \text{ m/s}^2$    **(b)** 35.0 m    **(c)** 8.75 m/s

---

## EXAMPLE 2.5    CAR CHASE

**GOAL** Solve a problem involving two objects, one moving at constant acceleration and the other at constant velocity.

**PROBLEM** A car traveling at a constant speed of 24.0 m/s passes a trooper hidden behind a billboard, as in Figure 2.17. One second after the speeding car passes the billboard, the trooper sets off in chase with a constant acceleration of 3.00 m/s$^2$. **(a)** How long does it take the trooper to overtake the speeding car? **(b)** How fast is the trooper going at that time?

**STRATEGY** Solving this problem involves two simultaneous kinematics equations of position: one for the trooper and the other for the car. Choose $t = 0$ to correspond to the time the trooper takes up the chase, when the car is at $x_{car} = 24.0$ m because of its head start

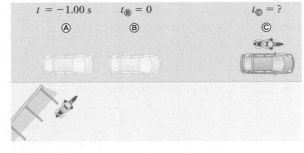

**Figure 2.17** (Example 2.5) A speeding car passes a hidden trooper. When does the trooper catch up to the car?

$(24.0 \text{ m/s} \times 1.00 \text{ s})$. The trooper catches up with the car when their positions are the same. This suggests setting $x_{trooper} = x_{car}$ and solving for time, which can then be used to find the trooper's speed in part **(b)**.

---

**SOLUTION**

**(a)** How long does it take the trooper to overtake the car?

Write the equation for the car's displacement:

$$\Delta x_{car} = x_{car} - x_0 = v_0 t + \tfrac{1}{2}a_{car}t^2$$

Take $x_0 = 24.0$ m, $v_0 = 24.0$ m/s, and $a_{car} = 0$. Solve for $x_{car}$:

$$x_{car} = x_0 + vt = 24.0 \text{ m} + (24.0 \text{ m/s})t$$

Write the equation for the trooper's position, taking $x_0 = 0$, $v_0 = 0$, and $a_{trooper} = 3.00$ m/s$^2$:

$$x_{trooper} = \tfrac{1}{2}a_{trooper}t^2 = \tfrac{1}{2}(3.00 \text{ m/s}^2)t^2 = (1.50 \text{ m/s}^2)t^2$$

Set $x_{trooper} = x_{car}$, and solve the quadratic equation. (The quadratic formula appears in Appendix A, Equation A.8.) Only the positive root is meaningful.

$$(1.50 \text{ m/s}^2)t^2 = 24.0 \text{ m} + (24.0 \text{ m/s})t$$

$$(1.50 \text{ m/s}^2)t^2 - (24.0 \text{ m/s})t - 24.0 \text{ m} = 0$$

$$t = \boxed{16.9 \text{ s}}$$

*(Continued)*

**(b)** Find the trooper's speed at that time.

Substitute the time into the trooper's velocity equation:

$$v_{\text{trooper}} = v_0 + a_{\text{trooper}}\, t = 0 + (3.00 \text{ m/s}^2)(16.9 \text{ s})$$
$$= \boxed{50.7 \text{ m/s}}$$

**REMARKS** The trooper, traveling about twice as fast as the car, must swerve or apply the brakes strongly to avoid a collision! This problem can also be solved graphically by plotting position vs. time for each vehicle on the same graph. The intersection of the two graphs corresponds to the time and position at which the trooper overtakes the car.

**QUESTION 2.5** The graphical solution corresponds to finding the intersection of what two types of curves in the *tx*-plane?

**EXERCISE 2.5** A motorist with an expired license tag is traveling at a constant 10.0 m/s down a street, and a police officer on a motorcycle, taking another 5.00 s to react, gives chase at an acceleration of 2.00 m/s$^2$. Find **(a)** the time required to catch the car and **(b)** the distance the officer travels while overtaking the motorist.

**ANSWERS** **(a)** 13.7 s    **(b)** 188 m

---

## EXAMPLE 2.6    RUNWAY LENGTH

**GOAL** Apply kinematics to horizontal motion with two phases.

**PROBLEM** A typical jetliner lands at a speed of $1.60 \times 10^2$ mi/h and decelerates at the rate of (10.0 mi/h)/s. If the plane travels at a constant speed of $1.60 \times 10^2$ mi/h for 1.00 s after landing before applying the brakes, what is the total displacement of the aircraft between touchdown on the runway and coming to rest?

**STRATEGY** See Figure 2.18. First, convert all quantities to SI units. The problem must be solved in two parts, or phases, corresponding to the initial coast after touchdown, followed by braking. Using the kinematic equations, find the displacement during each part and add the two displacements.

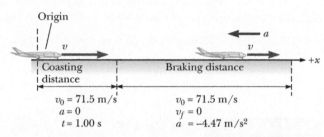

**Figure 2.18** (Example 2.6) Coasting and braking distances for a landing jetliner.

### SOLUTION

Convert units of speed and acceleration to SI:

$$v_0 = (1.60 \times 10^2 \text{ mi/h})\left(\frac{0.447 \text{ m/s}}{1.00 \text{ mi/h}}\right) = 71.5 \text{ m/s}$$

$$a = (-10.0 \text{ (mi/h)/s})\left(\frac{0.447 \text{ m/s}}{1.00 \text{ mi/h}}\right) = -4.47 \text{ m/s}^2$$

Taking $a = 0$, $v_0 = 71.5$ m/s, and $t = 1.00$ s, find the displacement while the plane is coasting:

$$\Delta x_{\text{coasting}} = v_0 t + \tfrac{1}{2}at^2 = (71.5 \text{ m/s})(1.00 \text{ s}) + 0 = 71.5 \text{ m}$$

Use the time-independent kinematic equation to find the displacement while the plane is braking.

$$v^2 = v_0^2 + 2\,a\Delta x_{\text{braking}}$$

Take $a = -4.47$ m/s$^2$ and $v_0 = 71.5$ m/s. The negative sign on $a$ means that the plane is slowing down.

$$\Delta x_{\text{braking}} = \frac{v^2 - v_0^2}{2a} = \frac{0 - (71.5 \text{ m/s})^2}{2.00(-4.47 \text{ m/s}^2)} = 572 \text{ m}$$

Sum the two results to find the total displacement:

$$\Delta x_{\text{coasting}} + \Delta x_{\text{braking}} = 71.5 \text{ m} + 572 \text{ m} = \boxed{644 \text{ m}}$$

**REMARKS** To find the displacement while braking, we could have used the two kinematics equations involving time, namely, $\Delta x = v_0 t + \tfrac{1}{2}at^2$ and $v = v_0 + at$, but because we weren't interested in time, the time-independent equation was easier to use.

**QUESTION 2.6** How would the answer change if the plane coasted for 2.00 s before the pilot applied the brakes?

**EXERCISE 2.6** A jet lands at 80.0 m/s, and the pilot applies the brakes 2.00 s after landing. Find the acceleration needed to stop the jet within $5.00 \times 10^2$ m after touchdown.

**ANSWER** $-9.41$ m/s$^2$

## EXAMPLE 2.7   THE ACELA: THE PORSCHE OF AMERICAN TRAINS

**GOAL**  Find accelerations and displacements from a velocity vs. time graph.

**PROBLEM**  The sleek high-speed electric train known as the Acela (pronounced "ah cel´ la") is currently in service on the Washington-New York-Boston run. The Acela consists of two power cars and six coaches and can carry 304 passengers at speeds up to 170 mi/h. To negotiate curves comfortably at high speeds, the train carriages tilt as much as 6° from the vertical, preventing passengers from being pushed to the side. A velocity vs. time graph for the Acela is shown in Figure 2.19a. **(a)** Describe the motion of the Acela. **(b)** Find the peak acceleration of the Acela in miles per hour per second ((mi/h)/s) as the train speeds up from 45 mi/h to 170 mi/h. **(c)** Find the train's displacement in miles between $t = 0$ and $t = 200$ s. **(d)** Find the average acceleration of the Acela and its displacement in miles in the interval from 200 s to 300 s. (The train has regenerative braking, which means that it feeds energy back into the utility lines each time it stops!)

**(e)** Find the total displacement in the interval from 0 to 400 s. *Note:* Assume that all given quantities and estimates are good to two significant figures. (Estimates by different individuals may vary and result in slightly different answers.)

**STRATEGY**  For part **(a)**, remember that the slope of the tangent line at any point of the velocity vs. time graph gives the acceleration at that time. To find the peak acceleration in part **(b)**, study the graph and locate the point at which the slope is steepest. In parts **(c)** through **(e)**, estimating the area under the curve gives the displacement during a given period, with areas below the time axis, as in part **(e)**, subtracted from the total. The average acceleration in part **(d)** can be obtained by substituting numbers taken from the graph into the definition of average acceleration, $\bar{a} = \Delta v/\Delta t$.

. . . . . . . . . . . . . . . . . . . . . . . . . . . . . . . . . . . . . . . . . . . . . . . . . . . . . . . . . . . . . . . . . . . . . . . . . . . . . . . . . . . . . .

**SOLUTION**
**(a)** Describe the motion.

From about −50 s to 50 s, the Acela cruises at a constant velocity in the +x-direction. Then the train accelerates in the +x-direction from 50 s to 200 s, reaching a top speed of about 170 mi/h, whereupon it brakes to rest at 350 s and reverses, steadily gaining speed in the −x-direction.

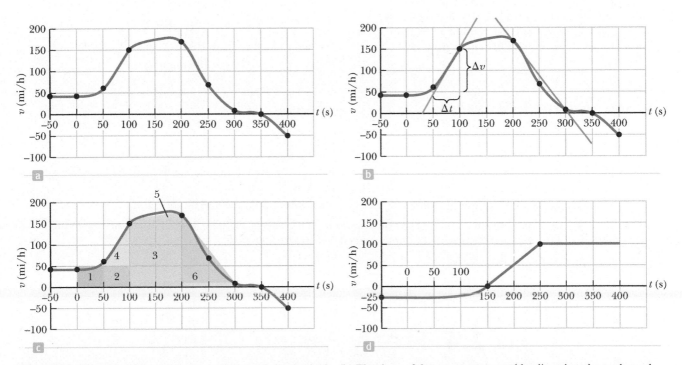

**Figure 2.19** (Example 2.7) (a) Velocity vs. time graph for the Acela. (b) The slope of the steepest tangent blue line gives the peak acceleration, and the slope of the green line is the average acceleration between 200 s and 300 s. (c) The area under the velocity vs. time graph in some time interval gives the displacement of the Acela in that time interval. (d) (Exercise 2.7).

(*Continued*)

**(b)** Find the peak acceleration.

Calculate the slope of the steepest tangent line, which connects the points (50 s, 50 mi/h) and (100 s, 150 mi/h) (the light blue line in Figure 2.19b):

$$a = \text{slope} = \frac{\Delta v}{\Delta t} = \frac{(1.5 \times 10^2 - 5.0 \times 10^1) \text{ mi/h}}{(1.0 \times 10^2 - 5.0 \times 10^1)\text{s}}$$

$$= 2.0 \text{ (mi/h)/s}$$

**(c)** Find the displacement between 0 s and 200 s.

Using triangles and rectangles, approximate the area in Figure 2.19c:

$$\Delta x_{0 \to 200\text{ s}} = \text{area}_1 + \text{area}_2 + \text{area}_3 + \text{area}_4 + \text{area}_5$$

$$\approx (5.0 \times 10^1 \text{ mi/h})(5.0 \times 10^1 \text{ s})$$

$$+ (5.0 \times 10^1 \text{ mi/h})(5.0 \times 10^1 \text{ s})$$

$$+ (1.6 \times 10^2 \text{ mi/h})(1.0 \times 10^2 \text{ s})$$

$$+ \tfrac{1}{2}(5.0 \times 10^1 \text{ s})(1.0 \times 10^2 \text{ mi/h})$$

$$+ \tfrac{1}{2}(1.0 \times 10^2 \text{ s})(1.7 \times 10^2 \text{ mi/h} - 1.6 \times 10^2 \text{ mi/h})$$

$$= 2.4 \times 10^4 \text{ (mi/h)s}$$

Convert units to miles by converting hours to seconds:

$$\Delta x_{0 \to 200\text{ s}} \approx 2.4 \times 10^4 \frac{\text{mi} \cdot \text{s}}{\text{h}}\left(\frac{1 \text{ h}}{3\ 600 \text{ s}}\right) = 6.7 \text{ mi}$$

**(d)** Find the average acceleration from 200 s to 300 s, and find the displacement.

The slope of the green line is the average acceleration from 200 s to 300 s (Fig. 2.19b):

$$\bar{a} = \text{slope} = \frac{\Delta v}{\Delta t} = \frac{(1.0 \times 10^1 - 1.7 \times 10^2) \text{ mi/h}}{1.0 \times 10^2 \text{ s}}$$

$$= -1.6 \text{ (mi/h)/s}$$

The displacement from 200 s to 300 s is equal to area$_6$, which is the area of a triangle plus the area of a very narrow rectangle beneath the triangle:

$$\Delta x_{200 \to 300\text{ s}} \approx \tfrac{1}{2}(1.0 \times 10^2 \text{ s})(1.7 \times 10^2 - 1.0 \times 10^1) \text{ mi/h}$$

$$+ (1.0 \times 10^1 \text{ mi/h})(1.0 \times 10^2 \text{ s})$$

$$= 9.0 \times 10^3 (\text{mi/h})(\text{s}) = 2.5 \text{ mi}$$

**(e)** Find the total displacement from 0 s to 400 s.

The total displacement is the sum of all the individual displacements. We still need to calculate the displacements for the time intervals from 300 s to 350 s and from 350 s to 400 s. The latter is negative because it's below the time axis.

$$\Delta x_{300 \to 350\text{ s}} \approx \tfrac{1}{2}(5.0 \times 10^1 \text{ s})(1.0 \times 10^1 \text{ mi/h})$$

$$= 2.5 \times 10^2 (\text{mi/h})(\text{s})$$

$$\Delta x_{350 \to 400\text{ s}} \approx \tfrac{1}{2}(5.0 \times 10^1 \text{ s})(-5.0 \times 10^1 \text{ mi/h})$$

$$= -1.3 \times 10^3 (\text{mi/h})(\text{s})$$

Find the total displacement by summing the parts:

$$\Delta x_{0 \to 400\text{ s}} \approx (2.4 \times 10^4 + 9.0 \times 10^3 + 2.5 \times 10^2$$

$$-1.3 \times 10^3)(\text{mi/h})(\text{s}) = 8.9 \text{ mi}$$

. . . . . . . . . . . . . . . . . . . . . . . . . . . . . . . . . . . . . . . . . . . . . . . . . . . . . . . . . . . . . . . . . . . . . . . . . . . . . . . . . . . . . . . . . . . . . .

**REMARKS** There are a number of ways to find the approximate area under a graph. Choice of technique is a personal preference.

**QUESTION 2.7** According to the graph in Figure 2.19a, at what different times is the acceleration zero?

**EXERCISE 2.7** Suppose the velocity vs. time graph of another train is given in Figure 2.19d. Find **(a)** the maximum instantaneous acceleration and **(b)** the total displacement in the interval from 0 s to $4.00 \times 10^2$ s.

**ANSWERS** **(a)** 1.0 (mi/h)/s   **(b)** 4.7 mi

# 2.4 Freely Falling Objects

When air resistance is negligible, all objects dropped under the influence of gravity near Earth's surface fall toward Earth with the same constant acceleration. This idea may seem obvious today, but it wasn't until about 1600 that it was accepted.

Prior to that time, the teachings of the great philosopher Aristotle (384–322 BC) had held that heavier objects fell faster than lighter ones.

According to legend, Galileo discovered the law of falling objects by observing that two different weights dropped simultaneously from the Leaning Tower of Pisa hit the ground at approximately the same time. Although it's unlikely that this particular experiment was carried out, we know that Galileo performed many systematic experiments with objects moving on inclined planes. In his experiments, he rolled balls down a slight incline and measured the distances they covered in successive time intervals. The purpose of the incline was to reduce the acceleration and enable Galileo to make accurate measurements of the intervals. (Some people refer to this experiment as "diluting gravity.") By gradually increasing the slope of the incline, he was finally able to draw mathematical conclusions about freely falling objects, because a falling ball is equivalent to a ball going down a vertical incline. Galileo's achievements in the science of mechanics paved the way for Newton in his development of the laws of motion, which we will study in Topic 4.

Try the following experiment: Drop a hammer and a feather simultaneously from the same height. The hammer hits the floor first because air drag has a greater effect on the much lighter feather. On August 2, 1971, this same experiment was conducted on the Moon by astronaut David Scott, and the hammer and feather fell with exactly the same acceleration, as expected, hitting the lunar surface at the same time. In the idealized case where air resistance is negligible, such motion is called *free fall*.

The expression *freely falling object* doesn't necessarily refer to an object dropped from rest. **A freely falling object is any object moving freely under the influence of gravity alone, regardless of its initial motion.** Objects thrown upward or downward and those released from rest are all considered freely falling.

We denote the magnitude of the **free-fall acceleration** by the symbol $g$. The value of $g$ decreases with increasing altitude, and varies slightly with latitude as well. At Earth's surface, the value of $g$ is approximately $9.80 \text{ m/s}^2$. Unless stated otherwise, we will use this value for $g$ in doing calculations. For quick estimates, use $g \approx 10 \text{ m/s}^2$.

If we neglect air resistance and assume that the free-fall acceleration doesn't vary with altitude over short vertical distances, then the motion of a freely falling object is the same as motion in one dimension under constant acceleration. This means that the kinematics equations developed in Section 2.3 can be applied. It's conventional to define "up" as the $+y$-direction and to use $y$ as the position variable. In that case, the acceleration is $a = -g = -9.80 \text{ m/s}^2$. In Topic 7, we study the variation in $g$ with altitude.

**GALILEO GALILEI**
**Italian Physicist and Astronomer (1564–1642)**

Galileo formulated the laws that govern the motion of objects in free fall. He also investigated the motion of an object on an inclined plane, established the concept of relative motion, invented the thermometer, and discovered that the motion of a swinging pendulum could be used to measure time intervals. After designing and constructing his own telescope, he discovered four of Jupiter's moons, found that our own Moon's surface is rough, discovered sunspots and the phases of Venus, and showed that the Milky Way consists of an enormous number of stars. Galileo publicly defended Nicolaus Copernicus's assertion that the Sun is at the center of the Universe (the heliocentric system). He published *Dialogue Concerning Two New World Systems* to support the Copernican model, a view the Church declared to be heretical. After being taken to Rome in 1633 on a charge of heresy, he was sentenced to life imprisonment and later was confined to his villa at Arcetri, near Florence, where he died in 1642.

## Quick Quiz

**2.6** A tennis player on serve tosses a ball straight up. While the ball is in free fall, does its acceleration (a) increase, (b) decrease, (c) increase and then decrease, (d) decrease and then increase, or (e) remain constant?

**2.7** As the tennis ball of Quick Quiz 2.6 travels through the air, does its speed (a) increase, (b) decrease, (c) decrease and then increase, (d) increase and then decrease, or (e) remain the same?

**2.8** A skydiver jumps out of a hovering helicopter. A few seconds later, another skydiver jumps out, so they both fall along the same vertical line relative to the helicopter. Assume both skydivers fall with the same acceleration. Does the vertical distance between them (a) increase, (b) decrease, or (c) stay the same? Does the difference in their velocities (d) increase, (e) decrease, or (f) stay the same?

**EXAMPLE 2.8    NOT A BAD THROW FOR A ROOKIE!**

**GOAL** Apply the kinematic equations to a freely falling object with a nonzero initial velocity.

**PROBLEM** A ball is thrown from the top of a building with an initial velocity of 20.0 m/s straight upward, at an initial height of 50.0 m above the ground. The ball just misses the edge of the roof on its way down, as shown in Figure 2.20. Determine **(a)** the time needed for the ball to reach its maximum height, **(b)** the maximum height, **(c)** the time needed for the ball to return to the height from which it was thrown and the velocity of the ball at that instant, **(d)** the time needed for the ball to reach the ground, and **(e)** the velocity and position of the ball at $t = 5.00$ s. Neglect air drag.

**STRATEGY** The diagram in Figure 2.20 establishes a coordinate system with $y_0 = 0$ at the level at which the ball is released from the thrower's hand, with $y$ positive upward. Write the velocity and position kinematic equations for the ball, and substitute the given information. All the answers come from these two equations by using simple algebra or by just substituting the time. In part **(a)**, for example, the ball comes to rest for an instant at its maximum height, so set $v = 0$ at this point and solve for time. Then substitute the time into the displacement equation, obtaining the maximum height.

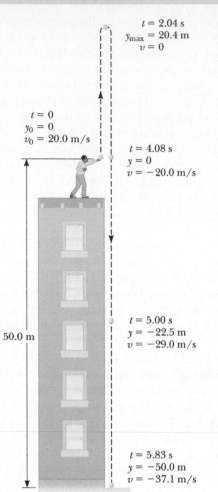

$t = 2.04$ s
$y_{max} = 20.4$ m
$v = 0$

$t = 0$
$y_0 = 0$
$v_0 = 20.0$ m/s

$t = 4.08$ s
$y = 0$
$v = -20.0$ m/s

50.0 m

$t = 5.00$ s
$y = -22.5$ m
$v = -29.0$ m/s

$t = 5.83$ s
$y = -50.0$ m
$v = -37.1$ m/s

**Figure 2.20** (Example 2.8) A ball is thrown upward with an initial velocity of $v_0 = +20.0$ m/s. Positions and velocities are given for several times.

**SOLUTION**

**(a)** Find the time when the ball reaches its maximum height.

Write the velocity and position kinematic equations:

$$v = at + v_0$$
$$\Delta y = y - y_0 = v_0 t + \tfrac{1}{2} a t^2$$

Substitute $a = -9.80$ m/s², $v_0 = 20.0$ m/s, and $y_0 = 0$ into the preceding two equations:

**(1)**   $v = (-9.80 \text{ m/s}^2)t + 20.0 \text{ m/s}$
**(2)**   $y = (20.0 \text{ m/s})t - (4.90 \text{ m/s}^2)t^2$

Substitute $v = 0$, the velocity at maximum height, into Equation (1) and solve for time:

$0 = (-9.80 \text{ m/s}^2)t + 20.0 \text{ m/s}$

$t = \dfrac{-20.0 \text{ m/s}}{-9.80 \text{ m/s}^2} = \boxed{2.04 \text{ s}}$

**(b)** Determine the ball's maximum height.

Substitute the time $t = 2.04$ s into Equation (2):

$y_{max} = (20.0 \text{ m/s})(2.04 \text{ s}) - (4.90 \text{ m/s}^2)(2.04 \text{ s})^2 = \boxed{20.4 \text{ m}}$

**(c)** Find the time the ball takes to return to its initial position, and find the velocity of the ball at that time.

Set $y = 0$ in Equation (2) and solve for $t$:

$0 = (20.0 \text{ m/s})t - (4.90 \text{ m/s}^2)t^2$
$\ \ = t(20.0 \text{ m/s} - (4.90 \text{ m/s}^2)t)$
$t = \boxed{4.08 \text{ s}}$

Substitute the time into Equation (1) to get the velocity:

$v = 20.0 \text{ m/s} + (-9.80 \text{ m/s}^2)(4.08 \text{ s}) = \boxed{-20.0 \text{ m/s}}$

**(d)** Find the time required for the ball to reach the ground.

In Equation (2), set $y = -50.0$ m:

$-50.0 \text{ m} = (20.0 \text{ m/s})t - (4.90 \text{ m/s}^2)t^2$

Apply the quadratic formula and take the positive root:

$t = \boxed{5.83 \text{ s}}$

**(e)** Find the velocity and position of the ball at $t = 5.00$ s.

Substitute values into Equations (1) and (2):

$v = (-9.80 \text{ m/s}^2)(5.00 \text{ s}) + 20.0 \text{ m/s} = \boxed{-29.0 \text{ m/s}}$

$y = (20.0 \text{ m/s})(5.00 \text{ s}) - (4.90 \text{ m/s}^2)(5.00 \text{ s})^2 = \boxed{-22.5 \text{ m}}$

**REMARKS** Notice how everything follows from the two kinematic equations. Once they are written down and the constants correctly identified as in Equations (1) and (2), the rest is relatively easy. If the ball were thrown downward, the initial velocity would have been negative.

**QUESTION 2.8** How would the answer to part **(b)**, the maximum height, change if the person throwing the ball was jumping upward at the instant he released the ball?

**EXERCISE 2.8** A projectile is launched straight up at 60.0 m/s from a height of 80.0 m, at the edge of a sheer cliff. The projectile falls, just missing the cliff and hitting the ground below. Find **(a)** the maximum height of the projectile above the point of firing, **(b)** the time it takes to hit the ground at the base of the cliff, and **(c)** its velocity at impact.

**ANSWERS** **(a)** 184 m    **(b)** 13.5 s    **(c)** $-72.3$ m/s

---

## EXAMPLE 2.9    MAXIMUM HEIGHT DERIVED

**GOAL** Find the maximum height of a thrown projectile using symbols.

**PROBLEM** Refer to Example 2.8. Use symbolic manipulation to find **(a)** the time $t_{max}$ it takes the ball to reach its maximum height and **(b)** an expression for the maximum height that doesn't depend on time. Answers should be expressed in terms of the quantities $v_0$, $g$, and $y_0$ only.

**STRATEGY** When the ball reaches its maximum height, its velocity is zero, so for part **(a)** solve the kinematics velocity equation for time $t$ and set $v = 0$. For part **(b)**, substitute the expression for time found in part **(a)** into the displacement equation, solving it for the maximum height.

**SOLUTION**

**(a)** Find the time it takes the ball to reach its maximum height.

Write the velocity kinematics equation:

$v = at + v_0$

Move $v_0$ to the left side of the equation:

$v - v_0 = at$

Divide both sides by $a$:

$\dfrac{v - v_0}{a} = \dfrac{at}{a} = t$

Turn the equation around so that $t$ is on the left and substitute $v = 0$, corresponding to the velocity at maximum height:

$(1) \quad t = \dfrac{-v_0}{a}$

Replace $t$ by $t_{max}$ and substitute $a = -g$:

$(2) \quad t_{max} = \dfrac{v_0}{g}$

**(b)** Find the maximum height.

Write the equation for the position $y$ at any time:

$y = y_0 + v_0 t + \frac{1}{2}at^2$

*(Continued)*

Substitute $t = -v_0/a$, which corresponds to the time it takes to reach $y_{max}$, the maximum height:

$$y_{max} = y_0 + v_0\left(\frac{-v_0}{a}\right) + \frac{1}{2}a\left(\frac{-v_0}{a}\right)^2$$

$$= y_0 - \frac{v_0^2}{a} + \frac{1}{2}\frac{v_0^2}{a}$$

Combine the last two terms and substitute $a = -g$:

$$(3) \quad y_{max} = y_0 + \frac{v_0^2}{2g}$$

. . . . . . . . . . . . . . . . . . . . . . . . . . . . . . . . . . . . . . . . . . . . . . . . . . . . . . . . . . . . . . . . . . . . . . . . . . . . . . . . . .

**REMARKS** Notice that $g = +9.8$ m/s², so the second term is positive overall. Equations (1)–(3) are much more useful than a numerical answer because the effect of changing one value can be seen immediately. For example, doubling the initial velocity $v_0$ quadruples the displacement above the point of release. Notice also that $y_{max}$ could be obtained more readily from the time-independent equation, $v^2 - v_0^2 = 2a\Delta y$.

**QUESTION 2.9** By what factor would the maximum displacement above the rooftop be increased if the building were transported to the Moon, where $a = -\frac{1}{6}g$?

**EXERCISE 2.9** (a) Using symbols, find the time $t_E$ it takes for a ball to reach the ground on Earth if released from rest at height $y_0$. (b) In terms of $t_E$, how much time $t_M$ would be required if the building were on Mars, where $a = -0.385g$?

**ANSWERS** (a) $t_E = \sqrt{\dfrac{2y_0}{g}}$  (b) $t_M = 1.61 t_E$

---

### EXAMPLE 2.10  A ROCKET GOES BALLISTIC

**GOAL** Solve a problem involving a powered ascent followed by free-fall motion.

**PROBLEM** A rocket moves straight upward, starting from rest with an acceleration of $+29.4$ m/s². It runs out of fuel at the end of 4.00 s and continues to coast upward, reaching a maximum height before falling back to Earth. **(a)** Find the rocket's velocity and position at the end of 4.00 s. **(b)** Find the maximum height the rocket reaches. **(c)** Find the velocity the instant before the rocket crashes on the ground.

**STRATEGY** Take $y = 0$ at the launch point and $y$ positive upward, as in Figure 2.21. The problem consists of two phases. In phase 1 the rocket has a net *upward* acceleration of 29.4 m/s², and we can use the kinematic equations with constant acceleration $a$ to find the height and velocity of the rocket at the end of phase 1, when the fuel is burned up. In phase 2 the rocket is in free fall and has an acceleration of $-9.80$ m/s², with initial velocity and position given by the results of phase 1. Apply the kinematic equations for free fall.

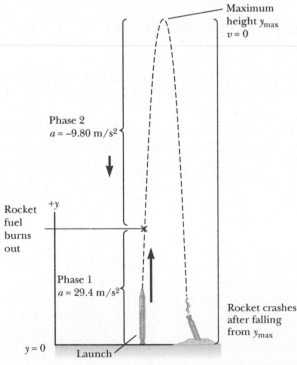

**Figure 2.21** (Example 2.10) Two linked phases of motion for a rocket that is launched, uses up its fuel, and crashes.

. . . . . . . . . . . . . . . . . . . . . . . . . . . . . . . . . . . . . . . . . . . . . . . . . . . . . . . . . . . . . . . . . . . . . . . . . . . . . . . . . .

### SOLUTION

**(a)** Phase 1: Find the rocket's velocity and position after 4.00 s.

Write the velocity and position kinematic equations:

$$(1) \quad v = v_0 + at$$

$$(2) \quad \Delta y = y - y_0 = v_0 t + \frac{1}{2}at^2$$

Adapt these equations to phase 1, substituting $a = 29.4$ m/s$^2$, $v_0 = 0$, and $y_0 = 0$:

**(3)** $v = (29.4 \text{ m/s}^2)t$

**(4)** $y = \frac{1}{2}(29.4 \text{ m/s}^2)t^2 = (14.7 \text{ m/s}^2)t^2$

Substitute $t = 4.00$ s into Equations (3) and (4) to find the rocket's velocity $v$ and position $y$ at the time of burnout. These will be called $v_b$ and $y_b$, respectively.

$v_b = 118$ m/s  and  $y_b = 235$ m

**(b)** Phase 2: Find the maximum height the rocket attains.

Adapt Equations (1) and (2) to phase 2, substituting $a = -9.8$ m/s$^2$, $v_0 = v_b = 118$ m/s, and $y_0 = y_b = 235$ m:

**(5)** $v = (-9.8 \text{ m/s}^2)t + 118$ m/s

**(6)** $y = 235 \text{ m} + (118 \text{ m/s})t - (4.90 \text{ m/s}^2)t^2$

Substitute $v = 0$ (the rocket's velocity at maximum height) in Equation (5) to get the time it takes the rocket to reach its maximum height:

$0 = (-9.8 \text{ m/s}^2)t + 118 \text{ m/s} \quad \rightarrow \quad t = \dfrac{118 \text{ m/s}}{9.80 \text{ m/s}^2} = 12.0$ s

Substitute $t = 12.0$ s into Equation (6) to find the rocket's maximum height:

$y_{\max} = 235 \text{ m} + (118 \text{ m/s})(12.0 \text{ s}) - (4.90 \text{ m/s}^2)(12.0 \text{ s})^2$

$= 945$ m

**(c)** Phase 2: Find the velocity of the rocket just prior to impact.

Find the time to impact by setting $y = 0$ in Equation (6) and using the quadratic formula:

$0 = 235 \text{ m} + (118 \text{ m/s})t - (4.90 \text{ m/s})t^2$

$t = 25.9$ s

Substitute this value of $t$ into Equation (5):

$v = (-9.80 \text{ m/s}^2)(25.9 \text{ s}) + 118 \text{ m/s} = -136$ m/s

**REMARKS** You may think that it is more natural to break this problem into three phases, with the second phase ending at the maximum height and the third phase a free fall from maximum height to the ground. Although this approach gives the correct answer, it's an unnecessary complication. Two phases are sufficient, one for each different acceleration.

**QUESTION 2.10** If, instead, some fuel remains, at what height should the engines be fired again to brake the rocket's fall and allow a perfectly soft landing? (Assume the same acceleration as during the initial ascent.)

**EXERCISE 2.10** An experimental rocket designed to land upright falls freely from a height of $2.00 \times 10^2$ m, starting at rest. At a height of 80.0 m, the rocket's engines start and provide constant upward acceleration until the rocket lands. What acceleration is required if the speed on touchdown is to be zero? (Neglect air resistance.)

**ANSWER** 14.7 m/s$^2$

---

## SUMMARY

### 2.1 Displacement, Velocity, and Acceleration

#### 2.1.1 Displacement

The **displacement** of an object moving along the $x$-axis is defined as the change in position of the object,

$$\Delta x \equiv x_f - x_i \qquad [2.1]$$

where $x_i$ is the initial position of the object and $x_f$ is its final position.

A **vector** quantity is characterized by both a magnitude and a direction. A **scalar** quantity has a magnitude only.

#### 2.1.2 Velocity

The **average speed** of an object is given by

$$\text{Average speed} \equiv \frac{\text{path length}}{\text{elapsed time}}$$

The **average velocity** $\bar{v}$ during a time interval $\Delta t$ is the displacement $\Delta x$ divided by $\Delta t$.

$$\bar{v} \equiv \frac{\Delta x}{\Delta t} = \frac{x_f - x_i}{t_f - t_i} \qquad [2.2]$$

The average velocity is equal to the slope of the straight line joining the initial and final points on a graph of the position of the object versus time.

The slope of the line tangent to the position versus time curve at some point is equal to the **instantaneous velocity** at that time. The **instantaneous speed** of an object is defined as the magnitude of the instantaneous velocity.

## 2.1.3 Acceleration

The **average acceleration** $\bar{a}$ of an object undergoing a change in velocity $\Delta v$ during a time interval $\Delta t$ is

$$\bar{a} \equiv \frac{\Delta v}{\Delta t} = \frac{v_f - v_i}{t_f - t_i} \qquad [2.4]$$

The **instantaneous acceleration** of an object at a certain time equals the slope of a velocity versus time graph at that instant.

## 2.3 One-Dimensional Motion with Constant Acceleration

The most useful equations that describe the motion of an object moving with constant acceleration along the $x$-axis are as follows:

$$v = v_0 + at \qquad [2.6]$$

$$\Delta x = v_0 t + \tfrac{1}{2}at^2 \qquad [2.9]$$

$$v^2 = v_0^2 + 2a\Delta x \qquad [2.10]$$

All problems can be solved with the first two equations alone, the last being convenient when time doesn't explicitly enter the problem. After the constants are properly identified, most problems reduce to one or two equations in as many unknowns.

## 2.4 Freely Falling Objects

An object falling in the presence of Earth's gravity exhibits a free-fall acceleration directed toward Earth's center. If air friction is neglected and if the altitude of the falling object is small compared with Earth's radius, then we can assume that the free-fall acceleration $g = 9.8$ m/s$^2$ is constant over the range of motion. Equations 2.6, 2.9, and 2.10 apply, with $a = -g$.

## CONCEPTUAL QUESTIONS

1. If the velocity of a particle is nonzero, can the particle's acceleration be zero? Explain.

2. If the velocity of a particle is zero, can the particle's acceleration be nonzero? Explain.

3. If a car is traveling eastward, can its acceleration be westward? Explain.

4. (a) Can the equations in Table 2.4 be used in a situation where the acceleration varies with time? (b) Can they be used when the acceleration is zero?

5. Two cars are moving in the same direction in parallel lanes along a highway. At some instant, car A is traveling faster than car B. Does that mean the acceleration of A is greater than that of B at that instant? (a) Yes. At any instant, a faster object always has a larger acceleration. (b) No. Acceleration only tells how an object's velocity is *changing* at some instant.

6. Figure CQ2.6 shows strobe photographs taken of a disk moving from left to right under different conditions. The time interval between images is constant. Taking the direction to the right to be positive, describe the motion of the disk in each case. For which case is (a) the acceleration positive? (b) the acceleration negative? (c) the velocity constant?

7. (a) Can the instantaneous velocity of an object at an instant of time ever be greater in magnitude than the average velocity over a time interval containing that instant? (b) Can it ever be less?

8. A ball is thrown vertically upward. (a) What are its velocity and acceleration when it reaches its maximum altitude? (b) What is the acceleration of the ball just before it hits the ground?

9. An object moves along the $x$-axis, its position given by $x(t) = 2t^2$. Which of the following cannot be obtained from a graph of $x$ vs. $t$? (a) The velocity at any instant (b) the acceleration at any instant (c) the displacement during some time interval (d) the average velocity during some time interval (e) the speed of the particle at any instant.

10. A ball is thrown straight up in the air. For which situation are both the instantaneous velocity and the acceleration zero? (a) On the way up (b) at the top of the flight path (c) on the way down (d) halfway up and halfway down (e) none of these.

11. A juggler throws a bowling pin straight up in the air. After the pin leaves his hand and while it is in the air, which statement is true? (a) The velocity of the pin is always in the same direction as its acceleration. (b) The velocity of the pin is never in the same direction as its acceleration. (c) The acceleration of the pin is zero. (d) The velocity of the pin is opposite its acceleration on the way up. (e) The velocity of the pin is in the same direction as its acceleration on the way up.

12. A racing car starts from rest and reaches a final speed $v$ in a time $t$. If the acceleration of the car is constant during this time, which of the following statements must be true? (a) The car travels a distance $vt$. (b) The average speed of the car is $v/2$. (c) The acceleration of the car is $v/t$. (d) The velocity of the car remains constant. (e) None of these.

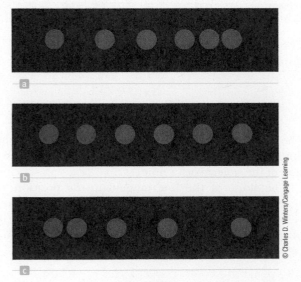

**Figure CQ2.6**

# PROBLEMS

*See the Preface for an explanation of the icons used in this problems set.*

## 2.1 Displacement, Velocity, and Acceleration

1. BIO The speed of a nerve impulse in the human body is about 100 m/s. If you accidentally stub your toe in the dark, estimate the time it takes the nerve impulse to travel to your brain.

2. Light travels at a speed of about $3 \times 10^8$ m/s. (a) How many miles does a pulse of light travel in a time interval of 0.1 s, which is about the blink of an eye? (b) Compare this distance to the diameter of Earth.

3. A person travels by car from one city to another with different constant speeds between pairs of cities. She drives for 30.0 min at 80.0 km/h, 12.0 min at 100 km/h, and 45.0 min at 40.0 km/h and spends 15.0 min eating lunch and buying gas. (a) Determine the average speed for the trip. (b) Determine the distance between the initial and final cities along the route.

4. A football player runs from his own goal line to the opposing team's goal line, returning to the fifty-yard line, all in 18.0 s. Calculate (a) his average speed, and (b) the magnitude of his average velocity.

5. Two boats start together and race across a 60-km-wide lake and back. Boat A goes across at 60 km/h and returns at 60 km/h. Boat B goes across at 30 km/h, and its crew, realizing how far behind it is getting, returns at 90 km/h. Turnaround times are negligible, and the boat that completes the round trip first wins. (a) Which boat wins and by how much? (Or is it a tie?) (b) What is the average velocity of the winning boat?

6. A graph of position versus time for a certain particle moving along the $x$-axis is shown in Figure P2.6. Find the average velocity in the time intervals from (a) 0 to 2.00 s, (b) 0 to 4.00 s, (c) 2.00 s to 4.00 s, (d) 4.00 s to 7.00 s, and (e) 0 to 8.00 s.

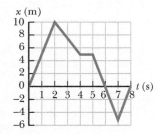

**Figure P2.6** (Problems 6 and 17)

7. V A motorist drives north for 35.0 minutes at 85.0 km/h and then stops for 15.0 minutes. He then continues north, traveling 130. km in 2.00 h. (a) What is his total displacement? (b) What is his average velocity?

8. A tennis player moves in a straight-line path as shown in Figure P2.8. Find her average velocity in the time intervals from (a) 0 to 1.0 s, (b) 0 to 4.0 s, (c) 1.0 s to 5.0 s, and (d) 0 to 5.0 s.

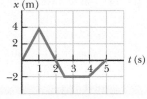

**Figure P2.8**

9. A jet plane has a takeoff speed of $v_{to} = 75$ m/s and can move along the runway at an average acceleration of 1.3 m/s$^2$. If the length of the runway is 2.5 km, will the plane be able to use this runway safely? Defend your answer.

10. Two cars travel in the same direction along a straight highway, one at a constant speed of 55 mi/h and the other at 70 mi/h. (a) Assuming they start at the same point, how much sooner does the faster car arrive at a destination 10 mi away? (b) How far must the faster car travel before it has a 15-min lead on the slower car?

11. The cheetah can reach a top speed of 114 km/h (71 mi/h). While chasing its prey in a short sprint, a cheetah starts from rest and runs 45 m in a straight line, reaching a final speed of 72 km/h. (a) Determine the cheetah's average acceleration during the short sprint, and (b) find its displacement at $t = 3.5$ s.

12. S An athlete swims the length $L$ of a pool in a time $t_1$ and makes the return trip to the starting position in a time $t_2$. If she is swimming initially in the positive $x$-direction, determine her average velocities symbolically in (a) the first half of the swim, (b) the second half of the swim, and (c) the round trip. (d) What is her average speed for the round trip?

13. T A person takes a trip, driving with a constant speed of 89.5 km/h, except for a 22.0-min rest stop. If the person's average speed is 77.8 km/h, (a) how much time is spent on the trip and (b) how far does the person travel?

14. A tortoise can run with a speed of 0.10 m/s, and a hare can run 20 times as fast. In a race, they both start at the same time, but the hare stops to rest for 2.0 minutes. The tortoise wins by a shell (20 cm). (a) How long does the race take? (b) What is the length of the race?

15. To qualify for the finals in a racing event, a race car must achieve an average speed of 250. km/h on a track with a total length of $1.60 \times 10^3$. If a particular car covers the first half of the track at an average speed of 230. km/h, what minimum average speed must it have in the second half of the event to qualify?

16. BIO A paper in the journal *Current Biology* tells of some jellyfish-like animals that attack their prey by launching stinging cells in one of the animal kingdom's fastest movements. High-speed photography showed the cells were accelerated from rest for 700. ns at $5.30 \times 10^7$ m/s$^2$. Calculate (a) the maximum speed reached by the cells and (b) the distance traveled during the acceleration.

17. A graph of position versus time for a certain particle moving along the $x$-axis is shown in Figure P2.6. Find the instantaneous velocity at the instants (a) $t = 1.00$ s, (b) $t = 3.00$ s, (c) $t = 4.50$ s, and (d) $t = 7.50$ s.

18. A race car moves such that its position fits the relationship

$$x = (5.0 \text{ m/s})t + (0.75 \text{ m/s}^3)t^3$$

where $x$ is measured in meters and $t$ in seconds. (a) Plot a graph of the car's position versus time. (b) Determine the instantaneous velocity of the car at $t = 4.0$ s, using time intervals of 0.40 s, 0.20 s, and 0.10 s. (c) Compare the average velocity during the first 4.0 s with the results of part (b).

19. Runner A is initially 4.0 mi west of a flagpole and is running with a constant velocity of 6.0 mi/h due east. Runner B is initially 3.0 mi east of the flagpole and is running with a constant velocity of 5.0 mi/h due west. How far are the runners from the flagpole when they meet?

20. **V** A particle starts from rest and accelerates as shown in Figure P2.20. Determine (a) the particle's speed at $t = 10.0$ s and at $t = 20.0$ s, and (b) the distance traveled in the first 20.0 s.

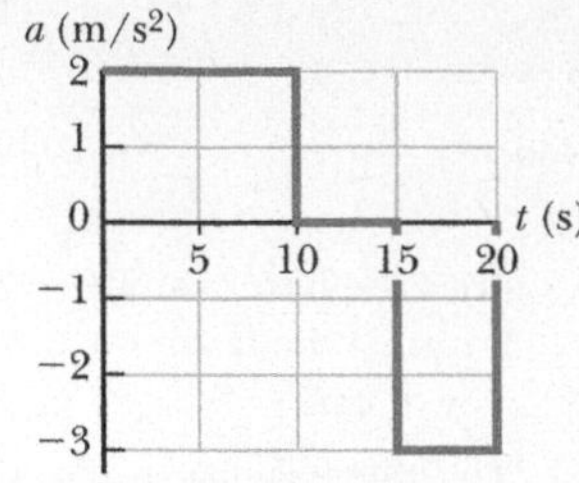

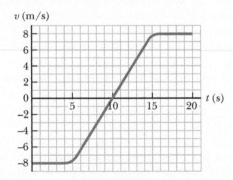

**Figure P2.20**

21. A 50.0-g Super Ball traveling at 25.0 m/s bounces off a brick wall and rebounds at 22.0 m/s. A high-speed camera records this event. If the ball is in contact with the wall for 3.50 ms, what is the magnitude of the average acceleration of the ball during this time interval?

22. **BIO** The average person passes out at an acceleration of $7g$ (that is, seven times the gravitational acceleration on Earth). Suppose a car is designed to accelerate at this rate. How much time would be required for the car to accelerate from rest to 60.0 miles per hour? (The car would need rocket boosters!)

23. **V** A certain car is capable of accelerating at a rate of 0.60 m/s². How long does it take for this car to go from a speed of 55 mi/h to a speed of 60 mi/h?

24. The velocity vs. time graph for an object moving along a straight path is shown in Figure P2.24. **(i)** Find the average acceleration of the object during the time intervals (a) 0 to 5.0 s, (b) 5.0 s to 15 s, and (c) 0 to 20 s. **(ii)** Find the instantaneous acceleration at (a) 2.0 s, (b) 10 s, and (c) 18 s.

**Figure P2.24**

25. A steam catapult launches a jet aircraft from the aircraft carrier *John C. Stennis*, giving it a speed of 175 mi/h in 2.50 s. (a) Find the average acceleration of the plane. (b) Assuming the acceleration is constant, find the distance the plane moves.

## 2.3 One-Dimensional Motion with Constant Acceleration

26. Solve Example 2.5, "Car Chase," by a graphical method. On the same graph, plot position versus time for the car and the trooper. From the intersection of the two curves, read the time at which the trooper overtakes the car.

27. **T** An object moving with uniform acceleration has a velocity of 12.0 cm/s in the positive $x$-direction when its $x$-coordinate is 3.00 cm. If its $x$-coordinate 2.00 s later is −5.00 cm, what is its acceleration?

28. **V** In 1865 Jules Verne proposed sending men to the Moon by firing a space capsule from a 220-m-long cannon with final speed of 10.97 km/h. What would have been the unrealistically large acceleration experienced by the space travelers during their launch? (A human can stand an acceleration of

$15g$ for a short time.) Compare your answer with the free-fall acceleration, 9.80 m/s².

29. A truck covers 40.0 m in 8.50 s while uniformly slowing down to a final velocity of 2.80 m/s. (a) Find the truck's original speed. (b) Find its acceleration.

30. **GP** A speedboat increases its speed uniformly from $v_i = 20.0$ m/s to $v_f = 30.0$ m/s in a distance of $2.00 \times 10^2$ m. (a) Draw a coordinate system for this situation and label the relevant quantities, including vectors. (b) For the given information, what single equation is most appropriate for finding the acceleration? (c) Solve the equation selected in part (b) symbolically for the boat's acceleration in terms of $v_f$, $v_i$, and $\Delta x$. (d) Substitute given values, obtaining that acceleration. (e) Find the time it takes the boat to travel the given distance.

31. A Cessna aircraft has a liftoff speed of 120. km/h. (a) What minimum constant acceleration does the aircraft require if it is to be airborne after a takeoff run of 240. m? (b) How long does it take the aircraft to become airborne?

32. An object moves with constant acceleration 4.00 m/s² and over a time interval reaches a final velocity of 12.0 m/s. (a) If its original velocity is 6.00 m/s, what is its displacement during the time interval? (b) What is the distance it travels during this interval? (c) If its original velocity is −6.00 m/s, what is its displacement during this interval? (d) What is the total distance it travels during the interval in part (c)?

33. **QC** In a test run, a certain car accelerates uniformly from zero to 24.0 m/s in 2.95 s. (a) What is the magnitude of the car's acceleration? (b) How long does it take the car to change its speed from 10.0 m/s to 20.0 m/s? (c) Will doubling the time always double the change in speed? Why?

34. **QC** A jet plane lands with a speed of 100 m/s and can accelerate at a maximum rate of −5.00 m/s² as it comes to rest. (a) From the instant the plane touches the runway, what is the minimum time needed before it can come to rest? (b) Can this plane land on a small tropical island airport where the runway is 0.800 km long?

35. **QC** Speedy Sue, driving at 30.0 m/s, enters a one-lane tunnel. She then observes a slow-moving van 155 m ahead traveling at 5.00 m/s. Sue applies her brakes but can accelerate only at −2.00 m/s² because the road is wet. Will there be a collision? State how you decide. If yes, determine how far into the tunnel and at what time the collision occurs. If no, determine the distance of closest approach between Sue's car and the van.

36. A record of travel along a straight path is as follows:

    1. Start from rest with a constant acceleration of 2.77 m/s² for 15.0 s.

    2. Maintain a constant velocity for the next 2.05 min.

    3. Apply a constant negative acceleration of −9.47 m/s² for 4.39 s.

    (a) What was the total displacement for the trip?

    (b) What were the average speeds for legs 1, 2, and 3 of the trip, as well as for the complete trip?

37. A train is traveling down a straight track at 20 m/s when the engineer applies the brakes, resulting in an acceleration of −1.0 m/s² as long as the train is in motion. How far does the train move during a 40-s time interval starting at the instant the brakes are applied?

38. A car accelerates uniformly from rest to a speed of 40.0 mi/h in 12.0 s. Find (a) the distance the car travels during this time and (b) the constant acceleration of the car.

39. A car starts from rest and travels for 5.0 s with a uniform acceleration of $+1.5 \text{ m/s}^2$. The driver then applies the brakes, causing a uniform acceleration of $-2.0 \text{ m/s}^2$. If the brakes are applied for 3.0 s, (a) how fast is the car going at the end of the braking period, and (b) how far has the car gone?

40. **S** A car starts from rest and travels for $t_1$ seconds with a uniform acceleration $a_1$. The driver then applies the brakes, causing a uniform acceleration $a_2$. If the brakes are applied for $t_2$ seconds, (a) how fast is the car going just before the beginning of the braking period? (b) How far does the car go before the driver begins to brake? (c) Using the answers to parts (a) and (b) as the initial velocity and position for the motion of the car during braking, what total distance does the car travel? Answers are in terms of the variables $a_1$, $a_2$, $t_1$, and $t_2$.

41. In the Daytona 500 auto race, a Ford Thunderbird and a Mercedes Benz are moving side by side down a straightaway at 71.5 m/s. The driver of the Thunderbird realizes that she must make a pit stop, and she smoothly slows to a stop over a distance of 250 m. She spends 5.00 s in the pit and then accelerates out, reaching her previous speed of 71.5 m/s after a distance of 350 m. At this point, how far has the Thunderbird fallen behind the Mercedes Benz, which has continued at a constant speed?

42. The kinematic equations can describe phenomena other than motion through space and time. Suppose $x$ represents a person's bank account balance. The units of $x$ would be dollars ($) , and velocity $v$ would give the rate at which the balance changes (in units of, for example, $/month). Acceleration would give the rate at which $v$ changes. Suppose a person begins with ten thousand dollars in the bank. Initial money management leads to no net change in the account balance so that $v_0 = 0$. Unfortunately, management worsens over time so that $a = -2.5 \times 10^2$ $/month$^2$. Assuming $a$ is constant, find the amount of time in months until the bank account is empty.

43. **T** A hockey player is standing on his skates on a frozen pond when an opposing player, moving with a uniform speed of 12 m/s, skates by with the puck. After 3.0 s, the first player makes up his mind to chase his opponent. If he accelerates uniformly at $4.0 \text{ m/s}^2$, (a) how long does it take him to catch his opponent, and (b) how far has he traveled in that time? (Assume the player with the puck remains in motion at constant speed.)

44. A train $4.00 \times 10^2$ m long is moving on a straight track with a speed of 82.4 km/h. The engineer applies the brakes at a crossing, and later the last car passes the crossing with a speed of 16.4 km/h. Assuming constant acceleration, determine how long the train blocked the crossing. Disregard the width of the crossing.

## 2.4 Freely Falling Objects

45. A ball is thrown vertically upward with a speed of 25.0 m/s. (a) How high does it rise? (b) How long does it take to reach its highest point? (c) How long does the ball take to hit the ground after it reaches its highest point? (d) What is its velocity when it returns to the level from which it started?

46. **V** A ball is thrown directly downward with an initial speed of 8.00 m/s, from a height of 30.0 m. After what time interval does it strike the ground?

47. A certain freely falling object, released from rest, requires 1.50 s to travel the last 30.0 m before it hits the ground. (a) Find the velocity of the object when it is 30.0 m above the ground. (b) Find the total distance the object travels during the fall.

48. **Q|C** An attacker at the base of a castle wall 3.65 m high throws a rock straight up with speed 7.40 m/s at a height of 1.55 m above the ground. (a) Will the rock reach the top of the wall? (b) If so, what is the rock's speed at the top? If not, what initial speed must the rock have to reach the top? (c) Find the change in the speed of a rock thrown straight down from the top of the wall at an initial speed of 7.40 m/s and moving between the same two points. (d) Does the change in speed of the downward-moving rock agree with the magnitude of the speed change of the rock moving upward between the same elevations? Explain physically why or why not.

49. **BIO** Traumatic brain injury such as concussion results when the head undergoes a very large acceleration. Generally, an acceleration less than $800 \text{ m/s}^2$ lasting for any length of time will not cause injury, whereas an acceleration greater than $1\,000 \text{ m/s}^2$ lasting for at least 1 ms will cause injury. Suppose a small child rolls off a bed that is 0.40 m above the floor. If the floor is hardwood, the child's head is brought to rest in approximately 2.0 mm. If the floor is carpeted, this stopping distance is increased to about 1.0 cm. Calculate the magnitude and duration of the deceleration in both cases, to determine the risk of injury. Assume the child remains horizontal during the fall to the floor. Note that a more complicated fall could result in a head velocity greater or less than the speed you calculate.

50. A small mailbag is released from a helicopter that is descending steadily at 1.50 m/s. After 2.00 s, (a) what is the speed of the mailbag, and (b) how far is it below the helicopter? (c) What are your answers to parts (a) and (b) if the helicopter is rising steadily at 1.50 m/s?

51. A tennis player tosses a tennis ball straight up and then catches it after 2.00 s at the same height as the point of release. (a) What is the acceleration of the ball while it is in flight? (b) What is the velocity of the ball when it reaches its maximum height? Find (c) the initial velocity of the ball and (d) the maximum height it reaches.

52. **S** A package is dropped from a helicopter that is descending steadily at a speed $v_0$. After $t$ seconds have elapsed, (a) what is the speed of the package in terms of $v_0$, $g$, and $t$? (b) What distance $d$ is it from the helicopter in terms of $g$ and $t$? (c) What are the answers to parts (a) and (b) if the helicopter is rising steadily at the same speed?

53. **Q|C** A model rocket is launched straight upward with an initial speed of 50.0 m/s. It accelerates with a constant upward acceleration of $2.00 \text{ m/s}^2$ until its engines stop at an altitude of 150. m. (a) What can you say about the motion of the rocket after its engines stop? (b) What is the maximum height reached by the rocket? (c) How long after liftoff does the rocket reach its maximum height? (d) How long is the rocket in the air?

54. **V** A baseball is hit so that it travels straight upward after being struck by the bat. A fan observes that it takes 3.00 s for the ball to reach its maximum height. Find (a) the ball's initial velocity and (b) the height it reaches.

## Additional Problems

**55.** A truck tractor pulls two trailers, one behind the other, at a constant speed of $1.00 \times 10^2$ km/h. It takes 0.600 s for the big rig to completely pass onto a bridge $4.00 \times 10^2$ m long. For what duration of time is all or part of the truck–trailer combination on the bridge?

**56.** **T** **BIO** Colonel John P. Stapp, USAF, participated in studying whether a jet pilot could survive emergency ejection. On March 19, 1954, he rode a rocket-propelled sled that moved down a track at a speed of 632 mi/h (see Fig. P2.56). He and the sled were safely brought to rest in 1.40 s. Determine in SI units (a) the negative acceleration he experienced and (b) the distance he traveled during this negative acceleration.

Courtesy U.S. Air Force

Courtesy of the U.S. Air Force

**Figure P2.56** (*left*) Col. John Stapp and his rocket sled are brought to rest in a very short time interval. (*right*) Stapp's face is contorted by the stress of rapid negative acceleration.

**57.** A bullet is fired through a board 10.0 cm thick in such a way that the bullet's line of motion is perpendicular to the face of the board. If the initial speed of the bullet is $4.00 \times 10^2$ m/s and it emerges from the other side of the board with a speed of $3.00 \times 10^2$ m/s, find (a) the acceleration of the bullet as it passes through the board and (b) the total time the bullet is in contact with the board.

**58.** A speedboat moving at 30.0 m/s approaches a no-wake buoy marker $1.00 \times 10^2$ m ahead. The pilot slows the boat with a constant acceleration of $-3.50$ m/s$^2$ by reducing the throttle. (a) How long does it take the boat to reach the buoy? (b) What is the velocity of the boat when it reaches the buoy?

**59.** A student throws a set of keys vertically upward to his fraternity brother, who is in a window 4.00 m above. The brother's outstretched hand catches the keys 1.50 s later. (a) With what initial velocity were the keys thrown? (b)? What was the velocity of the keys just before they were caught?

**60.** **BIO** Mature salmon swim upstream, returning to spawn at their birthplace. During the arduous trip they leap vertically upward over waterfalls as high as 3.6 m. With what minimum speed must a salmon launch itself into the air to clear a 3.6-m waterfall?

**61.** **V** **BIO** An insect called the froghopper (*Philaenus spumarius*) has been called the best jumper in the animal kingdom. This insect can accelerate at over $4.0 \times 10^3$ m/s$^2$ during a displacement of 2.0 mm as it straightens its specially equipped "jumping legs." (a) Assuming uniform acceleration, what is the insect's speed after it has accelerated through this short distance? (b) How long does it take to reach that speed? (c) How high could the insect jump if air resistance could be ignored? Note that the actual height obtained is about 0.70 m, so air resistance is important here.

**62.** **Q|C** An object is moving in the positive direction along the *x*-axis. Sketch plots of the object's position vs. time and velocity vs. time if (a) its speed is constant, (b) it's speeding up at a constant rate, and (c) it's slowing down at a constant rate.

**63.** **T** A ball is thrown upward from the ground with an initial speed of 25 m/s; at the same instant, another ball is dropped from a building 15 m high. After how long will the balls be at the same height?

**64.** **S** A player holds two baseballs a height $h$ above the ground. He throws one ball vertically upward at speed $v_0$ and the other vertically downward at the same speed. Calculate (a) the speed of each ball as it hits the ground and (b) the difference between their times of flight.

**65.** A ball thrown straight up into the air is found to be moving at 1.50 m/s after rising 2.00 m above its release point. Find the ball's initial speed.

**66.** **BIO** The thickest and strongest chamber in the human heart is the left ventricle, responsible during systole for pumping oxygenated blood through the aorta to rest of the body. Assume aortic blood starts from rest and accelerates at 22.5 m/s$^2$ to a peak speed of 1.05 m/s. (a) How far does the blood travel during this acceleration? (b) How much time is required for the blood to reach its peak speed?

**67.** **Q|C** Emily challenges her husband, David, to catch a $1 bill as follows. She holds the bill vertically as in Figure P2.67, with the center of the bill between David's index finger and thumb. David must catch the bill after Emily releases it without moving his hand downward. If his reaction time is 0.2 s, will he succeed? Explain your reasoning. (This challenge is a good trick you might want to try with your friends.)

© Cengage Learning/George Semple

**Figure P2.67**

**68.** A mountain climber stands at the top of a 50.0-m cliff that overhangs a calm pool of water. She throws two stones vertically downward 1.00 s apart and observes that they cause a single splash. The first stone had an initial velocity of $-2.00$ m/s. (a) How long after release of the first stone did the two stones hit the water? (b) What initial velocity must the second stone have had, given that they hit the water simultaneously? (c) What was the velocity of each stone at the instant it hit the water?

**69.** **Q|C** **S** One of Aesop's fables tells of a race between a tortoise and a hare. Suppose the overconfident hare takes a nap and wakes up to find the tortoise a distance $d$ ahead and a distance $L$ from the finish line. If the hare then begins running with constant speed $v_1$ and the tortoise continues crawling with constant speed $v_2$, it turns out that the tortoise wins the race if the distance $L$ is less than $(v_2/(v_1 - v_2))d$. Obtain this result by first writing expressions for the times taken by the hare and the tortoise to finish the race, and then noticing that to win, $t_{\text{tortoise}} < t_{\text{hare}}$. Assume $v_2 < v_1$.

**70.** In Bosnia, the ultimate test of a young man's courage used to be to jump off a 400-year-old bridge (destroyed in 1993; rebuilt in 2004) into the River Neretva, 23 m below the bridge. (a) How long did the jump last? (b) How fast was the jumper traveling upon impact with the river? (c) If the speed of sound in air is 340 m/s, how long after the jumper took off did a spectator on the bridge hear the splash?

**71.** A stuntman sitting on a tree limb wishes to drop vertically onto a horse galloping under the tree. The constant speed of the horse is 10.0 m/s, and the man is initially 3.00 m above the level of the saddle. (a) What must be the horizontal distance between the saddle and the limb when the man makes his move? (b) How long is he in the air?

# Motion in Two Dimensions

THE STUDY OF MOTION IN TOPIC 2 involved the concepts of displacement, velocity, and acceleration in one dimension. Topic 3 extends those concepts by applying vector techniques to a study of two-dimensional motion, including projectiles and relative motion.

## 3.1 Displacement, Velocity, and Acceleration in Two Dimensions

In one-dimensional motion, as discussed in Topic 2, the direction of a vector quantity such as a velocity or acceleration can be taken into account by specifying whether the quantity is positive or negative. The velocity of a rocket, for example, is positive if the rocket is going up and negative if it's going down. This simple solution is no longer available in two or three dimensions. Instead, we must make full use of the vector concept.

Consider an object moving through space as shown in Figure 3.1. When the object is at some point Ⓟ at time $t_i$, its position is described by the position vector $\vec{r}_i$, drawn from the origin to Ⓟ. When the object has moved to some other point Ⓠ at time $t_f$, its position vector is $\vec{r}_f$. From the vector diagram in Figure 3.1, the final position vector is the sum of the initial position vector and the displacement $\Delta\vec{r}$: $\vec{r}_f = \vec{r}_i + \Delta\vec{r}$. From this relationship, we obtain the following one:

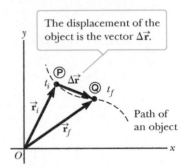

**Figure 3.1** An object moving along some curved path between points Ⓟ and Ⓠ. The displacement vector $\Delta\vec{r}$ is the difference in the position vectors: $\Delta\vec{r} = \vec{r}_f - \vec{r}_i$.

An object's **displacement** is defined as the change in its position vector, or

$$\Delta\vec{r} \equiv \vec{r}_f - \vec{r}_i \qquad [3.1]$$

**SI unit: meter (m)**

The displacement vector $\Delta\vec{r}$ has components $\Delta x$ and $\Delta y$ as shown in Figure 3.2 for an object that moves from location $(x_i, y_i)$ to $(x_f, y_f)$ in a time interval $\Delta t$. The $x$-component of the object's displacement is $\Delta x = x_f - x_i$ and the $y$-component is $\Delta y = y_f - y_i$.

We now present several generalizations of the definitions of velocity and acceleration given in Topic 2.

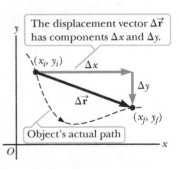

**Figure 3.2** The displacement vector has components $\Delta x = x_f - x_i$ and $\Delta y = y_f - y_i$.

An object's **average velocity** during a time interval $\Delta t$ is its displacement divided by $\Delta t$:

◀ Average velocity

$$\vec{v}_{av} \equiv \frac{\Delta\vec{r}}{\Delta t} \qquad [3.2]$$

**SI unit: meter per second (m/s)**

59

Because the displacement is a vector quantity and the time interval is a scalar quantity, we conclude that the average velocity is a *vector* quantity directed along $\Delta\vec{r}$. The x- and y-components of the average velocity are given by

$$v_{av,x} = \frac{\Delta x}{\Delta t} \text{ and } v_{av,y} = \frac{\Delta y}{\Delta t}$$

and express the rate at which an object's position is changing along the x- and y-axes, respectively. Note that the magnitude of the average velocity is just the distance between the endpoints divided by the elapsed time.

Instantaneous velocity ▶

An object's **instantaneous velocity** $\vec{v}$ is the limit of its average velocity as $\Delta t$ goes to zero:

$$\vec{v} \equiv \lim_{\Delta t \to 0} \frac{\Delta\vec{r}}{\Delta t} \qquad [3.3]$$

**SI unit: meter per second (m/s)**

The direction of the instantaneous velocity vector is along a line that is tangent to the object's path and in the direction of its motion.

Average acceleration ▶

An object's **average acceleration** during a time interval $\Delta t$ is the change in its velocity $\Delta\vec{v}$ divided by $\Delta t$, or

$$\vec{a}_{av} \equiv \frac{\Delta\vec{v}}{\Delta t} \qquad [3.4]$$

**SI unit: meter per second squared (m/s$^2$)**

Average acceleration has components given by

$$a_{av,x} = \frac{\Delta v_x}{\Delta t} \text{ and } a_{av,y} = \frac{\Delta v_y}{\Delta t}$$

where $v_x$ and $v_y$ are the x- and y-components of the velocity vector.

Instantaneous acceleration ▶

An object's **instantaneous acceleration** vector $\vec{a}$ is the limit of its average acceleration vector as $\Delta t$ goes to zero:

$$\vec{a} \equiv \lim_{\Delta t \to 0} \frac{\Delta\vec{v}}{\Delta t} \qquad [3.5]$$

**SI unit: meter per second squared (m/s$^2$)**

It's important to recognize that an object can accelerate in several ways. First, the magnitude of the velocity vector (the speed) may change with time. Second, the direction of the velocity vector may change with time, even though the speed is constant, as can happen along a curved path. Third, both the magnitude and the direction of the velocity vector may change at the same time.

### Quick Quiz

**3.1** Which of the following objects can't be accelerating? (a) An object moving with a constant speed; (b) an object moving with a constant velocity; or (c) an object moving along a curve.

**3.2** Consider the following controls in an automobile: gas pedal, brake, steering wheel. The controls in this list that can cause an acceleration of the car are (a) all three controls, (b) the gas pedal and the brake, (c) only the brake, or (d) only the gas pedal.

In Topic 2, a distinction was made between average velocity and average speed in one dimension. That distinction remains important for two-dimensional motion. Average velocity is a vector equal to the displacement divided by the time interval and depends *only* on the motion's endpoints but *not* on the actual path traveled between them. Average speed is a scalar equal to the actual path length traveled, including any retracing of steps or deviations from a straight line, divided by the elapsed time.

Figure 3.3 shows a woman jogging along the rectangular periphery of a small green space, going from point A to point B in 20.0 s. The path length is 30.0 m + 40.0 m = 70.0 m, so her average speed, equal to her path length divided by the elapsed time, is 70.0 m/20.0 s = 3.50 m/s. The magnitude of her average velocity, however, depends only on the distance between the endpoints, which is 50.0 m, so that $v_{av} = d/t = 50.0$ m/20.0 s = 2.50 m/s.

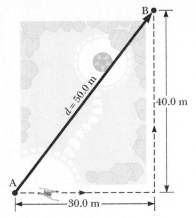

**Figure 3.3** A woman jogs along the dashed path from A to B in 20.0 s, traveling a path length of 30.0 m + 40.0 m = 70.0 m and a distance $d$ of 50.0 m.

### Quick Quiz

**3.3** A girl on a bicycle takes 15.0 s to ride half way around a circular track of radius 10.0 m (Figure 3.4). (a) What is the girl's average speed? (b) What is the magnitude of her average velocity?

**Figure 3.4** (Quick Quiz 3.3)

◀ Projectile motion

## 3.2 Two-Dimensional Motion

In Topic 2 we studied objects moving along straight-line paths, such as the x-axis. In this topic, we look at objects that move in both the x- and y-directions simultaneously under constant acceleration. An important special case of this two-dimensional motion is called **projectile motion.**

Anyone who has tossed any kind of object into the air has observed projectile motion. If the effects of air resistance and the rotation of Earth are neglected, the path of a projectile in Earth's gravity field is curved in the shape of a parabola, as shown in Figure 3.5.

The positive x-direction is horizontal and to the right, and the y-direction is vertical and positive upward. The most important experimental fact about projectile motion in two dimensions is that **the horizontal and vertical motions are completely independent of each other.** This means that motion in one direction has no effect on motion in the other direction. If a baseball is tossed in a parabolic path, as in Figure 3.5, the motion in the y-direction will look just like a ball tossed straight up under the influence of gravity. Figure 3.6 shows the effect

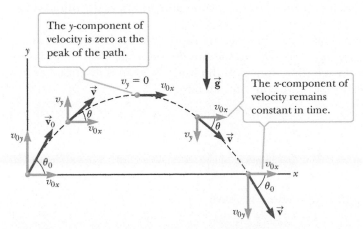

**Figure 3.5** The parabolic trajectory of a particle that leaves the origin with a velocity of $\vec{v}_0$. Note that $\vec{v}$ changes with time. However, the x-component of the velocity, $v_x$, remains constant in time, equal to its initial velocity, $v_{0x}$. Also, $v_y = 0$ at the peak of the trajectory, but the acceleration is always equal to the free-fall acceleration and acts vertically downward.

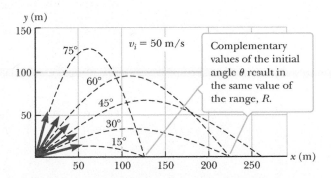

**Figure 3.6** A projectile launched from the origin with an initial speed of 50 m/s at various angles of projection.

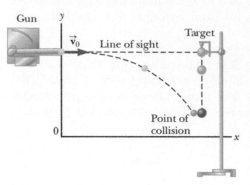

The velocity of the projectile (red arrows) changes in direction and magnitude, but its acceleration (purple arrows) remains constant.

© Charles D. Winters/Cengage Learning

**Figure 3.7** A ball is fired at a target at the same instant the target is released. Both fall vertically at the same rate and collide.

**Figure 3.8** Multiflash photograph of the projectile–target demonstration. If the gun is aimed directly at the target and is fired at the same instant the target begins to fall, the projectile will hit the target.

of various initial angles; note that complementary angles give the same horizontal range.

Figure 3.7 is an experiment illustrating the independence of horizontal and vertical motion. The gun is aimed directly at the target ball and fired at the instant the target is released. In the absence of gravity, the projectile would hit the target because the target wouldn't move. However, the projectile still hits the target in the presence of gravity. That means the projectile is falling through the same vertical displacement as the target despite its horizontal motion. The experiment also works when set up as in Figure 3.8, when the initial velocity has a vertical component.

In general, the equations of constant acceleration developed in Topic 2 follow separately for both the $x$-direction and the $y$-direction. An important difference is that the initial velocity now has two components, not just one as in that topic. We assume that at $t = 0$ the projectile leaves the origin with an initial velocity $\vec{v}_0$. If the velocity vector makes an angle $\theta_0$ with the horizontal, where $\theta_0$ is called the *projection angle*, then from the definitions of the cosine and sine functions and Figure 3.5 we have

$$v_{0x} = v_0 \cos \theta_0 \quad \text{and} \quad v_{0y} = v_0 \sin \theta_0$$

where $v_{0x}$ is the initial velocity (at $t = 0$) in the $x$-direction and $v_{0y}$ is the initial velocity in the $y$-direction.

Now, Equations 2.6, 2.9, and 2.10 developed in Topic 2 for motion with constant acceleration in one dimension carry over to the two-dimensional case; there is one set of three equations for each direction, with the initial velocities modified as just discussed. In the $x$-direction, with $a_x$ constant, we have

$$v_x = v_{0x} + a_x t \tag{3.6a}$$

$$\Delta x = v_{0x} t + \tfrac{1}{2} a_x t^2 \tag{3.6b}$$

$$v_x^2 = v_{0x}^2 + 2 a_x \Delta x \tag{3.6c}$$

where $v_{0x} = v_0 \cos \theta_0$. In the $y$-direction, we have

$$v_y = v_{0y} + a_y t \tag{3.7a}$$

$$\Delta y = v_{0y} t + \tfrac{1}{2} a_y t^2 \tag{3.7b}$$

$$v_y^2 = v_{0y}^2 + 2 a_y \Delta y \tag{3.7c}$$

**Tip 3.1** Acceleration at the Highest Point

The acceleration in the $y$-direction is *not* zero at the top of a projectile's trajectory. Only the $y$-component of the velocity is zero there. If the acceleration were zero, too, the projectile would never come down!

where $v_{0y} = v_0 \sin \theta_0$ and $a_y$ is constant. The object's speed $v$ can be calculated from the components of the velocity using the Pythagorean theorem:

$$v = \sqrt{v_x^2 + v_y^2}$$

The angle that the velocity vector makes with the $x$-axis is given by

$$\theta = \tan^{-1}\left(\frac{v_y}{v_x}\right)$$

This formula for $\theta$, as previously stated, must be used with care, because the inverse tangent function returns values only between $-90°$ and $+90°$. Adding $180°$ is necessary for vectors lying in the second or third quadrant.

The kinematic equations are easily adapted and simplified for projectiles close to the surface of the Earth. In this case, assuming air friction is negligible, the acceleration in the $x$-direction is 0 (because air resistance is neglected). **This means that $a_x = 0$, and the projectile's velocity component along the $x$-direction remains constant.** If the initial value of the velocity component in the $x$-direction is $v_{0x} = v_0 \cos \theta_0$, then this is also the value of $v_x$ at any later time, so

$$v_x = v_{0x} = v_0 \cos \theta_0 = \text{constant} \qquad \text{[3.8a]}$$

whereas the horizontal displacement is simply

$$\Delta x = v_{0x}t = (v_0 \cos \theta_0)t \qquad \text{[3.8b]}$$

For the motion in the $y$-direction, we make the substitution $a_y = -g$ and $v_{0y} = v_0 \sin \theta_0$ in Equations 3.7, giving

$$v_y = v_0 \sin \theta_0 - gt \qquad \text{[3.9a]}$$

$$\Delta y = (v_0 \sin \theta_0)t - \tfrac{1}{2}gt^2 \qquad \text{[3.9b]}$$

$$v_y^2 = (v_0 \sin \theta_0)^2 - 2g\,\Delta y \qquad \text{[3.9c]}$$

The important facts of projectile motion can be summarized as follows:

1. Provided air resistance is negligible, the horizontal component of the velocity $v_x$ remains constant because there is no horizontal component of acceleration.
2. The vertical component of the acceleration is equal to the free-fall acceleration $-g$.
3. The vertical component of the velocity $v_y$ and the displacement in the $y$-direction are identical to those of a freely falling body.
4. Projectile motion can be described as a superposition of two independent motions in the $x$- and $y$-directions.

## EXAMPLE 3.1    PROJECTILE MOTION WITH DIAGRAMS

**GOAL** Approximate answers in projectile motion using a motion diagram.

**PROBLEM** A ball is thrown so that its initial vertical and horizontal components of velocity are 40 m/s and 20 m/s, respectively. Use a motion diagram to estimate the ball's total time of flight and the distance it traverses before hitting the ground.

**STRATEGY** Use the diagram, estimating the acceleration of gravity as $-10$ m/s$^2$. By symmetry, the ball goes up and comes back down to the ground at the same $y$-velocity as when it left, except with opposite sign. With this fact and the fact that the acceleration of gravity decreases the velocity in the $y$-direction by 10 m/s every second, we can find the total time of flight and then the horizontal range.

*(Continued)*

## SOLUTION

In the motion diagram shown in Figure 3.9, the acceleration vectors are all the same, pointing downward with magnitude of nearly 10 m/s². By symmetry, we know that the ball will hit the ground at the same speed in the y-direction as when it was thrown, so the velocity in the y-direction goes from 40 m/s to −40 m/s in steps of −10 m/s every second; hence, approximately 8 seconds elapse during the motion.

The velocity vector constantly changes direction, but the horizontal velocity never changes because the acceleration in the horizontal direction is zero. Therefore, the displacement of the ball in the x-direction is given by Equation 3.8b, $\Delta x \approx v_{0x}t = (20 \text{ m/s})(8 \text{ s}) = 160 \text{ m}$.

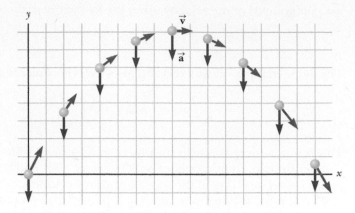

**Figure 3.9** (Example 3.1) Motion diagram for a projectile.

**REMARKS** This example emphasizes the independence of the x- and y-components in projectile motion problems.

**QUESTION 3.1** Neglecting air resistance, is the magnitude of the velocity vector at impact greater than, less than, or equal to the magnitude of the initial velocity vector? Why?

**EXERCISE 3.1** Estimate the maximum height in this same problem.

**ANSWER** 80 m

### Quick Quiz

**3.4** Suppose you are carrying a ball and running at constant velocity on level ground. You wish to throw the ball and catch it as it comes back down. Neglecting air resistance, should you (a) throw the ball at an angle of about 45° above the horizontal and maintain the same speed, (b) throw the ball straight up in the air and slow down to catch it, or (c) throw the ball straight up in the air and maintain the same speed?

**3.5** As a projectile moves in its parabolic path, where are the velocity and acceleration vectors perpendicular to each other? (a) Everywhere along the projectile's path, (b) at the peak of its path, (c) nowhere along its path, or (d) not enough information is given.

### PROBLEM-SOLVING STRATEGY

### Projectile Motion

1. **Select** a coordinate system and sketch the path of the projectile, including initial and final positions, velocities, and accelerations.
2. **Resolve** the initial velocity vector into x- and y-components.
3. **Treat** the horizontal motion and the vertical motion independently.
4. **Follow** the techniques for solving problems with constant velocity to analyze the horizontal motion of the projectile.
5. **Follow** the techniques for solving problems with constant acceleration to analyze the vertical motion of the projectile.

### EXAMPLE 3.2   STRANDED EXPLORERS

**GOAL** Solve a two-dimensional projectile motion problem in which an object has an initial horizontal velocity.

**PROBLEM** An Alaskan rescue plane drops a package of emergency rations to stranded hikers, as shown in Figure 3.10. The plane is traveling horizontally at 40.0 m/s at a height of $1.00 \times 10^2$ m above the ground. Neglect air resistance. **(a)** Where does the package strike the ground relative to the point at which it was released? **(b)** What are the horizontal and vertical components of the velocity of the package just before it hits the ground? **(c)** What is the angle of the impact?

**STRATEGY** Here, we're just taking Equations 3.8 and 3.9, filling in known quantities, and solving for the remaining unknown quantities. Sketch the problem using a coordinate system as in Figure 3.10. In part **(a)**, set the $y$-component of the displacement equations equal to $-1.00 \times 10^2$ m—the ground level where the package lands—and solve for the time it takes the package to reach the ground. Substitute this time into the displacement equation for the $x$-component to find the range. In part **(b)**, substitute the time found in part **(a)** into the velocity components. Notice that the initial velocity has only an $x$-component, which simplifies the math. Solving part **(c)** requires the inverse tangent function.

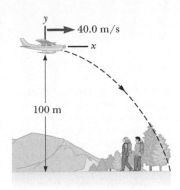

**Figure 3.10**  (Example 3.2) From the point of view of an observer on the ground, a package released from the rescue plane travels along the path shown.

## SOLUTION

**(a)** Find the range of the package.

Use Equation 3.9b to find the $y$-displacement:

$$\Delta y = y - y_0 = v_{0y}t - \tfrac{1}{2}gt^2$$

Substitute $y_0 = 0$ and $v_{0y} = 0$, and set $y = -1.00 \times 10^2$ m, the final vertical position of the package relative to the airplane. Solve for time:

$$y = -(4.90 \text{ m/s}^2)t^2 = -1.00 \times 10^2 \text{ m}$$
$$t = 4.52 \text{ s}$$

Use Equation 3.8b to find the $x$-displacement:

$$\Delta x = x - x_0 = v_{0x}t$$

Substitute $x_0 = 0$, $v_{0x} = 40.0$ m/s, and the time:

$$x = (40.0 \text{ m/s})(4.52 \text{ s}) = \boxed{181 \text{ m}}$$

**(b)** Find the components of the package's velocity at impact:

Find the $x$-component of the velocity at the time of impact:

$$v_x = v_0 \cos \theta = (40.0 \text{ m/s}) \cos 0° = \boxed{40.0 \text{ m/s}}$$

Find the $y$-component of the velocity at the time of impact:

$$v_y = v_0 \sin \theta - gt = 0 - (9.80 \text{ m/s}^2)(4.52 \text{ s}) = \boxed{-44.3 \text{ m/s}}$$

**(c)** Find the angle of the impact.

Write Equation 1.6 and substitute values:

$$\tan \theta = \frac{v_y}{v_x} = \frac{-44.3 \text{ m/s}}{40.0 \text{ m/s}} = -1.11$$

Apply the inverse tangent functions to both sides:

$$\theta = \tan^{-1}(-1.11) = \boxed{-48.0°}$$

**REMARKS** Notice how motion in the $x$-direction and motion in the $y$-direction are handled separately.

**QUESTION 3.2** Neglecting air resistance effects, what path does the package travel as observed by the pilot? Explain.

**EXERCISE 3.2** A bartender slides a beer mug at 1.50 m/s toward a customer at the end of a frictionless bar that is 1.20 m tall. The customer makes a grab for the mug and misses, and the mug sails off the end of the bar. **(a)** How far away from the end of the bar does the mug hit the floor? **(b)** What are the speed and direction of the mug at impact?

**ANSWERS** **(a)** 0.742 m   **(b)** 5.08 m/s, $\theta = -72.8°$

---

## EXAMPLE 3.3  THE LONG JUMP

**GOAL** Solve a two-dimensional projectile motion problem involving an object starting and ending at the same height.

**PROBLEM** A long jumper (Fig. 3.11) leaves the ground at an angle of 20.0° to the horizontal and at a speed of 11.0 m/s. **(a)** How long does it take for him to reach maximum height? **(b)** What is the maximum height? **(c)** How far does he jump? (Assume his motion is equivalent to that of a particle, disregarding the motion of his arms and legs.) **(d)** Use Equation 3.9c to find the maximum height he reaches.

(*Continued*)

**Figure 3.11** (Example 3.3) This multiple-exposure shot of a long jumper shows that in reality, the jumper's motion is not the equivalent of the motion of a particle. The center of mass of the jumper follows a parabola, but to extend the length of the jump before impact, the jumper pulls his feet up so he strikes the ground later than he otherwise would have.

iStockphoto.com/technotr

**STRATEGY** Again, we take the projectile equations, fill in the known quantities, and solve for the unknowns. At the maximum height, the velocity in the y-direction is zero, so setting Equation 3.9a equal to zero and solving gives the time it takes the jumper to reach his maximum height. By symmetry, given that his trajectory starts and ends at the same height, doubling this time gives the total time of the jump.

**SOLUTION**

**(a)** Find the time $t_{max}$ taken to reach maximum height.

Set $v_y = 0$ in Equation 3.9a and solve for $t_{max}$:

$$v_y = v_0 \sin \theta_0 - g t_{max} = 0$$

$$(1) \quad t_{max} = \frac{v_0 \sin \theta_0}{g}$$

$$= \frac{(11.0 \text{ m/s})(\sin 20.0°)}{9.80 \text{ m/s}^2} = \boxed{0.384 \text{ s}}$$

**(b)** Find the maximum height he reaches.

Substitute the time $t_{max}$ into the equation for the y-displacement, Equation 3.9b:

$$y_{max} = (v_0 \sin \theta_0) t_{max} - \tfrac{1}{2} g \, (t_{max})^2$$

$$y_{max} = (11.0 \text{ m/s})(\sin 20.0°)(0.384 \text{ s})$$
$$- \tfrac{1}{2}(9.80 \text{ m/s}^2)(0.384 \text{ s})^2$$

$$y_{max} = \boxed{0.722 \text{ m}}$$

**(c)** Find the horizontal distance he jumps.

First find the time for the jump, which is twice $t_{max}$:

$$t = 2t_{max} = 2(0.384 \text{ s}) = 0.768 \text{ s}$$

Substitute this result into the equation for the x-displacement:

$$(2) \quad \Delta x = (v_0 \cos \theta_0) t$$
$$= (11.0 \text{ m/s})(\cos 20.0°)(0.768 \text{ s}) = \boxed{7.94 \text{ m}}$$

**(d)** Use an alternate method to find the maximum height.

Use Equation 3.9c, solving for $\Delta y$:

$$v_y^2 - v_{0y}^2 = -2g \, \Delta y$$

$$\Delta y = \frac{v_y^2 - v_{0y}^2}{-2g}$$

Substitute $v_y = 0$ at maximum height, and the fact that $v_{0y} = (11.0 \text{ m/s}) \sin 20.0°$:

$$\Delta y = \frac{0 - [(11.0 \text{ m/s}) \sin 20.0°]^2}{-2(9.80 \text{ m/s}^2)} = \boxed{0.722 \text{ m}}$$

**REMARKS** Although modeling the long jumper's motion as that of a projectile is an oversimplification, the values obtained are reasonable.

**QUESTION 3.3** True or False: Because the $x$-component of the displacement doesn't depend explicitly on $g$, the horizontal distance traveled doesn't depend on the acceleration of gravity.

**EXERCISE 3.3** A grasshopper jumps a horizontal distance of 1.00 m from rest, with an initial velocity at a 45.0° angle with respect to the horizontal. Find **(a)** the initial speed of the grasshopper and **(b)** the maximum height reached.

**ANSWERS** **(a)** 3.13 m/s   **(b)** 0.250 m

---

### EXAMPLE 3.4   THE RANGE EQUATION

**GOAL** Find an equation for the maximum horizontal displacement of a projectile fired from ground level.

**PROBLEM** An athlete participates in a long-jump competition, leaping into the air with a velocity $v_0$ at an angle $\theta_0$ with the horizontal. Obtain an expression for the length of the jump in terms of $v_0$, $\theta_0$, and $g$.

**STRATEGY** Use the results of Example 3.3, eliminating the time $t$ from Equations (1) and (2).

**SOLUTION**

Use Equation (1) of Example 3.3 to find the time of flight, $t$:

$$t = 2t_{max} = \frac{2v_0 \sin \theta_0}{g}$$

Substitute that expression for $t$ into Equation (2) of Example 3.3:

$$\Delta x = (v_0 \cos \theta_0)t = (v_0 \cos \theta_0)\left(\frac{2v_0 \sin \theta_0}{g}\right)$$

Simplify:

$$\Delta x = \frac{2v_0^2 \cos \theta_0 \sin \theta_0}{g}$$

Substitute the identity $2 \cos \theta_0 \sin \theta_0 = \sin 2\theta_0$ to reduce the foregoing expression to a single trigonometric function:

$$(1) \quad \Delta x = \frac{v_0^2 \sin 2\theta_0}{g}$$

**REMARKS** The use of a trigonometric identity in the final step isn't necessary, but it makes Question 3.4 easier to answer.

**QUESTION 3.4** What angle $\theta_0$ produces the longest jump?

**EXERCISE 3.4** Obtain an expression for the athlete's maximum displacement in the vertical direction, $\Delta y_{max}$ in terms of $v_0$, $\theta_0$, and $g$.

**ANSWER** $\Delta y_{max} = \dfrac{v_0^2 \sin^2 \theta_0}{2g}$

---

### EXAMPLE 3.5   THAT'S QUITE AN ARM

**GOAL** Solve a two-dimensional kinematics problem with a nonhorizontal initial velocity, starting and ending at different heights.

**PROBLEM** A ball is thrown upward from the top of a building at an angle of 30.0° above the horizontal and with an initial speed of 20.0 m/s, as in Figure 3.12. The point of release is 45.0 m above the ground. **(a)** How long does it take for the ball to hit the ground? **(b)** Find the ball's speed at impact. **(c)** Find the horizontal range of the ball. Neglect air resistance.

**STRATEGY** Choose coordinates as in the figure, with the origin at the point of release. **(a)** Fill in the constants of Equation 3.9b for the $y$-displacement and set the displacement equal to $-45.0$ m, the $y$-displacement when the ball hits the ground. Using the quadratic formula, solve for the time. To solve part **(b)**, substitute the time from part **(a)** into the components of the velocity, and substitute the same time into the equation for the $x$-displacement to solve part **(c)**.

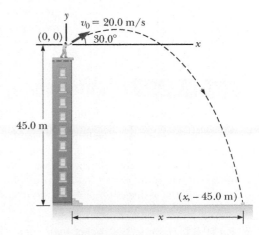

**Figure 3.12** (Example 3.5)

*(Continued)*

## SOLUTION

**(a)** Find the ball's time of flight.

Find the initial $x$- and $y$-components of the velocity:

$$v_{0x} = v_0 \cos \theta_0 = (20.0 \text{ m/s})(\cos 30.0°) = +17.3 \text{ m/s}$$
$$v_{0y} = v_0 \sin \theta_0 = (20.0 \text{ m/s})(\sin 30.0°) = +10.0 \text{ m/s}$$

Find the $y$-displacement, taking $y_0 = 0$, $y = -45.0$ m, and $v_{0y} = 10.0$ m/s:

$$\Delta y = y - y_0 = v_{0y}t - \tfrac{1}{2}gt^2$$
$$-45.0 \text{ m} = (10.0 \text{ m/s})t - (4.90 \text{ m/s}^2)t^2$$

Reorganize the equation into standard form and use the quadratic formula (see Appendix A) to find the positive root:

$$t = \boxed{4.22 \text{ s}}$$

**(b)** Find the ball's speed at impact.

Substitute the value of $t$ found in part (a) into Equation 3.9a to find the $y$-component of the velocity at impact:

$$v_y = v_{0y} - gt = 10.0 \text{ m/s} - (9.80 \text{ m/s}^2)(4.22 \text{ s})$$
$$= -31.4 \text{ m/s}$$

Use this value of $v_y$, the Pythagorean theorem, and the fact that $v_x = v_{0x} = 17.3$ m/s to find the speed of the ball at impact:

$$v = \sqrt{v_x^2 + v_y^2} = \sqrt{(17.3 \text{ m/s})^2 + (-31.4 \text{ m/s})^2}$$
$$= \boxed{35.9 \text{ m/s}}$$

**(c)** Find the horizontal range of the ball.

Substitute the time of flight into the range equation:

$$\Delta x = x - x_0 = (v_0 \cos \theta)t = (20.0 \text{ m/s})(\cos 30.0°)(4.22 \text{ s})$$
$$= \boxed{73.1 \text{ m}}$$

**REMARKS** The angle at which the ball is thrown affects the velocity vector throughout its subsequent motion, but doesn't affect the speed at a given height. This is a consequence of the conservation of energy, described in Topic 5.

**QUESTION 3.5** True or False: All other things being equal, if the ball is thrown at half the given speed it will travel half as far.

**EXERCISE 3.5** Suppose the ball is thrown from the same height as in the example at an angle of 30.0° below the horizontal. If it strikes the ground 57.0 m away, find **(a)** the time of flight, **(b)** the initial speed, and **(c)** the speed and the angle of the velocity vector with respect to the horizontal at impact. (*Hint:* For part **(a)**, use the equation for the $x$-displacement to eliminate $v_0 t$ from the equation for the $y$-displacement.)

**ANSWERS** **(a)** 1.57 s   **(b)** 41.9 m/s   **(c)** 51.3 m/s, −45.0°

## 3.2.1 Two-Dimensional Constant Acceleration

So far we have studied only problems in which an object with an initial velocity follows a trajectory determined by the acceleration of gravity alone. In the more general case, other agents, such as air drag, surface friction, or engines, can cause accelerations. These accelerations, taken together, form a vector quantity with components $a_x$ and $a_y$. When both components are constant, we can use Equations 3.6 and 3.7 to study the motion, as in the next example.

---

**EXAMPLE 3.6**   **THE ROCKET**

**GOAL** Solve a problem involving accelerations in two directions.

**PROBLEM** A jet plane traveling horizontally at $1.00 \times 10^2$ m/s drops a rocket from a considerable height. (See Fig. 3.13.) The rocket immediately fires its engines, accelerating at 20.0 m/s² in the $x$-direction while falling under the influence of gravity in the $y$-direction. When the rocket has fallen 1.00 km, find **(a)** its velocity in the $y$-direction, **(b)** its velocity in the $x$-direction, and **(c)** the magnitude and direction of its velocity. Neglect air drag and aerodynamic lift.

**STRATEGY** Because the rocket maintains a horizontal orientation (say, through gyroscopes), the $x$- and $y$-components of acceleration are independent of each other. Use the time-independent equation for the velocity in the $y$-direction to find the $y$-component

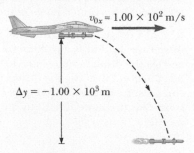

**Figure 3.13** (Example 3.6)

of the velocity after the rocket falls 1.00 km. Then calculate the time of the fall and use that time to find the velocity in the x-direction.

. . . . . . . . . . . . . . . . . . . . . . . . . . . . . . . . . . . . . . . . . . . . . . . . . . . . . . . . . . . . . . . . . . . . . . . . . . . .

**SOLUTION**

**(a)** Find the velocity in the y-direction.

Use Equation 3.9c:

$$v_y^2 = v_{0y}^2 - 2g\Delta y$$

Substitute $v_{0y} = 0$, $g = -9.80$ m/s$^2$, and $\Delta y = -1.00 \times 10^3$ m, and solve for $v_y$:

$$v_y^2 = 0 - 2(9.80 \text{ m/s}^2)(-1.00 \times 10^3 \text{ m})$$
$$v_y = \boxed{-1.40 \times 10^2 \text{ m/s}}$$

**(b)** Find the velocity in the x-direction.

Find the time it takes the rocket to drop $1.00 \times 10^3$ m, using the y-component of the velocity:

$$v_y = v_{0y} + a_y t$$
$$-1.40 \times 10^2 \text{ m/s} = 0 - (9.80 \text{ m/s}^2)t \quad \rightarrow \quad t = 14.3 \text{ s}$$

Substitute $t$, $v_{0x}$, and $a_x$ into Equation 3.6a to find the velocity in the x-direction:

$$v_x = v_{0x} + a_x t = 1.00 \times 10^2 \text{ m/s} + (20.0 \text{ m/s}^2)(14.3 \text{ s})$$
$$= \boxed{386 \text{ m/s}}$$

**(c)** Find the magnitude and direction of the velocity.

Find the magnitude using the Pythagorean theorem and the results of parts **(a)** and **(b)**:

$$v = \sqrt{v_x^2 + v_y^2} = \sqrt{(-1.40 \times 10^2 \text{ m/s})^2 + (386 \text{ m/s})^2}$$
$$= \boxed{411 \text{ m/s}}$$

Use the inverse tangent function to find the angle:

$$\theta = \tan^{-1}\left(\frac{v_y}{v_x}\right) = \tan^{-1}\left(\frac{-1.40 \times 10^2 \text{ m/s}}{386 \text{ m/s}}\right) = \boxed{-19.9°}$$

. . . . . . . . . . . . . . . . . . . . . . . . . . . . . . . . . . . . . . . . . . . . . . . . . . . . . . . . . . . . . . . . . . . . . . . . . . . .

**REMARKS** Notice the similarity: The kinematic equations for the x- and y-directions are handled in exactly the same way. Having a nonzero acceleration in the x-direction doesn't greatly increase the difficulty of the problem.

**QUESTION 3.6** True or False: Neglecting air friction and lift effects, a projectile with a horizontal acceleration always stays in the air longer than a projectile that is freely falling.

**EXERCISE 3.6** Suppose a rocket-propelled motorcycle is fired from rest horizontally across a canyon 1.00 km wide. **(a)** What minimum constant acceleration in the x-direction must be provided by the engines so the cycle crosses safely if the opposite side is 0.750 km lower than the starting point? **(b)** At what speed does the motorcycle land if it maintains this constant horizontal component of acceleration? Neglect air drag, but remember that gravity is still acting in the negative y-direction.

**ANSWERS** **(a)** 13.1 m/s$^2$  **(b)** 202 m/s

---

In a stunt similar to that described in Exercise 3.6, motorcycle daredevil Evel Knievel tried to vault across Hells Canyon, part of the Snake River system in Idaho, on his rocket-powered Harley-Davidson X-2 "Skycycle." He lost consciousness at takeoff and released a lever, prematurely deploying his parachute and falling short of the other side. He landed safely in the canyon.

# 3.3 Relative Velocity

Relative velocity is all about relating the measurements of two different observers, one moving with respect to the other. The measured velocity of an object depends on the velocity of the observer with respect to the object. On highways, for example, cars moving in the same direction are often moving at high speed relative to Earth, but relative to each other they hardly move at all. To an observer at rest at the side of the road, a car might be traveling at 60 mi/h, but to an observer in a truck traveling in the same direction at 50 mi/h, the car would appear to be traveling only 10 mi/h.

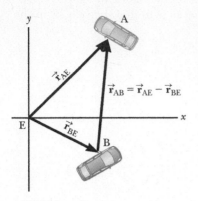

**Figure 3.14** The position of Car A relative to Car B can be found by vector subtraction. The rate of change of the resultant vector with respect to time is the relative velocity equation.

So measurements of velocity depend on the **reference frame** of the observer. Reference frames are just coordinate systems. Most of the time, we use a **stationary frame of reference** relative to Earth, but occasionally we use a **moving frame of reference** associated with a bus, car, or plane moving with constant velocity relative to Earth.

In two dimensions relative velocity calculations can be confusing, so a systematic approach is important and useful. Let E be an observer, assumed stationary with respect to Earth. Let two cars be labeled A and B, and introduce the following notation (see Fig. 3.14):

$\vec{r}_{AE}$ = the position of Car A as measured by E (in a coordinate system fixed with respect to Earth).

$\vec{r}_{BE}$ = the position of Car B as measured by E.

$\vec{r}_{AB}$ = the position of Car A as measured by an observer in Car B.

According to the preceding notation, the first letter tells us what the vector is pointing at and the second letter tells us where the position vector starts. The position vectors of Car A and Car B relative to E, $\vec{r}_{AE}$ and $\vec{r}_{BE}$, are given in the figure. How do we find $\vec{r}_{AB}$, the position of Car A as measured by an observer in Car B? We simply draw an arrow pointing from Car B to Car A, which can be obtained by subtracting $\vec{r}_{BE}$ from $\vec{r}_{AE}$:

$$\vec{r}_{AB} = \vec{r}_{AE} - \vec{r}_{BE} \qquad [3.10]$$

Now, the rate of change of these quantities with time gives us the relationship between the associated velocities:

$$\vec{v}_{AB} = \vec{v}_{AE} - \vec{v}_{BE} \qquad [3.11]$$

The coordinate system of observer E need not be fixed to Earth, although it often is. Take careful note of the pattern of subscripts; rather than memorize Equation 3.11, it's better to study the short derivation based on Figure 3.14. Note also that the equation doesn't work for observers traveling a sizable fraction of the speed of light, when Einstein's theory of special relativity comes into play.

### PROBLEM-SOLVING STRATEGY

**Relative Velocity**

1. **Label** each object involved (usually three) with a letter that reminds you of what it is (e.g., E for Earth).
2. **Look** through the problem for phrases such as "The velocity of A relative to B" and write the velocities as $\vec{v}_{AB}$. When a velocity is mentioned but it isn't explicitly stated as relative to something, it's almost always relative to Earth.
3. **Take** the three velocities you've found and assemble them into an equation just like Equation 3.11, with subscripts in an analogous order.
4. **There** will be two unknown components. Solve for them with the $x$- and $y$-components of the equation developed in step 3.

---

## EXAMPLE 3.7   PITCHING PRACTICE ON THE TRAIN

**GOAL** Solve a one-dimensional relative velocity problem.

**PROBLEM** A train is traveling with a speed of 15 m/s relative to Earth. A passenger standing at the rear of the train pitches a baseball with a speed of 15 m/s relative to the train off the back end, in the direction opposite the motion of the train. **(a)** What is the velocity of the baseball relative to Earth? **(b)** What is the velocity of the baseball relative to the Earth if thrown in the opposite direction at the same speed?

**STRATEGY** Solving these problems involves putting the proper subscripts on the velocities and arranging them as in Equation 3.11. In the first sentence of the problem statement, we are informed that the train travels at "15 m/s relative to Earth." This quantity is

$\vec{\mathbf{v}}_{TE}$, with T for train and E for Earth. The passenger throws the baseball at "15 m/s relative to the train," so this quantity is $\vec{\mathbf{v}}_{BT}$, where B stands for baseball. The second sentence asks for the velocity of the baseball relative to Earth, $\vec{\mathbf{v}}_{BE}$. The rest of the problem can be solved by identifying the correct components of the known quantities and solving for the unknowns, using an analog of Equation 3.11. Part **(b)** just requires a change of sign.

### SOLUTION

**(a)** What is the velocity of the baseball relative to the Earth?

Write the $x$-components of the known quantities:

$$(\vec{\mathbf{v}}_{TE})_x = +15 \text{ m/s}$$
$$(\vec{\mathbf{v}}_{BT})_x = -15 \text{ m/s}$$

Follow Equation 3.11:

$$\textbf{(1)} \quad (\vec{\mathbf{v}}_{BT})_x = (\vec{\mathbf{v}}_{BE})_x - (\vec{\mathbf{v}}_{TE})_x$$

Insert the given values and solve:

$$-15 \text{ m/s} = (\vec{\mathbf{v}}_{BE})_x - 15 \text{ m/s}$$
$$(\vec{\mathbf{v}}_{BE})_x = \boxed{0}$$

**(b)** What is the velocity of the baseball relative to the Earth if thrown in the opposite direction at the same speed?

Substitute $(\vec{\mathbf{v}}_{BT})_x = +15 \text{ m/s}$ into Equation (1):

$$+15 \text{ m/s} = (\vec{\mathbf{v}}_{BT})_x - 15 \text{ m/s}$$

Solve for $(\vec{\mathbf{v}}_{BT})_x$:

$$(\vec{\mathbf{v}}_{BT})_x = \boxed{+3.0 \times 10^1 \text{ m/s}}$$

**QUESTION 3.7** Describe the motion of the ball in part (a) as related by an observer on the ground.

**EXERCISE 3.7** A train is traveling at 27 m/s relative to Earth in the positive $x$-direction. A passenger standing on the ground throws a ball at 15 m/s relative to Earth in the same direction as the train's motion. **(a)** Find the speed of the ball relative to an observer on the train. **(b)** Repeat the exercise if the ball is thrown in the opposite direction.

**ANSWERS** **(a)** $-12$ m/s    **(b)** $-42$ m/s

---

### EXAMPLE 3.8    CROSSING A RIVER

**GOAL** Solve a simple two-dimensional relative motion problem.

**PROBLEM** The boat in Figure 3.15 is heading due north as it crosses a wide river with a velocity of 10.0 km/h relative to the water. The river has a uniform velocity of 5.00 km/h due east. Determine the magnitude and direction of the boat's velocity with respect to an observer on the riverbank.

**STRATEGY** Again, we look for key phrases. "The boat . . . (has) a velocity of 10.0 km/h relative to the water" gives $\vec{\mathbf{v}}_{BR}$. "The river has a uniform velocity of 5.00 km/h due east" gives $\vec{\mathbf{v}}_{RE}$, because this implies velocity with respect to Earth. The observer on the riverbank is in a reference frame at rest with respect to Earth. Because we're looking for the velocity of the boat with respect to that observer, this last velocity is designated $\vec{\mathbf{v}}_{BE}$. Take east to be the $+x$-direction, north the $+y$-direction.

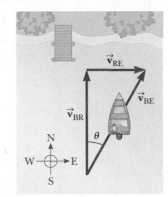

**Figure 3.15** (Example 3.8)

### SOLUTION

Arrange the three quantities into the proper relative velocity equation:

$$\vec{\mathbf{v}}_{BR} = \vec{\mathbf{v}}_{BE} - \vec{\mathbf{v}}_{RE}$$

Write the velocity vectors in terms of their components. For convenience, these are organized in the following table:

| Vector | $x$-Component (km/h) | $y$-Component (km/h) |
| --- | --- | --- |
| $\vec{\mathbf{v}}_{BR}$ | 0 | 10.0 |
| $\vec{\mathbf{v}}_{BE}$ | $v_x$ | $v_y$ |
| $\vec{\mathbf{v}}_{RE}$ | 5.00 | 0 |

(*Continued*)

Find the $x$-component of velocity:

$$0 = v_x - 5.00 \text{ km/h} \rightarrow v_x = 5.00 \text{ km/h}$$

Find the $y$-component of velocity:

$$10.0 \text{ km/h} = v_y - 0 \rightarrow v_y = 10.0 \text{ km/h}$$

Find the magnitude of $\vec{\mathbf{v}}_{\text{BE}}$:

$$v_{\text{BE}} = \sqrt{v_x^2 + v_y^2}$$
$$= \sqrt{(5.00 \text{ km/h})^2 + (10.0 \text{ km/h})^2} = \boxed{11.2 \text{ km/h}}$$

Find the direction of $\vec{\mathbf{v}}_{\text{BE}}$:

$$\theta = \tan^{-1}\left(\frac{v_x}{v_y}\right) = \tan^{-1}\left(\frac{5.00 \text{ m/s}}{10.0 \text{ m/s}}\right) = \boxed{26.6°}$$

. . . . . . . . . . . . . . . . . . . . . . . . . . . . . . . . . . . . . . . . . . . . . . . . . . . . . . . . . . . . . . . . . . . . . . . . . . . . . . . . . . . . . . . . . . . . .

**REMARKS** The boat travels at a speed of 11.2 km/h in the direction 26.6° east of north with respect to Earth.

**QUESTION 3.8** If the speed of the boat relative to the water is increased, what happens to the angle?

**EXERCISE 3.8** Suppose the river is flowing east at 3.00 m/s and the boat is traveling south at 4.00 m/s with respect to the river. Find the speed and direction of the boat relative to Earth.

**ANSWER** 5.00 m/s, 53.1° south of east

---

## EXAMPLE 3.9 | BUCKING THE CURRENT

**GOAL** Solve a complex two-dimensional relative motion problem.

**PROBLEM** If the skipper of the boat of Example 3.8 moves with the same speed of 10.0 km/h relative to the water but now wants to travel due north, as in Figure 3.16a, in what direction should he head? What is the speed of the boat, according to an observer on the shore? The river is flowing east at 5.00 km/h.

**STRATEGY** Proceed as in the previous example. In this situation, we must find the heading of the boat and its velocity with respect to the water, using the fact that the boat travels due north.

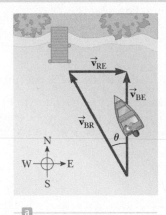
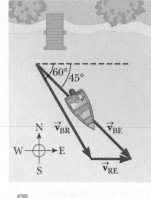

**Figure 3.16**
(a) (Example 3.9)
(b) (Exercise 3.9)

. . . . . . . . . . . . . . . . . . . . . . . . . . . . . . . . . . . . . . . . . . . . . . . . . . . . . . . . . . . . . . . . . . . . . . . . . . . . . . . . . . . . . . . . . . . . .

**SOLUTION**

Arrange the three quantities, as before:

$$\vec{\mathbf{v}}_{\text{BR}} = \vec{\mathbf{v}}_{\text{BE}} - \vec{\mathbf{v}}_{\text{RE}}$$

Organize a table of velocity components:

| Vector | $x$-Component (km/h) | $y$-Component (km/h) |
|---|---|---|
| $\vec{\mathbf{v}}_{\text{BR}}$ | $-(10.0 \text{ km/h}) \sin\theta$ | $(10.0 \text{ km/h}) \cos\theta$ |
| $\vec{\mathbf{v}}_{\text{BE}}$ | 0 | $v$ |
| $\vec{\mathbf{v}}_{\text{RE}}$ | 5.00 km/h | 0 |

The $x$-component of the relative velocity equation can be used to find $\theta$:

$$-(10.0 \text{ m/s}) \sin\theta = 0 - 5.00 \text{ km/h}$$
$$\sin\theta = \frac{5.00 \text{ km/h}}{10.0 \text{ km/h}} = \frac{1.00}{2.00}$$

Apply the inverse sine function and find $\theta$, which is the boat's heading, west of north:

$$\theta = \sin^{-1}\left(\frac{1.00}{2.00}\right) = \boxed{30.0°}$$

The $y$-component of the relative velocity equation can be used to find $v$:

$$(10.0 \text{ km/h}) \cos\theta = v \rightarrow v = \boxed{8.66 \text{ km/h}}$$

. . . . . . . . . . . . . . . . . . . . . . . . . . . . . . . . . . . . . . . . . . . . . . . . . . . . . . . . . . . . . . . . . . . . . . . . . . . . . . . . . . . . . . . . . . . . .

**REMARKS** From Figure 3.16, we see that this problem can be solved with the Pythagorean theorem, because the problem involves a right triangle: the boat's $x$-component of velocity exactly cancels the river's velocity. When this is not the case, a more general technique is necessary, as shown in the following exercise. Notice that in the $x$-component of the relative velocity equation a minus sign had to be included in the term $-(10.0 \text{ km/h})\sin\theta$ because the $x$-component of the boat's velocity with respect to the river is negative.

**QUESTION 3.9** The speeds in this example are the same as in Example 3.8. Why isn't the angle the same as before?

**EXERCISE 3.9** Suppose the river is moving east at 5.00 km/h and the boat is traveling 45.0° south of east with respect to Earth. Find **(a)** the speed of the boat with respect to Earth and **(b)** the speed of the boat with respect to the river if the boat's heading in the water is 60.0° south of east. (See Fig. 3.16b.) You will have to solve two equations with two unknowns. (As an alternative, the law of sines can be used.)

**ANSWERS (a)** 16.7 km/h   **(b)** 13.7 km/h

# SUMMARY

## 3.1 Displacement, Velocity, and Acceleration in Two Dimensions

The displacement of an object in two dimensions is defined as the change in the object's position vector:

$$\Delta \vec{r} \equiv \vec{r}_f - \vec{r}_i \qquad [3.1]$$

The average velocity of an object during the time interval $\Delta t$ is

$$\vec{v}_{av} \equiv \frac{\Delta \vec{r}}{\Delta t} \qquad [3.2]$$

Taking the limit of this expression as $\Delta t$ gets arbitrarily small gives the instantaneous velocity $\vec{v}$:

$$\vec{v} \equiv \lim_{\Delta t \to 0} \frac{\Delta \vec{r}}{\Delta t} \qquad [3.3]$$

The direction of the instantaneous velocity vector is along a line that is tangent to the path of the object and in the direction of its motion.

The average acceleration of an object with a velocity changing by $\Delta \vec{v}$ in the time interval $\Delta t$ is

$$\vec{a}_{av} \equiv \frac{\Delta \vec{v}}{\Delta t} \qquad [3.4]$$

Taking the limit of this expression as $\Delta t$ gets arbitrarily small gives the instantaneous acceleration vector $\vec{a}$:

$$\vec{a} \equiv \lim_{\Delta t \to 0} \frac{\Delta \vec{v}}{\Delta t} \qquad [3.5]$$

## 3.2 Two-Dimensional Motion

The general kinematic equations in two dimensions for objects with constant acceleration are, for the $x$-direction,

$$v_x = v_{0x} + a_x t \qquad [3.6a]$$

$$\Delta x = v_{0x} t + \tfrac{1}{2} a_x t^2 \qquad [3.6b]$$

$$v_x^2 = v_{0x}^2 + 2 a_x \Delta x \qquad [3.6c]$$

where $v_{0x} = v_0 \cos \theta_0$, and, for the $y$-direction,

$$v_y = v_{0y} + a_y t \qquad [3.7a]$$

$$\Delta y = v_{0y} t + \tfrac{1}{2} a_y t^2 \qquad [3.7b]$$

$$v_y^2 = v_{0y}^2 + 2 a_y \Delta y \qquad [3.7c]$$

where $v_{0y} = v_0 \sin \theta_0$. The speed $v$ of the object at any instant can be calculated from the components of velocity at that instant using the Pythagorean theorem:

$$v = \sqrt{v_x^2 + v_y^2}$$

The angle that the velocity vector makes with the $x$-axis is given by

$$\theta = \tan^{-1}\left(\frac{v_y}{v_x}\right)$$

The horizontal and vertical motions of a projectile are completely independent of each other (Fig. 3.17).

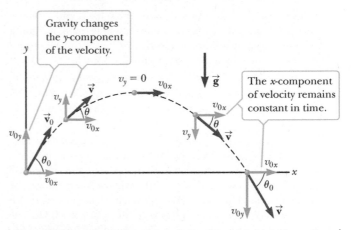

Gravity changes the $y$-component of the velocity.

The $x$-component of velocity remains constant in time.

**Figure 3.17** Gravity acts on the $y$-component of the velocity and has no effect on the $x$-component, illustrating the independence of horizontal and vertical projectile motion.

The kinematic equations are easily adapted and simplified for projectiles close to the surface of the Earth. The equations for the motion in the horizontal or $x$-direction are

$$v_x = v_{0x} = v_0 \cos \theta_0 = \text{constant} \qquad [3.8a]$$

$$\Delta x = v_{0x} t = (v_0 \cos \theta_0) t \qquad [3.8b]$$

while the equations for the motion in the vertical or $y$-direction are

$$v_y = v_0 \sin \theta_0 - g t \qquad [3.9a]$$

$$\Delta y = (v_0 \sin \theta_0) t - \tfrac{1}{2} g t^2 \qquad [3.9b]$$

$$v_y^2 = (v_0 \sin \theta_0)^2 - 2 g \Delta y \qquad [3.9c]$$

Problems are solved by algebraically manipulating one or more of these equations, which often reduces the system to two equations and two unknowns.

## 3.3 Relative Velocity

Let E be an observer, and B a second observer traveling with velocity $\vec{v}_{BE}$ as measured by E (Fig. 3.18). If E measures the velocity of an object A as $\vec{v}_{AE}$, then B will measure A's velocity as

$$\vec{v}_{AB} = \vec{v}_{AE} - \vec{v}_{BE} \qquad [3.11]$$

Equation 3.11 can be derived from Figure 3.18 by dividing the relative position equation by the $\Delta t$ and taking the limit as $\Delta t$ goes to zero.

Solving relative velocity problems involves identifying the velocities properly and labeling them correctly, substituting into Equation 3.11, and then solving for unknown quantities.

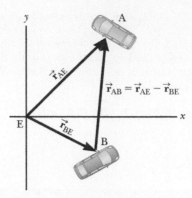

**Figure 3.18** The time rate of change of the difference of the two position vectors $\vec{r}_{AE}$ and $\vec{r}_{BE}$ gives the relative velocity equation, Equation 3.11.

## CONCEPTUAL QUESTIONS

1. As a projectile moves in its path, is there any point along the path where the velocity and acceleration vectors are (a) perpendicular to each other? (b) Parallel to each other?

2. Construct motion diagrams showing the velocity and acceleration of a projectile at several points along its path, assuming (a) the projectile is launched horizontally and (b) the projectile is launched at an angle $\theta$ with the horizontal.

3. Explain whether the following particles do or do not have an acceleration: (a) a particle moving in a straight line with constant speed and (b) a particle moving around a curve with constant speed.

4. A ball is projected horizontally from the top of a building. One second later, another ball is projected horizontally from the same point with the same velocity. (a) At what point in the motion will the balls be closest to each other? (b) Will the first ball always be traveling faster than the second? (c) What will be the time difference between them when the balls hit the ground? (d) Can the horizontal projection velocity of the second ball be changed so that the balls arrive at the ground at the same time?

5. A projectile is launched with speed $v_0$ at an angle $\theta_0$ above the horizon. Its motion is described in terms of position, velocity, and acceleration: $x, y, v_x, v_y, a_x, a_y$, respectively. (a) Which of those quantities are constant during the motion? (b) Which are equal to zero throughout the motion? Ignore any effects of air resistance.

6. Determine which of the following moving objects obey the equations of projectile motion developed in this topic. (a) A ball is thrown in an arbitrary direction. (b) A jet airplane crosses the sky with its engines thrusting the plane forward. (c) A rocket leaves the launch pad. (d) A rocket moves through the sky after its engines have failed. (e) A stone is thrown under water.

7. Two projectiles are thrown with the same initial speed, one at an angle $\theta$ with respect to the level ground and the other at angle $90° - \theta$. Both projectiles strike the ground at the same distance from the projection point. Are both projectiles in the air for the same length of time?

8. A ball is thrown upward in the air by a passenger on a train that is moving with constant velocity. (a) Describe the path of the ball as seen by the passenger. Describe the path as seen by a stationary observer outside the train. (b) How would these observations change if the train were accelerating along the track?

9. A projectile is launched at some angle to the horizontal with some initial speed $v_i$; air resistance is negligible. (a) Is the projectile a freely falling body? (b) What is its acceleration in the vertical direction? (c) What is its acceleration in the horizontal direction?

10. A baseball is thrown from the outfield toward the catcher. When the ball reaches its highest point, which statement is true? (a) Its velocity and its acceleration are both zero. (b) Its velocity is not zero, but its acceleration is zero. (c) Its velocity is perpendicular to its acceleration. (d) Its acceleration depends on the angle at which the ball was thrown. (e) None of statements (a) through (d) is true.

11. A student throws a heavy red ball horizontally from a balcony of a tall building with an initial speed $v_0$. At the same time, a second student drops a lighter blue ball from the same balcony. Neglecting air resistance, which statement is true? (a) The blue ball reaches the ground first. (b) The balls reach the ground at the same instant. (c) The red ball reaches the ground first. (d) Both balls hit the ground with the same speed. (e) None of statements (a) through (d) is true.

12. A boat is heading due east at speed $v$ when passengers onboard spot a dolphin swimming due north away from them, relative to their moving boat. Which of the following must be true of the dolphin's motion relative to a stationary observer floating in the water (choose one)? The dolphin is (a) heading south of east at a speed greater than $v$, (b) heading south of west at a speed less than $v$, (c) heading north of east at a speed greater than $v$, or (d) heading north of west at a speed less than $v$.

13. As an apple tree is transported by a truck moving to the right with a constant velocity, one of its apples shakes loose and falls toward the bed of the truck. Of the curves shown in Figure CQ3.13, (i) which best describes the path followed by the apple as seen by a stationary observer on the ground, who observes the truck moving from his left to his right? (ii) Which best describes the path as seen by an observer sitting in the truck?

**Figure CQ3.13**

## PROBLEMS

*See the Preface for an explanation of the icons used in this problems set.*

### 3.1 Displacement, Velocity, and Acceleration in Two Dimensions

1. An airplane in a holding pattern flies at constant altitude along a circular path of radius 3.50 km. If the airplane rounds half the circle in $1.50 \times 10^2$ s, determine the magnitude of its (a) displacement and (b) average velocity during that time. (c) What is the airplane's average speed during the same time interval?

2. A hiker walks 2.00 km north and then 3.00 km east, all in 2.50 hours. Calculate the magnitude and direction of the hiker's (a) displacement (in km) and (b) average velocity (in km/h) during those 2.50 hours. (c) What was her average speed during the same time interval?

3. A miniature quadcopter is located at $x_i = 2.00$ m and $y_i = 4.50$ m at $t = 0$ and moves with an average velocity having components $v_{av,x} = 1.50$ m/s and $v_{av,y} = -1.00$ m/s. What are the (a) x-coordinate and (b) y-coordinate of the quadcopter's position at $t = 2.00$ s?

4. An ant crawls on the floor along the curved path shown in Figure P3.4. The ant's positions and velocities are indicated for times $t_i = 0$ and $t_f = 5.00$ s. Determine the x- and y-components of the ant's (a) displacement, (b) average velocity, and (c) average acceleration between the two times.

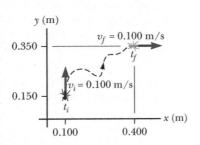

**Figure P3.4**

5. A car is traveling east at 25.0 m/s when it turns due north and accelerates to 35.0 m/s, all during a time of 6.00 s. Calculate the magnitude of the car's average acceleration.

6. A rabbit is moving in the positive x-direction at 2.00 m/s when it spots a predator and accelerates to a velocity of 12.0 m/s along the negative y-axis, all in 1.50 s. Determine (a) the x-component and (b) the y-component of the rabbit's acceleration.

### 3.2 Two-Dimensional Motion

7. **GP** A student stands at the edge of a cliff and throws a stone horizontally over the edge with a speed of 18.0 m/s. The cliff is 50.0 m above a flat, horizontal beach as shown in Figure P3.7. (a) What are the coordinates of the initial position of the stone? (b) What are the components of the initial velocity? (c) Write the equations for the x- and y-components of the velocity of the stone with time. (d) Write the equations for the position of the stone

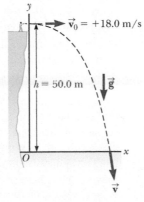

**Figure P3.7**

with time, using the coordinates in Figure P3.7. (e) How long after being released does the stone strike the beach below the cliff? (f) With what speed and angle of impact does the stone land?

8. One of the fastest recorded pitches in major league baseball, thrown by Tim Lincecum in 2009, was clocked at 101.0 mi/h (Fig. P3.8). If a pitch were thrown horizontally with this velocity, how far would the ball fall vertically by the time it reached home plate, 60.5 ft away?

**Figure P3.8** Tim Lincecum throws a baseball.

9. **V** The best leaper in the animal kingdom is the puma, which can jump to a height of 3.7 m when leaving the ground at an angle of 45°. With what speed must the animal leave the ground to reach that height?

10. **QC S** A rock is thrown upward from the level ground in such a way that the maximum height of its flight is equal to its horizontal range R. (a) At what angle $\theta$ is the rock thrown? (b) In terms of the original range R, what is the range $R_{max}$ the rock can attain if it is launched at the same speed but at the optimal angle for maximum range? (c) Would your answer to part (a) be different if the rock is thrown with the same speed on a different planet? Explain.

11. **T** A placekicker must kick a football from a point 36.0 m (about 40 yards) from the goal. Half the crowd hopes the ball will clear the crossbar, which is 3.05 m high. When kicked, the ball leaves the ground with a speed of 20.0 m/s at an angle of 53.0° to the horizontal. (a) By how much does the ball clear or fall short of clearing the crossbar? (b) Does the ball approach the crossbar while still rising or while falling?

12. **QC** The record distance in the sport of throwing cowpats is 81.1 m. This record toss was set by Steve Urner of the United States in 1981. Assuming the initial launch angle was 45° and neglecting air resistance, determine (a) the initial speed of the projectile and (b) the total time the projectile was in flight. (c) Qualitatively, how would the answers change if the launch angle were greater than 45°? Explain.

13. A brick is thrown upward from the top of a building at an angle of 25° to the horizontal and with an initial speed of 15 m/s. If the brick is in flight for 3.0 s, how tall is the building?

14. **GP** From the window of a building, a ball is tossed from a height $y_0$ above the ground with an initial velocity of 8.00 m/s

and angle of 20.0° below the horizontal. It strikes the ground 3.00 s later. (a) If the base of the building is taken to be the origin of the coordinates, with upward the positive $y$-direction, what are the initial coordinates of the ball? (b) With the positive $x$-direction chosen to be out the window, find the $x$- and $y$-components of the initial velocity. (c) Find the equations for the $x$- and $y$-components of the position as functions of time. (d) How far horizontally from the base of the building does the ball strike the ground? (e) Find the height from which the ball was thrown. (f) How long does it take the ball to reach a point 10.0 m below the level of launching?

15. A car is parked on a cliff overlooking the ocean on an incline that makes an angle of 24.0° below the horizontal. The negligent driver leaves the car in neutral, and the emergency brakes are defective. The car rolls from rest down the incline with a constant acceleration of 4.00 m/s² for a distance of 50.0 m to the edge of the cliff, which is 30.0 m above the ocean. Find (a) the car's position relative to the base of the cliff when the car lands in the ocean and (b) the length of time the car is in the air.

16. An artillery shell is fired with an initial velocity of 300 m/s at 55.0° above the horizontal. To clear an avalanche, it explodes on a mountainside 42.0 s after firing. What are the $x$- and $y$-coordinates of the shell where it explodes, relative to its firing point?

17. A projectile is launched with an initial speed of 60.0 m/s at an angle of 30.0° above the horizontal. The projectile lands on a hillside 4.00 s later. Neglect air friction. (a) What is the projectile's velocity at the highest point of its trajectory? (b) What is the straight-line distance from where the projectile was launched to where it hits its target?

18. **V** A fireman $d = 50.0$ m away from a burning building directs a stream of water from a ground-level fire hose at an angle of $\theta_i = 30.0°$ above the horizontal as shown in Figure P3.18. If the speed of the stream as it leaves the hose is $v_i = 40.0$ m/s, at what height will the stream of water strike the building?

**Figure P3.18**

19. A playground is on the flat roof of a city school, 6.00 m above the street below (Fig. P3.19). The vertical wall of the building is $h = 7.00$ m high, to form a 1-m-high railing around the playground. A ball has fallen to the street below, and a passerby returns it by launching it at an angle of $\theta = 53.0°$ above the horizontal at a point $d = 24.0$ m from the base of the building wall. The ball takes 2.20 s to reach a point vertically above the wall. (a) Find the speed at which the ball was launched. (b) Find the vertical distance by which the ball clears the wall. (c) Find the horizontal distance from the wall to the point on the roof where the ball lands.

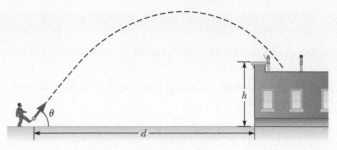

**Figure P3.19**

## 3.3 Relative Velocity

20. A cruise ship sails due north at 4.50 m/s while a Coast Guard patrol boat heads 45.0° north of west at 5.20 m/s. What are (a) the $x$-component and (b) $y$-component of the velocity of the cruise ship relative to the patrol boat?

21. Suppose a boat moves at 12.0 m/s relative to the water. If the boat is in a river with the current directed east at 2.50 m/s, what is the boat's speed relative to the ground when it is heading (a) east, with the current, and (b) west, against the current?

22. **T** A car travels due east with a speed of 50.0 km/h. Raindrops are falling at a constant speed vertically with respect to the Earth. The traces of the rain on the side windows of the car make an angle of 60.0° with the vertical. Find the velocity of the rain with respect to (a) the car and (b) the Earth.

23. **GP** A jet airliner moving initially at $3.00 \times 10^2$ mi/h due east enters a region where the wind is blowing $1.00 \times 10^2$ mi/h in a direction 30.0° north of east. (a) Find the components of the velocity of the jet airliner relative to the air, $\vec{v}_{JA}$. (b) Find the components of the velocity of the air relative to Earth, $\vec{v}_{AE}$. (c) Write an equation analogous to Equation 3.11 for the velocities $\vec{v}_{JA}$, $\vec{v}_{AE}$, and $\vec{v}_{JE}$. (d) What are the speed and direction of the aircraft relative to the ground?

24. A Coast Guard cutter detects an unidentified ship at a distance of 20.0 km in the direction 15.0° east of north. The ship is traveling at 26.0 km/h on a course at 40.0° east of north. The Coast Guard wishes to send a speedboat to intercept and investigate the vessel. (a) If the speedboat travels at 50.0 km/h, in what direction should it head? Express the direction as a compass bearing with respect to due north. (b) Find the time required for the cutter to intercept the ship.

25. **Q|C** A bolt drops from the ceiling of a moving train car that is accelerating northward at a rate of 2.50 m/s². (a) What is the acceleration of the bolt relative to the train car? (b) What is the acceleration of the bolt relative to the Earth? (c) Describe the trajectory of the bolt as seen by an observer fixed on the Earth.

26. **V** An airplane maintains a speed of 630 km/h relative to the air it is flying through, as it makes a trip to a city 750 km away to the north. (a) What time interval is required for the trip if the plane flies through a headwind blowing at 35.0 km/h toward the south? (b) What time interval is required if there is a tailwind with the same speed? (c) What time interval is required if there is a crosswind blowing at 35.0 km/h to the east relative to the ground?

27. **BIO** Suppose a chinook salmon needs to jump a waterfall that is 1.50 m high. If the fish starts from a distance 1.00 m from the base of the ledge over which the waterfall flows, (a) find the $x$- and $y$-components of the initial velocity the

salmon would need to just reach the ledge at the top of its trajectory. (b) Can the fish make this jump? (Note that a chinook salmon can jump out of the water with an initial speed of 6.26 m/s.)

28. A bomber is flying horizontally over level terrain at a speed of 275 m/s relative to the ground and at an altitude of 3.00 km. (a) The bombardier releases one bomb. How far does the bomb travel horizontally between its release and its impact on the ground? Ignore the effects of air resistance. (b) Firing from the people on the ground suddenly incapacitates the bombardier before he can call, "Bombs away!" Consequently, the pilot maintains the plane's original course, altitude, and speed through a storm of flak. Where is the plane relative to the bomb's point of impact when the bomb hits the ground? (c) The plane has a telescopic bombsight set so that the bomb hits the target seen in the sight at the moment of release. At what angle from the vertical was the bombsight set?

29. **Q|C** A river has a steady speed of 0.500 m/s. A student swims upstream a distance of 1.00 km and swims back to the starting point. (a) If the student can swim at a speed of 1.20 m/s in still water, how long does the trip take? (b) How much time is required in still water for the same length swim? (c) Intuitively, why does the swim take longer when there is a current?

30. **Q|C** **S** *This is a symbolic version of Problem 29.* A river has a steady speed of $v_s$. A student swims upstream a distance $d$ and back to the starting point. (a) If the student can swim at a speed of $v$ in still water, how much time $t_{up}$ does it take the student to swim upstream a distance $d$? Express the answer in terms of $d$, $v$, and $v_s$. (b) Using the same variables, how much time $t_{down}$ does it take to swim back downstream to the starting point? (c) Sum the answers found in parts (a) and (b) and show that the time $t_a$ required for the whole trip can be written as

$$t_a = \frac{2d/v}{1 - v_s^2/v^2}$$

(d) How much time $t_b$ does the trip take in still water?
(e) Which is larger, $t_a$ or $t_b$? Is it always larger?

## Additional Problems

31. **V** How long does it take an automobile traveling in the left lane of a highway at 60.0 km/h to overtake (become even with) another car that is traveling in the right lane at 40.0 km/h when the cars' front bumpers are initially 100 m apart?

32. **S** A moving walkway at an airport has a speed $v_1$ and a length $L$. A woman stands on the walkway as it moves from one end to the other, while a man in a hurry to reach his flight walks on the walkway with a speed of $v_2$ relative to the moving walkway. (a) How long does it take the woman to travel the distance $L$? (b) How long does it take the man to travel this distance?

33. A boy throws a baseball onto a roof and it rolls back down and off the roof with a speed of 3.75 m/s. If the roof is pitched at 35.0° below the horizon and the roof edge is 2.50 m above the ground, find (a) the time the baseball spends in the air, and (b) the horizontal distance from the roof edge to the point where the baseball lands on the ground.

34. **V** You can use any coordinate system you like to solve a projectile motion problem. To demonstrate the truth of this

statement, consider a ball thrown off the top of a building with a velocity $\vec{v}$ at an angle $\theta$ with respect to the horizontal. Let the building be 50.0 m tall, the initial horizontal velocity be 9.00 m/s, and the initial vertical velocity be 12.0 m/s. Choose your coordinates such that the positive y-axis is upward, the x-axis is to the right, and the origin is at the point where the ball is released. (a) With these choices, find the ball's maximum height above the ground and the time it takes to reach the maximum height. (b) Repeat your calculations choosing the origin at the base of the building.

35. **T** Towns A and B in Figure P3.35 are 80.0 km apart. A couple arranges to drive from town A and meet a couple driving from town B at the lake, L. The two couples leave simultaneously and drive for 2.50 h in the directions shown. Car 1 has a speed of 90.0 km/h. If the cars arrive simultaneously at the lake, what is the speed of car 2?

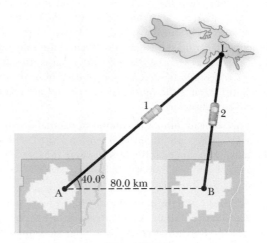

**Figure P3.35**

36. **V** **S** In a local diner, a customer slides an empty coffee cup down the counter for a refill. The cup slides off the counter and strikes the floor at distance $d$ from the base of the counter. If the height of the counter is $h$, (a) find an expression for the time $t$ it takes the cup to fall to the floor in terms of the variables $h$ and $g$. (b) With what speed does the mug leave the counter? Answer in terms of the variables $d$, $g$, and $h$. (c) In the same terms, what is the speed of the cup immediately before it hits the floor? (d) In terms of $h$ and $d$, what is the direction of the cup's velocity immediately before it hits the floor?

37. A father demonstrates projectile motion to his children by placing a pea on his fork's handle and rapidly depressing the curved tines, launching the pea to heights above the dining room table. Suppose the pea is launched at 8.25 m/s at an angle of 75.0° above the table. With what speed does the pea strike the ceiling 2.00 m above the table?

38. **V** Two canoeists in identical canoes exert the same effort paddling and hence maintain the same speed relative to the water. One paddles directly upstream (and moves upstream), whereas the other paddles directly downstream. With downstream as the positive direction, an observer on shore determines the velocities of the two canoes to be −1.2 m/s and +2.9 m/s, respectively. (a) What is the speed of the water relative to the shore? (b) What is the speed of each canoe relative to the water?

39. A rocket is launched at an angle of 53.0° above the horizontal with an initial speed of 100. m/s. The rocket moves for 3.00 s along its initial line of motion with an acceleration of 30.0 m/s². At this time, its engines fail and the rocket proceeds to move as a projectile. Find (a) the maximum altitude reached by the rocket, (b) its total time of flight, and (c) its horizontal range.

40. **Q|C** A farm truck travels due east with a constant speed of 9.50 m/s along a horizontal road. A boy riding in the back of the truck tosses a can of soda upward (Fig. P3.40) and catches it at the same location in the truck bed, but 16.0 m farther down the road. Ignore any effects of air resistance. (a) At what angle to the vertical does the boy throw the can, relative to the moving truck? (b) What is the can's initial speed relative to the truck? (c) What is the shape of the can's trajectory as seen by the boy? (d) What is the shape of the can's trajectory as seen by a stationary observer on the ground? (e) What is the initial velocity of the can, relative to the stationary observer?

**Figure P3.40**

41. (a) If a person can jump a maximum horizontal distance (by using a 45° projection angle) of 3.0 m on Earth, what would be his maximum range on the Moon, where the free-fall acceleration is $g/6$ and $g = 9.80$ m/s²? (b) Repeat for Mars, where the acceleration due to gravity is $0.38g$.

42. A ball is thrown straight upward and returns to the thrower's hand after 3.00 s in the air. A second ball thrown at an angle of 30.0° with the horizontal reaches the same maximum height as the first ball. (a) At what speed was the first ball thrown? (b) At what speed was the second ball thrown?

43. **T** A home run is hit in such a way that the baseball just clears a wall 21 m high, located 130 m from home plate. The ball is hit at an angle of 35° to the horizontal, and air resistance is negligible. Find (a) the initial speed of the ball, (b) the time it takes the ball to reach the wall, and (c) the velocity components and the speed of the ball when it reaches the wall. (Assume the ball is hit at a height of 1.0 m above the ground.)

44. A 2.00-m-tall basketball player is standing on the floor 10.0 m from the basket, as in Figure P3.44. If he shoots the ball at a 40.0° angle with the horizontal, at what initial speed must he throw the basketball so that it goes through the hoop without striking the backboard? The height of the basket is 3.05 m.

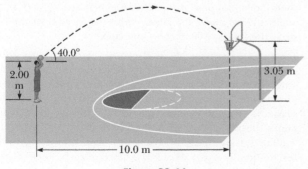

**Figure P3.44**

45. A quarterback throws a football toward a receiver with an initial speed of 20. m/s at an angle of 30.° above the horizontal. At that instant the receiver is 20. m from the quarterback. In (a) what direction and (b) with what constant speed should the receiver run in order to catch the football at the level at which it was thrown?

46. **S** The $x$- and $y$-coordinates of a projectile launched from the origin are $x = v_{0x}t$ and $y = v_{0y}t - \frac{1}{2}gt^2$. Solve the first of these equations for time $t$ and substitute into the second to show that the path of a projectile is a parabola with the form $y = ax + bx^2$, where $a$ and $b$ are constants.

47. **BIO** Spitting cobras can defend themselves by squeezing muscles around their venom glands to squirt venom at an attacker. Suppose a spitting cobra rears up to a height of 0.500 m above the ground and launches venom at 3.50 m/s, directed 50.0° above the horizon. Neglecting air resistance, find the horizontal distance traveled by the venom before it hits the ground.

48. **S** When baseball outfielders throw the ball, they usually allow it to take one bounce, on the theory that the ball arrives at its target sooner that way. Suppose that, after the bounce, the ball rebounds at the same angle $\theta$ that it had when it was released (as in Fig. P3.48), but loses half its speed. (a) Assuming that the ball is always thrown with the same initial speed, at what angle $\theta$ should the ball be thrown in order to go the same distance $D$ with one bounce as a ball thrown upward at 45.0° with no bounce? (b) Determine the ratio of the times for the one-bounce and no-bounce throws.

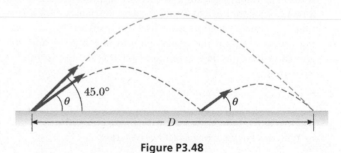

**Figure P3.48**

49. A hunter wishes to cross a river that is 1.5 km wide and flows with a speed of 5.0 km/h parallel to its banks. The hunter uses a small powerboat that moves at a maximum speed of 12 km/h with respect to the water. What is the minimum time necessary for crossing?

50. **BIO** Chinook salmon are able to move upstream faster by jumping out of the water periodically; this behavior is called *porpoising*. Suppose a salmon swimming in still water jumps out of the water with a speed of 6.26 m/s at an angle of 45°, sails through the air a distance $L$ before returning to the water, and then swims a distance $L$ underwater at a speed of 3.58 m/s before beginning another porpoising maneuver. Determine the average speed of the fish.

51. A daredevil is shot out of a cannon at 45.0° to the horizontal with an initial speed of 25.0 m/s. A net is positioned a horizontal distance of 50.0 m from the cannon. At what height above the cannon should the net be placed in order to catch the daredevil?

52. If raindrops are falling vertically at 7.50 m/s, what angle from the vertical do they make for a person jogging at 2.25 m/s?

**53.** **BIO** A celebrated Mark Twain story has motivated contestants in the Calaveras County Jumping Frog Jubilee, where frog jumps as long as 2.2 m have been recorded. If a frog jumps 2.2 m and the launch angle is 45°, find (a) the frog's launch speed and (b) the time the frog spends in the air. Ignore air resistance.

**54.** **QC** A landscape architect is planning an artificial waterfall in a city park. Water flowing at 0.750 m/s leaves the end of a horizontal channel at the top of a vertical wall $h = 2.35$ m high and falls into a pool (Fig. P3.54). (a) How far from the wall will the water land? Will the space behind the waterfall be wide enough for a pedestrian walkway? (b) To sell her plan to the city council, the architect wants to build a model to standard scale, one-twelfth actual size. How fast should the water flow in the channel in the model?

**Figure P3.54**

**55.** A golf ball with an initial speed of 50.0 m/s lands exactly 240 m downrange on a level course. (a) Neglecting air friction, what *two* projection angles would achieve this result? (b) What is the maximum height reached by the ball, using the two angles determined in part (a)?

**56.** **BIO** Antlion larvae lie in wait for prey at the bottom of a conical pit about 5.0 cm deep and 3.8 cm in radius. When a small insect ventures into the pit, it slides to the bottom and is seized by the antlion. If the prey attempts to escape, the antlion rapidly launches grains of sand at the prey, either knocking it down or causing a small avalanche that returns the prey to the bottom of the pit. Suppose an antlion launches grains of sand at an angle of 72° above the horizon. Find the launch speed $v_0$ required to hit a target at the top of the pit, 5.0 cm above and 3.8 cm to the right of the antlion.

**57.** One strategy in a snowball fight is to throw a snowball at a high angle over level ground. Then, while your opponent is watching that snowball, you throw a second one at a low angle timed to arrive before or at the same time as the first one. Assume both snowballs are thrown with a speed of 25.0 m/s. The first

is thrown at an angle of 70.0° with respect to the horizontal. (a) At what angle should the second snowball be thrown to arrive at the same point as the first? (b) How many seconds later should the second snowball be thrown after the first for both to arrive at the same time?

**58.** A football receiver running straight downfield at 5.50 m/s is 10.0 m in front of the quarterback when a pass is thrown downfield at 25.0° above the horizon (Fig. P3.58). If the receiver never changes speed and the ball is caught at the same height from which it was thrown, find (a) the football's initial speed, (b) the amount of time the football spends in the air, and (c) the distance between the quarterback and the receiver when the catch is made.

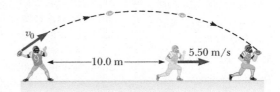

**Figure P3.58**

**59.** The determined Wile E. Coyote is out once more to try to capture the elusive roadrunner. The coyote wears a new pair of power roller skates, which provide a constant horizontal acceleration of 15.0 m/s², as shown in Figure P3.59. The coyote starts off at rest 70.0 m from the edge of a cliff at the instant the roadrunner zips by in the direction

**Figure P3.59**

of the cliff. (a) If the roadrunner moves with constant speed, find the minimum speed the roadrunner must have to reach the cliff before the coyote. (b) If the cliff is $1.00 \times 10^2$ m above the base of a canyon, find where the coyote lands in the canyon. (Assume his skates are still in operation when he is in "flight" and that his horizontal component of acceleration remains constant at 15.0 m/s².)

# TOPIC 4
# Newton's Laws of Motion

CLASSICAL MECHANICS DESCRIBES THE RELATIONSHIP between the motion of objects found in our everyday world and the forces acting on them. As long as the system under study doesn't involve objects comparable in size to an atom or traveling close to the speed of light, classical mechanics provides an excellent description of nature.

Topic 4 introduces Newton's three laws of motion and his law of gravity. The three laws are simple and sensible. The first law states that a force must be applied to an object in order to change its velocity. Changing an object's velocity means accelerating it, which implies a relationship between force and acceleration. This relationship, the second law, states that the net force on an object equals the object's mass times its acceleration. Finally, the third law says that whenever we push on something, it pushes back with equal force in the opposite direction. Those are the three laws in a nutshell.

Newton's three laws, together with his invention of calculus, opened avenues of inquiry and discovery that are used routinely today in virtually all areas of mathematics, science, engineering, and technology. Newton's theory of universal gravitation had a similar impact, starting a revolution in celestial mechanics and astronomy that continues to this day. With the advent of this theory, the orbits of all the planets could be calculated to high precision and the tides understood. The theory even led to the prediction of "dark stars," now called black holes, more than two centuries before any evidence for their existence was observed.[1] Newton's three laws of motion, together with his law of gravitation, are considered among the greatest achievements of the human mind.

## 4.1 Forces

A **force** is commonly imagined as a push or a pull on some object, perhaps rapidly, as when we hit a tennis ball with a racket. (See Fig. 4.1.) We can hit the ball at different speeds and direct it into different parts of the opponent's court. That means we can control the magnitude of the applied force and also its direction, so force is a vector quantity, just like velocity and acceleration.

If you pull on a spring (Fig. 4.2a), the spring stretches. If you pull hard enough on a wagon (Fig. 4.2b), the wagon moves. When you kick a football (Fig. 4.2c), it deforms briefly and is set in motion. These are all examples of **contact forces,** so named because they result from physical contact between two objects.

Contact forces give solidity to our physical environment and everything in it. The ground supports us, and walls block our progress. Friction forces both enable and impede motion. These properties derive primarily from electromagnetic forces, which are long range and enormously strong.

Atoms are composed of tightly bound nuclei of neutrons and protons surrounded by swarms of electrons. A nucleus is measured in femtometers (fm), where one femtometer is $10^{-15}$ m. A typical atom such as hydrogen, however, occupies a volume an Angstrom across, about $10^{-10}$ m. Roughly speaking, if the proton in a

**Figure 4.1** A tennis player applies a contact force to the ball with her racket, accelerating and directing the ball toward the open court.

---

[1] In 1783, John Michell combined Newton's theory of light and theory of gravitation, predicting the existence of "dark stars" from which light itself couldn't escape.

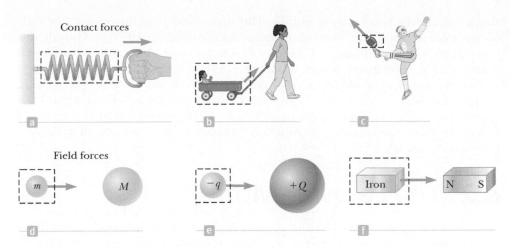

**Figure 4.2** Examples of forces applied to various objects. In each case, a force acts on the object surrounded by the dashed lines. Something in the environment external to the boxed area exerts the force.

hydrogen atom were the size of a ping pong ball, the electron would be the size of a bacterium, at most, and would orbit the nucleus at a distance of a kilometer.

The atomic electrons, via transfers or sharing in molecular orbits, or through polarization, are responsible for the electromagnetic forces forming chemical bonds and more extended structures, such as crystals or membranes and tissues, resulting in the macroscopic world of everyday life. The normal force (Section 4.3) that prevents us from falling toward the center of the Earth is due to an enormous number of electromagnetic interactions between the ground and the soles of our feet. We can't walk on water, because the hydrogen bonds between water molecules are too weak and random. If the water freezes to ice, the stronger crystalline structure supports our weight.

Other contact forces have similar origins. The tension force of a string depends on fibers, which in turn depend on microscopic and molecular structures ultimately created by electromagnetic forces. Friction forces result from microscopic irregularities of the areas in contact. Our macroscopic models of contact forces are approximations for the numerous, complex electromagnetic interactions at the molecular and atomic levels.

Another class of forces doesn't involve any direct physical contact. Early scientists, including Newton, were uneasy with the concept of forces that act between two disconnected objects. Nonetheless, Newton used this "action-at-a-distance" concept in his law of gravity, whereby a mass at one location, such as the Sun, affects the motion of a distant object such as Earth despite no evident physical connection between the two objects. To overcome the conceptual difficulty associated with action at a distance, Michael Faraday (1791–1867) introduced the concept of a *field*. The corresponding forces are called **field forces.** According to this approach, an object of mass $M$, such as the Sun, creates an invisible influence that stretches throughout space. A second object of mass $m$, such as Earth, interacts with the *field* of the Sun, not directly with the Sun itself. So the force of gravitational attraction between two objects, illustrated in Figure 4.2d, is an example of a field force. The force of gravity keeps objects bound to Earth and also gives rise to what we call the *weight* of those objects.

Another common example of a field force is the electric force that one electric charge exerts on another (Fig. 4.2e). A third example is the force exerted by a bar magnet on a piece of iron (Fig. 4.2f).

The known fundamental forces in nature are all field forces. These are, in order of decreasing strength: (1) the strong nuclear force between subatomic particles; (2) the electromagnetic forces between electric charges; (3) the weak nuclear force, which arises in certain radioactive decay processes; and (4) the gravitational force between objects. The strong force keeps the nucleus of an atom from flying apart due to the repulsive electric force of the protons. The weak force is involved in most radioactive processes and plays an important role in the nuclear reactions that generate the Sun's energy output. The strong and weak forces operate only on the nuclear scale, with a very short range on the order of $10^{-15}$ m. Outside this range they have no influence. Classical physics, however, deals only with gravitational and electromagnetic forces, which have infinite range.

Forces exerted on an object can change the object's shape. For example, striking a tennis ball with a racket, as in Figure 4.1, deforms the ball to some extent. Even objects we usually consider rigid and inflexible are deformed under the action of external forces. Often the deformations are permanent, as in the case of a collision between automobiles.

# 4.2 The Laws of Motion

Isaac Newton proposed the laws of motion both to correct previous misconceptions about motion and to offer a systematic method of calculating an object's motion due to forces exerted on it. This section focuses on each of his three laws in turn.

## 4.2.1 Newton's First Law

Consider a book lying on a table. Obviously, the book remains at rest if left alone. Now imagine pushing the book with a horizontal force great enough to overcome the force of friction between the book and the table, setting the book in motion. Because the magnitude of the applied force exceeds the magnitude of the friction force, the book accelerates. When the applied force is withdrawn (Fig. 4.3a), friction soon slows the book to a stop.

Now imagine pushing the book across a smooth, waxed floor. The book again comes to rest once the force is no longer applied, but not as quickly as before. Finally, if the book is moving on a horizontal frictionless surface (Fig. 4.3b), it continues to move in a straight line with constant velocity until it hits a wall or some other obstruction.

Before about 1600, scientists felt that the natural state of matter was the state of rest. Galileo, however, devised thought experiments—such as an object moving on a frictionless surface, as just described—and concluded that **it's not the nature of an object to stop once set in motion, but rather to continue in**

**Figure 4.3** The first law of motion. (a) A book moves at an initial velocity of $\vec{v}$ on a surface with friction. Because there is a friction force acting horizontally, the book slows to rest. (b) A book moves at velocity $\vec{v}$ on a frictionless surface. In the absence of a net force, the book keeps moving at velocity $\vec{v}$.

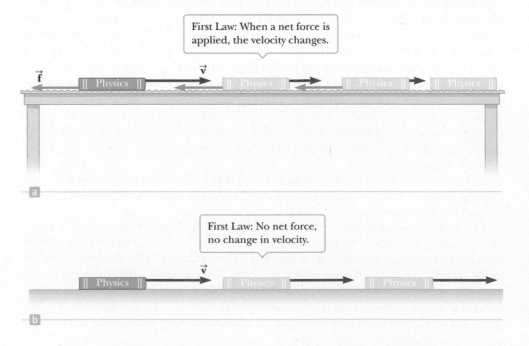

First Law: When a net force is applied, the velocity changes.

First Law: No net force, no change in velocity.

its original state of motion (Fig. 4.4). This approach was later formalized as Newton's first law of motion:

> An object moves with a velocity that is constant in magnitude and direction unless a nonzero net force acts on it.

◀ Newton's first law

The net force on an object is defined as the vector sum of all external forces exerted on the object. External forces come from the object's environment. If an object's velocity isn't changing in either magnitude or direction, then its acceleration and the net force acting on it must both be zero.

Internal forces originate within the object itself and can't change the object's velocity (although they can change the object's rate of rotation, as described in Topic 8). As a result, internal forces aren't included in Newton's first law. It's not really possible to "pull yourself up by your own bootstraps."

A consequence of the first law is the feasibility of space travel. After just a few moments of powerful thrust, the spacecraft coasts for months or years, its velocity only slowly changing with time under the relatively faint influence of the distant Sun and planets.

**Mass and Inertia** Imagine hitting a golf ball off a tee with a driver. If you're a good golfer, the ball will sail over two hundred yards down the fairway. Now imagine teeing up a bowling ball and striking it with the same club (an experiment we don't recommend). Your club would probably break, you might sprain your wrist, and the bowling ball, at best, would fall off the tee, take half a roll, and come to rest.

From this thought experiment, we conclude that although both balls resist changes in their state of motion, the bowling ball offers much more effective resistance. The tendency of an object to continue in its original state of motion is called **inertia.**

Although inertia is the tendency of an object to continue its motion in the absence of a force, **mass** is a measure of the object's resistance to changes in its motion due to a force. This kind of mass is often called *inertial mass* because it's associated with inertia. The greater the mass of a body, the less it accelerates under the action of a given applied force. The SI unit of mass is the kilogram. Mass is a scalar quantity that obeys the rules of ordinary arithmetic.

Inertia can be used to explain the operation of one type of seat belt mechanism. The purpose of the seat belt is to hold the passenger firmly in place relative to the car, to prevent serious injury in the event of an accident. Figure 4.5 illustrates how one type of shoulder harness operates. Under normal conditions, the ratchet turns freely to allow the harness to wind on or unwind from the pulley as the passenger moves. In an accident, the car undergoes a large acceleration and rapidly comes to rest. Because of its inertia, the large block under the seat continues to slide forward along the tracks. The pin connection between the block and the rod causes the rod to pivot about its center and engage the ratchet wheel. At this point, the ratchet wheel locks in place and the harness no longer unwinds.

## 4.2.2 Newton's Second Law

Newton's first law explains what happens to an object that has no net force acting on it: The object either remains at rest or continues moving in a straight line with constant speed. Newton's second law answers the question of what happens to an object that *does* have a net force acting on it.

Imagine pushing a block of ice across a frictionless horizontal surface. When you exert some horizontal force on the block, it moves with an acceleration of, say, $2 \text{ m/s}^2$. If you apply a force twice as large, the acceleration doubles to $4 \text{ m/s}^2$. Pushing three times as hard triples the acceleration, and so on. From such observations, we conclude that **the acceleration of an object is directly proportional to the net force acting on it.**

**Figure 4.4** Unless acted on by an external force, an object at rest will remain at rest and an object in motion will continue in motion with constant velocity. In this case, the wall of the building did not exert a large enough external force on the moving train to stop it.

**APPLICATION**
Seat Belts

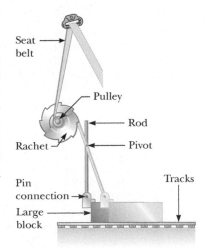

**Figure 4.5** A mechanical arrangement for an automobile seat belt.

**Tip 4.1** Forces Cause Changes in Motion

Motion can occur even in the absence of forces. Forces cause *changes* in motion.

Mass also affects acceleration. Suppose you stack identical blocks of ice on top of each other while pushing the stack with constant force. If the force applied to one block produces an acceleration of 2 m/s$^2$, then the acceleration drops to half that value, 1 m/s$^2$, when two blocks are pushed, to one-third the initial value when three blocks are pushed, and so on. We conclude that **the acceleration of an object is inversely proportional to its mass.** These observations are summarized in **Newton's second law:**

◀ Newton's second law

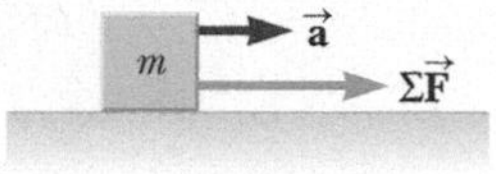

**Figure 4.6** The second law of motion. For the block of mass $m$, the net force $\sum \vec{\mathbf{F}}$ acting on the block equals the mass $m$ times the acceleration vector $\vec{\mathbf{a}}$.

The acceleration $\vec{\mathbf{a}}$ of an object is directly proportional to the net force acting on it and inversely proportional to its mass.

The constant of proportionality is equal to one, so in mathematical terms the preceding statement can be written

$$\vec{\mathbf{a}} = \frac{\sum \vec{\mathbf{F}}}{m}$$

where $\vec{\mathbf{a}}$ is the acceleration of the object, $m$ is its mass, and $\sum \vec{\mathbf{F}}$ is the vector sum of all forces acting on it. Multiplying through by $m$, we have

$$\sum \vec{\mathbf{F}} = m\vec{\mathbf{a}} \qquad [4.1]$$

Physicists commonly refer to this equation as '$F = ma$.' Figure 4.6 illustrates the relationship between the mass, acceleration, and the net force. The second law is a vector equation, equivalent to the following three component equations:

$$\sum F_x = ma_x \qquad \sum F_y = ma_y \qquad \sum F_z = ma_z \qquad [4.2]$$

When there is no net force on an object, its acceleration is zero, which means the velocity is constant.

**ISAAC NEWTON**
**English Physicist and Mathematician (1642–1727)**

Newton was one of the most brilliant scientists in history. Before he was 30, he formulated the basic concepts and laws of mechanics, discovered the law of universal gravitation, and invented the mathematical methods of calculus. As a consequence of his theories, Newton was able to explain the motions of the planets, the ebb and flow of the tides, and many special features of the motions of the Moon and Earth. He also interpreted many fundamental observations concerning the nature of light. His contributions to physical theories dominated scientific thought for two centuries and remain important today.

**Units of Force and Mass** The SI unit of force is the **newton.** When 1 newton of force acts on an object that has a mass of 1 kg, it produces an acceleration of 1 m/s$^2$ in the object. From this definition and Newton's second law, we see that the newton can be expressed in terms of the fundamental units of mass, length, and time as

$$1 \text{ N} \equiv 1 \text{ kg} \cdot \text{m/s}^2 \qquad [4.3]$$

In the U.S. customary system, the unit of force is the **pound.** The conversion from newtons to pounds is given by

$$1 \text{ N} = 0.225 \text{ lb} \qquad [4.4]$$

The units of mass, acceleration, and force in the SI and U.S. customary systems are summarized in Table 4.1.

**Tip 4.2** $m\vec{\mathbf{a}}$ Is Not a Force

Equation 4.1 does *not* say that the product $m\vec{\mathbf{a}}$ is a force. All forces exerted on an object are summed as vectors to generate the net force on the left side of the equation. This net force is then equated to the product of the mass and resulting acceleration of the object. Do *not* include an "$m\vec{\mathbf{a}}$ force" in your analysis.

*Quick Quiz*

**4.1** Which of the following statements are true? (a) An object can move even when no force acts on it. (b) If an object isn't moving, no external forces act on it. (c) If a single force acts on an object, the object accelerates. (d) If an object accelerates, at least one force is acting on it. (e) If an object isn't accelerating, no external force is acting on it. (f) If the net force acting on an object is in the positive $x$-direction, the object moves only in the positive $x$-direction.

**Table 4.1** Units of Mass, Acceleration, and Force

| System | Mass | Acceleration | Force |
| --- | --- | --- | --- |
| SI | kg | m/s$^2$ | N = kg·m/s$^2$ |
| U.S. customary | slug | ft/s$^2$ | lb = slug·ft/s$^2$ |

## EXAMPLE 4.1    AIRBOAT

**GOAL** Apply Newton's second law in one dimension, together with the equations of kinematics.

**PROBLEM** An airboat with mass $3.50 \times 10^2$ kg, including the passenger, has an engine that produces a net horizontal force of $7.70 \times 10^2$ N, after accounting for forces of resistance (see Fig. 4.7). **(a)** Find the acceleration of the airboat. **(b)** Starting from rest, how long does it take the airboat to reach a speed of 12.0 m/s? **(c)** After reaching that speed, the pilot turns off the engine and drifts to a stop over a distance of 50.0 m. Find the resistance force, assuming it's constant.

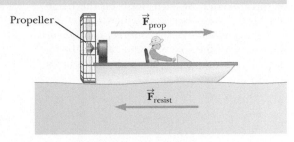

**Figure 4.7** (Example 4.1)

**STRATEGY** In part **(a)**, apply Newton's second law to find the acceleration, and in part **(b)**, use that acceleration in the one-dimensional kinematics equation for the velocity. When the engine is turned off in part **(c)**, only the resistance forces act on the boat in the x-direction, so the net acceleration can be found from $v^2 - v_0^2 = 2a\,\Delta x$. Then Newton's second law gives the resistance force.

. . . . . . . . . . . . . . . . . . . . . . . . . . . . . . . . . . . . . . . . . . . . . . . . . . . . . . . . . . . . . . . . . . . . . . . . . . . . . . . . . . . . . . . . . .

**SOLUTION**

**(a)** Find the acceleration of the airboat.

Apply Newton's second law and solve for the acceleration:

$$ma = F_{net} \quad \rightarrow \quad a = \frac{F_{net}}{m} = \frac{7.70 \times 10^2 \text{ N}}{3.50 \times 10^2 \text{ kg}}$$

$$= \boxed{2.20 \text{ m/s}^2}$$

**(b)** Find the time necessary to reach a speed of 12.0 m/s.

Apply the kinematics velocity equation:

$$v = at + v_0 = (2.20 \text{ m/s}^2)\,t = 12.0 \text{ m/s} \quad \rightarrow \quad t = \boxed{5.45 \text{ s}}$$

**(c)** Find the resistance force after the engine is turned off.

Using kinematics, find the net acceleration due to resistance forces:

$$v^2 - v_0^2 = 2a\,\Delta x$$
$$0 - (12.0 \text{ m/s})^2 = 2a(50.0 \text{ m}) \quad \rightarrow \quad a = -1.44 \text{ m/s}^2$$

Substitute the acceleration into Newton's second law, finding the resistance force:

$$F_{resist} = ma = (3.50 \times 10^2 \text{ kg})(-1.44 \text{ m/s}^2) = \boxed{-504 \text{ N}}$$

. . . . . . . . . . . . . . . . . . . . . . . . . . . . . . . . . . . . . . . . . . . . . . . . . . . . . . . . . . . . . . . . . . . . . . . .

**REMARKS** The propeller exerts a force on the air, pushing it backwards behind the boat. At the same time, the air exerts a force on the propellers and consequently on the airboat. Forces always come in pairs of this kind, which are formalized in the next section as Newton's third law of motion. The negative answer for the acceleration in part **(c)** means that the airboat is slowing down.

**QUESTION 4.1** What other forces act on the airboat? Describe them.

**EXERCISE 4.1** Suppose the pilot, starting again from rest, opens the throttle partway. At a constant acceleration, the airboat then covers a distance of 60.0 m in 10.0 s. Find the net force acting on the boat.

**ANSWER** $4.20 \times 10^2$ N

> **Tip 4.3** Newton's Second Law Is a *Vector* Equation
>
> In applying Newton's second law, add all of the forces on the object as vectors and then find the resultant vector acceleration by dividing by $m$. Don't find the individual magnitudes of the forces and add them like scalars.

## EXAMPLE 4.2    HORSES PULLING A BARGE

**GOAL** Apply Newton's second law in a two-dimensional problem.

**PROBLEM** Two horses are pulling a barge with mass $2.00 \times 10^3$ kg along a canal, as shown in Figure 4.8. The cable connected to the first horse makes an angle of $\theta_1 = 30.0°$ with respect to the direction of the canal, while the cable connected to the second horse makes an angle of $\theta_2 = -45.0°$. Find the initial acceleration of the barge, starting at rest, if each horse exerts a force of magnitude $6.00 \times 10^2$ N on the barge. Ignore forces of resistance on the barge.

*(Continued)*

**STRATEGY** Using trigonometry, find the vector force exerted by each horse on the barge. Add the x-components together to get the x-component of the resultant force, and then do the same with the y-components. Divide by the mass of the barge to get the accelerations in the x- and y-directions.

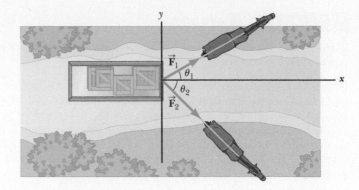

**Figure 4.8** (Example 4.2)

**SOLUTION**

Compute the x-components of the forces exerted by the horses.

$$F_{1x} = F_1 \cos \theta_1 = (6.00 \times 10^2 \text{ N}) \cos (30.0°) = 5.20 \times 10^2 \text{ N}$$

$$F_{2x} = F_2 \cos \theta_2 = (6.00 \times 10^2 \text{ N}) \cos (-45.0°) = 4.24 \times 10^2 \text{ N}$$

Find the total force in the x-direction by adding the x-components:

$$F_x = F_{1x} + F_{2x} = 5.20 \times 10^2 \text{ N} + 4.24 \times 10^2 \text{ N}$$

$$= 9.44 \times 10^2 \text{ N}$$

Compute the y-components of the forces exerted by the horses:

$$F_{1y} = F_1 \sin \theta_1 = (6.00 \times 10^2 \text{ N}) \sin 30.0° = 3.00 \times 10^2 \text{ N}$$

$$F_{2y} = F_2 \sin \theta_2 = (6.00 \times 10^2 \text{ N}) \sin (-45.0°)$$

$$= -4.24 \times 10^2 \text{ N}$$

Find the total force in the y-direction by adding the y-components:

$$F_y = F_{1y} + F_{2y} = 3.00 \times 10^2 \text{ N} - 4.24 \times 10^2 \text{ N}$$

$$= -1.24 \times 10^2 \text{ N}$$

Obtain the components of the acceleration by dividing each of the force components by the mass:

$$a_x = \frac{F_x}{m} = \frac{9.44 \times 10^2 \text{ N}}{2.00 \times 10^3 \text{ kg}} = 0.472 \text{ m/s}^2$$

$$a_y = \frac{F_y}{m} = \frac{-1.24 \times 10^2 \text{ N}}{2.00 \times 10^3 \text{ kg}} = -0.062 \ 0 \text{ m/s}^2$$

Calculate the magnitude of the acceleration:

$$a = \sqrt{a_x^2 + a_y^2} = \sqrt{(0.472 \text{ m/s}^2)^2 + (-0.062 \ 0 \text{ m/s}^2)^2}$$

$$= 0.476 \text{ m/s}^2$$

Calculate the direction of the acceleration using the tangent function:

$$\tan \theta = \frac{a_y}{a_x} = \frac{-0.062 \ 0 \text{ m/s}^2}{0.472 \text{ m/s}^2} = -0.131$$

$$\theta = \tan^{-1}(-0.131) = -7.46°$$

**REMARKS** Notice that the angle is in fourth quadrant, in the range of the inverse tangent function, so it is not necessary to add 180° to the answer. The horses exert a force on the barge through the tension in the cables, while the barge exerts an equal and opposite force on the horses, again through the cables. If that were not true, the horses would easily move forward, as if unburdened. This example is another illustration of forces acting in pairs.

**QUESTION 4.2** True or False: In general, the magnitude of the acceleration of an object is determined by the magnitudes of the forces acting on it.

**EXERCISE 4.2** Repeat Example 4.2, but assume the first horse pulls at a 40.0° angle, the second horse at −20.0°.

**ANSWER** 0.520 m/s², 10.0°

**The Gravitational Force** The **gravitational force** is the mutual force of attraction between any two objects in the Universe. Although the gravitational force can be very strong between very large objects, it's the weakest of the fundamental forces. A good demonstration of how weak it is can be carried out with a small balloon. Rubbing the balloon in your hair gives the balloon a tiny electric charge.

Through electric forces, the balloon then adheres to a wall, resisting the gravitational pull of the entire Earth!

In addition to contributing to the understanding of motion, Newton studied gravity extensively. **Newton's law of universal gravitation states that every particle in the Universe attracts every other particle with a force that is directly proportional to the product of the masses of the particles and inversely proportional to the square of the distance between them.** If the particles have masses $m_1$ and $m_2$ and are separated by a distance $r$, as in Figure 4.9, the magnitude of the gravitational force $F_g$ is

$$F_g = G\frac{m_1 m_2}{r^2} \qquad [4.5]$$

where $G = 6.67 \times 10^{-11}\ \text{N} \cdot \text{m}^2/\text{kg}^2$ is the **universal gravitation constant.** Technically, $m_1$ and $m_2$ are the *gravitational* masses of the two particles, involved in creating gravitational forces and conceivably distinct from the *inertial* mass found in the second law of motion. Experiment shows they can be taken as identical to high precision. (See Topic 26.) We examine the gravitational force in more detail in Topic 7.

## Weight

The magnitude of the gravitational force acting on an object of mass $m$ is called the **weight** $w$ of the object, given by

$$w = mg \qquad [4.6]$$

where $g$ is the acceleration of gravity.
**SI unit: newton (N)**

From Equation 4.5, an alternate definition of the weight of an object with mass $m$ can be written as

$$w = G\frac{M_E m}{r^2} \qquad [4.7]$$

where $M_E$ is the mass of Earth and $r$ is the distance from the object to Earth's center. If the object is at rest on Earth's surface, then $r$ is equal to Earth's radius $R_E$. Because $r^2$ is in the denominator of Equation 4.7, the weight decreases with increasing $r$. So the weight of an object on a mountaintop is less than the weight of the same object at sea level.

Comparing Equations 4.6 and 4.7, it follows that

$$g = G\frac{M_E}{r^2} \qquad [4.8]$$

Unlike mass, weight is not an inherent property of an object because it can take different values, depending on the value of $g$ in a given location (Fig. 4.10). If an object has a mass of 70.0 kg, for example, then its weight at a location where $g = 9.80\ \text{m/s}^2$ is $mg = 686$ N. In a high-altitude balloon, where $g$ might be $9.76\ \text{m/s}^2$, the object's weight would be 683 N. The value of $g$ also varies slightly due to the density of matter in a given locality. **In this text, unless stated otherwise, the value of $g$ will be understood to be $9.80\ \text{m/s}^2$, its value near the surface of the Earth.**

Equation 4.8 is a general result that can be used to calculate the acceleration of an object falling near the surface of any massive object if the more massive object's radius and mass are known. Using the values in Table 7.3 (p. 214), you should be able to show that $g_{\text{Sun}} = 274\ \text{m/s}^2$ and $g_{\text{Moon}} = 1.62\ \text{m/s}^2$. An important fact is that for spherical bodies, distances are calculated from the centers of the objects, a consequence of Gauss's law (explained in Topic 15), which holds for both gravitational and electric forces.

◄ Law of universal gravitation

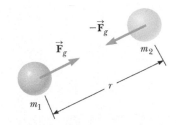

**Figure 4.9** The gravitational force between two particles is attractive.

NASA/Eugene Cernan

**Figure 4.10** The life-support unit strapped to the back of astronaut Harrison Schmitt weighed 300 lb on Earth and had a mass of 136 kg. During his training, a 50-lb mock-up with a mass of 23 kg was used. Although the mock-up had the same weight as the actual unit would have on the Moon, the smaller mass meant it also had a lower inertia. The weight of the unit is caused by the acceleration of the local gravity field, but the astronaut must also accelerate anything he's carrying in order to move it around. Consequently, the actual unit used on the Moon, with the same weight but greater inertia, was harder for the astronaut to handle than the mock-up unit on Earth.

**4.2** Which has greater value, a newton of gold on Earth or a newton of gold on the Moon? (a) The newton of gold on the Earth. (b) The newton of gold on the Moon. (c) The value is the same, regardless.

**4.3** Respond to each statement, true or false: (a) No force of gravity acts on an astronaut in an orbiting space station. (b) At three Earth radii from the center of Earth, the acceleration of gravity is one-ninth its surface value. (c) If two identical planets, each with surface gravity $g$ and volume $V$, coalesce into one planet with volume $2V$, the surface gravity of the new planet is $2g$. (d) One kilogram of gold would have greater value on Earth than on the Moon.

## EXAMPLE 4.3    FORCES OF DISTANT WORLDS

**GOAL** Calculate the magnitude of a gravitational force using Newton's law of gravitation.

**PROBLEM** (a) Find the gravitational force exerted by the Sun on a 70.0-kg man located at the Earth's equator at noon, when the man is closest to the Sun. (b) Calculate the gravitational force of the Sun on the man at midnight, when he is farthest from the Sun. (c) Calculate the difference in the acceleration due to the Sun between noon and midnight. (For values, see Table 7.3 on page 214.)

**STRATEGY** To obtain the distance of the Sun from the man at noon, subtract the Earth's radius from the solar distance. At midnight, add the Earth's radius. Retain enough digits so that rounding doesn't remove the small difference between the two answers. For part (c), subtract the answer for (b) from (a) and divide by the man's mass.

**SOLUTION**

(a) Find the gravitational force exerted by the Sun on the man at the Earth's equator at noon.

Write the law of gravitation, Equation 4.5, in terms of the distance from the Sun to the Earth, $r_S$, and Earth's radius, $R_E$:

$$(1) \quad F_{\text{Sun}}^{\text{noon}} = \frac{mM_SG}{r^2} = \frac{mM_SG}{(r_S - R_E)^2}$$

Substitute values into (1) and retain two extra digits:

$$F_{\text{Sun}}^{\text{noon}} = \frac{(70.0 \text{ kg})(1.991 \times 10^{30} \text{ kg})(6.67 \times 10^{-11} \text{ kg}^{-1}\text{m}^3/\text{s}^2)}{(1.496 \times 10^{11} \text{ m} - 6.38 \times 10^6 \text{ m})^2}$$

$$= 0.415\,40 \text{ N}$$

(b) Calculate the gravitational force of the Sun on the man at midnight.

Write the law of gravitation, adding Earth's radius this time:

$$(2) \quad F_{\text{Sun}}^{\text{mid}} = \frac{mM_SG}{r^2} = \frac{mM_SG}{(r_S + R_E)^2}$$

Substitute values into (2):

$$F_{\text{Sun}}^{\text{mid}} = \frac{(70.0 \text{ kg})(1.991 \times 10^{30} \text{ kg})(6.67 \times 10^{-11} \text{ kg}^{-1}\text{m}^3/\text{s}^2)}{(1.496 \times 10^{11} \text{ m} + 6.38 \times 10^6 \text{ m})^2}$$

$$= 0.415\,33 \text{ N}$$

(c) Calculate the difference in the man's solar acceleration between noon and midnight.

Write an expression for the difference in acceleration and substitute values:

$$a = \frac{F_{\text{Sun}}^{\text{noon}} - F_{\text{Sun}}^{\text{mid}}}{m} = \frac{0.415\,40 \text{ N} - 0.415\,33 \text{ N}}{70.0 \text{ kg}}$$

$$\cong 1 \times 10^{-6} \text{ m/s}^2$$

**REMARKS** The gravitational attraction between the Sun and objects on Earth is easily measurable and has been exploited in experiments to determine whether gravitational attraction depends on the composition of the object. The gravitational force on Earth due to the Moon is much weaker than the gravitational force on Earth due to the Sun. Paradoxically, the Moon's effect on the tides is over twice that of the Sun because the tides depend on *differences* in the gravitational force across the Earth, and those differences are greater for the Moon's gravitational force because the Moon is much closer to Earth than the Sun.

**QUESTION 4.3** Mars is about one and a half times as far from the Sun as Earth. Without doing an explicit calculation, estimate to one significant digit the gravitational force of the Sun on a 70.0-kg man standing on Mars.

**EXERCISE 4.3** Consider a location on the equator where the Moon is directly overhead in the middle of the day. **(a)** Find the gravitational force exerted by the Moon on a 70.0-kg man at noon. **(b)** Calculate the gravitational force of the Moon on the man at midnight, neglecting any motion of the Moon during this time. **(c)** Calculate the difference in the man's acceleration due to the Moon between noon and midnight. *Note:* The distance from the Earth to the Moon is $3.84 \times 10^8$ m. The mass of the Moon is $7.36 \times 10^{22}$ kg.

**ANSWERS** **(a)** $2.41 \times 10^{-3}$ N  **(b)** $2.25 \times 10^{-3}$ N  **(c)** $2.3 \times 10^{-6}$ m/s$^2$

---

**EXAMPLE 4.4    WEIGHT ON PLANET X**

**GOAL** Understand the effect of a planet's mass and radius on the weight of an object on the planet's surface.

**PROBLEM** An astronaut on a space mission lands on a planet with three times the mass and twice the radius of Earth. What is her weight $w_X$ on this planet as a multiple of her Earth weight $w_E$?

**STRATEGY** Write $M_X$ and $r_X$, the mass and radius of the planet, in terms of $M_E$ and $R_E$, the mass and radius of Earth, respectively, and substitute into the law of gravitation.

**SOLUTION**

From the statement of the problem, we have the following relationships:

$$M_X = 3M_E \quad r_X = 2R_E$$

Substitute the preceding expressions into Equation 4.5 and simplify, algebraically associating the terms giving the weight on Earth:

$$w_X = G\frac{M_X m}{r_X^2} = G\frac{3M_E m}{(2R_E)^2} = \frac{3}{4}G\frac{M_E m}{R_E^2} = \frac{3}{4}w_E$$

**REMARKS** This problem shows the interplay between a planet's mass and radius in determining the weight of objects on its surface. Although Jupiter has about three hundred times the mass of the Earth, the weight of an object at Jupiter's planetary radius is only a little over two and a half times the weight of the same object on Earth's surface.

**QUESTION 4.4** A volume of rock has a mass roughly three times a similar volume of ice. Suppose one world is made of ice whereas another world with the same radius is made of rock. If $g$ is the acceleration of gravity on the surface of the ice world, what is the approximate acceleration of gravity on the rock world?

**EXERCISE 4.4** An astronaut lands on Ganymede, a giant moon of Jupiter that is larger than the planet Mercury. Ganymede has one-fortieth the mass of Earth and two-fifths the radius. Find the weight of the astronaut standing on Ganymede in terms of his Earth weight $w_E$.

**ANSWER** $w_G = (5/32)w_E$

---

## 4.2.3 Newton's Third Law

Consider the task of driving a nail into a block of wood, for example, as illustrated in Figure 4.11a (page 90). To accelerate the nail and drive it into the block, the hammer must exert a net force on the nail. Newton recognized, however, that a single isolated force couldn't exist. Instead, **forces in nature always exist in pairs.** According to Newton, as the nail is driven into the block by the force exerted by the hammer, the hammer is slowed down and stopped by the force exerted by the nail.

Newton described such paired forces with his **third law:**

If object 1 and object 2 interact, the force $\vec{F}_{12}$ exerted by object 1 on object 2 is equal in magnitude but opposite in direction to the force $\vec{F}_{21}$ exerted by object 2 on object 1.

◄ Newton's third law

**Figure 4.11** Newton's third law. (a) The force exerted by the hammer on the nail is equal in magnitude and opposite in direction to the force exerted by the nail on the hammer. (b) The force $\vec{\mathbf{F}}_{12}$ exerted by object 1 on object 2 is equal in magnitude and opposite in direction to the force $\vec{\mathbf{F}}_{21}$ exerted by object 2 on object 1.

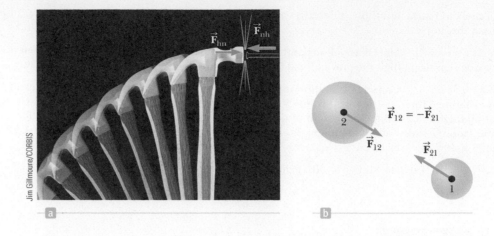

Jim Gillmoure/CORBIS

$$\vec{\mathbf{F}}_{12} = -\vec{\mathbf{F}}_{21}$$

**Tip 4.4** Action–Reaction Pairs

In applying Newton's third law, remember that an action and its reaction force always act on *different* objects. Two external forces acting on the same object, even if they are equal in magnitude and opposite in direction, *can't* be an action–reaction pair.

This law, which is illustrated in Figure 4.11b, states that **a single isolated force can't exist.** The force $\vec{\mathbf{F}}_{12}$ exerted by object 1 on object 2 is sometimes called the *action force,* and the force $\vec{\mathbf{F}}_{21}$ exerted by object 2 on object 1 is called the *reaction force.* In reality, either force can be labeled the action or reaction force. **The action force is equal in magnitude to the reaction force and opposite in direction. In all cases, the action and reaction forces act on different objects.** For example, the force acting on a freely falling projectile is the force of gravity exerted by Earth on the projectile, $\vec{\mathbf{F}}_g$, and the magnitude of this force is its weight *mg*. The reaction to force $\vec{\mathbf{F}}_g$ is the gravitational force exerted by the projectile on Earth, $\vec{\mathbf{F}}_g' = -\vec{\mathbf{F}}_g$. The reaction force $\vec{\mathbf{F}}_g'$ must accelerate the Earth towards the projectile, just as the action force $\vec{\mathbf{F}}_g$ accelerates the projectile towards Earth. Because Earth has such a large mass, however, its acceleration due to this reaction force is negligibly small.

Newton's third law constantly affects our activities in everyday life. Without it, no locomotion of any kind would be possible, whether on foot, on a bicycle, or in a motorized vehicle. When walking, for example, we exert a frictional force against the ground. The reaction force of the ground against our foot propels us forward. In the same way, the tires on a bicycle exert a frictional force against the ground, and the reaction of the ground pushes the bicycle forward. As we'll see shortly, friction plays a large role in such reaction forces.

**APPLICATION**

Helicopter Flight

For another example of Newton's third law, consider the helicopter. Most helicopters have a large set of blades rotating in a horizontal plane above the body of the vehicle and another, smaller set rotating in a vertical plane at the back. Other helicopters have two large sets of blades above the body rotating in opposite directions. Why do helicopters always have two sets of blades? In the first type of helicopter, the engine applies a force to the blades, causing them to change their rotational motion. According to Newton's third law, however, the blades must exert a force on the engine of equal magnitude and in the opposite direction. This force would cause the body of the helicopter to rotate in the direction opposite the blades. A rotating helicopter would be impossible to control, so a second set of blades is used. The small blades in the back provide a force opposite to that tending to rotate the body of the helicopter, keeping the body oriented in a stable position. In helicopters with two sets of large counterrotating blades, engines apply forces in opposite directions, so there is no net force rotating the helicopter.

As mentioned earlier, Earth exerts a gravitational force $\vec{\mathbf{F}}_g$ on any object. If the object is a monitor at rest on a table, as in Figure 4.12a, the reaction force to $\vec{\mathbf{F}}_g$ is the gravitational force the monitor exerts on the Earth, $\vec{\mathbf{F}}_g'$. The monitor doesn't accelerate downward because it's held up by the table. The table therefore exerts an upward force $\vec{\mathbf{n}}$, called the **normal force,** on the monitor. (*Normal,* a technical term from mathematics, means "perpendicular" in this context.) The normal force is an elastic force arising from the cohesion of matter and is electromagnetic in origin. It balances the gravitational force acting on the monitor, preventing the

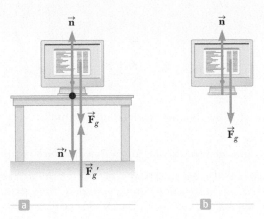

**Figure 4.12** When a monitor is sitting on a table, the forces acting on the monitor are the normal force $\vec{\mathbf{n}}$ exerted by the table and the force of gravity, $\vec{\mathbf{F}}_g$, as illustrated in (b). The reaction to $\vec{\mathbf{n}}$ is the force exerted by the monitor on the table, $\vec{\mathbf{n}}'$. The reaction to $\vec{\mathbf{F}}_g$ is the force exerted by the monitor on Earth, $\vec{\mathbf{F}}_g'$.

monitor from falling through the table, and can have any value needed, up to the point of breaking the table. The reaction to $\vec{\mathbf{n}}$ is the force exerted by the monitor on the table, $\vec{\mathbf{n}}'$. Therefore,

$$\vec{\mathbf{F}}_g = -\vec{\mathbf{F}}_g' \quad \text{and} \quad \vec{\mathbf{n}} = -\vec{\mathbf{n}}'$$

The forces $\vec{\mathbf{n}}$ and $\vec{\mathbf{n}}'$ both have the same magnitude as $\vec{\mathbf{F}}_g$. Note that the forces acting on the monitor are $\vec{\mathbf{F}}_g$ and $\vec{\mathbf{n}}$, as shown in Figure 4.12b. The two reaction forces, $\vec{\mathbf{F}}_g'$ and $\vec{\mathbf{n}}'$, are exerted by the monitor on objects other than the monitor. Remember that the two forces in an action–reaction pair always act on two different objects.

Because the monitor is not accelerating in any direction ($\vec{\mathbf{a}} = 0$), it follows from Newton's second law that $m\vec{\mathbf{a}} = 0 = \vec{\mathbf{F}}_g + \vec{\mathbf{n}}$. However, $F_g = -mg$, so $n = mg$, a useful result.

> **Tip 4.5** Equal and Opposite but Not a Reaction Force
>
> A common error in Figure 4.12b is to consider the normal force on the object to be the reaction force to the gravity force, because in this case these two forces are equal in magnitude and opposite in direction. That is impossible, however, because they act on the same object!

### Quick Quiz

**4.4** A small sports car collides head-on with a massive truck. The greater impact force (in magnitude) acts on (a) the car, (b) the truck, (c) neither, the force is the same on both. Which vehicle undergoes the greater magnitude acceleration? (d) the car, (e) the truck, (f) the accelerations are the same.

**APPLICATION**
Colliding Vehicles

## EXAMPLE 4.5    ACTION–REACTION AND THE ICE SKATERS

**GOAL** Illustrate Newton's third law of motion.

**PROBLEM** A man of mass $M = 75.0$ kg and woman of mass $m = 55.0$ kg stand facing each other on an ice rink, both wearing ice skates. The woman pushes the man with a horizontal force of $F = 85.0$ N in the positive $x$-direction. Assume the ice is frictionless. **(a)** What is the man's acceleration? **(b)** What is the reaction force acting on the woman? **(c)** Calculate the woman's acceleration.

**STRATEGY** Parts **(a)** and **(c)** are simple applications of the second law. An application of the third law solves part **(b)**.

. . . . . . . . . . . . . . . . . . . . . . . . . . . . . . . . . . . . . . . . . . . . . . . . . . . . . . . . . . . . . . . . . . . . . . . . . . . . .

**SOLUTION**

**(a)** What is the man's acceleration?

Write the second law for the man:

$$Ma_M = F$$

Solve for the man's acceleration and substitute values:

$$a_M = \frac{F}{M} = \frac{85.0 \text{ N}}{75.0 \text{ kg}} = \boxed{1.13 \text{ m/s}^2}$$

**(b)** What is the reaction force acting on the woman?

Apply Newton's third law of motion, finding that the reaction force $R$ acting on the woman has the same magnitude and opposite direction:

$$R = -F = \boxed{-85.0 \text{ N}}$$

*(Continued)*

**(c)** Calculate the woman's acceleration.

Write Newton's second law for the woman:

$$ma_W = R = -F$$

Solve for the woman's acceleration and substitute values:

$$a_W = \frac{-F}{m} = \frac{-85.0 \text{ N}}{55.0 \text{ kg}} = \boxed{-1.55 \text{ m/s}^2}$$

. . . . . . . . . . . . . . . . . . . . . . . . . . . . . . . . . . . . . . . . . . . . . . . . . . . . . . . . . . . . . . . . . . . . . . . . .

**REMARKS** Notice that the forces are equal and opposite each other, but the accelerations are not because the two masses differ from each other.

**QUESTION 4.5** Name two other forces acting on the man and the two reaction forces that are paired with them.

**EXERCISE 4.5** A space-walking astronaut of total mass 148 kg exerts a force of 265 N on a free-floating satellite of mass 635 kg, pushing it in the positive $x$-direction. **(a)** What is the reaction force exerted by the satellite on the astronaut? Calculate the accelerations of **(b)** the astronaut, and **(c)** the satellite.

**ANSWERS** **(a)** $-265$ N   **(b)** $-1.79$ m/s$^2$   **(c)** $0.417$ m/s$^2$

# 4.3 The Normal and Kinetic Friction Forces

In two-dimensional second law problems, the $y$-component is often used to determine the normal force acting on the object. This section introduces the most common four cases. Effectively, knowledge of these four cases reduces many complex, two-dimensional second law problems to one-dimensional problems, because the normal force is known in advance. Generally, the normal force is used in the $x$-component of the second law to determine the kinetic friction force acting on the object.

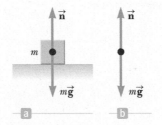

**Figure 4.13** (a) On a level surface, the normal force supports an object's weight. (b) Free-body diagram.

## 4.3.1 Case 1: The Normal Force on a Level Surface

Figure 4.13a shows a block at rest on a flat surface. The two forces acting on the block are the normal force, acting upward, and the gravity force, directed downward. Figure 4.13b is called the **free-body diagram** for the block, which, for clarity, consists of showing just the forces acting on the block. Free-body diagrams include *only* the forces acting directly on the object in question. Forces or reaction forces acting on different bodies aren't shown. For example, the reaction force to the gravity force acting on the block is the gravity force exerted by the block on the Earth, which doesn't appear in the block's free-body diagram. The $y$-component of the second law of motion, with $a_y = 0$, yields:

$$\sum F_y = ma_y$$
$$n - mg = 0$$
$$\boxed{n = mg}$$  [4.9]

The normal force in this case is equal to the weight of the object.

## 4.3.2 Case 2: The Normal Force on a Level Surface with an Applied Force

Figure 4.14a shows a block at rest on a flat surface. The three forces acting on the block are the normal force, directed upward; the gravity force, directed downward;

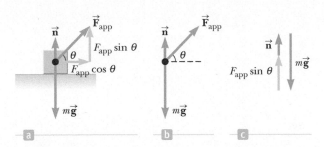

**Figure 4.14** (a) An applied force at a positive angle. (b) Free-body diagram. (c) When a force is applied at a positive angle, the weight is supported by the sum of the normal force and the $y$-component of the applied force.

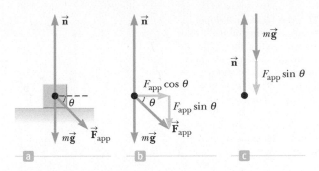

**Figure 4.15** (a) An applied force at a negative angle. (b) Free-body diagram. (c) The normal force must be equal in magnitude to the sum of the weight and the $y$-component of the applied force.

and an applied force, directed at a positive angle $\theta$. The $y$-component of the second law of motion yields:

$$\sum F_y = ma_y$$
$$n - mg + F_{app}\sin\theta = 0$$

That equation can easily be solved for the normal force $n$:

$$n = mg - F_{app}\sin\theta \qquad\qquad \text{[4.10]}$$

Notice that if the angle is positive, as in Figure 4.14a, then the sine of the angle is positive, and the $y$-component of the applied force supports some of the weight, reducing the normal force. That can be understood more easily by separating out the $y$-components, as in Figure 4.14c: the sum of the normal force and the $y$-component of the applied force equals the magnitude of the weight. If the angle is negative, however, then the sine of the angle is also negative, making a positive contribution to the normal force, as illustrated in Figure 4.15. The normal force must be larger, and in magnitude, equal to the sum of the weight and the $y$-component of the applied force.

### 4.3.3 Case 3: The Normal Force on a Level Surface Under Acceleration

Figure 4.16a shows a block on a flat surface that is under acceleration, such as in an elevator. The two forces acting on the block are the normal force, directed upward, and the gravity force, directed downward. An acceleration upward, however, will increase the magnitude of the normal force, because the normal force must not only compensate for gravity, but in addition, provide the acceleration. The $y$-component of the second law of motion yields:

$$\sum F_y = ma_y$$
$$n - mg = ma_y$$
$$n = ma_y + mg \qquad\qquad \text{[4.11]}$$

As can be seen in Figure 4.16b, the magnitude of the normal force vector must equal the sum of the magnitudes of the gravitational force and the inertial quantity, $ma$.

### 4.3.4 Case 4: The Normal Force on a Slope

A common variation on a second law problem is an object resting on a surface tilted at some constant angle. Although optional, in that circumstance a simplification of the problem can be achieved by selecting coordinates that are similarly

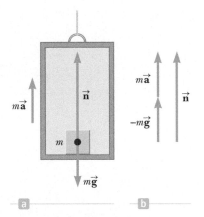

**Figure 4.16** (a) A block in an elevator accelerating upward. (b) When the acceleration is upward, the normal force must be equal in magnitude to the magnitudes of the weight, $mg$, and the inertia, $ma$.

**Figure 4.17** (a) A block on a slope, showing the forces acting on it. (b) In tilted coordinates, the gravity force splits into two components, one perpendicular to the slope and the other parallel to it. The component of the gravity force acting opposite the normal force is then given by $F_{y', \text{grav}} = -mg\cos\theta$, while the component acting down the slope is $F_{x', \text{grav}} = mg\sin\theta$.

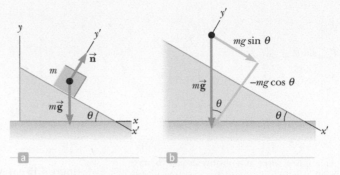

tilted, with the $x'$-axis running parallel to the slope and $y'$-axis perpendicular to the slope, as shown in Figure 4.17a. Using Figure 4.17b, the force due to gravity can then be broken into two components, $F_{x', \text{grav}} = mg\sin\theta$ and $F_{y', \text{grav}} = -mg\cos\theta$.

The second law for the $y'$-direction can be solved for the normal force:

$$\sum F_{y'} = ma_{y'}$$
$$n - mg\cos\theta = 0$$
$$n = mg\cos\theta \qquad [4.12]$$

The normal force on a slope is equal in magnitude to the component of the gravity force perpendicular to the slope. It's also useful to know that the force on the block directed down the slope due to gravity is given by

$$F_{x', \text{grav}} = mg\sin\theta \qquad [4.13]$$

When approaching these problems, it's often possible to immediately treat them as one dimensional, because all that is needed from the $y$-direction is the normal force.

### 4.3.5 The Normal Force and Atmospheric Pressure

The normal force, acting on an object such as a block on a level surface, opposes the gravity force drawing the block towards the center of the Earth. Technically, the normal force also opposes the downward pressure of the atmosphere on the top of the block. Because of microscopic imperfections, however, air infiltrates beneath even a very smooth object resting on a relatively smooth surface, effectively canceling out the downward pressure with an equal upward pressure. When air can be excluded from beneath an object, then atmospheric pressure acts only downward, and the object is held very firmly in place. This is the "suction-cup effect" that allows suction-cup darts to stick to windows. It works better when the cups are slightly damp and when the surface is extremely smooth, because under those conditions the air can be more effectively excluded from beneath the cup. In general, unless stated otherwise, it will always be assumed that pressure forces on objects cancel.

### 4.3.6 The Force of Kinetic Friction

*Friction* is a contact force that derives from the microscopic interactions between a body and its environment. Air friction affects vehicles from cars to rockets, and fluid friction the motion of ships or the passage of fluid in pipes. Friction is not always something that impedes motion; however, without friction, we wouldn't be able to walk or pick up and hold objects.

Friction is also caused when an object is in contact with a surface. Microscopically, tiny protrusions on the interacting surfaces interlock, and soften and break or bend when force is applied. *Kinetic friction* is friction that arises during motion. It depends on the materials the surfaces are made of, as well as how smooth they are, and how firmly they are in contact. The normal force is the measure of that firmness of contact, and it makes sense that the larger the normal force a surface exerts on an object, the larger the kinetic friction force. For many purposes, all the interactions between the two surfaces can be modeled as being proportional to the normal force.

The magnitude of the **kinetic friction force** $f_k$ acting on an object moving on a surface is given by

◄ Kinetic friction force

$$f_k = \mu_k n \qquad\qquad [4.14]$$

where $n$ is the normal force acting on the object and $\mu_k$ is called the **coefficient of kinetic friction** between the object and the surface. The force $f_k$ acts in the direction opposite that of the motion.

Calculating the kinetic friction force is easy: find the normal force and multiply it by the coefficient, $\mu_k$. That coefficient depends on the object and the surface and must be determined experimentally. Variations in the surface that are not apparent to the eye can alter the value of the coefficient as the object moves from one point to another. Further, the effect on the object can vary as a function of velocity. Such coefficients are nonetheless useful, although they are best regarded as averages over time and approximations. In this text, we'll assume a coefficient of friction is always constant for a given surface and object. Some representative values are given in Table 4.2, which also has values for the coefficient of static friction (Section 4.4).

A diagram illustrating the effect of a kinetic friction force on a block is given in Figure 4.18. In every case, this model of friction requires the calculation of the normal force, followed by multiplying by the kinetic friction coefficient, $\mu_k$, which is a constant derived from experiment. The normal force corresponding to Figure 4.18 can be taken from case 2 and is $n = mg - F_{app} \sin \theta$.

The $x$-component of the second law of motion can now be easily written down:

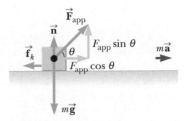

**Figure 4.18** Kinetic friction acts opposite an object's direction of motion.

$$ma_x = F_{app,x} - f_k = F_{app}\cos\theta - \mu_k n$$

$$ma_x = F_{app}\cos\theta - \mu_k(mg - F_{app}\sin\theta)$$

So calculating and using the kinetic friction force is no harder than calculating the normal force. In this case, the angle of the applied force is critical in determining the resultant acceleration of the block. As the angle increases from zero, the part of the applied force acting in the $x$-direction decreases, as do the normal and kinetic friction forces. Using calculus, that angle can be chosen so as to optimize the acceleration.

**Table 4.2** Coefficients of Friction[a]

|  | $\mu_s$ | $\mu_k$ |
|---|---|---|
| Steel on steel | 0.74 | 0.57 |
| Aluminum on steel | 0.61 | 0.47 |
| Copper on steel | 0.53 | 0.36 |
| Rubber on concrete | 1.0 | 0.8 |
| Wood on wood | 0.25–0.5 | 0.2 |
| Glass on glass | 0.94 | 0.4 |
| Waxed wood on wet snow | 0.14 | 0.1 |
| Waxed wood on dry snow | — | 0.04 |
| Metal on metal (lubricated) | 0.15 | 0.06 |
| Ice on ice | 0.1 | 0.03 |
| Teflon on Teflon | 0.04 | 0.04 |
| Synovial joints in humans | 0.01 | 0.003 |

[a]All values are approximate.

# 4.4 Static Friction Forces

When a horizontal force $\vec{F}$ is exerted against an object such as a trash can (Fig. 4.19a), the trash can typically doesn't move unless the force exceeds some critical value (Fig. 4.19b). That critical value is the **maximum static friction force**, $\vec{f}_{s,\max}$. The actual static friction force is always less than or equal to that maximum. As with the kinetic friction force, the static friction force arises from microscopic interactions between the object and the surface it rests upon.

Figure 4.19c illustrates graphically how static friction works. As the applied force gradually increases, so does the static friction force, acting in the opposite direction. When the applied force exceeds the maximum static friction force, the object begins to move, and kinetic friction takes over. In general, the kinetic friction coefficient is less than the static friction coefficient: $\mu_k < \mu_s$.

Static friction force ▶

The magnitude of the **static friction force** $f_s$ acting on an object at rest satisfies the inequality

$$0 \leq f_s \leq f_{s,\max} = \mu_s n \qquad [4.15]$$

where $n$ is the normal force and $\mu_s$ is the coefficient of the *maximum* static friction force between the object and the surface. The force $f_s$ acts in the direction opposite that of the impending motion (the motion that results if static friction is not sufficient to prevent it).

**Tip 4.6** The Static Friction Force Depends on Other Forces

The equation $f_{s,\max} = \mu_s n$ is used *only* to calculate the maximum *possible* static friction force. As long as that limit is not exceeded, the static friction force is actually the negative of the sum of all other forces acting on an unmoving object.

Unlike the kinetic friction force, the static friction force takes on any value between zero and its maximum value of $\mu_s n$, depending on the magnitude of the applied force. A common error is to substitute the maximum value routinely, when in fact that special situation usually doesn't hold. The static friction force self-adjusts, depending on the net force acting on an object. Such self-adjustment is illustrated in Figure 4.20.

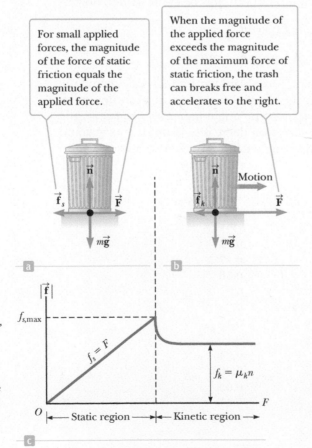

For small applied forces, the magnitude of the force of static friction equals the magnitude of the applied force.

When the magnitude of the applied force exceeds the magnitude of the maximum force of static friction, the trash can breaks free and accelerates to the right.

**Figure 4.19** (a) and (b) When pulling on a trash can, the direction of the force of friction ($\vec{f}_s$ in part (a) and $\vec{f}_k$ in part (b)) between the can and a rough surface is opposite the direction of the applied force $\vec{F}$. (c) A graph of the magnitude of the friction force versus applied force. Note that $f_{s,\max} > f_k$.

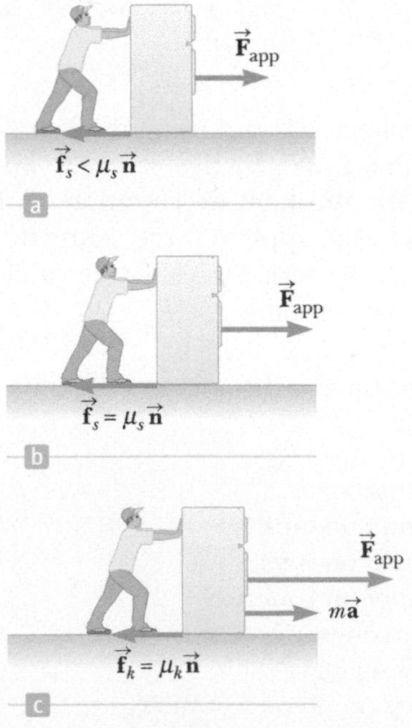

**Figure 4.20** As the applied force increases, the static friction force also increases, as in (a) and (b); however, the refrigerator doesn't move. (c) When the applied force surpasses the maximum static friction force, $f_{s,\max} = \mu_s n$, the refrigerator begins accelerating, while responding to a kinetic friction force.

## EXAMPLE 4.6    BLOCK ON A SLOPE

**GOAL** Calculate the static friction force and the maximum possible static friction force on a block on a sloping surface and distinguish between them.

**PROBLEM** As shown in the figure, a block having a mass of 4.00 kg rests on a slope that makes an angle of 30.0° with the horizontal. If the coefficient of static friction between the block and the surface it rests upon is 0.650, calculate **(a)** the normal force, **(b)** the maximum static friction force, and **(c)** the actual static friction force required to prevent the block from moving. **(d)** Will the block begin to move or remain at rest?

**STRATEGY** Calculate the normal force. The normal force is useful only for the calculation of the *maximum* static friction force. Use the second law to calculate the actual static friction force, and compare the two values.

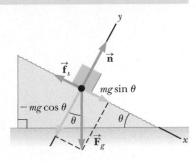

**Figure 4.21** (Example 4.6)

. . . . . . . . . . . . . . . . . . . . . . . . . . . . . . . . . . . . . . . . . . . . . . . . . . . . . . . . . . . . . . . . . . . . . . . . . . . . . . . . . . . . . . . . . . . . . . . . . . . . . . . . . . . . . . . . . . . . . . . . . . . . . . . . .

### SOLUTION

Three forces act on the block: the force of gravity, $F_g = mg$; the normal force $n$; and the static friction force, $f_s$. Choose coordinates so the positive $x$-axis is down the slope, whereas the positive $y$-axis is perpendicular to the slope.

**(a)** Calculate the normal force.

From Equation 4.12, the normal force is given by:

$$n = mg \cos\theta = (4.00 \text{ kg})(9.80 \text{ m/s}^2) \cos(30.0°)$$
$$n = \boxed{33.9 \text{ N}}$$

**(b)** Calculate the *maximum possible* static friction force.

Multiply the normal force by the static friction coefficient, $\mu_s$. This is the *maximum* force that static friction can exert on the block on this surface.

$$f_{s,max} = \mu_s n = (0.650)(33.9\text{N}) = 22.1\text{N}$$
$$f_{s,max} = \boxed{22.1\text{N}}$$

**(c)** Calculate the *actual static friction force* required to keep the block at rest.

Write the second law for the $x$-direction, down the slope.

$$ma_x = \Sigma F_x$$

Substitute $a_x = 0$, and expressions for the two forces acting parallel to the $x$-axis, the gravity force and static friction force:

$$0 = f_{x,\text{grav}} - f_s$$

Solve for the static friction force, $f_s$, and note the expression for $f_{x,\text{grav}}$ from Equation 4.13:

$$f_s = f_{x,\text{grav}} = mg \sin\theta$$
$$f_s = (4.00 \text{ kg})(9.80 \text{ m/s}^2)(\sin(30.0°)) = \boxed{19.6 \text{ N}}$$

**(d)** Determine whether the block will begin to move.

In this case, the actual, required static friction force, $f_s = 19.6$ N, is less than the maximum possible static friction force, $f_{s,max} = 22.1$ N, so the block remains at rest on the slope .

. . . . . . . . . . . . . . . . . . . . . . . . . . . . . . . . . . . . . . . . . . . . . . . . . . . . . . . . . . . . . . . . . . . . . . . . . . . . . . . . . . . . . . . . . . . . . . . . . . . . . . . . . . . . . . . . . . . . . . . . . . . . . . . . .

**REMARKS** As the angle of the slope increases, the magnitude of the static friction force decreases and the component of the gravitational force acting down the slope increases. When the angle of the slope exceeds a critical angle, the block will start to slide down the slope and kinetic friction will take over.

**QUESTION 4.6** How would the answer to part **(c)** change if the mass of the block were larger?

**EXERCISE** What angle must be exceeded so that the block will begin sliding? (*Hint:* Equate the expression for the static friction force to the maximum possible static friction force.)

**ANSWER** 33.0°

_____

### Quick Quiz

**4.5** If you press a book flat against a vertical wall with your hand, in what direction is the friction force exerted by the wall on the book? (a) downward (b) upward (c) out from the wall (d) into the wall.

(*Continued*)

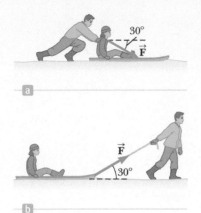

**Figure 4.22** (Quick Quiz 4.7)

**4.6** A crate is sitting in the center of a flatbed truck. As the truck accelerates to the east, the crate moves with it, and doesn't slide on the bed of the truck. In what direction is the friction force exerted by the bed of the truck on the crate? (a) To the west. (b) To the east. (c) There is no friction force because the crate isn't sliding.

**4.7** Suppose your friend is sitting on a sled and asks you to move her across a flat, horizontal field. You have a choice of (a) pushing her from behind by applying a force downward on her shoulders at 30° below the horizontal (Fig. 4.22a) or (b) attaching a rope to the front of the sled and pulling with a force at 30° above the horizontal (Fig. 4.22b). Which option would be easier and why?

# 4.5 Tension Forces

A tension force can be exerted by attaching a string or cable to an object and pulling it. The tension force acts along the direction of the cable and exerts a force both on the object and on the person exerting force on the cable, as illustrated in Figure 4.23. If the mass of the given cable is neglected, the tension is the same all along the cable. In that case the tension force on the car, $\vec{T}_C$, has the same magnitude as the tension force on the man, $\vec{T}_M$.

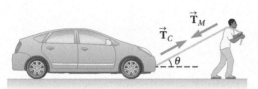

**Figure 4.23** A very strong man pulls a car via a tension force, but must lean forward to prevent an equal and opposite tension force from causing him to topple over backwards.

Tension can be understood as the force that would be read on a spring scale, if it were spliced into the cable. Like the normal force, tension forces have their origin in microscopic electromagnetic interactions that resist the separating of a long, thin string of matter when a force is applied to it. The string becomes taut, and through it the force can then be exerted on whatever object that is attached to the opposite end of the string. This section focuses on some common tension force contexts.

## 4.5.1 Case 1: Vertical Tension Forces on a Static Object

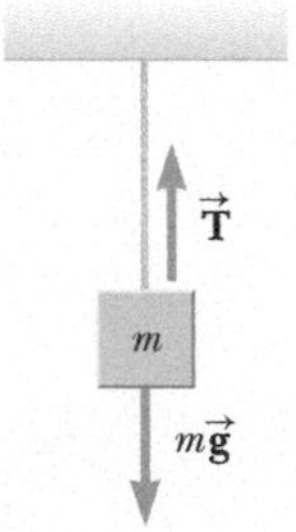

In this case, an object is dangling at the end of a vertical string, as in Figure 4.24. The object is in equilibrium so the acceleration is zero. The $y$-component of the second law gives an expression for the tension in the string:

$$\sum F_y = ma_y$$
$$T - mg = 0$$
$$\boxed{T = mg} \qquad [4.16]$$

**Figure 4.24** The tension in the string supports the block's weight.

The tension force therefore is equal to the supported weight.

## 4.5.2 Case 2: Vertical Tension Forces on an Accelerating Object

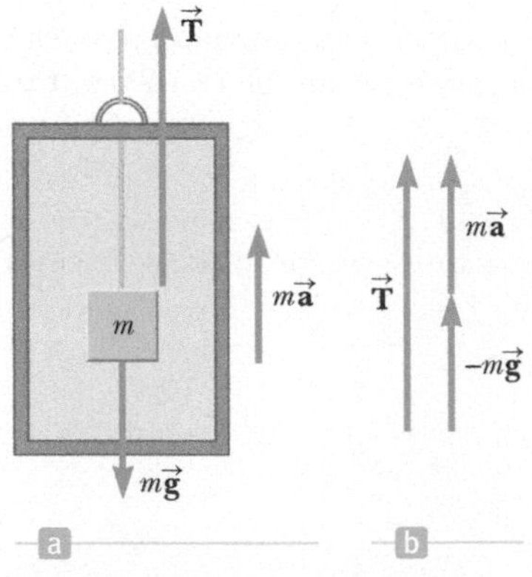

In this case, a string is attached to an object and put under such tension that it accelerates the object upward (Fig. 4.25). The tension force must support the weight and, in addition, provide the acceleration. The correct expression can be derived from the $y$-component of the second law:

$$ma_y = \sum F_y$$
$$ma_y = T - mg$$

Rearrange algebraically to obtain:

$$\boxed{T = ma_y + mg = m(a_y + g)} \qquad [4.17]$$

**Figure 4.25** (a) A block hangs by a string in an elevator accelerating upward. (b) The tension supplies the acceleration of the block while supporting it against gravity.

Accelerating upward increases the tension, whereas accelerating downward decreases the tension.

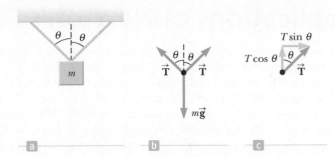

**Figure 4.26** (a) A block supported by two symmetric strings. (b) Free-body diagram. (c) Components of the tension, $\vec{\mathbf{T}}$.

### 4.5.3 Case 3: Two Tension Forces at Symmetric Angles

In this case, two tensions act on one object symmetrically, as illustrated in Figure 4.26. For this to work, naturally, the points of attachment must be chosen carefully, in which case the tensions will be equal. The $y$-component of each tension is given by $T \cos \theta$. The second law then gives an expression for the tension $T$:

$$ma_y = \sum F_y$$
$$ma_y = T \cos \theta + T \cos \theta - mg$$
$$0 = 2T \cos \theta - mg$$

Solve this expression algebraically for the tension $T$:

$$T = \frac{mg}{2 \cos\theta} \qquad [4.18]$$

Whether the expression involves a sine or a cosine depends on what angles are given in the problem.

### 4.5.4 Case 4: Two Tensions at Nonequal Angles

There are many contexts in which two tensions may act at nonequal angles, and the context in Figure 4.27 is one example. In such cases, it's necessary to develop two equations and two unknowns for the two tensions, $T_1$ and $T_2$.

Because the given angle $\theta$ is with respect to the vertical, the adjacent side of the triangle formed by the vector $\vec{\mathbf{T}}_2$ corresponds to the $y$-direction, requiring a cosine function. With $a_y = 0$, the $y$-component of the second law yields

$$ma_y = \sum F_y$$
$$0 = -mg + T_2 \cos \theta$$

which can easily be solved for the tension, $T_2$:

$$T_2 = \frac{mg}{\cos\theta}$$

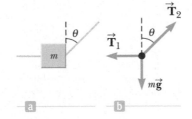

**Figure 4.27** (a) A block held by two strings. (b) Free-body diagram.

Again, the given angle $\theta$ is with respect to the vertical, so the opposite side of the triangle formed by the vector $\vec{\mathbf{T}}_2$ corresponds to the $x$-direction, requiring a sine function. With $a_x = 0$, the $x$-component of the second law can be solved for the tension $T_1$:

$$ma_x = \sum F_x$$
$$0 = -T_1 + T_2 \sin\theta$$
$$T_1 = T_2 \sin\theta = \left(\frac{mg}{\cos\theta}\right)\sin\theta$$
$$T_1 = mg \tan\theta$$

This illustrates how to determine tensions when two such tensions are involved in supporting an object. Using the two components of the second law, write two equations and then solve for the two unknowns.

**Figure 4.28** Newton's second law applied to a rope gives $T - T' = ma$. However, if $m = 0$, then $T = T'$. Thus, the tension in a massless rope is the same at all points in the rope.

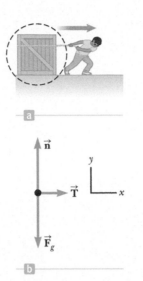

**Figure 4.29** (a) A crate being pulled to the right on a frictionless surface. (b) The free-body diagram that represents the forces exerted on the crate.

---

**Tip 4.7** Free-Body Diagrams

The *most important step* in solving a problem by means of Newton's second law is to draw the correct free-body diagram. Include only those forces that act directly on the object of interest.

---

**Tip 4.8** A Particle in Equilibrium

A zero net force on a particle does *not* mean that the particle isn't moving. It means that the particle isn't *accelerating*. If the particle has a nonzero initial velocity and is acted upon by a zero net force, it continues to move with the same velocity.

---

# 4.6 Applications of Newton's Laws

The foregoing four sections introduced several common forces in typical contexts. This section applies Newton's laws to objects under the influence of constant external forces. We assume that objects behave as particles, so we need not consider the possibility of rotational motion. We also neglect any friction effects and the masses of any ropes or strings involved. With these approximations, the magnitude of the tension exerted along a rope is the same at all points in the rope. This is illustrated by the rope in Figure 4.28, showing the forces $\vec{\mathbf{T}}$ and $\vec{\mathbf{T}}'$ acting on it. If the rope has mass $m$, then Newton's second law applied to the rope gives $T - T' = ma$. If the mass $m$ is taken to be negligible, however, as in the upcoming examples, then $T = T'$.

When we apply Newton's law to an object, we are interested only in those forces that act *on the object*. For example, in Figure 4.12b (page 91), the only external forces acting on the monitor are $\vec{\mathbf{n}}$ and $\vec{\mathbf{F}}_g$. The reactions to these forces, $\vec{\mathbf{n}}'$ and $\vec{\mathbf{F}}_g'$, act on the table and on Earth, respectively, and don't appear in Newton's second law applied to the monitor.

Consider a crate being pulled to the right on a frictionless, horizontal surface, as in Figure 4.29a. Suppose you wish to find the acceleration of the crate and the force the surface exerts on it. The horizontal force exerted on the crate acts through the rope. The force that the rope exerts on the crate is denoted by $\vec{\mathbf{T}}$ (because it's a tension force). The magnitude of $\vec{\mathbf{T}}$ is equal to the tension in the rope. A dashed circle is drawn around the crate in Figure 4.29a to emphasize the importance of isolating the crate from its surroundings.

**Because we are interested only in the motion of the crate, we must be able to identify all forces acting on it.** The free-body diagram in Figure 4.29b illustrates these forces. In addition to displaying the force $\vec{\mathbf{T}}$, the free-body diagram for the crate includes the force of gravity $\vec{\mathbf{F}}_g$ exerted by Earth and the normal force $\vec{\mathbf{n}}$ exerted by the floor. The construction of a correct free-body diagram is an essential step in applying Newton's laws. An incorrect diagram will most likely lead to incorrect answers!

The *reactions* to the forces we have listed—namely, the force exerted by the rope on the hand doing the pulling, the force exerted by the crate on Earth, and the force exerted by the crate on the floor—aren't included in the free-body diagram because they act on other objects and not on the crate. Consequently, they don't directly influence the crate's motion. Only forces acting directly on the crate are included.

Now let's apply Newton's second law to the crate. First, we choose an appropriate coordinate system. In this case it's convenient to use the one shown in Figure 4.29b, with the $x$-axis horizontal and the $y$-axis vertical. We can apply Newton's second law in the $x$-direction, $y$-direction, or both, depending on what we're asked to find in a problem. Newton's second law applied to the crate in the $x$- and $y$-directions yields the following two equations:

$$ma_x = T \qquad ma_y = n - mg = 0$$

From these equations, we find that the acceleration in the $x$-direction is constant, given by $a_x = T/m$, and that the normal force is given by $n = mg$. Because the acceleration is constant, the equations of kinematics can be applied to obtain further information about the velocity and displacement of the object.

## PROBLEM-SOLVING STRATEGY

### Newton's Second Law

*Problems involving Newton's second law can be very complex. The following protocol breaks the solution process down into smaller, intermediate goals:*

1. **Read** the problem carefully at least once.
2. **Draw** a picture of the system, identify the object of primary interest, and indicate forces with arrows.

3. **Label** each force in the picture in a way that will bring to mind what physical quantity the label stands for (e.g., $T$ for tension).

4. **Draw** a free-body diagram of the object of interest, based on the labeled picture. If additional objects are involved, draw separate free-body diagrams for them. Choose convenient coordinates for each object.

5. **Apply Newton's second law.** The $x$- and $y$-components of Newton's second law should be taken from the vector equation and written individually. This usually results in two equations and two unknowns.

6. **Solve** for the desired unknown quantity, and substitute the numbers.

In the special case of equilibrium, the foregoing process is simplified because the acceleration is zero.

## 4.6.1 Objects in Equilibrium

**Objects that are either at rest or moving with constant velocity are said to be in equilibrium.** Because $\vec{a} = 0$, Newton's second law applied to an object in equilibrium gives

$$\sum \vec{F} = 0 \qquad [4.19]$$

This statement signifies that the *vector* sum of all the forces (the net force) acting on an object in equilibrium is zero. Equation 4.19 is equivalent to the set of component equations given by

$$\sum F_x = 0 \quad \text{and} \quad \sum F_y = 0 \qquad [4.20]$$

We won't consider three-dimensional problems in this book, but the extension of Equation 4.20 to a three-dimensional problem can be made by adding a third equation: $\sum F_z = 0$.

**4.8** Consider the two situations shown in Figure 4.30, in which there is no acceleration. In both cases the men pull with a force of magnitude $F$. Is the reading on the scale in part (i) of the figure (a) greater than, (b) less than, or (c) equal to the reading in part (ii)?

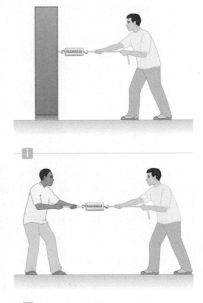

**Figure 4.30** (Quick Quiz 4.8) (i) A person pulls with a force of magnitude $F$ on a spring scale attached to a wall. (ii) Two people pull with forces of magnitude $F$ in opposite directions on a spring scale attached between two ropes.

---

**EXAMPLE 4.7 | A TRAFFIC LIGHT AT REST**

**GOAL** Use the second law in an equilibrium problem requiring two free-body diagrams.

**PROBLEM** A traffic light weighing $1.00 \times 10^2$ N hangs from a vertical cable tied to two other cables that are fastened to a support, as in Figure 4.31a. The upper cables make angles of $37.0°$ and $53.0°$ with the horizontal. Find the tension in each of the three cables.

**STRATEGY** There are three unknowns, so we need to generate three equations relating them, which can then be solved. One equation can be obtained by applying Newton's second law to the traffic light, which has forces in the $y$-direction only. Two more equations can be obtained by applying the second law to the knot joining the cables—one equation from the $x$-component and one equation from the $y$-component.

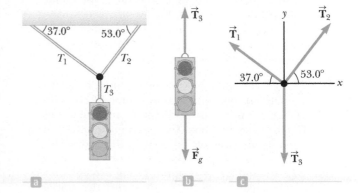

**Figure 4.31** (Example 4.7) (a) A traffic light suspended by cables. (b) The forces acting on the traffic light. (c) A free-body diagram for the knot joining the cables.

..................................................

**SOLUTION**

Find $T_3$ from Figure 4.31b, using the condition of equilibrium:

$$\sum F_y = 0 \;\rightarrow\; T_3 - F_g = 0$$
$$T_3 = F_g = 1.00 \times 10^2 \text{ N}$$

*(Continued)*

Using Figure 4.31c, resolve all three tension forces into components and construct a table for convenience:

| Force | x-Component | y-Component |
|---|---|---|
| $\vec{T}_1$ | $-T_1 \cos 37.0°$ | $T_1 \sin 37.0°$ |
| $\vec{T}_2$ | $T_2 \cos 53.0°$ | $T_2 \sin 53.0°$ |
| $\vec{T}_3$ | $0$ | $-1.00 \times 10^2$ N |

Apply the conditions for equilibrium to the knot, using the components in the table:

**(1)** $\sum F_x = -T_1 \cos 37.0° + T_2 \cos 53.0° = 0$

**(2)** $\sum F_y = T_1 \sin 37.0° + T_2 \sin 53.0° - 1.00 \times 10^2\,\text{N} = 0$

There are two equations and two remaining unknowns. Solve Equation (1) for $T_2$:

$$T_2 = T_1 \left(\frac{\cos 37.0°}{\cos 53.0°}\right) = T_1 \left(\frac{0.799}{0.602}\right) = 1.33 T_1$$

Substitute the result for $T_2$ into Equation (2):

$$T_1 \sin 37.0° + (1.33 T_1)(\sin 53.0°) - 1.00 \times 10^2\,\text{N} = 0$$

$$T_1 = \boxed{60.1\,\text{N}}$$

$$T_2 = 1.33 T_1 = 1.33(60.1\,\text{N}) = \boxed{79.9\,\text{N}}$$

. . . . . . . . . . . . . . . . . . . . . . . . . . . . . . . . . . . . . . . . . . . . . . . . . . . . . . . . . . . . . . . . . . . . . . . . . . . . . . . . . . . . . . . . . . . . .

**REMARKS** It's very easy to make sign errors in this kind of problem. One way to avoid them is to always measure the angle of a vector from the positive x-direction. The trigonometric functions of the angle will then automatically give the correct signs for the components. For example, $T_1$ makes an angle of $180° - 37° = 143°$ with respect to the positive x-axis, and its x-component, $T_1 \cos 143°$, is negative, as it should be.

**QUESTION 4.7** How would the answers change if a second traffic light were attached beneath the first?

**EXERCISE 4.7** Suppose the traffic light is hung so that the tensions $T_1$ and $T_2$ are both equal to 80.0 N. Find the new angles they make with respect to the x-axis. (By symmetry, these angles will be the same.)

**ANSWER** Both angles are 38.7°.

---

EXAMPLE 4.8 SLED ON A FRICTIONLESS HILL

**GOAL** Use the second law and the normal force in an equilibrium problem.

**PROBLEM** A sled is tied to a tree on a frictionless, snow-covered hill, as shown in Figure 4.32a. If the sled weighs 77.0 N, find the magnitude of the tension force $\vec{T}$ exerted by the rope on the sled and that of the normal force $\vec{n}$ exerted by the hill on the sled.

**STRATEGY** When an object is on a slope, it's convenient to use tilted coordinates, as in Figure 4.32b so that the normal force $\vec{n}$ is in the y-direction and the tension force $\vec{T}$ is in the x-direction. In the absence of friction, the hill exerts no force on the sled in the x-direction. Because the sled is at rest, the conditions for equilibrium, $\sum F_x = 0$ and $\sum F_y = 0$, apply, giving two equations for the two unknowns—the tension and the normal force.

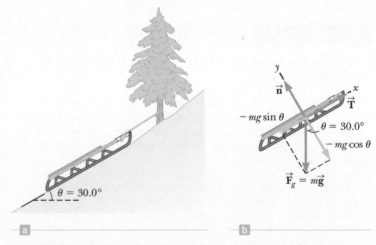

**Figure 4.32** (Example 4.8) (a) A sled tied to a tree on a frictionless hill. (b) A diagram of forces acting on the sled.

. . . . . . . . . . . . . . . . . . . . . . . . . . . . . . . . . . . . . . . . . . . . . . . . . . . . . . . . . . . . . . . . . . . . . . . . . . . . . . . . . . . . . . . . . . . . .

**SOLUTION**

Apply Newton's second law to the sled, with $\vec{a} = 0$:

$$\sum \vec{F} = \vec{T} + \vec{n} + \vec{F}_g = 0$$

*(Continued)*

Extract the $x$-component from this equation to find $T$. The $x$-component of the normal force is zero, and the sled's weight is given by $mg = 77.0$ N.

$$\sum F_x = T + 0 - mg \sin \theta = T - (77.0 \text{ N}) \sin 30.0° = 0$$

$$T = \boxed{38.5 \text{ N}}$$

Write the $y$-component of Newton's second law. The $y$-component of the tension is zero, so this equation will give the normal force.

$$\sum F_y = 0 + n - mg \cos \theta = n - (77.0 \text{ N})(\cos 30.0°) = 0$$

$$n = \boxed{66.7 \text{ N}}$$

. . . . . . . . . . . . . . . . . . . . . . . . . . . . . . . . . . . . . . . . . . . . . . . . . . . . . . . . . . . . .

**REMARKS** Unlike its value on a horizontal surface, $n$ is less than the weight of the sled when the sled is on the slope. This is because only part of the force of gravity (the $x$-component) is acting to pull the sled down the slope. The $y$-component of the force of gravity balances the normal force.

**QUESTION 4.8** Consider the same scenario on a hill with a steeper slope. Would the magnitude of the tension in the rope get larger, smaller, or remain the same as before? How would the normal force be affected?

**EXERCISE 4.8** Suppose a child of weight $w$ climbs onto the sled. If the tension force is measured to be 60.0 N, find the weight of the child and the magnitude of the normal force acting on the sled.

**ANSWERS** $w = 43.0$ N, $n = 104$ N

---

**Quick Quiz**

**4.9** For the woman being pulled forward on the toboggan in Figure 4.33, is the magnitude of the normal force exerted by the ground on the toboggan (a) equal to the total weight of the woman plus the toboggan, (b) greater than the total weight, (c) less than the total weight, or (d) possibly greater than or less than the total weight, depending on the size of the weight relative to the tension in the rope?

**Figure 4.33** (Quick Quiz 4.9)

## 4.6.2 Accelerating Objects and Newton's Second Law

When a net force acts on an object, the object accelerates, and we use Newton's second law to analyze the motion.

---

**EXAMPLE 4.9**    **MOVING A CRATE**

**GOAL** Apply the second law of motion for a system not in equilibrium, together with a kinematics equation.

**PROBLEM** The combined weight of the crate and dolly in Figure 4.34 is $3.00 \times 10^2$ N. If the man pulls on the rope with a constant force of 20.0 N, what is the acceleration of the system (crate plus dolly), and how far will it move in 2.00 s? Assume the system starts from rest and that there are no friction forces opposing the motion.

**STRATEGY** We can find the acceleration of the system from Newton's second law. Because the force exerted on the system is constant, its acceleration is constant. Therefore, we can apply a kinematics equation to find the distance traveled in 2.00 s.

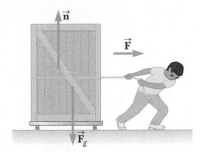

**Figure 4.34** (Example 4.9)

. . . . . . . . . . . . . . . . . . . . . . . . . . . . . . . . . . . . . . . . . . . . . . . . . . . . . . . . . . . . .

**SOLUTION**

Find the mass of the system from the definition of weight, $w = mg$:

$$m = \frac{w}{g} = \frac{3.00 \times 10^2 \text{ N}}{9.80 \text{ m/s}^2} = 30.6 \text{ kg}$$

Find the acceleration of the system from the second law:

$$a_x = \frac{F_x}{m} = \frac{20.0 \text{ N}}{30.6 \text{ kg}} = \boxed{0.654 \text{ m/s}^2}$$

Use kinematics to find the distance moved in 2.00 s, with $v_0 = 0$:

$$\Delta x = \tfrac{1}{2} a_x t^2 = \tfrac{1}{2} (0.654 \text{ m/s}^2)(2.00 \text{ s})^2 = \boxed{1.31 \text{ m}}$$

*(Continued)*

**REMARKS** Note that the constant applied force of 20.0 N is assumed to act on the system at all times during its motion. If the force were removed at some instant, the system would continue to move with constant velocity and hence zero acceleration. The rollers have an effect that was neglected, here.

**QUESTION 4.9** What effect does doubling the weight have on the acceleration and the displacement?

**EXERCISE 4.9** A man pulls a 50.0-kg box horizontally from rest while exerting a constant horizontal force, displacing the box 3.00 m in 2.00 s. Find the force the man exerts on the box. (Ignore friction.)

**ANSWER** 75.0 N

---

## EXAMPLE 4.10 THE RUNAWAY CAR

**GOAL** Apply the second law and kinematic equations to a problem involving an object moving on an incline.

**PROBLEM** **(a)** A car of mass $m$ is on an icy driveway inclined at an angle $\theta = 20.0°$, as in Figure 4.35a. Determine the acceleration of the car, assuming the incline is frictionless. **(b)** If the length of the driveway is 25.0 m and the car starts from rest at the top, how long does it take to travel to the bottom? **(c)** What is the car's speed at the bottom?

**STRATEGY** Choose tilted coordinates as in Figure 4.35b so that the normal force $\vec{\mathbf{n}}$ is in the positive $y$-direction, perpendicular to the driveway, and the positive $x$-axis is down the slope. The force of gravity $\vec{\mathbf{F}}_g$ then has an $x$-component, $mg \sin \theta$, and a $y$-component, $-mg \cos \theta$. The components of Newton's second law form a system of two equations and two unknowns for the acceleration down the slope, $a_x$, and the normal force. Parts **(b)** and **(c)** can be solved with the kinematics equations.

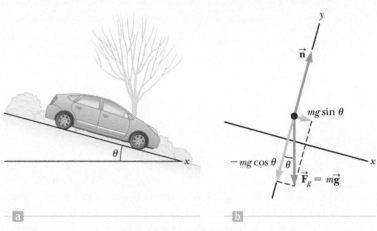

**Figure 4.35** (Example 4.10)

---

### SOLUTION

**(a)** Find the acceleration of the car.

Apply Newton's second law to the car:

$$m\vec{\mathbf{a}} = \sum \vec{\mathbf{F}} = \vec{\mathbf{F}}_g + \vec{\mathbf{n}}$$

Extract the $x$- and $y$-components from the second law:

(1)     $ma_x = \sum F_x = mg \sin \theta$

(2)     $0 = \sum F_y = -mg \cos \theta + n$

Divide Equation (1) by $m$ and substitute the given values:

$a_x = g \sin \theta = (9.80 \text{ m/s}^2) \sin 20.0° = \boxed{3.35 \text{ m/s}^2}$

**(b)** Find the time taken for the car to reach the bottom.

Use Equation 3.6b for displacement, with $v_{0x} = 0$:

$\Delta x = \frac{1}{2} a_x t^2 \quad \rightarrow \quad \frac{1}{2}(3.35 \text{ m/s}^2)t^2 = 25.0 \text{ m}$

$t = \boxed{3.86 \text{ s}}$

**(c)** Find the speed of the car at the bottom of the driveway.

Use Equation 3.6a for velocity, again with $v_{0x} = 0$:

$v_x = a_x t = (3.35 \text{ m/s}^2)(3.86 \text{ s}) = \boxed{12.9 \text{ m/s}}$

---

**REMARKS** Notice that the final answer for the acceleration depends only on $g$ and the angle $\theta$, not the mass. Equation (2), which gives the normal force, isn't useful here, but is essential when friction plays a role.

**QUESTION 4.10** If the car is parked on a more gentle slope, how will the time required for it to slide to the bottom of the hill be affected? Explain.

**EXERCISE 4.10** Suppose a hockey puck slides down a frictionless ramp with an acceleration of 5.00 m/s². **(a)** What angle does the ramp make with respect to the horizontal? **(b)** If the ramp has a length of 6.00 m, how long does it take the puck to reach the bottom? **(c)** Now suppose the mass of the puck is doubled. What's the puck's new acceleration down the ramp?

**ANSWER** **(a)** 30.7°  **(b)** 1.55 s  **(c)** unchanged, 5.00 m/s²

## EXAMPLE 4.11 | WEIGHING A FISH IN AN ELEVATOR

**GOAL** Explore the effect of acceleration on the apparent weight of an object.

**PROBLEM** A woman weighs a fish with a spring scale attached to the ceiling of an elevator, as shown in Figures 4.36a and 4.36b. While the elevator is at rest, she measures a weight of 40.0 N. **(a)** What weight does the scale read if the elevator accelerates upward at 2.00 m/s²? **(b)** What does the scale read if the elevator accelerates downward at 2.00 m/s², as in Figure 4.36b? **(c)** If the elevator cable breaks, what does the scale read?

**STRATEGY** Write down Newton's second law for the fish, including the force $\vec{T}$ exerted by the spring scale and the force of gravity, $m\vec{g}$. The scale doesn't measure the true weight, it measures the force $T$ that it exerts on the fish, so in each case solve for this force, which is the apparent weight as measured by the scale.

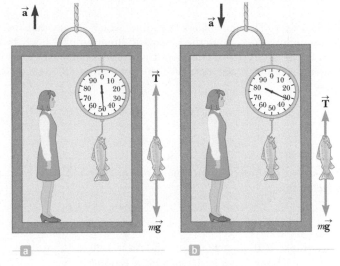

When the elevator accelerates upward, the spring scale reads a value greater than the weight of the fish.

When the elevator accelerates downward, the spring scale reads a value less than the weight of the fish.

**Figure 4.36** (Example 4.11)

### SOLUTION

**(a)** Find the scale reading as the elevator accelerates upward, as in Figure 4.36a.

Apply Newton's second law to the fish, taking upward as the positive direction:

$$ma = \sum F = T - mg$$

Solve for $T$:

$$T = ma + mg = m(a + g)$$

Find the mass of the fish from its weight of 40.0 N:

$$m = \frac{w}{g} = \frac{40.0\ \text{N}}{9.80\ \text{m/s}^2} = 4.08\ \text{kg}$$

Compute the value of $T$, substituting $a = +2.00$ m/s²:

$$T = m(a + g) = (4.08\ \text{kg})(2.00\ \text{m/s}^2 + 9.80\ \text{m/s}^2)$$
$$= \boxed{48.1\ \text{N}}$$

**(b)** Find the scale reading as the elevator accelerates downward, as in Figure 4.36b.

The analysis is the same, the only change being the acceleration, which is now negative: $a = -2.00$ m/s².

$$T = m(a + g) = (4.08\ \text{kg})(-2.00\ \text{m/s}^2 + 9.80\ \text{m/s}^2)$$
$$= \boxed{31.8\ \text{N}}$$

**(c)** Find the scale reading after the elevator cable breaks.

Now $a = -9.80$ m/s², the acceleration due to gravity:

$$T = m(a + g) = (4.08\ \text{kg})(-9.80\ \text{m/s}^2 + 9.80\ \text{m/s}^2)$$
$$= \boxed{0\ \text{N}}$$

**REMARKS** Notice how important it is to have correct signs in this problem! Accelerations can increase or decrease the apparent weight of an object. Astronauts experience very large changes in apparent weight, from several times normal weight during ascent to weightlessness in free fall.

**QUESTION 4.11** Starting from rest, an elevator accelerates upward, reaching and maintaining a constant velocity thereafter until it reaches the desired floor, when it begins to slow down. Describe the scale reading during this time.

**EXERCISE 4.11** Find the initial acceleration of a rocket if the astronauts on board experience eight times their normal weight during an initial vertical ascent. (*Hint:* In this exercise, the scale force is replaced by the normal force.)

**ANSWER** 68.6 m/s²

## EXAMPLE 4.12 | THE SLIDING HOCKEY PUCK

**GOAL** Apply the concept of kinetic friction.

**PROBLEM** The hockey puck in Figure 4.37, struck by a hockey stick, is given an initial speed of 20.0 m/s on a frozen pond. The puck remains on the ice and slides $1.20 \times 10^2$ m, slowing down steadily until it comes to rest. Determine the coefficient of kinetic friction between the puck and the ice.

**STRATEGY** The puck slows "steadily," which means that the acceleration is constant. Consequently, we can use the kinematic equation $v^2 = v_0^2 + 2a\Delta x$ to find $a$, the acceleration in the $x$-direction. The $x$- and $y$-components of Newton's second law then give two equations and two unknowns for the coefficient of kinetic friction, $\mu_k$, and the normal force $n$.

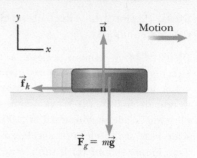

**Figure 4.37** (Example 4.12) *After* the puck is given an initial velocity to the right, the external forces acting on it are the force of gravity $\vec{F}_g$, the normal force $\vec{n}$, and the force of kinetic friction, $\vec{f}_k$.

### SOLUTION

Solve the time-independent kinematic equation for the acceleration $a$:

$$v^2 = v_0^2 + 2a\Delta x$$
$$a = \frac{v^2 - v_0^2}{2\Delta x}$$

Substitute $v = 0$, $v_0 = 20.0$ m/s, and $\Delta x = 1.20 \times 10^2$ m. Note the negative sign in the answer: $\vec{a}$ is opposite $\vec{v}$:

$$a = \frac{0 - (20.0 \text{ m/s})^2}{2(1.20 \times 10^2 \text{ m})} = -1.67 \text{ m/s}^2$$

Find the normal force from the $y$-component of the second law:

$$\sum F_y = n - F_g = n - mg = 0$$
$$n = mg$$

Obtain an expression for the force of kinetic friction, and substitute it into the $x$-component of the second law:

$$f_k = \mu_k n = \mu_k mg$$
$$ma = \sum F_x = -f_k = -\mu_k mg$$

Solve for $\mu_k$ and substitute values:

$$\mu_k = -\frac{a}{g} = \frac{1.67 \text{ m/s}^2}{9.80 \text{ m/s}^2} = \boxed{0.170}$$

**REMARKS** Notice how the problem breaks down into three parts: kinematics, Newton's second law in the $y$-direction, and then Newton's second law in the $x$-direction.

**QUESTION 4.12** How would the answer be affected if the puck were struck by an astronaut on a patch of ice on Mars, where the acceleration of gravity is $0.35g$, with all other given quantities remaining the same?

**EXERCISE 4.12** An experimental rocket plane lands on skids on a dry lake bed. If it's traveling at 80.0 m/s when it touches down, how far does it slide before coming to rest? Assume the coefficient of kinetic friction between the skids and the lake bed is 0.600.

**ANSWER** 544 m

# 4.7 Two-Body Problems

Newton's second law of motion also applies to systems of objects. Previously, an implicit simplification resulted from considering external forces on a single body, but pairs of bodies, if connected in some way, can influence each other.

Solving such two-body problems is a matter of writing the second law for both objects. In problems examined here, that can result in up to four equations and four unknown quantities. Symmetry can sometimes reduce the number of unknowns.

**EXAMPLE 4.13** ATWOOD'S MACHINE

**GOAL** Use the second law to solve a simple two-body problem symbolically.

**PROBLEM** Two objects of mass $m_1$ and $m_2$, with $m_2 > m_1$, are connected by a light, inextensible cord and hung over a frictionless pulley, as in Figure 4.38a. Both cord and pulley have negligible mass. Find the magnitude of the acceleration of the system and the tension in the cord.

**STRATEGY** The heavier mass, $m_2$, accelerates downward, in the negative $y$-direction. Because the cord can't be stretched, the accelerations of the two masses are equal in magnitude, but *opposite* in direction, so that $a_1$ is positive and $a_2$ is negative, and $a_2 = -a_1$. Each mass is acted on by a force of tension $\vec{T}$ in the upward direction and a force of gravity in the downward direction. Figure 4.38b shows free-body diagrams for the two masses. Newton's second law for each mass, together with the equation relating the accelerations, constitutes a set of three equations for the three unknowns—$a_1$, $a_2$, and $T$.

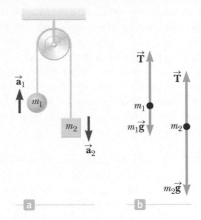

**Figure 4.38** (Example 4.13) Atwood's machine. (a) Two hanging objects connected by a light string that passes over a frictionless pulley. (b) Free-body diagrams for the objects.

**SOLUTION**

Apply the second law to each of the two objects individually:

**(1)** $m_1 a_1 = T - m_1 g$     **(2)** $m_2 a_2 = T - m_2 g$

Substitute $a_2 = -a_1$ into Equation (2) and multiply both sides by $-1$:

**(3)** $m_2 a_1 = -T + m_2 g$

Add Equations (1) and (3), and solve for $a_1$:

$(m_1 + m_2)a_1 = m_2 g - m_1 g$

$$a_1 = \left( \frac{m_2 - m_1}{m_1 + m_2} \right) g$$

Substitute this result into Equation (1) to find $T$:

$$T = \left( \frac{2 m_1 m_2}{m_1 + m_2} \right) g$$

**REMARKS** The acceleration of the second object is the same as that of the first, but negative. When $m_2$ gets very large compared with $m_1$, the acceleration of the system approaches $g$, as expected, because $m_2$ is falling nearly freely under the influence of gravity. Indeed, $m_2$ is only slightly restrained by the much lighter $m_1$.

**QUESTION 4.13** How could this simple machine be used to raise objects too heavy for a person to lift?

**EXERCISE 4.13** Suppose in the same Atwood setup another string is attached to the bottom of $m_1$ and a constant force $f$ is applied, retarding the upward motion of $m_1$. If $m_1 = 5.00$ kg and $m_2 = 10.00$ kg, what value of $f$ will reduce the acceleration of the system by 50%?

**ANSWER** 24.5 N

## 4.7.1 The System Approach

If two bodies are connected by an inextensible string, so that the bodies have a common acceleration, it's possible to use the system approach, which treat the bodies as a single object. The second law of motion for the system approach is:

$$\sum F_{\text{ext}} = \left( \sum m_i \right) a_{\text{sys}} \qquad \text{[4.21]}$$

where $F_{\text{ext}}$ are forces external to the system. Internal forces, such as a tension force holding the system together, do not appear. Internal forces must be found by applying the second law to individual bodies in the system. The system approach, naturally, can be extended to three or more bodies. Example 4.14 contrasts the two approaches to solving these problems.

## EXAMPLE 4.14  CONNECTED OBJECTS

**GOAL** Use both the general method and the system approach to solve a connected two-body problem involving gravity and friction.

**PROBLEM** (a) A block with mass $m_1 = 4.00$ kg and a ball with mass $m_2 = 7.00$ kg are connected by a light string that passes over a massless, frictionless pulley, as shown in Figure 4.39a. The coefficient of kinetic friction between the block and the surface is 0.300. Find the acceleration of the two objects and the tension in the string. (b) Check the answer for the acceleration by using the system approach.

**STRATEGY** Connected objects are handled by applying Newton's second law separately to each object. The force diagrams for the block and the ball are shown in Figure 4.39b, with the $+x$-direction to the right and the $+y$-direction upwards. The magnitude of the acceleration for both objects has the same value, $|a_1| = |a_2| = a$. The block with mass $m_1$ moves in the positive $x$-direction, and the ball with mass $m_2$ moves in the negative $y$-direction, so $a_1 = -a_2$. Using Newton's second law, we can develop two equations involving the unknowns $T$ and $a$ that can be solved simultaneously. In part (b), treat the two masses as a single object, with the gravity force on the ball increasing the combined object's speed and the friction force on the block retarding it. The tension forces then become internal and don't appear in the second law.

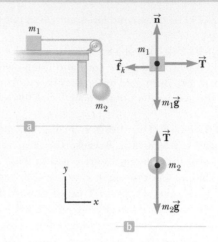

**Figure 4.39** (Example 4.14) (a) Two objects connected by a light string that passes over a frictionless pulley. (b) Force diagrams for the objects.

. . . . . . . . . . . . . . . . . . . . . . . . . . . . . . . . . . . . . . . . . . . . . . . . . . . . . . . . . . . . . . . . . . . . . . . . . . . . . . . . . . . . .

**SOLUTION**

(a) Find the acceleration of the objects and the tension in the string.

Write the components of Newton's second law for the block of mass $m_1$:

$$\sum F_x = T - f_k = m_1 a_1 \quad \text{and} \quad \sum F_y = n - m_1 g = 0$$

The equation for the $y$-component gives $n = m_1 g$. Substitute this value for $n$ and $f_k = \mu_k n$ into the equation for the $x$-component:

$$(1) \quad T - \mu_k m_1 g = m_1 a_1$$

Apply Newton's second law to the ball, recalling that $a_2 = -a_1$:

$$\sum F_y = T - m_2 g = m_2 a_2 = -m_2 a_1$$

$$(2) \quad T - m_2 g = -m_2 a_1$$

Subtract Equation (2) from Equation (1), eliminating $T$ and leaving an equation that can be solved for $a_1$:

$$m_2 g - \mu_k m_1 g = (m_1 + m_2) a_1$$

$$a_1 = \frac{m_2 g - \mu_k m_1 g}{m_1 + m_2}$$

Substitute the given values to obtain the acceleration:

$$a_1 = \frac{(7.00 \text{ kg})(9.80 \text{ m/s}^2) - (0.300)(4.00 \text{ kg})(9.80 \text{ m/s}^2)}{(4.00 \text{ kg} + 7.00 \text{ kg})}$$

$$= 5.17 \text{ m/s}^2$$

Substitute the value for $a_1$ into Equation (1) to find the tension $T$:

$$T = 32.4 \text{ N}$$

(b) Find the acceleration using the system approach, where the system consists of the two blocks.

Apply Newton's second law to the system and solve for $a$:

$$(m_1 + m_2) a = m_2 g - \mu_k n = m_2 g - \mu_k m_1 g$$

$$a = \frac{m_2 g - \mu_k m_1 g}{m_1 + m_2}$$

**REMARKS** Although the system approach appears quick and easy, it can be applied only in special cases and can't give any information about the internal forces, such as the tension. To find the tension, you must consider the free-body diagram of one of the blocks separately as was done in part **(a)** of Example 4.14.

**QUESTION 4.14** If mass $m_2$ is increased, does the acceleration of the system increase, decrease, or remain the same? Does the tension increase, decrease, or remain the same?

**EXERCISE 4.14** What if an additional mass is attached to the ball in Example 4.14? How large must this mass be to increase the downward acceleration by 50%? Why isn't it possible to add enough mass to double the acceleration?

**ANSWER** 14.0 kg. Doubling the acceleration to 10.3 m/s² isn't possible simply by suspending more mass because all objects, regardless of their mass, fall freely at 9.8 m/s² near Earth's surface.

---

## EXAMPLE 4.15 TWO BLOCKS AND A CORD

**GOAL** Apply Newton's second law and static friction to a two-body system.

**PROBLEM** A block of mass $m = 5.00$ kg rides on top of a second block of mass $M = 10.0$ kg. A person attaches a string to the bottom block and pulls the system horizontally across a frictionless surface, as in Figure 4.40a. Friction between the two blocks keeps the 5.00-kg block from slipping off. If the coefficient of static friction is 0.350, **(a)** what maximum force can be exerted by the string on the 10.0-kg block without causing the 5.00-kg block to slip? **(b)** Use the system approach to calculate the acceleration.

**STRATEGY** Draw a free-body diagram for each block. The static friction force causes the top block to move horizontally, and the maximum such force corresponds to $f_s = \mu_s n$. That same static friction retards the motion of the bottom block. As long as the top block isn't slipping, the acceleration of both blocks is the same. Write Newton's second law for each block, and eliminate the acceleration $a$ by substitution, solving for the tension $T$. Once the tension is known, use the system approach to calculate the acceleration.

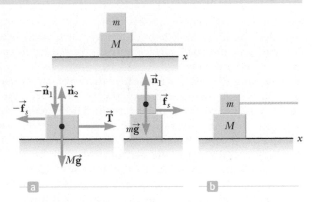

**Figure 4.40** (a) (Example 4.15) (b) (Exercise 4.15)

---

### SOLUTION

**(a)** Find the maximum force that can be exerted by the string.

Write the two components of Newton's second law for the top block:

$$x\text{-component:} \quad ma = \mu_s n_1$$
$$y\text{-component:} \quad 0 = n_1 - mg$$

Solve the $y$-component for $n_1$, substitute the result into the $x$-component, and then solve for $a$:

$$n_1 = mg \quad \rightarrow \quad ma = \mu_s mg \quad \rightarrow \quad a = \mu_s g$$

Write the $x$-component of Newton's second law for the bottom block:

$$(1) \quad Ma = -\mu_s mg + T$$

Substitute the expression for $a = \mu_s g$ into Equation (1) and solve for the tension $T$:

$$M\mu_s g = T - \mu_s mg \quad \rightarrow \quad T = (m + M)\mu_s g$$

Now evaluate to get the answer:

$$T = (5.00 \text{ kg} + 10.0 \text{ kg})(0.350)(9.80 \text{ m/s}^2) = \boxed{51.5 \text{ N}}$$

**(b)** Use the system approach to calculate the acceleration.

Write the second law for the $x$-component of the force on the system:

$$(m + M)a = T$$

Solve for the acceleration and substitute values:

$$a = \frac{T}{m + M} = \frac{51.5 \text{ N}}{5.00 \text{ kg} + 10.0 \text{ kg}} = \boxed{3.43 \text{ m/s}^2}$$

*(Continued)*

**REMARKS** Notice that the $y$-component for the 10.0-kg block wasn't needed because there was no friction between that block and the underlying surface. It's also interesting to note that the top block was accelerated by the force of static friction. The system acceleration could also have been calculated with $a = \mu_s g$. Does the result agree with the answer found by the system approach?

**QUESTION 4.15** What would happen if the tension force exceeded 51.5 N?

**EXERCISE 4.15** Suppose instead the string is attached to the top block in Example 4.15 (see Fig. 4.40b). Find the maximum force that can be exerted by the string on the block without causing the top block to slip.

**ANSWER** 25.7 N

---

## APPLYING PHYSICS 4.1    CARS AND FRICTION

Forces of friction are important in the analysis of the motion of cars and other wheeled vehicles. How do such forces both help and hinder the motion of a car?

**EXPLANATION** There are several types of friction forces to consider, the main ones being the force of friction between the tires and the road surface and the retarding force produced by air resistance.

Assuming the car is a four-wheel-drive vehicle of mass $m$, as each wheel turns to propel the car forward, the tire exerts a rearward force on the road. The reaction to this rearward force is a forward force $\vec{\mathbf{f}}$ exerted by the road on the tire (Fig. 4.41). If we assume the same forward force $\vec{\mathbf{f}}$ is exerted on each tire, the net forward force on the car is $4\vec{\mathbf{f}}$, and the car's acceleration is therefore $\vec{\mathbf{a}} = 4\vec{\mathbf{f}}/m$.

The friction between the moving car's wheels and the road is normally static friction, unless the car is skidding.

When the car is in motion, we must also consider the force of air resistance, $\vec{\mathbf{R}}$, which acts in the direction opposite the velocity of the car. The net force exerted on the car is therefore $4\vec{\mathbf{f}} - \vec{\mathbf{R}}$, so the car's acceleration is

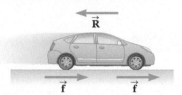

**Figure 4.41** (Applying Physics 4.1) The horizontal forces acting on the car are the *forward* forces $\vec{\mathbf{f}}$ exerted by the road on each tire and the force of air resistance $\vec{\mathbf{R}}$, which acts *opposite* the car's velocity. (The car's tires exert a rearward force on the road, not shown in the diagram.)

$\vec{\mathbf{a}} = (4\vec{\mathbf{f}} - \vec{\mathbf{R}})/m$. At normal driving speeds, the magnitude of $\vec{\mathbf{R}}$ is proportional to the first power of the speed, $R = bv$, where $b$ is a constant, so the force of air resistance increases with increasing speed. When $R$ is equal to $4f$, the acceleration is zero and the car moves at a constant speed. To minimize this resistive force, race cars often have very low profiles and streamlined contours. ∎

---

## APPLYING PHYSICS 4.2    AIR DRAG

Air resistance isn't always undesirable. What are some applications that depend on it?

**EXPLANATION** Consider a skydiver plunging through the air, as in Figure 4.42. Despite falling from a height of several thousand meters, she never exceeds a speed of around 120 miles per hour. This is because, aside from the downward force of gravity $m\vec{\mathbf{g}}$, there is also an upward force of air resistance, $\vec{\mathbf{R}}$. Before she reaches a final constant speed, the magnitude of $\vec{\mathbf{R}}$ is less than her weight. As her downward speed increases, the force of air resistance increases. The vector sum of the force of gravity and the force of air resistance gives a total force that decreases with time, so her acceleration decreases. Once the two forces balance each other, the net force is zero, so the acceleration is zero, and she reaches a **terminal speed**.

Terminal speed is generally still high enough to be fatal on impact, although there have been amazing stories of survival. In one case, a man fell flat on his back in a freshly plowed field and survived. (He did, however, break virtually every bone in his body.) In another case, a flight attendant survived a fall from thirty thousand feet into a snowbank. In

**Figure 4.42** (Applying Physics 4.2)

neither case would the person have had any chance of surviving without the effects of air drag.

Parachutes and paragliders create a much larger drag force due to their large area and can reduce the terminal speed to a few meters per second. Some sports enthusiasts have even developed special suits with wings, allowing a long glide to the ground. In each case, a larger cross-sectional area intercepts more air, creating greater air drag, so the terminal speed is lower.

Air drag is also important in space travel. Without it, returning to Earth would require a considerable amount of fuel. Air drag helps slow capsules and spaceships, and aerocapture techniques have been proposed for trips to other planets. These techniques significantly reduce fuel requirements by using air drag to reduce the speed of the spacecraft. ∎

## SUMMARY

### 4.1 Forces

There are four known fundamental forces of nature: (1) the strong nuclear force between subatomic particles; (2) the electromagnetic forces between electric charges; (3) the weak nuclear forces, which arise in certain radioactive decay processes; and (4) the gravitational force between objects. These are collectively called field forces. Classical physics deals only with the gravitational and electromagnetic forces.

Forces such as friction or the force of a bat hitting a ball are called contact forces. On a more fundamental level, contact forces have an electromagnetic nature.

### 4.2 The Laws of Motion

**Newton's first law** states that an object moves at constant velocity unless acted on by a force.

The tendency for an object to maintain its original state of motion is called **inertia. Mass** is the physical quantity that measures the resistance of an object to changes in its velocity.

**Newton's second law** states that the acceleration of an object is directly proportional to the net force acting on it and inversely proportional to its mass (Fig. 4.43). The net force acting on an object equals the product of its mass and acceleration:

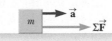

**Figure 4.43** A net force $\Sigma\vec{F}$ acting on a mass $m$ creates an acceleration proportional to the force and inversely proportional to the mass.

$$\sum \vec{F} = m\vec{a} \qquad [4.1]$$

Newton's universal law of gravitation is

$$F_g = G\frac{m_1 m_2}{r^2} \qquad [4.5]$$

(See Fig. 4.44.) The **weight** $w$ of an object is the magnitude of the force of gravity exerted on that object and is given by

$$w = mg \qquad [4.6]$$

where $g = F_g/m$ is the acceleration of gravity.

Solving problems with Newton's second law involves finding all the forces acting on a system and writing Equation 4.1 for the $x$-component and $y$-component separately. These two equations are then solved algebraically for the unknown quantities.

**Newton's third law** states that if two objects interact (Fig. 4.45), the force $\vec{F}_{12}$ exerted by object 1 on object 2 is

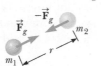

**Figure 4.44** The gravity force between any two objects is proportional to their masses and inversely proportional to the square of the distance between them.

equal in magnitude and opposite in direction to the force $\vec{F}_{21}$ exerted by object 2 on object 1:

$$\vec{F}_{12} = -\vec{F}_{21}$$

For example, the force of a hammer on a nail is equal in magnitude and opposite in direction to the force of the nail on the hammer (Fig. 4.45.) An isolated force can never occur in nature.

**Figure 4.45** Newton's third law in action: the hammer drives the nail forward into the wall, and the nail slows the head of the hammer down to rest with an equal and opposite force.

### 4.3 The Normal and Kinetic Friction Forces

The normal force is the response of a surface to contact with a given object. There are four common physical contexts and the normal force $n$ for those cases are:

**Case 1:** At rest on a level surface:

$$n = mg \qquad [4.9]$$

**Figure 4.46** The normal force is equal in magnitude and opposite in direction to the gravity force.

**Case 2:** Applied force at an angle:

$$n = mg - F_{app}\sin\theta \qquad [4.10]$$

**Figure 4.47** The normal force plus the magnitude of the $y$-component of the applied force is equal in magnitude to the gravity force.

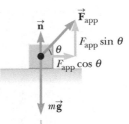

**Case 3:** Level surface, under acceleration:

$$n = ma_y + mg \qquad [4.11]$$

**Figure 4.48** The normal force provides the acceleration and compensates for the gravity force.

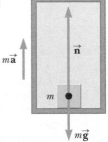

**Case 4:** Sloped surface:

$$n = mg\cos\theta \qquad [4.12]$$

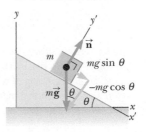

**Figure 4.49** The normal force decreases as the angle of the slope increases.

In Case 4, it's also useful to know that the component of the gravitational force acting down the slope is

$$F_{x',\text{grav}} = mg\sin\theta \qquad [4.13]$$

The magnitude of the **kinetic friction force** $f_k$ acting on an object moving on a surface is given by

$$f_k = \mu_k n \qquad [4.14]$$

## 4.4 Static Friction Forces

The magnitude of the **static friction force** $f_s$ acting on an object at rest satisfies the inequality

$$0 \le f_s \le f_{s,\max} = \mu_s n \qquad [4.15]$$

where $n$ is the normal force and $\mu_s$ is the coefficient of the *maximum* static friction force between the object and the surface.

Notice that only the *maximum* static friction force, $f_{s,\max}$, involves the use the static friction coefficient, $\mu_s$. In general, the static friction force $f_s$ is found by solving for it in the second law of motion for equilibrium.

## 4.5 Tension Forces

Tension forces are conveyed by the use of strings, cords, or cables. Finding their values involves using the second law for equilibrium and solving for the tensions. An applied force can also create a tension force. Unlike problems involving the normal force, there are numerous contexts. Some examples follow, the object considered a point particle:

**Case 1:** Vertical tension on a static object:

$$T = mg \qquad [4.16]$$

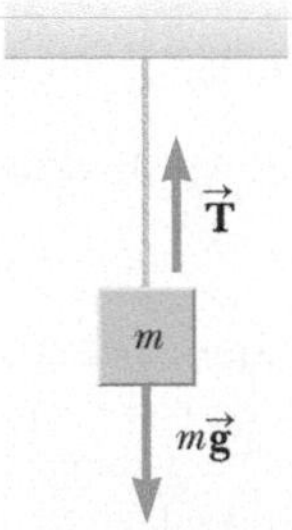

**Figure 4.50** The tension force equals the gravity force acting on the object.

**Case 2:** Vertical tension forces in an accelerating object:

$$T = m(a_y + g) \qquad [4.17]$$

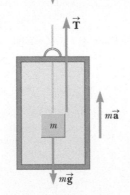

**Figure 4.51** The tension force must provide the acceleration as well as compensate for the gravity force.

**Case 3:** Two tension forces at symmetric angles:

$$T = \frac{mg}{2\cos\theta} \qquad [4.18]$$

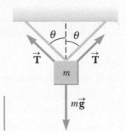

**Figure 4.52** The $y$-components of each tension force equal the gravity force.

When the angles are not equal, it's necessary to develop two equations and two unknowns using Newton's second law of motion for equilibrium.

## 4.6 Applications of Newton's Laws

An **object in equilibrium** has no net external force acting on it, and the second law, in component form, implies that $\Sigma F_x = 0$ and $\Sigma F_y = 0$ for such an object. These two equations are useful for solving problems in statics, in which the object is at rest or moving at constant velocity.

An object under acceleration requires the same two equations, but with the acceleration terms included: $\Sigma F_x = ma_x$ and $\Sigma F_y = ma_y$. When the acceleration is constant, the equations of kinematics can supplement Newton's second law.

## 4.7 Two-Body Problems

If two bodies are interacting with each other, the second law of motion must be developed for each individual body and simultaneously solve the resulting equations. If that interaction involves two objects connected by an inextensible string, so that the bodies have a common acceleration, it's possible to use the system approach. The second law of motion for the system approach is:

$$\sum F_{\text{ext}} = \left(\sum m_i\right) a_{\text{sys}} \qquad [4.21]$$

The system approach involves only forces external to the system, not internal forces between the bodies constituting the system. Internal forces must be found by applying the second law to individual bodies in the system.

## CONCEPTUAL QUESTIONS

1. A passenger sitting in the rear of a bus claims that she was injured as the driver slammed on the brakes, causing a suitcase to come flying toward her from the front of the bus. If you were the judge in this case, what disposition would you make? Explain.

2. A space explorer is moving through space far from any planet or star. He notices a large rock, taken as a specimen from an alien planet, floating around the cabin of the ship. Should he push it gently, or should he kick it toward the storage compartment? Explain.

3. (a) If gold were sold by weight, would you rather buy it in Denver or in Death Valley? (b) If it were sold by mass, in which of the two locations would you prefer to buy it? Why?

4. If you push on a heavy box that is at rest, you must exert some force to start its motion. Once the box is sliding, why does a smaller force maintain its motion?

5. A ball is held in a person's hand. (a) Identify all the external forces acting on the ball and the reaction to each. (b) If the ball is dropped, what force is exerted on it while it is falling? Identify the reaction force in this case. (Neglect air resistance.)

6. A weight lifter stands on a bathroom scale. (a) As she pumps a barbell up and down, what happens to the reading on the scale? (b) Suppose she is strong enough to actually *throw* the barbell upward. How does the reading on the scale vary now?

7. (a) What force causes an automobile to move? (b) A propeller-driven airplane? (c) A rowboat?

8. If only one force acts on an object, can it be in equilibrium? Explain.

9. In the motion picture *It Happened One Night* (Columbia Pictures, 1934), Clark Gable is standing inside a stationary bus in front of Claudette Colbert, who is seated. The bus suddenly starts moving forward and Clark falls into Claudette's lap. Why did this happen?

10. Analyze the motion of a rock dropped in water in terms of its speed and acceleration as it falls. Assume a resistive force is acting on the rock that increases as the velocity of the rock increases.

11. Identify the action–reaction pairs in the following situations: (a) a man takes a step, (b) a snowball hits a girl in the back, (c) a baseball player catches a ball, (d) a gust of wind strikes a window.

12. Draw a free-body diagram for each of the following objects: (a) a projectile in motion in the presence of air resistance, (b) a rocket leaving the launch pad with its engines operating, and (c) an athlete running along a horizontal track.

13. In a tug-of-war between two athletes, each pulls on the rope with a force of 200 N. What is the tension in the rope? If the rope doesn't move, what horizontal force does each athlete exert against the ground?

14. Suppose you are driving a car at a high speed. Why should you avoid slamming on your brakes when you want to stop in the shortest possible distance? (Newer cars have antilock brakes that avoid this problem.)

15. As a block slides down a frictionless incline, which of the following statements is true? (a) Both its speed and acceleration increase. (b) Its speed and acceleration remain constant. (c) Its speed increases and its acceleration remains constant. (d) Both its speed and acceleration decrease. (e) Its speed increases and its acceleration decreases.

16. A crate remains stationary after it has been placed on a ramp inclined at an angle with the horizontal. Which of the following statements must be true about the magnitude of the frictional force that acts on the crate? (a) It is larger than the weight of the crate. (b) It is at least equal to the weight of the crate. (c) It is equal to $\mu_s n$. (d) It is greater than the component of the gravitational force acting down the ramp. (e) It is equal to the component of the gravitational force acting down the ramp.

17. In Figure 4.4, a locomotive has broken through the wall of a train station. During the collision, what can be said about the force exerted by the locomotive on the wall? (a) The force exerted by the locomotive on the wall was larger than the force the wall could exert on the locomotive. (b) The force exerted by the locomotive on the wall was the same in magnitude as the force exerted by the wall on the locomotive. (c) The force exerted by the locomotive on the wall was less than the force exerted by the wall on the locomotive. (d) The wall cannot be said to "exert" a force; after all, it broke.

18. If an object is in equilibrium, which of the following statements is not true? (a) The speed of the object remains constant. (b) The acceleration of the object is zero. (c) The net force acting on the object is zero. (d) The object must be at rest. (e) The velocity is constant.

19. A truck loaded with sand accelerates along a highway. The driving force on the truck remains constant. What happens to the acceleration of the truck as its trailer leaks sand at a constant rate through a hole in its bottom? (a) It decreases at a steady rate. (b) It increases at a steady rate. (c) It increases and then decreases. (d) It decreases and then increases. (e) It remains constant.

20. A large crate of mass $m$ is placed on the back of a truck but not tied down. As the truck accelerates forward with an acceleration $a$, the crate remains at rest relative to the truck. What force causes the crate to accelerate forward? (a) the normal force (b) the force of gravity (c) the force of friction between the crate and the floor of the truck (d) the "$ma$" force (e) none of these

21. Which of the following statements are true? (a) An astronaut's weight is the same on the Moon as on Earth. (b) An astronaut's mass is the same on the International Space Station as it is on Earth. (c) Earth's gravity has no effect on astronauts inside the International Space Station. (d) An astronaut's mass is greater on Earth than on the Moon. (e) None of these statements are true.

22. A woman is standing on the Earth. In terms of magnitude, is her gravitational force on the Earth (a) equal to, (b) less than, or (c) greater than the Earth's gravitational force on her?

23. An exoplanet has twice the mass and half the radius of the Earth. Find the acceleration due to gravity on its surface, in terms of g, the acceleration of gravity at Earth's surface. (a) g (b) 0.5g (c) 2g (d) 4g (e) 8g.

24. Choose the best answer. A car traveling at constant speed has a net force of zero acting on it. (a) True (b) False (c) The answer depends on the motion.

## PROBLEMS

*See the Preface for an explanation of the icons used in this problems set.*

### 4.1 Forces

### 4.2 The Laws of Motion

1. The heaviest invertebrate is the giant squid, which is estimated to have a weight of about 2 tons spread out over its length of 70 feet. What is its weight in newtons?

2. A football punter accelerates a football from rest to a speed of 10 m/s during the time in which his toe is in contact with the ball (about 0.20 s). If the football has a mass of 0.50 kg, what average force does the punter exert on the ball?

3. A 6.0-kg object undergoes an acceleration of 2.0 m/s². (a) What is the magnitude of the resultant force acting on it? (b) If this same force is applied to a 4.0-kg object, what acceleration is produced?

4. **Q|C** One or more external forces are exerted on each object enclosed in a dashed box shown in Figure 4.2. Identify the reaction to each of these forces.

5. A bag of sugar weighs 5.00 lb on Earth. What would it weigh in newtons on the Moon, where the free-fall acceleration is one-sixth that on Earth? Repeat for Jupiter, where $g$ is 2.64 times that on Earth. Find the mass of the bag of sugar in kilograms at each of the three locations.

6. A freight train has a mass of $1.5 \times 10^7$ kg. If the locomotive can exert a constant pull of $7.5 \times 10^5$ N, how long does it take to increase the speed of the train from rest to 80 km/h?

7. Four forces act on an object, given by $\vec{A}$ = 40.0 N east, $\vec{B}$ = 50.0 north, $\vec{C}$ = 70.0 N west, and $\vec{D}$ = 90.0 N south. (a) What is the magnitude of the net force on the object? (b) What is the direction of the force?

8. **Q|C** Consider a solid metal sphere (S) a few centimeters in diameter and a feather (F). For each quantity in the list that follows, indicate whether the quantity is the same, greater, or lesser in the case of S or in that of F. Explain in each case why you gave the answer you did. Here is the list: (a) the gravitational force, (b) the time it will take to fall a given distance in air, (c) the time it will take to fall a given distance in vacuum, (d) the total force on the object when falling in vacuum.

9. **BIO** As a fish jumps vertically out of the water, assume that only two significant forces act on it: an upward force $F$ exerted by the tail fin and the downward force due to gravity. A record Chinook salmon has a length of 1.50 m and a mass of 61.0 kg. If this fish is moving upward at 3.00 m/s as its head first breaks the surface and has an upward speed of 6.00 m/s after two-thirds of its length has left the surface, assume constant acceleration and determine (a) the salmon's acceleration and (b) the magnitude of the force $F$ during this interval.

10. **V** A 5.0-g bullet leaves the muzzle of a rifle with a speed of 320 m/s. What force (assumed constant) is exerted on the bullet while it is traveling down the 0.82-m-long barrel of the rifle?

11. A boat moves through the water with two forces acting on it. One is a $2.00 \times 10^3$-N forward push by the water on the propeller, and the other is a $1.80 \times 10^3$-N resistive force due to the water around the bow. (a) What is the acceleration of the $1.00 \times 10^3$-kg boat? (b) If it starts from rest, how far will the boat move in 10.0 s? (c) What will its velocity be at the end of that time?

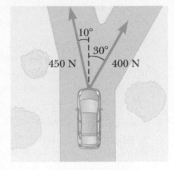

**Figure P4.12**

12. Two forces are applied to a car in an effort to move it, as shown in Figure P4.12. (a) What is the resultant vector of these two forces? (b) If the car has a mass of 3 000 kg, what acceleration does it have? Ignore friction.

13. A 970.-kg car starts from rest on a horizontal roadway and accelerates eastward for 5.00 s when it reaches a speed of 25.0 m/s. What is the average force exerted on the car during this time?

14. **S** An object of mass $m$ is dropped from the roof of a building of height $h$. While the object is falling, a wind blowing parallel to the face of the building exerts a constant horizontal force $F$ on the object. (a) How long does it take the object to strike the ground? Express the time $t$ in terms of $g$ and $h$. (b) Find an expression in terms of $m$ and $F$ for the acceleration $a_x$ of the object in the horizontal direction (taken as the positive $x$-direction). (c) How far is the object displaced horizontally before hitting the ground? Answer in terms of $m$, $g$, $F$, and $h$. (d) Find the magnitude of the object's acceleration while it is falling, using the variables $F$, $m$, and $g$.

15. After falling from rest from a height of 30.0 m, a 0.500-kg ball rebounds upward, reaching a height of 20.0 m. If the contact between ball and ground lasted 2.00 ms, what average force was exerted on the ball?

16. **T** The force exerted by the wind on the sails of a sailboat is 390 N north. The water exerts a force of 180 N east. If the boat (including its crew) has a mass of 270 kg, what are the magnitude and direction of its acceleration?

17. A force of 30.0 N is applied in the positive $x$-direction to a block of mass 8.00 kg, at rest on a frictionless surface. (a) What is the block's acceleration? (b) How fast is it going after 6.00 s?

18. What would be the acceleration of gravity at the surface of a world with twice Earth's mass and twice its radius?

### 4.3 The Normal and Kinetic Friction Forces

19. Calculate the magnitude of the normal force on a 15.0-kg block in the following circumstances: (a) The block is resting on a level surface. (b) The block is resting on a surface tilted up at a 30.0° angle with respect to the horizontal. (c) The block is resting on the floor of an elevator that is accelerating upwards at 3.00 m/s². (d) The block is on a level surface and a force of 125 N is exerted on it at an angle of 30.0° above the horizontal.

20. A horizontal force of 95.0 N is applied to a 60.0-kg crate on a rough, level surface. If the crate accelerates at 1.20 m/s², what is the magnitude of the force of kinetic friction acting on the crate?

21. A car of mass 875 kg is traveling 30.0 m/s when the driver applies the brakes, which lock the wheels. The car skids for 5.60 s in the positive $x$-direction before coming to rest. (a) What is the car's acceleration? (b) What magnitude force acted on the car during this time? (c) How far did the car travel?

22. A student of mass 60.0 kg, starting at rest, slides down a slide 20.0 m long, tilted at an angle of 30.0° with respect to the horizontal. If the coefficient of kinetic friction between the student and the slide is 0.120, find (a) the force of kinetic friction, (b) the acceleration, and (c) the speed she is traveling when she reaches the bottom of the slide.

23. A $1.00 \times 10^3$-N crate is being pushed across a level floor at a constant speed by a force $\vec{F}$ of $3.00 \times 10^2$ N at an angle of 20.0° below the horizontal, as shown in Figure P4.23a. (a) What is the coefficient of kinetic friction between the crate and the floor? (b) If the $3.00 \times 10^2$-N force is instead pulling the block at an angle of 20.0° above the horizontal, as shown in Figure P4.23b, what will be the acceleration of the crate? Assume that the coefficient of friction is the same as that found in part (a).

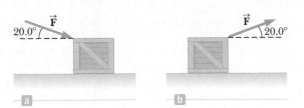

**Figure P4.23**

24. A block of mass $m = 5.8$ kg is pulled up a $\theta = 25°$ incline as in Figure P4.24 with a force of magnitude $F = 32$ N. (a) Find the acceleration of the block if the incline is frictionless. (b) Find the acceleration of the block if the coefficient of kinetic friction between the block and incline is 0.10.

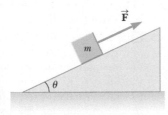

**Figure P4.24**

25. A rocket takes off from Earth's surface, accelerating straight up at 72.0 m/s². Calculate the normal force acting on an astronaut of mass 85.0 kg, including his space suit.

## 4.4 Static Friction Forces

26. A man exerts a horizontal force of 125 N on a crate with a mass of 30.0 kg. (a) If the crate doesn't move, what's the magnitude of the static friction force? (b) What is the minimum possible value of the coefficient of static friction between the crate and the floor?

27. A horse is harnessed to a sled having a mass of 236 kg, including supplies. The horse must exert a force exceeding 1 240 N at an angle of 35.0° in order to get the sled moving. Treat the sled as a point particle. (a) Calculate the normal force on the sled when the magnitude of the applied force is 1 240 N. (b) Find the coefficient of static friction between the sled and the ground beneath it. (c) Find the static friction force when the horse is exerting a force of $6.20 \times 10^2$ N on the sled at the same angle.

28. A block of mass 55.0 kg rests on a slope having an angle of elevation of 25.0°. If pushing downhill on the block with a force just exceeding 187 N and parallel to the slope is sufficient to cause the block to start moving, find the coefficient of static friction.

29. A dockworker loading crates on a ship finds that a 20.0-kg crate, initially at rest on a horizontal surface, requires a 75.0-N horizontal force to set it in motion. However, after the crate is in motion, a horizontal force of 60.0 N is required to keep it moving with a constant speed. Find the coefficients of static and kinetic friction between crate and floor.

30. Suppose the coefficient of static friction between a quarter and the back wall of a rocket car is 0.330. At what minimum rate would the car have to accelerate so that a quarter placed on the back wall would remain in place?

31. ⬛V⬛ The coefficient of static friction between the 3.00-kg crate and the 35.0° incline of Figure P4.31 is 0.300. What minimum force $\vec{F}$ must be applied to the crate perpendicular to the incline to prevent the crate from sliding down the incline?

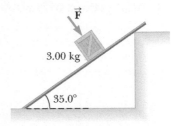

**Figure P4.31**

## 4.5 Tension Forces

32. Two identical strings making an angle of $\theta = 30.0°$ with respect to the vertical support a block of mass $m = 15.0$ kg (Fig. P4.32). What is the tension in each of the strings?

33. A 75-kg man standing on a scale in an elevator notes that as the elevator rises, the scale reads 825 N. What is the acceleration of the elevator?

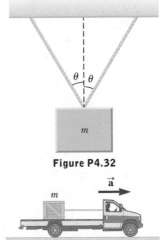

**Figure P4.32**

34. A crate of mass $m = 32$ kg rides on the bed of a truck attached by a cord to the back of the cab as in Figure P4.34. The cord can withstand a maximum tension of 68 N before breaking. Neglecting friction between the crate and truck bed, find the maximum acceleration the truck can have before the cord breaks.

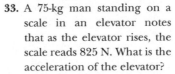

**Figure P4.34**

35. ⬛QC⬛ (a) Find the tension in each cable supporting the $6.00 \times 10^2$-N cat burglar in Figure P4.35. (b) Suppose the horizontal cable were reattached higher up on the wall. Would the tension in the other cables increase, decrease, or stay the same? Why?

36. The distance between two telephone poles is 50.0 m. When a 1.00-kg bird lands on the telephone wire midway between the poles, the wire sags 0.200 m. Draw a free-body diagram of the bird. How much tension does the bird produce in the wire? Ignore the weight of the wire.

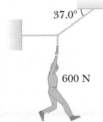

**Figure P4.35**

37. ⬛QC⬛ (a) An elevator of mass $m$ moving upward has two forces acting on it: the upward force of tension in the cable and the downward force due to gravity. When the elevator is

accelerating upward, which is greater, $T$ or $w$? (b) When the elevator is moving at a constant velocity upward, which is greater, $T$ or $w$? (c) When the elevator is moving upward, but the acceleration is downward, which is greater, $T$ or $w$? (d) Let the elevator have a mass of 1 500 kg and an upward acceleration of 2.5 m/s². Find $T$. Is your answer consistent with the answer to part (a)? (e) The elevator of part (d) now moves with a constant upward velocity of 10 m/s. Find $T$. Is your answer consistent with your answer to part (b)? (f) Having initially moved upward with a constant velocity, the elevator begins to accelerate downward at 1.50 m/s². Find $T$. Is your answer consistent with your answer to part (c)?

## 4.6 Applications of Newton's Laws

38. **BIO** A certain orthodontist uses a wire brace to align a patient's crooked tooth as in Figure P4.38. The tension in the wire is adjusted to have a magnitude of 18.0 N. Find the magnitude of the net force exerted by the wire on the crooked tooth.

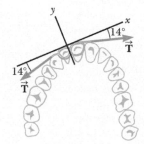

**Figure P4.38**

39. A 150-N bird feeder is supported by three cables as shown in Figure P4.39. Find the tension in each cable.

40. **BIO** The leg and cast in Figure P4.40 weigh 220 N ($w_1$). Determine the weight $w_2$ and the angle $\alpha$ needed so that no force is exerted on the hip joint by the leg plus the cast.

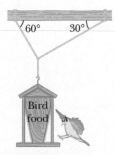

**Figure P4.39**

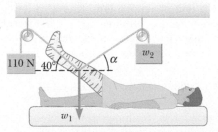

**Figure P4.40**

41. **V** A 276-kg glider is being pulled by a 1 950-kg jet along a horizontal runway with an acceleration of $\vec{a}$ = 2.20 m/s² to the right as in Figure P4.41. Find (a) the thrust provided by the jet's engines and (b) the magnitude of the tension in the cable connecting the jet and glider.

**Figure P4.41**

42. **V** A crate of mass 45.0 kg is being transported on the flatbed of a pickup truck. The coefficient of static friction between the crate and the truck's flatbed is 0.350, and the coefficient of kinetic friction is 0.320. (a) The truck accelerates forward on level ground. What is the maximum acceleration the truck can have so that the crate does not slide relative to the truck's flatbed? (b) The truck barely exceeds this acceleration and then moves with constant acceleration, with the crate sliding along its bed. What is the acceleration of the crate relative to the ground?

43. Consider a large truck carrying a heavy load, such as steel beams. A significant hazard for the driver is that the load may slide forward, crushing the cab, if the truck stops suddenly in an accident or even in braking. Assume, for example, a 10 000-kg load sits on the flatbed of a 20 000-kg truck moving at 12.0 m/s. Assume the load is not tied down to the truck and has a coefficient of static friction of 0.500 with the truck bed. (a) Calculate the minimum stopping distance for which the load will not slide forward relative to the truck. (b) Is any piece of data unnecessary for the solution?

44. A student decides to move a box of books into her dormitory room by pulling on a rope attached to the box. She pulls with a force of 80.0 N at an angle of 25.0° above the horizontal. The box has a mass of 25.0 kg, and the coefficient of kinetic friction between box and floor is 0.300. (a) Find the acceleration of the box. (b) The student now starts moving the box up a 10.0° incline, keeping her 80.0 N force directed at 25.0° above the line of the incline. If the coefficient of friction is unchanged, what is the new acceleration of the box?

45. **Q|C** An object falling under the pull of gravity is acted upon by a frictional force of air resistance. The magnitude of this force is approximately proportional to the speed of the object, which can be written as $f = bv$. Assume $b$ = 15 kg/s and $m$ = 50 kg. (a) What is the terminal speed the object reaches while falling? (b) Does your answer to part (a) depend on the initial speed of the object? Explain.

46. A 3.00-kg block starts from rest at the top of a 30.0° incline and slides 2.00 m down the incline in 1.50 s. Find (a) the acceleration of the block, (b) the coefficient of kinetic friction between the block and the incline, (c) the frictional force acting on the block, and (d) the speed of the block after it has slid 2.00 m.

47. To meet a U.S. Postal Service requirement, employees' footwear must have a coefficient of static friction of 0.500 or more on a specified tile surface. A typical athletic shoe has a coefficient of 0.800. In an emergency, what is the minimum time interval in which a person starting from rest can move 3.00 m on the tile surface if she is wearing (a) footwear meeting the Postal Service minimum and (b) a typical athletic shoe?

48. A block of mass 12.0 kg is sliding at an initial velocity of 8.00 m/s in the positive $x$-direction. The surface has a coefficient of kinetic friction of 0.300. (a) What is the force of kinetic friction acting on the block? (b) What is the block's acceleration? (c) How far will it slide before coming to rest?

49. **V** **BIO** The person in Figure P4.49 weighs 170. lb. Each crutch makes an angle of 22.0° with the vertical (as seen from the front). Half of the person's weight is supported by the crutches, the other half by the vertical forces exerted by the ground

on his feet. Assuming he is at rest and the force exerted by the ground on the crutches acts along the crutches, determine (a) the smallest possible coefficient of friction between crutches and ground and (b) the magnitude of the compression force supported by each crutch.

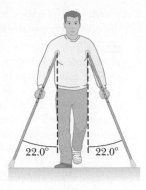

**Figure P4.49**

50. A car is traveling at 50.0 km/h on a flat highway. (a) If the coefficient of friction between road and tires on a rainy day is 0.100, what is the minimum distance in which the car will stop? (b) What is the stopping distance when the surface is dry and the coefficient of friction is 0.600?

51. **V** A 5.0-kg bucket of water is raised from a well by a rope. If the upward acceleration of the bucket is 3.0 m/s², find the force exerted by the rope on the bucket.

52. **S** A hockey puck struck by a hockey stick is given an initial speed $v_0$ in the positive $x$-direction. The coefficient of kinetic friction between the ice and the puck is $\mu_k$. (a) Obtain an expression for the acceleration of the puck. (b) Use the result of part (a) to obtain an expression for the distance $d$ the puck slides. The answer should be in terms of the variables $v_0$, $\mu_k$, and $g$ only.

53. **V** **BIO** A setup similar to the one shown in Figure P4.53 is often used in hospitals to support and apply a traction force to an injured leg. (a) Determine the force of tension in the rope supporting the leg. (b) What is the traction force exerted on the leg? Assume the traction force is horizontal.

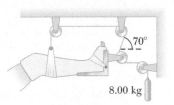

**Figure P4.53**

## 4.7 Two-Body Problems

54. An Atwood's machine (Fig. 4.38) consists of two masses: one of mass 3.00 kg and the other of mass 8.00 kg. When released from rest, what is the acceleration of the system?

55. A block of mass $m_1 = 16.0$ kg is on a frictionless table to the left of a second block of mass $m_2 = 24.0$ kg, attached by a horizontal string (Fig. P4.55). If a horizontal force of $1.20 \times 10^2$ N is exerted on the block $m_2$ in the positive $x$-direction, (a) use the system approach to find the acceleration of the two blocks. (b) What is the tension in the string connecting the blocks?

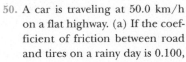

**Figure P4.55**

56. **S** Two blocks each of mass $m$ are fastened to the top of an elevator as in Figure P4.56. The elevator has an upward acceleration $a$. The strings have negligible mass. (a) Find the tensions $T_1$ and $T_2$ in the upper and lower strings in terms of $m$, $a$, and $g$. (b) Compare the

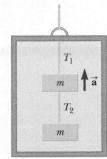

**Figure P4.56**
(Problems 56 and 71)

two tensions and determine which string would break first if $a$ is made sufficiently large. (c) What are the tensions if the cable supporting the elevator breaks?

57. **S** Two blocks of masses $m$ and $2m$ are held in equilibrium on a frictionless incline as in Figure P4.57. In terms of $m$ and $\theta$, find (a) the magnitude of the tension $T_1$ in the upper cord and (b) the magnitude of the tension $T_2$ in the lower cord connecting the two blocks.

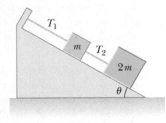

**Figure P4.57**

58. **V** The systems shown in Figure P4.58 are in equilibrium. If the spring scales are calibrated in newtons, what do they read? Ignore the masses of the pulleys and strings and assume the pulleys and the incline in Figure P4.58d are frictionless.

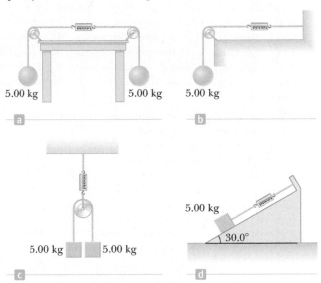

**Figure P4.58**

59. **T** Assume the three blocks portrayed in Figure P4.59 move on a frictionless surface and a 42-N force acts as shown on the 3.0-kg block. Determine (a) the acceleration given this system, (b) the tension in the cord connecting the 3.0-kg and the 1.0-kg blocks, and (c) the force exerted by the 1.0-kg block on the 2.0-kg block.

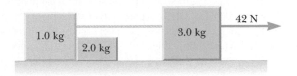

**Figure P4.59**

60. Two packing crates of masses 10.0 kg and 5.00 kg are connected by a light string that passes over a frictionless pulley as in Figure P4.60. The 5.00-kg crate lies on a smooth incline of angle 40.0°. Find (a) the acceleration of the 5.00-kg crate and (b) the tension in the string.

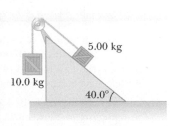

**Figure P4.60**

**61.** A $1.00 \times 10^3$ car is pulling a 300.-kg trailer. Together, the car and trailer have an acceleration of $2.15 \text{ m/s}^2$ in the positive $x$-direction. Neglecting frictional forces on the trailer, determine (a) the net force on the car, (b) the net force on the trailer, (c) the magnitude and direction of the force exerted by the trailer on the car, and (d) the resultant force exerted by the car on the road.

**62.** **GP** Two blocks of masses $m_1$ and $m_2$ ($m_1 > m_2$) are placed on a frictionless table in contact with each other. A horizontal force of magnitude $F$ is applied to the block of mass $m_1$ in

**Figure P4.62**

Figure P4.62. (a) If $P$ is the magnitude of the contact force between the blocks, draw the free-body diagrams for each block. (b) What is the net force on the system consisting of both blocks? (c) What is the net force acting on $m_1$? (d) What is the net force acting on $m_2$? (e) Write the $x$-component of Newton's second law for each block. (f) Solve the resulting system of two equations and two unknowns, expressing the acceleration $a$ and contact force $P$ in terms of the masses and force. (g) How would the answers change if the force had been applied to $m_2$ instead? (*Hint:* use symmetry; don't calculate!) Is the contact force larger, smaller, or the same in this case? Why?

**63.** **Q|C** In Figure P4.63, the light, taut, unstretchable cord B joins block 1 and the larger-mass block 2. Cord A exerts a force on block 1 to make it accelerate forward.

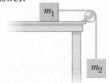

**Figure P4.63**

(a) How does the magnitude of the force exerted by cord A on block 1 compare with the magnitude of the force exerted by cord B on block 2? (b) How does the acceleration of block 1 compare with the acceleration of block 2? (c) Does cord B exert a force on block 1? Explain your answer.

**64.** **V** An object with mass $m_1 = 5.00$ kg rests on a frictionless horizontal table and is connected to a cable that passes over a pulley and is then fastened to a hanging object with mass $m_2 = 10.0$ kg, as shown in Figure P4.64. Find (a) the acceleration of each object and (b) the tension in the cable.

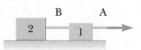

**Figure P4.64**
(Problems 64, 65, and 67)

**65.** **T** Objects with masses $m_1 = 10.0$ kg and $m_2 = 5.00$ kg are connected by a light string that passes over a frictionless pulley as in Figure P4.64. If, when the system starts from rest, $m_2$ falls 1.00 m in 1.20 s, determine the coefficient of kinetic friction between $m_1$ and the table.

**66.** Two objects with masses of 3.00 kg and 5.00 kg are connected by a light string that passes over a frictionless pulley, as in Figure P4.66. Determine (a) the tension in the string, (b) the acceleration of each object, and (c) the distance each object will move in the first second of motion if both objects start from rest.

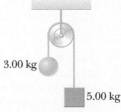

**Figure P4.66**

**67.** In Figure P4.64, $m_1 = 10.$ kg and $m_2 = 4.0$ kg. The coefficient of static friction between $m_1$ and the horizontal surface is 0.50, and the coefficient of kinetic friction is 0.30. (a) If the system is released from rest, what will its acceleration be? (b) If the

system is set in motion with $m_2$ moving downward, what will be the acceleration of the system?

**68.** **Q|C** **S** A block of mass $3m$ is placed on a frictionless horizontal surface, and a second block of mass $m$ is placed on top of the first block. The surfaces of the blocks are rough. A constant force of magnitude $F$ is

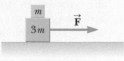

**Figure P4.68**

applied to the first block as in Figure P4.68. (a) Construct free-body diagrams for each block. (b) Identify the horizontal force that causes the block of mass $m$ to accelerate. (c) Assume that the upper block does not slip on the lower block, and find the acceleration of each block in terms of $m$ and $F$.

**69.** A 15.0-lb block rests on a horizontal floor. (a) What force does the floor exert on the block? (b) A rope is tied to the block and is run vertically over a pulley. The other end is attached to a free-hanging 10.0-lb object. What now is the force exerted by the floor on the 15.0-lb block? (c) If the 10.0-lb object in part (b) is replaced with a 20.0-lb object, what is the force exerted by the floor on the 15.0-lb block?

**70.** Objects of masses $m_1 = 4.00$ kg and $m_2 = 9.00$ kg are connected by a light string that passes over a frictionless pulley as in Figure P4.70. The object $m_1$ is held at rest on the floor, and $m_2$ rests on a fixed incline of $\theta = 40.0°$. The objects are released from rest, and $m_2$ slides 1.00 m down

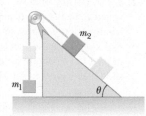

**Figure P4.70**

the incline in 4.00 s. Determine (a) the acceleration of each object, (b) the tension in the string, and (c) the coefficient of kinetic friction between $m_2$ and the incline.

**71.** **V** Two blocks each of mass $m = 3.50$ kg are fastened to the top of an elevator as in Figure P4.56. (a) If the elevator has an upward acceleration $a = 1.60 \text{ m/s}^2$, find the tensions $T_1$ and $T_2$ in the upper and lower strings. (b) If the strings can withstand a maximum tension of 85.0 N, what maximum acceleration can the elevator have before the upper string breaks?

## Additional Problems

**72.** As a protest against the umpire's calls, a baseball pitcher throws a ball straight up into the air at a speed of 20.0 m/s. In the process, he moves his hand through a distance of 1.50 m. If the ball has a mass of 0.150 kg, find the force he exerts on the ball to give it this upward speed.

**73.** **V** **Q|C** Three objects are connected on a table as shown in Figure P4.73. The coefficient of kinetic friction between the block of mass $m_2$ and the table is 0.350. The objects have masses

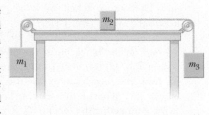

**Figure P4.73**

of $m_1 = 4.00$ kg, $m_2 = 1.00$ kg, and $m_3 = 2.00$ kg as shown, and the pulleys are frictionless. (a) Draw a diagram of the forces on each object. (b) Determine the acceleration of each object, including its direction. (c) Determine the tensions in the two cords. (d) If the tabletop were smooth, would the tensions increase, decrease, or remain the same? Explain.

74. (a) What is the minimum force of friction required to hold the system of Figure P4.74 in equilibrium? (b) What coefficient of static friction between the 100.-N block and the table ensures equilibrium? (c) If the coefficient of kinetic friction between the 100.-N block and the table is 0.250, what hanging weight should replace the 50.0-N weight to allow the system to move at a constant speed once it is set in motion?

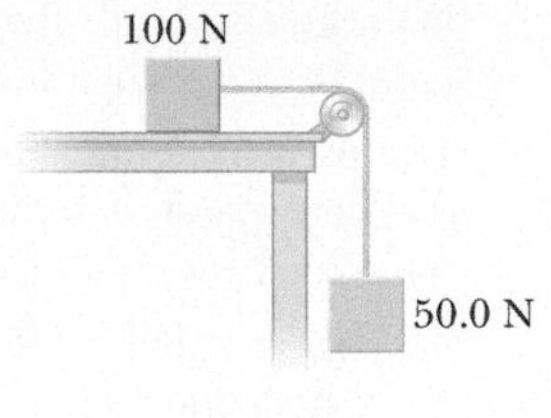

100 N

50.0 N

**Figure P4.74**

75. (a) What is the resultant force exerted by the two cables supporting the traffic light in Figure P4.75? (b) What is the weight of the light?

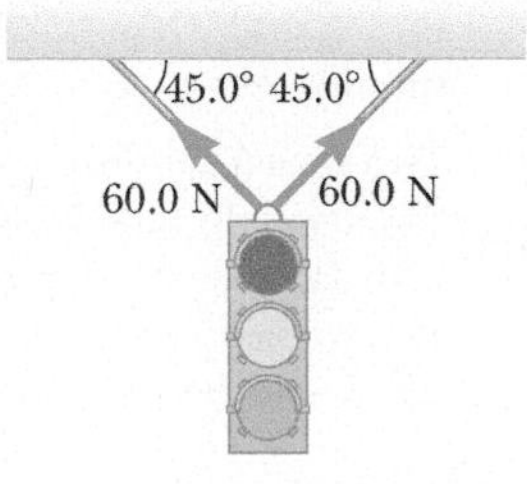

45.0° 45.0°

60.0 N  60.0 N

**Figure P4.75**

76. **V** A woman at an airport is towing her 20.0-kg suitcase at constant speed by pulling on a strap at an angle $\theta$ above the horizontal (Fig. 4.76). She pulls on the strap with a 35.0-N force, and the friction force on the suitcase is 20.0 N. (a) Draw a free-body diagram of the suitcase. (b) What angle does the strap make with the horizontal? (c) What is the magnitude of the normal force that the ground exerts on the suitcase?

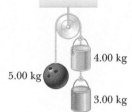

**Figure P4.76**

77. A boy coasts down a hill on a sled, reaching a level surface at the bottom with a speed of 7.00 m/s. If the coefficient of friction between the sled's runners and the snow is 0.050 0 and the boy and sled together weigh 600. N, how far does the sled travel on the level surface before coming to rest?

78. Three objects are connected by light strings as shown in Figure P4.78. The string connecting the 4.00-kg object and the 5.00-kg object passes over a light frictionless pulley. Determine (a) the acceleration of each object and (b) the tension in the two strings.

4.00 kg

5.00 kg

3.00 kg

**Figure P4.78**

79. **Q|C** A box rests on the back of a truck. The coefficient of static friction between the box and the bed of the truck is 0.300. (a) When the truck accelerates forward, what force accelerates the box? (b) Find the maximum acceleration the truck can have before the box slides.

80. A high diver of mass 70.0 kg steps off a board 10.0 m above the water and falls vertical to the water, starting from rest. If her downward motion is stopped 2.00 s after her feet first touch the water, what average upward force did the water exert on her?

81. **T** A frictionless plane is 10.0 m long and inclined at 35.0°. A sled starts at the bottom with an initial speed of 5.00 m/s up the incline. When the sled reaches the point at which it momentarily stops, a second sled is released from the top of the incline with an initial speed $v_i$. Both sleds reach the bottom of the incline at the same moment. (a) Determine the

distance that the first sled traveled up the incline. (b) Determine the initial speed of the second sled.

82. **V** **Q|C** *Measuring coefficients of friction* A coin is placed near one edge of a book lying on a table, and that edge of the book is lifted until the coin just slips down the incline as shown in Figure P4.82. The angle of the incline, $\theta_c$, called the critical angle, is measured. (a) Draw a free-body diagram for the coin when it is on the verge of slipping and identify all forces acting on it. Your free-body diagram should include a force of static friction acting up the incline. (b) Is the magnitude of the friction force equal to $\mu_s n$ for angles less than $\theta_c$? Explain. What can you definitely say about the magnitude of the friction force for any angle $\theta \le \theta_c$? (c) Show that the coefficient of static friction is given by $\mu_s = \tan \theta_c$. (d) Once the coin starts to slide down the incline, the angle can be adjusted to a new value $\theta_c' \le \theta_c$ such that the coin moves down the incline with constant speed. How does observation enable you to obtain the coefficient of kinetic friction?

Coin

**Figure P4.82**

83. A 2.00-kg aluminum block and a 6.00-kg copper block are connected by a light string over a frictionless pulley. The two blocks are allowed to move on a fixed steel block wedge (of angle $\theta = 30.0°$) as shown in Figure P4.83. Making use of Table 4.2, determine (a) the acceleration of the two blocks and (b) the tension in the string.

Aluminum

$m_1$

Copper

$m_2$

Steel

$\theta$

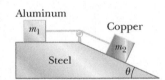

**Figure P4.83**

84. **T** On an airplane's takeoff, the combined action of the air around the engines and wings of an airplane exerts an 8 000-N force on the plane, directed upward at an angle of 65.0° above the horizontal. The plane rises with constant velocity in the vertical direction while continuing to accelerate in the horizontal direction. (a) What is the weight of the plane? (b) What is its horizontal acceleration?

85. Two boxes of fruit on a frictionless horizontal surface are connected by a light string as in Figure P4.85, where $m_1 = 10.0$ kg and $m_2 = 20.0$ kg. A force of 50.0 N is applied to the 20.0-kg box. (a) Determine the acceleration of each box and the tension in the string. (b) Repeat the problem for the case where the coefficient of kinetic friction between each box and the surface is 0.10.

$m_1$   $m_2$

$T$   Apples

Peaches   $\vec{F}$

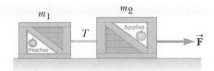

**Figure P4.85**

86. A sled weighing 60.0 N is pulled horizontally across snow so that the coefficient of kinetic friction between sled and snow is 0.100. A penguin weighing 70.0 N rides on

the sled, as in Figure P4.86. If the coefficient of static friction between penguin and sled is 0.700, find the maximum horizontal force that can be exerted on the sled before the penguin begins to slide off.

**Figure P4.86**

87. A car accelerates down a hill (Fig. P4.87), going from rest to 30.0 m/s in 6.00 s. During the acceleration, a toy ($m = 0.100$ kg) hangs by a string from the car's ceiling. The acceleration is such that the string remains perpendicular to the ceiling. Determine (a) the angle $\theta$ and (b) the tension in the string.

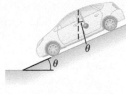

**Figure P4.87**

88. An inventive child wants to reach an apple in a tree without climbing the tree. Sitting in a chair connected to a rope that passes over a frictionless pulley (Fig. P4.88), the child pulls on the loose end of the rope with such a force that the spring scale reads 250 N. The child's true weight is 320 N, and the chair weighs 160 N. The child's feet are not touching the ground. (a) Show that

**Figure P4.88**

the acceleration of the system is *upward,* and find its magnitude. (b) Find the force the child exerts on the chair.

89. The parachute on a race car of weight 8 820 N opens at the end of a quarter-mile run when the car is traveling at 35.0 m/s. What total retarding force must be supplied by the parachute to stop the car in a distance of $1.00 \times 10^3$ m?

90. A fire helicopter carries a 620-kg bucket of water at the end of a 20.0-m-long cable. Flying back from a fire at a constant speed of 40.0 m/s, the cable makes an angle of 40.0° with respect to the vertical. Determine the force exerted by air resistance on the bucket.

91. The board sandwiched between two other boards in Figure P4.91 weighs 95.5 N. If the coefficient of friction between the boards is 0.663, what must be the magnitude of the compression forces (assumed to be horizontal) acting on both sides of the center board to keep it from slipping?

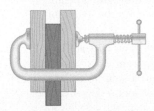

**Figure P4.91**

92. A 72-kg man stands on a spring scale in an elevator. Starting from rest, the elevator ascends, attaining its maximum speed of 1.2 m/s in 0.80 s. The elevator travels with this constant speed for 5.0 s, undergoes a uniform *negative* acceleration for 1.5 s, and then comes to rest. What does the spring scale register (a) before the elevator starts to move? (b) During the first 0.80 s of the elevator's ascent? (c) While the elevator is traveling at constant speed? (d) During the elevator's negative acceleration?

# Energy

*ENERGY* IS ONE OF THE MOST IMPORTANT CONCEPTS in the world of science. In everyday use energy is associated with the fuel needed for transportation and heating, with electricity for lights and appliances, and with the foods we consume. These associations, however, don't tell us what energy *is*, only what it *does*, and that producing it requires fuel. Our goal in Topic 5, therefore, is to develop a better understanding of energy and how to quantify it.

Energy is present in the Universe in a variety of forms, including mechanical, chemical, electromagnetic, and nuclear energy. Even the inert mass of everyday matter contains a very large amount of energy. Although energy can be transformed from one kind to another, all observations and experiments to date suggest that the total amount of energy in the Universe never changes. That's also true for an isolated system, which is a collection of objects that can exchange energy with each other, but not with the rest of the Universe. If one form of energy in an isolated system decreases, then another form of energy in the system must increase. For example, if the system consists of a motor connected to a battery, the battery converts chemical energy to electrical energy and the motor converts electrical energy to mechanical energy. Understanding how energy changes from one form to another is essential in all the sciences.

In this topic the focus is mainly on *mechanical energy,* which is the sum of *kinetic energy*, the energy associated with motion, and *potential energy*—the energy associated with relative position. Using an energy approach to solve certain problems is often much easier than using forces and Newton's three laws. These two very different approaches are linked through the concept of *work*.

## 5.1 Work

Work has a different meaning in physics than it does in everyday usage. In the physics definition, a physics textbook author does very little work typing away at a computer. A mason, by contrast, may do a lot of work laying concrete blocks. In physics, work is done only if an object is moved through some displacement while a force is applied to it. If either the force or displacement is doubled, the work is doubled. Double them both, and the work is quadrupled. Doing work involves applying a force to an object while moving it a given distance.

The definition for work $W$ might be taken as

$$W = Fd \qquad [5.1]$$

where $F$ is the magnitude of the force acting on the object and $d$ is the magnitude of the object's displacement. That definition, however, gives only the magnitude of work done on an object when the force is constant and parallel to the displacement, which must be along a line. A more sophisticated definition is required.

Figure 5.1 shows a block undergoing a displacement $\Delta \vec{\mathbf{x}}$ along a straight line while acted on by a constant force $\vec{\mathbf{F}}$ in the same direction. We have the following definition:

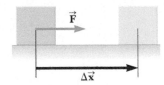

**Figure 5.1** A constant force $\vec{\mathbf{F}}$ in the same direction as the displacement, $\Delta\vec{\mathbf{x}}$, does work $F\Delta x$.

◀ Intuitive definition of work

121

The work $W$ done on an object by a constant force $\vec{F}$ during a linear displacement along the $x$-axis is

$$W = F_x \Delta x \qquad [5.2]$$

where $F_x$ is the $x$-component of the force $\vec{F}$ and $\Delta x = x_f - x_i$ is the object's displacement.

**SI unit: joule ( J ) = newton · meter (N · m) = kg · m²/s²**

Note that in one dimension, $\Delta x = x_f - x_i$ is a vector quantity, just as it was defined in Topic 2, not a magnitude as might be inferred from definitions of a vector and its magnitude in Topic 3. Therefore $\Delta x$ can be either positive or negative. Work as defined in Equation 5.2 is rigorous for displacements of any object along the $x$-axis while a constant force acts on it and, therefore, is suitable for many one-dimensional problems. Work is a positive number if $F_x$ and $\Delta x$ are both positive or both negative, in which case, as discussed in the next section, the work increases the mechanical energy of the object. If $F_x$ is positive and $\Delta x$ is negative, or vice versa, then the work is negative, and the object loses mechanical energy. The definition in Equation 5.2 works even when the constant force $\vec{F}$ is not parallel to the $x$-axis. Work is only done by the part of the force acting parallel to the object's direction of motion.

It's easy to see the difference between the physics definition and the everyday definition of work. The author exerts very little force on the keys of a keyboard, creating only small displacements, so relatively little physics work is done. The mason must exert much larger forces on concrete blocks and move them significant distances, and so performs a much greater amount of work. Even very tiring tasks, however, may not constitute work according to the physics definition. A truck driver, for example, may drive for several hours, but if he doesn't exert a force, then $F_x = 0$ in Equation 5.2 and he doesn't do any work. Similarly, a student pressing against a wall for hours in an isometric exercise also does no work, because the displacement in Equation 5.2, $\Delta x$, is zero.[1] Atlas, of Greek mythology, bore the world on his shoulders, but that, too, wouldn't qualify as work in the physics definition.

Work is a scalar quantity—a number rather than a vector—and consequently is easier to handle. No direction is associated with it. Further, work doesn't depend explicitly on time, which can be an advantage in problems involving only velocities and positions. Because the units of work are those of force and distance, the SI unit is the **newton-meter** (N · m). Another name for the newton-meter is the **joule ( J )** (rhymes with "pool"). The U.S. customary unit of work is the **foot-pound,** because distances are measured in feet and forces in pounds in that system.

A useful alternate definition relates the work done on an object to the angle the displacement makes with respect to the force. This definition exploits the triangle shown in Figure 5.2. The components of the vector $\vec{F}$ can be written as $F_x = F\cos\theta$ and $F_y = F\sin\theta$. However, only the $x$-component, which is parallel to the direction of motion, makes a nonzero contribution to the work done on the object.

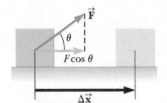

**Figure 5.2** A constant force $\vec{F}$ exerted at an angle $\theta$ with respect to the displacement, $\Delta\vec{x}$, does work $(F\cos\theta)\Delta x$.

The work $W$ done on an object by a constant force $\vec{F}$ during a linear displacement $\Delta\vec{x}$ is

$$W = (F\cos\theta)d \qquad [5.3]$$

where $d$ is the magnitude of the displacement and $\theta$ is the angle between the vectors $\vec{F}$ and $\Delta\vec{x}$.

**SI unit: joule ( J )**

[1]Actually, you do expend energy while doing isometric exercises because your muscles are continuously contracting and relaxing in the process. This internal muscular movement qualifies as work according to the physics definition.

The definition in Equation 5.3 can also be used more generally when the displacement is not specifically along the x-axis or any other axis.

In Figure 5.3, a man carries a bucket of water horizontally at constant velocity. The upward force exerted by the man's hand on the bucket is perpendicular to the direction of motion, so it does no work on the bucket. This can also be seen from Equation 5.3 because the angle between the force exerted by the hand and the direction of motion is 90°, giving cos 90° = 0 and W = 0. Similarly, the force of gravity does no work on the bucket.

Work always requires a system of more than just one object. A nail, for example, can't do work on itself, but a hammer can do work on the nail by driving it into a board. In general, an object may be moving under the influence of several external forces. In that case, the net work done on the object as it undergoes some displacement is just the sum of the amount of work done by each force.

Work can be either positive or negative. In the definition of work in Equation 5.3, F and d are magnitudes, which are never negative. Work is therefore positive or negative depending on whether cos θ is positive or negative. This, in turn, depends on the direction of $\vec{F}$ relative the direction of $\Delta\vec{x}$. When these vectors are pointing in the same direction, the angle between them is 0°, so cos 0° = +1 and the work is positive. For example, when a student lifts a box as in Figure 5.4, the work he does on the box is positive because the force he exerts on the box is upward, in the same direction as the displacement. In lowering the box slowly back down, however, the student still exerts an upward force on the box, but the motion of the box is downwards. Because the vectors $\vec{F}$ and $\Delta\vec{x}$ are now in opposite directions, the angle between them is 180°, and cos 180° = −1 and the work done by the student is negative. In general, when the part of $\vec{F}$ parallel to $\Delta\vec{x}$ points in the same direction as $\Delta\vec{x}$, the work is positive; otherwise, it's negative.

Because Equations 5.1–5.3 assume a force constant in both direction and magnitude, they are only special cases of a more general definition of work—that done by a varying force—treated briefly in Section 5.8.

**Figure 5.3** No work is done on a bucket when it is moved horizontally because the applied force $\vec{F}$ is perpendicular to the displacement.

**Tip 5.2** Work Is Done by Something, on Something Else

Work doesn't happen by itself. Work is done *by* something in the environment, *on* the object of interest.

### Quick Quiz

**5.1** In Figure 5.5 (a)–(d), a block moves to the right in the positive x-direction through the displacement $\Delta\vec{x}$ while under the influence of a force with the same magnitude $\vec{F}$. Which of the following is the correct order of the amount of work done by the force $\vec{F}$, from most positive to most negative? (a) d, c, a, b (b) c, a, b, d (c) c, a, d, b

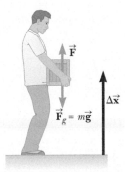

**Figure 5.4** The student does positive work when he lifts the box from the floor, because the applied force $\vec{F}$ is in the same direction as the displacement. When he lowers the box to the floor, he does negative work.

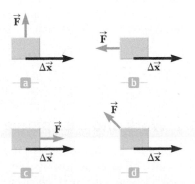

**Figure 5.5** (Quick Quiz 5.1) A force $\vec{F}$ is exerted on an object that undergoes a displacement $\Delta\vec{x}$ to the right. Both the magnitude of the force and the displacement are the same in all four cases.

## EXAMPLE 5.1   SLEDDING THROUGH THE YUKON

**GOAL** Apply the basic definitions of work done by a constant force.

**PROBLEM** A man returning from a successful ice fishing trip pulls a sled loaded with salmon. The total mass of the sled and salmon is 50.0 kg, and the man exerts a force of magnitude $1.20 \times 10^2$ N on the sled by pulling on the rope. **(a)** How much work does he do on the sled if the rope is horizontal to the ground ($\theta = 0°$ in Fig. 5.6) and he pulls the sled 5.00 m? **(b)** How much work does he do on the sled if $\theta = 30.0°$ and he pulls the sled the same distance? (Treat the sled as a point particle, so details such as the point of attachment of the rope make no difference.) **(c)** At a coordinate position of 12.4 m, the man lets up on the applied force. A friction force of 45.0 N between the ice and the sled brings the sled to rest at a coordinate position of 18.2 m. How much work does friction do on the sled?

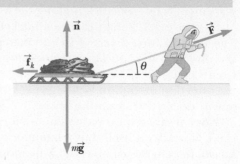

**Figure 5.6** (Examples 5.1 and 5.2) A man pulling a sled with a rope at an angle $\theta$ to the horizontal.

**STRATEGY** Substitute the given values of $F$ and $\Delta x$ into the basic equations for work, Equations 5.2 and 5.3.

. . . . . . . . . . . . . . . . . . . . . . . . . . . . . . . . . . . . . . . . . . . . . . . . . . . . . . . . . . . . . . . . . . . . . . . . . . . . . . . . . . . . . . . . . .

**SOLUTION**

**(a)** Find the work done when the force is horizontal.

Use Equation 5.2, substituting the given values:

$$W = F_x \Delta x = (1.20 \times 10^2 \text{ N})(5.00 \text{ m}) = \boxed{6.00 \times 10^2 \text{ J}}$$

**(b)** Find the work done when the force is exerted at a 30° angle.

Use Equation 5.3, again substituting the given values:

$$W = (F \cos \theta) d = (1.20 \times 10^2 \text{ N})(\cos 30°)(5.00 \text{ m})$$
$$= \boxed{5.20 \times 10^2 \text{ J}}$$

**(c)** How much work does a friction force of 45.0 N do on the sled as it travels from a coordinate position of 12.4 m to 18.2 m?

Use Equation 5.2, with $F_x$ replaced by $f_k$:

$$W_{\text{fric}} = F_x \Delta x = f_k(x_f - x_i)$$

Substitute $f_k = -45.0$ N and the initial and final coordinate positions into $x_i$ and $x_f$:

$$W_{\text{fric}} = (-45.0 \text{ N})(18.2 \text{ m} - 12.4 \text{ m}) = \boxed{-2.6 \times 10^2 \text{ J}}$$

. . . . . . . . . . . . . . . . . . . . . . . . . . . . . . . . . . . . . . . . . . . . . . . . . . . . . . . . . . . . . . . . . . . . . . . . . . . . . . . . . . . . . . . . . .

**REMARKS** The normal force $\vec{n}$, the gravitational force $m\vec{g}$, and the upward component of the applied force do *no* work on the sled because they're perpendicular to the displacement. The mass of the sled didn't come into play here, but it is important when the effects of friction must be calculated, as well as in the next section, where we introduce the work–energy theorem.

**QUESTION 5.1** How does the answer for the work done by the applied force change if the load is doubled? Explain.

**EXERCISE 5.1** Suppose the man is pushing the same 50.0-kg sled across level terrain with a force of 50.0 N. **(a)** If he does $4.00 \times 10^2$ J of work on the sled while exerting the force horizontally, through what distance must he have pushed it? **(b)** If he exerts the same force at an angle of 45.0° with respect to the horizontal and moves the sled through the same distance, how much work does he do on the sled?

**ANSWERS** **(a)** 8.00 m   **(b)** 283 J

---

## 5.1.1 Work and Dissipative Forces

*Frictional work* is extremely important in everyday life because doing almost any other kind of work is impossible without it. The man in the last example, for instance, depends on surface friction to pull his sled. Otherwise, the rope would slip in his hands and exert no force on the sled, while his feet slid out from underneath him and he fell flat on his face. Cars wouldn't work without friction, nor could conveyor belts, nor even our muscle tissue.

The work done by pushing or pulling an object is the application of a single force. Friction, on the other hand, is a complex process caused by numerous

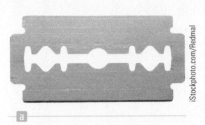

**Figure 5.7** The edge of a razor blade looks smooth to the eye, but under a microscope proves to have numerous irregularities.

microscopic interactions over the entire area of the surfaces in contact (Fig. 5.7). Consider a metal block sliding over a metal surface. Microscopic "teeth" in the block encounter equally microscopic irregularities in the underlying surface. Pressing against each other, the teeth deform, get hot, and weld to the opposite surface. Work must be done breaking these temporary bonds, and that comes at the expense of the energy of motion of the block, to be discussed in the next section. The energy lost by the block goes into heating both the block and its environment, with some energy converted to sound.

The friction force of two objects in contact and in relative motion to each other always dissipates energy in these relatively complex ways. For our purposes, the phrase "work done by friction" will denote the effect of these processes on mechanical energy alone.

## EXAMPLE 5.2    MORE SLEDDING

**GOAL** Calculate the work done by friction when an object is acted on by an applied force.

**PROBLEM** Suppose that in Example 5.1 the coefficient of kinetic friction between the loaded 50.0-kg sled and snow is 0.200. **(a)** The man again pulls the sled 5.00 m, exerting a force of $1.20 \times 10^2$ N at an angle of 0°. Find the work done on the sled by friction, and the net work. **(b)** Repeat the calculation if the applied force is exerted at an angle of 30.0° with the horizontal.

**STRATEGY** See Figure 5.6. The frictional work depends on the magnitude of the kinetic friction coefficient, the normal force, and the displacement. Use the $y$-component of Newton's second law to find the normal force $\vec{\mathbf{n}}$, calculate the work done by friction using the definitions, and sum with the result of Example 5.1(a) to obtain the net work on the sled. Part **(b)** is solved similarly, but the normal force is smaller because it has the help of the applied force $\vec{\mathbf{F}}_{\text{app}}$ in supporting the load.

......................................................................................................................................................

### SOLUTION

**(a)** Find the work done by friction on the sled and the net work, if the applied force is horizontal.

First, find the normal force from the $y$-component of Newton's second law, which involves only the normal force and the force of gravity:

$$\sum F_y = n - mg = 0 \quad \rightarrow \quad n = mg$$

Use the normal force to compute the work done by friction:

$$W_{\text{fric}} = -f_k \Delta x = -\mu_k n \Delta x = -\mu_k mg \Delta x$$
$$= -(0.200)(50.0 \text{ kg})(9.80 \text{ m/s}^2)(5.00 \text{ m})$$
$$= \boxed{-4.90 \times 10^2 \text{ J}}$$

Sum the frictional work with the work done by the applied force from Example 5.1 to get the net work (the normal and gravity forces are perpendicular to the displacement, so they don't contribute):

$$W_{\text{net}} = W_{\text{app}} + W_{\text{fric}} + W_n + W_g$$
$$= 6.00 \times 10^2 \text{ J} + (-4.90 \times 10^2 \text{ J}) + 0 + 0$$
$$= \boxed{1.10 \times 10^2 \text{ J}}$$

**(b)** Recalculate the frictional work and net work if the applied force is exerted at a 30.0° angle.

Find the normal force from the $y$-component of Newton's second law:

$$\sum F_y = n - mg + F_{\text{app}} \sin \theta = 0$$
$$n = mg - F_{\text{app}} \sin \theta$$

*(Continued)*

Use the normal force to calculate the work done by friction:

$$W_{\text{fric}} = -f_k \Delta x = -\mu_k n \Delta x = -\mu_k (mg - F_{\text{app}} \sin \theta)\, \Delta x$$
$$= -(0.200)(50.0 \text{ kg} \cdot 9.80 \text{ m/s}^2$$
$$-1.20 \times 10^2 \text{ N} \sin 30.0°)(5.00 \text{ m})$$
$$W_{\text{fric}} = \boxed{-4.30 \times 10^2 \text{ J}}$$

Sum this answer with the result of Example 5.1(b) to get the net work (again, the normal and gravity forces don't contribute):

$$W_{\text{net}} = W_{\text{app}} + W_{\text{fric}} + W_n + W_g$$
$$= 5.20 \times 10^2 \text{ J} - 4.30 \times 10^2 \text{ J} + 0 + 0 = \boxed{9.0 \times 10^1 \text{ J}}$$

. . . . . . . . . . . . . . . . . . . . . . . . . . . . . . . . . . . . . . . . . . . . . . . . . . . . . . . . . . . . . . . . . . . . . . . . . . . . . .

**REMARKS** The most important thing to notice here is that exerting the applied force at different angles can dramatically affect the work done on the sled. Pulling at the optimal angle (11.3° in this case) will result in the most net work for the same applied force.

**QUESTION 5.2** How does the net work change in each case if the displacement is doubled?

**EXERCISE 5.2** The man pushes the same 50.0-kg sled over level ground with a force of $1.75 \times 10^2$ N exerted horizontally, moving it a distance of 6.00 m over new terrain. **(a)** If the net work done on the sled is $1.50 \times 10^2$ J, find the coefficient of kinetic friction. **(b)** Repeat the exercise with the same data, finding the coefficient of kinetic friction, but assume the applied force is upwards at a 45.0° angle with the horizontal.

**ANSWERS** **(a)** 0.306   **(b)** 0.270

---

# 5.2 Kinetic Energy and the Work–Energy Theorem

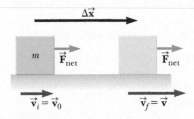

**Figure 5.8** An object undergoes a displacement and a change in velocity under the action of a constant net force $\vec{\mathbf{F}}_{\text{net}}$.

Solving problems using Newton's second law can be difficult if the forces involved are complicated. An alternative is to relate the speed of an object to the net work done on it by external forces. If the net work can be calculated for a given displacement, the change in the object's speed is easy to evaluate.

Figure 5.8 shows an object of mass $m$ moving to the right under the action of a constant net force $\vec{\mathbf{F}}_{\text{net}}$, also directed to the right. Because the force is constant, we know from Newton's second law that the object moves with constant acceleration $\vec{\mathbf{a}}$. If the object is displaced by $\Delta x$, the work done by $\vec{\mathbf{F}}_{\text{net}}$ on the object is

$$W_{\text{net}} = F_{\text{net}}\, \Delta x = (ma)\, \Delta x \qquad [5.4]$$

In Topic 2, we found that the following relationship holds when an object undergoes constant acceleration:

$$v^2 = v_0^2 + 2a\Delta x \qquad \text{or} \qquad a\Delta x = \frac{v^2 - v_0^2}{2}$$

We can substitute this expression into Equation 5.4 to get

$$W_{\text{net}} = m\left(\frac{v^2 - v_0^2}{2}\right)$$

or

$$W_{\text{net}} = \tfrac{1}{2}mv^2 - \tfrac{1}{2}mv_0^2 \qquad [5.5]$$

So the net work done on an object equals a change in a quantity of the form $\tfrac{1}{2}mv^2$. Because this term carries units of energy and involves the object's speed, it can be interpreted as energy associated with the object's motion, leading to the following definition:

The **kinetic energy** *KE* of an object of mass $m$ moving with a speed $v$ is

◄ Kinetic energy

$$KE \equiv \tfrac{1}{2}mv^2 \qquad [5.6]$$

**SI unit: joule (J) = kg · m²/s²**

Like work, kinetic energy is a scalar quantity. Using this definition in Equation 5.5, we arrive at an important result known as the **work–energy theorem:**

◄ Work–energy theorem

The net work done on an object is equal to the change in the object's kinetic energy:

$$W_{\text{net}} = KE_f - KE_i = \Delta KE \qquad [5.7]$$

where the change in the kinetic energy is due entirely to the object's change in speed.

The proviso on the speed is necessary because work that deforms or causes the object to warm up invalidates Equation 5.7, although under most circumstances it remains approximately correct. From that equation, a positive net work $W_{\text{net}}$ means that the final kinetic energy $KE_f$ is greater than the initial kinetic energy $KE_i$. This, in turn, means that the object's final speed is greater than its initial speed. So positive net work increases an object's speed, and negative net work decreases its speed.

We can also turn Equation 5.7 around and think of kinetic energy as the work a moving object can do in coming to rest. For example, suppose a hammer is on the verge of striking a nail, as in Figure 5.9. The moving hammer has kinetic energy and can therefore do work on the nail. The work done on the nail is $F\Delta x$, where $F$ is the average net force exerted on the nail and $\Delta x$ is the distance the nail is driven into the wall. That work, plus small amounts of energy carried away by heat and sound, is equal to the change in kinetic energy of the hammer, $\Delta KE$.

For convenience, the work–energy theorem was derived under the assumption that the net force acting on the object was constant. A more general derivation, using calculus, would show that Equation 5.7 is valid under all circumstances, including the application of a variable force.

**Figure 5.9** The moving hammer has kinetic energy and can do work on the nail, driving it into the wall.

### Quick Quiz

**5.2** A block slides at constant speed down a ramp while acted on by three forces: its weight, the normal force, and kinetic friction. Respond to each statement, true or false. (a) The combined net work done by all three forces on the block equals zero. (b) Each force does zero work on the block as it slides. (c) Each force does negative work on the block as it slides.

### APPLYING PHYSICS 5.1    SKID TO A STOP

Suppose a car traveling at a speed $v$ skids a distance $d$ after its brakes lock. Estimate how far it would skid if it were traveling at speed $2v$ when its brakes locked.

**EXPLANATION** Assume for simplicity that the force of kinetic friction between the car and the road surface is constant and the same at both speeds. From the work–energy theorem, the net force exerted on the car times the displacement of the car, $F_{\text{net}}\Delta x$, is equal in magnitude to its initial kinetic energy, $\tfrac{1}{2}mv^2$. When the speed is doubled, the kinetic energy of the car is quadrupled. So for a given applied friction force, the distance traveled must increase fourfold when the initial speed is doubled, and the estimated distance the car skids is $4d$. ■

## EXAMPLE 5.3   COLLISION ANALYSIS

**GOAL** Apply the work–energy theorem with a known force.

**PROBLEM** The driver of a $1.00 \times 10^3$ kg car traveling on the interstate at 35.0 m/s (nearly 80.0 mph) slams on his brakes to avoid hitting a second vehicle in front of him, which had come to rest because of congestion ahead (Fig. 5.10). After the brakes are applied, a constant kinetic friction force of magnitude $8.00 \times 10^3$ N acts on the car. Ignore air resistance. **(a)** At what minimum distance should the brakes be applied to avoid a collision with the other vehicle? **(b)** If the distance between the vehicles is initially only 30.0 m, at what speed would the collision occur?

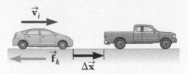

**Figure 5.10** (Example 5.3) A braking vehicle just prior to an accident.

**STRATEGY** Compute the net work, which involves just the kinetic friction, because the normal and gravity forces are perpendicular to the motion. Then set the net work equal to the change in kinetic energy. To get the minimum distance in part **(a)**, we take the final speed $v_f$ to be zero just as the braking vehicle reaches the rear of the vehicle at rest. Solve for the unknown, $\Delta x$. For part **(b)** proceed similarly, except that the unknown is the final velocity $v_f$.

**SOLUTION**

**(a)** Find the minimum necessary stopping distance.

Apply the work–energy theorem to the car:

$$W_{\text{net}} = \tfrac{1}{2}mv_f{}^2 - \tfrac{1}{2}mv_i^2$$

Substitute an expression for the frictional work and set $v_f = 0$:

$$-f_k \Delta x = 0 - \tfrac{1}{2}mv_i^2$$

Substitute $v_i = 35.0$ m/s, $f_k = 8.00 \times 10^3$ N, and $m = 1.00 \times 10^3$ kg. Solve for $\Delta x$:

$$-(8.00 \times 10^3 \text{ N})\,\Delta x = -\tfrac{1}{2}(1.00 \times 10^3 \text{ kg})(35.0 \text{ m/s})^2$$

$$\Delta x = \boxed{76.6 \text{ m}}$$

**(b)** At the given distance of 30.0 m, the car is too close to the other vehicle. Find the speed at impact.

Write down the work–energy theorem:

$$W_{\text{net}} = W_{\text{fric}} = -f_k \Delta x = \tfrac{1}{2}mv_f{}^2 - \tfrac{1}{2}mv_i^2$$

Multiply by $2/m$ and rearrange terms, solving for the final velocity $v_f$:

$$v_f^2 = v_i^2 - \frac{2}{m}f_k \Delta x$$

$$v_f^2 = (35.0 \text{ m/s})^2 - \left(\frac{2}{1.00 \times 10^3 \text{ kg}}\right)(8.00 \times 10^3 \text{ N})(30.0 \text{ m})$$

$$= 745 \text{ m}^2/\text{s}^2$$

$$v_f = \boxed{27.3 \text{ m/s}}$$

**REMARKS** This calculation illustrates how important it is to remain alert on the highway, allowing for an adequate stopping distance at all times. It takes about a second to react to the brake lights of the car in front of you. On a high-speed highway, your car may travel more than 30 m before you can engage the brakes. Bumper-to-bumper traffic at high speed, which often occurs on the highways near big cities, is extremely unsafe.

**QUESTION 5.3** Qualitatively, how would the answer for the final velocity change in part **(b)** if it's raining during the incident? Explain.

**EXERCISE 5.3** A police investigator measures straight skid marks 27.0 m long in an accident investigation. Assuming a friction force and car mass the same as in the previous problem, what was the minimum speed of the car when the brakes locked?

**ANSWER** 20.8 m/s

## 5.2.1 Conservative and Nonconservative Forces

It turns out there are two general kinds of forces. The first is called a **conservative force**. Gravity is probably the best example of a conservative force. To understand the origin of the name, think of a diver climbing to the top of a 10-meter platform. The diver has to do work against gravity in making the climb. Once at the top, however, she can recover the work as kinetic energy by taking a dive. Her speed just before hitting the water will give her a kinetic energy equal to the work she did against gravity in climbing to the top of the platform, minus the effect of some nonconservative forces, such as air drag and internal muscular friction.

A **nonconservative force** is generally dissipative, which means that it tends to randomly disperse the energy of bodies on which it acts. This dispersal of energy often takes the form of heat or sound. Kinetic friction and air drag are good examples. Propulsive forces, like the force exerted by a jet engine on a plane or by a propeller on a submarine, are also nonconservative.

Work done against a nonconservative force can't be easily recovered. Dragging objects over a rough surface requires work. When the man in Example 5.2 dragged the sled across terrain having a nonzero coefficient of friction, the net work was smaller than in the frictionless case. The missing energy went into warming the sled and its environment. As will be seen in the study of thermodynamics, such losses can't be avoided, nor all the energy recovered, so these forces are called nonconservative.

Another way to characterize conservative and nonconservative forces is to measure the work done by a force on an object traveling between two points along different paths. The work done by gravity on someone going down a frictionless slide, as in Figure 5.11, is the same as that done on someone diving into the water from the same height. This equality doesn't hold for nonconservative forces. For example, sliding a book directly from point Ⓐ to point Ⓓ in Figure 5.12 requires a certain amount of work against friction, but sliding the book along the three other legs of the square, from Ⓐ to Ⓑ, Ⓑ to Ⓒ, and finally Ⓒ to Ⓓ, requires three times as much work. This observation motivates the following definition of a conservative force:

**Figure 5.11** Because the gravity field is conservative, the diver regains as kinetic energy the work she did against gravity in climbing the ladder. Taking the frictionless slide gives the same result.

> A force is conservative if the work it does moving an object between two points is the same no matter what path is taken.

◀ Conservative force

Nonconservative forces, as we've seen, don't have this property. The work–energy theorem, Equation 5.7, can be rewritten in terms of the work done by conservative forces $W_c$ and the work done by nonconservative forces $W_{nc}$ because the net work is just the sum of these two:

$$W_{nc} + W_c = \Delta KE \qquad [5.8]$$

It turns out that conservative forces have another useful property: The work they do can be recast as something called **potential energy,** a quantity that depends only on the beginning and end points of a curve, not the path taken.

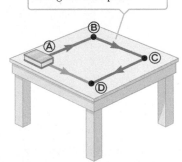

The work done in moving the book is greater along the rust-colored path than along the blue path.

**Figure 5.12** Because friction is a nonconservative force, a book pushed along the three segments Ⓐ–Ⓑ, Ⓑ–Ⓒ, and Ⓒ–Ⓓ requires three times the work as pushing the book directly from Ⓐ to Ⓓ.

# 5.3 Gravitational Potential Energy

An object with kinetic energy (energy of motion) can do work on another object, just like a moving hammer can drive a nail into a wall. A brick on a high shelf can also do work: it can fall off the shelf, accelerate downward, and hit a nail squarely, driving it into the floorboards. The brick is said to have **potential energy** associated with it, because from its location on the shelf it can potentially do work.

Potential energy is a property of a **system,** rather than of a single object, because it's due to the relative positions of interacting objects in the system, such as the position of the diver in Figure 5.11 relative to the Earth. In this topic we define a system as a collection of objects interacting via forces or other processes that are internal to the system. It turns out that potential energy is another way of looking at the work done by conservative forces.

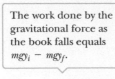

The work done by the gravitational force as the book falls equals $mgy_i - mgy_f$.

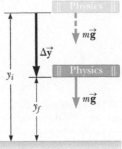

**Figure 5.13** A book of mass $m$ falls from a height $y_i$ to a height $y_f$.

## 5.3.1 Gravitational Work and Potential Energy

Using the work–energy theorem in problems involving gravitation requires computing the work done by gravity. For most trajectories—say, for a ball traversing a parabolic arc—finding the gravitational work done on the ball requires sophisticated techniques from calculus. Fortunately, for conservative fields there's a simple alternative: potential energy.

Gravity is a conservative force, and for every conservative force, a special expression called a potential energy function can be found. Evaluating that function at any two points in an object's path of motion and finding the difference will give the negative of the work done by that force between those two points. It's also advantageous that potential energy, like work and kinetic energy, is a scalar quantity.

Our first step is to find the work done by gravity on an object when it moves from one position to another. The negative of that work is the change in the gravitational potential energy of the system, and from that expression, we'll be able to identify the potential energy function.

In Figure 5.13, a book of mass $m$ falls from a height $y_i$ to a height $y_f$, where the positive $y$-coordinate represents position above the ground. We neglect the force of air friction, so the only force acting on the book is gravitation. How much work is done? The magnitude of the force is $mg$ and that of the displacement is $\Delta y = y_i - y_f$ (a positive number), while both $\vec{F}$ and $\Delta \vec{y}$ are pointing downwards, so the angle between them is zero. We apply the definition of work in Equation 5.3, with $d = y_i - y_f$:

$$W_g = Fd \cos \theta = mg(y_i - y_f) \cos 0° = -mg(y_f - y_i) \qquad [5.9]$$

Factoring out the minus sign was deliberate, to clarify the coming connection to potential energy. Equation 5.9 for gravitational work holds for any object, regardless of its trajectory in space, because the gravitational force is conservative. Now, $W_g$ will appear as the work done by gravity in the work–energy theorem. For the rest of this section, assume for simplicity that we are dealing only with systems involving gravity and nonconservative forces. Then Equation 5.8 can be written as

$$W_{\text{net}} = W_{nc} + W_g = \Delta KE$$

where $W_{nc}$ is the work done by the nonconservative forces. Substituting the expression for $W_g$ from Equation 5.9, we obtain

$$W_{nc} - mg(y_f - y_i) = \Delta KE \qquad [5.10a]$$

Next, we add $mg(y_f - y_i)$ to both sides:

$$W_{nc} = \Delta KE + mg(y_f - y_i) \qquad [5.10b]$$

Now, by definition, we'll make the connection between gravitational work and gravitational potential energy.

Gravitational potential ▶ energy

> The gravitational potential energy of a system consisting of Earth and an object of mass $m$ near Earth's surface is given by
>
> $$PE \equiv mgy \qquad [5.11]$$
>
> where $g$ is the acceleration of gravity and $y$ is the vertical position of the mass relative the surface of Earth (or some other reference point).
>
> **SI unit: joule (J)**

In this definition, $y = 0$ is usually taken to correspond to Earth's surface, but that is not strictly necessary, as discussed in the next subsection. It turns out that only *differences* in potential energy really matter.

So the gravitational potential energy associated with an object located near the surface of Earth is the object's weight $mg$ times its vertical position $y$ above Earth. From this *definition*, we have the relationship between gravitational work and gravitational potential energy:

$$W_g = -(PE_f - PE_i) = -(mgy_f - mgy_i) \qquad [5.12]$$

The work done by gravity is one and the same as the negative of the change in gravitational potential energy.

Finally, using the relationship in Equation 5.12 in Equation 5.10b, we obtain an extension of the work–energy theorem:

$$W_{nc} = (KE_f - KE_i) + (PE_f - PE_i) \qquad [5.13]$$

This equation says that the work done by nonconservative forces, $W_{nc}$, is equal to the change in the kinetic energy plus the change in the gravitational potential energy.

Equation 5.13 will turn out to be true in general, even when other conservative forces besides gravity are present. The work done by these additional conservative forces will again be recast as changes in potential energy and will appear on the right-hand side along with the expression for gravitational potential energy.

> **Tip 5.3** Potential Energy Takes Two
>
> Potential energy always takes a system of at least two interacting objects—for example, the Earth and a baseball interacting via the gravitational force.

### 5.3.2 Reference Levels for Gravitational Potential Energy

In solving problems involving gravitational potential energy, it's important to choose a location at which to set that energy equal to zero. Given the form of Equation 5.11, this is the same as choosing the place where $y = 0$. The choice is completely arbitrary because the important quantity is the *difference* in potential energy, and this difference will be the same regardless of the choice of zero level. However, once this position is chosen, it must remain fixed for a given problem.

While it's always possible to choose the surface of Earth as the reference position for zero potential energy, the statement of a problem will usually suggest a convenient position to use. As an example, consider a book at several possible locations, as in Figure 5.14. When the book is at Ⓐ, a natural zero level for potential energy is the surface of the desk. When the book is at Ⓑ, the floor might be a more convenient reference level. Finally, a location such as Ⓒ, where the book is held out a window, would suggest choosing the surface of Earth as the zero level of potential energy. The choice, however, makes no difference: Any of the three reference levels could be used as the zero level, regardless of whether the book is at Ⓐ, Ⓑ, or Ⓒ. Example 5.4 illustrates this important point.

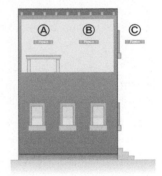

**Figure 5.14** Any reference level—the desktop, the floor of the room, or the ground outside the building—can be used to represent zero gravitational potential energy in the book–Earth system.

---

### EXAMPLE 5.4 | WAX YOUR SKIS

**GOAL** Calculate the change in gravitational potential energy for different choices of reference level.

**PROBLEM** A 60.0-kg skier is at the top of a slope, as shown in Figure 5.15. At the initial point Ⓐ, she is 10.0 m vertically above point Ⓑ. **(a)** Setting the zero level for gravitational potential energy at Ⓑ, find the gravitational potential energy of this system when the skier is at Ⓐ and then at Ⓑ. Finally, find the change in potential energy of the skier–Earth system as the skier goes from point Ⓐ to point Ⓑ. **(b)** Repeat this problem with the zero level at point Ⓐ. **(c)** Repeat again, with the zero level 2.00 m higher than point Ⓑ.

**STRATEGY** Follow the definition and be careful with signs. Ⓐ is the initial point, with gravitational potential energy $PE_i$, and Ⓑ is the final point, with gravitational potential energy $PE_f$. The location chosen for $y = 0$ is also the zero point for the potential energy, because $PE = mgy$.

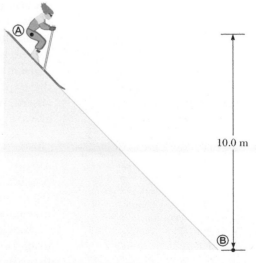

**Figure 5.15** (Example 5.4)

*(Continued)*

## SOLUTION

**(a)** Let $y = 0$ at Ⓑ. Calculate the potential energy at Ⓐ and at Ⓑ, and calculate the change in potential energy.

Find $PE_i$, the potential energy at Ⓐ, from Equation 5.11:

$$PE_i = mgy_i = (60.0 \text{ kg})(9.80 \text{ m/s}^2)(10.0 \text{ m}) = 5.88 \times 10^3 \text{ J}$$

$PE_f = 0$ at Ⓑ by choice. Find the difference in potential energy between Ⓐ and Ⓑ:

$$PE_f - PE_i = 0 - 5.88 \times 10^3 \text{ J} = \boxed{-5.88 \times 10^3 \text{ J}}$$

**(b)** Repeat the problem if $y = 0$ at Ⓐ, the new reference point, so that $PE = 0$ at Ⓐ.

Find $PE_f$, noting that point Ⓑ is now at $y = -10.0$ m:

$$PE_f = mgy_f = (60.0 \text{ kg})(9.80 \text{ m/s}^2)(-10.0 \text{ m})$$
$$= -5.88 \times 10^3 \text{ J}$$
$$PE_f - PE_i = -5.88 \times 10^3 \text{ J} - 0 = \boxed{-5.88 \times 10^3 \text{ J}}$$

**(c)** Repeat the problem, if $y = 0$ two meters above Ⓑ.

Find $PE_i$, the potential energy at Ⓐ:

$$PE_i = mgy_i = (60.0 \text{ kg})(9.80 \text{ m/s}^2)(8.00 \text{ m}) = 4.70 \times 10^3 \text{ J}$$

Find $PE_f$, the potential energy at Ⓑ:

$$PE_f = mgy_f = (60.0 \text{ kg})(9.80 \text{ m/s}^2)(-2.00 \text{ m})$$
$$= -1.18 \times 10^3 \text{ J}$$

Compute the change in potential energy:

$$PE_f - PE_i = -1.18 \times 10^3 \text{ J} - 4.70 \times 10^3 \text{ J}$$
$$= \boxed{-5.88 \times 10^3 \text{ J}}$$

**REMARKS** These calculations show that the change in the gravitational potential energy when the skier goes from the top of the slope to the bottom is $-5.88 \times 10^3$ J, *regardless of the zero level selected.*

**QUESTION 5.4** If the angle of the slope is increased, does the change in gravitational potential energy between two heights **(a)** increase, **(b)** decrease, **(c)** remain the same?

**EXERCISE 5.4** If the zero level for gravitational potential energy is selected to be midway down the slope, 5.00 m above point Ⓑ, find the initial potential energy, the final potential energy, and the change in potential energy as the skier goes from point Ⓐ to Ⓑ in Figure 5.15.

**ANSWER** 2.94 kJ, −2.94 kJ, −5.88 kJ

## 5.3.3 Gravity and the Conservation of Mechanical Energy

Conservation principles play a very important role in physics. **When a physical quantity is conserved the numeric value of the quantity remains the same throughout the physical process.** Although the form of the quantity may change in some way, **its final value is the same as its initial value.**

The kinetic energy $KE$ of an object falling only under the influence of gravity is constantly changing, as is the gravitational potential energy $PE$. Obviously, then, these quantities aren't conserved. Because all nonconservative forces are assumed absent, however, we can set $W_{nc} = 0$ in Equation 5.13. Rearranging the equation, we arrive at the following very interesting result:

$$KE_i + PE_i = KE_f + PE_f \qquad [5.14]$$

According to this equation, **the sum of the kinetic energy and the gravitational potential energy remains constant at all times and hence is a conserved quantity.** We denote the total mechanical energy by $E = KE + PE$, and say that **the total mechanical energy is conserved.**

To show how this concept works, think of tossing a rock off a cliff, ignoring the drag forces. As the rock falls, its speed increases, so its kinetic energy increases. As the rock approaches the ground, the potential energy of the rock–Earth system decreases. Whatever potential energy is lost as the rock moves downward appears

**Tip 5.4** Conservation Principles

There are many conservation laws like the conservation of mechanical energy in isolated systems, as in Equation 5.14. For example, momentum, angular momentum, and electric charge are all conserved quantities, as will be seen later. Conserved quantities may change form during physical interactions, but their sum total for a system never changes.

as kinetic energy, and Equation 5.14 says that in the absence of nonconservative forces like air drag, the trading of energy is exactly even. This is true for all conservative forces, not just gravity.

> In any isolated system of objects interacting only through conservative forces, the total mechanical energy, $E = KE + PE$, of the system remains the same at all times.

◄ Conservation of mechanical energy

If the force of gravity is the *only* force doing work within a system, then the principle of conservation of mechanical energy takes the form

$$\tfrac{1}{2}mv_i^2 + mgy_i = \tfrac{1}{2}mv_f^2 + mgy_f \qquad [5.15]$$

This form of the equation is particularly useful for solving problems explicitly involving only one mass and gravity. In that special case, which occurs commonly, notice that the mass cancels out of the equation. However, that is possible only because any change in kinetic energy of the Earth in response to the gravity field of the object of mass $m$ has been (rightfully) neglected. In general, there must be kinetic energy terms for each object in the system and gravitational potential energy terms for every pair of objects. Further terms have to be added when other conservative forces are present, as we'll soon see.

**Figure 5.16** (Quick Quiz 5.3) A student throws three identical balls from the top of a building, each at the same initial speed but at a different initial angle.

## Quick Quiz

**5.3** Three identical balls are thrown from the top of a building, all with the same initial speed. The first ball is thrown horizontally, the second at some angle above the horizontal, and the third at some angle below the horizontal, as in Figure 5.16. Neglecting air resistance, rank the speeds of the balls as they reach the ground, from fastest to slowest. (a) 1, 2, 3 (b) 2, 1, 3 (c) 3, 1, 2 (d) All three balls strike the ground at the same speed.

**5.4** Bob, of mass $m$, drops from a tree limb at the same time that Esther, also of mass $m$, begins her descent down a frictionless slide. If they both start at the same height above the ground, which of the following is true about their kinetic energies as they reach the ground?

(a) Bob's kinetic energy is greater than Esther's.
(b) Esther's kinetic energy is greater than Bob's.
(c) They have the same kinetic energy.
(d) The answer depends on the shape of the slide.

## PROBLEM-SOLVING STRATEGY

### Applying Conservation of Mechanical Energy

*Take the following steps when applying conservation of mechanical energy to problems involving gravity:*

1. **Define the system,** including all interacting bodies. Verify the absence of nonconservative forces.
2. **Choose a location for $y = 0$,** the zero point for gravitational potential energy.
3. **Select the body of interest and identify two points**—one point where you have given information and the other point where you want to find out something about the body of interest.
4. **Write down the conservation of energy equation,** Equation 5.15, for the system. **Identify the unknown quantity** of interest.
5. **Solve for the unknown quantity,** which is usually either a speed or a position, and substitute known values.

As previously stated, it's usually best to do the algebra with symbols rather than substituting known numbers first, because it's easier to check the symbols for possible errors. The exception is when a quantity is clearly zero, in which case immediate substitution greatly simplifies the ensuing algebra.

## EXAMPLE 5.5   PLATFORM DIVER

**GOAL** Use conservation of energy to calculate the speed of a body falling straight down in the presence of gravity.

**PROBLEM** A diver of mass $m$ drops from a board 10.0 m above the water's surface, as in Figure 5.17. Neglect air resistance. **(a)** Use conservation of mechanical energy to find his speed 5.00 m above the water's surface. **(b)** Find his speed as he hits the water.

**STRATEGY** Refer to the problem-solving strategy. Step 1: The system consists of the diver and Earth. As the diver falls, only the force of gravity acts on him (neglecting air drag), so the mechanical energy of the system is conserved, and we can use conservation of energy for both parts **(a)** and **(b)**. Step 2: Choose $y = 0$ for the water's surface. Step 3: In part **(a)**, $y = 10.0$ m and $y = 5.00$ m are the points of interest, while in part **(b)**, $y = 10.0$ m and $y = 0$ m are of interest.

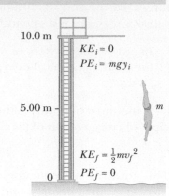

**Figure 5.17** (Example 5.5) The zero of gravitational potential energy is taken to be at the water's surface.

**SOLUTION**

**(a)** Find the diver's speed halfway down, at $y = 5.00$ m.

Step 4: Write the energy conservation equation and supply the proper terms:

$$KE_i + PE_i = KE_f + PE_f$$
$$\tfrac{1}{2}mv_i^2 + mgy_i = \tfrac{1}{2}mv_f^2 + mgy_f$$

Step 5: Substitute $v_i = 0$, cancel the mass $m$, and solve for $v_f$:

$$0 + gy_i = \tfrac{1}{2}v_f^2 + gy_f$$
$$v_f = \sqrt{2g(y_i - y_f)} = \sqrt{2(9.80 \text{ m/s}^2)(10.0 \text{ m} - 5.00 \text{ m})}$$
$$v_f = \boxed{9.90 \text{ m/s}}$$

**(b)** Find the diver's speed at the water's surface, $y = 0$.

Use the same procedure as in part **(a)**, taking $y_f = 0$:

$$0 + mgy_i = \tfrac{1}{2}mv_f^2 + 0$$
$$v_f = \sqrt{2gy_i} = \sqrt{2(9.80 \text{ m/s}^2)(10.0 \text{ m})} = \boxed{14.0 \text{ m/s}}$$

**REMARKS** Notice that the speed halfway down is not half the final speed. Another interesting point is that the final answer doesn't depend on the mass. That is really a consequence of neglecting the change in kinetic energy of Earth, which is valid when the mass of the object, the diver in this case, is much smaller than the mass of Earth. In reality, Earth also falls towards the diver, reducing the final speed, but the reduction is so minuscule it could never be measured.

**QUESTION 5.5** Qualitatively, how will the answers change if the diver takes a running dive off the end of the board?

**EXERCISE 5.5** Suppose the diver vaults off the springboard, leaving it with an initial speed of 3.50 m/s upward. Use energy conservation to find his speed when he strikes the water.

**ANSWER** 14.4 m/s

## EXAMPLE 5.6   THE JUMPING BUG

**GOAL** Use conservation of mechanical energy and concepts from ballistics in two dimensions to calculate a speed.

**PROBLEM** A powerful grasshopper launches itself at an angle of 45° above the horizontal and rises to a maximum height of 1.00 m during the leap. (See Fig. 5.18.) With what speed $v_i$ did it leave the ground? Neglect air resistance.

**STRATEGY** This problem can be solved with conservation of energy and the relation between the initial velocity and its $x$-component. Aside from the origin, the other point of interest is the maximum height $y = 1.00$ m, where the grasshopper has a velocity $v_x$ in the $x$-direction only. Energy conservation then gives one equation with two unknowns: the initial speed $v_i$ and speed at maximum height, $v_x$. Because there are no forces in the $x$-direction, however, $v_x$ is the same as the $x$-component of the initial velocity.

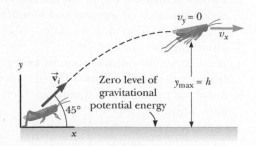

**Figure 5.18** (Example 5.6)

## SOLUTION

Use energy conservation:

$$\tfrac{1}{2}mv_i^2 + mgy_i = \tfrac{1}{2}mv_f^2 + mgy_f$$

Substitute $y_i = 0$, $v_f = v_x$, and $y_f = h$:

$$\tfrac{1}{2}mv_i^2 = \tfrac{1}{2}mv_x^2 + mgh$$

Multiply each side by $2/m$, obtaining one equation and two unknowns:

$$(1) \quad v_i^2 = v_x^2 + 2gh$$

Eliminate $v_x$ by substituting $v_x = v_i \cos 45°$ into Equation (1), solving for $v_i$, and substituting known values:

$$v_i^2 = (v_i \cos 45°)^2 + 2gh = \tfrac{1}{2}v_i^2 + 2gh$$
$$v_i = 2\sqrt{gh} = 2\sqrt{(9.80 \text{ m/s}^2)(1.00 \text{ m})} = \boxed{6.26 \text{ m/s}}$$

**REMARKS** The final answer is a surprisingly high value and illustrates how strong insects are relative to their size.

**QUESTION 5.6** All other given quantities remaining the same, how would the answer change if the initial angle were smaller? Why?

**EXERCISE 5.6** A catapult launches a rock at a 30.0° angle with respect to the horizontal. Find the maximum height attained if the speed of the rock at its highest point is 30.0 m/s.

**ANSWER** 15.3 m

# 5.4 Gravity and Nonconservative Forces

When nonconservative forces are involved along with gravitation, the full work–energy theorem must be used, often with techniques from Topic 4. Solving problems requires the basic procedure of the problem-solving strategy for conservation-of-energy problems in the previous section. The only difference lies in substituting Equation 5.13, the work–energy equation with potential energy, for Equation 5.15.

> **Tip 5.5** Don't Use Work Done by the Force of Gravity *and* Gravitational Potential Energy!
>
> Gravitational potential energy is just another way of including the work done by the force of gravity in the work–energy theorem. Don't use both of them in the same equation or you'll count it twice!

## EXAMPLE 5.7  DER STUKA!

**GOAL** Use the work–energy theorem with gravitational potential energy to calculate the work done by a nonconservative force.

**PROBLEM** Waterslides, which are nearly frictionless, can provide bored students with high-speed thrills (Fig. 5.19). One such slide, Der Stuka, named for the terrifying German dive bombers of World War II, is 72.0 feet high (21.9 m), is found at Six Flags in Dallas, Texas, and at Wet'n Wild in Orlando, Florida. **(a)** Determine the speed of a 60.0-kg woman at the bottom of such a slide, assuming no friction is present. **(b)** If the woman is clocked at 18.0 m/s at the bottom of the slide, find the work done on the woman by friction.

**STRATEGY** The system consists of the woman, Earth, and the slide. The normal force, always perpendicular to the displacement, does no work. Let $y = 0$ m represent the bottom of the slide. The two points of interest are $y = 0$ m and $y = 21.9$ m. Without friction, $W_{nc} = 0$, and we can apply conservation of mechanical energy, Equation 5.15. For part **(b)**, use Equation 5.13, substitute two velocities and heights, and solve for $W_{nc}$.

**Figure 5.19** (Example 5.7) If the slide is frictionless, the woman's speed at the bottom depends only on the height of the slide, not on the path it takes.

Wet'n Wild Orlando

(*Continued*)

## SOLUTION

**(a)** Find the woman's speed at the bottom of the slide, assuming no friction.

Write down Equation 5.15, for conservation of energy:

$$\tfrac{1}{2}mv_i^2 + mgy_i = \tfrac{1}{2}mv_f^2 + mgy_f$$

Insert the values $v_i = 0$ and $y_f = 0$:

$$0 + mgy_i = \tfrac{1}{2}mv_f^2 + 0$$

Solve for $v_f$ and substitute values for $g$ and $y_i$:

$$v_f = \sqrt{2gy_i} = \sqrt{2(9.80 \text{ m/s}^2)(21.9 \text{ m})} = \boxed{20.7 \text{ m/s}}$$

**(b)** Find the work done on the woman by friction if $v_f = 18.0$ m/s $< 20.7$ m/s.

Write Equation 5.13, substituting expressions for the kinetic and potential energies:

$$W_{nc} = (KE_f - KE_i) + (PE_f - PE_i)$$
$$= (\tfrac{1}{2}mv_f^2 - \tfrac{1}{2}mv_i^2) + (mgy_f - mgy_i)$$

Substitute $m = 60.0$ kg, $v_f = 18.0$ m/s, and $v_i = 0$, and solve for $W_{nc}$:

$$W_{nc} = [\tfrac{1}{2} \cdot 60.0 \text{ kg} \cdot (18.0 \text{ m/s})^2 - 0]$$
$$+ [0 - 60.0 \text{ kg} \cdot (9.80 \text{ m/s}^2) \cdot 21.9 \text{ m}]$$
$$W_{nc} = \boxed{-3.16 \times 10^3 \text{ J}}$$

**REMARKS** The speed found in part **(a)** is the same as if the woman fell vertically through a distance of 21.9 m, consistent with our intuition in Quick Quiz 5.4. The result of part **(b)** is negative because the system loses mechanical energy. Friction transforms part of the mechanical energy into thermal energy and mechanical waves, absorbed partly by the system and partly by the environment.

**QUESTION 5.7** If the slide were not frictionless, would the shape of the slide affect the final answer? Explain.

**EXERCISE 5.7** Suppose a slide similar to Der Stuka is 35.0 m high, but is a straight slope, inclined at 45.0° with respect to the horizontal. **(a)** Find the speed of a 60.0-kg woman at the bottom of the slide, assuming no friction. **(b)** If the woman has a speed of 20.0 m/s at the bottom, find the change in mechanical energy due to friction and **(c)** the magnitude of the force of friction, assumed constant.

**ANSWERS** **(a)** 26.2 m/s  **(b)** $-8.58 \times 10^3$ J  **(c)** 173 N

---

## EXAMPLE 5.8   HIT THE SKI SLOPES

**GOAL** Combine conservation of mechanical energy with the work–energy theorem involving friction on a horizontal surface.

**PROBLEM** A skier starts from rest at the top of a frictionless incline of height 20.0 m, as in Figure 5.20. At the bottom of the incline, the skier encounters a horizontal surface where the coefficient of kinetic friction between skis and snow is 0.210. **(a)** Find the skier's speed at the bottom. **(b)** How far does the skier travel on the horizontal surface before coming to rest? Neglect air resistance.

**STRATEGY** Going down the frictionless incline is physically no different than going down the slide of the previous example and is handled the same way, using conservation of mechanical energy to find the speed $v_\circledB$ at the bottom. On the flat, rough surface, use the work–energy theorem, Equation 5.13, with $W_{nc} = W_{\text{fric}} = -f_k d$, where $f_k$ is the magnitude of the force of friction and $d$ is the distance traveled on the horizontal surface before coming to rest.

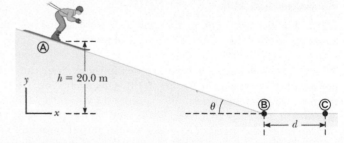

**Figure 5.20** (Example 5.8) The skier slides down the slope and onto a level surface, stopping after traveling a distance $d$ from the bottom of the hill.

## SOLUTION

**(a)** Find the skier's speed at the bottom.

Follow the procedure used in part **(a)** of the previous example as the skier moves from the top, point Ⓐ, to the bottom, point Ⓑ:

$$v_\circledB = \sqrt{2gh} = \sqrt{2(9.80 \text{ m/s}^2)(20.0 \text{ m})} = \boxed{19.8 \text{ m/s}}$$

**(b)** Find the distance traveled on the horizontal, rough surface.

Apply the work–energy theorem as the skier moves from Ⓑ to Ⓒ:

$$W_{net} = -f_k d = \Delta KE = \tfrac{1}{2} m v_{\textcircled{c}}^2 - \tfrac{1}{2} m v_{\textcircled{B}}^2$$

Substitute $v_{\textcircled{c}} = 0$ and $f_k = \mu_k n = \mu_k mg$:

$$-\mu_k mgd = -\tfrac{1}{2} m v_{\textcircled{B}}^2$$

Solve for $d$:

$$d = \frac{v_{\textcircled{B}}^2}{2\mu_k g} = \frac{(19.8 \text{ m/s})^2}{2(0.210)(9.80 \text{ m/s}^2)} = \boxed{95.2 \text{ m}}$$

.........................................................................................................................

**REMARKS** Substituting the symbolic expression $v_{\textcircled{B}} = \sqrt{2gh}$ into the equation for the distance $d$ shows that $d$ is linearly proportional to $h$: Doubling the height doubles the distance traveled.

**QUESTION 5.8** Give two reasons why skiers typically assume a crouching position down when going down a slope.

**EXERCISE 5.8** Find the horizontal distance the skier travels before coming to rest if the incline also has a coefficient of kinetic friction equal to 0.210. Assume that $\theta = 20.0°$.

**ANSWER** 40.3 m

---

# 5.5 Spring Potential Energy

Springs are important elements in modern technology. They are found in machines of all kinds, in watches, toys, cars, and trains. Springs will be introduced here, then studied in more detail in Topic 13.

Work done by an applied force in stretching or compressing a spring can be recovered by removing the applied force, so like gravity, the spring force is conservative, as long as losses through internal friction of the spring can be neglected. That means a potential energy function can be found and used in the work–energy theorem.

Figure 5.21a shows a spring in its equilibrium position, where the spring is neither compressed nor stretched. Pushing a block against the spring as in Figure 5.21b compresses it a distance $x$. Although $x$ appears to be merely a coordinate, for springs it also represents a displacement from the equilibrium position, which for our purposes will always be taken to be at $x = 0$. Experimentally, it turns out that doubling a given displacement requires twice the force, and tripling it takes three times the force. This means the force exerted by the spring, $F_s$, must be proportional to the displacement $x$, or

$$F_s = -kx \qquad [5.16]$$

where $k$ is a constant of proportionality, the *spring constant*, carrying units of newtons per meter. Equation 5.16 is called **Hooke's law,** after Sir Robert Hooke, who discovered the relationship. The force $F_s$ is often called a *restoring force* because the spring always exerts a force in a direction opposite the displacement of its end, tending to restore whatever is attached to the spring to its original position. For positive values of $x$, the force is negative, pointing back towards equilibrium at $x = 0$, and for negative $x$, the force is positive, again pointing towards $x = 0$. For a flexible spring, $k$ is a small number (about 100 N/m), whereas for a stiff spring $k$ is large (about 10 000 N/m). The value of the spring constant $k$ is determined by how the spring was formed, its material composition, and the thickness of the wire. The minus sign ensures that the spring force is always directed back towards the equilibrium point.

As in the case of gravitation, a potential energy, called the **elastic potential energy,** can be associated with the spring force. Elastic potential energy is another way of looking at the work done by a spring during motion because it is equal to

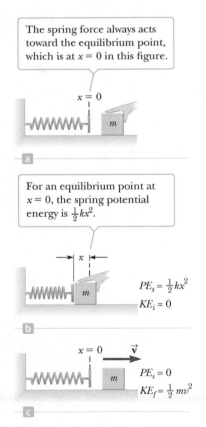

The spring force always acts toward the equilibrium point, which is at $x = 0$ in this figure.

$x = 0$

**a**

For an equilibrium point at $x = 0$, the spring potential energy is $\tfrac{1}{2} kx^2$.

$PE_s = \tfrac{1}{2} kx^2$
$KE_i = 0$

**b**

$x = 0$

$PE_s = 0$
$KE_f = \tfrac{1}{2} mv^2$

**c**

**Figure 5.21** (a) A spring at equilibrium, neither compressed nor stretched. (b) A block of mass $m$ on a frictionless surface is pushed against the spring. (c) When the block is released, the energy stored in the spring is transferred to the block in the form of kinetic energy.

the negative of the work done by the spring. It can also be considered stored energy arising from the work done to compress or stretch the spring.

Consider a horizontal spring and mass at the equilibrium position. We determine the work done by the spring when compressed by an applied force from equilibrium to a displacement $x$, as in Figure 5.21b. The spring force points in the direction opposite the motion, so we expect the work to be negative. When we studied the constant force of gravity near Earth's surface, we found the work done on an object by multiplying the gravitational force by the vertical displacement of the object. However, this procedure can't be used with a varying force such as the spring force. Instead, we use the average force, $\overline{F}$:

$$\overline{F} = \frac{F_0 + F_1}{2} = \frac{0 - kx}{2} = -\frac{kx}{2}$$

Therefore, the work *done by the spring force* is

$$W_s = \overline{F}x = -\tfrac{1}{2}kx^2$$

In general, when the spring is stretched or compressed from $x_i$ to $x_f$, the work done by the spring is

$$W_s = -(\tfrac{1}{2}kx_f^2 - \tfrac{1}{2}kx_i^2)$$

The work done by a spring can be included in the work–energy theorem. Assume Equation 5.13 now includes the work done by springs on the left-hand side. It then reads

$$W_{nc} - (\tfrac{1}{2}kx_f^2 - \tfrac{1}{2}kx_i^2) = \Delta KE + \Delta PE_g$$

where $PE_g$ is the gravitational potential energy.

Spring potential energy ▶ | The elastic potential energy associated with the spring force, $PE_s$, is given by

$$PE_s \equiv \tfrac{1}{2}kx^2 \qquad\qquad [5.17]$$

where $k$ is the spring constant and $x$ is the distance the spring is stretched or compressed from equilibrium.

**SI unit: joule (J)**

The quantity $PE_s$ is often called the *spring potential energy*, to differentiate it from other forms of elastic potential energy, such as that of rubber bands or piano strings, which have different mathematical forms.

Inserting this expression into the previous equation and rearranging gives the new form of the work–energy theorem, including both gravitational and elastic potential energy:

$$W_{nc} = (KE_f - KE_i) + (PE_{gf} - PE_{gi}) + (PE_{sf} - PE_{si}) \qquad\qquad [5.18]$$

where $W_{nc}$ is the work done by nonconservative forces, $KE$ is kinetic energy, $PE_g$ is gravitational potential energy, and $PE_s$ is the elastic potential energy. $PE$, formerly used to denote gravitational potential energy alone, will henceforth denote the total potential energy of a system, including potential energies due to all conservative forces acting on the system.

It's important to remember that the work done by gravity and springs in any given physical system is already included on the right-hand side of Equation 5.18 as potential energy and should not also be included on the left as work.

Figure 5.21c shows how the stored elastic potential energy can be recovered. When the block is released, the spring snaps back to its original length, and the stored elastic potential energy is converted to kinetic energy of the block. The elastic potential energy stored in the spring is zero when the spring is in the equilibrium position ($x = 0$). As given by Equation 5.17, potential energy is also stored in the spring when it's stretched. Further, the elastic potential energy is a maximum when the spring has reached its maximum compression or extension. Finally, because $PE_s$ is proportional to $x^2$, the potential energy is always positive when the spring is not in the equilibrium position.

In the absence of nonconservative forces, $W_{nc} = 0$, so the left-hand side of Equation 5.18 is zero, and an extended form for conservation of mechanical energy results:

$$(KE + PE_g + PE_s)_i = (KE + PE_g + PE_s)_f \qquad [5.19]$$

Problems involving springs, gravity, and other forces are handled in exactly the same way as described in the problem-solving strategy for conservation of mechanical energy, except that the equilibrium point of any spring in the problem must be defined in addition to the zero point for gravitational potential energy.

### Quick Quiz

**5.5** Calculate the elastic potential energy of a spring with spring constant $k = 225$ N/m that is (a) compressed and (b) stretched by $1.00 \times 10^{-2}$ m.

**5.6** True or False: The elastic potential energy of a stretched or compressed spring is always positive.

**5.7** Elastic potential energy depends on the spring constant and the distance the spring is stretched or compressed. By what factor does the elastic potential energy change if the spring's stretch is (a) doubled or (b) tripled?

---

### EXAMPLE 5.9   A HORIZONTAL SPRING

**GOAL** Use conservation of energy to calculate the speed of a block on a horizontal spring with and without friction.

**PROBLEM** A block with mass of 5.00 kg is attached to a horizontal spring with spring constant $k = 4.00 \times 10^2$ N/m, as in Figure 5.22. The surface the block rests upon is frictionless. If the block is pulled out to $x_i = 0.050\ 0$ m and released, **(a)** find the speed of the block when it first reaches the equilibrium point, **(b)** find the speed when $x = 0.025\ 0$ m, and **(c)** repeat part **(a)** if friction acts on the block, with coefficient $\mu_k = 0.150$.

**STRATEGY** In parts **(a)** and **(b)** there are no nonconservative forces, so conservation of energy, Equation 5.19, can be applied. In part **(c)** the definition of work and the work–energy theorem are needed to deal with the loss of mechanical energy due to friction.

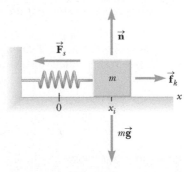

**Figure 5.22** (Example 5.9) A mass attached to a spring.

---

### SOLUTION

**(a)** Find the speed of the block at equilibrium point.

Start with Equation 5.19:

$$(KE + PE_g + PE_s)_i = (KE + PE_g + PE_s)_f$$

Substitute expressions for the block's kinetic energy and the potential energy, and set the gravity terms to zero:

$$(1) \quad \tfrac{1}{2}mv_i^2 + \tfrac{1}{2}kx_i^2 = \tfrac{1}{2}mv_f^2 + \tfrac{1}{2}kx_f^2$$

Substitute $v_i = 0$, $x_f = 0$, and multiply by $2/m$:

$$\frac{k}{m}x_i^2 = v_f^2$$

Solve for $v_f$ and substitute the given values:

$$v_f = \sqrt{\frac{k}{m}}\, x_i = \sqrt{\frac{4.00 \times 10^2\,\text{N/m}}{5.00\,\text{kg}}}\,(0.050\ 0\,\text{m})$$

$$= 0.447\ \text{m/s}$$

**(b)** Find the speed of the block at the halfway point.

Set $v_i = 0$ in Equation (1) and multiply by $2/m$:

$$\frac{kx_i^2}{m} = v_f^2 + \frac{kx_f^2}{m}$$

*(Continued)*

Solve for $v_f$ and substitute the given values:

$$v_f = \sqrt{\frac{k}{m}(x_i^2 - x_f^2)}$$

$$= \sqrt{\frac{4.00 \times 10^2 \text{ N/m}}{5.00 \text{ kg}}[(0.050\,0 \text{ m})^2 - (0.025\,0 \text{ m})^2]}$$

$$= \boxed{0.387 \text{ m/s}}$$

**(c)** Repeat part **(a)**, this time with friction.

Apply the work–energy theorem. The work done by the force of gravity and the normal force is zero because these forces are perpendicular to the motion.

$$W_{\text{fric}} = \tfrac{1}{2}mv_f^2 - \tfrac{1}{2}mv_i^2 + \tfrac{1}{2}kx_f^2 - \tfrac{1}{2}kx_i^2$$

Substitute $v_i = 0$, $x_f = 0$, and $W_{\text{fric}} = -\mu_k n x_i$:

$$-\mu_k n x_i = \tfrac{1}{2}mv_f^2 - \tfrac{1}{2}kx_i^2$$

Set $n = mg$ and solve for $v_f$:

$$\tfrac{1}{2}mv_f^2 = \tfrac{1}{2}kx_i^2 - \mu_k mgx_i$$

$$v_f = \sqrt{\frac{k}{m}x_i^2 - 2\mu_k g x_i}$$

$$v_f = \sqrt{\frac{4.00 \times 10^2 \text{ N/m}}{5.00 \text{ kg}}(0.050\,0 \text{ m})^2 - 2(0.150)(9.80 \text{ m/s}^2)(0.050\,0 \text{ m})}$$

$$v_f = \boxed{0.230 \text{ m/s}}$$

**REMARKS** Friction or drag from immersion in a fluid damps the motion of an object attached to a spring, eventually bringing the object to rest.

**QUESTION 5.9** In the case of friction, what percent of the mechanical energy was lost by the time the mass first reached the equilibrium point? [*Hint:* use the answers to parts **(a)** and **(c)**.]

**EXERCISE 5.9** Suppose the spring system in the last example starts at $x = 0$ and the attached object is given a kick to the right, so it has an initial speed of 0.600 m/s. **(a)** What distance from the origin does the object travel before coming to rest, assuming the surface is frictionless? **(b)** How does the answer change if the coefficient of kinetic friction is $\mu_k = 0.150$? (Use the quadratic formula.)

**ANSWERS** **(a)** 0.067 1 m   **(b)** 0.051 2 m

---

**EXAMPLE 5.10   CIRCUS ACROBAT**

**GOAL** Use conservation of mechanical energy to solve a one-dimensional problem involving gravitational potential energy and spring potential energy.

**PROBLEM** A 50.0-kg circus acrobat drops from a height of 2.00 m straight down onto a springboard with a force constant of $8.00 \times 10^3$ N/m, as in Figure 5.23. By what maximum distance does she compress the spring?

**STRATEGY** Nonconservative forces are absent, so conservation of mechanical energy can be applied. At the two points of interest, the acrobat's initial position and the point of maximum spring compression, her velocity is zero, so the kinetic energy terms will be zero. Choose $y = 0$ as the point of maximum compression, so the final gravitational potential energy is zero. This choice also means that the initial position of the acrobat is $y_i = h + d$, where $h$ is the acrobat's initial height above the platform and $d$ is the spring's maximum compression.

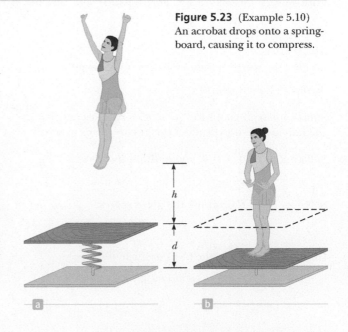

**Figure 5.23** (Example 5.10) An acrobat drops onto a springboard, causing it to compress.

## SOLUTION

Use conservation of mechanical energy:

**(1)**  $(KE + PE_g + PE_s)_i = (KE + PE_g + PE_s)_f$

The only nonzero terms are the initial gravitational potential energy and the final spring potential energy.

$0 + mg(h + d) + 0 = 0 + 0 + \frac{1}{2}kd^2$

$mg(h + d) = \frac{1}{2}kd^2$

Substitute the given quantities and rearrange the equation into standard quadratic form:

$(50.0 \text{ kg})(9.80 \text{ m/s}^2)(2.00 \text{ m} + d) = \frac{1}{2}(8.00 \times 10^3 \text{ N/m})\,d^2$

$d^2 - (0.123 \text{ m})d - 0.245 \text{ m}^2 = 0$

Solve with the quadratic formula (Equation A.8):

$d =$  $\boxed{0.560 \text{ m}}$

**REMARKS** The other solution, $d = -0.437$ m, can be rejected because $d$ was chosen to be a positive number at the outset. A change in the acrobat's center of mass, say, by crouching as she makes contact with the springboard, also affects the spring's compression, but that effect was neglected. Shock absorbers often involve springs, and this example illustrates how they work. The spring action of a shock absorber turns a dangerous jolt into a smooth deceleration, as excess kinetic energy is converted to spring potential energy.

**QUESTION 5.10** Is it possible for the acrobat to rebound to a height greater than her initial height? If so, how?

**EXERCISE 5.10** An 8.00-kg block drops straight down from a height of 1.00 m, striking a platform spring having force constant $1.00 \times 10^3$ N/m. Find the maximum compression of the spring.

**ANSWER** $d = 0.482$ m

---

## EXAMPLE 5.11   A BLOCK PROJECTED UP A FRICTIONLESS INCLINE

**GOAL** Use conservation of mechanical energy to solve a problem involving gravitational potential energy, spring potential energy, and a ramp.

**PROBLEM** A 0.500-kg block rests on a horizontal, frictionless surface as in Figure 5.24. The block is pressed back against a spring having a constant of $k = 625$ N/m, compressing the spring by 10.0 cm to point Ⓐ. Then the block is released. **(a)** Find the maximum distance $d$ the block travels up the frictionless incline if $\theta = 30.0°$. **(b)** How fast is the block going at half its maximum height?

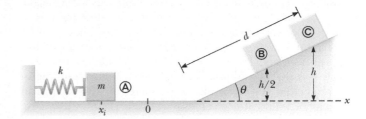

**Figure 5.24** (Example 5.11)

**STRATEGY** In the absence of other forces, conservation of mechanical energy applies to parts **(a)** and **(b)**. In part **(a)**, the block starts at rest and is also instantaneously at rest at the top of the ramp, so the kinetic energies at Ⓐ and Ⓒ are both zero. Note that the question asks for a distance $d$ along the ramp, not the height $h$. In part **(b)**, the system has both kinetic and gravitational potential energy at Ⓑ.

## SOLUTION

**(a)** Find the distance the block travels up the ramp.

Apply conservation of mechanical energy:

$\frac{1}{2}mv_i^2 + mgy_i + \frac{1}{2}kx_i^2 = \frac{1}{2}mv_f^2 + mgy_f + \frac{1}{2}kx_f^2$

Substitute $v_i = v_f = 0$, $y_i = 0$, $y_f = h = d \sin \theta$, and $x_f = 0$:

$\frac{1}{2}kx_i^2 = mgh = mgd \sin \theta$

Solve for the distance $d$ and insert the known values:

$d = \dfrac{\frac{1}{2}kx_i^2}{mg \sin \theta} = \dfrac{\frac{1}{2}(625 \text{ N/m})(-0.100 \text{ m})^2}{(0.500 \text{ kg})(9.80 \text{ m/s}^2) \sin (30.0°)}$

$= \boxed{1.28 \text{ m}}$

**(b)** Find the velocity at half the height, $h/2$. Note that $h = d \sin \theta = (1.28 \text{ m}) \sin 30.0° = 0.640$ m.

Use energy conservation again:

$\frac{1}{2}mv_i^2 + mgy_i + \frac{1}{2}kx_i^2 = \frac{1}{2}mv_f^2 + mgy_f + \frac{1}{2}kx_f^2$

(*Continued*)

Take $v_i = 0$, $y_i = 0$, $y_f = \frac{1}{2} h$, and $x_f = 0$, yielding

$$\frac{1}{2} kx_i^2 = \frac{1}{2} mv_f^2 + mg\left(\frac{1}{2}h\right)$$

Multiply by $2/m$ and solve for $v_f$:

$$\frac{k}{m} x_i^2 = v_f^2 + gh$$

$$v_f = \sqrt{\frac{k}{m} x_i^2 - gh}$$

$$= \sqrt{\left(\frac{625 \text{ N/m}}{0.500 \text{ kg}}\right)(-0.100 \text{ m})^2 - (9.80 \text{ m/s}^2)(0.640 \text{ m})}$$

$$v_f = \boxed{2.50 \text{ m/s}}$$

**REMARKS** Notice that it wasn't necessary to compute the velocity gained upon leaving the spring: only the mechanical energy at each of the two points of interest was required, where the block was at rest.

**QUESTION 5.11** A real spring will continue to vibrate slightly after the mass has left it. How would this affect the answer to part **(a)**, and why?

**EXERCISE 5.11** A 1.00-kg block is shot horizontally from a spring, as in the previous example, and travels 0.500 m up along a frictionless ramp before coming to rest and sliding back down. If the ramp makes an angle of 45.0° with respect to the horizontal, and the spring was originally compressed by 0.120 m, find the spring constant.

**ANSWER** 481 N/m

---

**APPLYING PHYSICS 5.2    ACCIDENT RECONSTRUCTION**

Sometimes people involved in automobile accidents make exaggerated claims of chronic pain due to subtle injuries to the neck or spinal column. The likelihood of injury can be determined by finding the change in velocity of a car during the accident. The larger the change in velocity, the more likely it is that the person suffered spinal injury resulting in chronic pain. How can reliable estimates for this change in velocity be found after the fact?

**EXPLANATION** The metal and plastic of an automobile acts much like a spring, absorbing the car's kinetic energy by flexing during a collision. When the magnitude of the difference in velocity of the two cars is under 5 mi/h, there is usually no visible damage, because bumpers are designed to absorb the impact and return to their original shape at such low speeds. At greater relative speeds there will be permanent damage to the vehicle. Despite the fact the structure of the car may not return to its original shape, a certain force per meter is still required to deform it, just as it takes a certain force per meter to compress a spring. The greater the original kinetic energy, the more the car is compressed during a collision, and the greater the damage. By using data obtained through crash tests, it's possible to obtain effective spring constants for all the different models of cars and determine reliable estimates of the change in velocity of a given vehicle during an accident. Medical research has established the likelihood of spinal injury for a given change in velocity, and the estimated velocity change can be used to help reduce insurance fraud. ■

# 5.6 Systems and Energy Conservation

Recall that the work–energy theorem can be written as

$$W_{nc} + W_c = \Delta KE$$

where $W_{nc}$ represents the work done by nonconservative forces and $W_c$ is the work done by conservative forces in a given physical context. As we have seen, any work done by conservative forces, such as gravity and springs, can be accounted for by changes in potential energy. The work–energy theorem can therefore be written in the following way:

$$W_{nc} = \Delta KE + \Delta PE = (KE_f - KE_i) + (PE_f - PE_i) \qquad [5.20]$$

where now, as previously stated, PE includes all potential energies. This equation is easily rearranged to

$$W_{nc} = (KE_f + PE_f) - (KE_i + PE_i) \qquad [5.21]$$

Recall, however, that the total mechanical energy is given by $E = KE + PE$. Making this substitution into Equation 5.21, we find that the work done on a system by all nonconservative forces is equal to the change in mechanical energy of that system:

$$W_{nc} = E_f - E_i = \Delta E \qquad [5.22]$$

If the mechanical energy is changing, it has to be going somewhere. The energy either leaves the system and goes into the surrounding environment, or it stays in the system and is converted into a nonmechanical form such as thermal energy.

A simple example is a block sliding along a rough surface. Friction creates thermal energy, absorbed partly by the block and partly by the surrounding environment. When the block warms up, something called *internal energy* increases. The internal energy of a system is related to its temperature, which in turn is a consequence of the activity of its parts, such as the motion of atoms in a gas or the vibration of atoms in a solid. (Internal energy will be studied in more detail in Topics 10–12.)

Energy can be transferred between a nonisolated system and its environment. If positive work is done on the system, energy is transferred from the environment to the system. If negative work is done on the system, energy is transferred from the system to the environment.

So far, we have encountered three methods of storing energy in a system: kinetic energy, potential energy, and internal energy. However, we've seen only one way of transferring energy into or out of a system: through work. Other methods will be studied in later topics but are summarized here:

- **Work**, in the mechanical sense of this topic transfers energy to a system by displacing it with an applied force.
- **Heat** is the process of transferring energy through microscopic collisions between atoms or molecules. For example, a metal spoon resting in a cup of coffee becomes hot because some of the kinetic energy of the molecules in the liquid coffee is transferred to the spoon as internal energy.
- **Mechanical waves** transfer energy by creating a disturbance that propagates through air or another medium. For example, energy in the form of sound leaves your stereo system through the loudspeakers and enters your ears to stimulate the hearing process. Other examples of mechanical waves are seismic waves and ocean waves.
- **Electrical transmission** transfers energy through electric currents. This is how energy enters your stereo system or any other electrical device.
- **Electromagnetic radiation** transfers energy in the form of electromagnetic waves such as light, microwaves, and radio waves. Examples of this method of transfer include cooking a potato in a microwave oven and light energy traveling from the Sun to Earth through space.

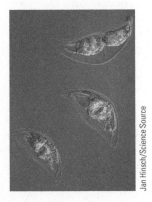

Jan Hinsch/Science Source

**Figure 5.25** This small plant, found in warm southern waters, exhibits bioluminescence, a process in which chemical energy is converted to light. The red areas are chlorophyll, which fluoresce when irradiated with blue light.

**BIO** APPLICATION

Flagellar Movement; Bioluminescence

## 5.6.1 Conservation of Energy in General

The most important feature of the energy approach is the idea that energy is conserved; it can't be created or destroyed, only transferred from one form into another. This is the principle of **conservation of energy.**

The principle of conservation of energy is not confined to physics. In biology, energy transformations take place in myriad ways inside all living organisms. One example is the transformation of chemical energy to mechanical energy that causes flagella to move and propel an organism. Some bacteria use chemical energy to produce light. (See Fig. 5.25.) Although the mechanisms that produce these light emissions are not well understood, living creatures often rely on this light for their existence. For example, certain fish have sacs beneath their eyes filled with light-emitting bacteria. The emitted light attracts creatures that become food for the fish.

**5.8** A book of mass $m$ is projected with a speed $v$ across a horizontal surface. The book slides until it stops due to the friction force between the book and the surface. The surface is now tilted 30°, and the book is projected up the surface with the same initial speed $v$. When the book has come to rest, how does the decrease in mechanical energy of the book–Earth system compare with that when the book slid over the horizontal surface? (a) It's the same. (b) It's larger on the tilted surface. (c) It's smaller on the tilted surface. (d) More information is needed.

## APPLYING PHYSICS 5.3    ASTEROID IMPACT!

An asteroid about 10 kilometers in diameter has been blamed for the extinction of the dinosaurs 65 million years ago. How can a relatively small object, which could fit inside a college campus, inflict such injury on the vast biosphere of Earth?

**EXPLANATION** Although such an asteroid is comparatively small, it travels at a very high speed relative to Earth, typically on the order of 40 000 m/s. A roughly spherical asteroid 10 kilometers in diameter and made mainly of rock has a mass of approximately 1 000 trillion kilograms—a mountain of matter. The kinetic energy of such an asteroid would be about $10^{24}$ J, or a trillion trillion joules. By contrast, the atomic bomb that devastated Hiroshima was equivalent to 15 kilotons of TNT, approximately $6 \times 10^{13}$ J of energy. On striking Earth, the asteroid's enormous kinetic energy would change into other forms, such as thermal energy, sound, and light, with a total energy release greater than ten billion Hiroshima explosions! Aside from the devastation in the immediate blast area and fires across a continent, gargantuan tidal waves would scour low-lying regions around the world and dust would block the Sun for decades.

For this reason, asteroid impacts represent a threat to life on Earth. Asteroids large enough to cause widespread extinction hit Earth only every 60 million years or so. Smaller asteroids, of sufficient size to cause serious damage

**Figure 5.26** Asteroid map of the inner solar system. The violet circles represent the orbits of the inner planets. Green dots stand for asteroids not considered dangerous to Earth; those that are considered threatening are represented by red dots.

to civilization on a global scale, are thought to strike every five to ten thousand years. There have been several near misses by such asteroids in the last century and even in the last decade. In 1907, a small asteroid or comet fragment struck Tunguska, Siberia, annihilating a region 60 kilometers across. Had it hit northern Europe, millions of people might have perished.

Figure 5.26 is an asteroid map of the inner solar system. More asteroids are being discovered every year. ■

# 5.7 Power

Power, the rate at which energy is transferred, is important in the design and use of practical devices, such as electrical appliances and engines of all kinds. The concept of power, however, is essential whenever a transfer of any kind of energy takes place. The issue is particularly interesting for living creatures because the maximum work per second, or power output, of an animal varies greatly with output duration. Power is defined as the rate of energy transfer with time:

Average power ▶ If an external force does work $W$ on an object in the time interval $\Delta t$, then the **average power** delivered to the object is the work done divided by the time interval, or

$$\overline{P} = \frac{W}{\Delta t} \qquad [5.23]$$

**SI unit: watt (W = J/s)**

It's sometimes useful to rewrite Equation 5.23 by substituting $W = F \Delta x$ and noticing that $\Delta x/\Delta t$ is the average velocity of the object during the time $\Delta t$:

$$\overline{P} = \frac{W}{\Delta t} = \frac{F \Delta x}{\Delta t} = F\overline{v} \qquad [5.24]$$

According to Equation 5.24, average power is a constant force times the average velocity. The force $F$ is the component of force in the direction of the average velocity. A more general definition, called the **instantaneous power,** can be written down with a little calculus and has the same form as Equation 5.24:

$$P = Fv \qquad\qquad [5.25]$$

◀ Instantaneous power

In Equation 5.25 both the force $F$ and the velocity $v$ must be parallel, but can change with time. The SI unit of power is the joule per second (J/s), also called the **watt,** named after James Watt:

$$1\ W = 1\ J/s = 1\ kg \cdot m^2/s^3 \qquad [5.26a]$$

The unit of power in the U.S. customary system is the horsepower (hp), where

$$1\ hp \equiv 550\ \frac{ft \cdot lb}{s} = 746\ W \qquad [5.26b]$$

The horsepower was first defined by Watt, who needed a large power unit to rate the power output of his new invention, the steam engine.

The watt is commonly used in electrical applications, but it can be used in other scientific areas as well. For example, European sports car engines are rated in kilowatts.

In electric power generation, it's customary to use the kilowatt-hour as a measure of energy. One kilowatt-hour (kWh) is the energy transferred in 1 h at the constant rate of 1 kW = 1 000 J/s. Therefore,

$$1\ kWh = (10^3\ W)(3\ 600\ s) = (10^3\ J/s)(3\ 600\ s) = 3.60 \times 10^6\ J$$

It's important to realize that a kilowatt-hour is a unit of energy, *not* power. When you pay your electric bill, you're buying energy, and that's why your bill lists a charge for electricity of about 10 cents/kWh. The amount of electricity used by an appliance can be calculated by multiplying its power rating (usually expressed in watts and valid only for normal household electrical circuits) by the length of time the appliance is operated. For example, an electric bulb rated at 100 W (= 0.100 kW) "consumes" $3.6 \times 10^5$ J of energy in 1 h.

**Tip 5.6** Watts the Difference?

Don't confuse the nonitalic symbol for watts, W, with the italic symbol $W$ for work. A watt is a unit, the same as joules per second. Work is a concept, carrying units of joules.

## EXAMPLE 5.12 POWER DELIVERED BY AN ELEVATOR MOTOR

**GOAL** Apply the force-times-velocity definition of power.

**PROBLEM** A $1.00 \times 10^3$-kg elevator car carries a maximum load of $8.00 \times 10^2$ kg. A constant frictional force of $4.00 \times 10^3$ N retards its motion upward, as in Figure 5.27. What minimum power, in kilowatts and in horsepower, must the motor deliver to lift the fully loaded elevator car at a constant speed of 3.00 m/s?

**STRATEGY** To solve this problem, we need to determine the force the elevator car's motor must deliver through the force of tension in the cable, $\vec{T}$. Substituting this force together with the given speed $v$ into $P = Fv$ gives the desired power. The tension in the cable, $T$, can be found with Newton's second law.

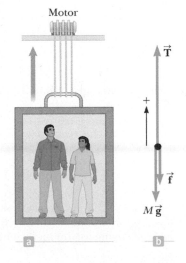

Motor

**Figure 5.27** (Example 5.12) (a) The motor exerts an upward force $\vec{T}$ on the elevator. A frictional force $\vec{f}$ and the force of gravity $M\vec{g}$ act downward. (b) The free-body diagram for the elevator car.

(*Continued*)

## SOLUTION

Apply Newton's second law to the elevator car:

$$\sum \vec{F} = m\vec{a}$$

The velocity is constant, so the acceleration is zero. The forces acting on the elevator car are the force of tension in the cable, $\vec{T}$, the friction $\vec{f}$, and gravity $M\vec{g}$, where $M$ is the mass of the elevator car.

$$\vec{T} + \vec{f} + M\vec{g} = 0$$

Write the equation in terms of its components:

$$T - f - Mg = 0$$

Solve this equation for the tension $T$ and evaluate it:

$$T = f + Mg$$
$$= 4.00 \times 10^3 \text{ N} + (1.80 \times 10^3 \text{ kg})(9.80 \text{ m/s}^2)$$
$$T = 2.16 \times 10^4 \text{ N}$$

Substitute this value of $T$ for $F$ in the power equation:

$$P = Fv = (2.16 \times 10^4 \text{ N})(3.00 \text{ m/s}) = \boxed{64.8 \text{ kW}}$$
$$P = 64.8 \text{ kW} = \boxed{86.9 \text{ hp}}$$

**REMARKS** The friction force acts to retard the motion, requiring more power. For a descending elevator car, the friction force can actually reduce the power requirement.

**QUESTION 5.12** In general, are the minimum power requirements of an elevator car ascending at constant velocity (a) greater than, (b) less than, or (c) equal to the minimum power requirements of an elevator car descending at constant velocity?

**EXERCISE 5.12** Suppose the same elevator car with the same load descends at 3.00 m/s. What minimum power is required? (Here, the motor removes energy from the elevator car by not allowing it to fall freely.)

**ANSWER** $4.09 \times 10^4 \text{ W} = 54.9 \text{ hp}$

---

## EXAMPLE 5.13 SHAMU SPRINT `BIO`

**GOAL** Calculate the average power needed to increase an object's kinetic energy.

**PROBLEM** Killer whales are known to reach 32 ft in length and have a mass of over 8 000 kg. They are also very quick, able to accelerate up to 30 mi/h in a matter of seconds. Disregarding the considerable drag force of water, calculate the average power a killer whale named Shamu with mass $8.00 \times 10^3$ kg would need to generate to reach a speed of 12.0 m/s in 6.00 s.

**STRATEGY** Find the change in kinetic energy of Shamu and use the work–energy theorem to obtain the minimum work Shamu has to do to effect this change. (Internal and external friction forces increase the necessary amount of energy.) Divide by the elapsed time to get the average power.

## SOLUTION

Calculate the change in Shamu's kinetic energy. By the work–energy theorem, this equals the minimum work Shamu must do:

$$\Delta KE = \tfrac{1}{2}mv_f^2 - \tfrac{1}{2}mv_i^2$$
$$= \tfrac{1}{2} \cdot 8.00 \times 10^3 \text{ kg} \cdot (12.0 \text{ m/s})^2 - 0$$
$$= 5.76 \times 10^5 \text{ J}$$

Divide by the elapsed time (Eq. 5.23), noting that $W = \Delta KE$:

$$\overline{P} = \frac{W}{\Delta t} = \frac{5.76 \times 10^5 \text{ J}}{6.00 \text{ s}} = \boxed{9.60 \times 10^4 \text{ W}}$$

**REMARKS** This is enough power to run a moderate-sized office building! The actual requirements are larger because of friction in the water and muscular tissues. Something similar can be done with gravitational potential energy, as the exercise illustrates.

**QUESTION 5.13** If Shamu could double his velocity in double the time, by what factor would the average power requirement change?

**EXERCISE 5.13** What minimum average power must a 35-kg human boy generate climbing up the stairs to the top of the Washington monument? The trip up the nearly 170-m-tall building takes him 10.0 minutes. Include only work done against gravity, ignoring biological inefficiency.

**ANSWER** 97 W

## EXAMPLE 5.14    SPEEDBOAT POWER

**GOAL**  Combine power, the work–energy theorem, and nonconservative forces with one-dimensional kinematics.

**PROBLEM**  **(a)** What average power would a $1.00 \times 10^3$-kg speedboat need to go from rest to 20.0 m/s in 5.00 s, assuming the water exerts a constant drag force of magnitude $f_d = 5.00 \times 10^2$ N and the acceleration is constant. **(b)** Find an expression for the instantaneous power in terms of the drag force $f_d$, the mass $m$, acceleration $a$, and time $t$.

**STRATEGY**  The power is provided by the engine, which creates a nonconservative force. Use the work–energy theorem together with the work done by the engine, $W_{engine}$, and the work done by the drag force, $W_{drag}$, on the left-hand side. Use one-dimensional kinematics to find the acceleration and then the displacement $\Delta x$. Solve the work–energy theorem for $W_{engine}$, and divide by the elapsed time to get the average power. For part **(b)**, use Newton's second law to obtain an equation for $F_E$, and then substitute into the definition of instantaneous power.

**SOLUTION**

**(a)** Write the work–energy theorem:

$$W_{net} = \Delta KE = \tfrac{1}{2}mv_f^2 - \tfrac{1}{2}mv_i^2$$

Fill in the two work terms and take $v_i = 0$:

**(1)**  $W_{engine} + W_{drag} = \tfrac{1}{2}mv_f^2$

To get the displacement $\Delta x$, first find the acceleration using the velocity equation of kinematics:

$$v_f = at + v_i \;\to\; v_f = at$$
$$20.0 \text{ m/s} = a(5.00 \text{ s}) \;\to\; a = 4.00 \text{ m/s}^2$$

Substitute $a$ into the time-independent kinematics equation and solve for $\Delta x$:

$$v_f^2 - v_i^2 = 2a\,\Delta x$$
$$(20.0 \text{ m/s})^2 - 0^2 = 2(4.00 \text{ m/s}^2)\,\Delta x$$
$$\Delta x = 50.0 \text{ m}$$

Now that we know $\Delta x$, we can find the mechanical energy lost due to the drag force:

$$W_{drag} = -f_d\,\Delta x = -(5.00 \times 10^2 \text{ N})(50.0 \text{ m}) = -2.50 \times 10^4 \text{ J}$$

Solve Equation (1) for $W_{engine}$:

$$W_{engine} = \tfrac{1}{2}mv_f^2 - W_{drag}$$
$$= \tfrac{1}{2}(1.00 \times 10^3 \text{ kg})(20.0 \text{ m/s})^2 - (-2.50 \times 10^4 \text{ J})$$
$$W_{engine} = 2.25 \times 10^5 \text{ J}$$

Compute the average power:

$$\overline{P} = \frac{W_{engine}}{\Delta t} = \frac{2.25 \times 10^5 \text{ J}}{5.00 \text{ s}} = 4.50 \times 10^4 \text{ W} = \boxed{60.3 \text{ hp}}$$

**(b)** Find a symbolic expression for the instantaneous power.

Use Newton's second law:

$$ma = F_E - f_d$$

Solve for the force exerted by the engine, $F_E$:

$$F_E = ma + f_d$$

Substitute the expression for $F_E$ and $v = at$ into Equation 5.25 to obtain the instantaneous power:

$$P = F_E v = (ma + f_d)(at)$$
$$\boxed{P = (ma^2 + af_d)t}$$

**REMARKS**  In fact, drag forces generally get larger with increasing speed.

**QUESTION 5.14**  How does the instantaneous power at the end of 5.00 s compare to the average power?

**EXERCISE 5.14**  What average power must be supplied to push a 5.00-kg block from rest to 10.0 m/s in 5.00 s when the coefficient of kinetic friction between the block and surface is 0.250? Assume the acceleration is uniform.

**ANSWER**  111 W

## 5.7.1 Energy and Power in a Vertical Jump  BIO

The stationary jump consists of two parts: extension and free flight.[2] In the extension phase the person jumps up from a crouch, straightening the legs and throwing up the arms; the free-flight phase occurs when the jumper leaves the ground.

---

[2]For more information on this topic, see E. J. Offenbacher, *American Journal of Physics* (38): 829, 1969.

**Figure 5.28** Extension and free flight in the vertical jump.

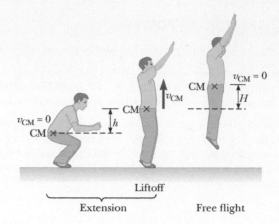

Because the body is an extended object and different parts move with different speeds, we describe the motion of the jumper in terms of the position and velocity of the **center of mass (CM)**, which is the point in the body at which all the mass may be considered to be concentrated. Figure 5.28 shows the position and velocity of the CM at different stages of the jump.

Using the principle of the conservation of mechanical energy, we can find $H$, the maximum increase in height of the CM, in terms of the velocity $v_{CM}$ of the CM at liftoff. Taking $PE_i$, the gravitational potential energy of the jumper–Earth system just as the jumper lifts off from the ground to be zero, and noting that the kinetic energy $KE_f$ of the jumper at the peak is zero, we have

$$PE_i + KE_i = PE_f + KE_f$$

$$\tfrac{1}{2}mv_{CM}^2 = mgH \quad \text{or} \quad H = \frac{v_{CM}^2}{2g}$$

We can estimate $v_{CM}$ by assuming that the acceleration of the CM is constant during the extension phase. If the depth of the crouch is $h$ and the time for extension is $\Delta t$, we find that $v_{CM} = 2\bar{v} = 2h/\Delta t$. Measurements on a group of male college students show typical values of $h = 0.40$ m and $\Delta t = 0.25$ s, the latter value being set by the fixed speed with which muscle can contract. Substituting, we obtain

$$v_{CM} = 2(0.40 \text{ m})/(0.25 \text{ s}) = 3.2 \text{ m/s}$$

and

$$H = \frac{v_{CM}^2}{2g} = \frac{(3.2 \text{ m/s})^2}{2(9.80 \text{ m/s}^2)} = 0.52 \text{ m}$$

Measurements on this same group of students found that $H$ was between 0.45 m and 0.61 m in all cases, confirming the basic validity of our simple calculation.

To relate the abstract concepts of energy, power, and efficiency to humans, it's interesting to calculate these values for the vertical jump. The kinetic energy given to the body in a jump is $KE = \tfrac{1}{2}mv_{CM}^2$, and for a person of mass 68 kg, the kinetic energy is

$$KE = \tfrac{1}{2}(68 \text{ kg})(3.2 \text{ m/s})^2 = 3.5 \times 10^2 \text{ J}$$

Although this may seem like a large expenditure of energy, we can make a simple calculation to show that jumping and exercise in general are not good ways to lose weight, in spite of their many health benefits. Because the muscles are at most 25% efficient at producing kinetic energy from chemical energy (muscles always produce a lot of internal energy and kinetic energy as well as work—that's why you perspire when you work out), they use up four times the 350 J (about 1 400 J) of chemical energy in one jump. This chemical energy ultimately comes from the food we eat, with energy content given in units of food

**BIO APPLICATION**
Diet vs. Exercise in Weight-loss Programs

calories and one food calorie equal to 4 200 J. So the total energy supplied by the body as internal energy and kinetic energy in a vertical jump is only about one-third of a food calorie!

Finally, it's interesting to calculate the average mechanical power that can be generated by the body in strenuous activity for brief periods. Here we find that

$$\overline{P} = \frac{KE}{\Delta t} = \frac{3.5 \times 10^2 \text{ J}}{0.25 \text{ s}} = 1.4 \times 10^3 \text{ W}$$

or $(1\ 400\text{ W})(1\text{ hp}/746\text{ W}) = 1.9$ hp. So humans can produce about 2 hp of mechanical power for periods on the order of seconds. Table 5.1 shows the maximum power outputs from humans for various periods while bicycling and rowing, activities in which it is possible to measure power output accurately.

**Table 5.1** Maximum Power Output from Humans over Various Periods BIO

| Power | Time |
|---|---|
| 2 hp, or 1 500 W | 6 s |
| 1 hp, or 750 W | 60 s |
| 0.35 hp, or 260 W | 35 min |
| 0.2 hp, or 150 W | 5 h |
| 0.1 hp, or 75 W (safe daily level) | 8 h |

# 5.8 Work Done by a Varying Force

Suppose an object is displaced along the $x$-axis under the action of a force $F_x$ that acts in the $x$-direction and varies with position, as shown in Figure 5.29. The object is displaced in the direction of increasing $x$ from $x = x_i$ to $x = x_f$. In such a situation, we can't use Equation 5.2 to calculate the work done by the force because this relationship applies only when $\vec{F}$ is constant in magnitude and direction. However, if we imagine that the object undergoes the *small* displacement $\Delta x$ shown in Figure 5.28a, then the $x$-component $F_x$ of the force is nearly constant over this interval and we can approximate the work done by the force for this small displacement as

$$W_1 \cong F_x \Delta x \qquad\qquad [5.27]$$

This quantity is just the area of the shaded rectangle in Figure 5.28a. If we imagine that the curve of $F_x$ versus $x$ is divided into a large number of such intervals, then the total work done for the displacement from $x_i$ to $x_f$ is approximately equal to the sum of the areas of a large number of small rectangles:

$$W \cong F_1 \Delta x_1 + F_2 \Delta x_2 + F_3 \Delta x_3 + \cdots \qquad\qquad [5.28]$$

Now imagine going through the same process with twice as many intervals, each half the size of the original $\Delta x$. The rectangles then have smaller widths and will better approximate the area under the curve. Continuing the process of increasing the number of intervals while allowing their size to approach zero,

**Figure 5.29** (a) The work done on a particle by the force component $F_x$ for the small displacement $\Delta x$ is approximately $F_x \Delta x$, the area of the shaded rectangle. (b) The width $\Delta x$ of each rectangle is shrunk to zero.

The sum of the areas of all the rectangles approximates the work done by the force $F_x$ on the particle during its displacement from $x_i$ to $x_f$.

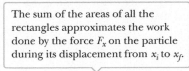
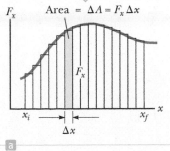

The area under the curve exactly equals the work done by the force $F_x$ on the particle during its displacement from $x_i$ to $x_f$.

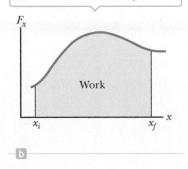

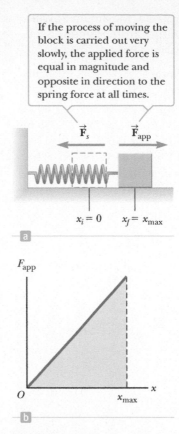

If the process of moving the block is carried out very slowly, the applied force is equal in magnitude and opposite in direction to the spring force at all times.

$\vec{F}_s$   $\vec{F}_{app}$

$x_i = 0$   $x_f = x_{max}$

a

$F_{app}$

$O$   $x_{max}$   $x$

b

**Figure 5.30** (a) A block being pulled from $x_i = 0$ to $x_f = x_{max}$ on a frictionless surface by a force $\vec{F}_{app}$. (b) A graph of $F_{app}$ vs. $x$.

the number of terms in the sum increases without limit, but the value of the sum approaches a definite value equal to the area under the curve bounded by $F_x$ and the x-axis in Figure 5.28b. In other words, **the work done by a variable force acting on an object that undergoes a displacement is equal to the area under the graph of $F_x$ versus x.**

A common physical system in which force varies with position consists of a block on a horizontal, frictionless surface connected to a spring, as discussed in Section 5.5. When the spring is stretched or compressed a small distance $x$ from its equilibrium position $x = 0$, it exerts a force on the block given by $F_x = -kx$, where $k$ is the force constant of the spring.

Now let's determine the work done by an external agent on the block as the spring is stretched *very slowly* from $x_i = 0$ to $x_f = x_{max}$, as in Figure 5.30a. This work can be easily calculated by noting that at any value of the displacement, Newton's third law tells us that the applied force $\vec{F}_{app}$ is equal in magnitude to the spring force $\vec{F}_s$ and acts in the opposite direction, so that $F_{app} = -(-kx) = kx$. A plot of $F_{app}$ versus $x$ is a straight line, as shown in Figure 5.30b. Therefore, the work done by this applied force in stretching the spring from $x = 0$ to $x = x_{max}$ is the area under the straight line in that figure, which in this case is the area of the shaded triangle:

$$W_{F_{app}} = \tfrac{1}{2}kx_{max}^2$$

During this same time the spring has done exactly the same amount of work, but that work is negative, because the spring force points in the direction opposite the motion. The potential energy of the system is exactly equal to the work done by the applied force and is the same sign, which is why potential energy is thought of as stored work.

---

## EXAMPLE 5.15 | WORK REQUIRED TO STRETCH A SPRING

**GOAL** Apply the graphical method of finding work.

**PROBLEM** One end of a horizontal spring ($k = 80.0$ N/m) is held fixed while an external force is applied to the free end, stretching it slowly from $x_{\circledA} = 0$ to $x_{\circledB} = 4.00$ cm. **(a)** Find the work done by the applied force on the spring. **(b)** Find the additional work done in stretching the spring from $x_{\circledB} = 4.00$ cm to $x_{\circledC} = 7.00$ cm.

**STRATEGY** For part **(a)**, simply find the area of the smaller triangle in Figure 5.31, using $A = \tfrac{1}{2}bh$, one-half the base times the height. For part **(b)**, the easiest way to find the additional work done from $x_{\circledB} = 4.00$ cm to $x_{\circledC} = 7.00$ cm is to find the area of the new, larger triangle and subtract the area of the smaller triangle.

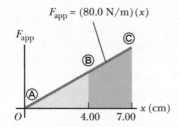

$F_{app} = (80.0 \text{ N/m})(x)$

$F_{app}$

$O$   $4.00$   $7.00$   $x$ (cm)

**Figure 5.31** (Example 5.15) A graph of the external force required to stretch a spring that obeys Hooke's law vs. the elongation of the spring.

---

### SOLUTION

**(a)** Find the work from $x_{\circledA} = 0$ cm to $x_{\circledB} = 4.00$ cm.

Compute the area of the smaller triangle:

$W = \tfrac{1}{2}kx_{\circledB}^2 = \tfrac{1}{2}(80.0 \text{ N/m})(0.040 \text{ m})^2 = \boxed{0.064\,0 \text{ J}}$

**(b)** Find the work from $x_{\circledB} = 4.00$ cm to $x_{\circledC} = 7.00$ cm.

Compute the area of the large triangle and subtract the area of the smaller triangle:

$W = \tfrac{1}{2}kx_{\circledC}^2 - \tfrac{1}{2}kx_{\circledB}^2$

$W = \tfrac{1}{2}(80.0 \text{ N/m})(0.070\,0 \text{ m})^2 - 0.064\,0 \text{ J}$

$= 0.196 \text{ J} - 0.064\,0 \text{ J}$

$= \boxed{0.132 \text{ J}}$

**REMARKS** Only simple geometries (rectangles and triangles) can be solved exactly with this method. More complex shapes require calculus or the square-counting technique in the next worked example.

**QUESTION 5.15** True or False: When stretching springs, half the displacement requires half as much work.

**EXERCISE 5.15** How much work is required to stretch this same spring from $x_i = 5.00$ cm to $x_f = 9.00$ cm?

**ANSWER** 0.224 J

---

## EXAMPLE 5.16 | ESTIMATING WORK BY COUNTING BOXES

**GOAL** Use the graphical method and counting boxes to estimate the work done by a force.

**PROBLEM** Suppose the force applied to stretch a thick piece of elastic changes with position as indicated in Figure 5.32a. Estimate the work done by the applied force.

**STRATEGY** To find the work, simply count the number of boxes underneath the curve and multiply that number by the area of each box. The curve will pass through the middle of some boxes, in which case only an estimated fractional part should be counted.

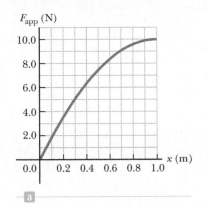
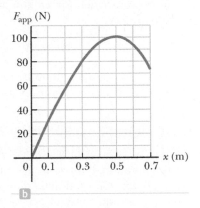

**Figure 5.32** (a) (Example 5.16) (b) (Exercise 5.16)

### SOLUTION

There are 62 complete or nearly complete boxes under the curve, 6 boxes that are about half under the curve, and a triangular area from $x = 0$ m to $x = 0.10$ m that is equivalent to 1 box, for a total of about 66 boxes. Because the area of each box is 0.10 J, the total work done is approximately $66 \times 0.10$ J $= 6.6$ J.

**REMARKS** Mathematically, there are a number of other methods for creating such estimates, all involving adding up regions approximating the area. To get a better estimate, make smaller boxes.

**QUESTION 5.16** In developing such an estimate, is it necessary for all boxes to have the same length and width?

**EXERCISE 5.16** Suppose the applied force necessary to pull the drawstring on a bow is given by Figure 5.32b. Find the approximate work done by counting boxes.

**ANSWER** About 50 J. (Individual answers may vary.)

---

## SUMMARY

### 5.1 Work

The work done on an object by a constant force is

$$W = (F \cos \theta) d \qquad [5.3]$$

where $F$ is the magnitude of the force, $d$ is the magnitude of the object's displacement, and $\theta$ is the angle between the direction of the force $\vec{F}$ and the displacement $\Delta \vec{x}$ (Fig. 5.33). Solving simple problems requires substituting values into this equation. More complex problems, such as those involving friction, often require using Newton's second law, $m\vec{a} = \vec{F}_{net}$, to determine forces.

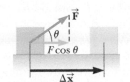

**Figure 5.33** A constant force $\vec{F}$ applied during a displacement $\Delta \vec{x}$ does work $(F \cos \theta) \Delta x$.

### 5.2 Kinetic Energy and the Work–Energy Theorem

The kinetic energy of a body with mass $m$ and speed $v$ is given by

$$KE \equiv \tfrac{1}{2} m v^2 \qquad [5.6]$$

The work–energy theorem states that the net work done on an object of mass $m$ is equal to the change in its kinetic energy (Fig. 5.34), or

$$W_{net} = KE_f - KE_i = \Delta KE \qquad [5.7]$$

Work and energy of any kind carry units of joules. Solving problems involves finding the work done by each force acting

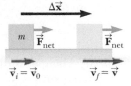

**Figure 5.34** Work done by a net force $\vec{F}_{net}$ on an object changes the object's velocity.

on the object and summing them up, which is $W_{net}$, followed by substituting known quantities into Equation 5.7, solving for the unknown quantity.

Conservative forces are special: Work done against them can be recovered—it's conserved. An example is gravity: The work done in lifting an object through a height is effectively stored in the gravity field and can be recovered in the kinetic energy of the object simply by letting it fall. Nonconservative forces, such as surface friction and drag, dissipate energy in a form that can't be readily recovered. To account for such forces, the work–energy theorem can be rewritten as

$$W_{nc} + W_c = \Delta KE \qquad [5.8]$$

where $W_{nc}$ is the work done by nonconservative forces and $W_c$ is the work done by conservative forces.

## 5.3 Gravitational Potential Energy
## 5.4 Gravity and Nonconservative Forces

The gravitational force is a conservative field. Gravitational potential energy is another way of accounting for gravitational work $W_g$:

$$W_g = -(PE_f - PE_i)$$
$$= -(mgy_f - mgy_i) \qquad [5.12]$$

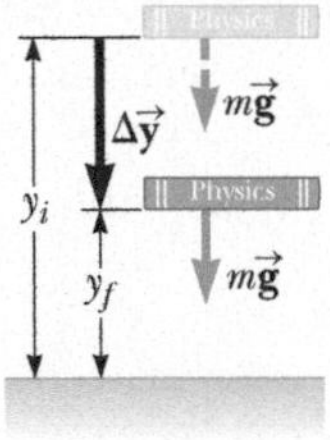

To find the change in gravitational potential energy as an object of mass $m$ moves between two points in a gravitational field (Fig. 5.35), substitute the values of the object's $y$-coordinates.

**Figure 5.35** The work done by the gravitational force as the book falls equals $mgy_i - mgy_f$.

The work–energy theorem can be generalized to include gravitational potential energy:

$$W_{nc} = (KE_f - KE_i) + (PE_f - PE_i) \qquad [5.13]$$

Gravitational work and gravitational potential energy should not both appear in the work–energy theorem at the same time, only one or the other, because they're equivalent. Setting the work due to nonconservative forces to zero and substituting the expressions for $KE$ and $PE$, a form of the conservation of mechanical energy with gravitation can be obtained:

$$\tfrac{1}{2}mv_i^2 + mgy_i = \tfrac{1}{2}mv_f^2 + mgy_f \qquad [5.15]$$

To solve problems with this equation, identify two points in the system—one where information is known and the other where information is desired. Substitute and solve for the unknown quantity.

The work done by other forces, as when frictional forces are present, isn't always zero. In that case, identify two points as before, calculate the work due to all other forces, and solve for the unknown in Equation 5.13.

## 5.5 Spring Potential Energy

The spring force is conservative, and its potential energy is given by

$$PE_s \equiv \tfrac{1}{2}kx^2 \qquad [5.17]$$

Spring potential energy can be put into the work–energy theorem, which then reads

$$W_{nc} = (KE_f - KE_i) + (PE_{gf} - PE_{gi}) + (PE_{sf} - PE_{si}) \qquad [5.18]$$

When nonconservative forces are absent, $W_{nc} = 0$ and mechanical energy is conserved.

## 5.6 Systems and Energy Conservation

The principle of the conservation of energy states that energy can't be created or destroyed. It can be transformed, but the total energy content of any isolated system is always constant. The same is true for the universe at large. The work done by all nonconservative forces acting on a system equals the change in the total mechanical energy of the system:

$$W_{nc} = (KE_f + PE_f) - (KE_i + PE_i) = E_f - E_i \qquad [5.21\text{–}5.22]$$

where $PE$ represents all potential energies present.

## 5.7 Power

Average power is the amount of energy transferred divided by the time taken for the transfer:

$$\overline{P} = \frac{W}{\Delta t} \qquad [5.23]$$

This expression can also be written

$$\overline{P} = F\overline{v} \qquad [5.24]$$

where $\overline{v}$ is the object's average velocity and $F$ is constant and parallel to $\overline{v}$. The instantaneous power is given by:

$$P = Fv \qquad [5.25]$$

where $F$ must be parallel to the velocity $v$ and both quantities can change with time. The unit of power is the watt ($W = J/s$). To solve simple problems, substitute given quantities into one of these equations. More difficult problems usually require finding the work done on the object using the work–energy theorem or the definition of work.

# CONCEPTUAL QUESTIONS

1. Consider a tug-of-war as in Figure CQ5.1, in which two teams pulling on a rope are evenly matched so that no motion takes place. Is work done on the rope? On the pullers? On the ground? Is work done on anything?

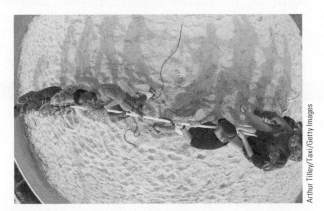

**Figure CQ5.1**

2. Choose the best answer. A car traveling at constant speed has a net work of zero done on it. (a) True (b) False (c) The answer depends on the motion.

3. (a) If the height of a playground slide is kept constant, will the length of the slide or whether it has bumps make any difference in the final speed of children playing on it? Assume that the slide is slick enough to be considered frictionless. (b) Repeat part (a), assuming that the slide is not frictionless.

4. (a) Can the kinetic energy of a system be negative? (b) Can the gravitational potential energy of a system be negative? Explain.

5. Two toboggans (with riders) of the same mass are at rest on top of a steep hill. As they slide down the hill, toboggan A takes a straight path down the slope while toboggan B winds back and forth along a more gently sloping path. Toboggan A makes the trip in half the time of toboggan B. (a) Compare the work done by gravity on each toboggan. (b) Compare the average power delivered by gravity on the two toboggans.

6. A bowling ball is suspended from the ceiling of a lecture hall by a strong cord. The ball is drawn away from its equilibrium position and released from rest at the tip of the demonstrator's nose, as shown in Figure CQ5.6. (a) If the demonstrator remains stationary, explain why the ball does not strike her on its return swing. (b) Would this demonstrator be safe if the ball were given a push from its starting position at her nose?

**Figure CQ5.6**

7. As a mass tied to the end of a string swings from its highest point down to its lowest point, it is acted on by three forces: gravity, tension, and air resistance. Which force does (a) positive work? (b) negative work? (c) zero work?

8. Discuss whether any work is being done by each of the following agents and, if so, whether the work is positive or negative: (a) a chicken scratching the ground, (b) a person studying, (c) a crane lifting a bucket of concrete, (d) the force of gravity on the bucket in part (c), (e) the leg muscles of a person in the act of sitting down.

9. When a punter kicks a football, is he doing any work on the ball while the toe of his foot is in contact with it? Is he doing any work on the ball after it loses contact with his toe? Are any forces doing work on the ball while it is in flight?

10. The driver of a car slams on her brakes to avoid colliding with a deer crossing the highway. What happens to the car's kinetic energy as it comes to rest?

11. A weight is connected to a spring that is suspended vertically from the ceiling. If the weight is displaced downward from its equilibrium position and released, it will oscillate up and down. (a) If air resistance is neglected, will the total mechanical energy of the system (weight plus Earth plus spring) be conserved? (b) How many forms of potential energy are there for this situation?

12. For each of the situations given, state whether frictional forces do positive, negative, or zero work on the italicized object. (a) A *plate* slides across a table and is brought to rest by friction. (b) A person pushes a *chair* at constant speed across a rough, horizontal surface. (c) A *box* is set down on a stationary conveyor belt. The conveyor belt is then turned on and the *box* begins to move, being carried along by the belt. (d) A person exerts a horizontal force on a *banana* at rest on a counter top. The *banana* remains at rest.

13. Suppose you are reshelving books in a library. As you lift a book from the floor and place it on the top shelf, two forces act on the book: your upward lifting force and the downward gravity force. (a) Is the mechanical work done by your lifting force positive, zero, or negative? (b) Is the work done by the gravity force positive, zero, or negative? (c) Is the sum of the works done by both forces positive, zero, or negative?

14. Two stones, one with twice the mass of the other, are thrown straight up and rise to the same height $h$. Compare their changes in gravitational potential energy (choose one): (a) They rise to the same height, so the stone with twice the mass has twice the change in gravitational potential energy. (b) They rise to the same height, so they have the same change in gravitational potential energy. (c) The answer depends on their speeds at height $h$.

15. An Earth satellite is in a circular orbit at an altitude of 500 km. Explain why the work done by the gravitational force acting on the satellite is zero. Using the work–energy theorem, what can you say about the speed of the satellite?

16. Mark and David are loading identical cement blocks onto David's pickup truck. Mark lifts his block straight up from the ground to the truck, whereas David slides his block up a ramp on massless, frictionless rollers. Which statement is true? (a) Mark does more work than David. (b) Mark and David do the same amount of work. (c) David does more work than Mark. (d) None of these statements is necessarily true because the angle of the incline is unknown. (e) None of these statements is necessarily true because the mass of one block is not given.

17. If the speed of a particle is doubled, what happens to its kinetic energy? (a) It becomes four times larger. (b) It becomes two times larger. (c) It becomes $\sqrt{2}$ times larger. (d) It is unchanged. (e) It becomes half as large.

18. A certain truck has twice the mass of a car. Both are moving at the same speed. If the kinetic energy of the truck is K, what is the kinetic energy of the car? (a) K/4 (b) K/2 (c) 0.71K (d) K (e) 2K

19. If the net work done on a particle is zero, which of the following statements must be true? (a) The velocity is zero. (b) The velocity is decreased. (c) The velocity is unchanged. (d) The speed is unchanged. (e) More information is needed.

20. A car accelerates uniformly from rest. Ignoring air friction, when does the car require the greatest power? (a) When the car first accelerates from rest, (b) just as the car reaches its maximum speed, (c) when the car reaches half its maximum speed. (d) The question is misleading because the power required is constant. (e) More information is needed.

# PROBLEMS

*See the Preface for an explanation of the icons used in this problems set.*

## 5.1 Work

1. A weight lifter lifts a 350-N set of weights from ground level to a position over his head, a vertical distance of 2.00 m. How much work does the weight lifter do, assuming he moves the weights at constant speed?

2. **V** In 1990 Walter Arfeuille of Belgium lifted a 281.5-kg object through a distance of 17.1 cm using only his teeth. (a) How much work did Arfeuille do on the object? (b) What magnitude force did he exert on the object during the lift, assuming the force was constant?

3. A cable exerts a constant upward tension of magnitude $1.25 \times 10^4$ N on a $1.00 \times 10^3$-kg elevator as it rises through a vertical distance of 2.00 m. (a) Find the work done by the tension force on the elevator. (b) Find the work done by the force of gravity on the elevator.

4. **Q|C** A shopper in a supermarket pushes a cart with a force of 35 N directed at an angle of 25° below the horizontal. The force is just sufficient to overcome various frictional forces, so the cart moves at constant speed. (a) Find the work done by the shopper as she moves down a 50.0-m length aisle. (b) What is the *net work* done on the cart? Why? (c) The shopper goes down the next aisle, pushing horizontally and maintaining the same speed as before. If the work done by frictional forces doesn't change, would the shopper's applied force be larger, smaller, or the same? What about the work done on the cart by the shopper?

5. **Q|C** Starting from rest, a 5.00-kg block slides 2.50 m down a rough 30.0° incline. The coefficient of kinetic friction between the block and the incline is $\mu_k = 0.436$. Determine (a) the work done by the force of gravity, (b) the work done by the friction force between block and incline, and (c) the work done by the normal force. (d) Qualitatively, how would the answers change if a shorter ramp at a steeper angle were used to span the same vertical height?

6. A horizontal force of 150 N is used to push a 40.0-kg packing crate a distance of 6.00 m on a rough horizontal surface. If the crate moves at constant speed, find (a) the work done by the 150-N force and (b) the coefficient of kinetic friction between the crate and surface.

7. A tension force of 175 N inclined at 20.0° above the horizontal is used to pull a 40.0-kg packing crate a distance of 6.00 m on a rough surface. If the crate moves at a constant speed, find (a) the work done by the tension force and (b) the coefficient of kinetic friction between the crate and surface.

8. **T** A block of mass $m = 2.50$ kg is pushed a distance $d = 2.20$ m along a frictionless horizontal table by a constant applied force of magnitude $F = 16.0$ N directed at an angle $\theta = 25.0°$ below the horizontal as shown in Figure P5.8. Determine the work done by (a) the applied force, (b) the normal force exerted by the table, (c) the force of gravity, and (d) the net force on the block.

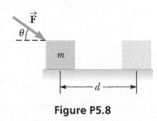

**Figure P5.8**

## 5.2 Kinetic Energy and the Work–Energy Theorem

9. A mechanic pushes a $2.50 \times 10^3$-kg car from rest to a speed of $v$, doing $5.00 \times 10^3$ J of work in the process. During this time, the car moves 25.0 m. Neglecting friction between car and road, find (a) $v$ and (b) the horizontal force exerted on the car.

10. A 7.00-kg bowling ball moves at 3.00 m/s. How fast must a 2.45-g Ping-Pong ball move so that the two balls have the same kinetic energy?

11. A 65.0-kg runner has a speed of 5.20 m/s at one instant during a long-distance event. (a) What is the runner's kinetic energy at this instant? (b) How much net work is required to double his speed?

12. **Q|C** A worker pushing a 35.0-kg wooden crate at a constant speed for 12.0 m along a wood floor does 350 J of work by applying a constant horizontal force of magnitude $F_0$ on the crate. (a) Determine the value of $F_0$. (b) If the worker now applies a force greater than $F_0$, describe the subsequent motion of the crate. (c) Describe what would happen to the crate if the applied force is less than $F_0$.

13. **V** A 70-kg base runner begins his slide into second base when he is moving at a speed of 4.0 m/s. The coefficient of friction between his clothes and Earth is 0.70. He slides so that his speed is zero just as he reaches the base. (a) How much mechanical energy is lost due to friction acting on the runner? (b) How far does he slide?

14. **BIO** A 62.0-kg cheetah accelerates from rest to its top speed of 32.0 m/s. (a) How much net work is required for the cheetah to reach its top speed? (b) One food Calorie equals 4 186 J. How many Calories of net work are required for the cheetah to reach its top speed? *Note:* Due to inefficiencies in converting chemical energy to mechanical energy, the amount calculated

here is only a fraction of the power that must be produced by the cheetah's body.

**15.** A 7.80-g bullet moving at 575 m/s penetrates a tree trunk to a depth of 5.50 cm. (a) Use work and energy considerations to find the average frictional force that stops the bullet. (b) Assuming the frictional force is constant, determine how much time elapses between the moment the bullet enters the tree and the moment it stops moving.

**16.** A 0.60-kg particle has a speed of 2.0 m/s at point $A$ and a kinetic energy of 7.5 J at point $B$. What is (a) its kinetic energy at $A$? (b) Its speed at point $B$? (c) The total work done on the particle as it moves from $A$ to $B$?

**17.** A large cruise ship of mass $6.50 \times 10^7$ kg has a speed of 12.0 m/s at some instant. (a) What is the ship's kinetic energy at this time? (b) How much work is required to stop it? (c) What is the magnitude of the constant force required to stop it as it undergoes a displacement of 2.50 km?

**18.** **V** A man pushing a crate of mass $m = 92.0$ kg at a speed of $v = 0.850$ m/s encounters a rough horizontal surface of length $\ell = 0.65$ m as in Figure P5.18. If the coefficient of kinetic friction between the crate and rough surface is 0.358 and he exerts a constant horizontal force of 275 N on the crate, find (a) the magnitude and direction of the net force on the crate while it is on the rough surface, (b) the net work done on the crate while it is on the rough surface, and (c) the speed of the crate when it reaches the end of the rough surface.

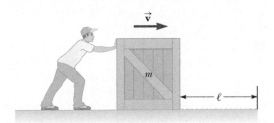

**Figure P5.18**

## 5.3 Gravitational Potential Energy

## 5.4 Gravity and Nonconservative Forces

## 5.5 Spring Potential Energy

**19.** A 0.20-kg stone is held 1.3 m above the top edge of a water well and then dropped into it. The well has a depth of 5.0 m. Taking $y = 0$ at the top edge of the well, what is the gravitational potential energy of the stone–Earth system (a) before the stone is released and (b) when it reaches the bottom of the well? (c) What is the change in gravitational potential energy of the system from release to reaching the bottom of the well?

**20.** When a 2.50-kg object is hung vertically on a certain light spring described by Hooke's law, the spring stretches 2.76 cm. (a) What is the force constant of the spring? (b) If the 2.50-kg object is removed, how far will the spring stretch if a 1.25-kg block is hung on it? (c) How much work must an external agent do to stretch the same spring 8.00 cm from its unstretched position?

**21.** A block of mass 3.00 kg is placed against a horizontal spring of constant $k = 875$ N/m and pushed so the spring compresses by 0.070 0 m. (a) What is the elastic potential energy of the

block–spring system? (b) If the block is now released and the surface is frictionless, calculate the block's speed after leaving the spring.

**22.** **GP** A 60.0-kg athlete leaps straight up into the air from a trampoline with an initial speed of 9.0 m/s. The goal of this problem is to find the maximum height she attains and her speed at half maximum height. (a) What are the interacting objects and how do they interact? (b) Select the height at which the athlete's speed is 9.0 m/s as $y = 0$. What is her kinetic energy at this point? What is the gravitational potential energy associated with the athlete? (c) What is her kinetic energy at maximum height? What is the gravitational potential energy associated with the athlete? (d) Write a general equation for energy conservation in this case and solve for the maximum height. Substitute and obtain a numerical answer. (e) Write the general equation for energy conservation and solve for the velocity at half the maximum height. Substitute and obtain a numerical answer.

**23.** **T** A $2.10 \times 10^3$-kg pile driver is used to drive a steel I-beam into the ground. The pile driver falls 5.00 m before coming into contact with the top of the beam, and it drives the beam 12.0 cm farther into the ground as it comes to rest. Using energy considerations, calculate the average force the beam exerts on the pile driver while the pile driver is brought to rest.

**24.** **S** Two blocks are connected by a light string that passes over two frictionless pulleys as in Figure P5.24. The block of mass $m_2$ is attached to a spring of force constant $k$ and $m_1 > m_2$. If the system is released from rest, and the spring is initially not stretched or compressed, find an expression for the maximum displacement $d$ of $m_2$.

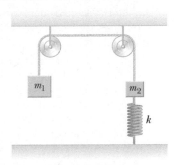

**Figure P5.24**

**25.** A daredevil on a motorcycle leaves the end of a ramp with a speed of 35.0 m/s as in Figure P5.25. If his speed is 33.0 m/s when he reaches the peak of the path, what is the maximum height that he reaches? Ignore friction and air resistance.

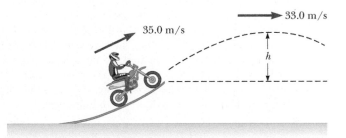

**Figure P5.25**

**26.** Truck suspensions often have "helper springs" that engage at high loads. One such arrangement is a leaf spring with a helper coil spring mounted on the axle, as shown in Figure P5.26. When the main leaf spring is compressed by distance $y_0$, the helper spring engages and then helps to support any additional load. Suppose the leaf spring constant is $5.25 \times 10^5$ N/m, the helper spring constant is $3.60 \times 10^5$ N/m, and $y_0 = 0.500$ m.

(a) What is the compression of the leaf spring for a load of $5.00 \times 10^5$ N? (b) How much work is done in compressing the springs?

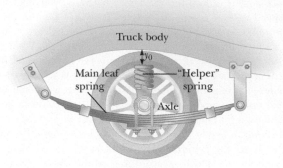

**Figure P5.26**

27. **BIO** The chin-up is one exercise that can be used to strengthen the biceps muscle. This muscle can exert a force of approximately $8.00 \times 10^2$ N as it contracts a distance of 7.5 cm in a 75-kg male.[3] (a) How much work can the biceps muscles (one in each arm) perform in a single contraction? (b) Compare this amount of work with the energy required to lift a 75-kg person 40. cm in performing a chin-up. (c) Do you think the biceps muscle is the only muscle involved in performing a chin-up?

28. **BIO** A flea is able to jump about 0.5 m. It has been said that if a flea were as big as a human, it would be able to jump over a 100-story building! When an animal jumps, it converts work done in contracting muscles into gravitational potential energy (with some steps in between). The maximum force exerted by a muscle is proportional to its cross-sectional area, and the work done by the muscle is this force times the length of contraction. If we magnified a flea by a factor of 1 000, the cross section of its muscle would increase by $1\,000^2$ and the length of contraction would increase by 1 000. How high would this "superflea" be able to jump? (Don't forget that the mass of the "superflea" increases as well.)

29. A 50.0-kg projectile is fired at an angle of 30.0° above the horizontal with an initial speed of $1.20 \times 10^2$ m/s from the top of a cliff 142 m above level ground, where the ground is taken to be $y = 0$. (a) What is the initial total mechanical energy of the projectile? (b) Suppose the projectile is traveling 85.0 m/s at its maximum height of $y = 427$ m. How much work has been done on the projectile by air friction? (c) What is the speed of the projectile immediately before it hits the ground if air friction does one and a half times as much work on the projectile when it is going down as it did when it was going up?

30. **S** A projectile of mass $m$ is fired horizontally with an initial speed of $v_0$ from a height of $h$ above a flat, desert surface. Neglecting air friction, at the instant before the projectile hits the ground, find the following in terms of $m$, $v_0$, $h$, and $g$: (a) the work done by the force of gravity on the projectile, (b) the change in kinetic energy of the projectile since it was fired, and (c) the final kinetic energy of the projectile. (d) Are any of the answers changed if the initial angle is changed?

[3] G. P. Pappas et al., "Nonuniform Shortening in the Biceps Brachii During Elbow Flexion," *Journal of Applied Physiology*, **92**(6): 2381–2389, 2002.

31. **GP** A horizontal spring attached to a wall has a force constant of 850 N/m. A block of mass 1.00 kg is attached to the spring and oscillates freely on a horizontal, frictionless surface as in Figure 5.22. The initial goal of this problem is to find the velocity at the equilibrium point after the block is released. (a) What objects constitute the system, and through what forces do they interact? (b) What are the two points of interest? (c) Find the energy stored in the spring when the mass is stretched 6.00 cm from equilibrium and again when the mass passes through equilibrium after being released from rest. (d) Write the conservation of energy equation for this situation and solve it for the speed of the mass as it passes equilibrium. Substitute to obtain a numerical value. (e) What is the speed at the halfway point? Why isn't it half the speed at equilibrium?

## 5.6 Systems and Energy Conservation

32. A 50.-kg pole vaulter running at 10. m/s vaults over the bar. Her speed when she is above the bar is 1.0 m/s. Neglect air resistance, as well as any energy absorbed by the pole, and determine her altitude as she crosses the bar.

33. **V** A child and a sled with a combined mass of 50.0 kg slide down a frictionless slope. If the sled starts from rest and has a speed of 3.00 m/s at the bottom, what is the height of the hill?

34. A 35.0-cm long spring is hung vertically from a ceiling and stretches to 41.5 cm when a 7.50-kg weight is hung from its free end. (a) Find the spring constant. (b) Find the length of the spring if the 7.50-kg weight is replaced with a 195-N weight.

35. **Q|C** A 0.250-kg block along a horizontal track has a speed of 1.50 m/s immediately before colliding with a light spring of force constant 4.60 N/m located at the end of the track. (a) What is the spring's maximum compression if the track is frictionless? (b) If the track is *not* frictionless, would the spring's maximum compression be greater than, less than, or equal to the value obtained in part (a)?

36. **V** A block of mass $m = 5.00$ kg is released from rest from point Ⓐ and slides on the frictionless track shown in Figure P5.36. Determine (a) the block's speed at points Ⓑ and Ⓒ and (b) the net work done by the gravitational force on the block as it moves from point from Ⓐ to Ⓒ.

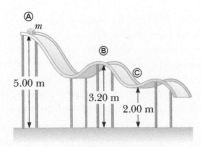

**Figure P5.36**

37. Tarzan swings on a 30.0-m-long vine initially inclined at an angle of 37.0° with the vertical. What is his speed at the bottom of the swing (a) if he starts from rest? (b) If he pushes off with a speed of 4.00 m/s?

38. **S** Two blocks are connected by a light string that passes over a frictionless pulley as in Figure P5.38. The system is

released from rest while $m_2$ is on the floor and $m_1$ is a distance $h$ above the floor. (a) Assuming $m_1 > m_2$, find an expression for the speed of $m_1$ just as it reaches the floor. (b) Taking $m_1$ = 6.5 kg, $m_2$ = 4.2 kg, and $h$ = 3.2 m, evaluate your answer to part (a), and (c) find the speed of each block when $m_1$ has fallen a distance of 1.6 m.

39. **T** The launching mechanism of a toy gun consists of a spring of unknown spring constant, as shown in Figure P5.39a. If the spring is compressed a distance of 0.120 m and the gun fired vertically as shown, the gun can launch a 20.0-g projectile from rest to a maximum height of 20.0 m above the starting point of the projectile. Neglecting all resistive forces, (a) describe the mechanical energy transformations that occur from the time the gun is fired until the projectile reaches its maximum height, (b) determine the spring constant, and (c) find the speed of the projectile as it moves through the equilibrium position of the spring (where $x = 0$), as shown in Figure P5.39b.

**Figure P5.38**

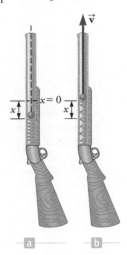

**Figure P5.39**

40. **GP** (a) A block with a mass $m$ is pulled along a horizontal surface for a distance $x$ by a constant force $\vec{F}$ at an angle $\theta$ with respect to the horizontal. The coefficient of kinetic friction between block and table is $\mu_k$. Is the force exerted by friction equal to $\mu_k mg$? If not, what is the force exerted by friction? (b) How much work is done by the friction force and by $\vec{F}$? (Don't forget the signs.) (c) Identify all the forces that do no work on the block. (d) Let $m$ = 2.00 kg, $x$ = 4.00 m, $\theta$ = 37.0°, $F$ = 15.0 N, and $\mu_k$ = 0.400, and find the answers to parts (a) and (b).

41. **Q|C** (a) A child slides down a water slide at an amusement park from an initial height $h$. The slide can be considered frictionless because of the water flowing down it. Can the equation for conservation of mechanical energy be used on the child? (b) Is the mass of the child a factor in determining his speed at the bottom of the slide? (c) The child drops straight down rather than following the curved ramp of the slide. In which case will he be traveling faster at ground level? (d) If friction is present, how would the conservation-of-energy equation be modified? (e) Find the maximum speed of the child when the slide is frictionless if the initial height of the slide is 12.0 m.

42. **Q|C** An airplane of mass $1.50 \times 10^4$ kg is moving at 60.0 m/s. The pilot then increases the engine's thrust to $7.50 \times 10^4$ N. The resistive force exerted by air on the airplane has a magnitude of $4.00 \times 10^4$ N. (a) Is the work done by the engine on the airplane equal to the change in the airplane's kinetic energy after it travels through some distance through the air? Is mechanical energy conserved? Explain. (b) Find the speed

of the airplane after it has traveled $5.00 \times 10^2$ m. Assume the airplane is in level flight throughout the motion.

43. **V** The system shown in Figure P5.43 is used to lift an object of mass $m$ = 76.0 kg. A constant downward force of magnitude $F$ is applied to the loose end of the rope such that the hanging object moves upward at constant speed. Neglecting the masses of the rope and pulleys, find (a) the required value of $F$, (b) the tensions $T_1$, $T_2$, and $T_3$, and (c) the work done by the applied force in raising the object a distance of 1.80 m.

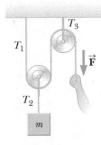

**Figure P5.43**

44. A 25.0-kg child on a 2.00-m-long swing is released from rest when the ropes of the swing make an angle of 30.0° with the vertical. (a) Neglecting friction, find the child's speed at the lowest position. (b) If the actual speed of the child at the lowest position is 2.00 m/s, what is the mechanical energy lost due to friction?

45. A $2.1 \times 10^3$-kg car starts from rest at the top of a 5.0-m-long driveway that is inclined at 20.0° with the horizontal. If an average friction force of $4.0 \times 10^3$ N impedes the motion, find the speed of the car at the bottom of the driveway.

46. **GP** **S** A child of mass $m$ starts from rest and slides without friction from a height $h$ along a curved waterslide (Fig. P5.46). She is launched from a height $h/5$ into the pool. (a) Is mechanical energy conserved? Why? (b) Give the gravitational potential energy associated with the child and her kinetic energy in terms of $mgh$ at the following positions: the top of the waterslide, the launching point, and the point where she lands in the pool. (c) Determine her initial speed $v_0$ at the launch point in terms of $g$ and $h$. (d) Determine her maximum airborne height $y_{max}$ in terms of $h$, $g$, and the horizontal speed at that height, $v_{0x}$ (e) Use the $x$-component of the answer to part (c) to eliminate $v_0$ from the answer to part (d), giving the height $y_{max}$ in terms of $g$, $h$, and the launch angle $\theta$. (f) Would your answers be the same if the waterslide were not frictionless? Explain.

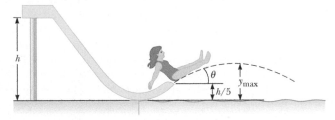

**Figure P5.46**

47. A skier starts from rest at the top of a hill that is inclined 10.5° with respect to the horizontal. The hillside is $2.00 \times 10^2$ m long, and the coefficient of friction between snow and skis is 0.075 0. At the bottom of the hill, the snow is level and the coefficient of friction is unchanged. How far does the skier glide along the horizontal portion of the snow before coming to rest?

48. In a circus performance, a monkey is strapped to a sled and both are given an initial speed of 4.0 m/s up a 20.0° inclined track. The combined mass of monkey and sled is 20. kg, and the coefficient of kinetic friction between sled and incline is 0.20. How far up the incline do the monkey and sled move?

**49.** An 80.0-kg skydiver jumps out of a balloon at an altitude of $1.00 \times 10^3$ m and opens the parachute at an altitude of 200.0 m. (a) Assuming that the total retarding force on the diver is constant at 50.0 N with the parachute closed and constant at $3.60 \times 10^3$ N with the parachute open, what is the speed of the diver when he lands on the ground? (b) Do you think the skydiver will get hurt? Explain. (c) At what height should the parachute be opened so that the final speed of the skydiver when he hits the ground is 5.00 m/s? (d) How realistic is the assumption that the total retarding force is constant? Explain.

## 5.7 Power

**50.** V A skier of mass 70.0 kg is pulled up a slope by a motor-driven cable. (a) How much work is required to pull him 60.0 m up a 30.0° slope (assumed frictionless) at a constant speed of 2.00 m/s? (b) What power (expressed in hp) must a motor have to perform this task?

**51.** What average mechanical power must a 70.0-kg mountain climber generate to climb to the summit of a hill of height 325 m in 45.0 min? *Note:* Due to inefficiencies in converting chemical energy to mechanical energy, the amount calculated here is only a fraction of the power that must be produced by the climber's body.

**52.** BIO While running, a person dissipates about 0.60 J of mechanical energy per step per kilogram of body mass. If a 60.-kg person develops a power of 70. W during a race, how fast is the person running? (Assume a running step is 1.5 m long.)

**53.** The electric motor of a model train accelerates the train from rest to 0.620 m/s in 21.0 ms. The total mass of the train is 875 g. Find the average power delivered to the train during its acceleration.

**54.** When an automobile moves with constant speed down a highway, most of the power developed by the engine is used to compensate for the mechanical energy loss due to frictional forces exerted on the car by the air and the road. If the power developed by an engine is 175 hp, estimate the total frictional force acting on the car when it is moving at a speed of 29 m/s. One horsepower equals 746 W.

**55.** BIO Under normal conditions the human heart converts about 13.0 J of chemical energy per second into 1.30 W of mechanical power as it pumps blood throughout the body. (a) Determine the number of Calories required to power the heart for one day, given that 1 Calorie equals 4 186 J. (b) Metabolizing 1.00 kg of fat can release about $9.00 \times 10^3$ Calories of energy. What mass of metabolized fat would power the heart for one day?

**56.** A certain rain cloud at an altitude of 1.75 km contains $3.20 \times 10^7$ kg of water vapor. How long would it take for a 2.70-kW pump to raise the same amount of water from Earth's surface to the cloud's position?

**57.** A $1.50 \times 10^3$-kg car starts from rest and accelerates uniformly to 18.0 m/s in 12.0 s. Assume that air resistance remains constant at $4.00 \times 10^2$ N during this time. Find (a) the average power developed by the engine and (b) the instantaneous power output of the engine at $t = 12.0$ s, just before the car stops accelerating.

**58.** A $6.50 \times 10^2$-kg elevator starts from rest and moves upward for 3.00 s with constant acceleration until it reaches its cruising speed, 1.75 m/s. (a) What is the average power of the elevator motor during this period? (b) How does this amount of power compare with its power during an upward trip with constant speed?

## 5.8 Work Done by a Varying Force

**59.** T The force acting on a particle varies as in Figure P5.59. Find the work done by the force as the particle moves (a) from $x = 0$ to $x = 8.00$ m, (b) from $x = 8.00$ m to $x = 10.0$ m, and (c) from $x = 0$ to $x = 10.0$ m.

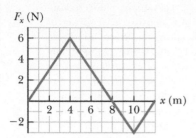

**Figure P5.59**

**60.** An object of mass 3.00 kg is subject to a force $F_x$ that varies with position as in Figure P5.60. Find the work done by the force on the object as it moves (a) from $x = 0$ to $x = 5.00$ m, (b) from $x = 5.00$ m to $x = 10.0$ m, and (c) from $x = 10.0$ m to $x = 15.0$ m. (d) If the object has a speed of 0.500 m/s at $x = 0$, find its speed at $x = 5.00$ m and its speed at $x = 15.0$ m.

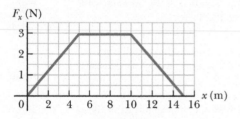

**Figure P5.60**

**61.** V The force acting on an object is given by $F_x = (8x - 16)$ N, where $x$ is in meters. (a) Make a plot of this force vs. $x$ from $x = 0$ to $x = 3.00$ m. (b) From your graph, find the net work done by the force as the object moves from $x = 0$ to $x = 3.00$ m.

## Additional Problems

**62.** An outfielder throws a 0.150-kg baseball at a speed of 40.0 m/s and an initial angle of 30.0°. What is the kinetic energy of the ball at the highest point of its motion?

**63.** A roller-coaster car of mass $1.50 \times 10^3$ kg is initially at the top of a rise at point Ⓐ. It then moves 35.0 m at an angle of 50.0° below the horizontal to a lower point Ⓑ. (a) Find both the potential energy of the system when the car is at points Ⓐ and Ⓑ and the change in potential energy as the car moves from point Ⓐ to point Ⓑ, assuming $y = 0$ at point Ⓑ. (b) Repeat part (a), this time choosing $y = 0$ at point Ⓒ, which is another 15.0 m down the same slope from point Ⓑ.

**64.** A ball of mass $m = 1.80$ kg is released from rest at a height $h = 65.0$ cm above a light vertical spring of force constant $k$ as in Figure P5.64a. The ball strikes the top of the spring and compresses it a distance $d = 9.00$ cm as in Figure P5.64b.

Neglecting any energy losses during the collision, find (a) the speed of the ball just as it touches the spring and (b) the force constant of the spring.

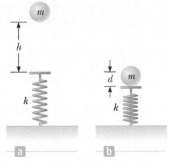

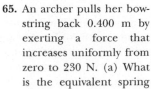

**Figure P5.64**

65. An archer pulls her bowstring back 0.400 m by exerting a force that increases uniformly from zero to 230 N. (a) What is the equivalent spring constant of the bow? (b) How much work does the archer do in pulling the bow?

66. A block of mass 12.0 kg slides from rest down a frictionless 35.0° incline and is stopped by a strong spring with $k = 3.00 \times 10^4$ N/m. The block slides 3.00 m from the point of release to the point where it comes to rest against the spring. When the block comes to rest, how far has the spring been compressed?

67. **BIO** (a) A 75-kg man steps out a window and falls (from rest) 1.0 m to a sidewalk. What is his speed just before his feet strike the pavement? (b) If the man falls with his knees and ankles locked, the only cushion for his fall is an approximately 0.50-cm give in the pads of his feet. Calculate the average force exerted on him by the ground during this 0.50 cm of travel. This average force is sufficient to cause damage to cartilage in the joints or to break bones.

68. **V** A toy gun uses a spring to project a 5.3-g soft rubber sphere horizontally. The spring constant is 8.0 N/m, the barrel of the gun is 15 cm long, and a constant frictional force of 0.032 N exists between barrel and projectile. With what speed does the projectile leave the barrel if the spring was compressed 5.0 cm for this launch?

69. Two objects ($m_1 = 5.00$ kg and $m_2 = 3.00$ kg) are connected by a light string passing over a light, frictionless pulley as in Figure P5.69. The 5.00-kg object is released from rest at a point $h = 4.00$ m above the table. (a) Determine the speed of each object when the two pass each other. (b) Determine the speed of each object at the moment the 5.00-kg object hits the table. (c) How much higher does the 3.00-kg object travel after the 5.00-kg object hits the table?

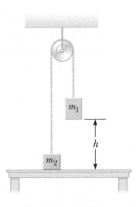

**Figure P5.69**

70. A 3.50-kN piano is lifted by three workers at constant speed to an apartment 25.0 m above the street using a pulley system fastened to the roof of the building. Each worker is able to deliver 165 W of power, and the pulley system is 75% efficient (so that 25% of the mechanical energy is lost due to friction in the pulley). Neglecting the mass of the pulley, find the time required to lift the piano from the street to the apartment.

71. A $2.00 \times 10^2$-g particle is released from rest at point $A$ on the inside of a smooth hemispherical bowl of radius $R = 30.0$ cm

(Fig. P5.71). Calculate (a) its gravitational potential energy at $A$ relative to $B$, (b) its kinetic energy at $B$, (c) its speed at $B$, (d) its potential energy at $C$ relative to $B$, and (e) its kinetic energy at $C$.

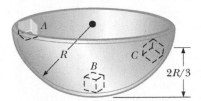

**Figure P5.71** Problems 71 and 72.

72. **Q|C** The particle described in Problem 71 (Fig. P5.71) is released from point $A$ at rest. Its speed at $B$ is 1.50 m/s. (a) What is its kinetic energy at $B$? (b) How much mechanical energy is lost as a result of friction as the particle goes from $A$ to $B$? (c) Is it possible to determine $\mu$ from these results in a simple manner? Explain.

73. **BIO** In terms of saving energy, bicycling and walking are far more efficient means of transportation than is travel by automobile. For example, when riding at 10.0 mi/h, a cyclist uses food energy at a rate of about 400 kcal/h above what he would use if he were merely sitting still. (In exercise physiology, power is often measured in kcal/h rather than in watts. Here, 1 kcal = 1 nutritionist's Calorie = 4 186 J.) Walking at 3.00 mi/h requires about 220 kcal/h. It is interesting to compare these values with the energy consumption required for travel by car. Gasoline yields about $1.30 \times 10^8$ J/gal. Find the fuel economy in equivalent miles per gallon for a person (a) walking and (b) bicycling.

74. **BIO** A 50.0-kg student evaluates a weight loss program by calculating the number of times she would need to climb a 12.0-m high flight of steps in order to lose one pound (0.45 kg) of fat. Metabolizing 1.00 kg of fat can release $3.77 \times 10^7$ J of chemical energy and the body can convert about 20.0% of this into mechanical energy. (The rest goes into internal energy.) (a) How much mechanical energy can the body produce from 0.450 kg of fat? (b) How many trips up the flight of steps are required for the student to lose 0.450 kg of fat? Ignore the relatively small amount of energy required to return down the stairs.

75. A ski jumper starts from rest 50.0 m above the ground on a frictionless track and flies off the track at an angle of 45.0° above the horizontal and at a height of 10.0 m above the level ground. Neglect air resistance. (a) What is her speed when she leaves the track? (b) What is the maximum altitude she attains after leaving the track? (c) Where does she land relative to the end of the track?

76. A 5.0-kg block is pushed 3.0 m up a vertical wall with constant speed by a constant force of magnitude $F$ applied at an angle of $\theta = 30°$ with the horizontal, as shown in Figure P5.76. If the coefficient of kinetic friction between block and wall is 0.30, determine the work done by (a) $\vec{F}$, (b) the force of gravity, and (c) the normal force between block and wall. (d) By how much does the gravitational potential energy increase during the block's motion?

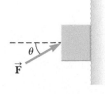

**Figure P5.76**

77. A child's pogo stick (Fig. P5.77) stores energy in a spring ($k = 2.50 \times 10^4$ N/m). At position Ⓐ ($x_1 = -0.100$ m), the spring compression is a maximum and the child is momentarily at rest. At position Ⓑ ($x = 0$), the spring is relaxed and the child is moving upward. At position Ⓒ, the child is again momentarily at rest at the top of the jump. Assuming that the combined mass of child and pogo stick is 25.0 kg, (a) calculate the total energy of the system if both potential energies are zero at $x = 0$, (b) determine $x_2$, (c) calculate the speed of the child at $x = 0$, (d) determine the value of $x$ for which the kinetic energy of the system is a maximum, and (e) obtain the child's maximum upward speed.

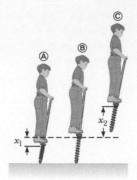

**Figure P5.77**

78. BIO A hummingbird hovers by exerting a downward force on the air equal, on average, to its weight. By Newton's third law, the air exerts an upward force of the same magnitude on the bird's wings. Find the average mechanical power delivered by a 3.00-g hummingbird while hovering if its wings beat 80.0 times per second through a stroke 3.50 cm long.

79. V In the dangerous "sport" of bungee jumping, a daring student jumps from a hot air balloon with a specially designed elastic cord attached to his waist. The unstretched length of the cord is 25.0 m, the student weighs $7.00 \times 10^2$ N, and the balloon is 36.0 m above the surface of a river below. Calculate the required force constant of the cord if the student is to stop safely 4.00 m above the river.

80. Apollo 14 astronaut Alan Shepard famously took two golf shots on the Moon where it's been estimated that an expertly hit shot could travel for 70.0 s through the Moon's reduced gravity, airless environment to a maximum range of 4.00 km (about 2.5 miles). Assuming such an expert shot has a launch angle of 45.0°, determine the golf ball's (a) kinetic energy as it leaves the club, and (b) maximum altitude in km above the lunar surface. Take the mass of a golf ball to be 0.045 0 kg and the Moon's gravitational acceleration to be $g_{\text{moon}} = 1.63$ m/s².

81. A truck travels uphill with constant velocity on a highway with a 7.0° slope. A 50.-kg package sits on the floor of the back of the truck and does not slide, due to a static frictional force. During an interval in which the truck travels 340 m, (a) what is the net work done on the package? What is the work done on the package by (b) the force of gravity, (c) the normal force, and (d) the friction force?

82. As a 75.0-kg man steps onto a bathroom scale, the spring inside the scale compresses by 0.650 mm. Excited to see that he has lost 2.50 kg since his previous weigh-in, the man jumps 0.300 m straight up into the air and lands directly on the scale. (a) What is the spring's maximum compression? (b) If the scale reads in kilograms, what reading does it give when the spring is at its maximum compression?

83. T A loaded ore car has a mass of $9.50 \times 10^2$ kg and rolls on rails with negligible friction. It starts from rest and is pulled up a mine shaft by a cable connected to a winch. The shaft is inclined at 30.0° above the horizontal. The car accelerates uniformly to a speed of 2.20 m/s in 12.0 s and then continues at constant speed. (a) What power must the winch motor provide when the car is moving at constant speed? (b) What maximum power must the motor provide? (c) What total energy transfers out of the motor by work by the time the car moves off the end of the track, which is of length 1 250 m?

84. A cat plays with a toy mouse suspended from a light string of length 1.25 m, rapidly batting the mouse so that it acquires a speed of 2.75 m/s while the string is still vertical. Use energy conservation to find the mouse's maximum height above its original position. (Assume the string always remains taut.)

85. Three objects with masses $m_1 = 5.00$ kg, $m_2 = 10.0$ kg, and $m_3 = 15.0$ kg, respectively, are attached by strings over frictionless pulleys as indicated in Figure P5.85. The horizontal surface exerts a force of friction of 30.0 N on $m_2$. If the system is released from rest, use energy concepts to find the speed of $m_3$ after it moves down 4.00 m.

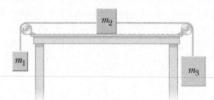

**Figure P5.85**

86. Two blocks, $A$ and $B$ (with mass 50.0 kg and $1.00 \times 10^2$ kg, respectively), are connected by a string, as shown in Figure P5.86. The pulley is frictionless and of negligible mass. The coefficient of kinetic friction between block $A$ and the incline is $\mu_k = 0.250$. Determine the change in the kinetic energy of block $A$ as it moves from Ⓒ to Ⓓ, a distance of 20.0 m up the incline (and block $B$ drops downward a distance of 20.0 m) if the system starts from rest.

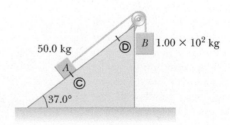

**Figure P5.86**

# Momentum, Impulse, and Collisions

**WHAT HAPPENS WHEN TWO AUTOMOBILES COLLIDE?** How does the impact affect the motion of each vehicle, and what basic physical principles determine the likelihood of serious injury? How do rockets work, and what mechanisms can be used to overcome the limitations imposed by exhaust speed? Why do we have to brace ourselves when firing small projectiles at high velocity? Finally, how can we use physics to improve our golf game?

To begin answering such questions, we introduce *momentum*. Intuitively, anyone or anything that has a lot of momentum is going to be hard to stop. In politics, the term is metaphorical. Physically, the more momentum an object has, the more force has to be applied to stop it in a given time. This concept leads to one of the most powerful principles in physics: *conservation of momentum*. Using this law, complex collision problems can be solved without knowing much about the forces involved during contact. We'll also be able to derive information about the average force delivered in an impact. With conservation of momentum, we'll have a better understanding of what choices to make when designing an automobile or a moon rocket, or when addressing a golf ball on a tee.

## 6.1 Momentum and Impulse

In physics, momentum has a precise definition. A slowly moving brontosaurus has a lot of momentum, but so does a little hot lead shot from the muzzle of a gun. We therefore expect that momentum will depend on an object's mass and velocity.

> The linear momentum $\vec{p}$ of an object of mass $m$ moving with velocity $\vec{v}$ is the product of its mass and velocity:
>
> $$\vec{p} \equiv m\vec{v} \qquad [6.1]$$
>
> **SI unit: kilogram meter per second (kg · m/s)**

◀ Linear momentum

Doubling either the mass or the velocity of an object doubles its momentum; doubling both quantities quadruples its momentum. Momentum is a vector quantity with the same direction as the object's velocity. Its components are given in two dimensions by

$$p_x = mv_x \qquad p_y = mv_y$$

where $p_x$ is the momentum of the object in the $x$-direction and $p_y$ its momentum in the $y$-direction.

The magnitude of the momentum $p$ of an object of mass $m$ can be related to its kinetic energy $KE$:

$$KE = \frac{p^2}{2m} \qquad [6.2]$$

This relationship is easy to prove using the definitions of kinetic energy and momentum (see Problem 6) and is valid for objects traveling at speeds much less than the speed of light. Equation 6.2 is useful in grasping the interplay between the two concepts, as illustrated in Quick Quiz 6.1.

*Quick Quiz*

**6.1** Two masses $m_1$ and $m_2$, with $m_1 < m_2$, have equal kinetic energy. How do the magnitudes of their momenta compare? (a) Not enough information is given. (b) $p_1 < p_2$ (c) $p_1 = p_2$ (d) $p_1 > p_2$.

Changing the momentum of an object requires the application of a force. This is, in fact, how Newton originally stated his second law of motion. Starting from the more common version of the second law, we have

$$\vec{\mathbf{F}}_{\text{net}} = m\vec{\mathbf{a}} = m\frac{\Delta\vec{\mathbf{v}}}{\Delta t} = \frac{\Delta(m\vec{\mathbf{v}})}{\Delta t}$$

where the mass $m$ and the forces are assumed constant. The quantity in parentheses is just the momentum, so we have the following result:

**The change in an object's momentum $\Delta\vec{\mathbf{p}}$ divided by the elapsed time $\Delta t$ equals the constant net force $\vec{\mathbf{F}}_{\text{net}}$ acting on the object:**

◀ Newton's second law and momentum

$$\frac{\Delta\vec{\mathbf{p}}}{\Delta t} = \frac{\text{change in momentum}}{\text{time interval}} = \vec{\mathbf{F}}_{\text{net}} \qquad [6.3]$$

This equation is also valid when the forces are not constant, provided the limit is taken as $\Delta t$ becomes infinitesimally small. Equation 6.3 says that if the net force on an object is zero, the object's momentum doesn't change. In other words, the linear momentum of an object is conserved when $\vec{\mathbf{F}}_{\text{net}} = 0$. Equation 6.3 also shows us that changing an object's momentum requires the continuous application of a force over a period of time $\Delta t$, leading to the definition of *impulse*:

If a constant force $\vec{\mathbf{F}}$ acts on an object, the **impulse $\vec{\mathbf{I}}$** delivered to the object over a time interval $\Delta t$ is given by

$$\vec{\mathbf{I}} \equiv \vec{\mathbf{F}}\,\Delta t \qquad [6.4]$$

**SI unit: kilogram meter per second (kg · m/s)**

Impulse is a vector quantity with the same direction as the constant force acting on the object. When a single constant force $\vec{\mathbf{F}}$ acts on an object, Equation 6.3 can be written as

◀ Impulse–momentum theorem

$$\vec{\mathbf{I}} = \vec{\mathbf{F}}\Delta t = \Delta\vec{\mathbf{p}} = m\vec{\mathbf{v}}_f - m\vec{\mathbf{v}}_i \qquad [6.5]$$

which is a special case of the **impulse–momentum theorem.** Equation 6.5 shows that **the impulse of the force acting on an object equals the change in momentum of that object.** That equality is true even if the force is not constant, as long as the time interval $\Delta t$ is taken to be arbitrarily small. (The proof of the general case requires concepts from calculus.)

In real-life situations, the force on an object is only rarely constant. For example, when a bat hits a baseball, the force increases sharply, reaches some maximum value, and then decreases just as rapidly. Figure 6.1a shows a typical graph of force versus time for such incidents. The force starts out small as the bat comes in contact with the ball, rises to a maximum value when they are firmly in contact, and then drops off as the ball leaves the bat. To analyze this rather complex interaction, it's useful to define an **average force $\vec{\mathbf{F}}_{\text{av}}$,** shown as the dashed line in Figure 6.1b. The average force is the constant force delivering the same impulse to the object in the time interval $\Delta t$ as the actual time-varying force. We can then write the impulse–momentum theorem as

$$\vec{\mathbf{F}}_{\text{av}}\,\Delta t = \Delta\vec{\mathbf{p}} \qquad [6.6]$$

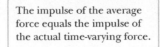

The impulse equals the area under the force vs. time curve.

The impulse of the average force equals the impulse of the actual time-varying force.

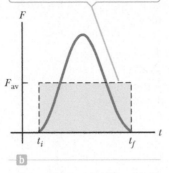

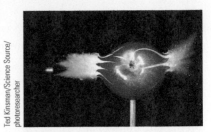

**Figure 6.2** An apple being pierced by a 30-caliber bullet traveling at a supersonic speed of 900 m/s. This collision was photographed with a microflash stroboscope using an exposure time of 0.33 $\mu$s. Shortly after the photograph was taken, the apple disintegrated completely. Note that the points of entry and exit of the bullet are visually explosive.

**Figure 6.1** (a) A net force acting on a particle may vary in time. (b) The value of the constant force $F_{av}$ (horizontal dashed line) is chosen so that the area of the rectangle $F_{av}\Delta t$ is the same as the area under the curve in (a).

**The magnitude of the impulse delivered by a force during the time interval $\Delta t$ is equal to the area under the force versus time graph as in Figure 6.1a or, equivalently, to $F_{av}\Delta t$ as shown in Figure 6.1b.** The brief collision between a bullet and an apple is illustrated in Figure 6.2.

---

**APPLYING PHYSICS 6.1**   **BOXING AND BRAIN INJURY**   BIO

Boxers in the nineteenth century used their bare fists. In modern boxing, fighters wear padded gloves. How do gloves protect the brain of the boxer from injury? Also, why do boxers often "roll with the punch"?

**EXPLANATION** The brain is immersed in a cushioning fluid inside the skull. If the head is struck suddenly by a bare fist, the skull accelerates rapidly. The brain matches this acceleration only because of the large impulsive force exerted by the skull on the brain. This large and sudden force (large $F_{av}$ and small $\Delta t$) can cause severe brain injury. Padded gloves extend the time $\Delta t$ over which the force is

applied to the head. For a given impulse $F_{av}\Delta t$, a glove results in a longer time interval than a bare fist, decreasing the average force. Because the average force is decreased, the acceleration of the skull is decreased, reducing (but not eliminating) the chance of brain injury. The same argument can be made for "rolling with the punch": If the head is held steady while being struck, the time interval over which the force is applied is relatively short and the average force is large. If the head is allowed to move in the same direction as the punch, the time interval is lengthened and the average force reduced. ■

---

**EXAMPLE 6.1**   **TEEING OFF**

**GOAL** Use the impulse–momentum theorem to estimate the average force exerted during an impact.

**PROBLEM** A golf ball with mass $5.0 \times 10^{-2}$ kg is struck with a club as in Figure 6.3. The force on the ball varies from zero when contact is made up to some maximum value (when the ball is maximally deformed) and then back to zero when the ball leaves the club, as in the graph of force vs. time in Figure 6.1. Assume that the ball leaves the club face with a velocity of 44 m/s. **(a)** Find the magnitude of the impulse due to the collision. **(b)** Estimate the duration of the collision and the average force acting on the ball.

**STRATEGY** In part **(a)**, use the fact that the impulse is equal to the change in momentum. The mass and the initial and final velocities are known, so this change can be computed. In part **(b)**, the average force is just the change in momentum computed in part **(a)** divided by an estimate of the duration of the collision. Estimate the distance the ball travels on the face of the club (about 2.0 cm, roughly the same as the radius of the ball). Divide this distance by the average velocity (half the final velocity) to get an estimate of the time of contact.

**Figure 6.3** (Example 6.1) During impact, the club head momentarily flattens the side of the golf ball.

*(Continued)*

## SOLUTION

**(a)** Find the impulse delivered to the ball.

The problem is essentially one dimensional. Note that $v_i = 0$, and calculate the change in momentum, which equals the impulse:

$$I = \Delta p = p_f - p_i = (5.0 \times 10^{-2} \text{ kg})(44 \text{ m/s}) - 0$$
$$= \boxed{2.2 \text{ kg} \cdot \text{m/s}}$$

**(b)** Estimate the duration of the collision and the average force acting on the ball.

Estimate the time interval of the collision, $\Delta t$, using the approximate displacement (radius of the ball) and its average speed (half the maximum speed):

$$\Delta t = \frac{\Delta x}{v_{av}} = \frac{2.0 \times 10^{-2} \text{ m}}{22 \text{ m/s}} = \boxed{9.1 \times 10^{-4} \text{ s}}$$

Estimate the average force from Equation 6.6:

$$F_{av} = \frac{\Delta p}{\Delta t} = \frac{2.2 \text{ kg} \cdot \text{m/s}}{9.1 \times 10^{-4} \text{ s}} = \boxed{2.4 \times 10^3 \text{ N}}$$

**REMARKS** This estimate shows just how large such contact forces can be. A good golfer achieves maximum momentum transfer by shifting weight from the back foot to the front foot, transmitting the body's momentum through the shaft and head of the club. This timing, involving a short movement of the hips, is more effective than a shot powered exclusively by the arms and shoulders. Following through with the swing ensures that the motion isn't slowed at the critical instant of impact.

**QUESTION 6.1** What average club speed would double the average force? (Assume the final velocity is unchanged.)

**EXERCISE 6.1** A 0.150-kg baseball, thrown with a speed of 40.0 m/s, is hit straight back at the pitcher with a speed of 50.0 m/s. **(a)** What is the magnitude of the impulse delivered by the bat to the baseball? **(b)** Find the magnitude of the average force exerted by the bat on the ball if the two are in contact for $2.00 \times 10^{-3}$ s.

**ANSWERS** **(a)** 13.5 kg · m/s   **(b)** 6.75 kN

---

### EXAMPLE 6.2   HOW GOOD ARE THE BUMPERS?

**GOAL** Find an impulse and estimate a force in a collision of a moving object with a stationary object.

**PROBLEM** In a crash test, a car of mass $1.50 \times 10^3$ kg collides with a wall and rebounds as in Figure 6.4a. The initial and final velocities of the car are $v_i = -15.0$ m/s and $v_f = 2.60$ m/s, respectively. If the collision lasts for 0.150 s, find **(a)** the impulse delivered to the car due to the collision and **(b)** the magnitude and direction of the average force exerted on the car.

**STRATEGY** This problem is similar to the previous example, except that the initial and final momenta are both nonzero. Find the momenta and substitute into the impulse–momentum theorem, Equation 6.6, solving for $F_{av}$.

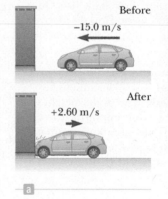

Before
−15.0 m/s

After
+2.60 m/s

a

b

**Figure 6.4** (Example 6.2) (a) This car's momentum changes as a result of its collision with the wall. (b) In a crash test (an inelastic collision), much of the car's initial kinetic energy is transformed into the energy it took to damage the vehicle.

TRL Ltd./Science Source

---

## SOLUTION

**(a)** Find the impulse delivered to the car.

Calculate the initial and final momenta of the car:

$$p_i = mv_i = (1.50 \times 10^3 \text{ kg})(-15.0 \text{ m/s})$$
$$= -2.25 \times 10^4 \text{ kg} \cdot \text{m/s}$$
$$p_f = mv_f = (1.50 \times 10^3 \text{ kg})(+2.60 \text{ m/s})$$
$$= +0.390 \times 10^4 \text{ kg} \cdot \text{m/s}$$

The impulse is just the difference between the final and initial momenta:

$$I = p_f - p_i$$
$$= +0.390 \times 10^4 \text{ kg} \cdot \text{m/s} - (-2.25 \times 10^4 \text{ kg} \cdot \text{m/s})$$
$$I = \boxed{2.64 \times 10^4 \text{ kg} \cdot \text{m/s}}$$

**(b)** Find the average force exerted on the car.

Apply Equation 6.6, the impulse–momentum theorem:

$$F_{av} = \frac{\Delta p}{\Delta t} = \frac{2.64 \times 10^4 \text{ kg} \cdot \text{m/s}}{0.150 \text{ s}} = \boxed{+1.76 \times 10^5 \text{ N}}$$

**REMARKS** If the car doesn't rebound off the wall, the average force exerted on the car is smaller than the value just calculated. With a final momentum of zero, the car undergoes a smaller change in momentum.

**QUESTION 6.2** When a person is involved in a car accident, why is the likelihood of injury greater in a head-on collision as opposed to being hit from behind? Answer using the concepts of relative velocity, momentum, and average force.

**EXERCISE 6.2** Suppose the car doesn't rebound off the wall, but the time interval of the collision remains at 0.150 s. In this case, the final velocity of the car is zero. Find the average force exerted on the car.

**ANSWER** $+1.50 \times 10^5 \, \text{N}$

## 6.1.1. Injury in Automobile Collisions

The main injuries that occur to a person hitting the interior of a car in a crash are brain damage, bone fracture, and trauma to the skin, blood vessels, and internal organs. Here, we compare the rather imprecisely known thresholds for human injury with typical forces and accelerations experienced in a car crash.

A force of about 90 kN (20 000 lb) compressing the tibia can cause fracture. Although the breaking force varies with the bone considered, we may take this value as the threshold force for fracture. It's well known that rapid acceleration of the head, even without skull fracture, can be fatal. Estimates show that head accelerations of $150g$ experienced for about 4 ms or $50g$ for 60 ms are fatal 50% of the time. Such injuries from rapid acceleration often result in nerve damage to the spinal cord where the nerves enter the base of the brain. The threshold for damage to skin, blood vessels, and internal organs may be estimated from whole-body impact data, where the force is uniformly distributed over the entire front surface area of 0.7 to 0.9 m². These data show that if the collision lasts for less than about 70 ms, a person will survive if the whole-body impact pressure (force per unit area) is less than $1.9 \times 10^5 \, \text{N/m}^2$ (28 lb/in.²). Death results in 50% of cases in which the whole-body impact pressure reaches $3.4 \times 10^5 \, \text{N/m}^2$ (50 lb/in.²).

Armed with the data above, we can estimate the forces and accelerations in a typical car crash and see how seat belts, air bags, and padded interiors can reduce the chance of death or serious injury in a collision. Consider a typical collision involving a 75-kg passenger not wearing a seat belt, traveling at 27 m/s (60 mi/h) who comes to rest in about 0.010 s after striking an unpadded dashboard. Using $F_{av} \Delta t = mv_f - mv_i$, we find that

$$F_{av} = \frac{mv_f - mv_i}{\Delta t} = \frac{0 - (75 \, \text{kg})(27 \, \text{m/s})}{0.010 \, \text{s}} = -2.0 \times 10^5 \, \text{N}$$

and

$$a = \left|\frac{\Delta v}{\Delta t}\right| = \frac{27 \, \text{m/s}}{0.010 \, \text{s}} = 2 \, 700 \, \text{m/s}^2 = \frac{2 \, 700 \, \text{m/s}^2}{9.8 \, \text{m/s}^2} g = 280g$$

If we assume the passenger crashes into the dashboard and windshield so that the head and chest, with a combined surface area of 0.5 m², experience the force, we find a whole-body pressure of

$$\frac{F_{av}}{A} = \frac{2.0 \times 10^5 \, \text{N}}{0.5 \, \text{m}^2} \cong 4 \times 10^5 \, \text{N/m}^2$$

We see that the force, the acceleration, and the whole-body pressure all *exceed* the threshold for fatality or broken bones and that an unprotected collision at 60 mi/h is almost certainly fatal.

What can be done to reduce or eliminate the chance of dying in a car crash? The most important factor is the collision time, or the time it takes the person to come to rest. If this time can be increased by 10 to 100 times the value of 0.01 s for a hard collision, the chances of survival in a car crash are much higher because the increase in $\Delta t$ makes the contact force 10 to 100 times smaller. Seat belts restrain people so that they come to rest in about the same amount of time it takes to stop the car, typically about 0.15 s. This increases the effective collision time by an order of magnitude. Figure 6.5 shows the measured force on a car versus time for a car crash.

**BIO APPLICATION**
Injury to Passengers in Car Collisions

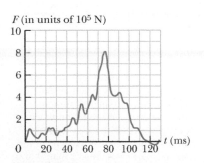

**Figure 6.5** Force on a car vs. time for a typical collision.

Air bags also increase the collision time, absorb energy from the body as they rapidly deflate, and spread the contact force over an area of the body of about 0.5 m², preventing penetration wounds and fractures. Air bags must deploy very rapidly (in less than 10 ms) in order to stop a human traveling at 27 m/s before he or she comes to rest against the steering column about 0.3 m away. To achieve this rapid deployment, accelerometers send a signal to discharge a bank of capacitors (devices that store electric charge), which then ignites an explosive, thereby filling the air bag with gas very quickly. The electrical charge for this ignition is stored in capacitors to ensure that the air bag deploys in the event of damage to the battery or the car's electrical system in a severe collision.

The important reduction in potentially fatal forces, accelerations, and pressures to tolerable levels by the simultaneous use of seat belts and air bags is summarized as follows. If a 75.0-kg person traveling 27.0 m/s is stopped by a seat belt in 0.150 s, the person experiences an average force of 13.5 kN, an average acceleration of 18.4$g$, and a whole-body pressure of $2.7 \times 10^4$ N/m² for a contact area of 0.50 m². These values are about one order of magnitude less than the values estimated earlier for an unprotected person and well below the thresholds for life-threatening injuries.

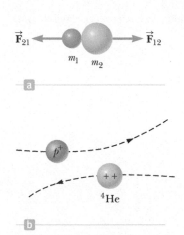

**Figure 6.6** (a) A collision between two objects resulting from direct contact. (b) A collision between two charged objects (in this case, a proton and a helium nucleus).

# 6.2 Conservation of Momentum

When a collision occurs in an isolated system, the total momentum of the system doesn't change with the passage of time. Instead, it remains constant both in magnitude and in direction. The momenta of the individual objects in the system may change, but the vector sum of *all* the momenta will not change. The total momentum is therefore said to be *conserved*. In this section, we will see how the laws of motion lead us to this important conservation law.

A collision may be the result of physical contact between two objects, as illustrated in Figure 6.6a. This is a common macroscopic event, as when a pair of billiard balls or a baseball and a bat strike each other. By contrast, because contact on a submicroscopic scale is hard to define accurately, the notion of *collision* must be generalized to that scale. Forces between two objects arise from the electrostatic interaction of the electrons in the surface atoms of the objects. As will be discussed in Topic 15, electric charges are either positive or negative. Charges with the same sign repel each other, while charges with opposite sign attract each other. To understand the distinction between macroscopic and microscopic collisions, consider the collision between two positive charges, as shown in Figure 6.6b. Because the two particles in the figure are both positively charged, they repel each other. During such a microscopic collision, particles need not touch in the normal sense in order to interact and transfer momentum.

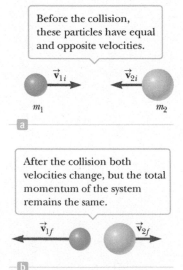

**Figure 6.7** Before and after a head-on collision between two particles. The momentum of each object changes during the collision, but the total momentum of the system is constant. Notice that the magnitude of the change of velocity of the lighter particle is greater than that of the heavier particle, which is true in general.

Figure 6.7 shows an isolated system of two particles before and after they collide. By "isolated," we mean that no external forces, such as the gravitational force or friction, act on the system. Before the collision, the velocities of the two particles are $\vec{\mathbf{v}}_{1i}$ and $\vec{\mathbf{v}}_{2i}$; after the collision, the velocities are $\vec{\mathbf{v}}_{1f}$ and $\vec{\mathbf{v}}_{2f}$. The impulse–momentum theorem applied to $m_1$ becomes

$$\vec{\mathbf{F}}_{21} \, \Delta t = m_1 \vec{\mathbf{v}}_{1f} - m_1 \vec{\mathbf{v}}_{1i}$$

Likewise, for $m_2$, we have

$$\vec{\mathbf{F}}_{12} \, \Delta t = m_2 \vec{\mathbf{v}}_{2f} - m_2 \vec{\mathbf{v}}_{2i}$$

where $\vec{\mathbf{F}}_{21}$ is the average force exerted by $m_2$ on $m_1$ during the collision and $\vec{\mathbf{F}}_{12}$ is the average force exerted by $m_1$ on $m_2$ during the collision, as in Figure 6.6a.

We use average values for $\vec{\mathbf{F}}_{21}$ and $\vec{\mathbf{F}}_{12}$ even though the actual forces may vary in time in a complicated way, as is the case in Figure 6.8. Newton's third law states that at all times these two forces are equal in magnitude and opposite in direction: $\vec{\mathbf{F}}_{21} = -\vec{\mathbf{F}}_{12}$. In addition, the two forces act over the same time interval. As a result, we have

$$\vec{\mathbf{F}}_{21} \, \Delta t = -\vec{\mathbf{F}}_{12} \, \Delta t$$

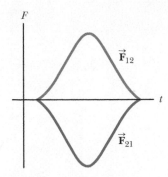

**Figure 6.8** Force as a function of time for the two colliding particles in Figures 6.6a and 6.7. Note that $\vec{\mathbf{F}}_{21} = -\vec{\mathbf{F}}_{12}$.

**Figure 6.9** Conservation of momentum is the principle behind a squid's propulsion system. It propels itself by expelling water at a high velocity.

Mike Severns/Stone/Getty Images

or

$$m_1\vec{\mathbf{v}}_{1f} - m_1\vec{\mathbf{v}}_{1i} = -(m_2\vec{\mathbf{v}}_{2f} - m_2\vec{\mathbf{v}}_{2i})$$

after substituting the expressions obtained for $\vec{\mathbf{F}}_{21}$ and $\vec{\mathbf{F}}_{12}$. This equation can be rearranged to give the following important result:

$$m_1\vec{\mathbf{v}}_{1i} + m_2\vec{\mathbf{v}}_{2i} = m_1\vec{\mathbf{v}}_{1f} + m_2\vec{\mathbf{v}}_{2f} \qquad [6.7]$$

This result is a special case of the law of **conservation of momentum** and is true of isolated systems containing any number of interacting objects (Fig. 6.9).

> When no net external force acts on a system, the total momentum of the system remains constant in time.

◀ **Conservation of momentum**

Defining the isolated system is an important feature of applying this conservation law. A cheerleader jumping upwards from rest might appear to violate conservation of momentum, because initially her momentum is zero and suddenly she's leaving the ground with velocity $\vec{\mathbf{v}}$. The flaw in this reasoning lies in the fact that the cheerleader isn't an isolated system. In jumping, she exerts a downward force on Earth, changing its momentum. This change in Earth's momentum isn't noticeable, however, because of Earth's gargantuan mass compared to the cheerleader's. When we define the system to be *the cheerleader and Earth,* momentum is conserved.

**BIO** **APPLICATION**

Conservation of Momentum and Squid Propulsion

Action and reaction, together with the accompanying exchange of momentum between two objects, is responsible for the phenomenon known as *recoil.* Everyone knows that throwing a baseball while standing straight up, without bracing one's feet against Earth, is a good way to fall over backwards. This reaction, an example of recoil, also happens when you fire a gun or shoot an arrow. Conservation of momentum provides a straightforward way to calculate such effects, as the next example shows.

## EXAMPLE 6.3   THE ARCHER

**GOAL** Calculate recoil velocity using conservation of momentum.

**PROBLEM** An archer stands at rest on frictionless ice; his total mass including his bow and quiver of arrows is 60.00 kg. (See Fig. 6.10.) **(a)** If the archer fires a 0.030 0-kg arrow horizontally at 50.0 m/s in the positive $x$-direction, what is his subsequent velocity across the ice? **(b)** He then fires a second identical arrow at the same speed relative to the ground but at an angle of 30.0° above the horizontal. Find his new speed. **(c)** Estimate the average normal force acting on the archer as the second arrow is accelerated by the bowstring. Assume a draw length of 0.800 m.

*(Continued)*

**STRATEGY** To solve part **(a)**, set up the conservation of momentum equation in the $x$-direction and solve for the final velocity of the archer. The system of the archer (including the bow) and the arrow is not isolated, because the gravitational and normal forces act on it. Those forces, however, are perpendicular to the motion of the system during the release of the arrow, and in addition are equal in magnitude and opposite in direction. Consequently, they produce no impulse during the arrow's release and conservation of momentum can be used. In part **(b)**, conservation of momentum can be applied again, neglecting the tiny effect of gravitation on the arrow during its release. This time there is a nonzero initial velocity. Part **(c)** requires using the impulse–momentum theorem and estimating the time, which can be carried out with simple ballistics.

**Figure 6.10** (Example 6.3) An archer fires an arrow horizontally to the right. Because he is standing on frictionless ice, he will begin to slide to the left across the ice.

**SOLUTION**

**(a)** Find the archer's subsequent velocity across the ice.

Write the conservation of momentum equation for the $x$-direction.

$$p_{ix} = p_{fx}$$

Let $m_1$ and $v_{1f}$ be the archer's mass and velocity after firing the arrow, respectively, and $m_2$ and $v_{2f}$ the arrow's mass and velocity. Both velocities are in the $x$-direction. Substitute $p_i = 0$ and expressions for the final momenta:

$$0 = m_1 v_{1f} + m_2 v_{2f}$$

Solve for $v_{1f}$ and substitute $m_1 = 59.97$ kg, $m_2 = 0.030\ 0$ kg, and $v_{2f} = 50.0$ m/s:

$$v_{1f} = -\frac{m_2}{m_1} v_{2f} = -\left(\frac{0.030\ 0 \text{ kg}}{59.97 \text{ kg}}\right)(50.0 \text{ m/s})$$

$$v_{1f} = \boxed{-0.025\ 0 \text{ m/s}}$$

**(b)** Calculate the archer's velocity after he fires a second arrow at an angle of 30.0° above the horizontal.

Write the $x$-component of the momentum equation with $m_1$ again the archer's mass after firing the first arrow as in part **(a)** and $m_2$ the mass of the next arrow:

$$m_1 v_{1i} = (m_1 - m_2)v_{1f} + m_2 v_{2f} \cos\theta$$

Solve for $v_{1f}$, the archer's final velocity, and substitute:

$$v_{1f} = \frac{m_1}{(m_1 - m_2)} v_{1i} - \frac{m_2}{(m_1 - m_2)} v_{2f} \cos\theta$$

$$= \left(\frac{59.97 \text{ kg}}{59.94 \text{ kg}}\right)(-0.025\ 0 \text{ m/s}) - \left(\frac{0.030\ 0 \text{ kg}}{59.94 \text{ kg}}\right)(50.0 \text{ m/s}) \cos(30.0°)$$

$$v_{1f} = \boxed{-0.046\ 7 \text{ m/s}}$$

**(c)** Estimate the average normal force acting on the archer as the arrow is accelerated by the bowstring.

Use kinematics in one dimension to estimate the acceleration of the arrow:

$$v^2 - v_0^2 = 2a\Delta x$$

Solve for the acceleration and substitute values setting $v = v_{2f}$, the final velocity of the arrow:

$$a = \frac{v_{2f}^2 - v_0^2}{2\Delta x} = \frac{(50.0 \text{ m/s})^2 - 0}{2(0.800 \text{ m})} = 1.56 \times 10^3 \text{ m/s}^2$$

Find the time the arrow is accelerated using $v = at + v_0$:

$$t = \frac{v_{2f} - v_0}{a} = \frac{50.0 \text{ m/s} - 0}{1.56 \times 10^3 \text{ m/s}^2} = 0.032\ 0 \text{ s}$$

Write the $y$-component of the impulse–momentum theorem:

$$F_{y,\text{av}} \, \Delta t = \Delta p_y$$

$$F_{y,\text{av}} = \frac{\Delta p_y}{\Delta t} = \frac{m_2 v_{2f} \sin \theta}{\Delta t}$$

$$F_{y,\text{av}} = \frac{(0.030\,0 \text{ kg})(50.0 \text{ m/s}) \sin (30.0°)}{0.032\,0 \text{ s}} = 23.4 \text{ N}$$

The average normal force is given by the archer's weight plus the reaction force $R$ of the arrow on the archer:

$$\sum F_y = n - mg - R = 0$$

$$n = mg + R = (59.94 \text{ kg})(9.80 \text{ m/s}^2) + (23.4 \text{ N}) = \boxed{6.11 \times 10^2 \text{ N}}$$

......................................................................................................

**REMARKS** The negative sign on $v_{1f}$ indicates that the archer is moving opposite the arrow's direction, in accordance with Newton's third law. Because the archer is much more massive than the arrow, his acceleration and velocity are much smaller than the acceleration and velocity of the arrow. A technical point: the second arrow was fired at the same velocity relative to the ground, but because the archer was moving backwards at the time, it was traveling slightly faster than the first arrow relative to the archer. Velocities must always be given relative to a frame of reference.

Notice that conservation of momentum was effective in leading to a solution in parts (a) and (b). The final answer for the normal force is only an average because the force exerted on the arrow is unlikely to be constant. If the ice really were frictionless, the archer would have trouble standing. In general the coefficient of static friction of ice is more than sufficient to prevent sliding in response to such small recoils.

**QUESTION 6.3** Would firing a heavier arrow necessarily increase the recoil velocity? Explain, using the result of Quick Quiz 6.1.

**EXERCISE 6.3** A 70.0-kg man and a 55.0-kg woman holding a 2.50-kg purse on ice skates stand facing each other. **(a)** If the woman pushes the man backwards so that his final speed is 1.50 m/s, with what average force did she push him, assuming they were in contact for 0.500 s? **(b)** What is the woman's recoil speed? **(c)** If she now throws her 2.50-kg purse at him at a 20.0° angle above the horizontal and at 4.20 m/s relative to the ground, what is her subsequent speed?

**ANSWERS** **(a)** $2.10 \times 10^2$ N    **(b)** 1.83 m/s    **(c)** 2.09 m/s

---

*Quick Quiz*

**6.2** A boy standing at one end of a floating raft that is stationary relative to the shore walks to the opposite end of the raft, away from the shore. As a consequence, the raft (a) remains stationary, (b) moves away from the shore, or (c) moves toward the shore. (*Hint:* Use conservation of momentum.)

# 6.3 Collisions in One Dimension

We have seen that for any type of collision, the total momentum of the system just before the collision equals the total momentum just after the collision as long as the system may be considered isolated. The total kinetic energy, on the other hand, is generally not conserved in a collision because some of the kinetic energy is converted to internal energy, sound energy, and the work needed to permanently deform the objects involved, such as cars in a car crash. **We define an inelastic collision as a collision in which momentum is conserved, but kinetic energy is not.** The collision of a rubber ball with a hard surface is inelastic, because some of the kinetic energy is lost when the ball is deformed during contact with the surface. **When two objects collide and stick together, the collision is called** *perfectly inelastic.* For example, if two pieces of putty collide, they stick together and move with some common velocity after the collision. If a meteorite collides head on with Earth, it becomes buried in Earth and the collision is considered perfectly inelastic. Only in very special circumstances is all the initial kinetic energy lost in a perfectly inelastic collision.

An elastic collision is defined as one in which both momentum and kinetic energy are conserved. Billiard ball collisions and the collisions of air molecules with the

**Tip 6.2** Inelastic vs. Perfectly Inelastic Collisions

If the colliding particles stick together, the collision is perfectly inelastic. If they bounce off each other (and kinetic energy is not conserved), the collision is inelastic.

**Tip 6.3** Momentum and Kinetic Energy in Collisions

The momentum of an isolated system is conserved in all collisions. However, the kinetic energy of an isolated system is conserved only when the collision is elastic.

walls of a container at ordinary temperatures are highly elastic. Macroscopic collisions such as those between billiard balls are only approximately elastic, because some loss of kinetic energy takes place—for example, in the clicking sound when two balls strike each other. Perfectly elastic collisions do occur, however, between atomic and subatomic particles. Elastic and perfectly inelastic collisions are *limiting* cases; most actual collisions fall into a range in between them.

**BIO APPLICATION**
Glaucoma Testing

As a practical application, an inelastic collision is used to detect glaucoma, a disease in which the pressure inside the eye builds up and leads to blindness by damaging the cells of the retina. In this application, medical professionals use a device called a *tonometer* to measure the pressure inside the eye. This device releases a puff of air against the outer surface of the eye and measures the speed of the air after reflection from the eye. At normal pressure, the eye is slightly spongy, and the pulse is reflected at low speed. As the pressure inside the eye increases, the outer surface becomes more rigid, and the speed of the reflected pulse increases. In this way, the speed of the reflected puff of air can measure the internal pressure of the eye.

We can summarize the types of collisions as follows:

Elastic collision ▶
Inelastic collision ▶

- In an elastic collision, both momentum and kinetic energy are conserved.
- In an inelastic collision, momentum is conserved but kinetic energy is not.
- In a *perfectly* inelastic collision, momentum is conserved, kinetic energy is not, and the two objects stick together after the collision, so their final velocities are the same.

In the remainder of this section, we will treat perfectly inelastic collisions and elastic collisions in one dimension.

*Quick Quiz*

**6.3** A car and a large truck traveling at the same speed collide head-on and stick together. Which vehicle undergoes the larger change in the magnitude of its momentum? (a) the car (b) the truck (c) the change in the magnitude of momentum is the same for both (d) impossible to determine without more information.

Before a perfectly inelastic collision the objects move independently.

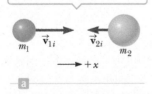

After the collision the objects remain in contact. System momentum *is* conserved, but system energy is *not* conserved.

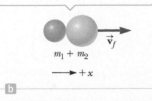

**Figure 6.11** (a) Before and (b) after a perfectly inelastic head-on collision between two objects.

### 6.3.1 Perfectly Inelastic Collisions

Consider two objects having masses $m_1$ and $m_2$ moving with known initial velocity components $v_{1i}$ and $v_{2i}$ along a straight line, as in Figure 6.11a. If the two objects collide head-on, stick together, and move with a common velocity component $v_f$ after the collision, then the collision is perfectly inelastic (Fig. 6.11b). Because the total momentum of the two-object isolated system before the collision equals the total momentum of the combined-object system after the collision, we can solve for the final velocity using conservation of momentum alone:

$$m_1 v_{1i} + m_2 v_{2i} = (m_1 + m_2) v_f \qquad [6.8]$$

$$v_f = \frac{m_1 v_{1i} + m_2 v_{2i}}{m_1 + m_2} \qquad [6.9]$$

It's important to notice that $v_{1i}$, $v_{2i}$, and $v_f$ represent the *x*-components of the velocity vectors, so care is needed in entering their known values, particularly with regard to signs. For example, in Figure 6.11a, $v_{1i}$ would have a positive value ($m_1$ moving to the right), whereas $v_{2i}$ would have a negative value ($m_2$ moving to the left). Once these values are entered, Equation 6.9 can be used to find the correct final velocity, as shown in Examples 6.4 and 6.5.

## EXAMPLE 6.4    A TRUCK VERSUS A COMPACT

**GOAL** Apply conservation of momentum to a one-dimensional inelastic collision.

**PROBLEM** A pickup truck with mass $1.80 \times 10^3$ kg is traveling eastbound at $+15.0$ m/s, while a compact car with mass $9.00 \times 10^2$ kg is traveling westbound at $-15.0$ m/s. (See Fig. 6.12.) The vehicles collide head-on, becoming entangled. **(a)** Find the speed of the entangled vehicles after the collision. **(b)** Find the change in the velocity of each vehicle. **(c)** Find the change in the kinetic energy of the system consisting of both vehicles.

**STRATEGY** The total momentum of the vehicles before the collision, $p_i$, equals the total momentum of the vehicles after the collision, $p_f$, if we ignore friction and assume the two vehicles form an isolated system. (This is called the "impulse approximation.") Solve the momentum conservation equation for the final velocity of the entangled vehicles. Once the velocities are in hand, the other parts can be solved by substitution.

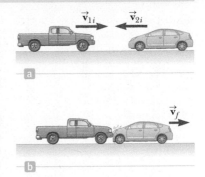

**Figure 6.12** (Example 6.4)

. . . . . . . . . . . . . . . . . . . . . . . . . . . . . . . . . . . . . . . . . . . . . . . . . . . . . . . . . . . . . . . . . .

## SOLUTION

**(a)** Find the final speed after collision.

Let $m_1$ and $v_{1i}$ represent the mass and initial velocity of the pickup truck, while $m_2$ and $v_{2i}$ pertain to the compact. Apply conservation of momentum:

$$p_i = p_f$$
$$m_1v_{1i} + m_2v_{2i} = (m_1 + m_2)v_f$$

Substitute the values and solve for the final velocity, $v_f$:

$$(1.80 \times 10^3 \text{ kg})(15.0 \text{ m/s}) + (9.00 \times 10^2 \text{ kg})(-15.0 \text{ m/s})$$
$$= (1.80 \times 10^3 \text{ kg} + 9.00 \times 10^2 \text{ kg})v_f$$
$$v_f = \boxed{+5.00 \text{ m/s}}$$

**(b)** Find the change in velocity for each vehicle.

Change in velocity of the pickup truck:

$$\Delta v_1 = v_f - v_{1i} = 5.00 \text{ m/s} - 15.0 \text{ m/s} = \boxed{-10.0 \text{ m/s}}$$

Change in velocity of the compact car:

$$\Delta v_2 = v_f - v_{2i} = 5.00 \text{ m/s} - (-15.0 \text{ m/s}) = \boxed{20.0 \text{ m/s}}$$

**(c)** Find the change in kinetic energy of the system.

Calculate the initial kinetic energy of the system:

$$KE_i = \tfrac{1}{2}m_1v_{1i}^2 + \tfrac{1}{2}m_2v_{2i}^2 = \tfrac{1}{2}(1.80 \times 10^3 \text{ kg})(15.0 \text{ m/s})^2$$
$$+\tfrac{1}{2}(9.00 \times 10^2 \text{ kg})(-15.0 \text{ m/s})^2$$
$$= 3.04 \times 10^5 \text{ J}$$

Calculate the final kinetic energy of the system and the change in kinetic energy, $\Delta KE$.

$$KE_f = \tfrac{1}{2}(m_1 + m_2)v_f^2$$
$$= \tfrac{1}{2}(1.80 \times 10^3 \text{ kg} + 9.00 \times 10^2 \text{ kg})(5.00 \text{ m/s})^2$$
$$= 3.38 \times 10^4 \text{ J}$$
$$\Delta KE = KE_f - KE_i = \boxed{-2.70 \times 10^5 \text{ J}}$$

. . . . . . . . . . . . . . . . . . . . . . . . . . . . . . . . . . . . . . . . . . . . . . . . . . . . . . . . . . . . . . . . . .

**REMARKS** During the collision, the system lost almost 90% of its kinetic energy. The change in velocity of the pickup truck was only 10.0 m/s, compared to twice that for the compact car. This example underscores perhaps the most important safety feature of any car: its mass. Injury is caused by a change in velocity, and the more massive vehicle undergoes a smaller velocity change in a typical accident.

**QUESTION 6.4** If the mass of both vehicles were doubled, how would the final velocity be affected? The change in kinetic energy?

**EXERCISE 6.4** Suppose the same two vehicles are both traveling eastward, the compact car leading the pickup truck.

The driver of the compact car slams on the brakes suddenly, slowing the vehicle to 6.00 m/s. If the pickup truck traveling at 18.0 m/s crashes into the compact car, find **(a)** the speed of the system right after the collision, assuming the two vehicles become entangled, **(b)** the change in velocity for both vehicles, and **(c)** the change in kinetic energy of the system, from the instant before impact (when the compact car is traveling at 6.00 m/s) to the instant right after the collision.

**ANSWERS** **(a)** 14.0 m/s   **(b)** pickup truck: $\Delta v_1 = -4.0$ m/s, compact car: $\Delta v_2 = 8.0$ m/s   **(c)** $-4.32 \times 10^4$ J

## EXAMPLE 6.5    THE BALLISTIC PENDULUM

**Figure 6.13** (Example 6.5)
(a) Diagram of a ballistic
pendulum. Note that $\vec{v}_{sys}$ is
the velocity of the system just
*after* the perfectly inelastic
collision. (b) Multiflash
photograph of a laboratory
ballistic pendulum.

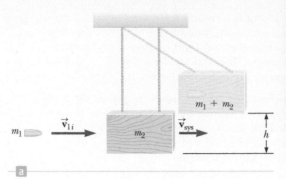

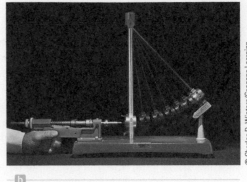

**GOAL**  Combine the concepts of conservation of energy and conservation of momentum in inelastic collisions.

**PROBLEM**  The ballistic pendulum (Fig. 6.13a) is a device used to measure the speed of a fast-moving projectile such as a bullet. The bullet is fired into a large block of wood suspended from some light wires. The bullet embeds in the block, and the entire system swings up to a height $h$. It is possible to obtain the initial speed of the bullet by measuring $h$ and the two masses. As an example of the technique, assume that the mass of the bullet, $m_1$, is 5.00 g, the mass of the pendulum, $m_2$, is 1.000 kg, and $h$ is 5.00 cm. **(a)** Find the velocity of the system after the bullet embeds in the block. **(b)** Calculate the initial speed of the bullet.

**STRATEGY**  Use conservation of energy to find the initial velocity of the block–bullet system, labeling it $v_{sys}$. Part **(b)** requires the conservation of momentum equation, which can be solved for the initial velocity of the bullet, $v_{1i}$.

### SOLUTION

**(a)** Find the velocity of the system after the bullet embeds in the block.

Apply conservation of energy to the block–bullet system after the collision:

$$(KE + PE)_{\text{after collision}} = (KE + PE)_{\text{top}}$$

Substitute expressions for the kinetic and potential energies. Note that both the potential energy at the bottom and the kinetic energy at the top are zero:

$$\tfrac{1}{2}(m_1 + m_2)v_{sys}^2 + 0 = 0 + (m_1 + m_2)gh$$

Solve for the final velocity of the block–bullet system, $v_{sys}$:

$$v_{sys}^2 = 2gh$$

$$v_{sys} = \sqrt{2gh} = \sqrt{2(9.80 \text{ m/s}^2)(5.00 \times 10^{-2} \text{ m})}$$

$$v_{sys} = \boxed{0.990 \text{ m/s}}$$

**(b)** Calculate the initial speed of the bullet.

Write the conservation of momentum equation and substitute expressions.

$$p_i = p_f$$

$$m_1 v_{1i} + m_2 v_{2i} = (m_1 + m_2)v_{sys}$$

Solve for the initial velocity of the bullet, and substitute values:

$$v_{1i} = \frac{(m_1 + m_2)v_{sys}}{m_1}$$

$$v_{1i} = \frac{(1.005 \text{ kg})(0.990 \text{ m/s})}{5.00 \times 10^{-3} \text{ kg}} = \boxed{199 \text{ m/s}}$$

**REMARKS**  Because the impact is inelastic, it would be incorrect to equate the initial kinetic energy of the incoming bullet to the final gravitational potential energy associated with the bullet–block combination. The energy isn't conserved!

**QUESTION 6.5**  List three ways mechanical energy can be lost from the system in this experiment.

**EXERCISE 6.5**  A bullet with mass 5.00 g is fired horizontally into a 2.000-kg block attached to a horizontal spring. The spring has a constant $6.00 \times 10^2$ N/m and reaches a maximum compression of 6.00 cm. **(a)** Find the initial speed of the bullet–block system. **(b)** Find the speed of the bullet.

**ANSWERS**  **(a)** 1.04 m/s    **(b)** 417 m/s

**6.4** An object of mass $m$ moves to the right with a speed $v$. It collides head-on with an object of mass $3m$ moving with speed $v/3$ in the opposite direction. If the two objects stick together, what is the speed of the combined object, of mass $4m$, after the collision?
(a) 0   (b) $v/2$   (c) $v$   (d) $2v$

**6.5** A skater is using very low-friction rollerblades. A friend throws a Frisbee to her, on the straight line along which she is coasting. Describe each of the following events as an elastic, an inelastic, or a perfectly inelastic collision between the skater and the Frisbee. (a) She catches the Frisbee and holds it. (b) She tries to catch the Frisbee, but it bounces off her hands and falls to the ground in front of her. (c) She catches the Frisbee and immediately throws it back with the same speed (relative to the ground) to her friend.

**6.6** In a perfectly inelastic one-dimensional collision between two objects, what initial condition alone is necessary so that *all* of the original kinetic energy of the system is gone after the collision? (a) The objects must have momenta with the same magnitude but opposite directions. (b) The objects must have the same mass. (c) The objects must have the same velocity. (d) The objects must have the same speed, with velocity vectors in opposite directions.

## 6.3.2 Elastic Collisions

Now consider two objects that undergo an **elastic head-on collision** (Fig. 6.14). In this situation, **both the momentum and the kinetic energy of the system of two objects are conserved.** We can write these conditions as

$$m_1 v_{1i} + m_2 v_{2i} = m_1 v_{1f} + m_2 v_{2f} \qquad [6.10]$$

and

$$\tfrac{1}{2}m_1 v_{1i}^2 + \tfrac{1}{2}m_2 v_{2i}^2 = \tfrac{1}{2}m_1 v_{1f}^2 + \tfrac{1}{2}m_2 v_{2f}^2 \qquad [6.11]$$

where $v$ is positive if an object moves to the right and negative if it moves to the left.

In a typical problem involving elastic collisions, there are two unknown quantities, and Equations 6.10 and 6.11 can be solved simultaneously to find them. These two equations are linear and quadratic, respectively. An alternate approach simplifies the quadratic equation to another linear equation, facilitating solution. Canceling the factor $\tfrac{1}{2}$ in Equation 6.11, we rewrite the equation as

$$m_1(v_{1i}^2 - v_{1f}^2) = m_2(v_{2f}^2 - v_{2i}^2)$$

Here we have moved the terms containing $m_1$ to one side of the equation and those containing $m_2$ to the other. Next, we factor both sides of the equation:

$$m_1(v_{1i} - v_{1f})(v_{1i} + v_{1f}) = m_2(v_{2f} - v_{2i})(v_{2f} + v_{2i}) \qquad [6.12]$$

Now we separate the terms containing $m_1$ and $m_2$ in the equation for the conservation of momentum (Eq. 6.10) to get

$$m_1(v_{1i} - v_{1f}) = m_2(v_{2f} - v_{2i}) \qquad [6.13]$$

Next, we divide Equation 6.12 by Equation 6.13, producing

$$v_{1i} + v_{1f} = v_{2f} + v_{2i} \qquad [6.14]$$

Gathering initial and final values on opposite sides of the equation gives

$$v_{1i} - v_{2i} = -(v_{1f} - v_{2f}) \qquad [6.15]$$

This equation, in combination with Equation 6.10, will be used to solve problems dealing with perfectly elastic head-on collisions. Equation 6.14 shows that the sum of the initial and final velocities for object 1 equals the sum of the initial and final velocities for object 2. According to Equation 6.15, the relative velocity of the two objects before the collision, $v_{1i} - v_{2i}$, equals the negative of the relative velocity of the two objects after the collision, $-(v_{1f} - v_{2f})$. To better understand

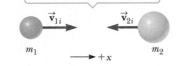

Before an elastic collision the two objects move independently.

$\vec{\mathbf{v}}_{1i}$     $\vec{\mathbf{v}}_{2i}$

$m_1$        $+x$        $m_2$

**a**

After the collision the object velocities change, but **both** the energy and momentum of the system are conserved.

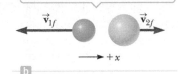

$\vec{\mathbf{v}}_{1f}$     $\vec{\mathbf{v}}_{2f}$

$+x$

**b**

**Figure 6.14** (a) Before and (b) after an elastic head-on collision between two hard spheres. Unlike an inelastic collision, both the total momentum and the total energy are conserved.

the equation, imagine that you are riding along on one of the objects. As you measure the velocity of the other object from your vantage point, you will be measuring the relative velocity of the two objects. In your view of the collision, the other object comes toward you and bounces off, leaving the collision with the same speed, but in the opposite direction. This is just what Equation 6.15 states.

---

**PROBLEM-SOLVING STRATEGY**

### One-Dimensional Collisions

*The following procedure is recommended for solving one-dimensional problems involving collisions between two objects:*

1. **Coordinates.** Choose a coordinate axis that lies along the direction of motion.
2. **Diagram.** Sketch the problem, representing the two objects as blocks and labeling velocity vectors and masses.
3. **Conservation of Momentum.** Write a general expression for the *total* momentum of the system of two objects *before* and *after* the collision, and equate the two, as in Equation 6.10. On the next line, fill in the known values.
4. **Conservation of Energy.** If the collision is elastic, write a general expression for the total energy before and after the collision, and equate the two quantities, as in Equation 6.11 or (preferably) Equation 6.15. Fill in the known values. (***Skip this step if the collision is *not* perfectly elastic.***)
5. **Solve** the equations simultaneously. Equations 6.10 and 6.15 form a system of two linear equations and two unknowns. If you have forgotten Equation 6.15, use Equation 6.11 instead.

---

Steps 1 and 2 of the problem-solving strategy are generally carried out in the process of sketching and labeling a diagram of the problem. This is clearly the case in our next example, which makes use of Figure 6.14. Other steps are pointed out as they are applied.

---

**EXAMPLE 6.6** **LET'S PLAY POOL**

**GOAL** Solve an elastic collision in one dimension.

**PROBLEM** Two billiard balls of identical mass move toward each other as in Figure 6.14, with the positive $x$-axis to the right (steps 1 and 2). Assume that the collision between them is perfectly elastic. If the initial velocities of the balls are $+30.0$ cm/s and $-20.0$ cm/s, what are the velocities of the balls after the collision? Assume friction and rotation are unimportant.

**STRATEGY** Solution of this problem is a matter of solving two equations, the conservation of momentum and conservation of energy equations, for two unknowns, the final velocities of the two balls. Instead of using Equation 6.11 for conservation of energy, use Equation 6.15, which is linear, hence easier to handle.

......................................................................................................................

**SOLUTION**

Write the conservation of momentum equation. Because $m_1 = m_2$, we can cancel the masses, then substitute $v_{1i} = +30.0$ cm/s and $v_{2i} = -20.0$ cm/s (Step 3).

$$m_1 v_{1i} + m_2 v_{2i} = m_1 v_{1f} + m_2 v_{2f}$$
$$30.0 \text{ cm/s} + (-20.0 \text{ cm/s}) = v_{1f} + v_{2f}$$
$$\textbf{(1)} \quad 10.0 \text{ cm/s} = v_{1f} + v_{2f}$$

Next, apply conservation of energy in the form of Equation 6.15 (Step 4):

$$\textbf{(2)} \quad v_{1i} - v_{2i} = -(v_{1f} - v_{2f})$$
$$30.0 \text{ cm/s} - (-20.0 \text{ cm/s}) = v_{2f} - v_{1f}$$
$$\textbf{(3)} \quad 50.0 \text{ cm/s} = v_{2f} - v_{1f}$$

Now solve Equations (1) and (3) simultaneously by adding them together (Step 5):

$$10.0 \text{ cm/s} + 50.0 \text{ cm/s} = (v_{1f} + v_{2f}) + (v_{2f} - v_{1f})$$
$$60.0 \text{ cm/s} = 2v_{2f} \quad \rightarrow \quad v_{2f} = \boxed{30.0 \text{ cm/s}}$$

Substitute the answer for $v_{2f}$ into Equation (1):

$$10.0 \text{ cm/s} = v_{1f} + 30.0 \text{ cm/s} \quad \rightarrow \quad v_{1f} = \boxed{-20.0 \text{ cm/s}}$$

......................................................................................................................

**REMARKS** Notice the balls exchanged velocities—almost as if they'd passed through each other. This is always the case when two objects of equal mass undergo an elastic head-on collision.

**QUESTION 6.6** In this example, is it possible to adjust the initial velocities of the balls so that both are at rest after the collision? Explain.

**EXERCISE 6.6** Find the final velocities of the two balls if the ball with initial velocity $v_{2i} = -20.0$ cm/s has a mass equal to one-half that of the ball with initial velocity $v_{1i} = +30.0$ cm/s.

**ANSWER** $v_{1f} = -3.33$ cm/s; $v_{2f} = +46.7$ cm/s

---

## EXAMPLE 6.7    TWO BLOCKS AND A SPRING

**GOAL** Solve an elastic collision involving spring potential energy.

**PROBLEM** A block of mass $m_1 = 1.60$ kg, initially moving to the right with a velocity of $+4.00$ m/s on a frictionless horizontal track, collides with a massless spring attached to a second block of mass $m_2 = 2.10$ kg moving to the left with a velocity of $-2.50$ m/s, as in Figure 6.15a. The spring has a spring constant of $6.00 \times 10^2$ N/m. **(a)** Determine the velocity of block 2 at the instant when block 1 is moving to the right with a velocity of $+3.00$ m/s, as in Figure 6.15b. **(b)** Find the compression of the spring at that time.

**STRATEGY** We identify the system as the two blocks and the spring. Write down the conservation of momentum equations, and solve for the final velocity of block 2, $v_{2f}$. Then use conservation of energy to find the compression of the spring at that time.

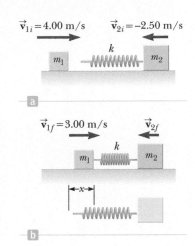

**Figure 6.15**
(Example 6.7)

---

### SOLUTION

**(a)** Find the velocity $v_{2f}$ when block 1 has velocity $+3.00$ m/s.

Write the conservation of momentum equation for the system and solve for $v_{2f}$:

$$\textbf{(1)} \quad m_1 v_{1i} + m_2 v_{2i} = m_1 v_{1f} + m_2 v_{2f}$$

$$v_{2f} = \frac{m_1 v_{1i} + m_2 v_{2i} - m_1 v_{1f}}{m_2}$$

$$= \frac{(1.60 \text{ kg})(4.00 \text{ m/s}) + (2.10 \text{ kg})(-2.50 \text{ m/s}) - (1.60 \text{ kg})(3.00 \text{ m/s})}{2.10 \text{ kg}}$$

$$v_{2f} = \boxed{-1.74 \text{ m/s}}$$

**(b)** Find the compression of the spring.

Use energy conservation for the system, noticing that potential energy is stored in the spring when it is compressed a distance $x$:

$$E_i = E_f$$
$$\tfrac{1}{2}m_1 v_{1i}^2 + \tfrac{1}{2}m_2 v_{2i}^2 + 0 = \tfrac{1}{2}m_1 v_{1f}^2 + \tfrac{1}{2}m_2 \, v_{2f}^2 + \tfrac{1}{2}kx^2$$

Substitute the given values and the result of part **(a)** into the preceding expression, solving for $x$:

$$x = \boxed{0.173 \text{ m}}$$

---

**REMARKS** The initial velocity component of block 2 is $-2.50$ m/s because the block is moving to the left. The negative value for $v_{2f}$ means that block 2 is still moving to the left at the instant under consideration.

**QUESTION 6.7** Is it possible for both blocks to come to rest while the spring is being compressed? Explain. *Hint:* Look at the momentum in Equation (1).

**EXERCISE 6.7** Find **(a)** the velocity of block 1 and **(b)** the compression of the spring at the instant that block 2 is at rest.

**ANSWERS** **(a)** 0.719 m/s to the right  **(b)** 0.251 m

# 6.4 Glancing Collisions

In Section 6.2 we showed that the total linear momentum of a system is conserved when the system is isolated (i.e., when no external forces act on the system). For a general collision of two objects in three-dimensional space, the conservation of momentum principle implies that the total momentum of the system in each direction is conserved. However, an important subset of collisions takes place in a plane. The game of billiards is a familiar example involving multiple collisions of objects moving on a two-dimensional surface. We restrict our attention to a single two-dimensional collision between two objects that takes place in a plane, and ignore any possible rotation. For such collisions, we obtain two component equations for the conservation of momentum:

$$m_1 v_{1ix} + m_2 v_{2ix} = m_1 v_{1fx} + m_2 v_{2fx}$$
$$m_1 v_{1iy} + m_2 v_{2iy} = m_1 v_{1fy} + m_2 v_{2fy}$$

We must use three subscripts in this general equation, to represent, respectively, (1) the object in question, and (2) the initial and final values of the components of velocity.

Now, consider a two-dimensional problem in which an object of mass $m_1$ collides with an object of mass $m_2$ that is initially at rest, as in Figure 6.16. After the collision, object 1 moves at an angle $\theta$ with respect to the horizontal, and object 2 moves at an angle $\phi$ with respect to the horizontal. This is called a *glancing* collision. Applying the law of conservation of momentum in component form, and noting that the initial $y$-component of momentum is zero, we have

$x$-component:  $\qquad m_1 v_{1i} + 0 = m_1 v_{1f} \cos\theta + m_2 v_{2f} \cos\phi$  [6.16]

$y$-component:  $\qquad 0 + 0 = m_1 v_{1f} \sin\theta + m_2 v_{2f} \sin\phi$  [6.17]

If the collision is elastic, we can write a third equation, for conservation of energy, in the form

$$\tfrac{1}{2} m_1 v_{1i}^2 = \tfrac{1}{2} m_1 v_{1f}^2 + \tfrac{1}{2} m_2 v_{2f}^2$$  [6.18]

If we know the initial velocity $v_{1i}$ and the masses, we are left with four unknowns ($v_{1f}$, $v_{2f}$, $\theta$, and $\phi$). Because we have only three equations, one of the four remaining quantities must be given in order to determine the motion after the collision from conservation principles alone.

If the collision is inelastic, the kinetic energy of the system is *not* conserved, and Equation 6.18 does *not* apply.

**Figure 6.16** A glancing collision between two objects.

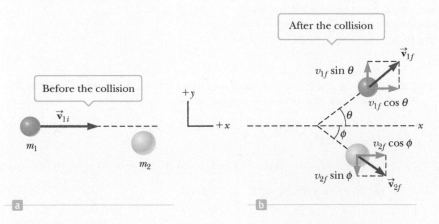

## PROBLEM-SOLVING STRATEGY

### Two-Dimensional Collisions

*To solve two-dimensional collisions, follow this procedure:*

1. **Coordinate Axes.** Use both $x$- and $y$-coordinates. It's convenient to have either the $x$-axis or the $y$-axis coincide with the direction of one of the initial velocities.
2. **Diagram.** Sketch the problem, labeling velocity vectors and masses.
3. **Conservation of Momentum.** Write a separate conservation of momentum equation for each of the $x$- and $y$-directions. In each case, the total initial momentum in a given direction equals the total final momentum in that direction.
4. **Conservation of Energy.** If the collision is elastic, write a general expression for the total energy before and after the collision, and equate the two expressions, as in Equation 6.11. Fill in the known values. (**Skip** this step if the collision is not **perfectly elastic.**) The energy equation can't be simplified as in the one-dimensional case, so a quadratic expression such as Equation 6.11 or 6.18 must be used when the collision is elastic.
5. **Solve** the equations simultaneously. There are two equations for inelastic collisions and three for elastic collisions.

---

## EXAMPLE 6.8    COLLISION AT AN INTERSECTION

**GOAL**  Analyze a two-dimensional inelastic collision.

**PROBLEM**  A car with mass $1.50 \times 10^3$ kg traveling east at a speed of 25.0 m/s collides at an intersection with a $2.50 \times 10^3$-kg pickup truck traveling north at a speed of 20.0 m/s, as shown in Figure 6.17. Find the magnitude and direction of the velocity of the wreckage immediately after the collision, assuming that the vehicles undergo a perfectly inelastic collision (that is, they stick together) and assuming that friction between the vehicles and the road can be neglected.

**STRATEGY**  Use conservation of momentum in two dimensions. (Kinetic energy is not conserved.) Choose coordinates as in Figure 6.17. Before the collision, the only object having momentum in the $x$-direction is the car, while the pickup truck carries all the momentum in the $y$-direction. After the totally inelastic collision, both vehicles move together at some common speed $v_f$ and angle $\theta$. Solve for these two unknowns, using the two components of the conservation of momentum equation.

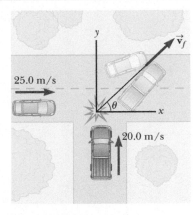

**Figure 6.17**  (Example 6.8) A top view of a perfectly inelastic collision between a car and a pickup truck.

. . . . . . . . . . . . . . . . . . . . . . . . . . . . . . . . . . . . . . . . . . . . . . . . . . . . . . . . . . . . . . . . . . . . . . . . . . . . . . . . . . .

### SOLUTION

Find the $x$-components of the initial and final total momenta:

$$\sum p_{xi} = m_{car}v_{car} = (1.50 \times 10^3 \text{ kg})(25.0 \text{ m/s})$$
$$= 3.75 \times 10^4 \text{ kg} \cdot \text{m/s}$$
$$\sum p_{xf} = (m_{car} + m_{truck})v_f \cos \theta = (4.00 \times 10^3 \text{ kg})v_f \cos \theta$$

Set the initial $x$-momentum equal to the final $x$-momentum:

**(1)**   $3.75 \times 10^4 \text{ kg} \cdot \text{m/s} = (4.00 \times 10^3 \text{ kg})v_f \cos \theta$

Find the $y$-components of the initial and final total momenta:

$$\sum p_{iy} = m_{truck}v_{truck} = (2.50 \times 10^3 \text{ kg})(20.0 \text{ m/s})$$
$$= 5.00 \times 10^4 \text{ kg} \cdot \text{m/s}$$
$$\sum p_{fy} = (m_{car} + m_{truck})v_f \sin \theta = (4.00 \times 10^3 \text{ kg})v_f \sin \theta$$

Set the initial $y$-momentum equal to the final $y$-momentum:

**(2)**   $5.00 \times 10^4 \text{ kg} \cdot \text{m/s} = (4.00 \times 10^3 \text{ kg})v_f \sin \theta$

*(Continued)*

Divide Equation (2) by Equation (1) and solve for $\theta$:

$$\tan\theta = \frac{5.00 \times 10^4 \text{ kg} \cdot \text{m/s}}{3.75 \times 10^4 \text{ kg} \cdot \text{m}} = 1.33$$

$$\theta = \boxed{53.1°}$$

Substitute this angle back into Equation (2) to find $v_f$:

$$v_f = \frac{5.00 \times 10^4 \text{ kg} \cdot \text{m/s}}{(4.00 \times 10^3 \text{ kg}) \sin 53.1°} = \boxed{15.6 \text{ m/s}}$$

**REMARKS** It's also possible to first find the $x$- and $y$-components $v_{fx}$ and $v_{fy}$ of the resultant velocity. The magnitude and direction of the resultant velocity can then be found with the Pythagorean theorem, $v_f = \sqrt{v_{fx}^2 + v_{fy}^2}$, and the inverse tangent function $\theta = \tan^{-1}(v_{fy}/v_{fx})$. Setting up this alternate approach is a simple matter of substituting $v_{fx} = v_f \cos\theta$ and $v_{fy} = v_f \sin\theta$ into Equations (1) and (2).

**QUESTION 6.8** If the car and truck had identical mass and speed, what would the resultant angle have been?

**EXERCISE 6.8** A 3.00-kg object initially moving in the positive $x$-direction with a velocity of $+5.00$ m/s collides with and sticks to a 2.00-kg object initially moving in the negative $y$-direction with a velocity of $-3.00$ m/s. Find the final components of velocity of the composite object.

**ANSWER** $v_{fx} = 3.00$ m/s; $v_{fy} = -1.20$ m/s

---

A rocket reaction chamber without a nozzle has reaction forces pushing equally in all directions, so no motion results.

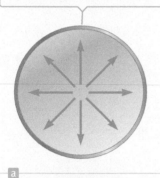

 a

An opening at the bottom of the chamber removes the downward reaction force, resulting in a net upward reaction force.

b

**Figure 6.18** A rocket reaction chamber containing a combusting gas works because it has a nozzle where gases can escape. The chamber wall acts on the expanding gas; the reaction force of the gas on the chamber wall pushes the rocket forward.

# 6.5 Rocket Propulsion

When ordinary vehicles such as cars and locomotives move, the driving force of the motion is friction. In the case of the car, this driving force is exerted by the road on the car, a reaction to the force exerted by the wheels against the road. Similarly, a locomotive "pushes" against the tracks; hence, the driving force is the reaction force exerted by the tracks on the locomotive. However, a rocket moving in space has no road or tracks to push against. How can it move forward?

In fact, reaction forces also propel a rocket. (You should review Newton's third law, discussed in Topic 4.) To illustrate this point, we model our rocket with a spherical chamber containing a combustible gas, as in Figure 6.18a. When an explosion occurs in the chamber, the hot gas expands and presses against all sides of the chamber, as indicated by the arrows. Because the sum of the forces exerted on the rocket is zero, it doesn't move. Now suppose a hole is drilled in the bottom of the chamber, as in Figure 6.18b. When the explosion occurs, the gas presses against the chamber in all directions, but can't press against anything at the hole, where it simply escapes into space. Adding the forces on the spherical chamber now results in a net force upward. Just as in the case of cars and locomotives, this is a reaction force. A car's wheels press against the ground, and the reaction force of the ground on the car pushes it forward. The wall of the rocket's combustion chamber exerts a force on the gas expanding against it. The reaction force of the gas on the wall then pushes the rocket upward.

In a now infamous 1920 article in *The New York Times*, rocket pioneer Robert Goddard was ridiculed for thinking that rockets would work in space, where, according to the *Times*, there was nothing to push against. The *Times* retracted, rather belatedly, during the first *Apollo* moon landing mission in 1969. The hot gases are not pushing against anything external, but against the rocket itself—and ironically, rockets actually work *better* in a vacuum. In an atmosphere, the gases have to do work against the outside air pressure to escape the combustion chamber, slowing the exhaust velocity and reducing the reaction force.

At the microscopic level, this process is complicated, but it can be simplified by applying conservation of momentum to the rocket and its ejected fuel. In principle, the solution is similar to that in Example 6.3, with the archer representing the rocket and the arrows the exhaust gases.

Suppose that at some time $t$, the momentum of the rocket plus the fuel is given by $(M + \Delta m)v$, where $\Delta m$ is an amount of fuel about to be burned (Fig. 6.19a).

This fuel is traveling at a speed $v$ relative to, say, Earth, just like the rest of the rocket. During a short time interval $\Delta t$, the rocket ejects fuel of mass $\Delta m$, and the rocket's speed increases to $v + \Delta v$ (Fig. 6.19b). If the fuel is ejected with exhaust speed $v_e$ *relative to the rocket*, the speed of the fuel relative to the Earth is $v - v_e$. Equating the total initial momentum of the system with the total final momentum, we have

$$(M + \Delta m)v = M(v + \Delta v) + \Delta m(v - v_e)$$

Simplifying this expression gives

$$M\Delta v = v_e \Delta m$$

The increase $\Delta m$ in the mass of the exhaust corresponds to an equal decrease in the mass of the rocket, so that $\Delta m = -\Delta M$. Using this fact, we have

$$M\Delta v = -v_e \Delta M \qquad [6.19]$$

This result, together with the methods of calculus, can be used to obtain the following equation:

$$v_f - v_i = v_e \ln\left(\frac{M_i}{M_f}\right) \qquad [6.20]$$

where $M_i$ is the initial mass of the rocket plus fuel and $M_f$ is the final mass of the rocket plus its remaining fuel. This is the basic expression for rocket propulsion; it tells us that the increase in velocity is proportional to the exhaust speed $v_e$ and to the natural logarithm of $M_i/M_f$. Because the maximum ratio of $M_i$ to $M_f$ for a single-stage rocket is about 10:1, the increase in speed can reach $v_e \ln 10 = 2.3v_e$ or about twice the exhaust speed! For best results, therefore, the exhaust speed should be as high as possible. Currently, typical rocket exhaust speeds are several kilometers per second.

The **thrust** on the rocket is defined as the force exerted on the rocket by the ejected exhaust gases. We can obtain an expression for the instantaneous thrust by dividing Equation 6.18 by $\Delta t$:

$$\text{Instantaneous thrust} = Ma = M\frac{\Delta v}{\Delta t} = \left| v_e \frac{\Delta M}{\Delta t} \right| \qquad [6.21] \qquad \blacktriangleleft \text{ Rocket thrust}$$

The absolute value signs are used for clarity: In Equation 6.19, $-\Delta M$ is a positive quantity (as is $v_e$, a speed). Here we see that the thrust increases as the exhaust velocity increases and as the rate of change of mass $\Delta M/\Delta t$ (the burn rate) increases.

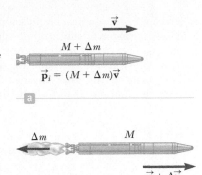

**Figure 6.19** Rocket propulsion. (a) The initial mass of the rocket and fuel is $M + \Delta m$ at a time $t$, and the rocket's speed is $v$. (b) At a time $t + \Delta t$, the rocket's mass has been reduced to $M$, and an amount of fuel $\Delta m$ has been ejected. The rocket's speed increases by an amount $\Delta v$.

---

### APPLYING PHYSICS 6.2    MULTISTAGE ROCKETS

The current maximum exhaust speed of $v_e = 4\,500$ m/s can be realized with rocket engines fueled with liquid hydrogen and liquid oxygen. But this means that the maximum speed attainable for a given rocket with a mass ratio of 10 is $v_e \ln 10 \simeq 10\,000$ m/s. To reach the Moon, however, requires a change in velocity of over 11 000 m/s. Further, this change must occur while working against gravity and atmospheric friction. How can that be managed without developing better engines?

**EXPLANATION** The answer is the multistage rocket. By dropping stages, the spacecraft becomes lighter, so that fuel burned later in the mission doesn't have to accelerate mass that no longer serves any purpose. Strap-on boosters, as used by the space shuttle and a number of other rockets, such as the *Titan 4* or Russian *Proton,* employ a similar method. The boosters are jettisoned after their fuel is exhausted, so the rocket is no longer burdened by their weight. ∎

---

### EXAMPLE 6.9    SINGLE STAGE TO ORBIT (SSTO)

**GOAL** Apply the velocity and thrust equations of a rocket.

**PROBLEM** A rocket has a total mass of $1.00 \times 10^5$ kg and a burnout mass of $1.00 \times 10^4$ kg, including engines, shell, and payload. The rocket blasts off from Earth and exhausts all its fuel in 4.00 min, burning the fuel at a steady rate with an exhaust velocity of $v_e = 4.50 \times 10^3$ m/s. **(a)** If air friction and gravity are neglected, what is the speed of the rocket at burnout? **(b)** What thrust does the engine develop at liftoff? **(c)** What is the initial acceleration of the rocket if gravity is not neglected? **(d)** Estimate the speed at burnout if gravity isn't neglected.

*(Continued)*

**STRATEGY** Although it sounds sophisticated, this problem is mainly a matter of substituting values into the appropriate equations. Part **(a)** requires substituting values into Equation 6.20 for the velocity. For part **(b)**, divide the change in the rocket's mass by the total time, getting $\Delta M/\Delta t$, then substitute into Equation 6.21 to find the thrust. **(c)** Using Newton's second law, the force of gravity, and the result of **(b)**, we can find the initial acceleration. For part **(d)**, the acceleration of gravity is approximately constant over the few kilometers involved, so the velocity found in part **(b)** will be reduced by roughly $\Delta v_g = -gt$. Add this loss to the result of part **(a)**.

## SOLUTION

**(a)** Calculate the velocity at burnout, ignoring gravity and air drag.

Substitute $v_i = 0$, $v_e = 4.50 \times 10^3$ m/s, $M_i = 1.00 \times 10^5$ kg, and $M_f = 1.00 \times 10^4$ kg into Equation 6.20:

$$v_f = v_i + v_e \ln\left(\frac{M_i}{M_f}\right)$$

$$= 0 + (4.5 \times 10^3 \text{ m/s}) \ln\left(\frac{1.00 \times 10^5 \text{ kg}}{1.00 \times 10^4 \text{ kg}}\right)$$

$$v_f = \boxed{1.04 \times 10^4 \text{ m/s}}$$

**(b)** Find the thrust at liftoff.

Compute the change in the rocket's mass:

$$\Delta M = M_f - M_i = 1.00 \times 10^4 \text{ kg} - 1.00 \times 10^5 \text{ kg}$$
$$= -9.00 \times 10^4 \text{ kg}$$

Calculate the rate at which rocket mass changes by dividing the change in mass by the time (where the time interval equals 4.00 min = $2.40 \times 10^2$ s):

$$\frac{\Delta M}{\Delta t} = \frac{-9.00 \times 10^4 \text{ kg}}{2.40 \times 10^2 \text{ s}} = -3.75 \times 10^2 \text{ kg/s}$$

Substitute this rate into Equation 6.21, obtaining the thrust:

$$\text{Thrust} = \left| v_e \frac{\Delta M}{\Delta t} \right| = (4.50 \times 10^3 \text{ m/s})(3.75 \times 10^2 \text{ kg/s})$$

$$= \boxed{1.69 \times 10^6 \text{ N}}$$

**(c)** Find the initial acceleration, including the gravity force.

Write Newton's second law, where $T$ stands for thrust, and solve for the acceleration $a$:

$$Ma = \sum F = T - Mg$$

$$a = \frac{T}{M} - g = \frac{1.69 \times 10^6 \text{ N}}{1.00 \times 10^5 \text{ kg}} - 9.80 \text{ m/s}^2$$

$$= \boxed{7.10 \text{ m/s}^2}$$

**(d)** Estimate the speed at burnout when gravity is not neglected.

Find the approximate loss of speed due to gravity:

$$\Delta v_g = -g\Delta t = -(9.80 \text{ m/s}^2)(2.40 \times 10^2 \text{ s})$$
$$= -2.35 \times 10^3 \text{ m/s}$$

Add this loss to the result of part **(a)**:

$$v_f = 1.04 \times 10^4 \text{ m/s} - 2.35 \times 10^3 \text{ m/s}$$
$$= \boxed{8.05 \times 10^3 \text{ m/s}}$$

**REMARKS** Even taking gravity into account, the speed is sufficient to attain orbit. Some additional boost may be required to overcome air drag.

**QUESTION 6.9** What initial normal force would be exerted on an astronaut of mass $m$ in a rocket traveling vertically upward with an acceleration $a$? Answer symbolically in terms of the positive quantities $m$, $g$, and $a$.

**EXERCISE 6.9** A spaceship with a mass of $5.00 \times 10^4$ kg is traveling at $6.00 \times 10^3$ m/s relative to a space station. What mass will the ship have after it fires its engines in order to reach a relative speed of $8.00 \times 10^3$ m/s, traveling in the same direction? Assume an exhaust velocity of $4.50 \times 10^3$ m/s.

**ANSWER** $3.21 \times 10^4$ kg

# SUMMARY

## 6.1 Momentum and Impulse

The **linear momentum** $\vec{p}$ of an object of mass $m$ moving with velocity $\vec{v}$ is defined as

$$\vec{p} \equiv m\vec{v} \qquad [6.1]$$

Momentum carries units of kg · m/s. The **impulse** $\vec{I}$ of a constant force $\vec{F}$ delivered to an object is equal to the product of the force and the time interval during which the force acts:

$$\vec{I} \equiv \vec{F}\Delta t \qquad [6.4]$$

These two concepts are unified in the **impulse–momentum theorem**, which states that the impulse of a constant force delivered to an object is equal to the change in momentum of the object:

$$\vec{I} = \vec{F}\Delta t = \Delta\vec{p} \equiv m\vec{v}_f - m\vec{v}_i \qquad [6.5]$$

Solving problems with this theorem often involves estimating speeds or contact times (or both), leading to an average force.

## 6.2 Conservation of Momentum

When no net external force acts on an isolated system, the total momentum of the system is constant. This principle is called **conservation of momentum.** In particular, if the isolated system consists of two objects undergoing a collision, the total momentum of the system is the same before and after the collision (Fig. 6.20). Conservation of momentum can be written mathematically for this case as

$$m_1\vec{v}_{1i} + m_2\vec{v}_{2i} = m_1\vec{v}_{1f} + m_2\vec{v}_{2f} \qquad [6.7]$$

**Figure 6.20** In an isolated system of two objects undergoing a collision, the total momentum of the system remains constant.

Collision and recoil problems typically require finding unknown velocities in one or two dimensions. Each vector component gives an equation, and the resulting equations are solved simultaneously.

## 6.3 Collisions in One Dimension

In an **inelastic collision,** the momentum of the system is conserved, but kinetic energy is not. In a **perfectly inelastic collision,** the colliding objects stick together. In an **elastic collision,** both the momentum and the kinetic energy of the system are conserved.

A one-dimensional **elastic collision** between two objects can be solved by using the conservation of momentum and conservation of energy equations:

$$m_1v_{1i} + m_2v_{2i} = m_1v_{1f} + m_2v_{2f} \qquad [6.10]$$

$$\tfrac{1}{2}m_1v_{1i}^2 + \tfrac{1}{2}m_2v_{2i}^2 = \tfrac{1}{2}m_1v_{1f}^2 + \tfrac{1}{2}m_2v_{2f}^2 \qquad [6.11]$$

The following equation, derived from Equations 6.10 and 6.11, is usually more convenient to use than the original conservation of energy equation:

$$v_{1i} - v_{2i} = -(v_{1f} - v_{2f}) \qquad [6.15]$$

These equations can be solved simultaneously for the unknown velocities. Energy is not conserved in **inelastic collisions,** so such problems must be solved with Equation 6.10 alone.

## 6.4 Glancing Collisions

In glancing collisions, conservation of momentum can be applied along two perpendicular directions: an $x$-axis and a $y$-axis. Problems can be solved by using the $x$- and $y$-components of Equation 6.7. Elastic two-dimensional collisions will usually require Equation 6.11 as well. (Equation 6.15 doesn't apply to two dimensions.) Generally, one of the two objects is taken to be traveling along the $x$-axis, undergoing a deflection at some angle $\theta$ after the collision. The final velocities and angles can be found with elementary trigonometry.

## 6.5 Rocket Propulsion

From conservation of momentum, the change in a rocket's velocity as it ejects fuel is given by

$$v_f - v_i = v_e \ln\left(\frac{M_i}{M_f}\right) \qquad [6.20]$$

where $M_i$ and $M_f$ are the rocket's initial and final masses, respectively, and $v_e$ is the fuel's exhaust speed relative to the rocket.

The thrust on a rocket depends on the fuel's exhaust speed and the rate at which fuel is ejected:

$$\text{Instantaneous thrust} = \left| v_e \frac{\Delta M}{\Delta t} \right| \qquad [6.21]$$

where $\Delta M/\Delta t$ is the rate at which the rocket's mass changes as fuel is burned.

## CONCEPTUAL QUESTIONS

1. A batter bunts a pitched baseball, blocking the ball without swinging. (a) Can the baseball deliver more kinetic energy to the bat and batter than the ball carries initially? (b) Can the baseball deliver more momentum to the bat and batter than the ball carries initially? Explain each of your answers.

2. If two objects collide and one is initially at rest, (a) is it possible for both to be at rest after the collision? (b) Is it possible for only one to be at rest after the collision? Explain.

3. Two carts on an air track have the same mass and speed and are traveling towards each other. If they collide and stick together, find (a) the total momentum and (b) total kinetic energy of the system. (c) Describe a different colliding system with this same final momentum and kinetic energy.

4. Two identical ice hockey pucks, labeled A and B, are sliding towards each other at speed $v$. Which one of the following statements is true concerning their momenta and kinetic energies?

   (a) $\vec{\mathbf{p}}_A = \vec{\mathbf{p}}_B$ and $KE_A = KE_B$    (b) $\vec{\mathbf{p}}_A = -\vec{\mathbf{p}}_B$ and $KE_A = -KE_B$

   (c) $\vec{\mathbf{p}}_A = -\vec{\mathbf{p}}_B$ and $KE_A = KE_B$   (d) $\vec{\mathbf{p}}_A = \vec{\mathbf{p}}_B$ and $KE_A = -KE_B$

5. A ball of clay of mass $m$ is thrown with a speed $v$ against a brick wall. The clay sticks to the wall and stops. Is the principle of conservation of momentum violated in this example?

6. A skater is standing still on a frictionless ice rink. Her friend throws a Frisbee straight to her. In which of the following cases is the largest momentum transferred to the skater? (a) The skater catches the Frisbee and holds onto it. (b) The skater catches the Frisbee momentarily, but then drops it vertically downward. (c) The skater catches the Frisbee, holds it momentarily, and throws it back to her friend.

7. A baseball is thrown from the outfield toward home plate. (a) True or False: Neglecting air resistance, the momentum of the baseball is conserved during its flight. (b) True or False: Neglecting air resistance, the momentum of the baseball–Earth system is conserved during the baseball's flight.

8. (a) If two automobiles collide, they usually do not stick together. Does this mean the collision is elastic? (b) Explain why a head-on collision is likely to be more dangerous than other types of collisions.

9. Your physical education teacher throws you a tennis ball at a certain velocity, and you catch it. You are now given the following choice: The teacher can throw you a medicine ball (which is much more massive than the tennis ball) with the same velocity, the same momentum, or the same kinetic energy as the tennis ball. Which option would you choose in order to make the easiest catch, and why?

10. Two carts move in the same direction along a frictionless air track, each acted on by the same constant force for a time interval $\Delta t$. Cart 2 has twice the mass of cart 1. Which one of the following statements is true? (a) Each cart has the same change in momentum. (b) Cart 1 has the greater change in momentum. (c) Cart 2 has the greater change in momentum. (d) The changes in momenta depend on the initial velocities.

11. For the situation described in the previous question, which cart experiences the greater change in kinetic energy? (a) Each cart has the same change in kinetic energy. (b) Cart 1 (c) Cart 2 (d) It's impossible to tell without knowing the initial velocities.

12. An air bag inflates when a collision occurs, protecting a passenger (the dummy in Figure CQ6.12) from serious injury. Why does the air bag soften the blow? Discuss the physics involved in this dramatic photograph.

**Figure CQ6.12**

13. At a bowling alley, two players each score a spare when their bowling balls make head-on, approximately elastic collisions at the same speed with identical pins. After the collisions, the pin hit by ball A moves much more quickly than the pin hit by ball B. Which ball has more mass?

14. An open box slides with constant speed across the frictionless surface of a frozen lake. If water from a rain shower falls vertically downward into it, does the box: (a) speed up, (b) slow down, or (c) continue to move with constant speed?

15. Does a larger net force exerted on an object always produce a larger change in the momentum of the object, compared to a smaller net force? Explain.

16. Does a larger net force always produce a larger change in kinetic energy than a smaller net force? Explain.

17. If two particles have equal momenta, are their kinetic energies equal? (a) yes, always (b) no, never (c) no, except when their masses are equal (d) no, except when their speeds are the same (e) yes, as long as they move along parallel lines.

18. Two particles of different mass start from rest. The same net force acts on both of them as they move over equal distances. How do their final kinetic energies compare? (a) The particle of larger mass has more kinetic energy. (b) The particle of smaller mass has more kinetic energy. (c) The particles have equal kinetic energies. (d) Either particle might have more kinetic energy.

## PROBLEMS

*See the Preface for an explanation of the icons used in this problems set.*

### 6.1 Momentum and Impulse

1. Calculate the magnitude of the linear momentum for the following cases: (a) a proton with mass equal to $1.67 \times 10^{-27}$ kg, moving with a speed of $5.00 \times 10^6$ m/s; (b) a 15.0-g bullet moving with a speed of 300 m/s; (c) a 75.0-kg sprinter running with a speed of 10.0 m/s; (d) the Earth (mass = $5.98 \times 10^{24}$ kg) moving with an orbital speed equal to $2.98 \times 10^4$ m/s.

**2.** A high-speed photograph of a club hitting a golf ball is shown in Figure 6.3. The club was in contact with a ball, initially at rest, for about 0.002 0 s. If the ball has a mass of 55 g and leaves the head of the club with a speed of $2.0 \times 10^2$ ft/s, find the average force exerted on the ball by the club.

**3.** A pitcher claims he can throw a 0.145-kg baseball with as much momentum as a 3.00-g bullet moving with a speed of $1.50 \times 10^3$ m/s. (a) What must the baseball's speed be if the pitcher's claim is valid? (b) Which has greater kinetic energy, the ball or the bullet?

**4.** A 0.280-kg volleyball approaches a player horizontally with a speed of 15.0 m/s. The player strikes the ball with her fist and causes the ball to move in the opposite direction with a speed of 22.0 m/s. (a) What impulse is delivered to the ball by the player? (b) If the player's fist is in contact with the ball for 0.060 0 s, find the magnitude of the average force exerted on the player's fist.

**5.** **QC** Drops of rain fall perpendicular to the roof of a parked car during a rainstorm. The drops strike the roof with a speed of 12 m/s, and the mass of rain per second striking the roof is 0.035 kg/s. (a) Assuming the drops come to rest after striking the roof, find the average force exerted by the rain on the roof. (b) If hailstones having the same mass as the raindrops fall on the roof at the same rate and with the same speed, how would the average force on the roof compare to that found in part (a)?

**6.** **S** Show that the kinetic energy of a particle of mass $m$ is related to the magnitude of the momentum $p$ of that particle by $KE = p^2/2m$. (*Note:* This expression is invalid for particles traveling at speeds near that of light.)

**7.** An object has a kinetic energy of 275 J and a momentum of magnitude 25.0 kg · m/s. Find the (a) speed and (b) mass of the object.

**8.** An estimated force vs. time curve for a baseball struck by a bat is shown in Figure P6.8. From this curve, determine (a) the impulse delivered to the ball and (b) the average force exerted on the ball.

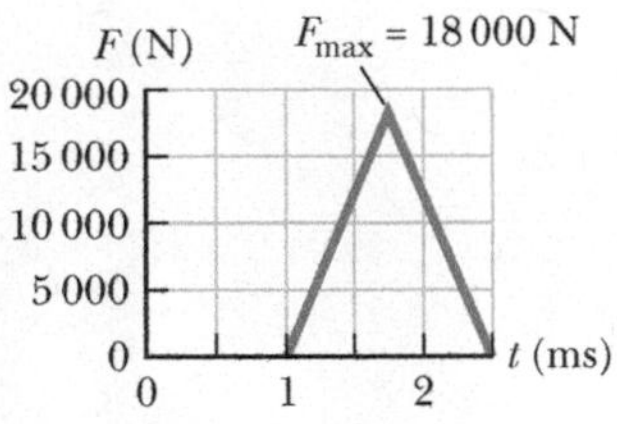

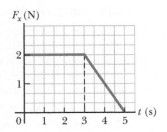

**Figure P6.8**

**9.** A soccer player takes a corner kick, lofting a stationary ball 35.0° above the horizon at 22.5 m/s. If the soccer ball has a mass of 0.425 kg and the player's foot is in contact with it for $5.00 \times 10^{-2}$ s, find (a) the x- and y-components of the soccer ball's change in momentum and (b) the magnitude of the average force exerted by the player's foot on the ball.

**10.** **QC** A man claims he can safely hold on to a 12.0-kg child in a head-on collision with a relative speed of 120-mi/h lasting for 0.10 s as long as he has his seat belt on. (a) Find the magnitude of the average force needed to hold onto the child. (b) Based on the result to part (a), is the man's claim valid? (c) What does the answer to this problem say about laws requiring the use of proper safety devices such as seat belts and special toddler seats?

**11.** A ball of mass 0.150 kg is dropped from rest from a height of 1.25 m. It rebounds from the floor to reach a height of 0.960 m. What impulse was given to the ball by the floor?

**12.** A tennis player receives a shot with the ball (0.060 0 kg) traveling horizontally at 50.0 m/s and returns the shot with the ball traveling horizontally at 40.0 m/s in the opposite direction. (a) What is the impulse delivered to the ball by the racket? (b) What work does the racket do on the ball?

**13.** A car is stopped for a traffic signal. When the light turns green, the car accelerates, increasing its speed from 0 to 5.20 m/s in 0.832 s. What are the magnitudes of (a) the linear impulse and (b) the average total force experienced by a 70.0-kg passenger in the car during the time the car accelerates?

**14.** **V** A 65.0-kg basketball player jumps vertically and leaves the floor with a velocity of 1.80 m/s upward. (a) What impulse does the player experience? (b) What force does the floor exert on the player before the jump? (c) What is the total average force exerted by the floor on the player if the player is in contact with the floor for 0.450 s during the jump?

**15.** The force shown in the force vs. time diagram in Figure P6.15 acts on a 1.5-kg object. Find (a) the impulse of the force, (b) the final velocity of the object if it is initially at rest, and (c) the final velocity of the object if it is initially moving along the x-axis with a velocity of −2.0 m/s.

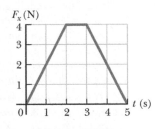

**Figure P6.15**

**16.** A force of magnitude $F_x$ acting in the x-direction on a 2.00-kg particle varies in time as shown in Figure P6.16. Find (a) the impulse of the force, (b) the final velocity of the particle if it is initially at rest, and (c) the final velocity of the particle if it is initially moving along the x-axis with a velocity of −2.00 m/s.

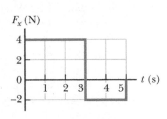

**Figure P6.16**

**17.** The forces shown in the force vs. time diagram in Figure P6.17 act on a 1.5-kg particle. Find (a) the impulse for the interval from $t = 0$ to $t = 3.0$ s and (b) the impulse for the interval from $t = 0$ to $t = 5.0$ s. If the forces act on a 1.5-kg particle that is initially at rest, find the particle's speed (c) at $t = 3.0$ s and (d) at $t = 5.0$ s.

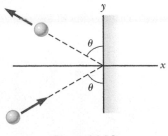

**Figure P6.17**

**18.** **V** A 3.00-kg steel ball strikes a massive wall at 10.0 m/s at an angle of $\theta = 60.0°$ with the plane of the wall. It bounces off the wall with the same speed and angle (Fig. P6.18). If the ball is in contact with the wall for 0.200 s, what is the average force exerted by the wall on the ball?

**Figure P6.18**

19. **T** The front 1.20 m of a 1 400-kg car is designed as a "crumple zone" that collapses to absorb the shock of a collision. If a car traveling 25.0 m/s stops uniformly in 1.20 m, (a) how long does the collision last, (b) what is the magnitude of the average force on the car, and (c) what is the acceleration of the car? Express the acceleration as a multiple of the acceleration of gravity.

20. **Q|C** A pitcher throws a 0.14-kg baseball toward the batter so that it crosses home plate horizontally and has a speed of 42 m/s just before it makes contact with the bat. The batter then hits the ball straight back at the pitcher with a speed of 48 m/s. Assume the ball travels along the same line leaving the bat as it followed before contacting the bat. (a) What is the magnitude of the impulse delivered by the bat to the baseball? (b) If the ball is in contact with the bat for 0.005 0 s, what is the magnitude of the average force exerted by the bat on the ball? (c) How does your answer to part (b) compare to the weight of the ball?

## 6.2 Conservation of Momentum

21. **V** High-speed stroboscopic photographs show that the head of a $2.00 \times 10^2$-g golf club is traveling at 55.0 m/s just before it strikes a 46.0-g golf ball at rest on a tee. After the collision, the club head travels (in the same direction) at 40.0 m/s. Find the speed of the golf ball just after impact.

22. A rifle with a weight of 30.0 N fires a 5.00-g bullet with a speed of $3.00 \times 10^2$ m/s. (a) Find the recoil speed of the rifle. (b) If a $7.00 \times 10^2$-N man holds the rifle firmly against his shoulder, find the recoil speed of the man and rifle.

23. A 45.0-kg girl is standing on a 150.-kg plank. The plank, originally at rest, is free to slide on a frozen lake, which is a flat, frictionless surface. The girl begins to walk along the plank at a constant velocity of 1.50 m/s to the right relative to the plank. (a) What is her velocity relative to the surface of the ice? (b) What is the velocity of the plank relative to the surface of the ice?

24. **S** *This is a symbolic version of Problem 23.* A girl of mass $m_G$ is standing on a plank of mass $m_P$. Both are originally at rest on a frozen lake that constitutes a frictionless, flat surface. The girl begins to walk along the plank at a constant velocity $v_{GP}$ to the right relative to the plank. (The subscript *GP* denotes the girl relative to plank.) (a) What is the velocity $v_{PI}$ of the plank relative to the surface of the ice? (b) What is the girl's velocity $v_{GI}$ relative to the ice surface?

25. **BIO** Squids are the fastest marine invertebrates, using a powerful set of muscles to take in and then eject water in a form of jet propulsion that can propel them to speeds of over 11.5 m/s. What speed would a stationary 1.50-kg squid achieve by ejecting 0.100 kg of water (not included in the squid's mass) at 3.25 m/s? Neglect other forces, including the drag force on the squid.

26. A 75-kg fisherman in a 125-kg boat throws a package of mass $m = 15$ kg horizontally toward the right with a speed of $v_i = 4.5$ m/s as in Figure P6.26. Neglecting water resistance, and

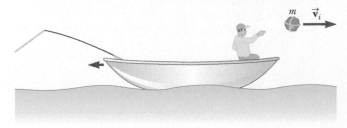

**Figure P6.26**

assuming the boat is at rest before the package is thrown, find the velocity of the boat after the package is thrown.

27. **V** A 65.0-kg person throws a 0.045 0-kg snowball forward with a ground speed of 30.0 m/s. A second person, with a mass of 60.0 kg, catches the snowball. Both people are on skates. The first person is initially moving forward with a speed of 2.50 m/s, and the second person is initially at rest. What are the velocities of the two people after the snowball is exchanged? Disregard friction between the skates and the ice.

28. Two objects of masses $m_1 = 0.56$ kg and $m_2 = 0.88$ kg are placed on a horizontal frictionless surface and a compressed spring of force constant $k = 280$ N/m is placed between them as in Figure P6.28a. Neglect the mass of the spring. The spring is not attached to either object and is compressed a distance of 9.8 cm. If the objects are released from rest, find the final velocity of each object as shown in Figure P6.28b.

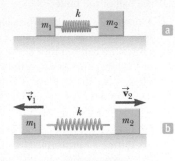

**Figure P6.28**

29. **Q|C** An astronaut in her space suit has a total mass of 87.0 kg, including suit and oxygen tank. Her tether line loses its attachment to her spacecraft while she's on a spacewalk. Initially at rest with respect to her spacecraft, she throws her 12.0-kg oxygen tank away from her spacecraft with a speed of 8.00 m/s to propel herself back toward it (Fig. P6.29). (a) Determine the maximum distance she can be from the craft and still return within 2.00 min (the amount of time the air in her helmet remains breathable). (b) Explain in terms of Newton's laws of motion why this strategy works.

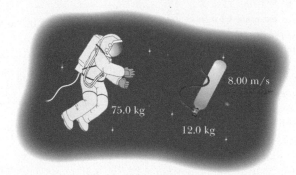

**Figure P6.29**

30. Three ice skaters meet at the center of a rink and each stands at rest facing the center, within arm's reach of the other two. On a signal, each skater pushes himself away from the other two across the frictionless ice. After the push, skater A with mass $m_A = 80.0$ kg moves in the negative *y*-direction at 3.50 m/s and skater B with mass $m_B = 75.0$ kg moves in the negative *x*-direction at 4.00 m/s. Find the *x*- and *y*-components of the 90.0-kg skater C's velocity after the push.

## 6.3 Collisions in One Dimension

## 6.4 Glancing Collisions

31. **GP** A man of mass $m_1 = 70.0$ kg is skating at $v_1 = 8.00$ m/s behind his wife of mass $m_2 = 50.0$ kg, who is skating at

$v_2 = 4.00$ m/s. Instead of passing her, he inadvertently collides with her. He grabs her around the waist, and they maintain their balance. (a) Sketch the problem with before-and-after diagrams, representing the skaters as blocks. (b) Is the collision best described as elastic, inelastic, or perfectly inelastic? Why? (c) Write the general equation for conservation of momentum in terms of $m_1$, $v_1$, $m_2$, $v_2$, and final velocity $v_f$. (d) Solve the momentum equation for $v_f$. (e) Substitute values, obtaining the numerical value for $v_f$, their speed after the collision.

32. An archer shoots an arrow toward a $3.00 \times 10^2$-g target that is sliding in her direction at a speed of 2.50 m/s on a smooth, slippery surface. The 22.5-g arrow is shot with a speed of 35.0 m/s and passes through the target, which is stopped by the impact. What is the speed of the arrow after passing through the target?

33. Gayle runs at a speed of 4.00 m/s and dives on a sled, initially at rest on the top of a frictionless, snow-covered hill. After she has descended a vertical distance of 5.00 m, her brother, who is initially at rest, hops on her back, and they continue down the hill together. What is their speed at the bottom of the hill if the total vertical drop is 15.0 m? Gayle's mass is 50.0 kg, the sled has a mass of 5.00 kg, and her brother has a mass of 30.0 kg.

34. **BIO** A 75.0-kg ice skater moving at 10.0 m/s crashes into a stationary skater of equal mass. After the collision, the two skaters move as a unit at 5.00 m/s. Suppose the average force a skater can experience without breaking a bone is 4 500 N. If the impact time is 0.100 s, does a bone break?

35. A railroad car of mass $2.00 \times 10^4$ kg moving at 3.00 m/s collides and couples with two coupled railroad cars, each of the same mass as the single car and moving in the same direction at 1.20 m/s. (a) What is the speed of the three coupled cars after the collision? (b) How much kinetic energy is lost in the collision?

36. **S** *This is a symbolic version of Problem 35.* A railroad car of mass $M$ moving at a speed $v_1$ collides and couples with two coupled railroad cars, each of the same mass $M$ and moving in the same direction at a speed $v_2$. (a) What is the speed $v_f$ of the three coupled cars after the collision in terms of $v_1$ and $v_2$? (b) How much kinetic energy is lost in the collision? Answer in terms of $M$, $v_1$, and $v_2$.

37. **S** Consider the ballistic pendulum device discussed in Example 6.5 and illustrated in Figure 6.13. (a) Determine the ratio of the momentum immediately after the collision to the momentum immediately before the collision. (b) Show that the ratio of the kinetic energy immediately after the collision to the kinetic energy immediately before the collision is $m_1/(m_1 + m_2)$.

38. A cue ball traveling at 4.00 m/s makes a glancing, elastic collision with a target ball of equal mass that is initially at rest. The target ball deflects the cue ball so that its subsequent motion makes an angle of 30.0° with respect to its original direction of travel. Find (a) the angle between the velocity vectors of the two balls after the collision and (b) the speed of each ball after the collision.

39. In a Broadway performance, an 80.0-kg actor swings from a 3.75-m-long cable that is horizontal when he starts. At the bottom of his arc, he picks up his 55.0-kg costar in an inelastic collision. What maximum height do they reach after their upward swing?

40. **V** Two shuffleboard disks of equal mass, one orange and the other green, are involved in a perfectly elastic glancing collision. The green disk is initially at rest and is struck by the orange disk moving initially to the right at 5.00 m/s as in Figure P6.40a. After the collision, the orange disk moves in a direction that makes an angle of 37.0° with the horizontal axis while the green disk makes an angle of 53.0° with this axis as in Figure P6.40b. Determine the speed of each disk after the collision.

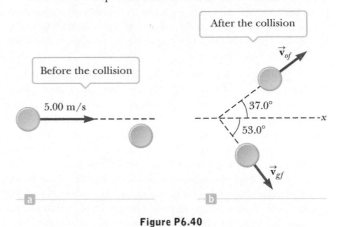

**Figure P6.40**

41. A 0.030-kg bullet is fired vertically at 200 m/s into a 0.15-kg baseball that is initially at rest. How high does the combined bullet and baseball rise after the collision, assuming the bullet embeds itself in the ball?

42. **T** An bullet of mass $m = 8.00$ g is fired into a block of mass $M = 250$ g that is initially at rest at the edge of a table of height $h = 1.00$ m (Fig. P6.42). The bullet remains in the block, and after the impact the block lands $d = 2.00$ m from the bottom of the table. Determine the initial speed of the bullet.

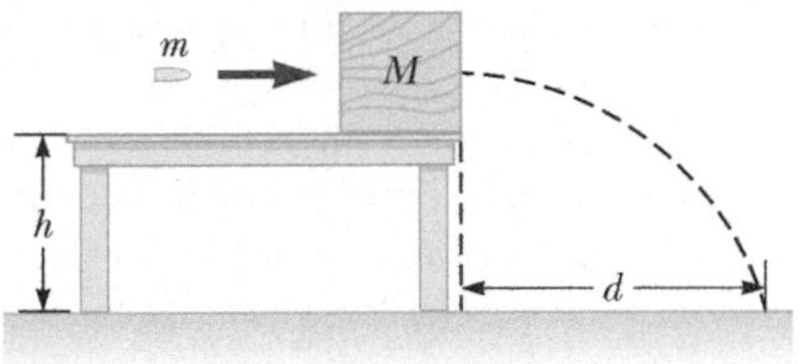

**Figure P6.42**

43. **V** A 12.0-g bullet is fired horizontally into a 100-g wooden block that is initially at rest on a frictionless horizontal surface and connected to a spring having spring constant 150 N/m. The bullet becomes embedded in the block. If the bullet–block system compresses the spring by a maximum of 80.0 cm, what was the speed of the bullet at impact with the block?

44. A 1200-kg car traveling initially with a speed of 25.0 m/s in an easterly direction crashes into the rear end of a 9 000-kg truck moving in the same direction at 20.0 m/s (Fig. P6.44). The velocity of the car right after the collision is 18.0 m/s to the east. (a) What is the velocity of the truck right after the collision? (b) How much mechanical energy is lost in the collision? Account for this loss in energy.

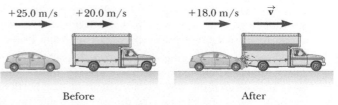

Before          After

**Figure P6.44**

45. **T** A tennis ball of mass 57.0 g is held just above a basketball of mass 590 g. With their centers vertically aligned, both balls are released from rest at the same time, falling through a distance of 1.20 m, as shown in Figure P6.45. (a) Find the magnitude of the basketball's velocity the instant before the basketball reaches the ground. (b) Assume that an elastic collision with the ground instantaneously reverses the velocity of the basketball so that it collides with the tennis ball just above it. To what height does the tennis ball rebound?

**Figure P6.45**

46. **GP** **QC** A space probe, initially at rest, undergoes an internal mechanical malfunction and breaks into three pieces. One piece of mass $m_1 = 48.0$ kg travels in the positive $x$-direction at 12.0 m/s, and a second piece of mass $m_2 = 62.0$ kg travels in the $xy$-plane at an angle of 105° at 15.0 m/s. The third piece has mass $m_3 = 112$ kg. (a) Sketch a diagram of the situation, labeling the different masses and their velocities. (b) Write the general expression for conservation of momentum in the $x$- and $y$-directions in terms of $m_1$, $m_2$, $m_3$, $v_1$, $v_2$, and $v_3$ and the sines and cosines of the angles, taking $\theta$ to be the unknown angle. (c) Calculate the final $x$-components of the momenta of $m_1$ and $m_2$. (d) Calculate the final $y$-components of the momenta of $m_1$ and $m_2$. (e) Substitute the known momentum components into the general equations of momentum for the $x$- and $y$-directions, along with the known mass $m_3$. (f) Solve the two momentum equations for $v_3 \cos \theta$ and $v_3 \sin \theta$, respectively, and use the identity $\cos^2 \theta + \sin^2 \theta = 1$ to obtain $v_3$. (g) Divide the equation for $v_3 \sin \theta$ by that for $v_3 \cos \theta$ to obtain $\tan \theta$, then obtain the angle by taking the inverse tangent of both sides. (h) In general, would three such pieces necessarily have to move in the same plane? Why?

47. **V** A 25.0-g object moving to the right at 20.0 cm/s overtakes and collides elastically with a 10.0-g object moving in the same direction at 15.0 cm/s. Find the velocity of each object after the collision.

48. A billiard ball rolling across a table at 1.50 m/s makes a head-on elastic collision with an identical ball. Find the speed of each ball after the collision (a) when the second ball is initially at rest, (b) when the second ball is moving toward the first at a speed of 1.00 m/s, and (c) when the second ball is moving away from the first at a speed of 1.00 m/s.

49. **QC** **T** A 90.0-kg fullback running east with a speed of 5.00 m/s is tackled by a 95.0-kg opponent running north with a speed of 3.00 m/s. (a) Why does the tackle constitute a perfectly inelastic collision? (b) Calculate the velocity of the players immediately after the tackle and (c) determine the mechanical energy that is lost as a result of the collision. (d) Where did the lost energy go?

50. Identical twins, each with mass 55.0 kg, are on ice skates and at rest on a frozen lake, which may be taken as frictionless. Twin A is carrying a backpack of mass 12.0 kg. She throws it horizontally at 3.00 m/s to Twin B. Neglecting any gravity effects, what are the subsequent speeds of Twin A and Twin B?

51. A $2.00 \times 10^3$-kg car moving east at 10.0 m/s collides with a $3.00 \times 10^3$-kg car moving north. The cars stick together and move as a unit after the collision, at an angle of 40.0° north of east and a speed of 5.22 m/s. Find the speed and direction of the $3.00 \times 10^3$-kg car before the collision.

52. Two automobiles of equal mass approach an intersection. One vehicle is traveling with velocity 13.0 m/s toward the east, and the other is traveling north with velocity $v_{2i}$. Neither driver sees the other. The vehicles collide in the intersection and stick together, leaving parallel skid marks at an angle of 55.0° north of east. The speed limit for both roads is 35 mi/h, and the driver of the northward-moving vehicle claims he was within the limit when the collision occurred. Is he telling the truth?

53. A billiard ball moving at 5.00 m/s strikes a stationary ball of the same mass. After the collision, the first ball moves at 4.33 m/s at an angle of 30.0° with respect to the original line of motion. (a) Find the velocity (magnitude and direction) of the second ball after collision. (b) Was the collision inelastic or elastic?

## 6.5 Rocket Propulsion

54. The Merlin rocket engines developed by SpaceX produce $8.01 \times 10^5$ N of instantaneous thrust with an exhaust speed of $3.05 \times 10^3$ m/s in vacuum. What mass of fuel does the engine burn each second?

55. One of the first ion engines on a commercial satellite used Xenon as a propellant and could eject the ionized gas at a rate of $3.03 \times 10^{-6}$ kg/s with an exhaust speed of $3.04 \times 10^4$ m/s. What instantaneous thrust could the engine provide?

56. NASA's Saturn V rockets that launched astronauts to the moon were powered by the strongest rocket engine ever developed, providing $6.77 \times 10^6$ N of thrust while burning fuel at a rate of $2.63 \times 10^3$ kg/s. Calculate the engine's exhaust speed.

57. **BIO** A typical person begins to lose consciousness if subjected to accelerations greater than about $5g$ (49.0 m/s²) for more than a few seconds. Suppose a $3.00 \times 10^4$-kg manned spaceship's engine has an exhaust speed of $2.50 \times 10^3$ m/s. What maximum burn rate $|\Delta M/\Delta t|$ could the engine reach before the ship's acceleration exceeded $5g$ and its human occupants began to lose consciousness?

58. A spaceship at rest relative to a nearby star in interplanetary space has a total mass of $2.50 \times 10^4$ kg. Its engines fire at $t = 0$, steadily burning fuel at 76.7 kg/s with an exhaust speed of $4.25 \times 10^3$ m/s. Calculate the spaceship's (a) acceleration at $t = 0$, (b) mass at $t = 125$ s, (c) acceleration at $t = 125$ s, and (d) speed at $t = 125$ s, relative to the same nearby star.

59. A spaceship's orbital maneuver requires a speed increase of $1.20 \times 10^3$ m/s. If its engine has an exhaust speed of $2.50 \times 10^3$ m/s, determine the required ratio $M_i/M_f$ of its initial mass to its final mass. (The difference $M_i - M_f$ equals the mass of the ejected fuel.)

## Additional Problems

60. **BIO** In research in cardiology and exercise physiology, it is often important to know the mass of blood pumped by a person's heart in one stroke. This information can be obtained by means of a *ballistocardiograph*. The instrument works as follows: The subject lies on a horizontal pallet floating on a film of air. Friction on the pallet is negligible. Initially, the momentum of the system is zero. When the heart beats, it expels a mass $m$ of blood into the aorta with speed $v$, and the body and platform move in the opposite direction with speed $V$. The speed of the

blood can be determined independently (e.g., by observing an ultrasound Doppler shift). Assume that the blood's speed is 50.0 cm/s in one typical trial. The mass of the subject plus the pallet is 54.0 kg. The pallet moves at a speed of $6.00 \times 10^{-5}$ m in 0.160 s after one heartbeat. Calculate the mass of blood that leaves the heart. Assume that the mass of blood is negligible compared with the total mass of the person. This simplified example illustrates the principle of ballistocardiography, but in practice a more sophisticated model of heart function is used.

61. **Q|C** Most of us know intuitively that in a head-on collision between a large dump truck and a subcompact car, you are better off being in the truck than in the car. Why is this? Many people imagine that the collision force exerted on the car is much greater than that exerted on the truck. To substantiate this view, they point out that the car is crushed, whereas the truck is only dented. This idea of unequal forces, of course, is false; Newton's third law tells us that both objects are acted upon by forces of the same magnitude. The truck suffers less damage because it is made of stronger metal. But what about the two drivers? Do they experience the same forces? To answer this question, suppose that each vehicle is initially moving at 8.00 m/s and that they undergo a perfectly inelastic head-on collision. Each driver has mass 80.0 kg. Including the masses of the drivers, the total masses of the vehicles are 800 kg for the car and $4.00 \times 10^3$ kg for the truck. If the collision time is 0.120 s, what force does the seat belt exert on each driver?

62. Consider a frictionless track as shown in Figure P6.62. A block of mass $m_1 = 5.00$ kg is released from Ⓐ. It makes a head-on elastic collision at Ⓑ with a block of mass $m_2 = 10.0$ kg that is initially at rest. Calculate the maximum height to which $m_1$ rises after the collision.

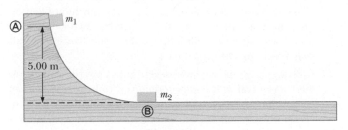

**Figure P6.62**

63. **T** A 2.0-g particle moving at 8.0 m/s makes a perfectly elastic head-on collision with a resting 1.0-g object. (a) Find the speed of each particle after the collision. (b) Find the speed of each particle after the collision if the stationary particle has a mass of 10 g. (c) Find the final kinetic energy of the incident 2.0-g particle in the situations described in parts (a) and (b). In which case does the incident particle lose more kinetic energy?

64. **S** A bullet of mass $m$ and speed $v$ passes completely through a pendulum bob of mass $M$ as shown in Figure P6.64. The bullet emerges with a speed of $v/2$. The pendulum bob is suspended by a stiff rod of length $\ell$ and negligible mass. What is the minimum value of $v$ such that the bob will barely swing through a complete vertical circle?

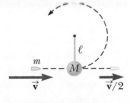

**Figure P6.64**

65. **Q|C** **S** An amateur skater of mass $M$ is trapped in the middle of an ice rink and is unable to return to the side where there is no ice. Every motion she makes causes her to slip on the ice and remain in the same spot. She decides to try to return to safety by removing her gloves of mass m and throwing them in the direction opposite the safe side. (a) She throws the gloves as hard as she can, and they leave her hand with a velocity $\vec{v}_{gloves}$. Explain whether or not she moves. If she does move, calculate her velocity $\vec{v}_{girl}$ relative to the Earth after she throws the gloves. (b) Discuss her motion from the point of view of the forces acting on her.

66. A 0.400-kg blue bead slides on a frictionless, curved wire, starting from rest at point Ⓐ in Figure P6.66, where $h = 1.50$ m. At point Ⓑ, the bead collides elastically with a 0.600-kg green bead at rest. Find the maximum height the green bead rises as it moves up the wire.

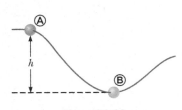

**Figure P6.66**

67. A 730-N man stands in the middle of a frozen pond of radius 5.0 m. He is unable to get to the other side because of a lack of friction between his shoes and the ice. To overcome this difficulty, he throws his 1.2-kg physics textbook horizontally toward the north shore at a speed of 5.0 m/s. How long does it take him to reach the south shore?

68. An unstable nucleus of mass $1.7 \times 10^{-26}$ kg, initially at rest at the origin of a coordinate system, disintegrates into three particles. One particle, having a mass of $m_1 = 5.0 \times 10^{-27}$ kg, moves in the positive y-direction with speed $v_1 = 6.0 \times 10^6$ m/s. Another particle, of mass $m_2 = 8.4 \times 10^{-27}$ kg, moves in the positive x-direction with speed $v_2 = 4.0 \times 10^6$ m/s. Find the magnitude and direction of the velocity of the third particle.

69. **S** Two blocks of masses $m_1$ and $m_2$ approach each other on a horizontal table with the same constant speed, $v_0$, as measured by a laboratory observer. The blocks undergo a perfectly elastic collision, and it is observed that $m_1$ stops but $m_2$ moves opposite its original motion with some constant speed, $v$. (a) Determine the ratio of the two masses, $m_1/m_2$. (b) What is the ratio of their speeds, $v/v_0$?

70. Two blocks of masses $m_1 = 2.00$ kg and $m_2 = 4.00$ kg are each released from rest at a height of $h = 5.00$ m on a frictionless track, as shown in Figure P6.70, and undergo an elastic head-on collision. (a) Determine the velocity of each block just before the collision. (b) Determine the velocity of each block immediately after the collision. (c) Determine the maximum heights to which $m_1$ and $m_2$ rise after the collision.

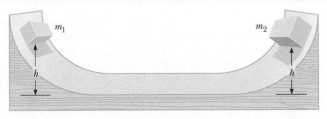

**Figure P6.70**

**71.** A block with mass $m_1 = 0.500$ kg is released from rest on a frictionless track at a distance $h_1 = 2.50$ m above the top of a table. It then collides elastically with an object having mass $m_2 = 1.00$ kg that is initially at rest on the table, as shown in Figure P6.71. (a) Determine the velocities of the two objects just after the collision. (b) How high up the track does the 0.500-kg object travel back after the collision? (c) How far away from the bottom of the table does the 1.00-kg object land, given that the height of the table is $h_2 = 2.00$ m? (d) How far away from the bottom of the table does the 0.500-kg object eventually land?

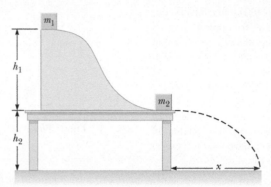

**Figure P6.71**

**72.** [S] Two objects of masses $m$ and $3m$ are moving toward each other along the $x$-axis with the same initial speed $v_0$. The object with mass $m$ is traveling to the left, and the object with mass $3m$ is traveling to the right. They undergo an elastic glancing collision such that $m$ is moving downward after the collision at right angles from its initial direction. (a) Find the final speeds of the two objects. (b) What is the angle $\theta$ at which the object with mass $3m$ is scattered?

**73.** A small block of mass $m_1 = 0.500$ kg is released from rest at the top of a curved wedge of mass $m_2 = 3.00$ kg, which sits on a frictionless horizontal surface as in Figure P6.73a. When the block leaves the wedge, its velocity is measured to be 4.00 m/s to the right, as in Figure P6.73b. (a) What is the velocity of the wedge after the block reaches the horizontal surface? (b) What is the height $h$ of the wedge?

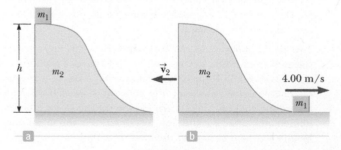

**Figure P6.73**

**74.** [S] A car of mass $m$ moving at a speed $v_1$ collides and couples with the back of a truck of mass $2m$ moving initially in the same direction as the car at a lower speed $v_2$. (a) What is the speed $v_f$ of the two vehicles immediately after the collision? (b) What is the change in kinetic energy of the car–truck system in the collision?

**75.** [Q|C] A cannon is rigidly attached to a carriage, which can move along horizontal rails, but is connected to a post by a large spring,

initially unstretched and with force constant $k = 2.00 \times 10^4$ N/m, as in Figure P6.75. The cannon fires a $2.00 \times 10^2$-kg projectile at a velocity of 125 m/s directed 45.0° above the horizontal. (a) If the mass of the cannon and its carriage is $5.00 \times 10^3$-kg, find the

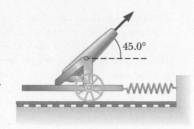

**Figure P6.75**

recoil speed of the cannon. (b) Determine the maximum extension of the spring. (c) Find the maximum force the spring exerts on the carriage. (d) Consider the system consisting of the cannon, the carriage, and the shell. Is the momentum of this system conserved during the firing? Why or why not?

**76.** [Q|C] [S] Two blocks collide on a frictionless surface. After the collision, the blocks stick together. Block A has a mass $M$ and is initially moving to the right at speed $v$. Block B has a mass $2M$ and is initially at rest. System C is composed of both blocks. (a) Draw a force diagram for each block at an instant *during* the collision. (b) Rank the magnitudes of the horizontal forces in your diagram. Explain your reasoning. (c) Calculate the change in momentum of block A, block B, and system C. (d) Is kinetic energy conserved in this collision? Explain your answer. (This problem is courtesy of Edward F. Redish. For more such problems, visit http://www.physics.umd.edu/perg.)

**77.** [Q|C] (a) A car traveling due east strikes a car traveling due north at an intersection, and the two move together as a unit. A property owner on the southeast corner of the intersection claims that his fence was torn down in the collision. Should he be awarded damages by the insurance company? Defend your answer. (b) Let the eastward-moving car have a mass of $1.30 \times 10^3$-kg and a speed of 30.0 km/h and the northward-moving car a mass of $1.10 \times 10^3$-kg and a speed of 20.0 km/h. Find the velocity after the collision. Are the results consistent with your answer to part (a)?

**78.** A 60-kg soccer player jumps vertically upwards and heads the 0.45-kg ball as it is descending vertically with a speed of 25 m/s. (a) If the player was moving upward with a speed of 4.0 m/s just before impact, what will be the speed of the ball immediately after the collision if the ball rebounds vertically upwards and the collision is elastic? (b) If the ball is in contact with the player's head for 20 ms, what is the average acceleration of the ball? (Note that the force of gravity may be ignored during the brief collision time.)

**79.** [Q|C] [S] A boy of mass $m_b$ and his girlfriend of mass $m_g$, both wearing ice skates, face each other at rest while standing on a frictionless ice rink. The boy pushes the girl, giving her a velocity $v_g$ toward the east. Assume that $m_b > m_g$. (a) Describe the subsequent motion of the boy. (b) Find expressions for the final kinetic energy of the girl and the final kinetic energy of the boy, and show that the girl has greater kinetic energy than the boy. (c) The boy and girl had zero kinetic energy before the boy pushed the girl, but ended up with kinetic energy after the event. How do you account for the appearance of mechanical energy?

**80.** A 20.0-kg toboggan with 70.0-kg driver is sliding down a frictionless chute directed 30.0° below the horizontal at 8.00 m/s when a 55.0-kg woman drops from a tree limb straight down behind the driver. If she drops through a vertical displacement of 2.00 m, what is the subsequent velocity of the toboggan immediately after impact?

81. **S** *Measuring the speed of a bullet.* A bullet of mass $m$ is fired horizontally into a wooden block of mass $M$ lying on a table. The bullet remains in the block after the collision. The coefficient of friction between the block and table is $\mu$, and the block slides a distance $d$ before stopping. Find the initial speed $v_0$ of the bullet in terms of $M$, $m$, $\mu$, $g$, and $d$.

82. A flying squid (family Ommastrephidae) is able to "jump" off the surface of the sea by taking water into its body cavity and then ejecting the water vertically downward. A 0.850-kg squid is able to eject 0.300 kg of water with a speed of 20.0 m/s. (a) What will be the speed of the squid immediately after ejecting the water? (b) How high in the air will the squid rise?

83. A 0.30-kg puck, initially at rest on a frictionless horizontal surface, is struck by a 0.20-kg puck that is initially moving along the x-axis with a velocity of 2.0 m/s. After the collision, the 0.20-kg puck has a speed of 1.0 m/s at an angle of $\theta = 53°$ to the positive x-axis. (a) Determine the velocity of the 0.30-kg puck after the collision. (b) Find the fraction of kinetic energy lost in the collision.

84. **Q|C** **S** A wooden block of mass $M$ rests on a table over a large hole as in Figure P6.84. A bullet of mass $m$ with an initial velocity $v_i$ is fired upward into the bottom of the block and remains in the block after the collision. The block and bullet rise to a maximum height of $h$. (a) Describe how you would find the initial velocity of the bullet using ideas you have learned in this topic. (b) Find an expression for the initial velocity of the bullet.

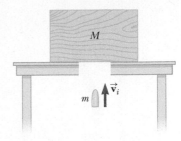

**Figure P6.84** Problems 84 and 85.

85. **V** **Q|C** A 1.25-kg wooden block rests on a table over a large hole as in Figure P6.84. A 5.00-g bullet with an initial velocity $v_i$ is fired upward into the bottom of the block and remains in the block after the collision. The block and bullet rise to a maximum height of 22.0 cm. (a) Describe how you would find the initial velocity of the bullet using ideas you have learned in this topic. (b) Calculate the initial velocity of the bullet from the information provided.

# Rotational Motion and Gravitation

**ROTATIONAL MOTION IS AN IMPORTANT PART** of everyday life. The rotation of the Earth creates the cycle of day and night, the rotation of wheels enables easy vehicular motion, and modern technology depends on circular motion in a variety of contexts, from the tiny gears in a Swiss watch to the operation of lathes and other machinery. The concepts of *angular velocity, angular acceleration,* and *centripetal acceleration* are central to understanding the motions of a diverse range of phenomena, from a car moving around a circular race-track to clusters of galaxies orbiting a common center.

Rotational motion, when combined with Newton's law of universal gravitation and his laws of motion, can also explain certain facts about space travel and satellite motion, such as where to place a satellite so it will remain fixed in position over the same spot on the Earth. The generalization of gravitational potential energy and energy conservation offers an easy route to such results as planetary escape speed. Finally, we present Kepler's three laws of planetary motion, which formed the foundation of Newton's approach to gravity.

## 7.1 Angular Velocity and Angular Acceleration

In the study of linear motion, the important concepts are *displacement* $\Delta x$, *velocity* $v$, and *acceleration* $a$. Each of these concepts has its analog in rotational motion: *angular displacement* $\Delta \theta$, *angular velocity* $\omega$, and *angular acceleration* $\alpha$.

The *radian,* a unit of angular measure, is essential to the understanding of these concepts. Recall that the distance $s$ around a circle is given by $s = 2\pi r$, where $r$ is the radius of the circle. Dividing both sides by $r$ results in $s/r = 2\pi$. This quantity is dimensionless because both $s$ and $r$ have dimensions of length, but the value $2\pi$ corresponds to an angular displacement around a circle. A half circle would give an answer of $\pi$, a quarter circle an answer of $\pi/2$. The numbers $2\pi$, $\pi$, and $\pi/2$ correspond to angles of 360°, 180°, and 90°, respectively, so a new unit of angular measure, the **radian,** can be introduced, with 180° = $\pi$ rad relating degrees to radians.

In general, the angle $\theta$ subtended by an arclength $s$ along a circle of radius $r$, measured in radians counterclockwise from the positive $x$-axis, is

$$\theta = \frac{s}{r} \qquad [7.1]$$

The angle $\theta$ in Equation 7.1 is actually an angular displacement from the positive $x$-axis, and $s$ the corresponding displacement along the circular arc, again measured from the positive $x$-axis. Figure 7.1 illustrates the size of 1 radian, which is approximately 57°. Converting from degrees to radians requires multiplying by the ratio ($\pi$ rad/180°). For example, 45° ($\pi$ rad/180°) = ($\pi/4$) rad.

Generally, angular quantities in physics must be expressed in radians. Be sure to set your calculator to radian mode; neglecting to do so is a common error.

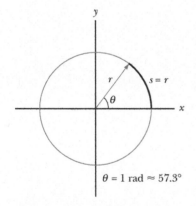

**Figure 7.1** For a circle of radius $r$, one radian is the angle subtended by an arclength equal to $r$.

Armed with the concept of radian measure, we can now discuss angular concepts in physics. Consider Figure 7.2a, a top view of a rotating compact disc. Such a disk is an example of a "rigid body," with each part of the body fixed in position relative to all other parts of the body. When a rigid body rotates through a given angle, all parts of the body rotate through the same angle at the same time. For the compact disc, the axis of rotation is at the center of the disc, *O*. A point *P* on the disc is at a distance *r* from the origin and moves about *O* in a circle of radius *r*. We set up a *fixed* reference line, as shown in Figure 7.2a, and assume that at time $t = 0$ the point *P* is on that reference line. After a time interval $\Delta t$ has elapsed, *P* has advanced to a new position (Fig. 7.2b). In this interval, the line *OP* has moved through the angle $\theta$ with respect to the reference line. The angle $\theta$, measured in radians, is called the **angular position** and is analogous to the linear position variable *x*. Likewise, *P* has moved an arclength *s* measured along the circumference of the circle.

In Figure 7.3, as a point on the rotating disc moves from Ⓐ to Ⓑ in a time $\Delta t$, it starts at an angle $\theta_i$ and ends at an angle $\theta_f$. The difference $\theta_f - \theta_i$ is called the **angular displacement.**

An object's **angular displacement**, $\Delta\theta$, is the difference in its final and initial angles:

$$\Delta\theta = \theta_f - \theta_i \qquad [7.2]$$

**SI unit: radian (rad)**

For example, if a point on a disc is at $\theta_i = 4$ rad and rotates to angular position $\theta_f = 7$ rad, the angular displacement is $\Delta\theta = \theta_f - \theta_i = 7$ rad $-$ 4 rad $= 3$ rad. Note that we use angular variables to describe the rotating disc because **each point on the disc undergoes the same angular displacement in any given time interval.**

Using the definition in Equation 7.2, Equation 7.1 can be written more generally as $\Delta\theta = \Delta s/r$, where $\Delta s$ is a displacement along the circular arc subtended by the angular displacement. Having defined angular displacements, it's natural to define an angular velocity:

The **average angular velocity** $\omega_{av}$ of a rotating rigid object during the time interval $\Delta t$ is the angular displacement $\Delta\theta$ divided by $\Delta t$:

$$\omega_{av} \equiv \frac{\theta_f - \theta_i}{t_f - t_i} = \frac{\Delta\theta}{\Delta t} \qquad [7.3]$$

**SI unit: radian per second (rad/s)**

For very short time intervals, the average angular velocity approaches the instantaneous angular velocity, just as in the linear case.

The **instantaneous angular velocity** $\omega$ of a rotating rigid object is the limit of the average velocity $\Delta\theta/\Delta t$ as the time interval $\Delta t$ approaches zero:

$$\omega \equiv \lim_{\Delta t \to 0} \frac{\Delta\theta}{\Delta t} \qquad [7.4]$$

**SI unit: radian per second (rad/s)**

We take $\omega$ to be positive when $\theta$ is increasing (counterclockwise motion) and negative when $\theta$ is decreasing (clockwise motion). When the angular velocity is constant, the instantaneous angular velocity is equal to the average angular velocity.

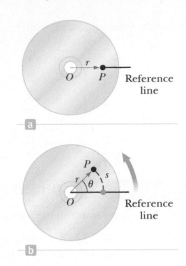

**Figure 7.2** (a) The point *P* on a rotating compact disc at $t = 0$. (b) As the disc rotates, *P* moves through an arclength *s*.

◀ Angular displacement

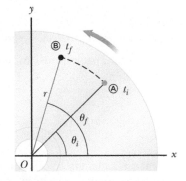

**Figure 7.3** As a point on the compact disc moves from Ⓐ to Ⓑ, the disc rotates through the angle $\Delta\theta = \theta_f - \theta_i$.

◀ Average angular velocity

**Tip 7.1** Remember the Radian

Equation 7.1 uses angles measured in *radians*. Angles expressed in terms of degrees must first be converted to radians. Also, be sure to check whether your calculator is in degree or radian mode when solving problems involving rotation.

◀ Instantaneous angular velocity

Instantaneous angular speed ▶   The **instantaneous angular speed** of an object is defined as the magnitude of the instantaneous angular velocity. Instantaneous angular speed (or simply, "angular speed") has no direction associated with it and carries no algebraic sign. For example, if one object is rotating counterclockwise with an angular velocity of $+5.0$ rad/s and another object is rotating clockwise with an angular velocity of $-5.0$ rad/s, both have an angular speed of 5.0 rad/s.

## EXAMPLE 7.1   WHIRLYBIRDS

**GOAL** Perform some elementary calculations with angular variables.

**PROBLEM** The rotor on a helicopter turns at an angular velocity of $3.20 \times 10^2$ revolutions per minute. (In this book, we sometimes use the abbreviation rpm, but in most cases we use rev/min.) **(a)** Express this angular velocity in radians per second. **(b)** If the rotor has a radius of 2.00 m, what arclength does the tip of the blade trace out in $3.00 \times 10^2$ s? **(c)** The pilot opens the throttle, and the angular velocity of the blade increases while rotating twenty-six times in 3.60 s. Calculate the average angular velocity during that time.

**STRATEGY** During one revolution, the rotor turns through an angle of $2\pi$ radians. Use this relationship as a conversion factor. For part **(b)**, first calculate the angular displacement in radians by multiplying the angular velocity by time. Part **(c)** is a simple application of Equation 7.3.

. . . . . . . . . . . . . . . . . . . . . . . . . . . . . . . . . . . . . . . . . . . . . . . . . . . . . . . . . . . . . . . . . . . . . . . . . . . . . . . . . . . . . . . . . . . .

**SOLUTION**

**(a)** Express this angular velocity in radians per second.

Apply the conversion factors 1 rev = $2\pi$ rad and 60.0 s = 1 min:

$$\omega = 3.20 \times 10^2 \, \frac{\text{rev}}{\text{min}}$$

$$= 3.20 \times 10^2 \, \frac{\text{rev}}{\text{min}} \left( \frac{2\pi \text{ rad}}{1 \text{ rev}} \right) \left( \frac{1.00 \text{ min}}{60.0 \text{ s}} \right)$$

$$= \boxed{33.5 \text{ rad/s}}$$

**(b)** Find the arclength traced out by the tip of the blade.

Multiply the angular velocity by the time to obtain the angular displacement:

$$\Delta\theta = \omega t = (33.5 \text{ rad/s})(3.00 \times 10^2 \text{ s}) = 1.01 \times 10^4 \text{ rad}$$

Multiply the angular displacement by the radius to get the arclength:

$$\Delta s = r\Delta\theta = (2.00 \text{ m})(1.01 \times 10^4 \text{ rad}) = \boxed{2.02 \times 10^4 \text{ m}}$$

**(c)** Calculate the average angular velocity of the blade while its angular velocity increases.

Apply Equation 7.3, noticing that

$$\Delta\theta = (26 \text{ rev})(2\pi \text{ rad/rev}) = 52\pi \text{ rad}:$$

$$\omega_{\text{av}} = \frac{\Delta\theta}{\Delta t} = \frac{52\pi \text{ rad}}{3.60 \text{ s}} = \boxed{45 \text{ rad/s}}$$

. . . . . . . . . . . . . . . . . . . . . . . . . . . . . . . . . . . . . . . . . . . . . . . . . . . . . . . . . . . . . . . . . . . . . . . . . . . . . . . . . . . . . . . . . . . .

**REMARKS** It's best to express angular velocity in radians per second. Consistent use of radian measure minimizes errors.

**QUESTION 7.1** Is it possible to express angular velocity in degrees per second? If so, what's the conversion factor from radians per second?

**EXERCISE 7.1** A Ferris wheel turns at a constant 185.0 revolutions per hour. **(a)** Express this rate of rotation in units of radians per second. **(b)** If the wheel has a radius of 12.0 m, what arclength does a passenger trace out during a ride lasting 5.00 min? **(c)** If the wheel then slows to rest in 9.72 s while making a quarter turn, calculate the magnitude of its average angular velocity during that time.

**ANSWERS** **(a)** 0.323 rad/s   **(b)** $1.16 \times 10^3$ m   **(c)** 0.162 rad/s

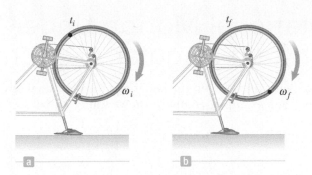

**Figure 7.4** An accelerating bicycle wheel rotates with (a) angular velocity $\omega_i$ at time $t_i$ and (b) angular velocity $\omega_f$ at time $t_f$.

## Quick Quiz

**7.1** A rigid body is rotating counterclockwise about a fixed axis. Each of the following pairs of quantities represents an initial angular position and a final angular position of the rigid body. Which of the sets can occur *only* if the rigid body rotates through more than 180°? (a) 3 rad, 6 rad; (b) −1 rad, 1 rad; (c) 1 rad, 5 rad.

**7.2** Suppose the change in angular position for each of the pairs of values in Quick Quiz 7.1 occurred in 1 s. Which choice represents the lowest average angular velocity?

Figure 7.4 shows a bicycle turned upside down so that a repair technician can work on the rear wheel. The bicycle pedals are turned so that at time $t_i$ the wheel has angular velocity $\omega_i$ (Fig. 7.4a), and at a later time $t_f$ it has angular velocity $\omega_f$ (Fig. 7.4b). Just as a changing velocity leads to the concept of an acceleration, a changing angular velocity leads to the concept of an angular acceleration.

An object's **average angular acceleration** $\alpha_{av}$ during the time interval $\Delta t$ is the change in its angular velocity $\Delta\omega$ divided by $\Delta t$:

$$\alpha_{av} \equiv \frac{\omega_f - \omega_i}{t_f - t_i} = \frac{\Delta\omega}{\Delta t} \qquad [7.5]$$

◀ Average angular acceleration

**SI unit: radian per second squared (rad/s$^2$)**

As with angular velocity, positive angular accelerations are in the counterclockwise direction, negative angular accelerations in the clockwise direction. If the angular velocity goes from 15 rad/s to 9.0 rad/s in 3.0 s, the average angular acceleration during that time interval is

$$\alpha_{av} = \frac{\Delta\omega}{\Delta t} = \frac{9.0 \text{ rad/s} - 15 \text{ rad/s}}{3.0 \text{ s}} = -2.0 \text{ rad/s}^2$$

The negative sign indicates that the angular acceleration is clockwise (although the angular velocity, still positive but slowing down, is in the counterclockwise direction). There is also an instantaneous version of angular acceleration:

The **instantaneous angular acceleration** $\alpha$ is the limit of the average angular acceleration $\Delta\omega/\Delta t$ as the time interval $\Delta t$ approaches zero:

$$\alpha = \lim_{\Delta t \to 0} \frac{\Delta\omega}{\Delta t} \qquad [7.6]$$

◀ Instantaneous angular acceleration

**SI unit: radian per second squared (rad/s$^2$)**

**When a rigid object rotates about a fixed axis, as does the bicycle wheel, every portion of the object has the same angular velocity and the same angular acceleration.** This fact is what makes these variables so useful for describing rotational motion. In contrast, the tangential velocity and acceleration of the object take different values that depend on the distance from a given point to the axis of rotation.

# 7.2 Rotational Motion Under Constant Angular Acceleration

A number of parallels exist between the equations for rotational motion and those for linear motion. For example, compare the defining equation for the average angular velocity,

$$\omega_{av} \equiv \frac{\theta_f - \theta_i}{t_f - t_i} = \frac{\Delta\theta}{\Delta t}$$

with that of the average linear velocity,

$$v_{av} \equiv \frac{x_f - x_i}{t_f - t_i} = \frac{\Delta x}{\Delta t}$$

In these equations, $\omega$ takes the place of $v$ and $\theta$ takes the place of $x$, so the equations differ only in the names of the variables. In the same way, every linear quantity we have encountered so far has a corresponding "twin" in rotational motion.

The procedure used in Section 2.3 to develop the kinematic equations for linear motion under constant acceleration can be used to derive a similar set of equations for rotational motion under constant angular acceleration. The resulting equations of rotational kinematics, along with the corresponding equations for linear motion, are as follows:

| Linear Motion with $a$ Constant (Variables: $x$ and $v$) | Rotational Motion About a Fixed Axis with $\alpha$ Constant (Variables: $\theta$ and $\omega$) | |
|:---:|:---:|:---:|
| $v = v_i + at$ | $\omega = \omega_i + \alpha t$ | [7.7] |
| $\Delta x = v_i t + \frac{1}{2}at^2$ | $\Delta\theta = \omega_i t + \frac{1}{2}\alpha t^2$ | [7.8] |
| $v^2 = v_i^2 + 2a\Delta x$ | $\omega^2 = \omega_i^2 + 2\alpha\Delta\theta$ | [7.9] |

Notice that every term in a given linear equation has a corresponding term in the analogous rotational equation.

## Quick Quiz

**7.3** Consider again the pairs of angular positions for the rigid object in Quick Quiz 7.1. If the object starts from rest at the initial angular position, moves counterclockwise with constant angular acceleration, and arrives at the final angular position with the same angular velocity in all three cases, for which choice is the angular acceleration the highest?

## EXAMPLE 7.2   A ROTATING WHEEL

**GOAL** Apply the rotational kinematic equations.

**PROBLEM** A wheel rotates with a constant angular acceleration of 3.50 rad/s². If the angular velocity of the wheel is 2.00 rad/s at $t = 0$, (a) through what angle does the wheel rotate between $t = 0$ and $t = 2.00$ s? Give your answer in radians and in revolutions. (b) What is the angular velocity of the wheel at $t = 2.00$ s? (c) What angular displacement (in revolutions) results while the angular velocity found in part (b) doubles?

**STRATEGY** The angular acceleration is constant, so this problem just requires substituting given values into Equations 7.7–7.9. Note that both the angular acceleration and initial angular velocity are positive, meaning the rotation is strictly in the counterclockwise (positive angular) direction.

. . . . . . . . . . . . . . . . . . . . . . . . . . . . . . . . . . . . . . . . . . . . . . . . . . . . . . . . . . . . . . . . . . . . . . . . . . . . . . . . . . . . . . . . . . . . . . . . . . . . . . . . . . . . . . . . . . . . . . . . . . . . . . .

**SOLUTION**

(a) Find the angular displacement after 2.00 s, in both radians and revolutions.

Use Equation 7.8, setting $\omega_i = 2.00$ rad/s, $\alpha = 3.5$ rad/s², and $t = 2.00$ s:

$$\Delta\theta = \omega_i t + \tfrac{1}{2}\alpha t^2$$
$$= (2.00 \text{ rad/s})(2.00 \text{ s}) + \tfrac{1}{2}(3.50 \text{ rad/s}^2)(2.00 \text{ s})^2$$
$$= \boxed{11.0 \text{ rad}}$$

Convert radians to revolutions.

$\Delta\theta = (11.0 \text{ rad})(1.00 \text{ rev}/2\pi \text{ rad}) = \boxed{1.75 \text{ rev}}$

**(b)** What is the angular velocity of the wheel at $t = 2.00$ s?

Substitute the same values into Equation 7.7:

$\omega = \omega_i + \alpha t = 2.00 \text{ rad/s} + (3.50 \text{ rad/s}^2)(2.00 \text{ s})$

$\qquad = \boxed{9.00 \text{ rad/s}}$

**(c)** What angular displacement (in revolutions) results during the time in which the angular velocity found in part **(b)** doubles?

Apply the time-independent rotational kinematics equation:

$\omega_f^2 - \omega_i^2 = 2\alpha\Delta\theta$

Substitute values, noting that $\omega_f = 2\omega_i$:

$(2 \times 9.00 \text{ rad/s})^2 - (9.00 \text{ rad/s})^2 = 2(3.50 \text{ rad/s}^2)\Delta\theta$

Solve for the angular displacement and convert to revolutions:

$\Delta\theta = (34.7 \text{ rad})(1 \text{ rev}/2\pi \text{ rad}) = \boxed{5.52 \text{ rev}}$

. . . . . . . . . . . . . . . . . . . . . . . . . . . . . . . . . . . . . . . . . . . . . . . . . . . . . . . . . . . . . . . . . . . . . . . . . . . . .

**REMARKS** The result of part **(b)** could also be obtained from Equation 7.9 and the results of part **(a)**.

**QUESTION 7.2** Suppose the radius of the wheel is doubled. Are the answers affected? If so, in what way?

**EXERCISE 7.2** **(a)** Find the angle through which the wheel rotates between $t = 2.00$ s and $t = 3.00$ s. **(b)** Find the angular velocity when $t = 3.00$ s. **(c)** What is the magnitude of the angular velocity two revolutions following $t = 3.00$ s?

**ANSWERS** **(a)** 10.8 rad    **(b)** 12.5 rad/s    **(c)** 15.6 rad/s

# 7.3 Tangential Velocity, Tangential Acceleration, and Centripetal Acceleration

## 7.3.1 Tangential Velocity and Acceleration

Angular variables are closely related to linear variables. Consider the arbitrarily shaped object in Figure 7.5 rotating about the $z$-axis through the point $O$. Assume the object rotates through the angle $\Delta\theta$, and hence point $P$ moves through the arclength $\Delta s$, in the interval $\Delta t$. We know from the defining equation for radian measure that

$$\Delta\theta = \frac{\Delta s}{r}$$

Dividing both sides of this equation by $\Delta t$, the time interval during which the rotation occurs, yields

$$\frac{\Delta\theta}{\Delta t} = \frac{1}{r}\frac{\Delta s}{\Delta t}$$

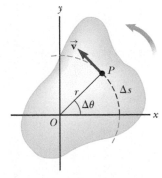

**Figure 7.5** Rotation of an object about an axis through $O$ (the $z$-axis) that is perpendicular to the plane of the figure. Note that a point $P$ on the object rotates in a circle of radius $r$ centered at $O$.

When $\Delta t$ is very small, the angle $\Delta\theta$ through which the object rotates is also small, and the ratio $\Delta\theta/\Delta t$ is close to the instantaneous angular velocity $\omega$. On the other side of the equation, similarly, the ratio $\Delta s/\Delta t$ approaches the instantaneous linear speed $v$ for small values of $\Delta t$. Hence, when $\Delta t$ gets arbitrarily small, the preceding equation is equivalent to

$$\omega = \frac{v}{r}$$

In Figure 7.5, the point $P$ traverses a distance $\Delta s$ along a circular arc during the time interval $\Delta t$ at a linear speed of $v$. The direction of $P$'s velocity vector $\vec{v}$ is *tangent to the circular path.* The angular component of $\vec{v}$ is $v = v_t$, called the **tangential velocity** of a particle moving in a circular path, written

Tangential velocity ▶

$$v_t = r\omega \qquad [7.10]$$

**The tangential velocity of a point on a rotating object equals the distance of that point from the axis of rotation multiplied by the angular velocity.** Equation 7.10 shows that the tangential velocity of a point on a rotating object increases as that point is moved outward from the center of rotation toward the rim, as expected; however, **every point on the rotating object has the same *angular* velocity.** Tangential speed is the magnitude of the tangential velocity, and is also called the *linear speed.*

Equation 7.10, derived using the defining equation for radian measure, is valid only when $\omega$ is measured in radians per unit time. Other measures of angular velocity, such as degrees per second and revolutions per second, shouldn't be used.

To find a second equation relating linear and angular quantities, refer again to Figure 7.5 and suppose the rotating object changes its angular speed by $\Delta\omega$ in the time interval $\Delta t$. At the end of this interval, the speed of a point on the object, such as $P$, has changed by the amount $\Delta v_t$. From Equation 7.10 we have

$$\Delta v_t = r\,\Delta\omega$$

Dividing by $\Delta t$ gives

$$\frac{\Delta v_t}{\Delta t} = r\frac{\Delta\omega}{\Delta t}$$

As the time interval $\Delta t$ is taken to be arbitrarily small, $\Delta\omega/\Delta t$ approaches the instantaneous angular acceleration. On the left-hand side of the equation, note that the ratio $\Delta v_t/\Delta t$ tends to the tangential acceleration of that point, given by

Tangential acceleration ▶

$$a_t = r\alpha \qquad [7.11]$$

**The tangential acceleration of a point on a rotating object equals the distance of that point from the axis of rotation multiplied by the angular acceleration.** Again, radian measure must be used for the angular acceleration term in this equation.

One last equation that relates linear quantities to angular quantities is derived in the next section.

### Quick Quiz

**7.4** Andrea and Chuck are riding on a merry-go-round. Andrea rides on a horse at the outer rim of the circular platform, twice as far from the center of the circular platform as Chuck, who rides on an inner horse. When the merry-go-round is rotating at a constant angular speed, Andrea's angular speed is (a) twice Chuck's (b) the same as Chuck's (c) half of Chuck's (d) impossible to determine.

**7.5** When the merry-go-round of Quick Quiz 7.4 is rotating at a constant angular speed, Andrea's tangential speed is (a) twice Chuck's (b) the same as Chuck's (c) half of Chuck's (d) impossible to determine.

### APPLYING PHYSICS 7.1    EUROPEAN SPACE AGENCY LAUNCH SITE

Why is the launch area for the European Space Agency in South America and not in Europe?

**EXPLANATION** Satellites are boosted into orbit on top of rockets, which provide the large tangential velocity necessary to achieve orbit. Due to its rotation, the surface of Earth is already traveling toward the east at a tangential velocity of nearly 1 700 m/s at the equator. This tangential velocity is steadily reduced farther north because the distance to the axis of rotation is decreasing. It finally goes to zero at the North Pole. Launching eastward from the equator gives the satellite a starting initial tangential velocity of 1 700 m/s, whereas a European launch provides roughly half that value (depending on the exact latitude). ■

## EXAMPLE 7.3 COMPACT DISCS

**GOAL** Apply the rotational kinematics equations in tandem with tangential acceleration and speed.

**PROBLEM** A compact disc (CD) rotates from rest up to an angular velocity of $-31.4$ rad/s in a time of 0.892 s. **(a)** What is the angular acceleration of the disc, assuming the angular acceleration is uniform? **(b)** Through what angle does the disc turn while coming up to speed? **(c)** If the radius of the disc is 4.45 cm, find the tangential velocity of a microbe riding on the rim of the disc when $t = 0.892$ s. **(d)** What is the magnitude of the tangential acceleration of the microbe at the given time?

**STRATEGY** We can solve parts (a) and (b) by applying the kinematic equations for angular velocity and angular displacement (Eqs. 7.7 and 7.8). Multiplying the radius by the angular acceleration yields the tangential acceleration at the rim, whereas multiplying the radius by the angular velocity gives the tangential velocity at that point.

. . . . . . . . . . . . . . . . . . . . . . . . . . . . . . . . . . . . . . . . . . . . . . . . . . . . . . . . . . . . . . . . . . . . . . . . . . . . . . . . . . . . . . . . . . .

### SOLUTION

**(a)** Find the angular acceleration of the disc.

Apply the angular velocity equation $\omega = \omega_i + \alpha t$, taking $\omega_i = 0$ at $t = 0$:

$$\alpha = \frac{\omega}{t} = \frac{-31.4 \text{ rad/s}}{0.892 \text{ s}} = \boxed{-35.2 \text{ rad/s}^2}$$

**(b)** Through what angle does the disc turn?

Use Equation 7.8 for angular displacement, with $t = 0.892$ s and $\omega_i = 0$:

$$\Delta\theta = \omega_i t + \tfrac{1}{2}\alpha t^2 = \tfrac{1}{2}(-35.2 \text{ rad/s}^2)(0.892 \text{ s})^2 = \boxed{-14.0 \text{ rad}}$$

**(c)** Find the final tangential velocity of a microbe at $r = 4.45$ cm.

Substitute into Equation 7.10:

$$v_t = r\omega = (0.044\,5 \text{ m})(-31.4 \text{ rad/s}) = \boxed{-1.40 \text{ m/s}}$$

**(d)** Find the tangential acceleration of the microbe at $r = 4.45$ cm.

Substitute into Equation 7.11:

$$a_t = r\alpha = (0.044\,5 \text{ m})(-35.2 \text{ rad/s}^2) = \boxed{-1.57 \text{ m/s}^2}$$

. . . . . . . . . . . . . . . . . . . . . . . . . . . . . . . . . . . . . . . . . . . . . . . . . . . . . . . . . . . . . . . . . . . . . . . . . . . . . . . . . . . . . . . . . . .

**REMARKS** Because $2\pi$ rad = 1 rev, the angular displacement in part **(b)** corresponds to 2.23 revolutions in the clockwise direction. In general, dividing the number of radians by 6 gives a rough approximation to the number of revolutions, because $2\pi \sim 6$.

**QUESTION 7.3** If the angular acceleration were doubled for the same duration, by what factor would the angular displacement change? Why is the answer true in this case but not in general?

**EXERCISE 7.3** **(a)** What are the angular velocity and angular displacement of the disc 0.300 s after it begins to rotate? **(b)** Find the tangential velocity at the rim at this time.

**ANSWERS** **(a)** $-10.6$ rad/s; $-1.58$ rad   **(b)** $-0.472$ m/s

Before MP3s and streaming became the mediums of choice for recorded music, compact discs (CDs), and phonographs were popular. There are similarities and differences between the rotational motion of phonograph records and that of CDs. A phonograph record rotates at a constant angular speed. Popular angular speeds were $33\tfrac{1}{3}$ rev/min for long-playing albums (hence the nickname "LP"), 45 rev/min for "singles," and 78 rev/min used in very early recordings. At the outer edge of the record, the pickup needle (stylus) moves over the vinyl material at a faster tangential speed than when the needle is close to the center of the record. As a result, the sound information is compressed into a smaller length of track near the center of the record than near the outer edge.

CDs, on the other hand, are designed so that the disc moves under the laser pickup at a constant tangential speed. Because the pickup moves radially as it follows the tracks of information, the angular speed of the CD must vary according to the radial position of the laser. Because the tangential speed is fixed, the information density (per length of track) anywhere on the disc is the same. Example 7.4 demonstrates numerical calculations for both CDs and phonograph records.

**APPLICATION**
Phonograph Records and Compact Discs

## EXAMPLE 7.4   TRACK LENGTH OF A COMPACT DISC

**GOAL** Relate angular to linear variables.

**PROBLEM** In a compact disc player, as the read head moves out from the center of the disc, the angular speed of the disc changes so that the linear speed at the position of the head remains at a constant value of about 1.3 m/s. (a) Find the angular speed of a CD of radius 6.00 cm when the read head is at $r = 2.0$ cm and again at $r = 5.6$ cm. (b) An old-fashioned record player rotates at a constant angular speed, so the linear speed of the record groove moving under the detector (the stylus) changes. Find the linear speed of a 45.0-rpm record at points 2.0 cm and 5.6 cm from the center. (c) In both the CDs and phonograph records, information is recorded in a continuous spiral track. Calculate the total length of the track for a CD designed to play for 1.0 h.

**STRATEGY** This problem is just a matter of substituting numbers into the appropriate equations. Part (a) requires relating angular and linear speed with Equation 7.10, $v_t = r\omega$, solving for $\omega$, and substituting given values. In part (b), convert from rev/min to rad/s and substitute straight into Equation 7.10 to obtain the linear speeds. In part (c), linear speed multiplied by time gives the total distance.

- - - - - - - - - - - - - - - - - - - - - - - - - - - - - - - - - - - - - - - - - - - - - - - - - - - - - - - - - - - -

**SOLUTION**

(a) Find the angular speed of the disc when the read head is at $r = 2.0$ cm and $r = 5.6$ cm.

Solve $v_t = r\omega$ for $\omega$ and calculate the angular speed at $r = 2.0$ cm:

$$\omega = \frac{v_t}{r} = \frac{1.3 \text{ m/s}}{2.0 \times 10^{-2} \text{ m}} = \boxed{65 \text{ rad/s}}$$

Likewise, find the angular speed at $r = 5.6$ cm:

$$\omega = \frac{v_t}{r} = \frac{1.3 \text{ m/s}}{5.6 \times 10^{-2} \text{ m}} = \boxed{23 \text{ rad/s}}$$

(b) Find the linear speed in m/s of a 45.0-rpm record at points 2.0 cm and 5.6 cm from the center.

Convert rev/min to rad/s:

$$45.0 \frac{\text{rev}}{\text{min}} = 45.0 \frac{\text{rev}}{\text{min}} \left( \frac{2\pi \text{ rad}}{\text{rev}} \right) \left( \frac{1.00 \text{ min}}{60.0 \text{ s}} \right) = 4.71 \frac{\text{rad}}{\text{s}}$$

Calculate the linear speed at $r = 2.0$ cm:

$$v_t = r\omega = (2.0 \times 10^{-2} \text{ m})(4.71 \text{ rad/s}) = \boxed{0.094 \text{ m/s}}$$

Calculate the linear speed at $r = 5.6$ cm:

$$v_t = r\omega = (5.6 \times 10^{-2} \text{ m})(4.71 \text{ rad/s}) = \boxed{0.26 \text{ m/s}}$$

(c) Calculate the total length of the track for a CD designed to play for 1.0 h.

Multiply the linear speed of the read head by the time in seconds:

$$d = v_t t = (1.3 \text{ m/s})(3\,600 \text{ s}) = \boxed{4\,700 \text{ m}}$$

- - - - - - - - - - - - - - - - - - - - - - - - - - - - - - - - - - - - - - - - - - - - - - - - - - - - - - - - - - - -

**REMARKS** Notice that for the record player in part (b), even though the angular speed is constant at all points along a radial line, the tangential speed steadily increases with increasing $r$. The calculation for a CD in part (c) is easy only because the linear (tangential) speed is constant. It would be considerably more difficult for a record player, where the tangential speed depends on the distance from the center.

**QUESTION 7.4** What is the angular acceleration of a record player while it's playing a song? Can a CD player have the same angular acceleration as a record player? Explain.

**EXERCISE 7.4** Compute the linear speed on a record playing at $33\frac{1}{3}$ revolutions per minute (a) at $r = 2.00$ cm and (b) at $r = 5.60$ cm.

**ANSWERS** (a) 0.069 8 m/s   (b) 0.195 m/s

---

## 7.3.2 Centripetal Acceleration

Figure 7.6a shows a car moving in a circular path with *constant linear speed v*. **Even though the car moves at a constant speed, it still has an acceleration.** To understand this, consider the defining equation for average acceleration:

$$\vec{a}_{av} = \frac{\vec{v}_f - \vec{v}_i}{t_f - t_i} \qquad [7.12]$$

The numerator represents the difference between the velocity vectors $\vec{\mathbf{v}}_f$ and $\vec{\mathbf{v}}_i$. These vectors may have the same *magnitude,* corresponding to the same speed, but if they have different *directions,* their difference can't equal zero. The direction of the car's velocity as it moves in the circular path is continually changing, as shown in Figure 7.6b. For circular motion at constant speed, the acceleration vector always points toward the center of the circle. Such an acceleration is called a **centripetal** (center-seeking) **acceleration.** Its magnitude is given by

$$a_c = \frac{v^2}{r} \qquad [7.13]$$

To derive Equation 7.13, consider Figure 7.7a. An object is first at point Ⓐ with velocity $\vec{\mathbf{v}}_i$ at time $t_i$ and then at point Ⓑ with velocity $\vec{\mathbf{v}}_f$ at a later time $t_f$. We assume $\vec{\mathbf{v}}_i$ and $\vec{\mathbf{v}}_f$ differ only in direction; their magnitudes are the same ($v_i = v_f = v$). To calculate the acceleration, we begin with Equation 7.12,

$$\vec{\mathbf{a}}_{\text{av}} = \frac{\vec{\mathbf{v}}_f - \vec{\mathbf{v}}_i}{t_f - t_i} = \frac{\Delta \vec{\mathbf{v}}}{\Delta t} \qquad [7.14]$$

where $\Delta \vec{\mathbf{v}} = \vec{\mathbf{v}}_f - \vec{\mathbf{v}}_i$ is the change in velocity. When $\Delta t$ is very small, $\Delta s$ and $\Delta \theta$ are also very small. In Figure 7.7b $\vec{\mathbf{v}}_f$ is almost parallel to $\vec{\mathbf{v}}_i$, and the vector $\Delta \vec{\mathbf{v}}$ is approximately perpendicular to them, pointing toward the center of the circle. In the limiting case when $\Delta t$ becomes vanishingly small, $\Delta \vec{\mathbf{v}}$ points exactly toward the center of the circle, and the average acceleration $\vec{\mathbf{a}}_{\text{av}}$ becomes the instantaneous acceleration $\vec{\mathbf{a}}$. From Equation 7.14, $\vec{\mathbf{a}}$ and $\Delta \vec{\mathbf{v}}$ point in the same direction (in this limit), so the instantaneous acceleration points to the center of the circle.

The triangle in Figure 7.7a, which has sides $\Delta s$ and $r$, is similar to the one formed by the vectors in Figure 7.7b, so the ratios of their sides are equal:

$$\frac{\Delta v}{v} = \frac{\Delta s}{r}$$

or

$$\Delta v = \frac{v}{r} \Delta s \qquad [7.15]$$

Substituting the result of Equation 7.15 into $a_{\text{av}} = \Delta v / \Delta t$ gives

$$a_{\text{av}} = \frac{v}{r} \frac{\Delta s}{\Delta t} \qquad [7.16]$$

But $\Delta s$ is the distance traveled along the arc of the circle in time $\Delta t$, and in the limiting case when $\Delta t$ becomes very small, $\Delta s / \Delta t$ approaches the instantaneous value of the tangential speed, $v$. At the same time, the average acceleration $a_{\text{av}}$ approaches $a_c$, the instantaneous centripetal acceleration, so Equation 7.16 reduces to

$$a_c = \frac{v^2}{r}$$

Because the tangential velocity is related to the angular velocity through the relation $v_t = r\omega$ (Eq. 7.10), an alternate form of Equation 7.13 is

$$a_c = \frac{r^2 \omega^2}{r} = r\omega^2 \qquad [7.17]$$

Dimensionally, $[r] = \text{L}$ and $[\omega] = 1/\text{T}$, so the units of centripetal acceleration are $\text{L}/\text{T}^2$, as they should be. This is a geometric result relating the centripetal acceleration to the angular speed, but physically an acceleration can occur only if some force is present. For example, if a car travels in a circle on flat ground, the force of static friction between the tires and the ground provides the necessary centripetal force.

Note that $a_c$ in Equations 7.13 and 7.17 represents only the *magnitude* of the centripetal acceleration. The acceleration itself is always directed toward the center of rotation.

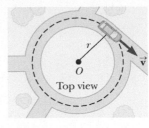
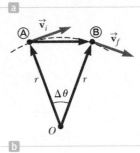

ⓐ

ⓑ

**Figure 7.6** (a) Circular motion of a car moving with constant speed. (b) As the car moves along the circular path from Ⓐ to Ⓑ, the direction of its velocity vector changes, so the car undergoes a centripetal acceleration.

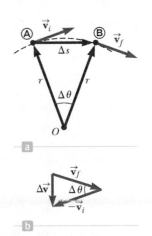

ⓐ

ⓑ

**Figure 7.7** (a) As the particle moves from Ⓐ to Ⓑ, the direction of its velocity vector changes from $\vec{\mathbf{v}}_i$ to $\vec{\mathbf{v}}_f$. (b) The construction for determining the direction of the change in velocity $\Delta \vec{\mathbf{v}}$, which is toward the center of the circle.

The foregoing derivations concern circular motion at constant speed. When an object moves in a circle but is speeding up or slowing down, a tangential component of acceleration, $a_t = r\alpha$, is also present. Because the tangential and centripetal components of acceleration are perpendicular to each other, we can find the magnitude of the **total acceleration** with the Pythagorean theorem:

Total acceleration ▶

$$a = \sqrt{a_t^2 + a_c^2}$$                                    [7.18]

### Quick Quiz

**7.6** A racetrack is constructed such that two arcs of radius 80 m at Ⓐ and 40 m at Ⓑ are joined by two stretches of straight track as in Figure 7.8. In a particular trial run, a driver travels at a constant speed of 50 m/s for one complete lap.
  1. The ratio of the tangential acceleration at Ⓐ to that at Ⓑ is
     (a) $\frac{1}{2}$ (b) $\frac{1}{4}$ (c) 2 (d) 4 (e) The tangential acceleration is zero at both points.
  2. The ratio of the centripetal acceleration at Ⓐ to that at Ⓑ is
     (a) $\frac{1}{2}$ (b) $\frac{1}{4}$ (c) 2 (d) 4 (e) The centripetal acceleration is zero at both points.
  3. The angular speed is greatest at
     (a) Ⓐ (b) Ⓑ (c) It is equal at both Ⓐ and Ⓑ.

**7.7** An object moves in a circular path with constant speed $v$. Which of the following statements is true concerning the object? (a) Its velocity is constant, but its acceleration is changing. (b) Its acceleration is constant, but its velocity is changing. (c) Both its velocity and acceleration are changing. (d) Its velocity and acceleration remain constant.

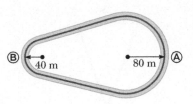

**Figure 7.8** (Quick Quiz 7.6)

---

## EXAMPLE 7.5  AT THE RACETRACK

**GOAL** Apply the concepts of centripetal acceleration and tangential velocity.

**PROBLEM** A race car accelerates uniformly from a speed of 40.0 m/s to a speed of 60.0 m/s in 5.00 s while traveling counterclockwise around a circular track of radius $4.00 \times 10^2$ m. When the car reaches a speed of 50.0 m/s, calculate (a) the magnitude of the car's centripetal acceleration, (b) the angular velocity, (c) the magnitude of the tangential acceleration, and (d) the magnitude of the total acceleration.

**STRATEGY** Substitute values into the definitions of centripetal acceleration (Eq. 7.13), tangential velocity (Eq. 7.10), and total acceleration (Eq. 7.18). Dividing the change in tangential velocity by the time yields the tangential acceleration.

· · · · · · · · · · · · · · · · · · · · · · · · · · · · · · · · · · · · · · · · · · · · · · · · · · · · · · · · · · · · · · · · · · · · · · · · · ·

**SOLUTION**

(a) Calculate the magnitude of the centripetal acceleration when $v = 50.0$ m/s.

Substitute into Equation 7.13:

$$a_c = \frac{v^2}{r} = \frac{(50.0 \text{ m/s})^2}{4.00 \times 10^2 \text{ m}} = \boxed{6.25 \text{ m/s}^2}$$

(b) Calculate the angular velocity.

Solve Equation 7.10 for $\omega$ and substitute:

$$\omega = \frac{v}{r} = \frac{50.0 \text{ m/s}}{4.00 \times 10^2 \text{ m}} = \boxed{0.125 \text{ rad/s}}$$

(c) Calculate the magnitude of the tangential acceleration.

Divide the change in tangential velocity by the time:

$$a_t = \frac{v_f - v_i}{\Delta t} = \frac{60.0 \text{ m/s} - 40.0 \text{ m/s}}{5.00 \text{ s}} = \boxed{4.00 \text{ m/s}^2}$$

(d) Calculate the magnitude of the total acceleration.

Substitute into Equation 7.18:

$$a = \sqrt{a_t^2 + a_c^2} = \sqrt{(4.00 \text{ m/s}^2)^2 + (6.25 \text{ m/s}^2)^2}$$

$$a = \boxed{7.42 \text{ m/s}^2}$$

· · · · · · · · · · · · · · · · · · · · · · · · · · · · · · · · · · · · · · · · · · · · · · · · · · · · · · · · · · · · · · · · · · · · · · · · · ·

**REMARKS** We can also find the centripetal acceleration by substituting the derived value of $\omega$ into Equation 7.17.

**QUESTION 7.5** If the force causing the centripetal acceleration suddenly vanished, would the car (a) slide away along a radius, (b) proceed along a line tangent to the circular motion, or (c) proceed at an angle intermediate between the tangent and radius?

**EXERCISE 7.5** Suppose the race car now slows down uniformly from 60.0 m/s to 30.0 m/s in 4.50 s to avoid an accident, while still traversing a circular path $4.00 \times 10^2$ m in radius. Calculate the car's **(a)** centripetal acceleration, **(b)** angular velocity, **(c)** tangential acceleration, and **(d)** total acceleration when the speed is 40.0 m/s.

**ANSWERS** **(a)** $4.00 \text{ m/s}^2$   **(b)** $0.100 \text{ rad/s}$   **(c)** $-6.67 \text{ m/s}^2$   **(d)** $7.78 \text{ m/s}^2$

# 7.4 Newton's Second Law for Uniform Circular Motion

Newton's second law of motion can be applied to problems involving circular motion. In the case of uniform circular motion, that law will feature the centripetal acceleration and a variety of radial forces that are either directed towards or away from the center of the circular motion. Just as acceleration and force are vector quantities, so are centripetal acceleration and the radial forces that appear in the statement of Newton's second law for uniform circular motion.

After discussing the concepts and sign conventions, Newton's second law for uniform circular motion will be applied in some elementary physical contexts.

## 7.4.1 Forces Causing Centripetal Acceleration

An object can have a centripetal acceleration *only* if some external force acts on it. For a ball whirling in a circle at the end of a string, that force is the tension in the string. In the case of a car moving on a flat circular track, the force is friction between the car and track. A satellite in circular orbit around Earth has a centripetal acceleration due to the gravitational force between the satellite and Earth.

Some books use the term "centripetal force," which can give the mistaken impression that it is a new force of nature. This is not the case: The adjective "centripetal" in "centripetal force" simply means that the force in question acts toward a center. The force of tension in the string of a yo-yo whirling in a vertical circle is an example of a centripetal force, as is the force of gravity on a satellite circling the Earth.

Consider a puck of mass $m$ that is tied to a string of length $r$ and is being whirled at constant speed in a horizontal circular path, as illustrated in Figure 7.9. Its weight is supported by a frictionless table. Why does the puck move in a circle? Because of its inertia, the tendency of the puck is to move in a straight line; however, the string prevents motion along a straight line by exerting a radial force on the puck—a tension force—that makes it follow the circular path. The tension $\vec{\mathbf{T}}$ is directed along the string toward *the center of the circle*, as shown in the figure.

In general, converting Newton's second law to polar coordinates yields an equation relating the net centripetal force, $F_c$, which is the sum of the radial components of all forces acting on a given object, to the centripetal acceleration. The *magnitude* of the **net** centripetal force equals the mass times the magnitude of the centripetal acceleration:

$$F_c = ma_c = m\frac{v^2}{r} \qquad [7.19]$$

A net force causing a centripetal acceleration acts toward the center of the circular path and effects a change in the direction of the velocity vector. If that force should vanish, the object would immediately leave its circular path and move along a straight line tangent to the circle at the point where the force vanished.

Centrifugal ('center-fleeing') forces also exist, such as the force between two particles with the same sign charge (see Topic 15). The normal force that prevents an object from falling toward the center of the Earth is another example of

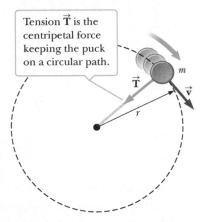

Tension $\vec{\mathbf{T}}$ is the centripetal force keeping the puck on a circular path.

**Figure 7.9** A puck attached to a string of length $r$ rotates in a horizontal plane at constant speed.

**Tip 7.2** Centripetal Force Is a *Type* of Force, Not a Force in Itself!

"Centripetal force" is a classification that includes forces acting toward a central point, like the horizontal component of the string tension on a tetherball or gravity on a satellite. A centripetal force must be *supplied* by some actual, physical force.

a centrifugal force. Sometimes an insufficient centripetal force is mistaken for the presence of a centrifugal force (see section 7.4.2 Fictitious Forces, page 206).

A radial force is a vector and has a direction. The second law for uniform circular motion involves forces that are directed either towards the center of a circle or away from it. **A force acting towards the center of the circle is by convention negative**. Examples include the gravity force on a satellite or the string tension of a whirling yo-yo. **A force acting away from the center of the circle is positive**. Examples include the normal force on a car traveling over the circular crest of a hill or the force of repulsion between like electric charges. Similarly, **the centripetal acceleration is negative because it acts towards the center of the circle**.

Newton's second law for uniform circular motion, written as a vector, therefore reads

$$-m\frac{v^2}{r} = \sum F_r \qquad [7.20]$$

where the forces $F_r$ are the radial forces acting on the mass $m$, positive if the force is away from the center of the circle, negative if the force is towards the center of the circle. Centripetal, or center-seeking, forces have negative radial components, whereas centrifugal, or center-fleeing, forces have positive radial components.

Newton's second law for uniform circular motion is illustrated in Applying Physics 7.2 and Examples 7.6–7.8.

---

**APPLYING PHYSICS 7.2    ARTIFICIAL GRAVITY**

Astronauts spending lengthy periods of time in space experience a number of negative effects due to weightlessness, such as weakening of muscle tissue and loss of calcium in bones. These effects may make it very difficult for them to return to their usual environment on Earth. How could artificial gravity be generated in space to overcome such complications?

**SOLUTION** A rotating cylindrical space station creates an environment of artificial gravity. The normal force of the rigid walls provides the centripetal force, which keeps the astronauts moving in a circle (Fig. 7.10). To an astronaut, the normal force can't be easily distinguished from a gravitational force as long as the radius of the station is large compared with the astronaut's height. (Otherwise, there are unpleasant inner ear effects.) This same principle is used in certain amusement park rides in which passengers are pressed against the inside of a rotating cylinder as it tilts in various directions. The visionary physicist Gerard O'Neill proposed creating a giant space colony a kilometer in radius that rotates slowly, creating Earth-normal artificial gravity

for the inhabitants in its interior. These inside-out artificial worlds could enable safe transport on a several-thousand-year journey to another star system. ∎

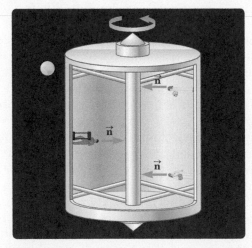

**Figure 7.10** Artificial gravity inside a spinning cylinder is provided by the normal force.

---

**PROBLEM-SOLVING STRATEGY**

### Forces That Cause Centripetal Acceleration
*Use the following steps in dealing with centripetal accelerations and the forces that produce them:*

1. **Draw a free-body diagram** of the object under consideration, labeling all forces that act on it.
2. **Choose a coordinate system** that has one axis perpendicular to the circular path followed by the object (the radial direction) and one axis tangent to the circular path (the tangential, or angular, direction). The normal direction, perpendicular to the plane of motion, is also often needed.

3. **Find the net force $F_c$ toward the center** of the circular path, $F_c = \Sigma F_r$, where $\Sigma F_r$ is the sum of the radial components of the forces. This net radial force causes the centripetal acceleration.
4. **Use Newton's second law for the radial, tangential, and normal directions,** as required, writing $\Sigma F_r = ma_c$, $\Sigma F_t = ma_t$, and $\Sigma F_n = ma_n$. Remember that the magnitude of the centripetal acceleration for uniform circular motion can always be written $a_c = v_t^2/r$.
5. **Solve** for the unknown quantities.

## EXAMPLE 7.6   BUCKLE UP FOR SAFETY

**GOAL** Calculate the frictional force that causes an object to have a centripetal acceleration.

**PROBLEM** A car travels at a constant speed of 30.0 mi/h (13.4 m/s) on a level circular turn of radius 50.0 m, as shown in the bird's-eye view in Figure 7.11a. What minimum coefficient of static friction, $\mu_s$, between the tires and roadway will allow the car to make the circular turn without sliding?

**STRATEGY** In the car's free-body diagram (Fig. 7.11b) the normal direction is vertical and the tangential direction is into the page (Step 2). Use Newton's second law. The net force acting on the car in the radial direction is the force of static friction toward the center of the circular path, which causes the car to have a centripetal acceleration. Calculating the maximum static friction force requires the normal force, obtained from the normal component of the second law.

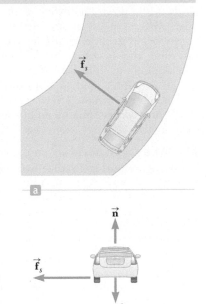

**Figure 7.11** (Example 7.6) (a) The centripetal force is provided by the force of static friction, which is directed radially toward the center of the circular path. (b) Gravity, the normal force, and the static friction force act on the car.

## SOLUTION

(Steps 3, 4) Write the components of Newton's second law. The radial component involves only the maximum static friction force, $f_{s,\text{max}}$:

$$-m\frac{v^2}{r} = -f_{s,\text{max}} = -\mu_s n$$

In the vertical component of the second law, the gravity force and the normal force are in equilibrium:

$$n - mg = 0 \quad \rightarrow \quad n = mg$$

(Step 5) Substitute the expression for $n$ into the first equation and solve for $\mu_s$:

$$-m\frac{v^2}{r} = -\mu_s mg$$

$$\mu_s = \frac{v^2}{rg} = \frac{(13.4 \text{ m/s})^2}{(50.0 \text{ m})(9.80 \text{ m/s}^2)} = \boxed{0.366}$$

**REMARKS** The value of $\mu_s$ for rubber on dry concrete is very close to 1, so the car can negotiate the curve with ease. If the road were wet or icy, however, the value for $\mu_s$ could be 0.2 or lower. Under such conditions, the radial force provided by static friction wouldn't be great enough to keep the car on the circular path, and it would slide off on a tangent, leaving the roadway.

**QUESTION 7.6** If the static friction coefficient were increased, would the maximum safe speed be reduced, increased, or remain the same?

**EXERCISE 7.6** At what maximum speed can a car negotiate a turn on a wet road with coefficient of static friction 0.230 without sliding out of control? The radius of the turn is 25.0 m.

**ANSWER** 7.51 m/s

## EXAMPLE 7.7 DAYTONA INTERNATIONAL SPEEDWAY

**GOAL** Solve a centripetal force problem involving two dimensions.

**PROBLEM** The Daytona International Speedway in Daytona Beach, Florida, is famous for its races, especially the Daytona 500, held every February. Both of its courses feature four-story, 31.0° banked curves, with maximum radius of 316 m. If a car negotiates the curve too slowly, it tends to slip down the incline of the turn, whereas if it's going too fast, it may begin to slide up the incline. **(a)** Find the necessary centripetal acceleration on this banked curve so the car won't tend to slip down or slide up the incline. (Neglect friction.) **(b)** Calculate the speed of the race car.

**APPLICATION**
Banked Roadways

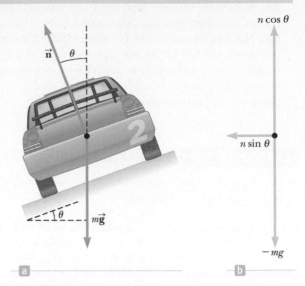

**Figure 7.12** (Example 7.7) As the car rounds a curve banked at an angle $\theta$, the centripetal force that keeps it on a circular path is supplied by the radial component of the normal force. Friction also contributes, although is neglected in this example. The car is moving forward, into the page. (a) Force diagram for the car. (b) Components of the forces.

**STRATEGY** Two forces act on the race car: the force of gravity and the normal force $\vec{\mathbf{n}}$. (See Fig. 7.12.) Use Newton's second law in the upward and radial directions to find the centripetal acceleration $a_c$. Solving $a_c = -v^2/r$ for $v$ then gives the race car's speed.

......................................................................................................

**SOLUTION**

**(a)** Find the centripetal acceleration.

Write Newton's second law for the car:

$$m\vec{\mathbf{a}} = \sum \vec{\mathbf{F}} = \vec{\mathbf{n}} + m\vec{\mathbf{g}}$$

Use the $y$-component of Newton's second law to solve for the normal force $n$:

$$n \cos \theta - mg = 0$$
$$n = \frac{mg}{\cos \theta}$$

Obtain an expression for the horizontal component of $\vec{\mathbf{n}}$, which is the centripetal force $F_c$ in this example:

$$F_c = -n \sin \theta = -\frac{mg \sin \theta}{\cos \theta} = -mg \tan \theta$$

Substitute this expression for $F_c$ into the radial component of Newton's second law and divide by $m$ to get the centripetal acceleration:

$$ma_c = F_c$$
$$a_c = \frac{F_c}{m} = -\frac{mg \tan \theta}{m} = -g \tan \theta$$
$$a_c = -(9.80 \text{ m/s}^2)(\tan 31.0°) = \boxed{-5.89 \text{ m/s}^2}$$

**(b)** Find the speed of the race car.

Apply Equation 7.13:

$$-\frac{v^2}{r} = a_c$$
$$v = \sqrt{-ra_c} = \sqrt{-(316 \text{ m})(-5.89 \text{ m/s}^2)} = \boxed{43.1 \text{ m/s}^2}$$

......................................................................................................

**REMARKS** In fact, both banking and friction assist in keeping the race car on the track.

**QUESTION 7.7** What three physical quantities determine the minimum and maximum safe speeds on a banked racetrack?

**EXERCISE 7.7** A racetrack is to have a banked curve with radius of 245 m. What should be the angle of the bank if the normal force alone is to allow safe travel around the curve at 58.0 m/s?

**ANSWER** 54.5°

## EXAMPLE 7.8    RIDING THE TRACKS

**GOAL** Combine centripetal force with conservation of energy. Derive results symbolically.

**PROBLEM** Figure 7.13a shows a roller-coaster car moving around a circular loop of radius $R$. **(a)** What speed must the car have at the top of the loop so that it will just make it over the top without any assistance from the track? **(b)** What speed will the car subsequently have at the bottom of the loop? **(c)** What will be the normal force on a passenger at the bottom of the loop if the loop has a radius of 10.0 m?

**STRATEGY** This problem requires Newton's second law and centripetal acceleration to find an expression for the car's speed at the top of the loop, followed by conservation of energy to find its speed at the bottom. If the car just makes it over the top, the force $\vec{\mathbf{n}}$ must become zero there, so the only force exerted on the car at that point is the force of gravity, $m\vec{\mathbf{g}}$. At the bottom of the loop, the normal force acts up toward the center and the gravity force acts down, away from the center. The difference of these two is the centripetal force. The normal force can then be calculated from Newton's second law.

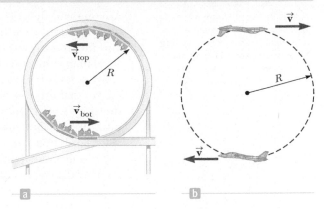

**Figure 7.13** (a) (Example 7.8) A roller coaster traveling around a nearly circular track. (b) (Exercise 7.8) A jet executing a vertical loop.

## SOLUTION

**(a)** Find the speed at the top of the loop.

Write Newton's second law for the car:

$$(1) \quad m\vec{\mathbf{a}}_c = \vec{\mathbf{n}} + m\vec{\mathbf{g}}$$

At the top of the loop, set $n = 0$. The force of gravity acts toward the center and provides the centripetal acceleration $a_c = -v^2/R$:

$$-m\frac{v_{\text{top}}^2}{R} = -mg$$

Solve the foregoing equation for $v_{\text{top}}$:

$$v_{\text{top}} = \boxed{\sqrt{gR}}$$

**(b)** Find the speed at the bottom of the loop.

Apply conservation of mechanical energy to find the total mechanical energy at the top of the loop:

$$E_{\text{top}} = \tfrac{1}{2}mv_{\text{top}}^2 + mgh = \tfrac{1}{2}mgR + mg(2R) = 2.5mgR$$

Find the total mechanical energy at the bottom of the loop:

$$E_{\text{bot}} = \tfrac{1}{2}mv_{\text{bot}}^2$$

Energy is conserved, so these two energies may be equated and solved for $v_{\text{bot}}$:

$$\tfrac{1}{2}mv_{\text{bot}}^2 = 2.5mgR$$

$$v_{\text{bot}} = \boxed{\sqrt{5gR}}$$

**(c)** Find the normal force on a passenger at the bottom. (This is the passenger's perceived weight.)

Use Equation (1). The net centripetal force is $mg - n$. (The normal force acts toward the center of the circle, in the negative radial direction. The car's weight acts away from the center, in the positive radial direction.)

$$-m\frac{v_{\text{bot}}^2}{R} = mg - n$$

Solve for $n$:

$$n = mg + m\frac{v_{\text{bot}}^2}{R} = mg + m\frac{5gR}{R} = \boxed{6mg}$$

**REMARKS** The final answer for $n$ shows that the rider experiences a force six times normal weight at the bottom of the loop! Astronauts experience a similar force during space launches.

**QUESTION 7.8** Suppose the car subsequently goes over a rise with the same radius of curvature and at the same speed as part **(a)**. What is the normal force in this case?

**EXERCISE 7.8** A jet traveling at a constant speed of $1.20 \times 10^2$ m/s executes a vertical loop with a radius of $5.00 \times 10^2$ m. (See Fig. 7.13b.) Find the magnitude of the force of the seat on a 70.0-kg pilot at **(a)** the top and **(b)** the bottom of the loop.

**ANSWERS** **(a)** $1.33 \times 10^3$ N    **(b)** $2.70 \times 10^3$ N

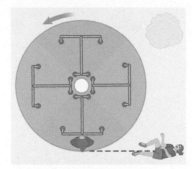

**Figure 7.14** A fun-loving student loses her grip and falls along a line tangent to the rim of the merry-go-round.

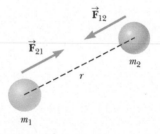

**Figure 7.15** The gravitational force between two particles is attractive and acts along the line joining the particles. Note that according to Newton's third law, $\vec{F}_{12} = -\vec{F}_{21}$.

**Table 7.1** Free-Fall Acceleration $g$ at Various Altitudes

| Altitude (km)[a] | $g$ (m/s²) |
|---|---|
| 1 000 | 7.33 |
| 2 000 | 5.68 |
| 3 000 | 4.53 |
| 4 000 | 3.70 |
| 5 000 | 3.08 |
| 6 000 | 2.60 |
| 7 000 | 2.23 |
| 8 000 | 1.93 |
| 9 000 | 1.69 |
| 10 000 | 1.49 |
| 50 000 | 0.13 |

[a]All figures are distances above Earth's surface.

## 7.4.2 Fictitious Forces

Anyone who has ridden a merry-go-round as a child (or as a fun-loving grown-up) has experienced what feels like a "center-fleeing" force. Holding onto the railing and moving toward the center feels like a walk up a steep hill.

Actually, this so-called centrifugal force is *fictitious.* In reality, the rider is exerting a centripetal force on her body with her hand and arm muscles. In addition, a smaller centripetal force is exerted by the static friction between her feet and the platform. If the rider's grip slipped, she wouldn't be flung radially away; rather, she would go off on a straight line, tangent to the point in space where she let go of the railing. The rider lands at a point that is farther away from the center, but not by "fleeing the center" along a radial line. Instead, she travels perpendicular to a radial line, traversing an angular displacement while increasing her radial displacement. (See Fig. 7.14.)

# 7.5 Newtonian Gravitation

Prior to 1686, a great deal of data had been collected on the motions of the Moon and planets, but no one had a clear understanding of the forces affecting them. In that year, Isaac Newton provided the key that unlocked the secrets of the heavens. He knew from the first law that a net force had to be acting on the Moon. If it were not, the Moon would move in a straight-line path rather than in its almost circular orbit around Earth. Newton reasoned that it was the same kind of force that attracted objects—such as apples—close to the surface of the Earth. He called it the force of gravity.

In 1687 Newton published his work on the law of universal gravitation:

> If two particles with masses $m_1$ and $m_2$ are separated by a distance $r$, a gravitational force $F$ acts along a line joining them, with magnitude given by
>
> $$F = G\frac{m_1 m_2}{r^2}$$ [7.21]
>
> where $G = 6.673 \times 10^{-11}$ kg⁻¹ · m³ · s⁻² is a constant of proportionality called the **constant of universal gravitation.** The gravitational force is always attractive.

This force law is an example of an **inverse-square law,** in that it varies as one over the square of the distance between particles. From Newton's third law, we know that the force exerted by $m_1$ on $m_2$, designated $\vec{F}_{12}$ in Figure 7.15, is equal in magnitude but opposite in direction to the force $\vec{F}_{21}$ exerted by $m_2$ on $m_1$, forming an action–reaction pair.

Another important fact is that **the gravitational force exerted by a uniform sphere on a particle outside the sphere is the same as the force exerted if the entire mass of the sphere were concentrated at its center.** This is called Gauss' law, after the German mathematician and astronomer Karl Friedrich Gauss, and is also true of electric fields, which we will encounter in Topic 15. Gauss' law is a mathematical result, true because the force falls off as an inverse square of the separation between the particles.

Near the surface of the Earth, the expression $F = mg$ is valid. As shown in Table 7.1, however, the free-fall acceleration $g$ varies considerably with altitude above the Earth.

> ### Quick Quiz
>
> **7.8** A ball is falling toward the ground. Which of the following statements are false? (a) The force that the ball exerts on Earth is equal in magnitude to the force that Earth exerts on the ball. (b) The ball undergoes the same acceleration as Earth. (c) The magnitude of the force the Earth exerts on the ball is greater than the magnitude of the force the ball exerts on the Earth.

**7.9** A planet has two moons with identical mass. Moon 1 is in a circular orbit of radius $r$. Moon 2 is in a circular orbit of radius $2r$. The magnitude of the gravitational force exerted by the planet on Moon 2 is (a) four times as large (b) twice as large (c) the same (d) half as large (e) one-fourth as large as the gravitational force exerted by the planet on Moon 1.

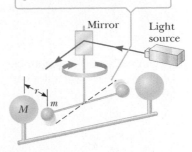

Gravity forces cause the rod to rotate away from its original position (the dashed line).

### 7.5.1 Measurement of the Gravitational Constant

The gravitational constant $G$ in Equation 7.21 was first measured in an important experiment by Henry Cavendish in 1798. His apparatus consisted of two small spheres, each of mass $m$, fixed to the ends of a light horizontal rod suspended by a thin metal wire, as in Figure 7.16. Two large spheres, each of mass $M$, were placed near the smaller spheres. The attractive force between the smaller and larger spheres caused the rod to rotate in a horizontal plane and the wire to twist. The angle through which the suspended rod rotated was measured with a light beam reflected from a mirror attached to the vertical suspension. (Such a moving spot of light is an effective technique for amplifying motion.) The experiment was carefully repeated with different masses at various separations. In addition to providing a value for $G$, the results showed that the force is attractive, proportional to the product $mM$, and inversely proportional to the square of the distance $r$. Modern forms of such experiments are carried out regularly today in an effort to determine $G$ with greater precision.

**Figure 7.16** A schematic diagram of the Cavendish apparatus for measuring $G$. The smaller spheres of mass $m$ are attracted to the large spheres of mass $M$, and the rod rotates through a small angle. A light beam reflected from a mirror on the rotating apparatus measures the angle of rotation.

---

### EXAMPLE 7.9 | BILLIARDS, ANYONE?

**GOAL** Use vectors to find the net gravitational force on an object.

**PROBLEM** (a) Three 0.300-kg billiard balls are placed on a table at the corners of a right triangle, as shown from overhead in Figure 7.17. Find the net gravitational force on the cue ball (designated as $m_1$) resulting from the forces exerted by the other two balls. (b) Find the components of the gravitational force of $m_2$ on $m_3$.

**STRATEGY** (a) To find the net gravitational force on the cue ball of mass $m_1$, we first calculate the force $\vec{F}_{21}$ exerted by $m_2$ on $m_1$. This force is the $y$-component of the net force acting on $m_1$. Then we find the force $\vec{F}_{31}$ exerted by $m_3$ on $m_1$, which is the $x$-component of the net force acting on $m_1$. With these two components, we can find the magnitude and direction of the net force on the cue ball. (b) In this case, we must use trigonometry to find the components of the force $\vec{F}_{23}$.

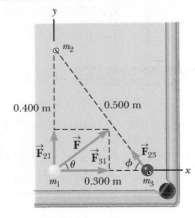

**Figure 7.17** (Example 7.9)

- - - - - - - - - - - - - - - - - - - - - - - - - - - - - - - - - - - - - - - - - - -

**SOLUTION**

**(a)** Find the net gravitational force on the cue ball.

Find the magnitude of the force $\vec{F}_{21}$ exerted by $m_2$ on $m_1$ using the law of gravitation, Equation 7.21:

$$F_{21} = G\frac{m_2 m_1}{r_{21}^2} = (6.67 \times 10^{-11}\,\text{N} \cdot \text{m}^2/\text{kg}^2)\frac{(0.300\,\text{kg})(0.300\,\text{kg})}{(0.400\,\text{m})^2}$$

$$F_{21} = 3.75 \times 10^{-11}\,\text{N}$$

Find the magnitude of the force $\vec{F}_{31}$ exerted by $m_3$ on $m_1$, again using Newton's law of gravity:

$$F_{31} = G\frac{m_3 m_1}{r_{31}^2} = (6.67 \times 10^{-11}\,\text{N} \cdot \text{m}^2/\text{kg}^2)\frac{(0.300\,\text{kg})(0.300\,\text{kg})}{(0.300\,\text{m})^2}$$

$$F_{31} = 6.67 \times 10^{-11}\,\text{N}$$

The net force has components $F_x = F_{31}$ and $F_y = F_{21}$. Compute the magnitude of this net force:

$$F = \sqrt{F_x^2 + F_y^2} = \sqrt{(6.67)^2 + (3.75)^2} \times 10^{-11}\,\text{N}$$

$$= 7.65 \times 10^{-11}\,\text{N}$$

Use the inverse tangent to obtain the direction of $\vec{F}$:

$$\theta = \tan^{-1}\left(\frac{F_y}{F_x}\right) = \tan^{-1}\left(\frac{3.75 \times 10^{-11}\,\text{N}}{6.67 \times 10^{-11}\,\text{N}}\right) = 29.3°$$

*(Continued)*

**(b)** Find the components of the force of $m_2$ on $m_3$.

First, compute the magnitude of $\vec{F}_{23}$:

$$F_{23} = G\frac{m_2 m_1}{r_{23}^2}$$

$$= (6.67 \times 10^{-11}\ kg^{-1}m^3s^{-2})\frac{(0.300\ kg)(0.300\ kg)}{(0.500\ m)^2}$$

$$= 2.40 \times 10^{-11}\ N$$

To obtain the $x$- and $y$-components of $F_{23}$, we need $\cos\phi$ and $\sin\phi$. Use the sides of the large triangle in Figure 7.17:

$$\cos\phi = \frac{adj}{hyp} = \frac{0.300\ m}{0.500\ m} = 0.600$$

$$\sin\phi = \frac{opp}{hyp} = \frac{0.400\ m}{0.500\ m} = 0.800$$

Compute the components of $\vec{F}_{23}$. A minus sign must be supplied for the $x$-component because it's in the negative $x$-direction.

$$F_{23x} = -F_{23}\cos\phi = -(2.40 \times 10^{-11}\ N)(0.600)$$

$$= -1.44 \times 10^{-11}\ N$$

$$F_{23y} = F_{23}\sin\phi = (2.40 \times 10^{-11}\ N)(0.800) = 1.92 \times 10^{-11}\ N$$

**REMARKS**  Notice how small the gravity forces are between everyday objects. Nonetheless, such forces can be measured directly with torsion balances.

**QUESTION 7.9**  Is the gravity force a significant factor in a game of billiards? Explain.

**EXERCISE 7.9**  Find the magnitude and direction of the force exerted by $m_1$ and $m_3$ on $m_2$.

**ANSWERS**  $5.85 \times 10^{-11}\ N$, $-75.7°$

---

**EXAMPLE 7.10  CERES**

**GOAL**  Relate Newton's universal law of gravity to $mg$ and show how $g$ changes with position.

**PROBLEM**  An astronaut standing on the surface of Ceres, the largest asteroid, drops a rock from a height of 10.0 m. It takes 8.06 s to hit the ground. **(a)** Calculate the acceleration of gravity on Ceres. **(b)** Find the mass of Ceres, given that the radius of Ceres is $R_C = 5.10 \times 10^2$ km. **(c)** Calculate the gravitational acceleration 50.0 km from the surface of Ceres.

**STRATEGY**  Part **(a)** is a review of one-dimensional kinematics. In part **(b)** the weight of an object, $w = mg$, is the same as the magnitude of the force given by the universal law of gravity. Solve for the unknown mass of Ceres, after which the answer for **(c)** can be found by substitution into the universal law of gravity, Equation 7.21.

**SOLUTION**

**(a)** Calculate the acceleration of gravity, $g_C$, on Ceres. Apply the kinematics equation of displacement to the falling rock:

$$(1)\quad \Delta x = \tfrac{1}{2}at^2 + v_0 t$$

Substitute $\Delta x = -10.0$ m, $v_0 = 0$, $a = -g_C$, and $t = 8.06$ s, and solve for the gravitational acceleration on Ceres, $g_C$:

$$-10.0\ m = -\tfrac{1}{2}g_C(8.06\ s)^2 \quad \rightarrow \quad g_C = 0.308\ m/s^2$$

**(b)** Find the mass of Ceres.

Equate the weight of the rock on Ceres to the gravitational force acting on the rock:

$$mg_C = G\frac{M_C m}{R_C^{\,2}}$$

Solve for the mass of Ceres, $M_C$:

$$M_C = \frac{g_C R_C^{\,2}}{G} = 1.20 \times 10^{21}\ kg$$

**(c)** Calculate the acceleration of gravity at a height of 50.0 km above the surface of Ceres.

Equate the weight at 50.0 km to the gravitational force:

$$mg_C' = G\frac{mM_C}{r^2}$$

Cancel $m$, then substitute $r = 5.60 \times 10^5$ m and the mass of Ceres:

$$g'_C = G \frac{M_C}{r^2}$$

$$= (6.67 \times 10^{-11} \text{ kg}^{-1}\text{m}^3\text{s}^{-2}) \frac{1.20 \times 10^{21} \text{ kg}}{(5.60 \times 10^5 \text{ m})^2}$$

$$= 0.255 \text{ m/s}^2$$

**REMARKS** This is the standard method of finding the mass of a planetary body: study the motion of a falling (or orbiting) object.

**QUESTION 7.10** Give two reasons Equation (1) could not be used for every asteroid as it is used in part **(a)**.

**EXERCISE 7.10** An object takes 2.40 s to fall 5.00 m on a certain planet. **(a)** Find the acceleration due to gravity on the planet. **(b)** Find the planet's mass if its radius is 5 250 km.

**ANSWERS** **(a)** 1.74 m/s$^2$  **(b)** $7.19 \times 10^{23}$ kg

## 7.5.2 Gravitational Potential Energy Revisited

In Topic 5 we introduced the concept of gravitational potential energy and found that the potential energy associated with an object could be calculated from the equation $PE = mgh$, where $h$ is the height of the object above or below some reference level. This equation, however, is valid *only* when the object is near Earth's surface. For objects high above Earth's surface, such as a satellite, an alternative must be used because $g$ varies with distance from the surface, as shown in Table 7.1.

The gravitational potential energy associated with an object of mass $m$ at a distance $r$ from the center of Earth is

$$PE = -G \frac{M_E m}{r} \qquad [7.22]$$

where $M_E$ is the mass of Earth and $r \geq R_E$ (the radius of Earth).
**SI units: Joules (J)**

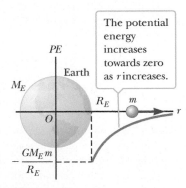

**Figure 7.18** As a mass $m$ moves radially away from the Earth, the potential energy of the Earth-mass system, which is $PE = -G(M_E m/R_E)$ at Earth's surface, increases toward a limit of zero as the mass $m$ travels away from Earth, as shown in the graph.

As before, gravitational potential energy is a property of a *system*, in this case the object of mass $m$ and Earth. Equation 7.22, illustrated in Figure 7.18, is valid for the special case where the zero level for potential energy is at an infinite distance from the center of Earth. Recall that the gravitational potential energy associated with an object is nothing more than the negative of the work done by the force of gravity in moving the object. If an object falls under the force of gravity from a great distance (effectively infinity), the change in gravitational potential energy is negative, which corresponds to a positive amount of gravitational work done on the system. This positive work is equal to the (also positive) change in kinetic energy, as the next example shows.

### EXAMPLE 7.11    A NEAR-EARTH ASTEROID

**GOAL** Use gravitational potential energy to calculate the work done by gravity on a falling object.

**PROBLEM** An asteroid with mass $m = 1.00 \times 10^9$ kg comes from deep space, effectively from infinity, and falls toward Earth. **(a)** Find the change in potential energy when it reaches a point $4.00 \times 10^8$ m from the center of the Earth (just beyond the orbital radius of the Moon). In addition, find the work done by the force of gravity. **(b)** Calculate the asteroid's speed at that point, assuming it was initially at rest when it was arbitrarily far away. **(c)** How much work would have to be done on the asteroid by some other agent so the asteroid would be traveling at only half the speed found in **(b)** at the same point?

**STRATEGY** Part **(a)** requires simple substitution into the definition of gravitational potential energy. To find the work done by the force of gravity, recall that the work done on an object by a conservative force is just the negative of the change in potential energy. Part **(b)** can be solved with conservation of energy, and part **(c)** is an application of the work–energy theorem.

*(Continued)*

## SOLUTION

**(a)** Find the change in potential energy and the work done by the force of gravity.

Apply Equation 7.22:

$$\Delta PE = PE_f - PE_i = -\frac{GM_E m}{r_f} - \left(-\frac{GM_E m}{r_i}\right)$$

$$= GM_E m \left(-\frac{1}{r_f} + \frac{1}{r_i}\right)$$

Substitute known quantities. The asteroid's initial position is effectively infinity, so $1/r_i$ is zero:

$$\Delta PE = (6.67 \times 10^{-11} \text{ kg}^{-1} \text{ m}^3/\text{s}^2)(5.98 \times 10^{24} \text{ kg})$$

$$\times (1.00 \times 10^9 \text{ kg})\left(-\frac{1}{4.00 \times 10^8 \text{ m}} + 0\right)$$

$$\Delta PE = \boxed{-9.97 \times 10^{14} \text{ J}}$$

Compute the work done by the force of gravity:

$$W_{grav} = -\Delta PE = \boxed{9.97 \times 10^{14} \text{ J}}$$

**(b)** Find the speed of the asteroid when it reaches $r_f = 4.00 \times 10^8$ m.

Use conservation of energy:

$$\Delta KE + \Delta PE = 0$$
$$(\tfrac{1}{2}mv^2 - 0) - 9.97 \times 10^{14} \text{ J} = 0$$
$$v = \boxed{1.41 \times 10^3 \text{ m/s}}$$

**(c)** Find the work needed to reduce the speed to $7.05 \times 10^2$ m/s (half the value just found) at this point.

Apply the work–energy theorem:

$$W = \Delta KE + \Delta PE$$

The change in potential energy remains the same as in part **(a)**, but substitute only half the speed in the kinetic-energy term:

$$W = (\tfrac{1}{2}mv^2 - 0) - 9.97 \times 10^{14} \text{ J}$$

$$W = \tfrac{1}{2}(1.00 \times 10^9 \text{ kg})(7.05 \times 10^2 \text{ m/s})^2 - 9.97 \times 10^{14} \text{ J}$$

$$= \boxed{-7.48 \times 10^{14} \text{ J}}$$

**REMARKS** The amount of work calculated in part **(c)** is negative because an external agent must exert a force against the direction of motion of the asteroid. It would take a thruster with a megawatt of output about 24 years to slow down the asteroid to half its original speed. An asteroid endangering Earth need not be slowed that much: A small change in its speed, if applied early enough, will cause it to miss Earth. Timeliness of the applied thrust, however, is important. By the time an astronaut on the asteroid can look over his shoulder and see the Earth, it's already far too late, despite how these scenarios play out in Hollywood. Last-minute rescues won't work!

**QUESTION 7.11** As the asteroid approaches Earth, does the gravitational potential energy associated with the asteroid–Earth system **(a)** increase, **(b)** decrease, **(c)** remain the same?

**EXERCISE 7.11** Suppose the asteroid starts from rest at a great distance (effectively infinity), falling toward Earth. How much work would have to be done on the asteroid to slow it to 425 m/s by the time it reached a distance of $2.00 \times 10^8$ m from Earth?

**ANSWER** $-1.90 \times 10^{15}$ J

---

## APPLYING PHYSICS 7.3    WHY IS THE SUN HOT?

**EXPLANATION** The Sun formed when particles in a cloud of gas coalesced, due to gravitational attraction, into a massive astronomical object. Before this occurred, the particles in the cloud were widely scattered, representing a large amount of gravitational potential energy. As the particles fell closer together, their kinetic energy increased, but the gravitational potential energy of the system decreased, as required by the conservation of energy. With further slow collapse, the cloud became more dense and the average kinetic energy of the particles increased. This kinetic energy is the internal energy of the cloud, which is proportional to the temperature. If enough particles come together, the temperature can rise to a point at which nuclear fusion occurs and the ball of gas becomes a star. Otherwise, the temperature may rise, but not enough to ignite fusion reactions, and the object becomes a brown dwarf (a failed star) or a planet. ∎

On inspecting Equation 7.22, some may wonder what happened to $mgh$, the gravitational potential energy expression introduced in Topic 5. That expression is still valid when $h$ is small compared with Earth's radius. To see this, we write the change in potential energy as an object is raised from the ground to height $h$, using the general form for gravitational potential energy (see Fig. 7.19):

$$PE_2 - PE_1 = -G\frac{M_E m}{(R_E + h)} - \left(-G\frac{M_E m}{R_E}\right)$$

$$= -GM_E m\left[\frac{1}{(R_E + h)} - \frac{1}{R_E}\right]$$

After finding a common denominator and applying some algebra, we obtain

$$PE_2 - PE_1 = \frac{GM_E mh}{R_E(R_E + h)}$$

When the height $h$ is very small compared with $R_E$, $h$ can be dropped from the second factor in the denominator, yielding

$$\frac{1}{R_E(R_E + h)} \cong \frac{1}{R_E^2}$$

Substituting this into the previous expression, we have

$$PE_2 - PE_1 \cong \frac{GM_E}{R_E^2}\, mh$$

Now recall from Topic 4 that the free-fall acceleration at the surface of Earth is given by $g = GM_E/R_E^2$, giving

$$PE_2 - PE_1 \cong mgh$$

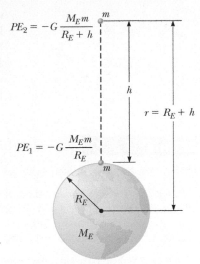

**Figure 7.19** Relating the general form of gravitational potential energy to $mgh$. The height $h$ is greatly exaggerated for clarity.

### 7.5.3 Escape Speed

If an object is projected upward from Earth's surface with a large enough speed, it can soar off into space and never return. This speed is called Earth's **escape speed.** (It is also commonly called the *escape velocity,* but in fact is more properly a speed.)

Earth's escape speed can be found by applying conservation of energy. Suppose an object of mass $m$ is projected vertically upward from Earth's surface with an initial speed $v_i$. The initial mechanical energy (kinetic plus potential energy) of the object–Earth system is given by

$$KE_i + PE_i = \tfrac{1}{2}mv_i^2 - \frac{GM_E m}{R_E}$$

We neglect air resistance and assume the initial speed is just large enough to allow the object to reach infinity with a speed of zero. This value of $v_i$ is the escape speed $v_{esc}$. When the object is at an infinite distance from Earth, its kinetic energy is zero because $v_f = 0$, and the gravitational potential energy is also zero because $1/r$ goes to zero as $r$ goes to infinity. Hence the total mechanical energy is zero, and the law of conservation of energy gives

$$\tfrac{1}{2}mv_{esc}^2 - \frac{GM_E m}{R_E} = 0$$

so that

$$v_{esc} = \sqrt{\frac{2GM_E}{R_E}} \qquad\qquad [7.23]$$

The escape speed for Earth is about 11.2 km/s, which corresponds to about 25 000 mi/h. (See Example 7.12.) Note that the expression for $v_{esc}$ doesn't depend on the mass of the object projected from Earth, so a spacecraft has the same escape speed as a molecule. Escape speeds for the planets, the Moon, and the Sun are listed in Table 7.2. Escape speed and temperature determine to a large extent whether a

**Table 7.2** Escape Speeds for the Planets and the Moon

| Planet | $v_{esc}$ (km/s) |
|---|---|
| Mercury | 4.3 |
| Venus | 10.3 |
| Earth | 11.2 |
| Moon | 2.3 |
| Mars | 5.0 |
| Jupiter | 60.0 |
| Saturn | 36.0 |
| Uranus | 22.0 |
| Neptune | 24.0 |
| Pluto[a] | 1.1 |

[a]In August 2006, the International Astronomical Union adopted a definition of a planet that separates Pluto from the other eight planets. Pluto is now defined as a "dwarf planet" (like the asteroid Ceres).

world has an atmosphere and, if so, what the constituents of the atmosphere are. Planets with low escape speeds, such as Mercury, generally don't have atmospheres because the average speed of gas molecules is close to the escape speed. Venus has a very thick atmosphere, but it's almost entirely carbon dioxide, a heavy gas. The atmosphere of Earth has very little hydrogen or helium but has retained the much heavier nitrogen and oxygen molecules.

## EXAMPLE 7.12 | FROM THE EARTH TO THE MOON

**GOAL** Apply conservation of energy with the general form of Newton's universal law of gravity.

**PROBLEM** In Jules Verne's classic novel *From the Earth to the Moon,* a giant cannon dug into the Earth in Florida fired a spacecraft all the way to the Moon. **(a)** If the spacecraft leaves the cannon at escape speed, at what speed is it moving when $1.50 \times 10^5$ km from the center of Earth? Neglect any friction effects. **(b)** Approximately what constant acceleration is needed to propel the spacecraft to escape speed through a cannon bore 1.00 km long?

**STRATEGY** For part **(a)**, use conservation of energy and solve for the final speed $v_f$. Part **(b)** is an application of the time-independent kinematic equation: solve for the acceleration $a$.

**SOLUTION**

**(a)** Find the speed at $r = 1.50 \times 10^5$ km.

Apply conservation of energy:

$$\tfrac{1}{2}mv_i^2 - \frac{GM_E m}{R_E} = \tfrac{1}{2}mv_f^2 - \frac{GM_E m}{r_f}$$

Multiply by $2/m$ and rearrange, solving for $v_f^2$. Then substitute known values and take the square root.

$$v_f^2 = v_i^2 + \frac{2GM_E}{r_f} - \frac{2GM_E}{R_E} = v_i^2 + 2GM_E\left(\frac{1}{r_f} - \frac{1}{R_E}\right)$$

$$v_f^2 = (1.12 \times 10^4 \text{ m/s})^2 + 2(6.67 \times 10^{-11} \text{ kg}^{-1}\text{m}^3\text{s}^{-2})$$

$$\times (5.98 \times 10^{24} \text{ kg})\left(\frac{1}{1.50 \times 10^8 \text{ m}} - \frac{1}{6.38 \times 10^6 \text{ m}}\right)$$

$$v_f = 2.39 \times 10^3 \text{ m/s}$$

**(b)** Find the acceleration through the cannon bore, assuming it's constant.

Use the time-independent kinematics equation:

$$v^2 - v_0^2 = 2a\Delta x$$

$$(1.12 \times 10^4 \text{ m/s})^2 - 0 = 2a(1.00 \times 10^3 \text{ m})$$

$$a = 6.27 \times 10^4 \text{ m/s}^2$$

**REMARKS** This result corresponds to an acceleration of over 6 000 times the free-fall acceleration on Earth. Such a huge acceleration is far beyond what the human body can tolerate.

**QUESTION 7.12** Suppose the spacecraft managed to go into an elliptical orbit around Earth, with a nearest point (perigee) and farthest point (apogee). At which point is the kinetic energy of the spacecraft higher, and why?

**EXERCISE 7.12** Using the data in Table 7.3 (see page 214), find **(a)** the escape speed from the surface of Mars and **(b)** the speed of a space vehicle when it is $1.25 \times 10^7$ m from the center of Mars if it leaves the surface at the escape speed.

**ANSWERS** **(a)** $5.04 \times 10^3$ m/s  **(b)** $2.62 \times 10^3$ m/s

## 7.5.4 Kepler's Laws

The movements of the planets, stars, and other celestial bodies have been observed for thousands of years. In early history scientists regarded Earth as the center of the Universe. This **geocentric model** was developed extensively by the Greek astronomer Claudius Ptolemy in the second century AD and was accepted for the next 1 400 years. In 1543 Polish astronomer Nicolaus Copernicus (1473–1543) showed that Earth and the other planets revolve in circular orbits around the Sun (the **heliocentric model**).

Danish astronomer Tycho Brahe (pronounced "brah" or "brah´ huh"; 1546–1601) made accurate astronomical measurements over a period of 20 years, providing the data for the currently accepted model of the solar system. Brahe's precise observations of the planets and 777 stars were carried out with nothing more elaborate than a large sextant and compass; the telescope had not yet been invented.

German astronomer Johannes Kepler, who was Brahe's assistant, acquired Brahe's astronomical data and spent about 16 years trying to deduce a mathematical model for the motions of the planets. After many laborious calculations, he found that Brahe's precise data on the motion of Mars about the Sun provided the answer. Kepler's analysis first showed that the concept of circular orbits about the Sun had to be abandoned. He eventually discovered that the orbit of Mars could be accurately described by an ellipse with the Sun at one focus. He then generalized this analysis to include the motions of all planets. The complete analysis is summarized in three statements known as **Kepler's laws:**

1. All planets move in elliptical orbits with the Sun at one of the focal points.
2. A line drawn from the Sun to any planet sweeps out equal areas in equal time intervals.
3. The square of the orbital period of any planet is proportional to the cube of the average distance from the planet to the Sun.

◀ Kepler's laws

Newton later demonstrated that these laws are consequences of the gravitational force that exists between any two objects. Newton's law of universal gravitation, together with his laws of motion, provides the basis for a full mathematical description of the motions of planets and satellites.

**Kepler's First Law**  The first law arises as a natural consequence of the inverse-square nature of Newton's law of gravitation. Any object bound to another by a force that varies as $1/r^2$ will move in an elliptical orbit. As shown in Figure 7.20a, an ellipse is a curve drawn so that the sum of the distances from any point on the curve to two internal points called *focal points* or *foci* (singular, *focus*) is always the same. The semimajor axis $a$ is half the length of the line that goes across the ellipse and contains both foci. For the Sun–planet configuration (Fig. 7.20b), the Sun is at one focus and the other focus is empty. Because the orbit is an ellipse, the distance from the Sun to the planet continuously changes.

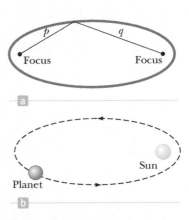

**Figure 7.20**  (a) The sum $p + q$ is the same for every point on the ellipse. (b) In the Solar System, the Sun is at one focus of the elliptical orbit of each planet and the other focus is empty.

**Kepler's Second Law**  Kepler's second law states that a line drawn from the Sun to any planet sweeps out equal areas in equal time intervals. Consider a planet in an elliptical orbit about the Sun, as in Figure 7.21. In a given period $\Delta t$, the planet moves from point Ⓐ to point Ⓑ. The planet moves more slowly on that side of the orbit because it's farther away from the sun. On the opposite side of its orbit, the planet moves from point Ⓒ to point Ⓓ in the same amount of time, $\Delta t$, moving faster because it's closer to the sun. Kepler's second law says that any two wedges formed as in Figure 7.21 will always have the same area. As we will see in Topic 8, Kepler's second law is related to a physical principle known as conservation of angular momentum.

**Kepler's Third Law**  The derivation of Kepler's third law is simple enough to carry out for the special case of a circular orbit. Consider a planet of mass $M_p$ moving around the Sun, which has a mass of $M_S$, in a circular orbit. Because the orbit is circular, the planet moves at a constant speed $v$. Newton's second law, his law of gravitation, and centripetal acceleration then give the following equation:

$$M_p a_c = \frac{M_p v^2}{r} = \frac{GM_S M_p}{r^2}$$

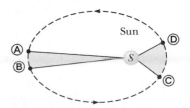

**Figure 7.21**  The two areas swept out by the planet in its elliptical orbit about the Sun are equal if the time interval between points Ⓐ and Ⓑ is equal to the time interval between points Ⓒ and Ⓓ.

The speed $v$ of the planet in its orbit is equal to the circumference of the orbit divided by the time required for one revolution, $T$, called the **period** of the planet, so $v = 2\pi r/T$. Substituting, the preceding expression becomes

$$\frac{GM_S}{r^2} = \frac{(2\pi r/T)^2}{r}$$

Kepler's third law ▶

$$T^2 = \left(\frac{4\pi^2}{GM_S}\right)r^3 = K_S r^3 \qquad [7.24]$$

where $K_S$ is a constant given by

$$K_S = \frac{4\pi^2}{GM_S} = 2.97 \times 10^{-19}\ \text{s}^2/\text{m}^3$$

Equation 7.24 is Kepler's third law for a circular orbit. The orbits of most of the planets are very nearly circular. Comets and asteroids, however, usually have elliptical orbits. For these orbits, the radius $r$ must be replaced with $a$, the semimajor axis—half the longest distance across the elliptical orbit. (This is also the average distance of the comet or asteroid from the Sun.) A more detailed calculation shows that $K_S$ actually depends on the sum of both the mass of a given planet and the Sun's mass. The masses of the planets, however, are negligible compared with the Sun's mass; hence can be neglected, meaning Equation 7.24 is valid for any planet in the Sun's family. If we consider the orbit of a satellite such as the Moon around Earth, then the constant has a different value, with the mass of the Sun replaced by the mass of Earth. In that case, $K_E$ equals $4\pi^2/GM_E$.

The mass of the Sun can be determined from Kepler's third law because the constant $K_S$ in Equation 7.24 includes the mass of the Sun and the other variables and constants can be easily measured. The value of this constant can be found by substituting the values of a planet's period and orbital radius and solving for $K_S$. The mass of the Sun is then

$$M_S = \frac{4\pi^2}{GK_S}$$

This same process can be used to calculate the mass of Earth (by considering the period and orbital radius of the Moon) and the mass of other planets in the solar system that have satellites.

The last column in Table 7.3 confirms that $T^2/r^3$ is very nearly constant. When time is measured in Earth years and the semimajor axis in astronomical units (1 AU = the distance from Earth to the Sun), Kepler's law takes the following simple form:

$$T^2 = a^3$$

**Table 7.3** Useful Planetary Data

| Body | Mass (kg) | Mean Radius (m) | Period (s) | Mean Distance from Sun (m) | $\frac{T^2}{r^3}\ 10^{-19}\left(\frac{\text{s}^2}{\text{m}^3}\right)$ |
|---|---|---|---|---|---|
| Mercury | $3.18 \times 10^{23}$ | $2.43 \times 10^{6}$ | $7.60 \times 10^{6}$ | $5.79 \times 10^{10}$ | 2.97 |
| Venus | $4.88 \times 10^{24}$ | $6.06 \times 10^{6}$ | $1.94 \times 10^{7}$ | $1.08 \times 10^{11}$ | 2.99 |
| Earth | $5.98 \times 10^{24}$ | $6.38 \times 10^{6}$ | $3.156 \times 10^{7}$ | $1.496 \times 10^{11}$ | 2.97 |
| Mars | $6.42 \times 10^{23}$ | $3.37 \times 10^{6}$ | $5.94 \times 10^{7}$ | $2.28 \times 10^{11}$ | 2.98 |
| Jupiter | $1.90 \times 10^{27}$ | $6.99 \times 10^{7}$ | $3.74 \times 10^{8}$ | $7.78 \times 10^{11}$ | 2.97 |
| Saturn | $5.68 \times 10^{26}$ | $5.85 \times 10^{7}$ | $9.35 \times 10^{8}$ | $1.43 \times 10^{12}$ | 2.99 |
| Uranus | $8.68 \times 10^{25}$ | $2.33 \times 10^{7}$ | $2.64 \times 10^{9}$ | $2.87 \times 10^{12}$ | 2.95 |
| Neptune | $1.03 \times 10^{26}$ | $2.21 \times 10^{7}$ | $5.22 \times 10^{9}$ | $4.50 \times 10^{12}$ | 2.99 |
| Pluto[a] | $1.27 \times 10^{23}$ | $1.14 \times 10^{6}$ | $7.82 \times 10^{9}$ | $5.91 \times 10^{12}$ | 2.96 |
| Moon | $7.36 \times 10^{22}$ | $1.74 \times 10^{6}$ | — | — | — |
| Sun | $1.991 \times 10^{30}$ | $6.96 \times 10^{8}$ | — | — | — |

[a]In August 2006, the International Astronomical Union adopted a definition of a planet that separates Pluto from the other eight planets. Pluto is now defined as a "dwarf planet" like the asteroid Ceres.

This equation can be easily checked: Earth has a semimajor axis of one astronomical unit (by definition), and it takes one year to circle the Sun. This equation, of course, is valid only for the Sun and its planets, asteroids, and comets.

Quick Quiz

**7.10** Suppose an asteroid has a semimajor axis of 4 AU. How long does it take the asteroid to go around the Sun? (a) 2 years (b) 4 years (c) 6 years (d) 8 years

## EXAMPLE 7.13   GEOSYNCHRONOUS ORBIT AND TELECOMMUNICATIONS SATELLITES

**GOAL** Apply Kepler's third law to an Earth satellite.

**PROBLEM** From a telecommunications point of view, it's advantageous for satellites to remain at the same location relative to a location on Earth. This can occur only if the satellite's orbital period is the same as the Earth's period of rotation, approximately 24.0 h. **(a)** At what distance from the center of the Earth can this geosynchronous orbit be found? **(b)** What's the orbital speed of the satellite?

**STRATEGY** This problem can be solved with the same method that was used to derive a special case of Kepler's third law, with Earth's mass replacing the Sun's mass. There's no need to repeat the analysis; just replace the Sun's mass with Earth's mass in Kepler's third law, substitute the period $T$ (converted to seconds), and solve for $r$. For part **(b)**, find the circumference of the circular orbit and divide by the elapsed time.

### SOLUTION

**(a)** Find the distance $r$ to geosynchronous orbit.

Apply Kepler's third law:

$$T^2 = \left(\frac{4\pi^2}{GM_E}\right)r^3$$

Substitute the period in seconds, $T = 86\,400$ s, the gravity constant $G = 6.67 \times 10^{-11}$ kg$^{-1}$ m$^3$/s$^2$, and the mass of the Earth, $M_E = 5.98 \times 10^{24}$ kg. Solve for $r$:

$$r = 4.23 \times 10^7 \text{ m}$$

**(b)** Find the orbital speed.

Divide the distance traveled during one orbit by the period:

$$v = \frac{d}{T} = \frac{2\pi r}{T} = \frac{2\pi(4.23 \times 10^7 \text{ m})}{8.64 \times 10^4 \text{ s}} = 3.08 \times 10^3 \text{ m/s}$$

**REMARKS** Earth's motion around the Sun was neglected; that requires using Earth's "sidereal" period (about four minutes shorter). Notice that Earth's mass could be found by substituting the Moon's distance and period into this form of Kepler's third law.

**QUESTION 7.13** If the satellite was placed in an orbit three times as far away, about how long would it take to orbit the Earth once? Answer in days, rounding to one digit.

**EXERCISE 7.13** Mars rotates on its axis once every 1.02 days (almost the same as Earth does). **(a)** Find the distance from the center of Mars at which a satellite would remain in one spot over the Martian surface. **(b)** Find the speed of the satellite.

**ANSWERS** **(a)** $2.03 \times 10^7$ m   **(b)** $1.45 \times 10^3$ m/s

## SUMMARY

### 7.1 Angular Velocity and Angular Acceleration

The **average angular velocity** $\omega_{av}$ of a rigid object is defined as the ratio of the angular displacement $\Delta\theta$ to the time interval $\Delta t$, or

$$\omega_{av} \equiv \frac{\theta_f - \theta_i}{t_f - t_i} = \frac{\Delta\theta}{\Delta t} \quad [7.3]$$

where $\omega_{av}$ is in radians per second (rad/s).

The **average angular acceleration** $\alpha_{av}$ of a rotating object is defined as the ratio of the change in angular velocity $\Delta\omega$ to the time interval $\Delta t$, or

$$\alpha_{av} \equiv \frac{\omega_f - \omega_i}{t_f - t_i} = \frac{\Delta\omega}{\Delta t} \quad [7.5]$$

where $\alpha_{av}$ is in radians per second per second (rad/s$^2$).

## 7.2 Rotational Motion Under Constant Angular Acceleration

If an object undergoes rotational motion about a fixed axis under a constant angular acceleration $\alpha$, its motion can be described with the following set of equations:

$$\omega = \omega_i + \alpha t \qquad [7.7]$$

$$\Delta\theta = \omega_i t + \tfrac{1}{2}\alpha t^2 \qquad [7.8]$$

$$\omega^2 = \omega_i^2 + 2\alpha\,\Delta\theta \qquad [7.9]$$

Problems are solved as in one-dimensional kinematics.

## 7.3 Tangential Velocity, Tangential Acceleration, and Centripetal Acceleration

When an object rotates about a fixed axis, the angular velocity and angular acceleration are related to the tangential velocity and tangential acceleration through the relationships

$$v_t = r\omega \qquad [7.10]$$

and

$$a_t = r\alpha \qquad [7.11]$$

Any object moving in a circular path has an acceleration directed toward the center of the circular path, called a **centripetal acceleration.** Its magnitude is given by

$$a_c = \frac{v^2}{r} = r\omega^2 \qquad [7.13, 7.17]$$

Any object moving in a circular path must have a net force exerted on it that is directed toward the center of the path. Some examples of forces that cause centripetal acceleration are the force of gravity (as in the motion of a satellite) and the force of tension in a string.

## 7.4 Newton's Second Law for Uniform Circular Motion

The second law for uniform circular motion involves forces that are directed either towards the center of a circle or away from it. A force acting towards the center of the circle is by convention negative and Newton's second law for uniform circular motion therefore reads

$$-m\frac{v^2}{r} = \sum F_r \qquad [7.20]$$

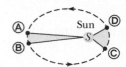

## 7.5 Newtonian Gravitation

**Newton's law of universal gravitation** states that every particle in the Universe attracts every other particle with a force (Fig. 7.22) that

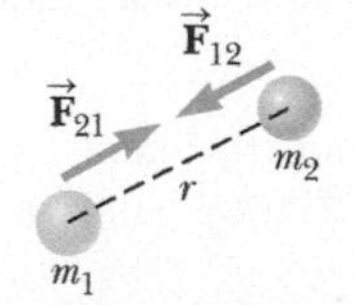

**Figure 7.22** The gravitational force is attractive and acts along the line joining the particles.

is directly proportional to the product of their masses and inversely proportional to the square of the distance $r$ between them:

$$F = G\frac{m_1 m_2}{r^2} \qquad [7.21]$$

where $G = 6.673 \times 10^{-11}$ N $\cdot$ m$^2$/kg$^2$ is the **constant of universal gravitation.** A general expression for gravitational potential energy (Fig. 7.23) is

$$PE = -G\frac{M_E m}{r} \qquad [7.22]$$

This expression reduces to $PE = mgh$ close to the surface of Earth and holds for other worlds through replacement of the mass $M_E$. Problems such as finding the escape velocity from Earth can be solved by using Equation 7.22 in the conservation of energy equation. Kepler derived the following three laws of planetary motion:

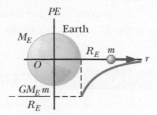

**Figure 7.23** The gravitational potential energy increases towards zero as $r$ increases.

1. All planets move in elliptical orbits with the Sun at one of the focal points (Fig. 7.24).

| **Figure 7.24** Kepler's first law. |

2. A line drawn from the Sun to any planet sweeps out equal areas in equal time intervals (Fig. 7.25).

| **Figure 7.25** Kepler's second law. |

3. The square of the orbital period of a planet is proportional to the cube of the average distance from the planet to the Sun:

$$T^2 = \left(\frac{4\pi^2}{GM_S}\right)r^3 \quad | \text{ Kepler's third law. } | \qquad [7.24]$$

The third law can be applied to any large body and its system of satellites by replacing the Sun's mass with the body's mass. In particular, it can be used to determine the mass of the central body once the average distance to a satellite and its period are known.

## CONCEPTUAL QUESTIONS

1. A disk rotates about an axis through its center. Point $A$ is located on its rim and point $B$ is located exactly halfway between the center and the rim. What is the ratio of (a) the angular velocity $\omega_A$ to that of $\omega_B$, and (b) the tangential velocity $v_A$ to that of $v_B$?

2. Suppose an alien civilization has a space station in circular orbit around its home planet. The station's orbital radius is twice the planet's radius. (a) If an alien astronaut has weight $w$ just before launch from the surface, will she be weightless when she reaches the station and floats inside of it? (b) If not,

what will be the ratio of her weight in orbit to her weight on the planet's surface?

3. If a car's wheels are replaced with wheels of greater diameter, will the reading of the speedometer change? Explain.

4. Objects moving along a circular path have a centripetal acceleration provided by a net force directed towards the center. Identify the force(s) providing the centripetal acceleration in each of these cases: (a) a planet in circular orbit around its sun; (b) a car going around an unbanked, circular turn; (c) a rock tied to a string and swung in a vertical circle, as it passes through its highest point; and (d) a dry sock in a clothes dryer as it spins in a horizontal circle.

5. A pendulum consists of a small object called a bob hanging from a light cord of fixed length, with the top end of the cord fixed, as represented in Figure CQ7.5. The bob moves without friction, swinging equally high on both sides. It moves from its turning point A through point B and reaches its maximum speed at point C. (a) At what point does the bob have non-zero radial acceleration and zero tangential acceleration? What is the direction of its total acceleration at this point? (b) At what point does the bob have nonzero tangential acceleration and zero radial acceleration? What is the direction of its total acceleration at this point? (c) At what point does the bob have both nonzero tangential and radial acceleration? What is the direction of its total acceleration at this point?

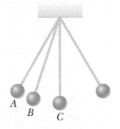

**Figure CQ7.5**

6. Because of Earth's rotation about its axis, you weigh slightly less at the equator than at the poles. Explain.

7. It has been suggested that rotating cylinders about 10 miles long and 5 miles in diameter be placed in space for colonies.

The purpose of their rotation is to simulate gravity for the inhabitants. Explain the concept behind this proposal.

8. Describe the path of a moving object in the event that the object's acceleration is constant in magnitude at all times and (a) perpendicular to its velocity; (b) parallel to its velocity.

9. A pail of water can be whirled in a vertical circular path such that no water is spilled. Why does the water remain in the pail, even when the pail is upside down above your head?

10. A car of mass $m$ follows a truck of mass $2m$ around a circular turn. Both vehicles move at speed $v$. (a) What is the ratio of the truck's net centripetal force to the car's net centripetal force? (b) At what new speed $v_{truck}$ will the net centripetal force acting on the truck equal the net centripetal force acting on the car still moving at the original speed $v$?

11. Is it possible for a car to move in a circular path in such a way that it has a tangential acceleration but no centripetal acceleration?

12. A child is practicing for a BMX race. His speed remains constant as he goes counterclockwise around a level track with two nearly straight sections and two nearly semicircular sections, as shown in the aerial view of Figure CQ7.12. (a) What are the directions of his velocity at points A, B, and C? For each point, choose one: north, south, east, west, or nonexistent. (b) What are the directions of his acceleration at points A, B, and C?

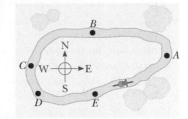

**Figure CQ7.12**

13. An object executes circular motion with constant speed whenever a net force of constant magnitude acts perpendicular to the velocity. What happens to the speed if the force is not perpendicular to the velocity?

# PROBLEMS

*See the Preface for an explanation of the icons used in this problems set.*

## 7.1 Angular Velocity and Angular Acceleration

1. Convert (a) 47.0° to radians, (b) 12.0 rad to revolutions, and (c) 75.0 rpm to rad/s.

2. A bicycle tire is spinning clockwise at 2.50 rad/s. During a time period $\Delta t = 1.25$ s, the tire is stopped and spun in the opposite (counterclockwise) direction, also at 2.50 rad/s. Calculate (a) the change in the tire's angular velocity $\Delta\omega$ and (b) the tire's average angular acceleration $\alpha_{av}$.

3. The tires on a new compact car have a diameter of 2.0 ft and are warranted for 60 000 miles. (a) Determine the angle (in radians) through which one of these tires will rotate during the warranty period. (b) How many revolutions of the tire are equivalent to your answer in part (a)?

4. **Q C** A potter's wheel moves uniformly from rest to an angular velocity of 1.00 rev/s in 30.0 s. (a) Find its angular acceleration in radians per second per second. (b) Would doubling the angular acceleration during the given period have doubled final angular velocity?

## 7.2 Rotational Motion Under Constant Angular Acceleration

## 7.3 Tangential Velocity, Tangential Acceleration, and Centripetal Acceleration

5. A dentist's drill starts from rest. After 3.20 s of constant angular acceleration, it turns at a rate of $2.51 \times 10^4$ rev/min. (a) Find the drill's angular acceleration. (b) Determine the angle (in radians) through which the drill rotates during this period.

6. **V** A centrifuge in a medical laboratory rotates at an angular velocity of 3 600 rev/min. When switched off, it rotates through 50.0 revolutions before coming to rest. Find the constant angular acceleration (in rad/s²) of the centrifuge.

7. A bicyclist starting at rest produces a constant angular acceleration of 1.60 rad/s² for wheels that are 38.0 cm in radius. (a) What is the bicycle's linear acceleration? (b) What is the angular speed of the wheels when the bicyclist reaches 11.0 m/s? (c) How many radians have the wheels turned through in that time? (d) How far has the bicycle traveled?

**8.** **Q|C** A bicycle is turned upside down while its owner repairs a flat tire. A friend spins the other wheel and observes that drops of water fly off tangentially. She measures the heights reached by drops moving vertically (Fig. P7.8). A drop that breaks loose from the tire on one turn rises vertically 54.0 cm above the tangent point. A drop that breaks loose on the next turn rises 51.0 cm above the tangent point. The radius of the wheel is 0.381 m. (a) Why does the first drop rise higher than the second drop? (b) Neglecting air friction and using only the observed heights and the radius of the wheel, find the wheel's angular acceleration (assuming it to be constant).

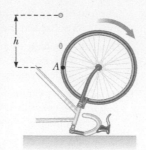

**Figure P7.8**
Problems 8 and 69.

**9.** **V** The diameters of the main rotor and tail rotor of a single-engine helicopter are 7.60 m and 1.02 m, respectively. The respective rotational speeds are 450 rev/min and 4 138 rev/min. Calculate the speeds of the tips of both rotors. Compare these speeds with the speed of sound, 343 m/s.

**10.** The tub of a washer goes into its spin-dry cycle, starting from rest and reaching an angular speed of 5.0 rev/s in 8.0 s. At this point, the person doing the laundry opens the lid, and a safety switch turns off the washer. The tub slows to rest in 12.0 s. Through how many revolutions does the tub turn during the entire 20-s interval? Assume constant angular acceleration while it is starting and stopping.

**11.** A car initially traveling at 29.0 m/s undergoes a constant negative acceleration of magnitude 1.75 m/s² after its brakes are applied. (a) How many revolutions does each tire make before the car comes to a stop, assuming the car does not skid and the tires have radii of 0.330 m? (b) What is the angular speed of the wheels when the car has traveled half the total distance?

**12.** A 45.0-cm diameter disk rotates with a constant angular acceleration of 2.50 rad/s². It starts from rest at $t = 0$, and a line drawn from the center of the disk to a point $P$ on the rim of the disk makes an angle of 57.3° with the positive $x$-axis at this time. At $t = 2.30$ s, find (a) the angular speed of the wheel, (b) the linear speed and tangential acceleration of $P$, and (c) the position of $P$ (in degrees, with respect to the positive $x$-axis).

**13.** **T** A rotating wheel requires 3.00 s to rotate 37.0 revolutions. Its angular velocity at the end of the 3.00-s interval is 98.0 rad/s. What is the constant angular acceleration (in rad/s²) of the wheel?

**14.** An electric motor rotating a workshop grinding wheel at a rate of $1.00 \times 10^2$ rev/min is switched off. Assume the wheel has a constant negative angular acceleration of magnitude 2.00 rad/s². (a) How long does it take for the grinding wheel to stop? (b) Through how many radians has the wheel turned during the interval found in part (a)?

**15.** **V** A car initially traveling eastward turns north by traveling in a circular path at uniform speed as shown in Figure P7.15. The length of

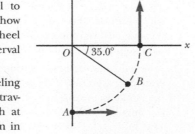

**Figure P7.15**

the arc $ABC$ is 235 m, and the car completes the turn in 36.0 s. (a) Determine the car's speed. (b) What is the magnitude and direction of the acceleration when the car is at point $B$?

**16.** It has been suggested that rotating cylinders about 10 mi long and 5.0 mi in diameter be placed in space and used as colonies. What angular speed must such a cylinder have so that the centripetal acceleration at its surface equals the free-fall acceleration on Earth?

**17.** (a) What is the tangential acceleration of a bug on the rim of a 10.0-in.-diameter disk if the disk accelerates uniformly from rest to an angular velocity of 78.0 rev/min in 3.00 s? (b) When the disk is at its final speed, what is the tangential velocity of the bug? One second after the bug starts from rest, what are its (c) tangential acceleration, (d) centripetal acceleration, and (e) total acceleration?

## 7.4 Newton's Second Law for Uniform Circular Motion

**18.** An adventurous archeologist ($m = 85.0$ kg) tries to cross a river by swinging from a vine. The vine is 10.0 m long, and his speed at the bottom of the swing is 8.00 m/s. The archeologist doesn't know that the vine has a breaking strength of 1 000 N. Does he make it across the river without falling in?

**19.** **Q|C** One end of a cord is fixed and a small 0.500-kg object is attached to the other end, where it swings in a section of a vertical circle of radius 2.00 m, as shown in Figure P7.19. When $\theta = 20.0°$, the speed of the object is 8.00 m/s. At this instant, find (a) the tension in the string, (b) the tangential and radial components of acceleration, and (c) the total acceleration. (d) Is your answer changed if the object is swinging down toward its lowest point instead of swinging up? (e) Explain your answer to part (d).

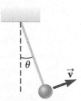

**Figure P7.19**

**20.** **BIO** Human centrifuges are used to train military pilots and astronauts in preparation for high-$g$ maneuvers. A trained, fit person wearing a $g$-suit can withstand accelerations up to about $9g$ (88.2 m/s²) without losing consciousness. (a) If a human centrifuge has a radius of 4.50 m, what angular speed results in a centripetal acceleration of $9g$? (b) What linear speed would a person in the centrifuge have at this acceleration?

**21.** A 55.0-kg ice skater is moving at 4.00 m/s when she grabs the loose end of a rope, the opposite end of which is tied to a pole. She then moves in a circle of radius 0.800 m around the pole. (a) Determine the force exerted by the horizontal rope on her arms. (b) Compare this force with her weight.

**22.** **T** A 40.0-kg child swings in a swing supported by two chains, each 3.00 m long. The tension in each chain at the lowest point is 350 N. Find (a) the child's speed at the lowest point and (b) the force exerted by the seat on the child at the lowest point. (Ignore the mass of the seat.)

**23.** A certain light truck can go around a flat curve having a radius of 150 m with a maximum speed of 32.0 m/s. With what maximum speed can it go around a curve having a radius of 75.0 m?

**24.** **BIO** A sample of blood is placed in a centrifuge of radius 15.0 cm. The mass of a red blood cell is $3.0 \times 10^{-16}$ kg, and the magnitude of the force acting on it as it settles out of the plasma is $4.0 \times 10^{-11}$ N. At how many revolutions per second should the centrifuge be operated?

**25.** A 50.0-kg child stands at the rim of a merry-go-round of radius 2.00 m, rotating with an angular speed of 3.00 rad/s. (a) What is the magnitude of the child's centripetal acceleration? (b) What is the magnitude of the minimum force between her feet and the floor of the carousel that is required to keep her in the circular path? (c) What minimum coefficient of static friction is required? Is the answer you found reasonable? In other words, is she likely to stay on the merry-go-round?

**26.** **GP** A space habitat for a long space voyage consists of two cabins each connected by a cable to a central hub as shown in Figure P7.26. The cabins are set spinning around the hub axis, which is connected to the rest of the spacecraft to generate artificial gravity. (a) What forces are acting on an astronaut in one of the cabins? (b) Write Newton's second law for an astronaut lying on the "floor" of one of the habitats, relating the astronaut's mass $m$, his velocity $v$, his radial distance from the hub $r$, and the normal force $n$. (c) What would $n$ have to equal if the 60.0-kg astronaut is to experience half his normal Earth weight? (d) Calculate the necessary tangential speed of the habitat from Newton's second law. (e) Calculate the angular speed from the tangential speed. (f) Calculate the period of rotation from the angular speed. (g) If the astronaut stands up, will his head be moving faster, slower, or at the same speed as his feet? Why? Calculate the tangential speed at the top of his head if he is 1.80 m tall.

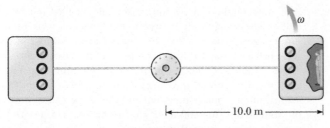

**Figure P7.26**

**27.** An air puck of mass $m_1 = 0.25$ kg is tied to a string and allowed to revolve in a circle of radius $R = 1.0$ m on a frictionless horizontal table. The other end of the string passes through a hole in the center of the table, and a mass of $m_2 = 1.0$ kg is tied to it (Fig. P7.27). The suspended mass remains in equilibrium while the puck on the tabletop revolves. (a) What is the tension in the string? (b) What is the horizontal force acting on the puck? (c) What is the speed of the puck?

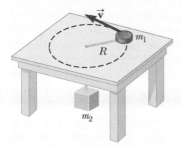

**Figure P7.27**

**28.** **S** A snowboarder drops from rest into a halfpipe of radius $R$ and slides down its frictionless surface to the bottom (Fig. P7.28). Show that (a) the snowboarder's speed at the bottom of the halfpipe is $v = \sqrt{2gR}$ (*Hint:* Use conservation of energy), (b) the snowboarder's centripetal acceleration at the bottom is $a_c = 2g$, and (c) the normal force on the snowboarder at the bottom of the halfpipe has magnitude $3mg$ (*Hint:* Use Newton's second law of motion).

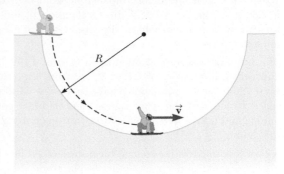

**Figure P7.28**

**29.** **V** **QC** A woman places her briefcase on the backseat of her car. As she drives to work, the car negotiates an unbanked curve in the road that can be regarded as an arc of a circle of radius 62.0 m. While on the curve, the speed of the car is 15.0 m/s at the instant the briefcase starts to slide across the backseat toward the side of the car. (a) What force causes the centripetal acceleration of the briefcase when it is stationary relative to the car? Under what condition does the briefcase begin to move relative to the car? (b) What is the coefficient of static friction between the briefcase and seat surface?

**30.** **QC** A pail of water is rotated in a vertical circle of radius 1.00 m. (a) What two external forces act on the water in the pail? (b) Which of the two forces is most important in causing the water to move in a circle? (c) What is the pail's minimum speed at the top of the circle if no water is to spill out? (d) If the pail with the speed found in part (c) were to suddenly disappear at the top of the circle, describe the subsequent motion of the water. Would it differ from the motion of a projectile?

**31.** **T** A 40.0-kg child takes a ride on a Ferris wheel that rotates four times each minute and has a diameter of 18.0 m. (a) What is the centripetal acceleration of the child? (b) What force (magnitude and direction) does the seat exert on the child at the lowest point of the ride? (c) What force does the seat exert on the child at the highest point of the ride? (d) What force does the seat exert on the child when the child is halfway between the top and bottom?

**32.** **V** A roller-coaster vehicle has a mass of 500 kg when fully loaded with passengers (Fig. P7.32). (a) If the vehicle has a speed of 20.0 m/s at point Ⓐ, what is the force of the track on the vehicle at this point? (b) What is the maximum speed the vehicle can have at point Ⓑ for gravity to hold it on the track?

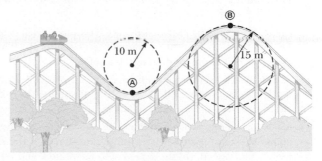

**Figure P7.32**

## 7.5 Newtonian Gravitation

33. (a) Find the magnitude of the gravitational force between a planet with mass $7.50 \times 10^{24}$ kg and its moon, with mass $2.70 \times 10^{22}$ kg, if the average distance between their centers is $2.80 \times 10^8$ m. (b) What is the acceleration of the moon towards the planet? (c) What is the acceleration of the planet towards the moon?

34. The International Space Station has a mass of $4.19 \times 10^5$ kg and orbits at a radius of $6.79 \times 10^6$ m from the center of Earth. Find (a) the gravitational force exerted by Earth on the space station, (b) the space station's gravitational potential energy, and (c) the weight of an 80.0-kg astronaut living inside the station.

35. A coordinate system (in meters) is constructed on the surface of a pool table, and three objects are placed on the table as follows: a 2.0-kg object at the origin of the coordinate system, a 3.0-kg object at (0, 2.0), and a 4.0-kg object at (4.0, 0). Find the resultant gravitational force exerted by the other two objects on the object at the origin.

36. After the Sun exhausts its nuclear fuel, its ultimate fate may be to collapse to a *white dwarf* state. In this state, it would have approximately the same mass as it has now, but its radius would be equal to the radius of Earth. Calculate (a) the average density of the white dwarf, (b) the surface free-fall acceleration, and (c) the gravitational potential energy associated with a 1.00-kg object at the surface of the white dwarf.

37. V Objects with masses of 200. kg and 500. kg are separated by 0.400 m. (a) Find the net gravitational force exerted by these objects on a 50.0-kg object placed midway between them. (b) At what position (other than infinitely remote ones) can the 50.0-kg object be placed so as to experience a net force of zero?

38. Use the data of Table 7.3 to find the point between Earth and the Sun at which an object can be placed so that the net gravitational force exerted by Earth and the Sun on that object is zero.

39. A projectile is fired straight upward from the Earth's surface at the South Pole with an initial speed equal to one third the escape speed. (a) Ignoring air resistance, determine how far from the center of the Earth the projectile travels before stopping momentarily. (b) What is the altitude of the projectile at this instant?

40. Two objects attract each other with a gravitational force of magnitude $1.00 \times 10^{-8}$ N when separated by 20.0 cm. If the total mass of the objects is 5.00 kg, what is the mass of each?

41. T A satellite is in a circular orbit around the Earth at an altitude of $2.80 \times 10^6$ m. Find (a) the period of the orbit, (b) the speed of the satellite, and (c) the acceleration of the satellite. *Hint:* Modify Equation 7.23 so it is suitable for objects orbiting the Earth rather than the Sun.

42. An artificial satellite circling the Earth completes each orbit in 110 minutes. (a) Find the altitude of the satellite. (b) What is the value of $g$ at the location of this satellite?

43. A satellite of Mars, called Phobos, has an orbital radius of $9.4 \times 10^6$ m and a period of $2.8 \times 10^4$ s. Assuming the orbit is circular, determine the mass of Mars.

44. A 600-kg satellite is in a circular orbit about Earth at a height above Earth equal to Earth's mean radius. Find (a) the satellite's orbital speed, (b) the period of its revolution, and (c) the gravitational force acting on it.

45. A comet has a period of 76.3 years and moves in an elliptical orbit in which its perihelion (closest approach to the Sun) is 0.610 AU. Find (a) the semimajor axis of the comet and (b) an estimate of the comet's maximum distance from the Sun, both in astronomical units.

## Additional Problems

46. V A synchronous satellite, which always remains above the same point on a planet's equator, is put in circular orbit around Jupiter to study that planet's famous red spot. Jupiter rotates once every 9.84 h. Use the data of Table 7.3 to find the altitude of the satellite.

47. (a) One of the moons of Jupiter, named Io, has an orbital radius of $4.22 \times 10^8$ m and a period of 1.77 days. Assuming the orbit is circular, calculate the mass of Jupiter. (b) The largest moon of Jupiter, named Ganymede, has an orbital radius of $1.07 \times 10^9$ m and a period of 7.16 days. Calculate the mass of Jupiter from this data. (c) Are your results to parts (a) and (b) consistent? Explain.

48. Neutron stars are extremely dense objects that are formed from the remnants of supernova explosions. Many rotate very rapidly. Suppose the mass of a certain spherical neutron star is twice the mass of the Sun and its radius is 10.0 km. Determine the greatest possible angular speed the neutron star can have so that the matter at its surface on the equator is just held in orbit by the gravitational force.

49. BIO One method of pitching a softball is called the "windmill" delivery method, in which the pitcher's arm rotates through approximately 360° in a vertical plane before the 198-gram ball is released at the lowest point of the circular motion. An experienced pitcher can throw a ball with a speed of 98.0 mi/h. Assume the angular acceleration is uniform throughout the pitching motion and take the distance between the softball and the shoulder joint to be 74.2 cm. (a) Determine the angular speed of the arm in rev/s at the instant of release. (b) Find the value of the angular acceleration in rev/s$^2$ and the radial and tangential acceleration of the ball just before it is released. (c) Determine the force exerted on the ball by the pitcher's hand (both radial and tangential components) just before it is released.

50. V A digital audio compact disc (CD) carries data along a continuous spiral track from the inner circumference of the disc to the outside edge. Each bit occupies 0.6 $\mu$m of the track. A CD player turns the disc to carry the track counterclockwise above a lens at a constant speed of 1.30 m/s. Find the required angular speed (a) at the beginning of the recording, where the spiral has a radius of 2.30 cm, and (b) at the end of the recording, where the spiral has a radius of 5.80 cm. (c) A full-length recording lasts for 74 min, 33 s. Find the average angular acceleration of the disc. (d) Assuming the acceleration is constant, find the total angular displacement of the disc as it plays. (e) Find the total length of the track.

51. An athlete swings a 5.00-kg ball horizontally on the end of a rope. The ball moves in a circle of radius 0.800 m at an angular speed of 0.500 rev/s. What are (a) the tangential speed of the ball and (b) its centripetal acceleration? (c) If the maximum tension the rope can withstand before breaking is 100. N, what is the maximum tangential speed the ball can have?

52. BIO The dung beetle is known as one of the strongest animals for its size, often forming balls of dung up to 10 times their own

mass and rolling them to locations where they can be buried and stored as food. A typical dung ball formed by the species *K. nigroaeneus* has a radius of 2.00 cm and is rolled by the beetle at 6.25 cm/s. (a) What is the rolling ball's angular speed? (b) How many full rotations are required if the beetle rolls the ball a distance of 1.00 m?

53. The Solar Maximum Mission Satellite was placed in a circular orbit about 150 mi above Earth. Determine (a) the orbital speed of the satellite and (b) the time required for one complete revolution.

54. A 0.400-kg pendulum bob passes through the lowest part of its path at a speed of 3.00 m/s. (a) What is the tension in the pendulum cable at this point if the pendulum is 80.0 cm long? (b) When the pendulum reaches its highest point, what angle does the cable make with the vertical? (c) What is the tension in the pendulum cable when the pendulum reaches its highest point?

55. **Q|C** A car moves at speed $v$ across a bridge made in the shape of a circular arc of radius $r$. (a) Find an expression for the normal force acting on the car when it is at the top of the arc. (b) At what minimum speed will the normal force become zero (causing the occupants of the car to seem weightless) if $r = 30.0$ m?

56. **BIO** Keratinocytes are the most common cells in the skin's outer layer. As these approximately circular cells migrate across a wound during the healing process, they roll in a way that reduces the frictional forces impeding their motion. (a) Given a cell body diameter of $1.00 \times 10^{-5}$ m (10 $\mu$m), what minimum angular speed would be required to produce the observed linear speed of $1.67 \times 10^{-7}$ m/s (10 $\mu$m/min)? (b) How many complete revolutions would be required for the cell to roll a distance of $5.00 \times 10^{-3}$ m? (Because of slipping as the cells roll, averages of observed angular speeds and the number of complete revolutions are about three times these minimum values.)

57. **Q|C** Because of Earth's rotation about its axis, a point on the equator has a centripetal acceleration of 0.034 0 m/s², whereas a point at the poles has no centripetal acceleration. (a) Show that, at the equator, the gravitational force on an object (the object's true weight) must exceed the object's apparent weight. (b) What are the apparent weights of a 75.0-kg person at the equator and at the poles? (Assume Earth is a uniform sphere and take $g = 9.800$ m/s².)

58. **Q|C** A roller coaster travels in a circular path. (a) Identify the forces on a passenger at the top of the circular loop that cause centripetal acceleration. Show the direction of all forces in a sketch. (b) Identify the forces on the passenger at the bottom of the loop that produce centripetal acceleration. Show these in a sketch. (c) Based on your answers to parts (a) and (b), at what point, top or bottom, should the seat exert the greatest force on the passenger? (d) Assume the speed of the roller coaster is 4.00 m/s at the top of the loop of radius 8.00 m. Find the force exerted by the seat on a 70.0-kg passenger at the top of the loop. Then, assume the speed remains the same at the bottom of the loop and find the force exerted by the seat on the passenger at this point. Are your answers consistent with your choice of answers for parts (a) and (b)?

59. In Robert Heinlein's *The Moon Is a Harsh Mistress,* the colonial inhabitants of the Moon threaten to launch rocks down onto Earth if they are not given independence (or at least representation). Assuming a gun could launch a rock of mass $m$ at twice the lunar escape speed, calculate the speed of the rock as it enters Earth's atmosphere.

60. **T** A model airplane of mass 0.750 kg flies with a speed of 35.0 m/s in a horizontal circle at the end of a 60.0-m control wire as shown in Figure P7.60a. The forces exerted on the airplane are shown in Figure P7.60b; the tension in the control wire, $\theta = 20.0°$ inward from the vertical. Compute the tension in the wire, assuming the wire makes a constant angle of $\theta = 20.0°$ with the horizontal.

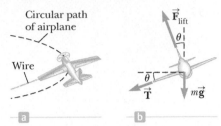

**Figure P7.60**

61. In a home laundry dryer, a cylindrical tub containing wet clothes is rotated steadily about a horizontal axis, as shown in Figure P7.61. So that the clothes will dry uniformly, they are made to tumble. The rate of rotation of the smooth-walled tub is chosen so that a small piece of cloth will lose contact with the tub when the cloth is at an angle of $\theta = 68.0°$ above the horizontal. If the radius of the tub is $r = 0.330$ m, what rate of revolution is needed in revolutions per second?

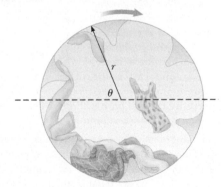

**Figure P7.61**

62. Casting of molten metal is important in many industrial processes. *Centrifugal casting* is used for manufacturing pipes, bearings, and many other structures. A cylindrical enclosure is rotated rapidly and steadily about a horizontal axis, as in Figure P7.62. Molten metal is poured into the rotating cylinder and then cooled, forming the finished product. Turning the cylinder at a high rotation rate forces the solidifying metal strongly to the outside. Any bubbles are displaced toward the axis so that unwanted voids will not be present in the casting.

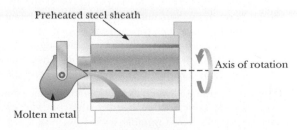

**Figure P7.62**

Suppose a copper sleeve of inner radius 2.10 cm and outer radius 2.20 cm is to be cast. To eliminate bubbles and give high structural integrity, the centripetal acceleration of each bit of metal should be 100$g$. What rate of rotation is required? State the answer in revolutions per minute.

63. **S** A skier starts at rest at the top of a large hemispherical hill (Fig. P7.63). Neglecting friction, show that the skier will leave the hill and become airborne at a distance $h = R/3$ below the top of the hill. *Hint:* At this point, the normal force goes to zero.

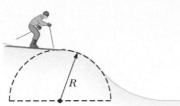

**Figure P7.63**

64. **V** A stuntman whose mass is 70 kg swings from the end of a 4.0-m-long rope along the arc of a vertical circle. Assuming he starts from rest when the rope is horizontal, find the tensions in the rope that are required to make him follow his circular path (a) at the beginning of his motion, (b) at a height of 1.5 m above the bottom of the circular arc, and (c) at the bottom of the arc.

65. Suppose a 1 800-kg car passes over a bump in a roadway that follows the arc of a circle of radius 20.4 m, as in Figure P7.65. (a) What force does the road exert on the car as the car passes the highest point of the bump if the car travels at 8.94 m/s? (b) What is the maximum speed the car can have without losing contact with the road as it passes this highest point?

**Figure P7.65**

66. **Q|C** The pilot of an airplane executes a constant-speed loop-the-loop maneuver in a vertical circle as in Figure 7.13b. The speed of the airplane is $2.00 \times 10^2$ m/s, and the radius of the circle is $3.20 \times 10^3$ m. (a) What is the pilot's apparent weight at the lowest point of the circle if his true weight is 712 N? (b) What is his apparent weight at the highest point of the circle? (c) Describe how the pilot could experience weightlessness if both the radius and the speed can be varied. *Note:* His apparent weight is equal to the magnitude of the force exerted by the seat on his body. Under what conditions does this occur? (d) What speed would have resulted in the pilot experiencing weightlessness at the top of the loop?

67. **Q|C** A minimum-energy orbit to an outer planet consists of putting a spacecraft on an elliptical trajectory with the departure planet corresponding to the perihelion of the ellipse, or closest point to the Sun, and the arrival planet corresponding to the aphelion of the ellipse, or farthest point from the Sun. (a) Use Kepler's third law to calculate how long it would take to go from Earth to Mars on such an orbit. (Answer in years.) (b) Can such an orbit be undertaken at any time? Explain.

68. **Q|C** A coin rests 15.0 cm from the center of a turntable. The coefficient of static friction between the coin and turntable surface is 0.350. The turntable starts from rest at $t = 0$ and rotates with a constant angular acceleration of 0.730 rad/s².

(a) Once the turntable starts to rotate, what force causes the centripetal acceleration when the coin is stationary relative to the turntable? Under what condition does the coin begin to move relative to the turntable? (b) After what period of time will the coin start to slip on the turntable?

69. A 4.00-kg object is attached to a vertical rod by two strings as shown in Figure P7.69. The object rotates in a horizontal circle at constant speed 6.00 m/s. Find the tension in (a) the upper string and (b) the lower string.

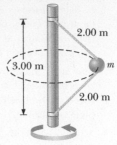

**Figure P7.69**

70. **Q|C** A 0.275-kg object is swung in a *vertical* circular path on a string 0.850 m long as in Figure P7.70. (a) What are the forces acting on the ball at any point along this path? (b) Draw free-body diagrams for the ball when it is at the bottom of the circle and when it is at the top. (c) If its speed is 5.20 m/s at the top of the circle, what is the tension in the string there? (d) If the string breaks when its tension exceeds 22.5 N, what is the maximum speed the object can have at the bottom before the string breaks?

**Figure P7.70**

71. (a) A luggage carousel at an airport has the form of a section of a large cone, steadily rotating about its vertical axis. Its metallic surface slopes downward toward the outside, making an angle of 20.0° with the horizontal. A 30.0-kg piece of luggage is placed on the carousel, 7.46 m from the axis of rotation. The travel bag goes around once in 38.0 s. Calculate the force of static friction between the bag and the carousel. (b) The drive motor is shifted to turn the carousel at a higher constant rate of rotation, and the piece of luggage is bumped to a position 7.94 m from the axis of rotation. The bag is on the verge of slipping as it goes around once every 34.0 s. Calculate the coefficient of static friction between the bag and the carousel.

72. **BIO** The maximum lift force on a bat is proportional to the square of its flying speed $v$. For the hoary bat (*Lasiurus cinereus*), the magnitude of the lift force is given by

$$F_L \leq (0.018 \text{ N} \cdot \text{s}^2/\text{m}^2)v^2$$

The bat can fly in a horizontal circle by "banking" its wings at an angle $\theta$, as shown in Figure P7.72. In this situation, the magnitude of the vertical component of the lift force must equal the bat's weight. The horizontal component of the

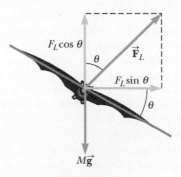

**Figure P7.72**

force provides the centripetal acceleration. (a) What is the minimum speed that the bat can have if its mass is 0.031 kg? (b) If the maximum speed of the bat is 10 m/s, what is the maximum banking angle that allows the bat to stay in a horizontal plane? (c) What is the radius of the circle of its flight when the bat flies at its maximum speed? (d) Can the bat turn with a smaller radius by flying more slowly?

73. In a popular amusement park ride, a rotating cylinder of radius 3.00 m is set in rotation at an angular speed of 5.00 rad/s, as in Figure P7.73. The floor then drops away, leaving the riders suspended against the wall in a vertical position. What minimum coefficient of friction between a rider's clothing and the wall is needed to keep the rider from slipping? *Hint:* Recall that the magnitude of the maximum force of static friction is equal to $\mu_s n$, where $n$ is the normal force—in this case, the force causing the centripetal acceleration.

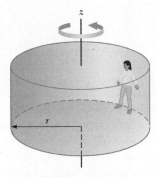

**Figure P7.73**

74. A massless spring of constant $k = 78.4$ N/m is fixed on the left side of a level track. A block of mass $m = 0.50$ kg is pressed against the spring and compresses it a distance $d$, as in Figure P7.74. The block (initially at rest) is then released and travels toward a circular loop-the-loop of radius $R = 1.5$ m. The entire track and the loop-the-loop are frictionless, except for the section of track between points $A$ and $B$. Given that the coefficient of kinetic friction between the block and the track along $AB$ is $\mu_k = 0.30$ and that the length of $AB$ is 2.5 m, determine the minimum compression $d$ of the spring that enables the block to just make it through the loop-the-loop at point $C$. *Hint:* The force exerted by the track on the block will be zero if the block barely makes it through the loop-the-loop.

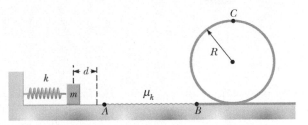

**Figure P7.74**

75. A 0.50-kg ball that is tied to the end of a 1.5-m light cord is revolved in a horizontal plane, with the cord making a 30° angle with the vertical. (See Fig. P7.75.) (a) Determine the ball's speed. (b) If, instead, the ball is revolved so that its speed is 4.0 m/s, what angle does the cord make with the vertical? (c) If the cord can withstand a maximum tension of 9.8 N, what is the highest speed at which the ball can move?

**Figure P7.75**

# Rotational Equilibrium and Dynamics

**IN THE STUDY OF LINEAR MOTION,** objects were treated as point particles without structure. It didn't matter *where* a force was applied, only *whether* it was applied or not.

The reality is that the point of application of a force *does* matter. In football, for example, if the ball carrier is tackled near his midriff, he might carry the tackler several yards before falling. If tackled well below the waistline, however, his center of mass rotates toward the ground, and he can be brought down immediately. Tennis provides another good example. If a tennis ball is struck with a strong horizontal force acting through its center of mass, it may travel a long distance before hitting the ground, far out-of-bounds. Instead, the same force applied in an upward, glancing stroke will impart topspin to the ball, which can cause it to land in the opponent's court.

The concepts of rotational equilibrium and rotational dynamics are also important in other disciplines. For example, students of architecture benefit from understanding the forces that act on buildings, and biology students should understand the forces at work in muscles and on bones and joints. These forces create torques, which tell us how the forces affect an object's equilibrium and rate of rotation.

We will find that an object remains in a state of uniform rotational motion unless acted on by a net torque. That principle is the equivalent of Newton's first law. Further, the angular acceleration of an object is proportional to the net torque acting on it, which is the analog of Newton's second law. A net torque acting on an object causes a change in its rotational energy.

Finally, torques applied to an object through a given time interval can change the object's angular momentum. In the absence of external torques, angular momentum is conserved, a property that explains some of the mysterious and formidable properties of pulsars, remnants of supernova explosions that rotate at equatorial speeds approaching that of light.

## 8.1 Torque

Forces cause accelerations; *torques* cause angular accelerations. There is a definite relationship, however, between the two concepts.

Figure 8.1 depicts an overhead view of a door hinged at point *O*. From this viewpoint, the door is free to rotate around an axis perpendicular to the page and passing through *O*. If a force $\vec{\mathbf{F}}$ is applied to the door, there are three factors that determine the effectiveness of the force in opening the door: the *magnitude* of the force, the *position* of application of the force, and the *angle* at which it is applied.

For simplicity, we restrict our discussion to position and force vectors lying in a plane. When the applied force $\vec{\mathbf{F}}$ is perpendicular to the outer edge of the door, as in Figure 8.1, the door rotates counterclockwise with constant angular acceleration. The same perpendicular force applied at a point nearer the hinge results in a smaller angular acceleration. In general, a larger radial distance *r* between the applied force and the axis of rotation results in a larger angular acceleration. Similarly, a larger applied force will also result in a larger angular acceleration. These considerations motivate the basic definition of **torque** for the special case of forces perpendicular to the position vector:

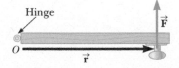

**Figure 8.1** A bird's-eye view of a door hinged at *O*, with a force applied perpendicular to the door.

Let $\vec{\mathbf{F}}$ be a force acting on an object, and let $\vec{\mathbf{r}}$ be a position vector from a chosen point $O$ to the point of application of the force, with $\vec{\mathbf{F}}$ perpendicular to $\vec{\mathbf{r}}$. The magnitude of the torque $\vec{\tau}$ exerted by the force $\vec{\mathbf{F}}$ is given by

$$\tau = rF \qquad [8.1]$$

where $r$ is the length of the position vector and $F$ is the magnitude of the force.

**SI unit: Newton-meter (N · m)**

◀ **Basic definition of torque**

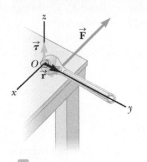

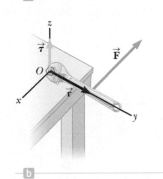

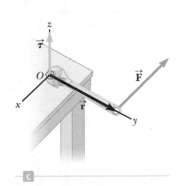

The vectors $\vec{\mathbf{r}}$ and $\vec{\mathbf{F}}$ lie in a plane. Figure 8.2 illustrates how the point of the force's application affects the magnitude of the torque. As discussed in detail shortly in conjunction with Figure 8.6, the torque $\vec{\tau}$ is then perpendicular to this plane. The point $O$ is usually chosen to coincide with the axis the object is rotating around, such as the hinge of a door or hub of a merry-go-round. (Other choices are possible as well.) In addition, we consider only forces acting in the plane perpendicular to the axis of rotation. This criterion excludes, for example, a force with upward component on a merry-go-round railing, which can't affect the merry-go-round's rotation.

Under these conditions, an object can rotate around the chosen axis in one of two directions. By convention, counterclockwise is taken to be the positive direction, clockwise the negative direction. When an applied force causes an object to rotate counterclockwise, the torque on the object is positive. When the force causes the object to rotate clockwise, the torque on the object is negative. When two or more torques act on an object at rest, the torques are added. If the net torque isn't zero, the object starts rotating at an ever-increasing rate. If the net torque is zero, the object's rate of rotation doesn't change. These considerations lead to the rotational analog of the first law: **the rate of rotation of an object doesn't change, unless the object is acted on by a net torque.**

**Figure 8.2** As the force is applied farther out along the wrench, the magnitude of the torque increases.

## EXAMPLE 8.1    BATTLE OF THE REVOLVING DOOR

**GOAL** Apply the basic definition of torque.

**PROBLEM** Two disgruntled businesspeople are trying to use a revolving door, which is initially at rest (see Fig. 8.3.) The woman on the left exerts a force of 625 N perpendicular to the door and 1.20 m from the hub's center, while the man on the right exerts a force of $8.50 \times 10^2$ N perpendicular to the door and 0.800 m from the hub's center. Find the net torque on the revolving door.

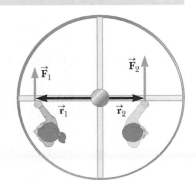

**Figure 8.3** (Example 8.1)

**STRATEGY** Calculate the individual torques on the door using the definition of torque, Equation 8.1, and then sum to find the net torque on the door. The woman exerts a negative torque, the man a positive torque. Their positions of application also differ.

**SOLUTION**

Calculate the torque exerted by the woman. A negative sign must be supplied because $\vec{\mathbf{F}}_1$, if unopposed, would cause a clockwise rotation:

$$\tau_1 = -r_1 F_1 = -(1.20 \text{ m})(625 \text{ N}) = -7.50 \times 10^2 \text{ N} \cdot \text{m}$$

*(Continued)*

Calculate the torque exerted by the man. The torque is positive because $\vec{\mathbf{F}}_2$, if unopposed, would cause a counterclockwise rotation:

$$\tau_2 = r_2 F_2 = (0.800 \text{ m})(8.50 \times 10^2 \text{ N}) = 6.80 \times 10^2 \text{ N} \cdot \text{m}$$

Sum the torques to find the net torque on the door:

$$\tau_{\text{net}} = \tau_1 + \tau_2 = \boxed{-7.0 \times 10^1 \text{ N} \cdot \text{m}}$$

..........................................................................................

**REMARKS** The negative result here means that the net torque will produce a clockwise rotation.

**QUESTION 8.1** Suppose the woman were pushing at a position 0.400 m closer to the hub. What direction would the revolving door rotate?

**EXERCISE 8.1** A businessman enters the same revolving door on the right, pushing with 576 N of force directed perpendicular to the door and 0.700 m from the hub, while a boy exerts a force of 365 N perpendicular to the door, 1.25 m to the left of the hub. Find **(a)** the torques exerted by each person and **(b)** the net torque on the door.

**ANSWERS** **(a)** $\tau_{\text{boy}} = -456 \text{ N} \cdot \text{m}$, $\tau_{\text{man}} = 403 \text{ N} \cdot \text{m}$   **(b)** $\tau_{\text{net}} = -53 \text{ N} \cdot \text{m}$

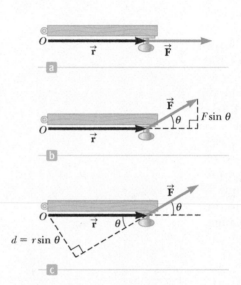

**Figure 8.4** (a) A force $\vec{\mathbf{F}}$ acting at an angle $\theta = 0°$ exerts zero torque about the pivot $O$. (b) The part of the force perpendicular to the door, $F \sin \theta$, exerts torque $rF \sin \theta$ about $O$. (c) An alternate interpretation of torque in terms of a *lever arm* $d = r \sin \theta$.

General definition of torque ▶

The applied force isn't always perpendicular to the position vector $\vec{\mathbf{r}}$. Suppose the force $\vec{\mathbf{F}}$ exerted on a door is directed away from the axis, as in Figure 8.4a, say, by someone's grasping the doorknob and pushing to the right. Exerting the force in this direction couldn't possibly open the door. However, if the applied force acts at an angle to the door as in Figure 8.4b, the component of the force *perpendicular* to the door will cause it to rotate. This figure shows that the component of the force perpendicular to the door is $F \sin \theta$, where $\theta$ is the angle between the position vector $\vec{\mathbf{r}}$ and the force $\vec{\mathbf{F}}$. When the force is directed away from the axis, $\theta = 0°$, $\sin (0°) = 0$, and $F \sin (0°) = 0$. When the force is directed toward the axis, $\theta = 180°$ and $F \sin (180°) = 0$. The maximum absolute value of $F \sin \theta$ is attained only when $\vec{\mathbf{F}}$ is perpendicular to $\vec{\mathbf{r}}$—that is, when $\theta = 90°$ or $\theta = 270°$. These considerations motivate a more general definition of torque:

Let $\vec{\mathbf{F}}$ be a force acting on an object, and let $\vec{\mathbf{r}}$ be a position vector from a chosen point $O$ to the point of application of the force. The magnitude of the torque $\vec{\boldsymbol{\tau}}$ exerted by the force $\vec{\mathbf{F}}$ is

$$\tau = rF \sin \theta \qquad [8.2]$$

where $r$ is the length of the position vector, $F$ the magnitude of the force, and $\theta$ the angle between $\vec{\mathbf{r}}$ and $\vec{\mathbf{F}}$.

**SI unit: Newton-meter (N · m)**

Again, the vectors $\vec{\mathbf{r}}$ and $\vec{\mathbf{F}}$ lie in a plane, and for our purposes, the chosen point $O$ will usually correspond to an axis of rotation perpendicular to the plane. Figure 8.5 illustrates how the magnitude of the torque exerted by a wrench increases as the angle between the position vector and the force vector increases at 90°, where the torque is a maximum.

A second way of understanding the $\sin \theta$ factor is to associate it with the magnitude $r$ of the position vector $\vec{\mathbf{r}}$. The quantity $d = r \sin \theta$ is called the **lever arm**, which is the perpendicular distance from the axis of rotation to a line drawn along the direction of the force. This alternate interpretation is illustrated in Figure 8.4c.

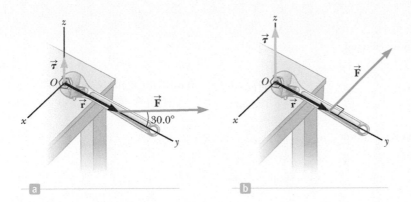

**Figure 8.5** As the angle between the position vector and force vector increases in parts (a)–(b), the torque exerted by the wrench increases.

It's important to remember that **the value of $\tau$ depends on the chosen axis of rotation.** Torques can be computed around any axis, regardless of whether there is some actual, physical rotation axis present. Once the point is chosen, however, it must be used consistently throughout a given problem.

Torque is a vector perpendicular to the plane determined by the position and force vectors, as illustrated in Figure 8.6. The direction can be determined by the *right-hand rule*:

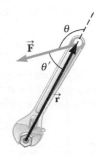

1. Point the fingers of your right hand in the direction of $\vec{r}$.
2. Curl your fingers toward the direction of vector $\vec{F}$.
3. Your thumb then points approximately in the direction of the torque, in this case out of the page.

**Figure 8.6** The right-hand rule: Point the fingers of your right hand along $\vec{r}$ and curl them in the direction of $\vec{F}$. Your thumb then points in the direction of the torque (out of the page, in this case). Note that either $\theta$ or $\theta'$ can be used in the definition of torque.

Notice the two choices of angle in Figure 8.6. The angle $\theta$ is the actual angle between the directions of the two vectors. The angle $\theta'$ is literally "between" the two vectors. Which angle is correct? Because $\sin\theta = \sin(180° - \theta) = \sin(180°)\cos\theta - \sin\theta\cos(180°) = 0 - \sin\theta \cdot (-1) = \sin\theta$, either angle is correct. Problems used in this book will be confined to objects rotating around an axis perpendicular to the plane containing $\vec{r}$ and $\vec{F}$, so if these vectors are in the plane of the page, the torque will always point either into or out of the page, parallel to the axis of rotation. If your right thumb is pointed in the direction of a torque, your fingers curl naturally in the direction of rotation that the torque would produce on an object at rest.

## EXAMPLE 8.2   THE SWINGING DOOR

**GOAL** Apply the more general definition of torque.

**PROBLEM** (a) A man applies a force of $F = 3.00 \times 10^2$ N at an angle of 60.0° to the door of Figure 8.7a, 2.00 m from well-oiled hinges. Find the torque on the door, choosing the position of the hinges as the axis of rotation. (b) Suppose a wedge is placed 1.50 m from the hinges on the other side of the door. What minimum force must the wedge exert so that the force applied in part (a) won't open the door?

**STRATEGY** Part (a) can be solved by substitution into the general torque equation. In part (b) the hinges, the wedge, and the applied force all exert torques on the door. The door doesn't open, so the sum of these torques must be zero, a condition that can be used to find the wedge force.

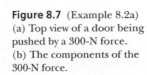

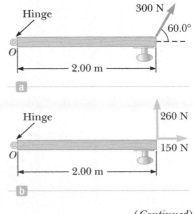

**Figure 8.7** (Example 8.2a) (a) Top view of a door being pushed by a 300-N force. (b) The components of the 300-N force.

*(Continued)*

## SOLUTION

(a) Compute the torque due to the applied force exerted at 60.0°.

Substitute into the general torque equation:

$$\tau_F = rF\sin\theta = (2.00\text{ m})(3.00 \times 10^2\text{ N})\sin 60.0°$$
$$= (2.00\text{ m})(2.60 \times 10^2\text{ N}) = \boxed{5.20 \times 10^2\text{ N}\cdot\text{m}}$$

(b) Calculate the force exerted by the wedge on the other side of the door.

Set the sum of the torques equal to zero:

$$\tau_{\text{hinge}} + \tau_{\text{wedge}} + \tau_F = 0$$

The hinge force provides no torque because it acts at the axis ($r = 0$). The wedge force acts at an angle of $-90.0°$, opposite the upward 260 N component.

$$0 + F_{\text{wedge}}(1.50\text{ m})\sin(-90.0°) + 5.20 \times 10^2\text{ N}\cdot\text{m} = 0$$
$$F_{\text{wedge}} = \boxed{347\text{ N}}$$

**REMARKS** Notice that the angle from the position vector to the wedge force is $-90°$. That's because, starting at the position vector, it's necessary to go 90° clockwise (the negative angular direction) to get to the force vector. Measuring the angle that way automatically supplies the correct sign for the torque term and is consistent with the right-hand rule. Alternately, the magnitude of the torque can be found and the correct sign chosen based on physical intuition. Figure 8.7b illustrates the fact that the component of the force perpendicular to the lever arm causes the torque.

**QUESTION 8.2** To make the wedge more effective in keeping the door closed, should it be placed closer to the hinge or to the doorknob?

**EXERCISE 8.2** A man ties one end of a strong rope 8.00 m long to the bumper of his truck, 0.500 m from the ground, and the other end to a vertical tree trunk at a height of 3.00 m. He uses the truck to create a tension of $8.00 \times 10^2$ N in the rope. Compute the magnitude of the torque on the tree due to the tension in the rope, with the base of the tree acting as the reference point.

**ANSWER** $2.28 \times 10^3$ N $\cdot$ m

# 8.2 Center of Mass and Its Motion

## 8.2.1 Center of Mass

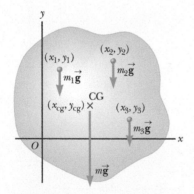

**Figure 8.8** The net gravitational torque on an object is zero if computed around the center of gravity. The object will balance if supported at that point (or at any point along a vertical line above or below that point).

Consider an object of arbitrary shape lying in the $xy$-plane, as in Figure 8.8. The object is divided into a large number of very small particles of weight $m_1g$, $m_2g$, $m_3g$, ... having coordinates $(x_1, y_1)$, $(x_2, y_2)$, $(x_3, y_3)$, .... If the object is free to rotate around the origin, each particle contributes a torque about the origin that is equal to its weight multiplied by its lever arm. For example, the torque due to the weight $m_1g$ is $m_1gx_1$, and so forth.

We wish to locate the point of application of the single force of magnitude $w = F_g = Mg$ (the total weight of the object), where the effect on the rotation of the object is the same as that of the individual particles. That point is called the object's **center of gravity**. Equating the torque exerted by $w$ at the center of gravity to the sum of the torques acting on the individual particles gives

$$(m_1g + m_2g + m_3g + \cdots)x_{\text{cg}} = m_1gx_1 + m_2gx_2 + m_3gx_3 + \cdots$$

We assume that $g$ is the same everywhere in the object (which is true for all objects we will encounter). Then the $g$ factors in the preceding equation cancel, resulting in

$$x_{\text{cg}} = \frac{m_1x_1 + m_2x_2 + m_3x_3 + \cdots}{m_1 + m_2 + m_3 + \cdots} = \frac{\sum m_i x_i}{\sum m_i} \qquad [8.3a]$$

where $x_{\text{cg}}$ is the $x$-coordinate of the center of gravity. Similarly, the $y$-coordinate and $z$-coordinate of the center of gravity of the system can be found from

$$y_{\text{cg}} = \frac{\sum m_i y_i}{\sum m_i} \qquad [8.3b]$$

and

$$z_{cg} = \frac{\sum m_i z_i}{\sum m_i}$$

[8.3c]

These three equations are identical to the equations for a similar concept called **center of mass.** The center of mass and center of gravity of an object are exactly the same when $g$ doesn't vary significantly over the object. In this topic, the concepts of center of gravity and center of mass will be used interchangeably.

It's often possible to guess the location of the center of mass. **The center of mass of a homogeneous, symmetric body must lie on the axis of symmetry.** For example, the center of mass of a homogeneous rod lies midway between the ends of the rod, and the center of mass of a homogeneous sphere or a homogeneous cube lies at the geometric center of the object. The center of mass of an irregularly shaped object, such as a wrench, can be determined experimentally by suspending the wrench from two different arbitrary points (Fig. 8.9). The wrench is first hung from point $A$, and a vertical line $AB$ (which can be established with a plumb bob) is drawn when the wrench is in equilibrium. The wrench is then hung from point $C$, and a second vertical line $CD$ is drawn. The center of mass coincides with the intersection of these two lines. In fact, if the wrench is hung freely from any point, the center of mass always lies straight below the point of support, so the vertical line through that point must pass through the center of mass.

Several examples in Section 8.3 involve homogeneous, symmetric objects where the centers of mass coincide with their geometric centers. A rigid object in a uniform gravitational field can be balanced by a single force equal in magnitude to the weight of the object, as long as the force is directed upward through the object's center of mass.

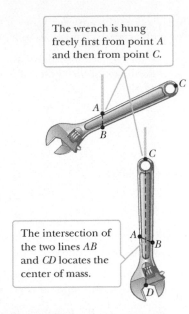

The wrench is hung freely first from point $A$ and then from point $C$.

The intersection of the two lines $AB$ and $CD$ locates the center of mass.

**Figure 8.9** An experimental technique for determining the center of mass of a wrench.

## EXAMPLE 8.3 WHERE IS THE CENTER OF MASS?

**GOAL** Find the center of mass of a system of objects.

**PROBLEM** (a) Three objects are located in a coordinate system as shown in Figure 8.10a. Find the center of mass. (b) How does the answer change if the object on the left is displaced upward by 1.00 m and the object on the right is displaced downward by 0.500 m (Fig. 8.10b)? Treat the objects as point particles.

**STRATEGY** The $y$-coordinate and $z$-coordinate of the center of mass in part (a) are both zero because all the objects are on the $x$-axis. We can find the $x$-coordinate of the center of mass using Equation 8.3a. Part (b) requires Equation 8.3b.

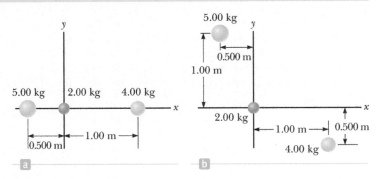

**Figure 8.10** (Example 8.3) Locating the center of mass of a system of three particles.

## SOLUTION

**(a)** Find the center of mass of the system in Figure 8.10a.

Apply Equation 8.3a to the system of three objects:

(1) $x_{cg} = \dfrac{\sum m_i x_i}{\sum m_i} = \dfrac{m_1 x_1 + m_2 x_2 + m_3 x_3}{m_1 + m_2 + m_3}$

Compute the numerator of Equation (1):

$$\sum m_i x_i = m_1 x_1 + m_2 x_2 + m_3 x_3$$
$$= (5.00 \text{ kg})(-0.500 \text{ m}) + (2.00 \text{ kg})(0 \text{ m}) + (4.00 \text{ kg})(1.00 \text{ m})$$
$$= 1.50 \text{ kg} \cdot \text{m}$$

Substitute the denominator, $\sum m_i = 11.0$ kg, and the numerator into Equation (1).

$$x_{cg} = \frac{1.50 \text{ kg} \cdot \text{m}}{11.0 \text{ kg}} = \boxed{0.136 \text{ m}}$$

*(Continued)*

**(b)** How does the answer change if the positions of the objects are changed as in Figure 8.10b?

Because the x-coordinates have not been changed, the x-coordinate of the center of mass is also unchanged:

$$x_{cg} = \boxed{0.136 \text{ m}}$$

Write Equation 8.3b:

$$y_{cg} = \frac{\sum m_i y_i}{\sum m_i} = \frac{m_1 y_1 + m_2 y_2 + m_3 y_3}{m_1 + m_2 + m_3}$$

Substitute values:

$$y_{cg} = \frac{(5.00 \text{ kg})(1.00 \text{ m}) + (2.00 \text{ kg})(0 \text{ m}) + (4.00 \text{ kg})(-0.500 \text{ m})}{5.00 \text{ kg} + 2.00 \text{ kg} + 4.00 \text{ kg}}$$

$$y_{cg} = \boxed{0.273 \text{ m}}$$

**REMARKS** Notice that translating objects in the y-direction doesn't change the x-coordinate of the center of mass. The three components of the center of mass are each independent of the other two coordinates.

**QUESTION 8.3** If 1.00 kg is added to the masses on the left and right in Figure 8.10a, does the center of mass **(a)** move to the left, **(b)** move to the right, or **(c)** remain in the same position?

**EXERCISE 8.3** If a fourth particle of mass 2.00 kg is placed at (0, 0.25 m) in Figure 8.10a, find the x- and y-coordinates of the center of mass for this system of four particles.

**ANSWER** $x_{cg} = 0.115$ m; $y_{cg} = 0.038\ 5$ m

---

**EXAMPLE 8.4   LOCATING YOUR LAB PARTNER'S CENTER OF GRAVITY   BIO**

**GOAL** Use torque to find a center of gravity.

**PROBLEM** In this example we show how to find the location of a person's center of gravity. Suppose your lab partner has a height $L$ of 173 cm (5 ft, 8 in.) and a weight $w$ of 715 N (160 lb). You can determine the position of his center of gravity by having him stretch out on a uniform board supported at one end by a scale, as shown in Figure 8.11. If the board's weight $w_b$ is 49 N and the scale reading $F$ is $3.50 \times 10^2$ N, find the distance of your lab partner's center of gravity from the left end of the board.

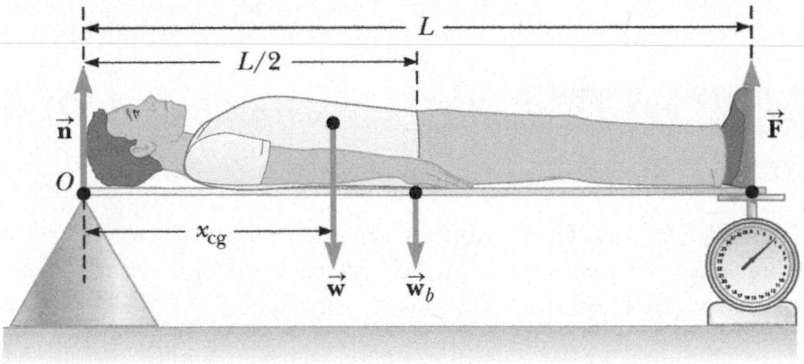

**STRATEGY** To find the position $x_{cg}$ of the center of gravity, compute the torques using an axis through $O$. There is no torque due to the normal force $\vec{\mathbf{n}}$ because its moment arm is zero about an axis through $O$. Because your lab partner is laying at rest on a stationary board, the sum of the torques must equal zero. Use this condition to solve for $x_{cg}$.

**Figure 8.11** (Example 8.4) Determining your lab partner's center of gravity.

**SOLUTION**

Set the sum of the torques equal to zero:

$$\sum \tau_i = \tau_n + \tau_w + \tau_{w_b} + \tau_F = 0$$

Substitute expressions for the torques:

$$0 - wx_{cg} - w_b(L/2) + FL = 0$$

Solve for $x_{cg}$ and substitute known values:

$$x_{cg} = \frac{FL - w_b(L/2)}{w}$$

$$= \frac{(350 \text{ N})(173 \text{ cm}) - (49 \text{ N})(86.5 \text{ cm})}{715 \text{ N}} = \boxed{79 \text{ cm}}$$

**REMARKS** The given information is sufficient only to determine the x-coordinate of the center of gravity. The other two coordinates can be estimated, based on the body's symmetry.

**QUESTION 8.4** What would happen if a support is placed exactly at $x = 79$ cm followed by the removal of the supports at the subject's head and feet?

**EXERCISE 8.4** Suppose a 416-kg alligator of length 3.5 m is stretched out on a board of the same length weighing 65 N. If the board is supported on the ends as in Figure 8.11 and the scale reads 1 880 N, find the *x*-component of the alligator's center of gravity.

**ANSWER** 1.59 m

## EXAMPLE 8.5    TUG OF WAR AT THE ICE FISHING HOLE

**GOAL** Use center of mass concepts to determine the motion of an isolated two-body system.

**PROBLEM** Bob and Sherry are lying on the ice, a fishing hole of radius 1.00 m cut in the ice halfway between them. A rope of length 10.0 m lies between them, and they both grip it and begin pulling, as in Figure 8.12a. Bob has mass of $m_B = 85.0$ kg and Sherry has mass of $m_S = 48.0$ kg, so Sherry reaches the hole, first. Where is Bob at that time? Assume the hole is centered on the origin and that Bob and Sherry start at $x_B = 5.00$ m and $x_S = -5.00$ m, respectively. Neglect forces of friction.

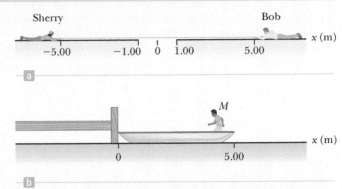

**Figure 8.12** (Example 8.5) (a) Bob and Sherry engage in a tug-of-war. (b) (Exercise 8.5)

**STRATEGY** The system consists of Bob and Sherry, and because there are no external forces acting horizontally and vertical forces sum to zero, the acceleration of the center of mass is zero. As they pull on the rope, they both get closer to the fishing hole; however, the center of mass doesn't change position. First, calculate the center of mass when Bob and Sherry are in their initial positions. Sherry reaches the hole first, by momentum conservation, because she weighs less. Solve Equation 8.3a for Bob's position at that time.

### SOLUTION

Calculate the center of mass using the initial positions:

$$x_{cm} = \frac{m_S x_S + m_B x_B}{m_S + m_B} = \frac{(48.0 \text{ kg})(-5.00 \text{ m}) + (85.0 \text{ kg})(5.00 \text{ m})}{48.0 \text{ kg} + 85.0 \text{ kg}} = 1.39 \text{ m}$$

Find Bob's position using the center of mass equation and the fact that Sherry reaches the hole first:

$$x_{cm} = 1.39 \text{ m} = \frac{m_S x_S + m_B x_B}{m_S + m_B} = \frac{(48.0 \text{ kg})(-1.00 \text{ m}) + (85.0 \text{ kg})x_B}{133 \text{ kg}}$$

Solve the expression for $x_B$:

$$x_B = 2.74 \text{ m}$$

**REMARKS** So when Sherry reaches the edge of the fishing hole, Bob is still nearly two meters away from the edge on his side of the hole.

**QUESTION 8.5** How do the speeds $v_S$ of Sherry and $v_B$ of Bob compare during the motion?

**EXERCISE 8.5** A man of mass $M = 75.0$ kg is standing in a canoe of mass 40.0 kg that is 5.00 m long, as in Figure 8.12b. The far end of the canoe is next to a dock. From a position 0.500 m from his end of the canoe, he walks to the same position at the other end of the canoe. (a) Find the center of mass of the canoe–man system, taking the end of the dock as the origin. (b) Neglecting drag forces, how far is he from the dock? (*Hint:* the final location of the canoe's center of mass will be 2.00 m farther from the dock than the man's final position, which is unknown.)

**ANSWERS** (a) 3.80 m  (b) 3.10 m

The center of mass (or of gravity) of an extended object can lie outside the object, as the next example shows.

## EXAMPLE 8.6    SYSTEM OF RODS

**GOAL** Find the center of mass of a continuous system of particles.

**PROBLEM** A rod system with a uniform linear density of 5.00 kg/m is cast into an irregular, upside-down T-shape as in Figure 8.13a, with rod 1 the horizontal rod and rod 2 the vertical rod. Neglect the diameter of the rods. (a) Find the

*(Continued)*

mass of each rod. **(b)** Determine the center of mass of each rod. **(c)** Calculate the center of mass of the system in the $xy$-plane.

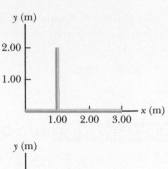

**STRATEGY** For part **(a)**, the mass of each rod is given by $m = \lambda L$, where $\lambda$ is the mass per unit length and $L$ is the length. The center of mass of each of the rods, part **(b)**, can be found by symmetry: it's in the middle of the rod. In this case, inspection suffices, but otherwise the midpoint formula could be applied. Finally, the center of mass of the system is the same as if the mass of each rod were concentrated at that rod's center of mass, in which case Equations 8.3a and 8.3b can be applied.

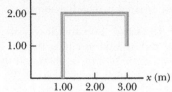

**Figure 8.13** (Example 8.6)
(a) A system of uniform rods.
(b) (Exercise 8.5)

**SOLUTION**

**(a)** Find the mass of the each rod.

Use the linear density and substitute, finding the mass of the horizontal rod, $m_1$:

$$m_1 = \lambda L_1 = (5.00 \text{ kg/m})(3.00 \text{ m}) = \boxed{15.0 \text{ kg}}$$

Do the same for the mass of the vertical rod, $m_2$:

$$m_2 = \lambda L_2 = (5.00 \text{ kg/m})(2.00 \text{ m}) = \boxed{10.0 \text{ kg}}$$

**(b)** Find the center of mass of each rod.

Because each rod has uniform density, the center of mass is in the center of each rod:

CM, horizontal rod: $(x_1, y_1) = \boxed{(1.50, 0) \text{ m}}$

CM, vertical rod: $(x_2, y_2) = \boxed{(1.00, 1.00) \text{ m}}$

**(c)** Find the center of mass of the system in the $xy$-plane.

Apply Equations 8.3a and 8.3b:

$$x_{cm} = \frac{m_1 x_1 + m_2 x_2}{m_1 + m_2} = \frac{(15.0 \text{ kg})(1.50 \text{ m}) + (10.0 \text{ kg})(1.00 \text{ m})}{15.0 \text{ kg} + 10.0 \text{ kg}} = \boxed{1.30 \text{ m}}$$

$$y_{cm} = \frac{m_1 y_1 + m_2 y_2}{m_1 + m_2} = \frac{(15.0 \text{ kg})(0 \text{ m}) + (10.0 \text{ kg})(1.00 \text{ m})}{15.0 \text{ kg} + 10.0 \text{ kg}} = \boxed{0.400 \text{ m}}$$

**REMARKS** The center of mass is outside the system at the position calculated. Note that knowledge of the linear density was not really required because it essentially occurs in both the numerator and denominator, and would cancel out. See the exercise.

**QUESTION 8.6** Suppose the vertical rod were replaced with one having a greater linear density than the horizontal rod. Would the $x$-component of the system's center of mass move to the left, to the right, or remain unchanged? Would the $y$-component move up, down, or remain unchanged?

**EXERCISE 8.6** Assume a system of linear rods takes the form of Figure 8.13b. Find the coordinates of the center of mass, assuming the rods have uniform linear density.

**ANSWER** $(1.80, 1.50)$ m

## 8.2.2 Center of Mass Motion

Newton's second law can be applied to a system. The total mass $M_{tot}$ times the acceleration of the center of mass $\vec{a}_{cm}$ equals the sum of the forces:

$$M_{tot}\vec{a}_{cm} = \sum \vec{F}_i = \sum \vec{F}_{ext} + \sum \vec{F}_{int} \qquad [8.4]$$

where $\vec{F}_{ext}$ are external forces acting on the system and $\vec{F}_{int}$ are internal forces acting on the system. All internal forces come as action–reaction pairs, however, and because the vectors in those pairs have equal magnitudes and opposite directions, they cancel out. Therefore, the second law for the center of mass may be written without internal forces as

$$M_{tot}\vec{a}_{cm} = \sum \vec{F}_{ext} \qquad [8.5]$$

A good application of center of mass motion includes simple ballistics. In both one dimension and in two dimensions, those equations are exactly the same as for single particles. For motion along a line, for example, the kinematics equations for motion are

$$x_{cm} = \frac{1}{2}a_{cm}t^2 + v_{0,cm}t + x_{0,cm} \qquad [8.6]$$

$$v_{cm} = a_{cm}t + v_{0,cm} \qquad [8.7]$$

and

$$v_{cm}^2 - v_{0,cm}^2 = 2a_{cm}\Delta x_{cm} \qquad [8.8]$$

The two dimensional equations can be written similarly.

## EXAMPLE 8.7    EXPLODING ROCKET

**GOAL**  Apply center of mass motion in a gravity field.

**PROBLEM**  A skyrocket of mass 3.00 kg is at an elevation of 30.0 m traveling 20.0 m/s straight upward when it explodes into two pieces, with mass $m_1 = 1.00$ kg and mass $m_2 = 2.00$ kg, respectively. Mass $m_1$ goes straight down, hitting the ground 1.00 s later. At what height is mass $m_2$ when $m_1$ hits the ground? Take the x-direction to be vertically upward.

**STRATEGY**  This problem looks difficult but in fact can be solved in one line with center of mass motion! All that need be done is write down the center of mass position equation for kinematics, Equation 8.6, and equate it to the definition of the center of mass, Equation 8.3a.

**SOLUTION**

Equate the right sides of Equations 8.3a and Equation 8.6:

$$x_{cm} = \frac{m_1 x_1 + m_2 x_2}{m_1 + m_2} = \frac{1}{2}a_{cm}t^2 + v_{0,cm}t + x_{0,cm}$$

Substitute values for $m_1$, $m_2$, $a_{cm}$, $v_{0,cm}$, $x_{0,cm}$, disregarding significant figures and units for clarity:

$$\frac{x_1 + 2x_2}{3} = -4.9t^2 + 20t + 30 \qquad (1)$$

Now set $x_1 = 0$ and $t = 1.00$ s and solve for $x_2$:

$$\frac{2x_2}{3} = 45.1 \quad \rightarrow \quad x_2 = \boxed{67.7\,m}$$

**REMARKS**  Note that Equation (1) is valid only until mass $m_1$ strikes the ground, because after that time $m_1$ is no longer in free fall.

**QUESTION 8.7**  Suppose that mass $m_1$ hit the ground in less than 1.00 s. Would the answer for $x_2$ be larger, smaller, or the same?

**EXERCISE 8.7**  From the top of a building 75.0 m tall, a red ball of mass 1.00 kg is shot upward at 10.0 m/s. A blue ball of mass 3.00 kg is dropped at the same instant. Find (a) the initial velocity of the center of mass of the system and (b) the position of the system's center of mass after 2.00 s.

**ANSWERS**  (a) 2.50 m/s  (b) 60.4 m

# 8.3 Torque and the Two Conditions for Equilibrium

An object in mechanical equilibrium (Fig. 8.14) must satisfy the following two conditions:

1. **The net external force must be zero:** $\sum \vec{F} = 0$

2. **The net external torque must be zero:** $\sum \vec{\tau} = 0$

The first condition is a statement of translational equilibrium: The sum of all forces acting on the object must be zero, so the object has no translational acceleration, $\vec{a} = 0$. The second condition is a statement of rotational equilibrium: The sum of all torques on the object must be zero, so the object has no angular acceleration, $\vec{\alpha} = 0$. For an object to be in equilibrium, it must move through space at a constant speed and rotate at a constant angular speed.

Because we can choose any location for calculating torques, it's usually best to select an axis that will make at least one torque equal to zero, just to simplify the net torque equation. The following general procedure is recommended for solving problems that involve objects in equilibrium.

**Figure 8.14** This large balanced rock at the Garden of the Gods in Colorado Springs, Colorado, is in mechanical equilibrium.

**Tip 8.1 Specify Your Axis**

Choose the axis of rotation and use that axis exclusively throughout a given problem. The axis need not correspond to a physical axle or pivot point. Any convenient point will do.

**Tip 8.2** Rotary Motion Under Zero Torque

If a net torque of zero is exerted on an object, it will continue to rotate at a constant angular speed—which need not be zero. However, zero torque *does* imply that the angular acceleration is zero.

### PROBLEM-SOLVING STRATEGY

**Objects in Equilibrium**

1. **Diagram the system.** Include coordinates and choose a convenient rotation axis for computing the net torque on the object.
2. **Draw a force diagram** of the object of interest, showing all external forces acting on it. For systems with more than one object, draw a *separate* diagram for each object. (Most problems will have a single object of interest.)
3. **Apply $\sum \tau_i = 0$, the second condition of equilibrium.** This condition yields a single equation for each object of interest. If the axis of rotation has been carefully chosen, the equation often has only one unknown and can be solved immediately.
4. **Apply $\sum F_x = 0$ and $\sum F_y = 0$, the first condition of equilibrium.** This yields two more equations per object of interest.
5. **Solve the system of equations.** For each object, the two conditions of equilibrium yield three equations, usually with three unknowns. Solve by substitution.

### EXAMPLE 8.8   BALANCING ACT

**GOAL** Apply the conditions of equilibrium and illustrate the use of different axes for calculating the net torque on an object.

**PROBLEM** A woman of mass $m = 55.0$ kg sits on the left end of a seesaw—a plank of length $L = 4.00$ m, pivoted in the middle as in Figure 8.15. **(a)** First compute the torques on the seesaw about an axis that passes through the pivot point. Where should a man of mass $M = 75.0$ kg sit if the system (seesaw plus man and woman) is to be balanced? **(b)** Find the normal force exerted by the pivot if the plank has a mass of $m_{pl} = 12.0$ kg. **(c)** Repeat part **(a)**, but this time compute the torques about an axis through the left end of the plank.

**STRATEGY** In part **(a)**, apply the second condition of equilibrium, $\Sigma \tau = 0$, computing torques around the pivot point. The mass of the plank forming the seesaw is distributed evenly on either side of the pivot point, so the torque exerted by gravity on the plank, $\tau_{plank}$, can be computed as if all the plank's mass is concentrated at the pivot point. Then $\tau_{plank}$ is zero, as is the torque

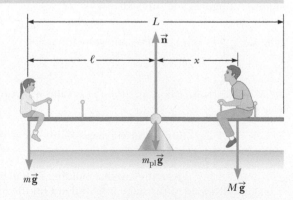

**Figure 8.15** (Example 8.8) The system consists of two people and a seesaw. Because the sum of the forces and the sum of the torques acting on the system are both zero, the system is said to be in equilibrium.

exerted by the pivot, because their lever arms are zero. In part **(b)** the first condition of equilibrium, $\sum \vec{F} = 0$, must be applied. Part **(c)** is a repeat of part **(a)** showing that choice of a different axis yields the same answer.

## SOLUTION

**(a)** Where should the man sit to balance the seesaw?

Apply the second condition of equilibrium to the plank by setting the sum of the torques equal to zero:

$$\tau_{\text{pivot}} + \tau_{\text{plank}} + \tau_{\text{man}} + \tau_{\text{woman}} = 0$$

The first two torques are zero. Let $x$ represent the man's distance from the pivot. The woman is at a distance $\ell = L/2$ from the pivot.

$$0 + 0 - Mgx + mg(L/2) = 0$$

Solve this equation for $x$ and evaluate it:

$$x = \frac{m(L/2)}{M} = \frac{(55.0\ \text{kg})(2.00\ \text{m})}{75.0\ \text{kg}} = \boxed{1.47\ \text{m}}$$

**(b)** Find the normal force $n$ exerted by the pivot on the seesaw.

Apply for first condition of equilibrium to the plank, solving the resulting equation for the unknown normal force, $n$:

$$-Mg - mg - m_{\text{pl}}g + n = 0$$
$$n = (M + m + m_{\text{pl}})g$$
$$= (75.0\ \text{kg} + 55.0\ \text{kg} + 12.0\ \text{kg})(9.80\ \text{m/s}^2)$$
$$n = \boxed{1.39 \times 10^3\ \text{N}}$$

**(c)** Repeat part **(a)**, choosing a new axis through the left end of the plank.

Compute the torques using this axis, and set their sum equal to zero. Now the pivot and gravity forces on the plank result in nonzero torques.

$$\tau_{\text{man}} + \tau_{\text{woman}} + \tau_{\text{plank}} + \tau_{\text{pivot}} = 0$$
$$-Mg(L/2 + x) + mg(0) - m_{\text{pl}}g(L/2) + n(L/2) = 0$$

Substitute all known quantities:

$$-(75.0\ \text{kg})(9.80\ \text{m/s}^2)(2.00\ \text{m} + x) + 0$$
$$- (12.0\ \text{kg})(9.80\ \text{m/s}^2)(2.00\ \text{m}) + n(2.00\ \text{m}) = 0$$
$$-(1.47 \times 10^3\ \text{N} \cdot \text{m}) - (735\ \text{N})x - (235\ \text{N} \cdot \text{m})$$
$$+ (2.00\ \text{m})n = 0$$

Solve for $x$, substituting the normal force found in part **(b)**:

$$x = \boxed{1.46\ \text{m}}$$

**REMARKS** The answers for $x$ in parts (a) and (c) agree except for a small rounding discrepancy. That illustrates how choosing a different axis leads to the same solution.

**QUESTION 8.8** What happens if the woman now leans backwards?

**EXERCISE 8.8** Suppose a 30.0-kg child sits 1.50 m to the left of center on the same seesaw. A second child sits at the end on the opposite side, and the system is balanced. **(a)** Find the mass of the second child. **(b)** Find the normal force acting at the pivot point.

**ANSWERS** **(a)** 22.5 kg    **(b)** 632 N

---

## EXAMPLE 8.9    A WEIGHTED FOREARM    BIO

**GOAL** Apply the equilibrium conditions to the human body.

**PROBLEM** A 50.0-N (11-lb) bowling ball is held in a person's hand with the forearm horizontal, as in Figure 8.16a. The biceps muscle is attached 0.030 0 m from the joint, and the ball is 0.350 m from the joint. Find the upward force $\vec{F}$ exerted

(Continued)

by the biceps on the forearm (the ulna) and the downward force $\vec{\mathbf{R}}$ exerted by the humerus on the forearm, acting at the joint. Neglect the weight of the forearm and slight deviation from the vertical of the biceps.

**STRATEGY** The forces acting on the forearm are equivalent to those acting on a bar of length 0.350 m, as shown in Figure 8.16b. Choose the usual x- and y-coordinates as shown and the axis at O on the left end. (This completes Steps 1 and 2.) Use the conditions of equilibrium to generate equations for the unknowns, and solve.

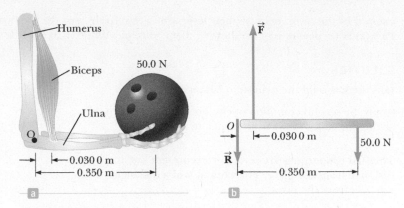

**Figure 8.16** (Example 8.9) (a) A weight held with the forearm horizontal. (b) The mechanical model for the system.

**SOLUTION**

Apply the second condition for equilibrium (Step 3) and solve for the upward force F:

$$\sum \tau_i = \tau_R + \tau_F + \tau_{BB} = 0$$
$$R(0) + F(0.030\ 0\ \text{m}) - (50.0\ \text{N})(0.350\ \text{m}) = 0$$
$$F = \boxed{583\ \text{N}\ (131\ \text{lb})}$$

Apply the first condition for equilibrium (Step 4) and solve (Step 5) for the downward force R:

$$\sum F_y = F - R - 50.0\ \text{N} = 0$$
$$R = F - 50.0\ \text{N} = 583\ \text{N} - 50\ \text{N} = \boxed{533\ \text{N}\ (120\ \text{lb})}$$

**REMARKS** The magnitude of the force supplied by the biceps must be about ten times as large as the bowling ball it is supporting!

**QUESTION 8.9** Suppose the biceps were surgically reattached three centimeters farther toward the person's hand. If the same bowling ball were again held in the person's hand, how would the force required of the biceps be affected? Explain.

**EXERCISE 8.9** Suppose you wanted to limit the force acting on your joint to a maximum value of $8.00 \times 10^2$ N. **(a)** Under these circumstances, what maximum weight would you attempt to lift? **(b)** What force would your biceps apply while lifting this weight?

**ANSWERS** **(a)** 75.0 N **(b)** 875 N

---

## EXAMPLE 8.10 DON'T CLIMB THE LADDER

**GOAL** Apply the two conditions of equilibrium.

**PROBLEM** A uniform ladder 10.0 m long and weighing 50.0 N rests against a frictionless vertical wall as in Figure 8.17a. If the ladder is just on the verge of slipping when it makes a 50.0° angle with the ground, find the coefficient of static friction between the ladder and ground.

**STRATEGY** Figure 8.17b is the force diagram for the ladder. The first condition of equilibrium, $\Sigma \vec{\mathbf{F}}_i = 0$, gives two equations for three unknowns: the magnitudes of the static friction force f and the normal force n, both acting on the base of the ladder, and the magnitude of the force of the wall, P, acting on the top of the ladder. The second condition of equilibrium, $\Sigma \tau_i = 0$, gives a third equation (for P), so all three quantities can be found. The definition of static friction then allows computation of the coefficient of static friction.

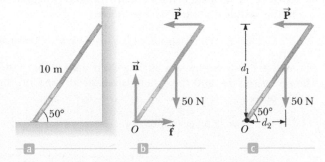

**Figure 8.17** (Example 8.10) (a) A ladder leaning against a frictionless wall. (b) A force diagram of the ladder. (c) Lever arms for the force of gravity and $\vec{\mathbf{P}}$.

**SOLUTION**

Apply the first condition of equilibrium to the ladder:

**(1)** $\Sigma F_x = f - P = 0 \rightarrow f = P$

**(2)** $\Sigma F_y = n - 50.0\ \text{N} = 0 \rightarrow n = 50.0\ \text{N}$

Apply the second condition of equilibrium, computing torques around the base of the ladder, with $\tau_{grav}$ standing for the torque due to the ladder's 50.0-N weight:

$$\Sigma\tau_i = \tau_f + \tau_n + \tau_{grav} + \tau_P = 0$$

The torques due to friction and the normal force are zero about $O$ because their moment arms are zero. (Moment arms can be found from Fig. 8.17c.)

$$0 + 0 - (50.0\ \text{N})(5.00\ \text{m})\sin 40.0° + P(10.0\ \text{m})\sin 50.0° = 0$$
$$P = 21.0\ \text{N}$$

From Equation (1), we now have $f = P = 21.0$ N. The ladder is on the verge of slipping, so write an expression for the maximum force of static friction and solve for $\mu_s$:

$$21.0\ \text{N} = f = f_{s,max} = \mu_s n = \mu_s(50.0\ \text{N})$$

$$\mu_s = \frac{21.0\ \text{N}}{50.0\ \text{N}} = \boxed{0.420}$$

. . . . . . . . . . . . . . . . . . . . . . . . . . . . . . . . . . . . . . . . . . . . . . . . . . . . . . . . . . . . . . . . . . . . . . . . . . . . . .

**REMARKS** Note that torques were computed around an axis through the bottom of the ladder so that only $\vec{P}$ and the force of gravity contributed nonzero torques. This choice of axis reduces the complexity of the torque equation, often resulting in an equation with only one unknown.

**QUESTION 8.10** If a 50.0-N monkey hangs from the middle rung, would the coefficient of static friction be **(a)** doubled, **(b)** halved, or **(c)** unchanged?

**EXERCISE 8.10** If the coefficient of static friction is 0.360, and the same ladder makes a 60.0° angle with respect to the horizontal, how far along the length of the ladder can a 70.0-kg painter climb before the ladder begins to slip?

**ANSWER** 6.33 m

---

## EXAMPLE 8.11    WALKING A HORIZONTAL BEAM

**GOAL** Apply the two conditions of equilibrium.

**PROBLEM** A uniform horizontal beam 5.00 m long and weighing $3.00 \times 10^2$ N is attached to a wall by a pin connection that allows the beam to rotate. Its far end is supported by a cable that makes an angle of 53.0° with the horizontal (Fig. 8.18a). If a person weighing $6.00 \times 10^2$ N stands 1.50 m from the wall, find the magnitude of the tension $\vec{T}$ in the cable and the components of the force $\vec{R}$ exerted by the wall on the beam.

**STRATEGY** See Figure 8.18a–c (Steps 1 and 2). The second condition of equilibrium, $\Sigma\tau_i = 0$, with torques computed around the pin, can be solved for the tension $T$ in the cable. The first condition of equilibrium, $\Sigma\vec{F}_i = 0$, gives two equations and two unknowns for the two components of the force exerted by the wall, $R_x$ and $R_y$.

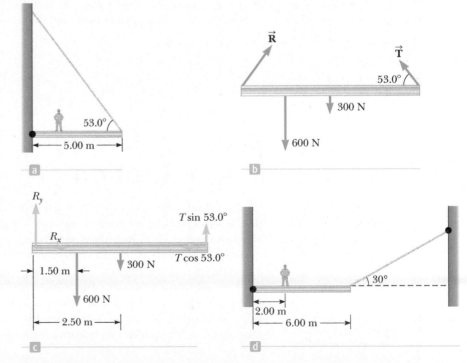

**Figure 8.18** (Example 8.11) (a) A uniform beam attached to a wall and supported by a cable. (b) A force diagram for the beam. (c) The component form of the force diagram. (d) (Exercise 8.11)

*(Continued)*

## SOLUTION

From Figure 8.18, the forces causing torques are the wall force $\vec{R}$, the gravity forces on the beam and the man, $w_B$ and $w_M$, and the tension force $\vec{T}$. Apply the condition of rotational equilibrium (Step 3):

$$\sum \tau_i = \tau_R + \tau_B + \tau_M + \tau_T = 0$$

Compute torques around the pin at $O$, so $\tau_R = 0$ (zero moment arm). The torque due to the beam's weight acts at the beam's center of gravity.

$$\sum \tau_i = 0 - w_B(L/2) - w_M(1.50 \text{ m}) + TL \sin (53°) = 0$$

Substitute $L = 5.00$ m and the weights, solving for $T$:

$$-(3.00 \times 10^2 \text{ N})(2.50 \text{ m}) - (6.00 \times 10^2 \text{ N})(1.50 \text{ m})$$
$$+ (T \sin 53.0°)(5.00 \text{ m}) = 0$$
$$T = \boxed{413 \text{ N}}$$

Now apply the first condition of equilibrium to the beam (Step 4):

(1) $\sum F_x = R_x - T \cos 53.0° = 0$

(2) $\sum F_y = R_y - w_B - w_M + T \sin 53.0° = 0$

Substituting the value of $T$ found in the previous step and the weights, obtain the components of $\vec{R}$ (Step 5):

$R_x = \boxed{249 \text{ N}}$   $R_y = \boxed{5.70 \times 10^2 \text{ N}}$

**REMARKS** Even if we selected some other axis for the torque equation, the solution would be the same. For example, if the axis were to pass through the center of gravity of the beam, the torque equation would involve both $T$ and $R_y$. Together with Equations (1) and (2), however, the unknowns could still be found—a good exercise. In both Example 8.9 and Example 8.11, notice the steps of the Problem-Solving Strategy could be carried out in the explicit recommended order.

**QUESTION 8.11** What happens to the tension in the cable if the man in Figure 8.18a moves farther away from the wall?

**EXERCISE 8.11** A person with mass 55.0 kg stands 2.00 m away from the wall on a uniform 6.00-m beam, as shown in Figure 8.18d. The mass of the beam is 40.0 kg. Find the hinge force components and the tension in the wire.

**ANSWERS** $T = 751$ N, $R_x = -6.50 \times 10^2$ N, $R_y = 556$ N

# 8.4 The Rotational Second Law of Motion

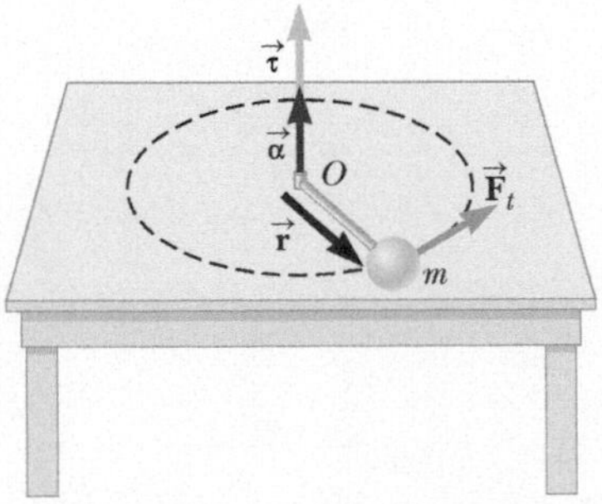

**Figure 8.19** An object of mass $m$ attached to a light rod of length $r$ moves in a circular path on a frictionless horizontal surface while a tangential force $\vec{F}_t$ acts on it. The acceleration and torque vectors are perpendicular to both the radial and the force vectors.

When a rigid object is subject to a net torque, it undergoes an angular acceleration that is directly proportional to the net torque. This result, which is analogous to Newton's second law, is derived as follows.

The system shown in Figure 8.19 consists of an object of mass $m$ connected to a very light rod of length $r$. The rod is pivoted at the point $O$, and its movement is confined to rotation on a frictionless *horizontal* table. Assume that a force $F_t$ acts perpendicular to the rod and hence is tangent to the circular path of the object. Because there is no force to oppose this tangential force, the object undergoes a tangential acceleration $a_t$ in accordance with Newton's second law:

$$F_t = ma_t$$

Multiply both sides of this equation by $r$:

$$F_t r = mra_t$$

Substituting the equation $a_t = r\alpha$ relating tangential and angular acceleration into the above expression gives

$$F_t r = mr^2 \alpha \qquad [8.9]$$

The left side of Equation 8.9 is the torque acting on the object about its axis of rotation, so we can rewrite it as

$$\tau = mr^2 \alpha \qquad [8.10]$$

Equation 8.10 shows that the torque on the object is proportional to the angular acceleration of the object, where the constant of proportionality $mr^2$ is called the **moment of inertia** of the object of mass $m$. (Because the rod is very light, its moment of inertia can be neglected.)

As discussed in Section 8.1, torque $\vec{\tau}$ is a vector perpendicular to a radial vector $\vec{r}$ and a force vector $\vec{F}$. Given Equation 8.10, the angular acceleration $\vec{\alpha}$ points in the same direction as $\vec{\tau}$, as illustrated in Figure 8.19. A force exerted clockwise yields a negative torque, whereas a force exerted counterclockwise results in a positive torque. In a similar way, the angular velocity vector, $\vec{\omega}$, points in either the same direction as $\vec{\alpha}$, or in the opposite direction.

### Quick Quiz

**8.1** Using a screwdriver, you try to remove a screw from a piece of furniture, but can't get it to turn. To increase the chances of success, you should use a screwdriver that (a) is longer, (b) is shorter, (c) has a narrower handle, or (d) has a wider handle.

## 8.4.1 Torque on a Rotating Object

Consider a solid disk rotating about its axis as in Figure 8.20a. The disk consists of many particles at various distances from the axis of rotation. (See Fig. 8.20b.) The torque on each one of these particles is given by Equation 8.10. The *net* torque on the disk is given by the sum of the individual torques on all the particles:

$$\Sigma \tau = (\Sigma\, mr^2)\alpha \qquad [8.11]$$

Because the disk is rigid, all of its particles have the *same* angular acceleration, so $\alpha$ is not involved in the sum. If the masses and distances of the particles are labeled with subscripts as in Figure 8.20b, then

$$\Sigma\, mr^2 = m_1 r_1^2 + m_2 r_2^2 + m_3 r_3^2 + \cdots$$

This quantity is the moment of inertia, $I$, of the whole body:

$$I \equiv \Sigma\, mr^2 \qquad [8.12]$$

◀ Moment of inertia

The moment of inertia has the SI units $\text{kg} \cdot \text{m}^2$. Using this result in Equation 8.11, we see that the net torque on a rigid body rotating about a fixed axis is given by

$$\Sigma \tau = I\alpha \qquad [8.13]$$

◀ Rotational analog of Newton's second law

**Figure 8.20** (a) A solid disk rotating about its axis. (b) The disk consists of many particles, all with the same angular acceleration.

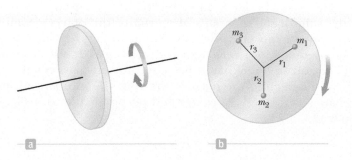

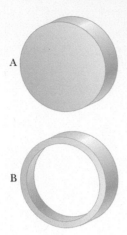

**Figure 8.21** (Quick Quiz 8.3)

Equation 8.13 says that **the angular acceleration of an extended rigid object is proportional to the net torque acting on it.** This equation is the rotational analog of Newton's second law of motion, with torque replacing force, moment of inertia replacing mass, and angular acceleration replacing linear acceleration. Although the moment of inertia of an object is related to its mass, there is an important difference between them. The mass $m$ depends only on the quantity of matter in an object, whereas the moment of inertia, $I$, depends on both the quantity of matter and its distribution (through the $r^2$ term in $I = \Sigma mr^2$) in the rigid object.

### Quick Quiz

**8.2** A constant net torque is applied to an object. Which one of the following will *not* be constant? (a) angular acceleration, (b) angular velocity, (c) moment of inertia, or (d) center of gravity.

**8.3** The two rigid objects shown in Figure 8.21 have the same mass, radius, and angular speed, each spinning around an axis through the center of its circular shape. If the same braking torque is applied to each, which takes longer to stop? (a) A (b) B (c) more information is needed

**APPLICATION**

**Bicycle Gears**

The gear system on a bicycle provides an easily visible example of the relationship between torque and angular acceleration. Consider first a five-speed gear system in which the drive chain can be adjusted to wrap around any of five gears attached to the back wheel (Fig. 8.22). The gears, with different radii, are concentric with the wheel hub. When the cyclist begins pedaling from rest, the chain is attached to the largest gear. Because it has the largest radius, this gear provides the largest torque to the drive wheel. A large torque is required initially, because the bicycle starts from rest. As the bicycle rolls faster, the tangential speed of the chain increases, eventually becoming too fast for the cyclist to maintain by pushing the pedals. The chain is then moved to a gear with a smaller radius, so the chain has a smaller tangential speed that the cyclist can more easily maintain. This gear doesn't provide as much torque as the first, but the cyclist needs to accelerate only to a somewhat higher speed. This process continues as the bicycle moves faster and faster and the cyclist shifts through all five gears. The fifth gear supplies the lowest torque, but now the main function of that torque is to counter the frictional torque from the rolling tires, which tends to reduce the speed of the bicycle. The small radius of the fifth gear allows the cyclist to keep up with the chain's movement by pushing the pedals.

**Figure 8.22** The drive wheel and gears of a bicycle.

© Cengage Learning/George Semple

A 15-speed bicycle has the same gear structure on the drive wheel, but has three gears on the sprocket connected to the pedals. By combining different positions of the chain on the rear gears and the sprocket gears, 15 different torques are available.

## 8.4.2 More on the Moment of Inertia

As we have seen, a small object (or a particle) has a moment of inertia equal to $mr^2$ about some axis. The moment of inertia of a *composite* object about some axis is just the sum of the moments of inertia of the object's components. For example, suppose a majorette twirls a baton as in Figure 8.23. Assume that the baton can be modeled as a very light rod of length $2\ell$ with a heavy object at each end. (The rod of a real baton has a significant mass relative to its ends.) Because we are neglecting the mass of the rod, the moment of inertia of the baton about an axis through its center and perpendicular to its length is given by Equation 8.12:

$$I = \Sigma mr^2$$

**Figure 8.23** A baton of length $2\ell$ and mass $2m$. (The mass of the connecting rod is neglected.) The moment of inertia about the axis through the baton's center and perpendicular to its length is $2m\ell^2$.

Because this system consists of two objects with equal masses equidistant from the axis of rotation, $r = \ell$ for each object, and the sum is

$$I = \Sigma mr^2 = m\ell^2 + m\ell^2 = 2m\ell^2$$

If the mass of the rod were not neglected, we would have to include its moment of inertia to find the total moment of inertia of the baton.

We pointed out earlier that $I$ is the rotational counterpart of $m$. However, there are some important distinctions between the two. For example, mass is an intrinsic property of an object that doesn't change, whereas **the moment of inertia of a system depends on how the mass is distributed and on the location of the axis of rotation.** Example 8.12 illustrates this point.

## EXAMPLE 8.12 | THE BATON TWIRLER

**GOAL** Calculate a moment of inertia.

**PROBLEM** In an effort to be the star of the halftime show, a majorette twirls an unusual baton made up of four balls fastened to the ends of very light rods (Fig. 8.24). Each rod is 1.0 m long. **(a)** Find the moment of inertia of the baton about an axis perpendicular to the page and passing through the point where the rods cross. **(b)** The majorette tries spinning her strange baton about the axis $OO'$, as shown in Figure 8.25 on page 242. Calculate the moment of inertia of the baton about this axis.

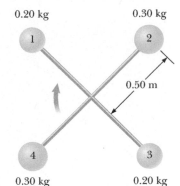

**Figure 8.24** (Example 8.12a) Four balls connected to light rods rotating in the plane of the page.

**STRATEGY** In Figure 8.24, all four balls contribute to the moment of inertia, whereas in Figure 8.25, with the new axis, only the two balls on the left and right contribute. Technically, the balls on the top and bottom in Figure 8.25 still make a small contribution because they're not really point particles. However, their contributions can be neglected because the distance from the axis of rotation of the balls on the horizontal rod is much greater than the radii of the balls on the vertical rod.

· · · · · · · · · · · · · · · · · · · · · · · · · · · · · · · · · · · · · · · · · · · · · · · · · · · · · · · · · · · · · · · · · · · · · · · · ·

### SOLUTION

**(a)** Calculate the moment of inertia of the baton when oriented as in Figure 8.24.

Apply Equation 8.12, neglecting the mass of the connecting rods:

$$I = \Sigma mr^2 = m_1 r_1^2 + m_2 r_2^2 + m_3 r_3^2 + m_4 r_4^2$$
$$= (0.20 \text{ kg})(0.50 \text{ m})^2 + (0.30 \text{ kg})(0.50 \text{ m})^2$$
$$+ (0.20 \text{ kg})(0.50 \text{ m})^2 + (0.30 \text{ kg})(0.50 \text{ m})^2$$
$$I = \boxed{0.25 \text{ kg} \cdot \text{m}^2}$$

**(b)** Calculate the moment of inertia of the baton when oriented as in Figure 8.25.

Apply Equation 8.12 again, neglecting the radii of the 0.20-kg balls.

$$I = \Sigma mr^2 = m_1 r_1^2 + m_2 r_2^2 + m_3 r_3^2 + m_4 r_4^2$$
$$= (0.20 \text{ kg})(0)^2 + (0.30 \text{ kg})(0.50 \text{ m})^2 + (0.20 \text{ kg})(0)^2$$
$$+ (0.30 \text{ kg})(0.50 \text{ m})^2$$
$$I = \boxed{0.15 \text{ kg} \cdot \text{m}^2}$$

· · · · · · · · · · · · · · · · · · · · · · · · · · · · · · · · · · · · · · · · · · · · · · · · · · · · · · · · · · · · · · · · · · · · · · · · ·

**REMARKS** The moment of inertia is smaller in part (b) because in this configuration the 0.20-kg balls are essentially located on the axis of rotation.

*(Continued)*

**QUESTION 8.12** If one of the rods is lengthened, which one would cause the larger change in the moment of inertia, the rod connecting balls one and three or the rod connecting balls two and four?

**EXERCISE 8.12** Yet another bizarre baton is created by taking four identical balls, each with mass 0.300 kg, and fixing them as before, except that one of the rods has a length of 1.00 m and the other has a length of 1.50 m. Calculate the moment of inertia of this baton **(a)** when oriented as in Figure 8.24; **(b)** when oriented as in Figure 8.25, with the shorter rod vertical; and **(c)** when oriented as in Figure 8.25, but with longer rod vertical.

**ANSWERS** **(a)** 0.488 kg · m$^2$  **(b)** 0.338 kg · m$^2$
**(c)** 0.150 kg · m$^2$

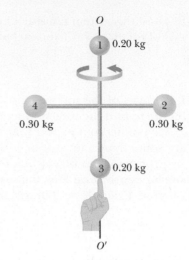

**Figure 8.25** (Example 8.12b) A double baton rotating about the axis $OO'$.

### 8.4.3 Calculation of Moments of Inertia for Extended Objects

The method used for calculating moments of inertia in Example 8.12 is simple when only a few small objects rotate about an axis. When the object is an extended one, such as a sphere, a cylinder, or a cone, techniques of calculus are often required, unless some simplifying symmetry is present. One such extended object amenable to a simple solution is a hoop rotating about an axis perpendicular to its plane and passing through its center, as shown in Figure 8.26. (A bicycle tire, for example, would approximately fit into this category.)

To evaluate the moment of inertia of the hoop, we can still use the equation $I = \Sigma mr^2$ and imagine that the mass of the hoop $M$ is divided into $n$ small segments having masses $m_1, m_2, m_3, \cdots, m_n$, as in Figure 8.26, with $M = m_1 + m_2 + m_3 + \cdots + m_n$. This approach is just an extension of the baton problem described in the preceding examples, except that now we have a large number of small masses in rotation instead of only four.

We can express the sum for $I$ as

$$I = \Sigma mr^2 = m_1 r_1^2 + m_2 r_2^2 + m_3 r_3^2 + \cdots + m_n r_n^2$$

All of the segments around the hoop are at the *same distance $R$* from the axis of rotation, so we can drop the subscripts on the distances and factor out $R^2$ to obtain

$$I = (m_1 + m_2 + m_3 + \cdots + m_n) R^2 = MR^2 \qquad [8.14]$$

This expression can be used for the moment of inertia of any ring-shaped object rotating about an axis through its center and perpendicular to its plane. Note that the result is strictly valid only if the thickness of the ring is small relative to its inner radius.

The hoop we selected as an example is unique in that we were able to find an expression for its moment of inertia by using only simple algebra. Unfortunately, for most extended objects the calculation is much more difficult because the mass elements are not all located at the same distance from the axis, so the methods of integral calculus are required. The moments of inertia for some other common shapes are given without proof in Table 8.1. You can use this table as needed to determine the moment of inertia of a body having any one of the listed shapes.

If mass elements in an object are redistributed parallel to the axis of rotation, the moment of inertia of the object doesn't change. Consequently, the expression $I = MR^2$ can be used equally well to find the axial moment of inertia of an embroidery hoop or of a long sewer pipe. Likewise, a door turning on its hinges is described by the same moment-of-inertia expression as that tabulated for a long, thin rod rotating about an axis through its end.

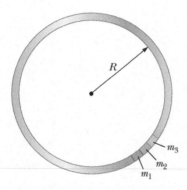

**Figure 8.26** A uniform hoop can be divided into a large number of small segments that are equidistant from the center of the hoop.

**Tip 8.3** No Single Moment of Inertia

Moment of inertia is analogous to mass, but there are major differences. Mass is an inherent property of an object. The moment of inertia of an object depends on the shape of the object, its mass, and the choice of rotation axis.

**Table 8.1** Moments of Inertia for Various Rigid Objects of Uniform Composition

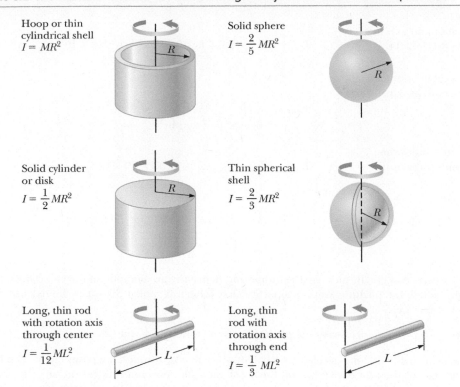

Hoop or thin cylindrical shell
$I = MR^2$

Solid sphere
$I = \frac{2}{5} MR^2$

Solid cylinder or disk
$I = \frac{1}{2} MR^2$

Thin spherical shell
$I = \frac{2}{3} MR^2$

Long, thin rod with rotation axis through center
$I = \frac{1}{12} ML^2$

Long, thin rod with rotation axis through end
$I = \frac{1}{3} ML^2$

---

## EXAMPLE 8.13 | WARMING UP

**GOAL** Find a moment of inertia and apply the rotational analog of Newton's second law.

**PROBLEM** A baseball player loosening up his arm before a game tosses a 0.150-kg baseball, using only the rotation of his forearm to accelerate the ball (Fig. 8.27). The forearm has a mass of 1.50 kg and the length from the elbow to the ball's center is 0.350 m. The ball starts at rest and is released with a speed of 30.0 m/s in 0.300 s. **(a)** Find the constant angular acceleration of the arm and ball. **(b)** Calculate the moment of inertia of the system consisting of the forearm and ball. **(c)** Find the torque exerted on the system that results in the angular acceleration found in part **(a)**.

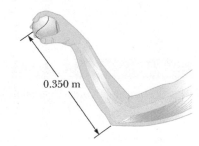

**Figure 8.27** (Example 8.13) A ball being tossed by a pitcher. The forearm is used to accelerate the ball.

0.350 m

**STRATEGY** The angular acceleration can be found with rotational kinematic equations, while the moment of inertia of the system can be obtained by summing the separate moments of inertia of the ball and forearm. The ball is treated as a point particle. Multiplying these two results together gives the torque.

. . . . . . . . . . . . . . . . . . . . . . . . . . . . . . . . . . . . . . . . . . . . . . . . . . . . . . . . . . . . . . . . . . . . . . . . . . . . . . . . . . . . . . . . . . . . . . . .

### SOLUTION

**(a)** Find the angular acceleration of the ball.

The angular acceleration is constant, so use the angular velocity kinematic equation with $\omega_i = 0$:

$$\omega = \omega_i + \alpha t \quad \rightarrow \quad \alpha = \frac{\omega}{t}$$

The ball accelerates along a circular arc with radius given by the length of the forearm. Solve $v = r\omega$ for $\omega$ and substitute:

$$\alpha = \frac{\omega}{t} = \frac{v}{rt} = \frac{30.0 \text{ m/s}}{(0.350 \text{ m})(0.300 \text{ s})} = \boxed{286 \text{ rad/s}^2}$$

*(Continued)*

**(b)** Find the moment of inertia of the system (forearm plus ball).

Find the moment of inertia of the ball about an axis that passes through the elbow, perpendicular to the arm:

$$I_{ball} = mr^2 = (0.150 \text{ kg})(0.350 \text{ m})^2 = 1.84 \times 10^{-2} \text{ kg} \cdot \text{m}^2$$

Obtain the moment of inertia of the forearm, modeled as a rod rotating about an axis through one end, by consulting Table 8.1:

$$I_{forearm} = \tfrac{1}{3} ML^2 = \tfrac{1}{3}(1.50 \text{ kg})(0.350 \text{ m})^2$$
$$= 6.13 \times 10^{-2} \text{ kg} \cdot \text{m}^2$$

Sum the individual moments of inertia to obtain the moment of inertia of the system (ball plus forearm):

$$I_{system} = I_{ball} + I_{forearm} = \boxed{7.97 \times 10^{-2} \text{ kg} \cdot \text{m}^2}$$

**(c)** Find the torque exerted on the system.

Apply Equation 8.13, using the results of parts **(a)** and **(b)**:

$$\tau = I_{system}\alpha = (7.97 \times 10^{-2} \text{ kg} \cdot \text{m}^2)(286 \text{ rad/s}^2)$$
$$= \boxed{22.8 \text{ N} \cdot \text{m}}$$

**REMARKS** Notice that having a long forearm can greatly increase the torque and hence the acceleration of the ball. This is one reason it's advantageous for a pitcher to be tall: the pitching arm is proportionately longer. A similar advantage holds in tennis, where taller players can usually deliver faster serves.

**QUESTION 8.13** Why do pitchers step forward when delivering the pitch? Why is the timing important?

**EXERCISE 8.13** A catapult with a radial arm 4.00 m long accelerates a ball of mass 20.0 kg through a quarter circle. The ball leaves the apparatus at 45.0 m/s. If the mass of the arm is 25.0 kg and the acceleration is constant, find **(a)** the angular acceleration, **(b)** the moment of inertia of the arm and ball, and **(c)** the net torque exerted on the ball and arm. *Hint:* Use the time-independent rotational kinematics equation to find the angular acceleration, rather than the angular velocity equation.

**ANSWERS** **(a)** 40.3 rad/s² **(b)** 453 kg · m² **(c)** 1.83 × 10⁴ N · m

---

## EXAMPLE 8.14  THE FALLING BUCKET

**GOAL** Combine Newton's second law with its rotational analog.

**PROBLEM** A solid, uniform, frictionless cylindrical reel of mass $M = 3.00$ kg and radius $R = 0.400$ m is used to draw water from a well (Fig. 8.28a). A bucket of mass $m = 2.00$ kg is attached to a cord that is wrapped around the cylinder. **(a)** Find the tension $T$ in the cord and acceleration $a$ of the bucket. **(b)** If the bucket starts from rest at the top of the well and falls for 3.00 s before hitting the water, how far does it fall?

**STRATEGY** This problem involves three equations and three unknowns. The three equations are Newton's second law applied to the bucket, $ma = \Sigma F_i$; the rotational version of the second law applied to the cylinder, $I\alpha = \Sigma \tau_i$; and the relationship between linear and angular acceleration, $a = r\alpha$, which connects the dynamics of the bucket and cylinder. The three unknowns are the acceleration $a$ of the bucket, the angular acceleration $a$ of the cylinder, and the tension $T$ in the rope. Assemble the terms of the three equations and solve for the three unknowns by substitution. Part **(b)** is a review of kinematics.

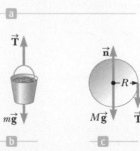

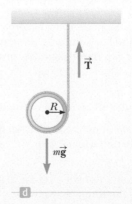

**Figure 8.28** (Example 8.14) (a) A water bucket attached to a rope passing over a frictionless reel. (b) A force diagram for the bucket. (c) The tension produces a torque on the cylinder about its axis of rotation. (d) A falling cylinder (Exercise 8.14).

## SOLUTION

**(a)** Find the tension in the cord and the acceleration of the bucket.

Apply Newton's second law to the bucket in Figure 8.28b. There are two forces: the tension $\vec{T}$ acting upward and gravity $m\vec{g}$ acting downward.

**(1)**  $ma = -mg + T$

Apply $\tau = I\alpha$ to the cylinder in Figure 8.28c:

$\sum \tau = I\alpha = \frac{1}{2}MR^2\alpha$   (solid cylinder)

Notice the angular acceleration is clockwise, so the torque is negative. The normal and gravity forces have zero moment arm and don't contribute any torque.

**(2)**  $-TR = \frac{1}{2}MR^2\alpha$

Solve for $T$ and substitute $\alpha = a/R$ (notice that both $\alpha$ and $a$ are negative):

**(3)**  $T = -\frac{1}{2}MR\alpha = -\frac{1}{2}Ma$

Substitute the expression for $T$ in Equation (3) into Equation (1), and solve for the acceleration:

$ma = -mg - \frac{1}{2}Ma \quad \rightarrow \quad a = -\dfrac{mg}{m + \frac{1}{2}M}$

Substitute the values for $m$, $M$, and $g$, getting $a$, then substitute $a$ into Equation (3) to get $T$:

$a = \boxed{-5.60 \text{ m/s}^2} \qquad T = \boxed{8.40 \text{ N}}$

**(b)** Find the distance the bucket falls in 3.00 s.

Apply the displacement kinematic equation for constant acceleration, with $t = 3.00$ s and $v_0 = 0$:

$\Delta y = v_0 t + \frac{1}{2}at^2 = -\frac{1}{2}(5.60 \text{ m/s}^2)(3.00 \text{ s})^2 = \boxed{-25.2 \text{ m}}$

**REMARKS** Proper handling of signs is very important in these problems. All such signs should be chosen initially and checked mathematically and physically. In this problem, for example, both the angular acceleration $\alpha$ and the acceleration $a$ are negative, so $\alpha = a/R$ applies. If the rope had been wound the other way on the cylinder, causing counterclockwise rotation, the torque would have been positive, and the relationship would have been $\alpha = -a/R$, with the double negative making the right-hand side positive, just like the left-hand side.

**QUESTION 8.14** How would the acceleration and tension change if most of the reel's mass were at its rim?

**EXERCISE 8.14** A hollow cylinder of mass 0.100 kg and radius 4.00 cm has a string wrapped several times around it, as in Figure 8.28d. If the string is attached to a rigid support and the cylinder allowed to drop from rest, find **(a)** the acceleration of the cylinder and **(b)** the speed of the cylinder when a meter of string has unwound off of it.

**ANSWERS** **(a)** $-4.90 \text{ m/s}^2$  **(b)** 3.13 m/s

## 8.4.4 Extending the System Approach

Notice that Example 8.14 required solving three equations and three unknowns to find the unknown acceleration. That's a lengthy task. In many contexts, including this one, the solution can be considerably simplified by using an extension of the system approach.

In the system approach, all masses in the system accelerate at the same rate under external forces. The masses are connected by internal forces, such as tensions, which can be disregarded because they come in equal and opposite pairs. When pulleys or other rotating objects are involved, the additional requirement is that the cables causing them to rotate must not slip, so that the tangential acceleration at the point of contact is the same as the common acceleration of the masses in the system. Each rotating object then contributes an additional effective mass of $I/r^2$ to the total mass (or inertia) of the system. For simplicity, external torques (for example, applied with a crank) will not be considered, although they can be included if the conditions aren't violated. The extended system approach, in the absence of external torques, can then be written as follows:

$$\left( \sum m_i + \sum \frac{I_i}{r_i^2} \right) a = \sum F_{\text{ext}}$$

The context of Example 8.14 involves a falling bucket of mass $m$ and a rotating, solid cylindrical reel of mass $M$ that can be treated as a disk. The only external force accelerating the system is the gravity force on the bucket. Newton's second law for the extended system approach can therefore be written:

$$\left(m + \frac{I}{R^2}\right)a = -mg$$

Substituting the moment of inertia for a disk, $I = (1/2)MR^2$, gives the same result for the acceleration as in Example 8.14:

$$\left(m + \frac{\frac{1}{2}MR^2}{R^2}\right)a = -mg$$

$$a = -\frac{mg}{m + \frac{1}{2}M}$$

So in problems involving the rotational second law of motion, it's clear that both translational inertia and rotational inertia combine to affect the extent to which a force can accelerate the system. Knowing this can greatly simplify some problems involving both the linear and rotational forms of Newton's second law of motion.

# 8.5 Rotational Kinetic Energy

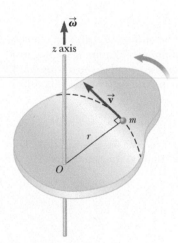

**Figure 8.29** A rigid plate rotates about the $z$-axis with angular speed $v$. The kinetic energy of a particle of mass $m$ is $\frac{1}{2}mv^2$. The total kinetic energy of the plate is $\frac{1}{2}I\omega^2$.

In Topic 5, we defined the kinetic energy of a particle moving through space with a speed $v$ as the quantity $\frac{1}{2}mv^2$. Analogously, **an object rotating about some axis with an angular speed $\omega$ has rotational kinetic energy given by $\frac{1}{2}I\omega^2$.** To prove this, consider an object in the shape of a thin, rigid plate rotating around some axis perpendicular to its plane, as in Figure 8.29. The plate consists of many small particles, each of mass $m$. All these particles rotate in circular paths around the axis. If $r$ is the distance of one of the particles from the axis of rotation, the speed of that particle is $v = r\omega$. Because the *total* kinetic energy of the plate's rotation is the sum of all the kinetic energies associated with its particles, we have

$$KE_r = \sum \left(\tfrac{1}{2}mv^2\right) = \sum \left(\tfrac{1}{2}mr^2\omega^2\right) = \tfrac{1}{2}\left(\sum mr^2\right)\omega^2$$

In the last step, the $\omega^2$ term is factored out because it's the same for every particle. Now, the quantity in parentheses on the right is the moment of inertia of the plate in the limit as the particles become vanishingly small, so

$$KE_r = \tfrac{1}{2}I\omega^2 \qquad\qquad [8.15]$$

where $I = \sum mr^2$ is the moment of inertia of the plate.

A system such as a bowling ball rolling down a ramp is described by three types of energy: **gravitational potential energy $PE_g$, translational kinetic energy $KE_t$,** and **rotational kinetic energy $KE_r$.** All these forms of energy, plus the potential energies of any other conservative forces, must be included in our equation for the conservation of mechanical energy of an isolated system:

◄ **Conservation of mechanical energy**

$$(KE_t + KE_r + PE)_i = (KE_t + KE_r + PE)_f \qquad\qquad [8.16]$$

where $i$ and $f$ refer to initial and final values, respectively, and $PE$ includes the potential energies of all conservative forces in a given problem. This relation is true *only* if we ignore dissipative forces such as friction. Otherwise, it's necessary to resort to a generalization of the work–energy theorem:

◄ **Work–energy theorem including rotational energy**

$$W_{nc} = \Delta KE_t + \Delta KE_r + \Delta PE \qquad\qquad [8.17]$$

## PROBLEM-SOLVING STRATEGY

### Energy Methods and Rotation

1. **Choose two points of interest,** one where all necessary information is known, and the other where information is desired.
2. **Identify** the conservative and nonconservative forces acting on the system being analyzed.
3. **Write the general work–energy theorem,** Equation 8.17, or Equation 8.16 if all forces are conservative.
4. **Substitute general expressions** for the terms in the equation.
5. **Use** $v = r\omega$ to eliminate either $\omega$ or $v$ from the equation.
6. **Solve** for the unknown.

## EXAMPLE 8.15  A BALL ROLLING DOWN AN INCLINE

**GOAL**  Combine gravitational, translational, and rotational energy.

**PROBLEM**  A uniform, solid ball of mass $M$ and radius $R$ starts from rest at a height of $h = 2.00$ m and rolls down a $\theta = 30.0°$ slope, as in Figure 8.30. What is the linear speed of the ball when it leaves the incline? Assume that the ball rolls without slipping.

**STRATEGY**  The two points of interest are the top and bottom of the incline, with the bottom acting as the zero point of gravitational potential energy. As the ball rolls down the ramp, gravitational potential energy is converted into both translational and rotational kinetic energy without dissipation, so conservation of mechanical energy can be applied with the use of Equation 8.16.

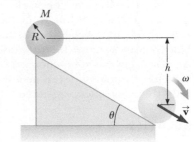

**Figure 8.30**  (Example 8.15) A ball starts from rest at the top of an incline and rolls to the bottom without slipping.

### SOLUTION

Apply conservation of energy with $PE = PE_g$, the potential energy associated with gravity:

$$(KE_t + KE_r + PE_g)_i = (KE_t + KE_r + PE_g)_f$$

Substitute the appropriate general expressions, noting that $(KE_t)_i = (KE_r)_i = 0$ and $(PE_g)_f = 0$ (obtain the moment of inertia of a ball from Table 8.1):

$$0 + 0 + Mgh = \tfrac{1}{2}Mv^2 + \tfrac{1}{2}(\tfrac{2}{5}MR^2)\omega^2 + 0$$

The ball rolls without slipping, so $R\omega = v$, the "no-slip condition," can be applied:

$$Mgh = \tfrac{1}{2}Mv^2 + \tfrac{1}{5}Mv^2 = \tfrac{7}{10}Mv^2$$

Solve for $v$, noting that $M$ cancels.

$$v = \sqrt{\frac{10gh}{7}} = \sqrt{\frac{10(9.80 \text{ m/s}^2)(2.00 \text{ m})}{7}} = \boxed{5.29 \text{ m/s}}$$

**REMARKS**  Notice the translational speed is less than that of a block sliding down a frictionless slope, $v = \sqrt{2gh}$. That's because some of the original potential energy must go to increasing the rotational kinetic energy.

**QUESTION 8.15**  Rank from fastest to slowest: **(a)** a solid ball rolling down a ramp without slipping, **(b)** a cylinder rolling down the same ramp without slipping, **(c)** a block sliding down a frictionless ramp with the same height and slope.

**EXERCISE 8.15**  Repeat this example for a solid cylinder of the same mass and radius as the ball and released from the same height. In a race between the two objects on the incline, which one would win?

**ANSWER**  $v = \sqrt{4gh/3} = 5.11$ m/s; the ball would win

### Quick Quiz

**8.4**  Two spheres, one hollow and one solid, are rotating with the same angular speed around an axis through their centers. Both spheres have the same mass and radius. Which sphere, if either, has the higher rotational kinetic energy? (a) The hollow sphere. (b) The solid sphere. (c) They have the same kinetic energy.

## EXAMPLE 8.16    BLOCKS AND PULLEY

**GOAL** Solve a system requiring rotation concepts and the work–energy theorem.

**PROBLEM** Two blocks with masses $m_1 = 5.00$ kg and $m_2 = 7.00$ kg are attached by a string as in Figure 8.31a, over a pulley with mass $M = 2.00$ kg. The pulley, which turns on a frictionless axle, is a hollow cylinder with radius 0.050 0 m over which the string moves without slipping. The horizontal surface has coefficient of kinetic friction 0.350. Find the speed of the system when the block of mass $m_2$ has dropped 2.00 m.

**STRATEGY** This problem can be solved with the extension of the work–energy theorem, Equation 8.15b. If the block of mass $m_2$ falls from height $h$ to 0, then the block of mass $m_1$ moves the same distance, $\Delta x = h$. Apply the work–energy theorem, solve for $v$, and substitute. Kinetic friction is the sole nonconservative force.

**Figure 8.31** (a) (Example 8.16) $\vec{T}_1$ and $\vec{T}_2$ exert torques on the pulley. (b) (Exercise 8.16)

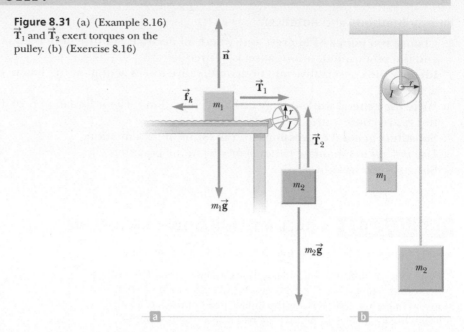

### SOLUTION

Apply the work–energy theorem, with $PE = PE_g$, the potential energy associated with gravity:

$$W_{nc} = \Delta KE_t + \Delta KE_r + \Delta PE_g$$

Substitute the frictional work for $W_{nc}$, kinetic energy changes for the two blocks, the rotational kinetic energy change for the pulley, and the potential energy change for the second block:

$$-\mu_k n\, \Delta x = -\mu_k(m_1 g)\, \Delta x = \left(\tfrac{1}{2}m_1 v^2 - 0\right) + \left(\tfrac{1}{2}m_2 v^2 - 0\right)$$
$$+ \left(\tfrac{1}{2}I\omega^2 - 0\right) + (0 - m_2 g h)$$

Substitute $\Delta x = h$, and write $I$ as $(I/r^2)r^2$:

$$-\mu_k(m_1 g)h = \tfrac{1}{2}m_1 v^2 + \tfrac{1}{2}m_2 v^2 + \tfrac{1}{2}\left(\frac{I}{r^2}\right)r^2\omega^2 - m_2 g h$$

For a hoop, $I = Mr^2$ so $(I/r^2) = M$. Substitute this quantity and $v = r\omega$:

$$-\mu_k(m_1 g)h = \tfrac{1}{2}m_1 v^2 + \tfrac{1}{2}m_2 v^2 + \tfrac{1}{2}Mv^2 - m_2 g h$$

Solve for $v$:

$$m_2 g h - \mu_k(m_1 g)h = \tfrac{1}{2}m_1 v^2 + \tfrac{1}{2}m_2 v^2 + \tfrac{1}{2}Mv^2$$
$$= \tfrac{1}{2}(m_1 + m_2 + M)v^2$$
$$v = \sqrt{\frac{2gh(m_2 - \mu_k m_1)}{m_1 + m_2 + M}}$$

Substitute $m_1 = 5.00$ kg, $m_2 = 7.00$ kg, $M = 2.00$ kg, $g = 9.80$ m/s², $h = 2.00$ m, and $\mu_k = 0.350$:

$$v = \boxed{3.83 \text{ m/s}}$$

**REMARKS** In the expression for the speed $v$, the mass $m_1$ of the first block and the mass $M$ of the pulley all appear in the denominator, reducing the speed, as they should. In the numerator, $m_2$ is positive while the friction term is negative. Both assertions are reasonable because the force of gravity on $m_2$ increases the speed of the system while the force of friction on $m_1$ slows it down. This problem can also be solved with Newton's second law together with $\tau = I\alpha$, a good exercise.

**QUESTION 8.16** How would increasing the radius of the pulley affect the final answer? Assume the angles of the cables are unchanged and the mass is the same as before.

**EXERCISE 8.16** Two blocks with masses $m_1 = 2.00$ kg and $m_2 = 9.00$ kg are attached over a pulley with mass $M = 3.00$ kg, hanging straight down as in Atwood's machine (Fig. 8.31b). The pulley is a solid cylinder with radius 0.050 0 m, and there is some friction in the axle. The system is released from rest, and the string moves without slipping over the pulley. If the larger mass is traveling at a speed of 2.50 m/s when it has dropped 1.00 m, how much mechanical energy was lost due to friction in the pulley's axle?

**ANSWER** 29.5 J

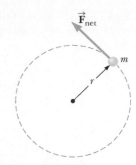

# 8.6 Angular Momentum

In Figure 8.32, an object of mass $m$ rotates in a circular path of radius $r$, acted on by a net force, $\vec{F}_{net}$. The resulting net torque on the object increases its angular speed from the value $\omega_0$ to the value $\omega$ in a time interval $\Delta t$. Therefore, we can write

$$\sum \tau = I\alpha = I\frac{\Delta \omega}{\Delta t} = I\left(\frac{\omega - \omega_0}{\Delta t}\right) = \frac{I\omega - I\omega_0}{\Delta t}$$

If we define the product

$$L \equiv I\omega \qquad [8.18]$$

as the **angular momentum** of the object, then we can write

$$\sum \tau = \frac{\text{change in angular momentum}}{\text{time interval}} = \frac{\Delta L}{\Delta t} \qquad [8.19]$$

**Figure 8.32** An object of mass $m$ rotating in a circular path under the action of a constant torque.

Equation 8.19 is the rotational analog of Newton's second law, which can be written in the form $F = \Delta p/\Delta t$ and states that **the net torque acting on an object is equal to the time rate of change of the object's angular momentum.** Recall that this equation also parallels the impulse–momentum theorem.

When the net external torque ($\sum \tau$) acting on a system is zero, Equation 8.19 gives $\Delta L/\Delta t = 0$, which says that the time rate of change of the system's angular momentum is zero. We then have the following important result:

Let $L_i$ and $L_f$ be the angular momenta of a system at two different times, and suppose there is no net external torque, so $\sum \tau = 0$. Then

$$L_i = L_f \qquad [8.20]$$

and angular momentum is said to be *conserved.*

◀ Conservation of angular momentum

Equation 8.20 gives us a third conservation law to add to our list: **conservation of angular momentum.** We can now state that **the mechanical energy, linear momentum, and angular momentum of an isolated system all remain constant.**

If the moment of inertia of an isolated rotating system changes, the system's angular speed will change (Fig. 8.33). Conservation of angular momentum then requires that

$$I_i\omega_i = I_f\omega_f \qquad \text{if} \qquad \sum \tau = 0 \qquad [8.21]$$

Note that conservation of angular momentum applies to macroscopic objects such as planets and people, as well as to atoms and molecules. There are many examples of conservation of angular momentum; one of the most dramatic is that of a figure skater spinning in the finale of his act. In Figure 8.34a, the skater has pulled his arms and legs close to his body, reducing their distance from his axis of rotation and hence also reducing his moment of inertia. By conservation of angular momentum, a reduction in his moment of inertia must increase his angular speed. Coming out of the spin in Figure 8.34b, he needs to reduce his angular speed, so he extends his arms and legs again, increasing his moment of inertia and thereby slowing his rotation.

Similarly, when a diver or an acrobat wishes to make several somersaults (Fig. 8.33), she pulls her hands and feet close to the trunk of her body to rotate at a greater angular speed. In this case, the external force due to gravity acts through her center of gravity and hence exerts no torque about her axis of rotation, so the

**Figure 8.33** Tightly curling her body, a diver decreases her moment of inertia, increasing her angular speed.

Action Plus/Stone/Getty Images

**APPLICATION**
Figure Skating

**APPLICATION**
Aerial Somersaults

**Figure 8.34** Evgeni Plushenko varies his moment of inertia to change his angular speed.

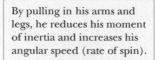

By pulling in his arms and legs, he reduces his moment of inertia and increases his angular speed (rate of spin).

Upon landing, extending his arms and legs increases his moment of inertia and helps slow his spin.

Clive Rose/Getty Images

Al Bello/Getty Images

APPLICATION

Rotating Neutron Stars

angular momentum about her center of gravity is conserved. For example, when a diver wishes to double her angular speed, she must reduce her moment of inertia to half its initial value.

An interesting astrophysical example of conservation of angular momentum occurs when a massive star, at the end of its lifetime, uses up all its fuel and collapses under the influence of gravitational forces, causing a gigantic outburst of energy called a supernova. The best-studied example of a remnant of a supernova explosion is the Crab Nebula, a chaotic, expanding mass of gas (Fig. 8.35). In a supernova, part of the star's mass is ejected into space, where it eventually condenses into new stars and planets. Most of what is left behind typically collapses into a **neutron star**—an extremely dense sphere of matter with a diameter of about 10 km, greatly reduced from the $10^6$-km diameter of the original star and containing a large fraction of the star's original mass. In a neutron star, pressures become so great that atomic electrons combine with protons, becoming neutrons. As the moment of inertia of the system decreases during the collapse, the star's rotational speed increases. More than 700 rapidly rotating neutron stars have been identified since their first discovery in 1967, with periods of rotation ranging from a millisecond to several seconds. The neutron star is an amazing system—an object with a mass greater than the Sun, fitting comfortably within the space of a small county and rotating so fast that the tangential speed of the surface approaches a sizable fraction of the speed of light!

Max Planck Institute for Astronomy, Heidelberg, Germany

Smithsonian Institution/Science Source/ Photo Researcher Inc.

Smithsonian Institution/Science Source/ Photo Researcher Inc.

**Figure 8.35** (a) The Crab Nebula in the constellation Taurus. This nebula is the remnant of a supernova seen on Earth in AD 1054. It is located some 6 300 light-years away and is approximately 6 light-years in diameter, still expanding outward. A pulsar deep inside the nebula flashes 30 times every second. (b) Pulsar off. (c) Pulsar on.

**8.5** A horizontal disk with moment of inertia $I_1$ rotates with angular speed $\omega_1$ about a vertical frictionless axle. A second horizontal disk having moment of inertia $I_2$ drops onto the first, initially not rotating but sharing the same axis as the first disk. Because their surfaces are rough, the two disks eventually reach the same angular speed $\omega$. The ratio $\omega/\omega_1$ is equal to (a) $I_1/I_2$  (b) $I_2/I_1$  (c) $I_1/(I_1 + I_2)$  (d) $I_2/(I_1 + I_2)$

**8.6** If global warming continues, it's likely that some ice from the polar ice caps of the Earth will melt and the water will be distributed closer to the equator. If this occurs, would the length of the day (one rotation) (a) increase, (b) decrease, or (c) remain the same?

---

### EXAMPLE 8.17 | THE SPINNING STOOL

**GOAL** Apply conservation of angular momentum to a simple system.

**PROBLEM** A student sits on a pivoted stool while holding a pair of weights. (See Fig. 8.36.) The stool is free to rotate about a vertical axis with negligible friction. The moment of inertia of student, weights, and stool is $2.25\ \text{kg} \cdot \text{m}^2$. The student is set in rotation with arms outstretched, making one complete turn every 1.26 s, arms outstretched. **(a)** What is the initial angular speed of the system? **(b)** As he rotates, he pulls the weights inward so that the new moment of inertia of the system (student, objects, and stool) becomes $1.80\ \text{kg} \cdot \text{m}^2$. What is the new angular speed of the system? **(c)** Find the work done by the student on the system while pulling in the weights. (Ignore energy lost through dissipation in his muscles.)

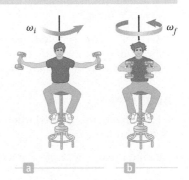

**STRATEGY** **(a)** The angular speed can be obtained from the frequency, which is the inverse of the period. **(b)** There are no external torques acting on the system, so the new angular speed can be found with the principle of conservation of angular momentum. **(c)** The work done on the system during this process is the same as the system's change in rotational kinetic energy.

**Figure 8.36** (Example 8.17)
(a) The student is given an initial angular speed while holding two weights out. (b) The angular speed increases as the student draws the weights inwards.

---

### SOLUTION

**(a)** Find the initial angular speed of the system.

Invert the period to get the frequency, and multiply by $2\pi$:

$$\omega_i = 2\pi f = 2\pi/T = \boxed{4.99\ \text{rad/s}}$$

**(b)** After he pulls the weights in, what's the system's new angular speed?

Equate the initial and final angular momenta of the system:

$$\textbf{(1)}\quad L_i = L_f \;\rightarrow\; I_i\omega_i = I_f\omega_f$$

Substitute and solve for the final angular speed $\omega_f$:

$$\textbf{(2)}\quad (2.25\ \text{kg} \cdot \text{m}^2)(4.99\ \text{rad/s}) = (1.80\ \text{kg} \cdot \text{m}^2)\omega_f$$

$$\omega_f = \boxed{6.24\ \text{rad/s}}$$

**(c)** Find the work the student does on the system.

Apply the work–energy theorem:

$$W_{\text{student}} = \Delta K_r = \tfrac{1}{2}I_f\omega_f^2 - \tfrac{1}{2}I_i\omega_i^2$$

$$= \tfrac{1}{2}(1.80\ \text{kg} \cdot \text{m}^2)(6.24\ \text{rad/s})^2$$

$$- \tfrac{1}{2}(2.25\ \text{kg} \cdot \text{m}^2)(4.99\ \text{rad/s})^2$$

$$W_{\text{student}} = \boxed{7.03\ \text{J}}$$

---

**REMARKS** Although the angular momentum of the system is conserved, mechanical energy is not conserved because the student does work on the system.

**QUESTION 8.17** If the student suddenly releases the weights, will his angular speed increase, decrease, or remain the same?

**EXERCISE 8.17** A star with an initial radius of $1.0 \times 10^8$ m and period of 30.0 days collapses suddenly to a radius of $1.0 \times 10^4$ m. **(a)** Find the period of rotation after collapse. **(b)** Find the work done by gravity during the collapse if the mass of the star is $2.0 \times 10^{30}$ kg. **(c)** What is the speed of an indestructible person standing on the equator of the collapsed star? (Neglect any relativistic or thermal effects, and assume the star is spherical before and after it collapses.)

**ANSWERS** **(a)** $2.6 \times 10^{-2}$ s  **(b)** $2.3 \times 10^{42}$ J  **(c)** $2.4 \times 10^6$ m/s

## EXAMPLE 8.18   THE MERRY-GO-ROUND

**GOAL** Apply conservation of angular momentum while combining two moments of inertia.

**PROBLEM** A merry-go-round modeled as a disk of mass $M = 1.00 \times 10^2$ kg and radius $R = 2.00$ m is rotating in a horizontal plane about a frictionless vertical axle (Fig. 8.37 is an overhead view of the system). **(a)** After a student with mass $m = 60.0$ kg jumps on the rim of the merry-go-round, the system's angular speed decreases to 2.00 rad/s. If the student walks slowly from the edge toward the center, find the angular speed of the system when she reaches a point 0.500 m from the center. **(b)** Find the change in the system's rotational kinetic energy caused by her movement to $r = 0.500$ m. **(c)** Find the work done on the student as she walks to $r = 0.500$ m.

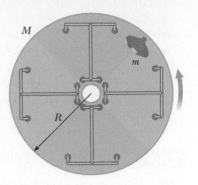

**Figure 8.37** (Example 8.18) As the student walks toward the center of the merry-go-round, the moment of inertia $I$ of the system becomes smaller. Because angular momentum is conserved and $L = I\omega$, the angular speed must increase.

**STRATEGY** This problem can be solved with conservation of angular momentum by equating the system's initial angular momentum when the student stands at the rim to the angular momentum when the student has reached $r = 0.500$ m. The key is to find the different moments of inertia.

. . . . . . . . . . . . . . . . . . . . . . . . . . . . . . . . . . . . . . . . . . . . . . . . . . . . . . . . . . . . . . . . . . . . . . . . . . . . . . . . . . . . . .

**SOLUTION**

**(a)** Find the angular speed when the student reaches a point 0.500 m from the center.

Calculate the moment of inertia of the disk, $I_D$:

$$I_D = \tfrac{1}{2}MR^2 = \tfrac{1}{2}(1.00 \times 10^2 \text{ kg})(2.00 \text{ m})^2$$
$$= 2.00 \times 10^2 \text{ kg} \cdot \text{m}^2$$

Calculate the initial moment of inertia of the student. This is the same as the moment of inertia of a mass a distance $R$ from the axis:

$$I_{Si} = mR^2 = (60.0 \text{ kg})(2.00 \text{ m})^2 = 2.40 \times 10^2 \text{ kg} \cdot \text{m}^2$$

Sum the two moments of inertia and multiply by the initial angular speed to find $L_i$, the initial angular momentum of the system:

$$L_i = (I_D + I_{Si})\omega_i$$
$$= (2.00 \times 10^2 \text{ kg} \cdot \text{m}^2 + 2.40 \times 10^2 \text{ kg} \cdot \text{m}^2)(2.00 \text{ rad/s})$$
$$= 8.80 \times 10^2 \text{ kg} \cdot \text{m}^2/\text{s}$$

Calculate the student's final moment of inertia, $I_{Sf}$, when she is 0.500 m from the center:

$$I_{Sf} = mr_f^2 = (60.0 \text{ kg})(0.50 \text{ m})^2 = 15.0 \text{ kg} \cdot \text{m}^2$$

The moment of inertia of the platform is unchanged. Add it to the student's final moment of inertia, and multiply by the unknown final angular speed to find $L_f$:

$$L_f = (I_D + I_{Sf})\omega_f = (2.00 \times 10^2 \text{ kg} \cdot \text{m}^2 + 15.0 \text{ kg} \cdot \text{m}^2)\omega_f$$
$$= (2.15 \times 10^2 \text{ kg} \cdot \text{m}^2)\omega_f$$

Equate the initial and final angular momenta and solve for the final angular speed of the system:

$$L_i = L_f$$
$$(8.80 \times 10^2 \text{ kg} \cdot \text{m}^2/\text{s}) = (2.15 \times 10^2 \text{ kg} \cdot \text{m}^2)\omega_f$$
$$\omega_f = \boxed{4.09 \text{ rad/s}}$$

**(b)** Find the change in the rotational kinetic energy of the system.

Calculate the initial kinetic energy of the system:

$$KE_i = \tfrac{1}{2}I_i\omega_i^2 = \tfrac{1}{2}(4.40 \times 10^2 \text{ kg} \cdot \text{m}^2)(2.00 \text{ rad/s})^2$$
$$= 8.80 \times 10^2 \text{ J}$$

Calculate the final kinetic energy of the system:

$$KE_f = \tfrac{1}{2}I_f\,\omega_f^2 = \tfrac{1}{2}(215 \text{ kg} \cdot \text{m}^2)(4.09 \text{ rad/s})^2 = 1.80 \times 10^3 \text{ J}$$

Calculate the change in kinetic energy of the system:

$$KE_f - KE_i = \boxed{920 \text{ J}}$$

**(c)** Find the work done on the student.

The student undergoes a change in kinetic energy that equals the work done on her. Apply the work–energy theorem:

$$W = \Delta KE_{\text{student}} = \tfrac{1}{2}I_{Sf}\omega_f^2 - \tfrac{1}{2}I_{Si}\omega_i^2$$
$$= \tfrac{1}{2}(15.0 \text{ kg} \cdot \text{m}^2)(4.09 \text{ rad/s})^2$$
$$\quad - \tfrac{1}{2}(2.40 \times 10^2 \text{ kg} \cdot \text{m}^2)(2.00 \text{ rad/s})^2$$
$$W = \boxed{-355 \text{ J}}$$

**REMARKS** The angular momentum is unchanged by internal forces; however, the kinetic energy increases because the student must perform positive work to walk toward the center of the platform.

**QUESTION 8.18** Is energy conservation violated in this example? Explain why there is a positive net change in mechanical energy. What is the origin of this energy?

**EXERCISE 8.18** **(a)** Find the angular speed of the merry-go-round before the student jumped on, assuming the student didn't transfer any momentum or energy as she jumped on the merry-go-round. **(b)** By how much did the kinetic energy of the system change when the student jumped on? Notice that energy is lost in this process, as should be expected, since it is essentially a perfectly inelastic collision.

**ANSWERS** **(a)** 4.40 rad/s    **(b)** $KE_f - KE_i = -1.06 \times 10^3$ J

## SUMMARY

### 8.1 Torque

Let $\vec{F}$ be a force acting on an object, and let $\vec{r}$ be a position vector from a chosen point $O$ to the point of application of the force. Then the magnitude of the torque $\vec{\tau}$ of the force $\vec{F}$ is given by

$$\tau = rF \sin \theta \qquad [8.2]$$

where $r$ is the length of the position vector, $F$ the magnitude of the force, and u the angle between $\vec{F}$ and $\vec{r}$ (Fig. 8.38).

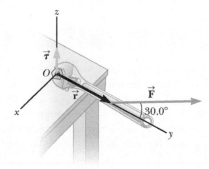

**Figure 8.38** The torque at O depends on the distance to the point of application of the force $\vec{F}$ and the force's magnitude and direction.

The quantity $d = r \sin \theta$ is called the *lever arm* of the force (Fig. 8.39).

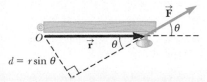

**Figure 8.39** An alternate interpretation of torque involves the concept of a lever arm $d = r \sin \theta$ that is perpendicular to the force.

### 8.2 Center of Mass and Its Motion

The torque of an object's weight can be calculated using a single force of magnitude $w = F_g = Mg$ (the total weight of the object) applied at the object's center of gravity. The $x$-, $y$-, and $z$-of an object's center of gravity are given by

$$x_{cg} = \frac{\sum m_i x_i}{\sum m_i} \; ; \; y_{cg} = \frac{\sum m_i y_i}{\sum m_i} \; ; \; z_{cg} = \frac{\sum m_i z_i}{\sum m_i} \quad [8.3]$$

The center of mass and center of gravity are exactly the same when $g$ doesn't vary significantly over the object, which is the case for all objects in this topic.

### 8.3 Torque and the Two Conditions for Equilibrium

An object in mechanical equilibrium must satisfy the following two conditions:

1. The net external force must be zero: $\sum \vec{F} = 0$.
2. The net external torque must be zero: $\sum \vec{\tau} = 0$.

These two conditions, used in solving problems involving rotation in a plane—result in three equations and three unknowns—two from the first condition (corresponding to the $x$- and $y$-components of the force) and one from the second condition, on torques. These equations must be solved simultaneously.

### 8.4 The Rotational Second Law of Motion

The **moment of inertia** of a group of particles is

$$I \equiv \Sigma mr^2 \qquad [8.12]$$

If a rigid object free to rotate about a fixed axis has a net external torque $\Sigma \tau$ acting on it, then the object undergoes an angular acceleration $a$, where

$$\Sigma \tau = I\alpha \qquad [8.13]$$

This equation is the rotational equivalent of the second law of motion.

Problems are solved by using Equation 8.13 together with Newton's second law and solving the resulting equations simultaneously. The relation $a = r\alpha$ is often key in relating the translational equations to the rotational equations.

## 8.5 Rotational Kinetic Energy

If a rigid object rotates about a fixed axis with angular speed $\omega$, its **rotational kinetic energy** is

$$KE_r = \tfrac{1}{2}I\omega^2 \qquad [8.15]$$

where $I$ is the moment of inertia of the object around the axis of rotation.

A system involving rotation is described by three types of energy: potential energy $PE$, translational kinetic energy $KE_t$, and rotational kinetic energy $KE_r$ (Fig. 8.40). All these forms of energy must be included in the equation for conservation of mechanical energy for an isolated system:

$$(KE_t + KE_r + PE)_i = (KE_t + KE_r + PE)_f \qquad [8.16]$$

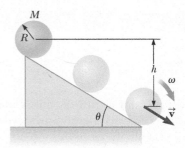

**Figure 8.40** A ball rolling down an incline converts potential energy to translational and rotational kinetic energy.

where $i$ and $f$ refer to initial and final values, respectively. When nonconservative forces are present, it's necessary to use a generalization of the work–energy theorem:

$$W_{nc} = \Delta KE_t + \Delta KE_r + \Delta PE \qquad [8.17]$$

## 8.6 Angular Momentum

The **angular momentum** of a rotating object is given by

$$L \equiv I\omega \qquad [8.18]$$

Angular momentum is related to torque in the following equation:

$$\sum \tau = \frac{\text{change in angular momentum}}{\text{time interval}} = \frac{\Delta L}{\Delta t} \qquad [8.19]$$

If the net external torque acting on a system is zero, the total angular momentum of the system is constant,

$$L_i = L_f \qquad [8.20]$$

and is said to be conserved. Solving problems usually involves substituting into the expression

$$I_i\omega_i = I_f\omega_f \qquad [8.21]$$

and solving for the unknown.

## CONCEPTUAL QUESTIONS

1. Why can't you put your heels firmly against a wall and then bend over without falling?

2. Two point masses are the same distance $R$ from an axis of rotation and have moments of inertia $I_A$ and $I_B$. (a) If $I_B = 4I_A$, what is the ratio $m_B/m_A$ of the two masses? (b) At what distance from the axis of rotation should mass A be placed so that $I_B = I_A$?

3. If you see an object rotating, is there necessarily a net torque acting on it?

4. (a) Is it possible to calculate the torque acting on a rigid object without specifying an origin? (b) Is the torque independent of the location of the origin?

5. Why does a long pole help a tightrope walker stay balanced?

6. A person stands a distance $R$ from a door's hinges and pushes with a force $F$ directed perpendicular to its surface. By what factor does the applied torque change if the person's position and force change to (a) $2R$ and $2F$, (b) $2R$ and $F$, (c) $R$ and $F/2$, (d) $R/2$ and $F/2$?

7. Orbiting spacecraft contain internal gyroscopes that are used to control their orientation. (a) Apply the principle of conservation of angular momentum to determine the direction a spacecraft will rotate if an internal gyroscope begins to rotate in the counterclockwise direction. (b) How many mutually perpendicular gyroscopes with fixed axes of rotation are required to have full control over the spacecraft's orientation?

8. If you toss a textbook into the air, rotating it each time about one of the three axes perpendicular to it, you will find that it

will not rotate smoothly about one of those axes. (Try placing a strong rubber band around the book before the toss so that it will stay closed.) The book's rotation is stable about those axes having the largest and smallest moments of inertia, but unstable about the axis of intermediate moment. Try this on your own to find the axis that has this intermediate moment of inertia.

9. Stars originate as large bodies of slowly rotating gas. Because of gravity, these clumps of gas slowly decrease in size. What happens to the angular speed of a star as it shrinks? Explain.

10. An object is acted on by a single nonzero force of magnitude $F$. (a) Is it possible for the object to have zero acceleration $a$? (b) Is it possible for the object to have zero angular acceleration $\alpha$? (c) Is it possible for the object to be in mechanical equilibrium?

11. In a tape recorder, the tape is pulled past the read–write heads at a constant speed by the drive mechanism. Consider the reel from which the tape is pulled: As the tape is pulled off, the radius of the roll of remaining tape decreases. (a) How does the torque on the reel change with time? (b) If the tape mechanism is suddenly turned on so that the tape is quickly pulled with a large force, is the tape more likely to break when pulled from a nearly full reel or from a nearly empty reel?

12. (a) Give an example in which the net force acting on an object is zero, yet the net torque is nonzero. (b) Give an example in which the net torque acting on an object is zero, yet the net force is nonzero.

13. Gravity is an example of a central force that acts along the line connecting two spherical masses. As a planet orbits its sun, (a) how much torque does the sun's gravitational force exert on the planet? (b) What is the change in the planet's orbital angular momentum?

14. A cat usually lands on its feet regardless of the position from which it is dropped. A slow-motion film of a cat falling shows that the upper half of its body twists in one direction while the lower half twists in the opposite direction. (See Fig. CQ8.14.) Why does this type of rotation occur?

15. A solid disk and a hoop are simultaneously released from rest at the top of an incline and roll down without slipping. Which object reaches the bottom first? (a) The one that has the largest mass arrives first. (b) The one that has the largest radius arrives

first. (c) The hoop arrives first. (d) The disk arrives first. (e) The hoop and the disk arrive at the same time.

16. A mouse is initially at rest on a horizontal turntable mounted on a frictionless, vertical axle. As the mouse begins to walk clockwise around the perimeter, which of the following statements *must* be true of the turntable? (a) It also turns clockwise. (b) It turns counterclockwise with the same angular velocity as the mouse. (c) It remains stationary. (d) It turns counterclockwise because angular momentum is conserved. (e) It turns clockwise because mechanical energy is conserved.

17. The cars in a soapbox derby have no engines; they simply coast downhill. Which of the following design criteria is best from a competitive point of view? The car's wheels should (a) have large moments of inertia, (b) be massive, (c) be hoop-like wheels rather than solid disks, (d) be large wheels rather than small wheels, or (e) have small moments of inertia.

**Figure CQ8.14**

# PROBLEMS

*See the Preface for an explanation of the icons used in this problems set.*

## 8.1 Torque

1. A man opens a 1.00-m wide door by pushing on it with a force of 50.0 N directed perpendicular to its surface. What magnitude of torque does he apply about an axis through the hinges if the force is applied (a) at the center of the door? (b) at the edge farthest from the hinges?

2. A worker applies a torque to a nut with a wrench 0.500 m long. Because of the cramped space, she must exert a force upward at an angle of 60.0° with respect to a line from the nut through the end of the wrench. If the force she exerts has magnitude 80.0 N, what magnitude torque does she apply to the nut?

3. [V] The fishing pole in Figure P8.3 makes an angle of 20.0° with the horizontal. What is the magnitude of the torque exerted by the fish about an axis perpendicular to the page and passing through the angler's hand if the fish pulls on the fishing line with a force $\vec{\mathbf{F}} = 1.00 \times 10^2$ N at an angle 37.0° below the horizontal? The force is applied at a point 2.00 m from the angler's hands.

4. [T] Find the net torque on the wheel in Figure P8.4 about the axle through $O$ perpendicular to the page, taking $a = 10.0$ cm and $b = 25.0$ cm.

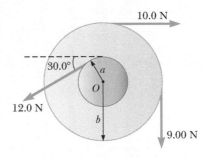

**Figure P8.4**

5. Calculate the net torque (magnitude and direction) on the beam in Figure P8.5 about (a) an axis through $O$ perpendicular to the page and (b) an axis through $C$ perpendicular to the page.

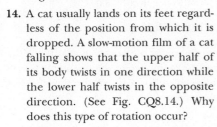

**Figure P8.3**

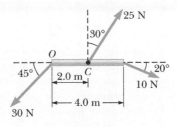

**Figure P8.5**

**6.** **BIO** A dental bracket exerts a horizontal force of 80.0 N on a tooth at point $B$ in Figure P8.6. What is the torque on the root of the tooth about point $A$?

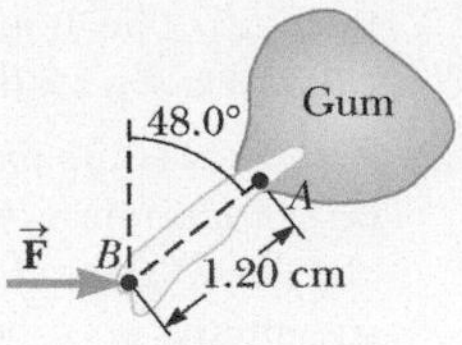

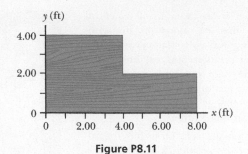

**Figure P8.6**

**7.** **QC** A simple pendulum consists of a small object of mass 3.0 kg hanging at the end of a 2.0-m-long light string that is connected to a pivot point. (a) Calculate the magnitude of the torque (due to the force of gravity) about this pivot point when the string makes a 5.0° angle with the vertical. (b) Does the torque increase or decrease as the angle increases? Explain.

## 8.2 Center of Mass and Its Motion

**8.** **V** Consider the following mass distribution, where $x$- and $y$-coordinates are given in meters: 5.0 kg at (0.0, 0.0) m, 3.0 kg at (0.0, 4.0) m, and 4.0 kg at (3.0, 0.0) m. Where should a fourth object of 8.0 kg be placed so that the center of mass of the four-object arrangement will be at (0.0, 0.0) m?

**9.** Two bowling balls are at rest on top of a uniform wooden plank with their centers of mass located as in Figure P8.9. The plank has a mass of 5.00 kg and is 1.00 m long. Find the horizontal distance from the left end of the plank to the center of mass of the plank–bowling balls system.

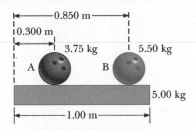

**Figure P8.9**

**10.** Three solid, uniform boxes are aligned as in Figure P8.10. Find the $x$- and $y$-coordinates of the center of mass of the three boxes, measured from the bottom left corner of box A.

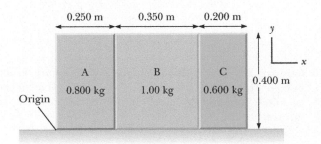

**Figure P8.10**

**11.** Find the $x$- and $y$-coordinates of the center of gravity of a 4.00-ft by 8.00-ft uniform sheet of plywood with the upper right quadrant removed as shown in Figure P8.11. *Hint:* The mass of any segment of the plywood sheet is proportional to the area of that segment.

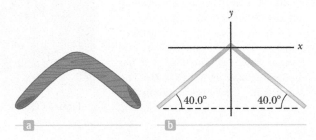

**Figure P8.11**

**12.** Find the $x$- and $y$-coordinates of the center of gravity for the boomerang in Figure P8.12a, modeling the boomerang as in Figure P8.12b, where each uniform leg of the model has a length of 0.300 m and a mass of 0.250 kg. (*Note:* Treat the legs like thin rods.)

**Figure P8.12**

**13.** A block of mass $m = 1.50$ kg is at rest on a ramp of mass $M = 4.50$ kg which, in turn, is at rest on a frictionless horizontal surface (Fig. P8.13a). The block and the ramp are aligned so that each has its center of mass located at $x = 0$. When released, the block slides down the ramp to the left and the ramp, also free to slide on the frictionless surface, slides to the right as in Figure P8.13b. Calculate $x_{ramp}$, the distance the ramp has moved to the right, when $x_{block} = -0.300$ m.

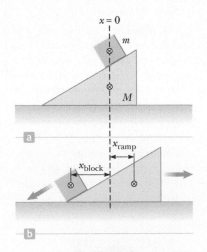

**Figure P8.13**

**14.** The Xanthar mothership locks onto an enemy cruiser with its tractor beam (Fig. P8.14); each ship is at rest in deep space with no propulsion following a devastating battle. The mothership is at $x = 0$ when its tractor beams are first engaged, a distance $d = 215$ xiles from the cruiser. Determine the x-position in xiles of the two spacecraft when the tractor beam has pulled them together. Model each spacecraft as a point particle with the mothership of mass $M = 185$ xons and the cruiser of mass $m = 20.0$ xons.

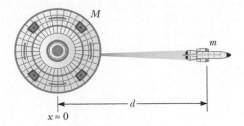

**Figure P8.14**

15. A hiker inspects a tree frog sitting on a small stick in his hand. Suddenly startled, the hiker drops the stick from rest at a height of 1.85 m above the ground and, at the same instant, the frog leaps vertically upward, pushing the stick down so that it hits the ground 0.450 s later. Find the height of the frog at the instant the stick hits the ground if the frog and the stick have masses of 7.25 g and 4.50 g, respectively. (*Hint:* Find the center-of-mass height at $t = 0.450$ s for the frog–stick system and then use the definition of center of mass to solve for the frog's height.)

16. Spectators watch a bicycle stunt rider travel off the end of a 60.0° ramp, rise to the top of his trajectory and, at that instant, suddenly push his bike away from him so that he falls vertically straight down, reaching the ground 0.550 s later. How far from the rider does the bicycle land if the rider has mass $M = 72.0$ kg and the bike has mass $m = 12.0$ kg? Neglect air resistance and assume the ground is level.

## 8.3 Torque and the Two Conditions for Equilibrium

17. **BIO** The arm in Figure P8.17 weighs 41.5 N. The force of gravity acting on the arm acts through point $A$. Determine the magnitudes of the tension force $\vec{F}_t$ in the deltoid muscle and the force $\vec{F}_s$ exerted by the shoulder on the humerus (upper-arm bone) to hold the arm in the position shown.

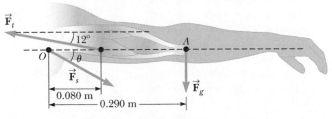

**Figure P8.17**

18. A uniform 35.0-kg beam of length $\ell = 5.00$ m is supported by a vertical rope located $d = 1.20$ m from its left end as in Figure P8.18. The right end of the beam is supported by a vertical column. Find (a) the tension in the rope and (b) the force that the column exerts on the right end of the beam.

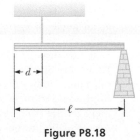

**Figure P8.18**

19. **BIO** A cook holds a 2.00-kg carton of milk at arm's length (Fig. P8.19). What force $\vec{F}_B$ must be exerted by the biceps muscle? (Ignore the weight of the forearm.)

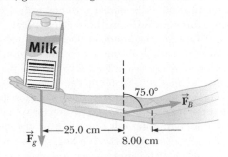

**Figure P8.19**

20. A meter stick is found to balance at the 49.7-cm mark when placed on a fulcrum. When a 50.0-gram mass is attached at the 10.0-cm mark, the fulcrum must be moved to the 39.2-cm mark for balance. What is the mass of the meter stick?

21. **BIO** In exercise physiology studies, it is sometimes important to determine the location of a person's center of gravity. This can be done with the arrangement shown in Figure P8.21. A light plank rests on two scales that read $F_{g1} = 380.$ N and $F_{g2} = 320.$ N. The scales are separated by a distance of 2.00 m. How far from the woman's feet is her center of gravity?

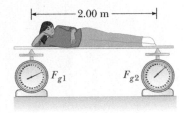

**Figure P8.21**

22. **GP** A beam resting on two pivots has a length of $L = 6.00$ m and mass $M = 90.0$ kg. The pivot under the left end exerts a normal force $n_1$ on the beam, and the second pivot placed a distance $\ell = 4.00$ m from the left end exerts a normal force $n_2$. A woman of mass $m = 55.0$ kg steps onto the left end of the beam and begins walking to the right as in Figure P8.22. The goal is to find the woman's position when the beam begins to tip. (a) Sketch a free-body diagram, labeling the gravitational and normal forces acting on the beam and placing the woman $x$ meters to the right of the first pivot, which is the origin. (b) Where is the woman when the normal force $n_1$ is the greatest? (c) What is $n_1$ when the beam is about to tip? (d) Use the force equation of equilibrium to find the value of $n_2$ when the beam is about to tip. (e) Using the result of part (c) and the torque equilibrium equation, with torques computed around the second pivot point, find the woman's position when the beam is about to tip. (f) Check the answer to part (e) by computing torques around the first pivot point. Except for possible slight differences due to rounding, is the answer the same?

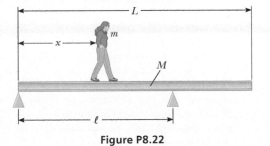

**Figure P8.22**

23. **BIO** A person bending forward to lift a load "with his back" (Fig. P8.23a) rather than "with his knees" can be injured by large forces exerted on the muscles and vertebrae. The spine pivots mainly at the fifth lumbar vertebra, with the principal supporting force provided by the erector spinalis muscle in the back. To see the magnitude of the forces involved, and to understand why back problems are common among humans, consider the model shown in Figure P8.23b of a person bending forward to lift a 200.-N object. The spine and upper body are represented as a uniform horizontal rod of weight 350. N, pivoted at the base of the spine. The erector spinalis muscle, attached at a point two-thirds of the way up the spine, maintains the position of the back. The angle between the spine and this muscle is 12.0°. Find (a) the tension in the back muscle and (b) the compressional force in the spine.

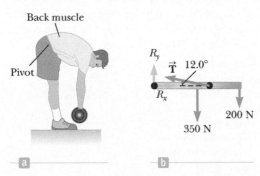

Figure P8.23

24. **BIO** When a person stands on tiptoe (a strenuous position), the position of the foot is as shown in Figure P8.24a. The total gravitational force on the body, $\vec{F}_g$, is supported by the force $\vec{n}$ exerted by the floor on the toes of one foot. A mechanical model of the situation is shown in Figure P8.24b, where $\vec{T}$ is the force exerted by the Achilles tendon on the foot and $\vec{R}$ is the force exerted by the tibia on the foot. Find the values of $T$, $R$, and $\theta$ when $F_g = n = 700.$ N.

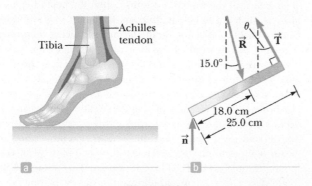

Figure P8.24

25. **T** A 500.-N uniform rectangular sign 4.00 m wide and 3.00 m high is suspended from a horizontal, 6.00-m-long, uniform, 100.-N rod as indicated in Figure P8.25. The left end of the rod is supported by a hinge, and the right end is supported by a thin cable making a 30.0° angle with the vertical. (a) Find the tension $T$ in the cable. (b) Find the horizontal and vertical components of force exerted on the left end of the rod by the hinge.

Figure P8.25

26. A window washer is standing on a scaffold supported by a vertical rope at each end. The scaffold weighs 200 N and is 3.00 m long. What is the tension in each rope when the 700-N worker stands 1.00 m from one end?

27. **V** A uniform plank of length 2.00 m and mass 30.0 kg is supported by three ropes, as indicated by the blue vectors in Figure P8.27. Find the tension in each rope when a 700.-N person is $d = 0.500$ m from the left end.

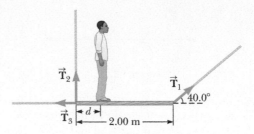

Figure P8.27

28. A hungry bear weighing 700. N walks out on a beam in an attempt to retrieve a basket of goodies hanging at the end of the beam (Fig. P8.28). The beam is uniform, weighs 200. N, and is 6.00 m long, and it is supported by a wire at an angle of $\theta = 60.0°$. The basket weighs 80.0 N. (a) Draw a force diagram for the beam. (b) When the bear is at $x = 1.00$ m, find the tension in the wire supporting the beam and the components of the force exerted by the wall on the left end of the beam. (c) If the wire can withstand a maximum tension of 900. N, what is the maximum distance the bear can walk before the wire breaks?

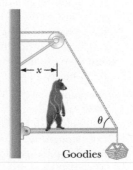

Figure P8.28

29. **S** Figure P8.29 shows a uniform beam of mass $m$ pivoted at its lower end, with a horizontal spring attached between its top end and a vertical wall. The beam makes an angle $\theta$ with the horizontal. Find expressions for (a) the distance $d$ the spring is stretched from equilibrium and (b) the components of the force exerted by the pivot on the beam.

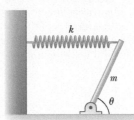

Figure P8.29

30. **GP** A strut of length $L = 3.00$ m and mass $m = 16.0$ kg is held by a cable at an angle of $\theta = 30.0°$ with respect to the horizontal as shown in Figure P8.30. (a) Sketch a force diagram, indicating all the forces and their placement on the strut. (b) Why is the hinge a good place to use for calculating torques? (c) Write the condition for rotational equilibrium symbolically, calculating the torques around the hinge. (d) Use the torque equation to calculate the tension

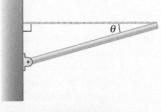

Figure P8.30

in the cable. (e) Write the $x$- and $y$-components of Newton's second law for equilibrium. (f) Use the force equation to find the $x$- and $y$-components of the force on the hinge. (g) Assuming the strut position is to remain the same, would it be advantageous to attach the cable higher up on the wall? Explain the benefit in terms of the force on the hinge and cable tension.

31. **S** A refrigerator of width $w$ and height $h$ rests on a rough incline as in Figure P8.31. Find an expression for the maximum value $\theta$ can have before the refrigerator tips over. Note, the contact point between the refrigerator and incline shifts as $\theta$ increases and treat the refrigerator as a uniform box.

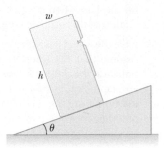

**Figure P8.31**

32. **S** Write the necessary equations of equilibrium of the object shown in Figure P8.32. Take the origin of the torque equation about an axis perpendicular to the page through the point $O$.

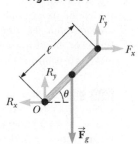

**Figure P8.32**

33. **V** **BIO** The chewing muscle, the masseter, is one of the strongest in the human body. It is attached to the mandible (lower jawbone) as shown in Figure P8.33a. The jawbone is pivoted about a socket just in front of the auditory canal. The forces acting on the jawbone are equivalent to those acting on the curved bar in Figure P8.33b. $\vec{F}_C$ is the force exerted by the food being chewed against the jawbone, $\vec{T}$ is the force of tension in the masseter, and $\vec{R}$ is the force exerted by the socket on the mandible. Find $\vec{T}$ and $\vec{R}$ for a person who bites down on a piece of steak with a force of 50.0 N.

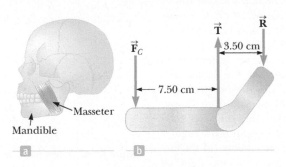

**Figure P8.33**

34. A 1 200-N uniform boom at $\phi = 65°$ to the horizontal is supported by a cable at an angle $\theta = 25.0°$ to the horizontal as shown in Figure P8.34. The boom is pivoted at the bottom, and an object of weight $w = 2\,000$ N hangs from its top. Find (a) the tension in the support cable and

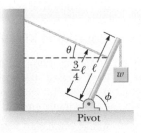

**Figure P8.34**

(b) the components of the reaction force exerted by the pivot on the boom.

35. **BIO** The large quadriceps muscle in the upper leg terminates at its lower end in a tendon attached to the upper end of the tibia (Fig. P8.35a). The forces on the lower leg when the leg is extended are modeled as in Figure P8.35b, where $\vec{T}$ is the force of tension in the tendon, $\vec{w}$ is the force of gravity acting on the lower leg, and $\vec{F}$ is the force of gravity acting on the foot. Find $\vec{T}$ when the tendon is at an angle of 25.0° with the tibia, assuming that $w = 30.0$ N, $F = 12.5$ N, and the leg is extended at an angle $\theta$ of 40.0° with the vertical. Assume that the center

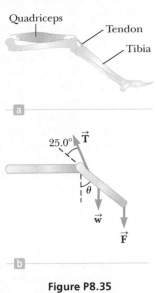

**Figure P8.35**

of gravity of the lower leg is at its center and that the tendon attaches to the lower leg at a point one-fifth of the way down the leg.

36. **V** One end of a uniform 4.0-m-long rod of weight $w$ is supported by a cable at an angle of $\theta = 37°$ with the rod. The other end rests against a wall, where it is held by friction. (See Fig. P8.36.) The coefficient of static friction between the wall and the rod is $\mu_s = 0.50$. Determine the minimum distance $x$ from point $A$ at

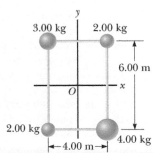

**Figure P8.36**

which an additional weight $w$ (the same as the weight of the rod) can be hung without causing the rod to slip at point $A$.

## 8.4 The Rotational Second Law of Motion

37. Four objects are held in position at the corners of a rectangle by light rods as shown in Figure P8.37. Find the moment of inertia of the system about (a) the $x$-axis, (b) the $y$-axis, and (c) an axis through $O$ and perpendicular to the page.

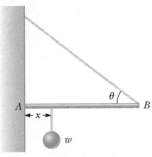

**Figure P8.37** Problems 37 and 38.

38. If the system shown in Figure P8.37 is set in rotation about each of the axes mentioned in Problem 37, find the torque that will produce an angular acceleration of 1.50 rad/s² in each case.

39. A large grinding wheel in the shape of a solid cylinder of radius 0.330 m is free to rotate on a frictionless, vertical axle. A constant tangential force of 250. N applied to its edge causes the wheel to have an angular acceleration of 0.940 rad/s².

(a) What is the moment of inertia of the wheel? (b) What is the mass of the wheel? (c) If the wheel starts from rest, what is its angular velocity after 5.00 s have elapsed, assuming the force is acting during that time?

40. **GP** An oversized yo-yo is made from two identical solid disks each of mass $M = 2.00$ kg and radius $R = 10.0$ cm. The two disks are joined by a solid cylinder of radius $r = 4.00$ cm and mass $m = 1.00$ kg as in Figure P8.40. Take the center of the cylinder as the axis of the system, with positive torques directed to the left along this axis. All torques and angular variables are to be calculated around this axis. Light string is wrapped around the cylinder, and the system is then allowed to drop from rest. (a) What is the moment of inertia of the system? Give a symbolic answer. (b) What torque does gravity exert on the system with respect to the given axis? (c) Take downward as the negative coordinate direction. As depicted in Figure P8.40, is the torque exerted by the tension positive or negative? Is the angular acceleration positive or negative? What about the translational acceleration? (d) Write an equation for the angular acceleration $\alpha$ in terms of the translational acceleration $a$ and radius $r$. (Watch the sign!) (e) Write Newton's second law for the system in terms of $m$, $M$, $a$, $T$, and $g$. (f) Write Newton's second law for rotation in terms of $I$, $\alpha$, $T$, and $r$. (g) Eliminate $\alpha$ from the rotational second law with the expression found in part (d) and find a symbolic expression for the acceleration $a$ in terms of $m$, $M$, $g$, $r$, and $R$. (h) What is the numeric value for the system's acceleration? (i) What is the tension in the string? (j) How long does it take the system to drop 1.00 m from rest?

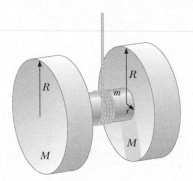

**Figure P8.40**

41. An approximate model for a ceiling fan consists of a cylindrical disk with four thin rods extending from the disk's center, as in Figure P8.41. The disk has mass 2.50 kg and radius 0.200 m. Each rod has mass 0.850 kg and is 0.750 m long. (a) Find the ceiling fan's moment of

**Figure P8.41**

inertia about a vertical axis through the disk's center. (b) Friction exerts a constant torque of magnitude 0.115 N · m on the fan as it rotates. Find the magnitude of the constant torque provided by the fan's motor if the fan starts from rest and takes 15.0 s and 18.5 full revolutions to reach its maximum speed.

42. **V** A potter's wheel having a radius of 0.50 m and a moment of inertia of 12 kg · m² is rotating freely at 50 rev/min. The potter can stop the wheel in 6.0 s by pressing a wet rag against the rim and exerting a radially inward force of 70 N. Find the

effective coefficient of kinetic friction between the wheel and the wet rag.

43. A model airplane with mass 0.750 kg is tethered by a wire so that it flies in a circle 30.0 m in radius. The airplane engine provides a net thrust of 0.800 N perpendicular to the tethering wire. (a) Find the torque the net thrust produces about the center of the circle. (b) Find the angular acceleration of the airplane when it is in level flight. (c) Find the linear acceleration of the airplane tangent to its flight path.

44. A bicycle wheel has a diameter of 64.0 cm and a mass of 1.80 kg. Assume that the wheel is a hoop with all the mass concentrated on the outside radius. The bicycle is placed on a stationary stand, and a resistive force of 120 N is applied tangent to the rim of the tire. (a) What force must be applied by a chain passing over a 9.00-cm-diameter sprocket to give the wheel an acceleration of 4.50 rad/s²? (b) What force is required if you shift to a 5.60-cm-diameter sprocket?

45. **T** A 150.-kg merry-go-round in the shape of a uniform, solid, horizontal disk of radius 1.50 m is set in motion by wrapping a rope about the rim of the disk and pulling on the rope. What constant force must be exerted on the rope to bring the merry-go-round from rest to an angular speed of 0.500 rev/s in 2.00 s?

46. **Q|C** **S** An Atwood's machine consists of blocks of masses $m_1 = 10.0$ kg and $m_2 = 20.0$ kg attached by a cord running over a pulley as in Figure P8.46. The pulley is a solid cylinder with mass $M = 8.00$ kg and radius $r = 0.200$ m. The block of mass $m_2$ is allowed to drop, and the cord turns the pulley without slipping. (a) Why must the tension $T_2$ be greater than the tension $T_1$? (b) What is the acceleration of the system, assuming the pulley axis is frictionless? (c) Find the tensions $T_1$ and $T_2$.

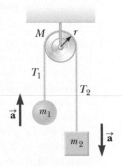

**Figure P8.46**

47. The uniform thin rod in Figure P8.47 has mass $M = 3.50$ kg and length $L = 1.00$ m and is free to rotate on a frictionless pin. At the instant the rod is released from rest in the horizontal position, find the magnitude of (a) the rod's angular acceleration, (b) the tangential acceleration of the rod's center of mass, and (c) the tangential acceleration

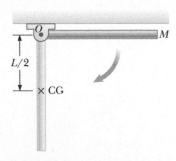

**Figure P8.47** Problems 47 and 86.

of the rod's free end.

## 8.5 Rotational Kinetic Energy

48. A 2.50-kg solid, uniform disk rolls without slipping across a level surface, translating at 3.75 m/s. If the disk's radius is 0.100 m, find its (a) translational kinetic energy and (b) rotational kinetic energy.

49. A horizontal 800.-N merry-go-round of radius 1.50 m is started from rest by a constant horizontal force of 50.0 N applied

tangentially to the merry-go-round. Find the kinetic energy of the merry-go-round after 3.00 s. (Assume it is a solid cylinder.)

50. **Q|C** Four objects—a hoop, a solid cylinder, a solid sphere, and a thin, spherical shell—each have a mass of 4.80 kg and a radius of 0.230 m. (a) Find the moment of inertia for each object as it rotates about the axes shown in Table 8.1. (b) Suppose each object is rolled down a ramp. Rank the translational speed of each object from highest to lowest. (c) Rank the objects' rotational kinetic energies from highest to lowest as the objects roll down the ramp.

51. A light rod of length $\ell = 1.00$ m rotates about an axis perpendicular to its length and passing through its center as in Figure P8.51. Two particles of masses $m_1 = 4.00$ kg and $m_2 = 3.00$ kg are connected to the ends of the rod. (a) Neglecting the mass of the rod, what is the system's kinetic energy when its angular speed is 2.50 rad/s? (b) Repeat the problem, assuming the mass of the rod is taken to be 2.00 kg.

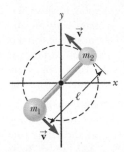

**Figure P8.51** Problems 51 and 65.

52. A 240-N sphere 0.20 m in radius rolls without slipping 6.0 m down a ramp that is inclined at 37° with the horizontal. What is the angular speed of the sphere at the bottom of the slope if it starts from rest?

53. A solid, uniform disk of radius 0.250 m and mass 55.0 kg rolls down a ramp of length 4.50 m that makes an angle of 15.0° with the horizontal. The disk starts from rest from the top of the ramp. Find (a) the speed of the disk's center of mass when it reaches the bottom of the ramp and (b) the angular speed of the disk at the bottom of the ramp.

54. A car is designed to get its energy from a rotating solid-disk flywheel with a radius of 2.00 m and a mass of $5.00 \times 10^2$ kg. Before a trip, the flywheel is attached to an electric motor, which brings the flywheel's rotational speed up to $5.00 \times 10^3$ rev/min. (a) Find the kinetic energy stored in the flywheel. (b) If the flywheel is to supply energy to the car as a 10.0-hp motor would, find the length of time the car could run before the flywheel would have to be brought back up to speed.

55. The top in Figure P8.55 has a moment of inertia of $4.00 \times 10^{-4}$ kg · m$^2$ and is initially at rest. It is free to rotate about a stationary axis $AA'$. A string wrapped around a peg along the axis of the top is pulled in such a manner as to maintain a constant tension of 5.57 N in the string. If the string does not slip while wound around the peg, what is the angular speed of the top after 80.0 cm of string has been pulled off the peg? *Hint:* Consider the work that is done.

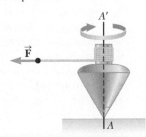

**Figure P8.55**

56. A constant torque of 25.0 N · m is applied to a grindstone whose moment of inertia is 0.130 kg · m$^2$. Using energy principles and neglecting friction, find the angular speed after the grindstone has made 15.0 revolutions. *Hint:* The angular

equivalent of $W_{net} = F\Delta x = \frac{1}{2}mv_f^2 - \frac{1}{2}mv_i^2$ is $W_{net} = \tau\Delta\theta = \frac{1}{2}I\omega_f^2 - \frac{1}{2}I\omega_i^2$. You should convince yourself that this last relationship is correct.

57. A 10.0-kg cylinder rolls without slipping on a rough surface. At an instant when its center of gravity has a speed of 10.0 m/s, determine (a) the translational kinetic energy of its center of gravity, (b) the rotational kinetic energy about its center of gravity, and (c) its total kinetic energy.

58. **V** Use conservation of energy to determine the angular speed of the spool shown in Figure P8.58 after the 3.00-kg bucket has fallen 4.00 m, starting from rest. The light string attached to the bucket is wrapped around the spool and does not slip as it unwinds.

**Figure P8.58**

59. A 2.00-kg solid, uniform ball of radius 0.100 m is released from rest at point A in Figure P8.59, its center of gravity a distance of 1.50 m above the ground. The ball rolls without slipping to the bottom of an incline and back up to point B where it is launched vertically into the air. The ball rises to its maximum height $h_{max}$ at point C. At point B, find the ball's (a) translational speed $v_B$ and (b) rotational speed $\omega_B$. At point C, find the ball's (c) rotational speed $\omega_C$ and (d) maximum height $h_{max}$ of its center of gravity.

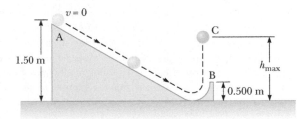

**Figure P8.59**

## 8.6 Angular Momentum

60. Each of the following objects has a radius of 0.180 m and a mass of 2.40 kg, and each rotates about an axis through its center (as in Table 8.1) with an angular speed of 35.0 rad/s. Find the magnitude of the angular momentum of each object. (a) a hoop (b) a solid cylinder (c) a solid sphere (d) a hollow spherical shell

61. A metal hoop lies on a horizontal table, free to rotate about a fixed vertical axis through its center while a constant tangential force applied to its edge exerts a torque of magnitude $1.25 \times 10^{-2}$ N · m for 2.00 s. (a) Calculate the magnitude of the hoop's change in angular momentum. (b) Find the change in the hoop's angular speed if its mass and radius are 0.250 kg and 0.100 m, respectively.

62. A disk of mass $m$ is spinning freely at 6.00 rad/s when a second identical disk, initially not spinning, is dropped onto it so that their axes coincide. In a short time the two disks are corotating. (a) What is the angular speed of the new system? (b) If a third such disk is dropped on the first two, find the final angular speed of the system.

63. (a) Calculate the angular momentum of Earth that arises from its spinning motion on its axis, treating Earth as a uniform solid sphere. (b) Calculate the angular momentum of Earth that arises from its orbital motion about the Sun, treating Earth as a point particle.

64. Q|C A 0.005 00-kg bullet traveling horizontally with a speed of $1.00 \times 10^3$ m/s enters an 18.0-kg door, embedding itself 10.0 cm from the side opposite the hinges as in Figure P8.64. The 1.00-m-wide door is free to swing on its hinges. (a) Before it hits the door, does the bullet have angular momentum relative the door's axis of rotation? Explain. (b) Is mechanical energy conserved in this collision? Answer without doing a calculation. (c) At what angular speed does the door swing open immediately after the collision? (The door has the same moment of inertia as a rod with axis at one end.) (d) Calculate the energy of the door–bullet system and determine whether it is less than or equal to the kinetic energy of the bullet before the collision.

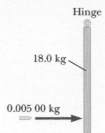

**Figure P8.64.** An overhead view of a bullet striking a door.

Hinge

18.0 kg

0.005 00 kg

65. A light rigid rod of length $\ell = 1.00$ m rotates about an axis perpendicular to its length and through its center, as shown in Figure P8.51. Two particles of masses $m_1 = 4.00$ kg and $m_2 = 3.00$ kg are connected to the ends of the rod. What is the angular momentum of the system if the speed of each particle is 5.00 m/s? (Neglect the rod's mass.)

66. Halley's comet moves about the Sun in an elliptical orbit, with its closest approach to the Sun being 0.59 AU and its greatest distance being 35 AU (1 AU is the Earth–Sun distance). If the comet's speed at closest approach is 54 km/s, what is its speed when it is farthest from the Sun? You may neglect any change in the comet's mass and assume that its angular momentum about the Sun is conserved.

67. A student holds a spinning bicycle wheel while sitting motionless on a stool that is free to rotate about a vertical axis through its center (Fig. P8.67). The wheel spins with an angular speed of 17.5 rad/s and its initial angular momentum is directed up. The wheel's moment of inertia is 0.150 kg · m² and the moment of inertia for the student plus stool is 3.00 kg · m². (a) Find the student's final angular speed after he turns the wheel over so that it spins at the same speed but with its angular momentum directed down. (b) Will the student's final angular momentum be directed up or down?

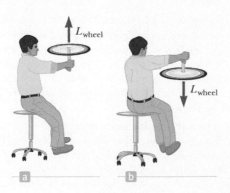

$L_{wheel}$

$L_{wheel}$

a

b

**Figure P8.67**

68. T A 60.0-kg woman stands at the rim of a horizontal turntable having a moment of inertia of 500 kg · m² and a radius of 2.00 m. The turntable is initially at rest and is free to rotate about a frictionless, vertical axle through its center. The woman then starts walking around the rim clockwise (as viewed from above the system) at a constant speed of 1.50 m/s relative to Earth. (a) In what direction and with what angular speed does the turntable rotate? (b) How much work does the woman do to set herself and the turntable into motion?

69. A solid, horizontal cylinder of mass 10.0 kg and radius 1.00 m rotates with an angular speed of 7.00 rad/s about a fixed vertical axis through its center. A 0.250-kg piece of putty is dropped vertically onto the cylinder at a point 0.900 m from the center of rotation and sticks to the cylinder. Determine the final angular speed of the system.

70. A student sits on a rotating stool holding two 3.0-kg objects. When his arms are extended horizontally, the objects are 1.0 m from the axis of rotation and he rotates with an angular speed of 0.75 rad/s. The moment of inertia of the student plus stool is 3.0 kg · m² and is assumed to be constant. The student then pulls in the objects horizontally to 0.30 m from the rotation axis. (a) Find the new angular speed of the student. (b) Find the kinetic energy of the student before and after the objects are pulled in.

71. V The puck in Figure P8.71 has a mass of 0.120 kg. Its original distance from the center of rotation is 40.0 cm, and it moves with a speed of 80.0 cm/s. The string is pulled downward 15.0 cm through the hole in the frictionless table. Determine the work done on the puck. *Hint:* Consider the change in kinetic energy of the puck.

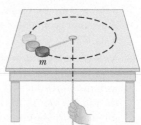

$m$

**Figure P8.71**

72. A space station shaped like a giant wheel has a radius of 100 m and a moment of inertia of $5.00 \times 10^8$ kg · m². A crew of 150 lives on the rim, and the station is rotating so that the crew experiences an apparent acceleration of 1$g$ (Fig. P8.72). When 100 people move to the center of the station for a union meeting, the angular speed changes. What apparent acceleration is experienced by the managers remaining at the rim? Assume the average mass of a crew member is 65.0 kg.

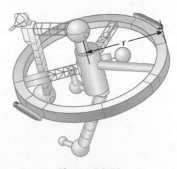

$r$

**Figure P8.72**

73. Q|C S A cylinder with moment of inertia $I_1$ rotates with angular velocity $\omega_0$ about a frictionless vertical axle. A second cylinder, with moment of inertia $I_2$, initially not rotating, drops onto the first cylinder (Fig. P8.73). Because the surfaces are rough, the two cylinders eventually reach the same angular speed $\omega$. (a) Calculate $\omega$. (b) Show that kinetic energy is

lost in this situation, and calculate the ratio of the final to the initial kinetic energy.

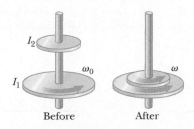

Before          After

**Figure P8.73**

74. **T** A particle of mass 0.400 kg is attached to the 100-cm mark of a meter stick of mass 0.100 kg. The meter stick rotates on a horizontal, frictionless table with an angular speed of 4.00 rad/s. Calculate the angular momentum of the system when the stick is pivoted about an axis (a) perpendicular to the table through the 50.0-cm mark and (b) perpendicular to the table through the 0-cm mark.

## Additional Problems

75. **GP** A typical propeller of a turbine used to generate electricity from the wind consists of three blades as in Figure P8.75. Each blade has a length of $L = 35$ m and a mass of $m = 420$ kg. The propeller rotates at the rate of 25 rev/min. (a) Convert the angular speed of the propeller to units of rad/s. Find

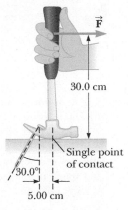

**Figure P8.75**

(b) the moment of inertia of the propeller about the axis of rotation and (c) the total kinetic energy of the propeller.

76. Figure P8.76 shows a clawhammer as it is being used to pull a nail out of a horizontal board. If a force of magnitude 150 N is exerted horizontally as shown, find (a) the force exerted by the hammer claws on the nail and (b) the force exerted by the surface at the point of contact with the hammer head. Assume that the force the hammer exerts on the nail is parallel to the nail and perpendicular to the position vector from the point of contact.

**Figure P8.76**

77. **QC** A 40.0-kg child stands at one end of a 70.0-kg boat that is 4.00 m long (Fig. P8.77). The boat is initially 3.00 m from the pier. The child notices a turtle on a rock beyond the far end of the boat and proceeds to walk to that end to catch the turtle. (a) Neglecting friction between the boat and water, describe the motion of the system (child plus boat). (b) Where will the child be relative to the pier when he reaches the far end of the boat? (c) Will he catch the turtle? (Assume that he can reach out 1.00 m from the end of the boat.)

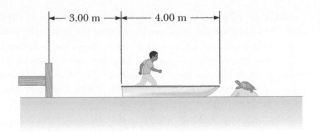

**Figure P8.77**

78. An object of mass $M = 12.0$ kg is attached to a cord that is wrapped around a wheel of radius $r = 10.0$ cm (Fig. P8.78). The acceleration of the object down the frictionless incline is measured to be $a = 2.00$ m/s$^2$ and the incline makes an angle $\theta = 37.0°$ with the horizontal. Assuming the axle of the wheel to be frictionless, determine (a) the tension in the rope, (b) the moment of inertia of the wheel, and (c) the angular speed of the wheel 2.00 s after it begins rotating, starting from rest.

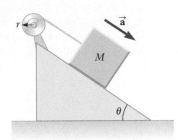

**Figure P8.78**

79. A uniform ladder of length $L$ and weight $w$ is leaning against a vertical wall. The coefficient of static friction between the ladder and the floor is the same as that between the ladder and the wall. If this coefficient of static friction is $\mu_s = 0.500$, determine the smallest angle the ladder can make with the floor without slipping.

80. Two astronauts (Fig. P8.80), each having a mass of 75.0 kg, are connected by a 10.0-m rope of negligible mass. They are isolated in space, moving in circles around the point halfway between them at a speed of 5.00 m/s. Treating the astronauts as particles, calculate (a) the magnitude of the angular momentum and (b) the rotational energy of the system. By pulling on the rope, the astronauts shorten the distance between them to 5.00 m. (c) What is the new angular momentum of the system? (d) What are their new speeds? (e) What is the new rotational energy of the system? (f) How much work is done by the astronauts in shortening the rope?

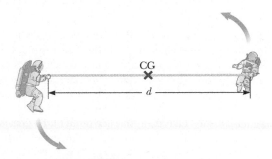

**Figure P8.80** Problems 80 and 81.

81. **S** *This is a symbolic version of problem 80.* Two astronauts (Fig. P8.80), each having a mass $M$, are connected by a rope of length $d$ having negligible mass. They are isolated in space, moving in circles around the point halfway between them at a speed $v$. (a) Calculate the magnitude of the angular

momentum of the system by treating the astronauts as particles. (b) Calculate the rotational energy of the system. By pulling on the rope, the astronauts shorten the distance between them to $d/2$. (c) What is the new angular momentum of the system? (d) What are their new speeds? (e) What is the new rotational energy of the system? (f) How much work is done by the astronauts in shortening the rope?

82. **V** Two window washers, Bob and Joe, are on a 3.00-m-long, 345-N scaffold supported by two cables attached to its ends. Bob weighs 750 N and stands 1.00 m from the left end, as shown in Figure P8.82. Two meters from the left end is the 500-N washing equipment. Joe is 0.500 m from the right end and weighs 1 000 N. Given that the scaffold is in rotational and translational equilibrium, what are the forces on each cable?

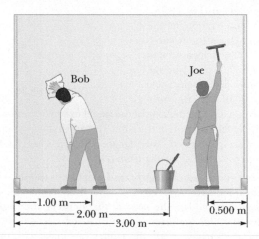

**Figure P8.82**

83. A 2.35-kg uniform bar of length $\ell = 1.30$ m is held in a horizontal position by three vertical springs as in Figure P8.83. The two lower springs are compressed and exert *upward* forces on the bar of magnitude $F_1 = 6.80$ N and $F_2 = 9.50$ N, respectively. Find (a) the force $F_s$ exerted by the top spring on the bar,

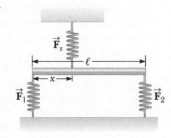

**Figure P8.83**

and (b) the location $x$ of the upper spring that will keep the bar in equilibrium.

84. **QC S** A light rod of length $2L$ is free to rotate in a *vertical* plane about a frictionless pivot through its center. A particle of mass $m_1$ is attached at one end of the rod, and a mass $m_2$ is at the opposite end, where $m_1 > m_2$. The system is released from rest in the vertical position shown in Figure P8.84a, and at some later time, the system is rotating in the position shown in Figure P8.84b. Take the reference point of the gravitational potential energy to be at the pivot. (a) Find an expression for the system's total mechanical energy in the vertical position. (b) Find an expression for the total mechanical energy in the rotated position shown in Figure P8.84b. (c) Using the fact that the mechanical energy of the system is conserved, how would you determine the angular speed $\omega$ of the system in the rotated position? (d) Find the magnitude of the torque on the system in the vertical position and in the rotated position.

Is the torque constant? Explain what these results imply regarding the angular momentum of the system. (e) Find an expression for the magnitude of the angular acceleration of the system in the rotated position. Does your result make sense when the rod is horizontal? When it is vertical? Explain.

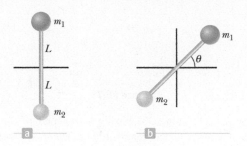

**Figure P8.84**

85. **BIO** Many aspects of a gymnast's motion can be modeled by representing the gymnast by four segments consisting of arms, torso (including the head), thighs, and lower legs, as in Figure P8.85. Figure P8.85b shows arrows of lengths $r_{cg}$ locating the center of gravity of each segment. Use the data below and the coordinate system shown in Figure P8.85b to locate the center of gravity of the gymnast shown in Figure P8.85a. Masses for the arms, thighs, and legs include both appendages.

| Segment | Mass (kg) | Length (m) | $r_{cg}$ (m) |
|---|---|---|---|
| Arms | 6.87 | 0.548 | 0.239 |
| Torso | 33.57 | 0.601 | 0.337 |
| Thighs | 14.07 | 0.374 | 0.151 |
| Legs | 7.54 | 0.350 | 0.227 |

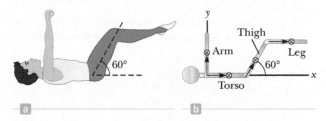

**Figure P8.85**

86. **S** A uniform thin rod of length $L$ and mass $M$ is free to rotate on a frictionless pin passing through one end (Fig. P8.47). The rod is released from rest in the horizontal position. (a) What is the speed of its center of gravity when the rod reaches its lowest position? (b) What is the tangential speed of the lowest point on the rod when it is in the vertical position?

87. **S** A uniform solid cylinder of mass $M$ and radius $R$ rotates on a frictionless horizontal axle (Fig. P8.87). Two objects with equal masses $m$ hang from light cords wrapped around the cylinder. If the system is released from rest, find (a) the tension in each cord and (b) the acceleration of each object after the objects have descended a distance $h$.

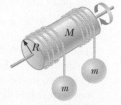

**Figure P8.87**

88. **Q|C** **S** A painter climbs a ladder leaning against a smooth wall. At a certain height, the ladder is on the verge of slipping. (a) Explain why the force exerted by the vertical wall on the ladder is horizontal. (b) If the ladder of length $L$ leans at an angle $\theta$ with the horizontal, what is the lever arm for this horizontal force with the axis of rotation taken at the base of the ladder? (c) If the ladder is uniform, what is the lever arm for the force of gravity acting on the ladder? (d) Let the mass of the painter be 80 kg, $L = 4.0$ m, the ladder's mass be 30 kg, $\theta = 53°$, and the coefficient of friction between ground and ladder be 0.45. Find the maximum distance the painter can climb up the ladder.

89. A *war-wolf*, or *trebuchet*, is a device used during the Middle Ages to throw rocks at castles and now sometimes used to fling pumpkins and pianos. A simple trebuchet is shown in Figure P8.89. Model it as a stiff rod of negligible mass 3.00 m long and joining particles of mass $m_1 = 0.120$ kg and $m_2 = 60.0$ kg at its ends. It can turn on a frictionless horizontal axle perpendicular to the rod and 14.0 cm from the particle of larger mass. The rod is released from rest in a horizontal orientation. Find the maximum speed that the object of smaller mass attains.

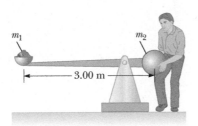

**Figure P8.89**

90. A string is wrapped around a uniform cylinder of mass $M$ and radius $R$. The cylinder is released from rest with the string vertical and its top end tied to a fixed bar (Fig. P8.90). Show that (a) the tension in the string is one-third the weight of the cylinder, (b) the magnitude of the acceleration of the center of gravity is $2g/3$, and (c) the speed of the center of gravity is $(4gh/3)^{1/2}$ after the cylinder has descended through distance $h$. Verify your answer to part (c) with the energy approach.

**Figure P8.90**

91. **V** **BIO** **The Iron Cross** When a gymnast weighing 750 N executes the iron cross as in Figure P8.91a, the primary muscles involved in supporting this position are the latissimus dorsi ("lats") and the pectoralis major ("pecs"). The rings exert an upward force on the arms and support the weight of the gymnast. The force exerted by the shoulder joint on the arm is labeled $\vec{\mathbf{F}}_s$, while the two muscles exert a total force $\vec{\mathbf{F}}_m$ on the arm. Estimate the magnitude of the force $\vec{\mathbf{F}}_m$. Note that one ring supports half the weight of the gymnast, which is 375 N as indicated in Figure P8.91b. Assume that the force $\vec{\mathbf{F}}_m$ acts at an angle of 45° below the horizontal at a distance of 4.0 cm from the shoulder joint. In your estimate, take the distance from the shoulder joint to the hand to be $L = 70$ cm and ignore the weight of the arm.

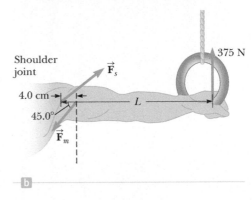

**Figure P8.91**

92. **BIO** In an emergency situation, a person with a broken forearm ties a strap from his hand to clip on his shoulder as in Figure P8.92. His 1.60-kg forearm remains in a horizontal position and the strap makes an angle of $\theta = 50.0°$ with the horizontal. Assume the forearm is uniform, has a length of $\ell = 0.320$ m, assume the biceps muscle is relaxed, and ignore the mass and length of the hand. Find (a) the tension in the strap and (b) the components of the reaction force exerted by the humerus on the forearm.

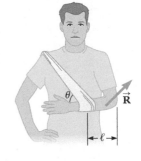

**Figure P8.92**

93. An object of mass $m_1 = 4.00$ kg is connected by a light cord to an object of mass $m_2 = 3.00$ kg on a frictionless surface (Fig. P8.93). The pulley rotates about a frictionless axle and has a moment of inertia of 0.500 kg · m² and a radius of 0.300 m. Assuming that the cord does not slip on the pulley, find (a) the acceleration of the two masses and (b) the tensions $T_1$ and $T_2$.

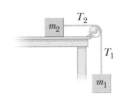

**Figure P8.93**

94. **Q|C** A 10.0-kg monkey climbs a uniform ladder with weight $w = 1.20 \times 10^2$ N and length $L = 3.00$ m as shown in Figure P8.94. The ladder rests against the wall at an angle of $\theta = 60.0°$. The upper and lower ends of the ladder rest on frictionless surfaces, with the lower end fastened to the wall by a horizontal rope that is frayed and that can

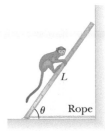

**Figure P8.94**

support a maximum tension of only 80.0 N. (a) Draw a force diagram for the ladder. (b) Find the normal force exerted by the bottom of the ladder. (c) Find the tension in the rope when the monkey is two-thirds of the way up the ladder. (d) Find the maximum distance $d$ that the monkey can climb up the ladder before the rope breaks. (e) If the horizontal surface were rough and the rope were removed, how would your analysis of the problem be changed and what other information would you need to answer parts (c) and (d)?

95. A 3.2-kg sphere is suspended by a cord that passes over a 1.8-kg pulley of radius 3.8 cm. The cord is attached to a spring whose force constant is $k = 86$ N/m as in Figure P8.95. Assume the pulley is a solid disk. (a) If the sphere is released from rest with the spring unstretched, what distance does the sphere fall through before stopping? (b) Find the speed of the sphere after it has fallen 25 cm.

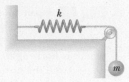

**Figure P8.95**

# Fluids and Solids

THERE ARE FOUR KNOWN STATES OF MATTER: solids, liquids, gases, and plasmas. In the Universe at large, plasmas—systems of charged particles interacting electromagnetically—are the most common. In our environment on Earth, solids, liquids, and gases predominate.

An understanding of the fundamental properties of these different states of matter is important in all the sciences, in engineering, and in medicine. Forces put stresses on solids, and stresses can strain, deform, and break those solids, whether they are steel beams or bones. Fluids under pressure can perform work or carry nutrients and essential solutes, like the blood flowing through our arteries and veins. Flowing gases cause pressure differences that can lift a massive cargo plane or the roof off a house in a hurricane. High-temperature plasmas created in fusion reactors may someday allow humankind to harness the energy source of the Sun.

The study of any one of these states of matter is itself a vast discipline. Here, we'll introduce basic properties of solids and liquids, the latter including some properties of gases. In addition, we'll take a brief look at surface tension, viscosity, osmosis, and diffusion.

## 9.1 States of Matter

Matter is normally classified as being in one of three states: **solid, liquid,** or **gas.** Often this classification system is extended to include a fourth state of matter, called a **plasma.**

Everyday experience tells us that a solid has a definite volume and shape. A brick, for example, maintains its familiar shape and size day in and day out.

A liquid has a definite volume but no definite shape. When you fill the tank on a lawn mower, the gasoline changes its shape from that of the original container to that of the tank on the mower, but the original volume is unchanged. A gas differs from solids and liquids in that it has neither definite volume nor definite shape. Because gas can flow, however, it shares many properties with liquids.

All matter consists of some distribution of atoms or molecules. The atoms in a solid, held together by forces that are mainly electrical, are located at specific positions with respect to one another and vibrate about those positions. At low temperatures, the vibrating motion is slight and the atoms can be considered essentially fixed. As energy is added to the material, the amplitude of the vibrations increases. A vibrating atom can be viewed as being bound in its equilibrium position by springs attached to neighboring atoms. A collection of such atoms and imaginary springs is shown in Figure 9.1. We can picture applied external forces as compressing these tiny internal springs. When the external forces are removed, the solid tends to return to its original shape and size. Consequently, a solid is said to have *elasticity.*

Solids can be classified as either crystalline or amorphous. In a **crystalline solid,** the atoms have an ordered structure (Fig. 9.2). For example, in the sodium chloride crystal (common table salt), sodium and chlorine atoms occupy alternate corners of a cube, as in Figure 9.3a. In an **amorphous solid,** such as glass, the atoms are arranged almost randomly, as in Figure 9.3b.

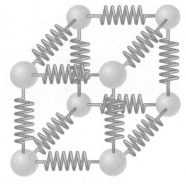

**Figure 9.1** A model of a portion of a solid. The atoms (spheres) are imagined as being attached to each other by springs, which represent the elastic nature of the interatomic forces. A solid consists of trillions of segments like this, with springs connecting all of them.

**Figure 9.2** Crystals of natural quartz ($SiO_2$), one of the most common minerals on Earth. Quartz crystals are used to make special lenses and prisms and are employed in certain electronic applications.

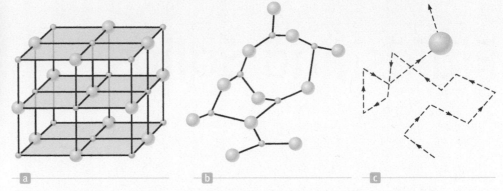

**Figure 9.3** (a) The NaCl structure, with the $Na^+$ (gray) and $Cl^-$ (green) ions at alternate corners of a cube. (b) In an amorphous solid, the atoms are arranged randomly. (c) Erratic motion of a molecule in a liquid.

For any given substance, the liquid state exists at a higher temperature than the solid state. The intermolecular forces in a liquid aren't strong enough to keep the molecules in fixed positions, and they wander through the liquid in random fashion (Fig. 9.3c). Solids and liquids both have the property that when an attempt is made to compress them, strong repulsive atomic forces act internally to resist the compression.

In the gaseous state, molecules are in constant random motion and exert only weak forces on each other. The average distance between the molecules of a gas is quite large compared with the size of the molecules. Occasionally, the molecules collide with each other, but most of the time they move as nearly free, noninteracting particles. As a result, unlike solids and liquids, gases can be easily compressed. We'll say more about gases in subsequent topics.

When a gas is heated to high temperature, many of the electrons surrounding each atom are freed from the nucleus. The resulting system is a collection of free, electrically charged particles—negatively charged electrons and positively charged ions. Such a highly ionized state of matter containing equal amounts of positive and negative charges is called a **plasma.** Unlike a neutral gas, the long-range electric and magnetic forces allow the constituents of a plasma to interact with each other. Plasmas are found inside stars and in accretion disks around black holes, for example, and are far more common than the solid, liquid, and gaseous states because there are far more stars around than any other form of celestial matter.

Normal matter, however, may constitute less than 5% of all matter in the Universe. Observations of the last several years point to the existence of an invisible **dark matter,** which affects the motion of stars orbiting the centers of galaxies. Dark matter may comprise nearly 25% of the matter in the Universe, several times larger than the amount of normal matter. Finally, the rapid acceleration of the expansion of the Universe may be driven by an even more mysterious form of matter, called **dark energy,** which may account for over 70% of all matter in the Universe.

## 9.2 Density and Pressure

Equal masses of aluminum and gold have an important physical difference: The aluminum takes up over seven times as much space as the gold. Although the reasons for the difference lie at the atomic and nuclear levels, a simple measure of this difference is the concept of *density*.

Density ▶

The **density** $\rho$ of an object having uniform composition is its mass $M$ divided by its volume $V$:

$$\rho \equiv \frac{M}{V} \qquad [9.1]$$

**SI unit: kilogram per meter cubed ($kg/m^3$)**

**Table 9.1** Densities of Some Common Substances

| Substance | $\rho$ (kg/m³)ᵃ | Substance | $\rho$ (kg/m³)ᵃ |
|---|---|---|---|
| Ice | $0.917 \times 10^3$ | Water | $1.00 \times 10^3$ |
| Aluminum | $2.70 \times 10^3$ | Glycerin | $1.26 \times 10^3$ |
| Iron | $7.86 \times 10^3$ | Ethyl alcohol | $0.806 \times 10^3$ |
| Copper | $8.92 \times 10^3$ | Benzene | $0.879 \times 10^3$ |
| Silver | $10.5 \times 10^3$ | Mercury | $13.6 \times 10^3$ |
| Lead | $11.3 \times 10^3$ | Air | $1.29$ |
| Gold | $19.3 \times 10^3$ | Oxygen | $1.43$ |
| Platinum | $21.4 \times 10^3$ | Hydrogen | $8.99 \times 10^{-2}$ |
| Uranium | $18.7 \times 10^3$ | Helium | $1.79 \times 10^{-1}$ |

ᵃAll values are at standard atmospheric temperature and pressure (STP), defined as 0°C (273 K) and 1 atm $(1.013 \times 10^5$ Pa). To convert to grams per cubic centimeter, multiply by $10^{-3}$.

For an object with nonuniform composition, Equation 9.1 defines an average density. The most common units used for density are kilograms per cubic meter in the SI system and grams per cubic centimeter in the cgs system. Table 9.1 lists the densities of some substances. The densities of most liquids and solids vary slightly with changes in temperature and pressure; the densities of gases vary greatly with such changes. Under normal conditions, the densities of solids and liquids are about 1 000 times greater than the densities of gases. This difference implies that the average spacing between molecules in a gas under such conditions is about ten times greater than in a solid or liquid.

The **specific gravity** of a substance is the ratio of its density to the density of water at 4°C, which is $1.0 \times 10^3$ kg/m³. (The size of the kilogram was originally defined to make the density of water $1.0 \times 10^3$ kg/m³ at 4°C.) By definition, specific gravity is a dimensionless quantity. For example, if the specific gravity of a substance is 3.0, its density is $3.0(1.0 \times 10^3$ kg/m³$) = 3.0 \times 10^3$ kg/m³.

## Quick Quiz

**9.1** Suppose you have one cubic meter of gold, two cubic meters of silver, and six cubic meters of aluminum. Rank them by mass, from smallest to largest. (a) gold, aluminum, silver (b) gold, silver, aluminum (c) aluminum, gold, silver (d) silver, aluminum, gold

The force exerted by a fluid on an object is always perpendicular to the surfaces of the object, as shown in Figure 9.4a.

The pressure at a specific point in a fluid can be measured with the device pictured in Figure 9.4b: an evacuated cylinder enclosing a light piston connected to a

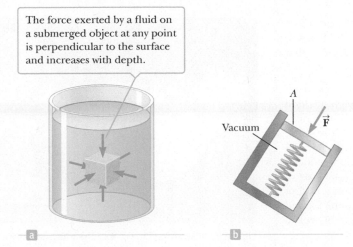

The force exerted by a fluid on a submerged object at any point is perpendicular to the surface and increases with depth.

$A$

Vacuum

$\vec{\mathbf{F}}$

a

b

**Figure 9.4** (a) The force exerted by a fluid on the surfaces of a submerged object. (b) A simple device for measuring pressure in a fluid.

**Tip 9.1** Force and Pressure

*Force is a vector* and *pressure is a scalar.* There is no direction associated with pressure, but the direction of the force associated with the pressure is perpendicular to the surface of interest.

spring that has been previously calibrated with known weights. As the device is submerged in a fluid, the fluid presses down on the top of the piston and compresses the spring until the inward force exerted by the fluid is balanced by the outward force exerted by the spring. Let $F$ be the magnitude of the force on the piston and $A$ the area of the top surface of the piston. Notice that the force that compresses the spring is spread out over the entire area, motivating our formal definition of pressure:

Pressure ▶

If $F$ is the magnitude of a force exerted perpendicular to a given surface of area $A$, then the average pressure $P$ is the force divided by the area:

$$P \equiv \frac{F}{A} \qquad [9.2]$$

**SI unit: pascal (Pa = N/m$^2$)**

**Figure 9.5** Snowshoes prevent the person from sinking into the soft snow because the force on the snow is spread over a larger area, reducing the pressure on the snow's surface.

Pressure can change from point to point, which is why the pressure in Equation 9.2 is called an average. Because pressure is defined as force per unit area, it has units of pascals (newtons per square meter). The English customary unit for pressure is the pound per inch squared. Atmospheric pressure at sea level is 14.7 lb/in.$^2$, which in SI units is $1.01 \times 10^5$ Pa.

As we see from Equation 9.2, the effect of a given force depends critically on the area to which it's applied. A 700-N man can stand on a vinyl-covered floor in regular street shoes without damaging the surface, but if he wears golf shoes, the metal cleats protruding from the soles can do considerable damage to the floor. With the cleats, the same force is concentrated into a smaller area, greatly elevating the pressure in those areas, resulting in a greater likelihood of exceeding the ultimate strength of the floor material.

Snowshoes use the same principle (Fig. 9.5). The snow exerts an upward normal force on the shoes to support the person's weight. According to Newton's third law, this upward force is accompanied by a downward force exerted by the shoes on the snow. If the person is wearing snowshoes, that force is distributed over the very large area of each snowshoe, so that the pressure at any given point is relatively low and the person doesn't penetrate very deeply into the snow.

---

## APPLYING PHYSICS 9.1    BED OF NAILS TRICK

After an exciting but exhausting lecture, a physics professor stretches out for a nap on a bed of nails, as in Figure 9.6, suffering no injury and only moderate discomfort. How is that possible?

**EXPLANATION** If you try to support your entire weight on a single nail, the pressure on your body is your weight divided by the very small area of the end of the nail. The resulting pressure is large enough to penetrate the skin. If you distribute your weight over several hundred nails, however, as demonstrated by the professor, the pressure is considerably reduced because the area that supports your weight is the total area of all nails in contact with your body. (Why is lying on a bed of nails more comfortable than sitting on the same bed? Extend the logic to show that it would be more uncomfortable yet to stand on a bed of nails without shoes.) ■

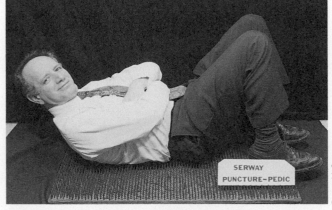

**Figure 9.6** (Applying Physics 9.1) Does anyone have a pillow?

## EXAMPLE 9.1  PRESSURE AND WEIGHT OF WATER

**GOAL** Relate density, pressure, and weight.

**PROBLEM** (a) Calculate the weight of a cylindrical column of water with height $h = 40.0$ m and radius $r = 1.00$ m. (See Fig. 9.7.) (b) Calculate the force exerted by air on a disk of radius 1.00 m at the water's surface. (c) What pressure at a depth of 40.0 m supports the water column?

**STRATEGY** For part (a), calculate the volume and multiply by the density to get the mass of water, then multiply the mass by $g$ to get the weight. Part (b) requires substitution into the definition of pressure. Adding the results of parts (a) and (b) and dividing by the area gives the pressure of water at the bottom of the column.

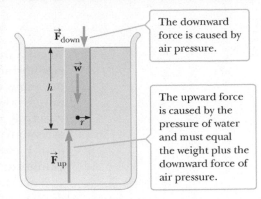

The downward force is caused by air pressure.

The upward force is caused by the pressure of water and must equal the weight plus the downward force of air pressure.

**Figure 9.7** (Example 9.1)

### SOLUTION

(a) Calculate the weight of a cylindrical column of water with height 40.0 m and radius 1.00 m.

Calculate the volume of the cylinder:

$$V = \pi r^2 h = \pi (1.00 \text{ m})^2 (40.0 \text{ m}) = 126 \text{ m}^3$$

Multiply the volume by the density of water to obtain the mass of water in the cylinder:

$$m = \rho V = (1.00 \times 10^3 \text{ kg/m}^3)(126 \text{ m}^3) = 1.26 \times 10^5 \text{ kg}$$

Multiply the mass by the acceleration of gravity $g$ to obtain the weight $w$:

$$w = mg = (1.26 \times 10^5 \text{ kg})(9.80 \text{ m/s}^2) = \boxed{1.23 \times 10^6 \text{ N}}$$

(b) Calculate the force exerted by air on a disk of radius 1.00 m at the surface of the lake.

Write the equation for pressure:

$$P = \frac{F}{A}$$

Solve the pressure equation for the force and substitute $A = \pi r^2$:

$$F = PA = P\pi r^2$$

Substitute values:

$$F = (1.01 \times 10^5 \text{ Pa})\pi (1.00 \text{ m})^2 = \boxed{3.17 \times 10^5 \text{ N}}$$

(c) What pressure at a depth of 40.0 m supports the water column?

Write Newton's second law for the water column:

$$-F_{\text{down}} - w + F_{\text{up}} = 0$$

Solve for the upward force:

$$F_{\text{up}} = F_{\text{down}} + w = (3.17 \times 10^5 \text{ N}) + (1.23 \times 10^6 \text{ N}) = 1.55 \times 10^6 \text{ N}$$

Divide the force by the area to obtain the required pressure:

$$P = \frac{F_{\text{up}}}{A} = \frac{1.55 \times 10^6 \text{ N}}{\pi (1.00 \text{ m})^2} = \boxed{4.93 \times 10^5 \text{ Pa}}$$

**REMARKS** Notice that the pressure at a given depth is related to the sum of the weight of the water and the force exerted by the air pressure at the water's surface. Water at a depth of 40.0 m must push upward to maintain the column in equilibrium. Notice also the important role of density in determining the pressure at a given depth.

**QUESTION 9.1** A giant oil storage facility contains oil to a depth of 40.0 m. How does the pressure at the bottom of the tank compare to the pressure at a depth of 40.0 m in water? Explain.

**EXERCISE 9.1** A large rectangular tub is filled to a depth of 2.60 m with olive oil, which has density 915 kg/m³. If the tub has length 5.00 m and width 3.00 m, calculate (a) the weight of the olive oil, (b) the force of air pressure on the surface of the oil, and (c) the pressure exerted upward by the bottom of the tub.

**ANSWERS** (a) $3.50 \times 10^5$ N  (b) $1.52 \times 10^6$ N  (c) $1.25 \times 10^5$ Pa

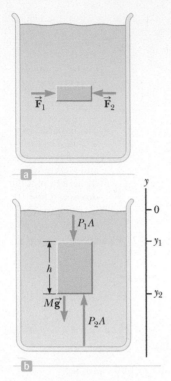

**Figure 9.8** (a) In a static fluid, all points at the same depth are at the same pressure, so the force $\vec{\mathbf{F}}_1$ must equal the force $\vec{\mathbf{F}}_2$. (b) Because the volume of the shaded fluid isn't sinking or rising, the net force on it must equal zero.

# 9.3 Variation of Pressure with Depth

When a fluid is at rest in a container, **all portions of the fluid must be in static equilibrium**—at rest with respect to the observer. Furthermore, **all points at the same depth must be at the same pressure.** If this were not the case, fluid would flow from the higher pressure region to the lower pressure region. For example, consider the small block of fluid shown in Figure 9.8a. If the pressure were greater on the left side of the block than on the right, $\vec{\mathbf{F}}_1$ would be greater than $\vec{\mathbf{F}}_2$, and the block would accelerate to the right and thus would not be in equilibrium.

Next, let's examine the fluid contained within the volume indicated by the darker region in Figure 9.8b. This region has cross-sectional area $A$ and extends from position $y_1$ to position $y_2$ below the surface of the liquid. Three external forces act on this volume of fluid: the force of gravity, $Mg$; the upward force $P_2A$ exerted by the liquid below it; and a downward force $P_1A$ exerted by the fluid above it. Because the given volume of fluid is in equilibrium, these forces must add to zero, so we get

$$P_2A - P_1A - Mg = 0 \qquad [9.3]$$

From the definition of density, we have

$$M = \rho V = \rho A(y_1 - y_2) \qquad [9.4]$$

Substituting Equation 9.4 into Equation 9.3, canceling the area $A$, and rearranging terms, we get

$$P_2 = P_1 + \rho g(y_1 - y_2) \qquad [9.5]$$

Notice that $(y_1 - y_2)$ is positive, because $y_2 < y_1$. The force $P_2A$ is greater than the force $P_1A$ by exactly the weight of water between the two points. This is the same principle experienced by the person at the bottom of a pileup in football or rugby.

Atmospheric pressure is also caused by a piling up of fluid—in this case, the fluid is the gas of the atmosphere. The weight of all the air from sea level to the edge of space results in an atmospheric pressure of $P_0 = 1.013 \times 10^5$ Pa (equivalent to 14.7 lb/in.$^2$) at sea level. This result can be adapted to find the pressure $P$ at any depth $h = (y_1 - y_2) = (0 - y_2)$ below the surface of the water:

$$P = P_0 + \rho g h \qquad [9.6]$$

According to Equation 9.6, **the pressure $P$ at a depth $h$ below the surface of a liquid open to the atmosphere is greater than atmospheric pressure by the amount $\rho g h$.** Moreover, the pressure isn't affected by the shape of the vessel, as shown in Figure 9.9. Equation 9.6 is often called the *equation of hydrostatic equilibrium*. (Similar, related equations also go by that name.)

© Cengage Learning/Charles D. Winters

**Figure 9.9** This photograph illustrates the fact that the pressure in a liquid is the same at all points lying at the same elevation. Note that the shape of the vessel does not affect the pressure.

### Quick Quiz

**9.2** The pressure at the bottom of a glass filled with water ($\rho = 1\ 000$ kg/m$^3$) is $P$. The water is poured out and the glass is filled with ethyl alcohol ($\rho = 806$ kg/m$^3$). The pressure at the bottom of the glass is now (a) smaller than $P$ (b) equal to $P$ (c) larger than $P$ (d) indeterminate.

## EXAMPLE 9.2 | OIL AND WATER

**GOAL** Calculate pressures created by layers of different fluids.

**PROBLEM** In a huge oil tanker, salt water has flooded an oil tank to a depth of $h_2 = 5.00$ m. On top of the water is a layer of oil $h_1 = 8.00$ m deep, as in the cross-sectional view of the tank in Figure 9.10. The oil has a density of $0.700$ g/cm$^3$. Find the pressure at the bottom of the tank. (Take 1 025 kg/m$^3$ as the density of salt water.)

**STRATEGY** Equation 9.6 must be used twice. First, use it to calculate the pressure $P_1$ at the bottom of the oil layer. Then use this pressure in place of $P_0$ in Equation 9.6 and calculate the pressure $P_{bot}$ at the bottom of the water layer.

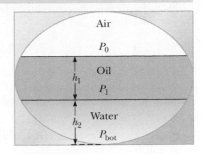

**Figure 9.10** (Example 9.2)

. . . . . . . . . . . . . . . . . . . . . . . . . . . . . . . . . . . . . . . . . . . . . . . . . . . . . . . . . . . . . . . . . . . . . . . . . . . . . . . . . . .

### SOLUTION

Use Equation 9.6 to calculate the pressure at the bottom of the oil layer:

(1) $P_1 = P_0 + \rho g h_1$

$= 1.01 \times 10^5 \text{ Pa}$

$\quad + (7.00 \times 10^2 \text{ kg/m}^3)(9.80 \text{ m/s}^2)(8.00 \text{ m})$

$P_1 = 1.56 \times 10^5 \text{ Pa}$

Now adapt Equation 9.6 to the new starting pressure, and use it to calculate the pressure at the bottom of the water layer:

(2) $P_{bot} = P_1 + \rho g h_2$

$= 1.56 \times 10^5 \text{ Pa}$

$\quad + (1.025 \times 10^3 \text{ kg/m}^3)(9.80 \text{ m/s}^2)(5.00 \text{ m})$

$P_{bot} = \boxed{2.06 \times 10^5 \text{ Pa}}$

. . . . . . . . . . . . . . . . . . . . . . . . . . . . . . . . . . . . . . . . . . . . . . . . . . . . . . . . . . . . . . . . . . . . . . . . . . . . . . . . . . .

**REMARKS** The weight of the atmosphere results in $P_0$ at the surface of the oil layer. Then the weight of the oil and the weight of the water combine to create the pressure at the bottom.

**QUESTION 9.2** Why does air pressure decrease with increasing altitude?

**EXERCISE 9.2** Calculate the pressure on the top lid of a chest buried under 4.00 m of mud with density equal to $1.75 \times 10^3$ kg/m$^3$ at the bottom of a 10.0-m-deep lake.

**ANSWER** $2.68 \times 10^5$ Pa

## EXAMPLE 9.3 | A PAIN IN THE EAR  BIO

**GOAL** Calculate a pressure difference at a given depth and estimate a force.

**PROBLEM** Estimate the net force exerted on your eardrum due to the water above when you are swimming at the bottom of a pool that is 5.0 m deep.

**STRATEGY** Use Equation 9.6 to find the pressure difference across the eardrum at the given depth. The air inside the ear is generally at atmospheric pressure. Estimate the eardrum's surface area, then use the definition of pressure to get the net force exerted on the eardrum.

. . . . . . . . . . . . . . . . . . . . . . . . . . . . . . . . . . . . . . . . . . . . . . . . . . . . . . . . . . . . . . . . . . . . . . . . . . . . . . . . . . .

### SOLUTION

Use Equation 9.6 to calculate the difference between the water pressure at the depth $h$ and the pressure inside the ear:

$\Delta P = P - P_0 = \rho g h$

$= (1.00 \times 10^3 \text{ kg/m}^3)(9.80 \text{ m/s}^2)(5.0 \text{ m})$

$= 4.9 \times 10^4 \text{ Pa}$

*(Continued)*

Multiply by area $A$ to get the net force on the eardrum associated with this pressure difference, estimating the area of the eardrum as 1 cm$^2$.

$$F_{net} = A\Delta P \approx (1 \times 10^{-4}\ m^2)\ (4.9 \times 10^4\ Pa) \approx \boxed{5\ N}$$

**REMARKS** Because a force on the eardrum of this magnitude is uncomfortable, swimmers often "pop their ears" by swallowing or expanding their jaws while underwater, an action that pushes air from the lungs into the middle ear. Using this technique equalizes the pressure on the two sides of the eardrum and relieves the discomfort.

**QUESTION 9.3** Why do water containers and gas cans often have a second, smaller cap opposite the spout?

**EXERCISE 9.3** An airplane takes off at sea level and climbs to a height of 425 m. Estimate the net outward force on a passenger's eardrum assuming the density of air is approximately constant at 1.3 kg/m$^3$ and that the inner ear pressure hasn't been equalized.

**ANSWER** 0.54 N

Because the pressure in a fluid depends on depth and on the value of $P_0$, any increase in pressure at the surface must be transmitted to every point in the fluid. This was first recognized by the French scientist Blaise Pascal (1623–1662) and is called **Pascal's principle:**

A change in pressure applied to an enclosed fluid is transmitted undiminished to every point of the fluid and to the walls of the container.

**APPLICATION**
Hydraulic Lifts

An important application of Pascal's principle is the hydraulic press (Fig. 9.11a). A downward force $\vec{F}_1$ is applied to a small piston of area $A_1$. The pressure is transmitted through a fluid to a larger piston of area $A_2$. As the pistons move and the fluids in the left and right cylinders change their relative heights, there are slight differences in the pressures at the input and output pistons. Neglecting these small differences, the fluid pressure on each of the pistons may be taken to be the same; $P_1 = P_2$. From the definition of pressure, it then follows that $F_1/A_1 = F_2/A_2$. Therefore, the magnitude of the force $\vec{F}_2$ is larger than the magnitude of $\vec{F}_1$ by the factor $A_2/A_1$. That's why a large load, such as a car, can be moved on the large piston by a much smaller force on the smaller piston. Hydraulic brakes, car lifts, hydraulic jacks, forklifts, and other machines make use of this principle.

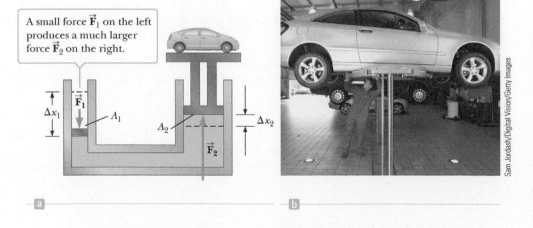

**Figure 9.11** (a) In a hydraulic press, an increase of pressure in the smaller area $A_1$ is transmitted to the larger area $A_2$. Because force equals pressure times area, the force $\vec{F}_2$ is larger than $\vec{F}_1$ by a factor of $A_2/A_1$. (b) A vehicle under repair is supported by a hydraulic lift in a garage.

A small force $\vec{F}_1$ on the left produces a much larger force $\vec{F}_2$ on the right.

Sam Jordash/Digital Vision/Getty Images

## EXAMPLE 9.4    THE CAR LIFT

**GOAL** Apply Pascal's principle to a car lift, and show that the input work is the same as the output work.

**PROBLEM** In a car lift used in a service station, compressed air exerts a force on a small piston of circular cross section having a radius of $r_1 = 5.00$ cm. This pressure is transmitted by an incompressible liquid to a second piston of radius $r_2 = 15.0$ cm. **(a)** What force must the compressed air exert on the small piston in order to lift a car weighing 13 300 N? Neglect the weights of the pistons. **(b)** What air pressure will produce a force of that magnitude? **(c)** Show that the work done by the input and output pistons is the same.

**STRATEGY** Substitute into Pascal's principle in part **(a)**, while recognizing that the magnitude of the output force, $F_2$, must be equal to the car's weight in order to support it. Use the definition of pressure in part **(b)**. In part **(c)**, use $W = F \Delta x$ to find the ratio $W_1/W_2$, showing that it must equal 1. This requires combining Pascal's principle with the fact that the input and output pistons move through the same volume.

**SOLUTION**

**(a)** Find the necessary force on the small piston.

Substitute known values into Pascal's principle, using $A = \pi r^2$ for the area of each piston:

$$F_1 = \left(\frac{A_1}{A_2}\right)F_2 = \frac{\pi r_1^2}{\pi r_2^2}F_2$$

$$= \frac{\pi(5.00 \times 10^{-2}\,\text{m})^2}{\pi(15.0 \times 10^{-2}\,\text{m})^2}(1.33 \times 10^4\,\text{N})$$

$$= 1.48 \times 10^3\,\text{N}$$

**(b)** Find the air pressure producing $F_1$.

Substitute into the definition of pressure:

$$P = \frac{F_1}{A_1} = \frac{1.48 \times 10^3\,\text{N}}{\pi(5.00 \times 10^{-2}\,\text{m})^2} = 1.88 \times 10^5\,\text{Pa}$$

**(c)** Show that the work done by the input and output pistons is the same.

First equate the volumes, and solve for the ratio of $A_2$ to $A_1$:

$$V_1 = V_2 \quad \rightarrow \quad A_1 \Delta x_1 = A_2 \Delta x_2$$

$$\frac{A_2}{A_1} = \frac{\Delta x_1}{\Delta x_2}$$

Now use Pascal's principle to get a relationship for $F_1/F_2$:

$$\frac{F_1}{A_1} = \frac{F_2}{A_2} \quad \rightarrow \quad \frac{F_1}{F_2} = \frac{A_1}{A_2}$$

Evaluate the work ratio, substituting the preceding two results:

$$\frac{W_1}{W_2} = \frac{F_1 \Delta x_1}{F_2 \Delta x_2} = \left(\frac{F_1}{F_2}\right)\left(\frac{\Delta x_1}{\Delta x_2}\right) = \left(\frac{A_1}{A_2}\right)\left(\frac{A_2}{A_1}\right) = 1$$

$$W_1 = W_2$$

**REMARKS** In this problem, we didn't address the effect of possible differences in the heights of the pistons. If the column of fluid is higher in the small piston, the fluid weight assists in supporting the car, reducing the necessary applied force. If the column of fluid is higher in the large piston, both the car and the extra fluid must be supported, so additional applied force is required.

**QUESTION 9.4** True or False: If the radius of the output piston is doubled, the output force increases by a factor of 4.

**EXERCISE 9.4** A hydraulic lift has pistons with diameters 8.00 cm and 36.0 cm, respectively. If a force of 825 N is exerted at the input piston, what maximum mass can be lifted by the output piston?

**ANSWER** $1.70 \times 10^3$ kg

A corollary to the statement that pressure in a fluid increases with depth is that water always seeks its own level. This means that if a vessel is filled with water, then regardless of the vessel's shape the surface of the water is perfectly flat and at the same height at all points. The ancient Egyptians used this fact to make the pyramids level. Devise a scheme showing how this could be done.

**EXPLANATION** There are many ways it could be done, but Figure 9.12 shows the scheme used by the Egyptians. The builders cut grooves in the base of the pyramid as in (a) and partially filled the grooves with water. The height of the water was marked as in (b), and the rock was chiseled down to the mark,

as in (c). Finally, the groove was filled with crushed rock and gravel, as in (d). ∎

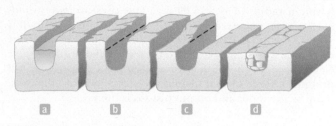

**Figure 9.12** (Applying Physics 9.2)

# 9.4 Pressure Measurements

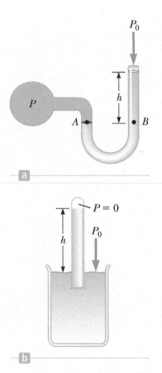

**Figure 9.13** Two devices for measuring pressure: (a) an open-tube manometer and (b) a mercury barometer.

A simple device for measuring pressure is the open-tube manometer (Fig. 9.13a). One end of a U-shaped tube containing a liquid is open to the atmosphere, and the other end is connected to a system of unknown pressure $P$. The pressure at point $B$ equals $P_0 + \rho gh$, where $\rho$ is the density of the fluid. The pressure at $B$, however, equals the pressure at $A$, which is also the unknown pressure $P$. We conclude that $P = P_0 + \rho gh$.

The pressure $P$ is called the **absolute pressure,** and $P - P_0$ is called the **gauge pressure.** If $P$ in the system is greater than atmospheric pressure, $h$ is positive. If $P$ is less than atmospheric pressure (a partial vacuum), $h$ is negative, meaning that the right-hand column in Figure 9.13a is lower than the left-hand column.

Another instrument used to measure pressure is the **barometer** (Fig. 9.13b), invented by Evangelista Torricelli (1608–1647). A long tube closed at one end is filled with mercury and then inverted into a dish of mercury. The closed end of the tube is nearly a vacuum, so its pressure can be taken to be zero. It follows that $P_0 = \rho gh$, where $\rho$ is the density of the mercury and $h$ is the height of the mercury column. Note that the barometer measures the pressure of the atmosphere, whereas the manometer measures pressure in an enclosed fluid.

**One atmosphere** of pressure is defined to be the pressure equivalent of a column of mercury that is exactly 0.76 m in height at 0°C with $g = 9.806\ 65\ \text{m/s}^2$. At this temperature, mercury has a density of $13.595 \times 10^3\ \text{kg/m}^3$; therefore,

$$P_0 = \rho gh = (13.595 \times 10^3\ \text{kg/m}^3)(9.806\ 65\ \text{m/s}^2)(0.760\ 0\ \text{m})$$

$$= 1.013 \times 10^5\ \text{Pa} = 1\ \text{atm}$$

It's interesting to note that the force of the atmosphere on our bodies (assuming a body area of 2 000 in.$^2$) is extremely large, on the order of 30 000 lbs! If it were not for the fluids permeating our tissues and body cavities, our bodies would collapse. The fluids provide equal and opposite forces. In the upper atmosphere or in space, sudden decompression can lead to serious injury and death. Air retained in the lungs can damage the tiny alveolar sacs, and intestinal gas can even rupture internal organs.

**BIO APPLICATION**
Decompression and Injury to the Lungs

*Quick Quiz*

**9.3** Several common barometers are built using a variety of fluids. For which fluid will the column of fluid in the barometer be the highest? (Refer to Table 9.1.)
(a) mercury (b) water (c) ethyl alcohol (d) benzene

## 9.4.1 Blood Pressure Measurements

**BIO** APPLICATION
Measuring Blood Pressure

A specialized manometer (called a sphygmomanometer) is often used to measure blood pressure. In this application, a rubber bulb forces air into a cuff wrapped tightly around the upper arm and simultaneously into a manometer, as in Figure 9.14. The pressure in the cuff is increased until the flow of blood through the brachial artery in the arm is stopped. A valve on the bulb is then opened, and the measurer listens with a stethoscope to the artery at a point just below the cuff. When the pressure in the cuff and brachial artery is just below the maximum value produced by the heart (the systolic pressure), the artery opens momentarily on each beat of the heart. At this point, the velocity of the blood is high and turbulent, and the flow is noisy and can be heard with the stethoscope. The manometer is calibrated to read the pressure in millimeters of mercury, and the value obtained is about 120 mm for a normal heart. Values of 130 mm or above are considered high, and medication to lower the blood pressure is often prescribed for such patients. As the pressure in the cuff is lowered further, intermittent sounds are still heard until the pressure falls just below the minimum heart pressure (the diastolic pressure). At this point, continuous sounds are heard. In the normal heart, this transition occurs at about 80 mm of mercury, and values above 90 require medical intervention. Blood pressure readings are usually expressed as the ratio of the systolic pressure to the diastolic pressure, which is 120/80 for a healthy heart.

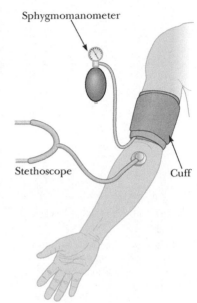

Figure 9.14 A sphygmomanometer can be used to measure blood pressure.

### Quick Quiz

**9.4** Blood pressure is normally measured with the cuff of the sphygmomanometer around the arm. Suppose the blood pressure is measured with the cuff around the calf of the leg of a standing person. Would the reading of the blood pressure be (a) the same here as it is for the arm, (b) greater than it is for the arm, or (c) less than it is for the arm?

---

**APPLYING PHYSICS 9.3    BALLPOINT PENS**

In a ballpoint pen, ink moves down a tube to the tip, where it is spread on a sheet of paper by a rolling stainless steel ball. Near the top of the ink cartridge, there is a small hole open to the atmosphere. If you seal this hole, you will find that the pen no longer functions. Use your knowledge of how a barometer works to explain this behavior.

**EXPLANATION** If the hole was sealed or if it was not present, the pressure of the air above the ink would decrease as the

ink was used. Consequently, atmospheric pressure exerted against the ink at the bottom of the cartridge would prevent some of the ink from flowing out. The hole allows the pressure above the ink to remain at atmospheric pressure. Why does a ballpoint pen seem to run out of ink when you write on a vertical surface? ■

---

# 9.5 Buoyant Forces and Archimedes' Principle

A fundamental principle affecting objects submerged in fluids was discovered by Greek mathematician and natural philosopher Archimedes. **Archimedes' principle** can be stated as follows:

> Any object completely or partially submerged in a fluid is buoyed up by a force with magnitude equal to the weight of the fluid displaced by the object.

◀ Archimedes' principle

Many historians attribute the concept of buoyancy to Archimedes' "bathtub epiphany," when he noticed an apparent change in his weight upon lowering

**Figure 9.15** (a) The arrows indicate forces on the sphere of fluid due to pressure, larger on the underside because pressure increases with depth. (b) The buoyant force, which is caused by the *surrounding* fluid, is the same on any object of the same volume, including this cannon ball.

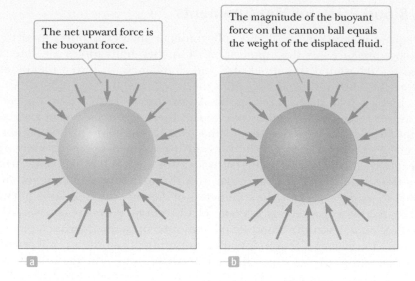

The net upward force is the buoyant force.

The magnitude of the buoyant force on the cannon ball equals the weight of the displaced fluid.

**ARCHIMEDES**
**Greek mathematician, physicist, and engineer (287–212 BC)**

Archimedes was probably the greatest scientist of antiquity. According to legend, King Hieron asked him to determine whether the king's crown was pure gold or a gold alloy. Archimedes allegedly arrived at a solution when bathing, noticing a partial loss of weight on lowering himself into the water. He was so excited that he reportedly ran naked through the streets of Syracuse shouting "Eureka!", which is Greek for "I have found it!"

**Tip 9.2** Buoyant Force Is Exerted by the Fluid

The buoyant force on an object is exerted by the fluid and is the same, regardless of the density of the object. Objects more dense than the fluid sink; objects less dense rise.

himself into a tub of water. As will be seen in Example 9.5, buoyancy yields a method of determining density.

Everyone has experienced Archimedes' principle. It's relatively easy, for example, to lift someone if you're both standing in a swimming pool, whereas lifting that same individual on dry land may be a difficult task. Water provides partial support to any object placed in it. We often say that an object placed in a fluid is buoyed up by the fluid, so we call this upward force the **buoyant force.**

The buoyant force is *not* a mysterious new force that arises in fluids. In fact, the physical cause of the buoyant force is the pressure difference between the upper and lower sides of the object. In Figure 9.15a, the fluid inside the indicated sphere, colored darker blue, is pressed on all sides by the surrounding fluid. Arrows indicate the forces arising from the pressure. Because pressure increases with depth, the arrows on the underside are larger than those on top. Adding them all up, the horizontal components cancel, but there is a net force upward. This force, due to differences in pressure, is the buoyant force $\vec{\mathbf{B}}$. The sphere of water neither rises nor falls, so the vector sum of the buoyant force and the force of gravity on the sphere of fluid must be zero, and it follows that $B = Mg$, where $M$ is the mass of the fluid. The buoyant force, therefore, is equal in magnitude to the weight of the displaced fluid.

Replacing the shaded fluid with a cannon ball of the same volume, as in Figure 9.15b, changes only the mass on which the pressure acts, so the buoyant force is the same: $B = Mg$, where $M$ is the mass of the displaced fluid, *not* the mass of the cannon ball. The force of gravity on the heavier ball is greater than it was on the fluid, so the cannon ball sinks.

Archimedes' principle can also be obtained from Equation 9.3, relating pressure and depth, using Figure 9.8b. Horizontal forces from the pressure cancel, but in the vertical direction $P_2A$ acts upward on the bottom of the block of fluid, and $P_1A$ and the gravity force on the fluid, $Mg$, act downward, giving

$$B = P_2A - P_1A = Mg \qquad [9.7a]$$

where the buoyancy force has been identified as the result of differences in pressure and is equal in magnitude to the weight of the displaced fluid. This buoyancy force remains the same regardless of the material occupying the volume in question because it's due to the *surrounding* fluid. Using the definition of density, Equation 9.7a becomes

$$B = \rho_{\text{fluid}} V_{\text{fluid}} g \qquad [9.7b]$$

Figure 9.16 Hot-air balloons. Because hot air is less dense than cold air, there is a net upward force on the balloons.

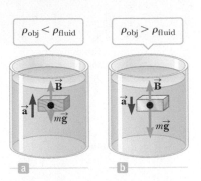

Figure 9.17 (a) A totally submerged object that is less dense than the fluid in which it is submerged is acted upon by a net upward force. (b) A totally submerged object that is denser than the fluid sinks.

where $\rho_{fluid}$ is the density of the fluid and $V_{fluid}$ is the volume of the displaced fluid. This result applies equally to all shapes because any irregular shape can be approximated by a large number of infinitesimal cubes.

It's instructive to compare the forces on a totally submerged object with those on a floating object.

**Case I: A Totally Submerged Object.** When an object is *totally* submerged in a fluid of density $\rho_{fluid}$, the upward buoyant force acting on the object has a magnitude of $B = \rho_{fluid}V_{obj}g$, where $V_{obj}$ is the volume of the object. If the object has density $\rho_{obj}$, the downward gravitational force acting on the object has a magnitude equal to $w = mg = \rho_{obj}V_{obj}g$, and the net force on it is $B - w = (\rho_{fluid} - \rho_{obj})V_{obj}g$. Therefore, if the density of the object is *less* than the density of the fluid, the net force exerted on the object is *positive* (upward) and the object accelerates *upward*, as in Figures 9.16 and 9.17a. If the density of the object is *greater* than the density of the fluid, as in Figure 9.17b, the net force is *negative* and the object accelerates *downward*.

Figure 9.18 Most of the volume of this iceberg is beneath the water. Can you determine what fraction of the total volume is under water?

**Case II: A Floating Object.** Now consider a partially submerged object in static equilibrium floating in a fluid, as in Figure 9.18. In this case, the upward buoyant force is balanced by the downward force of gravity acting on the object (Fig. 9.19). If $V_{fluid}$ is the volume of the fluid displaced by the object (which corresponds to the volume of the part of the object beneath the fluid level), then the magnitude of the buoyant force is given by $B = \rho_{fluid}V_{fluid}g$. Because the weight of the object is $w = mg = \rho_{obj}V_{obj}g$, and because $w = B$, it follows that $\rho_{fluid}V_{fluid}g = \rho_{obj}V_{obj}g$, or

$$\frac{\rho_{obj}}{\rho_{fluid}} = \frac{V_{fluid}}{V_{obj}}$$   [9.8]

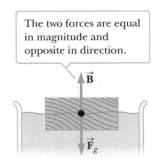

The two forces are equal in magnitude and opposite in direction.

Figure 9.19 An object floating on the surface of a fluid is acted upon by two forces: the gravitational force $\vec{F}_g$ and the buoyant force $\vec{B}$.

Equation 9.8 neglects the buoyant force of the air, which is slight because the density of air is only 1.29 kg/m³ at sea level.

Under normal circumstances, the average density of a fish is slightly greater than the density of water, so a fish would sink if it didn't have a mechanism for adjusting its density. By changing the size of an internal swim bladder, fish maintain neutral buoyancy as they swim to various depths.

The human brain is immersed in a fluid (the cerebrospinal fluid) of density 1 007 kg/m³, which is slightly less than the average density of the brain, 1040 kg/m³. Consequently, most of the weight of the brain is supported by the buoyant force of the surrounding fluid. In some clinical procedures, a portion of this fluid must be removed for diagnostic purposes. During such procedures, the nerves and blood vessels in the brain are placed under great strain, which in turn can cause extreme discomfort and pain. Great care must be exercised with such patients until the initial volume of brain fluid has been restored by the body.

When service station attendants check the antifreeze in your car or the condition of your battery, they often use devices that apply Archimedes' principle.

**BIO APPLICATION**
Buoyancy Control in Fish

**BIO APPLICATION**
Cerebrospinal Fluid

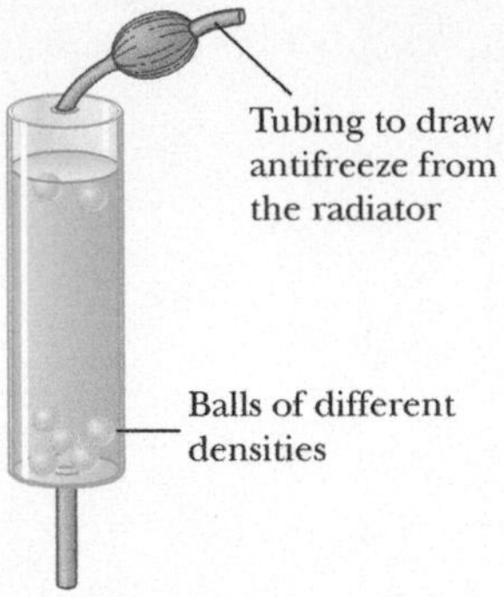

Tubing to draw antifreeze from the radiator

Balls of different densities

**Figure 9.20** The number of balls that float in this device is a measure of the density of the antifreeze solution in a vehicle's radiator and, consequently, a measure of the temperature at which freezing will occur.

**Figure 9.21** The orange ball in the plastic tube inside the battery serves as an indicator of whether the battery is (a) charged or (b) discharged.

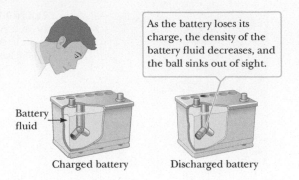

As the battery loses its charge, the density of the battery fluid decreases, and the ball sinks out of sight.

Battery fluid

Charged battery        Discharged battery

Figure 9.20 shows a common device that is used to check the antifreeze in a car radiator. The small balls in the enclosed tube vary in density so that all of them float when the tube is filled with pure water, none float in pure antifreeze, one floats in a 5% mixture, two in a 10% mixture, and so forth. The number of balls that float is a measure of the percentage of antifreeze in the mixture, which in turn is used to determine the lowest temperature the mixture can withstand without freezing.

Similarly, the degree of charge in some car batteries can be determined with a so-called magic-dot process that is built into the battery (Fig. 9.21). Inside a viewing port in the top of the battery, the appearance of an orange dot indicates that the battery is sufficiently charged; a black dot indicates that the battery has lost its charge. If the battery has sufficient charge, the density of the battery fluid is high enough to cause the orange ball to float. As the battery loses its charge, the density of the battery fluid decreases and the ball sinks beneath the surface of the fluid, making the dot appear black.

**APPLICATION**

Checking the Battery Charge

### Quick Quiz

**9.5** Atmospheric pressure varies from day to day. The level of a floating ship on a high-pressure day is (a) higher, (b) lower, or (c) no different than on a low-pressure day.

**9.6** The density of lead is greater than iron, and both metals are denser than water. Is the buoyant force on a solid lead object (a) greater than, (b) equal to, or (c) less than the buoyant force acting on a solid iron object of the same dimensions?

---

### EXAMPLE 9.5 A RED-TAG SPECIAL ON CROWNS

**GOAL** Apply Archimedes' principle to a submerged object.

**PROBLEM** A bargain hunter purchases a "gold" crown at a flea market. After she gets home, she hangs it from a scale and finds its weight to be 7.84 N (Fig. 9.22a). She then weighs the crown while it is immersed in water, as in Figure 9.22b, and now the scale reads 6.86 N. Is the crown made of pure gold?

**STRATEGY** The goal is to find the density of the crown and compare it to the density of gold. We already have the weight of the crown in air, so we can get the mass by dividing by the acceleration of gravity. If we can find the volume of the crown, we can obtain the desired density by dividing the mass by this volume.

**Figure 9.22** (Example 9.5) (a) When the crown is suspended in air, the scale reads $T_{air} = mg$, the crown's true weight. (b) When the crown is immersed in water, the buoyant force $\vec{B}$ reduces the scale reading by the magnitude of the buoyant force, $T_{water} = mg - B$.

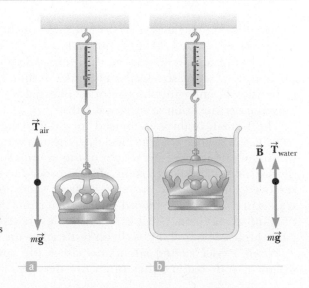

When the crown is fully immersed, the displaced water is equal to the volume of the crown. This same volume is used in calculating the buoyant force. So our strategy is as follows: (1) Apply Newton's second law to the crown, both in the water and in the air to find the buoyant force. (2) Use the buoyant force to find the crown's volume. (3) Divide the crown's scale weight in air by the acceleration of gravity to get the mass, then by the volume to get the density of the crown.

## SOLUTION

Apply Newton's second law to the crown when it's weighed in air. There are two forces on the crown—gravity $m\vec{g}$ and $\vec{T}_{air}$, the force exerted by the scale on the crown, with magnitude equal to the reading on the scale.

(1)     $T_{air} - mg = 0$

When the crown is immersed in water, the scale force is $\vec{T}_{water}$, with magnitude equal to the scale reading, and there is an upward buoyant force $\vec{B}$ and the force of gravity.

(2)     $T_{water} - mg + B = 0$

Solve Equation (1) for $mg$, substitute into Equation (2), and solve for the buoyant force, which equals the difference in scale readings:

$T_{water} - T_{air} + B = 0$
$B = T_{air} - T_{water} = 7.84\text{ N} - 6.86\text{ N} = 0.980\text{ N}$

Find the volume of the displaced water, using the fact that the magnitude of the buoyant force equals the weight of the displaced water:

$B = \rho_{water}\, g V_{water} = 0.980\text{ N}$

$V_{water} = \dfrac{0.980\text{ N}}{g\rho_{water}} = \dfrac{0.980\text{ N}}{(9.80\text{ m/s}^2)(1.00 \times 10^3\text{ kg/m}^3)}$

$= 1.00 \times 10^{-4}\text{ m}^3$

The crown is totally submerged, so $V_{crown} = V_{water}$. From Equation (1), the mass is the crown's weight in air, $T_{air}$, divided by $g$:

$m = \dfrac{T_{air}}{g} = \dfrac{7.84\text{ N}}{9.80\text{ m/s}^2} = 0.800\text{ kg}$

Find the density of the crown:

$\rho_{crown} = \dfrac{m}{V_{crown}} = \dfrac{0.800\text{ kg}}{1.00 \times 10^{-4}\text{ m}^3} = \boxed{8.00 \times 10^3\text{ kg/m}^3}$

**REMARKS** Because the density of gold is $19.3 \times 10^3$ kg/m$^3$, the crown is either hollow, made of an alloy, or both. Despite the mathematical complexity, it is certainly conceivable that this was the method that occurred to Archimedes. Conceptually, it's a matter of realizing (or guessing) that equal weights of gold and a silver–gold alloy would have different scale readings when immersed in water because their densities and hence their volumes are different, leading to differing buoyant forces.

**QUESTION 9.5** True or False: The magnitude of the buoyant force on a completely submerged object depends on the object's density.

**EXERCISE 9.5** The weight of a metal bracelet is measured to be 0.100 00 N in air and 0.092 00 N when immersed in water. Find its density.

**ANSWER** $1.25 \times 10^4$ kg/m$^3$

---

## EXAMPLE 9.6    FLOATING DOWN THE RIVER

**GOAL** Apply Archimedes' principle to a partially submerged object.

**PROBLEM** A raft is constructed of wood having a density of $6.00 \times 10^2$ kg/m$^3$. Its surface area is 5.70 m$^2$, and its volume is 0.60 m$^3$. When the raft is placed in fresh water as in Figure 9.23, to what depth $h$ is the bottom of the raft submerged?

**STRATEGY** There are two forces acting on the raft: the buoyant force of magnitude $B$, acting upward, and the force of gravity, acting downward. Because the raft is in equilibrium, the sum of these forces is zero. The buoyant force depends on the submerged volume $V_{water} = Ah$. Set up Newton's second law and solve for $h$, the depth reached by the bottom of the raft.

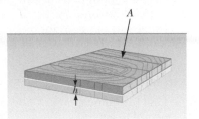

**Figure 9.23** (Example 9.6) A raft partially submerged in water.

*(Continued)*

## SOLUTION

Apply Newton's second law to the raft, which is in equilibrium:

$$B - m_{\text{raft}}g = 0 \quad \rightarrow \quad B = m_{\text{raft}}g$$

The volume of the raft submerged in water is given by $V_{\text{water}} = Ah$. The magnitude of the buoyant force is equal to the weight of this displaced volume of water:

$$B = m_{\text{water}}g = (\rho_{\text{water}}V_{\text{water}})g = (\rho_{\text{water}}Ah)g$$

Now rewrite the gravity force on the raft using the raft's density and volume:

$$m_{\text{raft}}g = (\rho_{\text{raft}}V_{\text{raft}})g$$

Substitute these two expressions into Newton's second law, $B = m_{\text{raft}}g$, and solve for $h$ (note that $g$ cancels):

$$(\rho_{\text{water}}Ah)g = (\rho_{\text{raft}}V_{\text{raft}})g$$

$$h = \frac{\rho_{\text{raft}}V_{\text{raft}}}{\rho_{\text{water}}A}$$

$$= \frac{(6.00 \times 10^2 \text{ kg/m}^3)(0.600 \text{ m}^3)}{(1.00 \times 10^3 \text{ kg/m}^3)(5.70 \text{ m}^2)}$$

$$= \boxed{0.0632 \text{ m}}$$

**REMARKS** How low the raft rides in the water depends on the density of the raft. The same is true of the human body: Fat is less dense than muscle and bone, so those with a higher percentage of body fat float better.

**QUESTION 9.6** If the raft is placed in salt water, which has a density greater than fresh water, would the value of $h$ (a) decrease, (b) increase, or (c) not change?

**EXERCISE 9.6** Calculate how much of an iceberg is beneath the surface of the ocean, given that the density of ice is $917 \text{ kg/m}^3$ and salt water has density $1\,025 \text{ kg/m}^3$.

**ANSWER** 89.5%

---

## EXAMPLE 9.7 | FLOATING IN TWO FLUIDS

**GOAL** Apply Archimedes' principle to an object floating in a fluid having two layers with different densities.

**PROBLEM** A $1.00 \times 10^3$-kg cube of aluminum is placed in a tank. Water is then added to the tank until half the cube is immersed. **(a)** What is the normal force on the cube? (See Fig. 9.24a.) **(b)** Mercury is now slowly poured into the tank until the normal force on the cube goes to zero. (See Fig. 9.24b.) How deep is the layer of mercury? Assume a very thin layer of fluid is underneath the block in both parts of Figure 9.24, due to imperfections between the surfaces in contact.

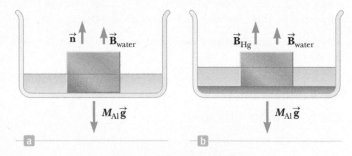

**Figure 9.24** (Example 9.7)

**STRATEGY** Both parts of this problem involve applications of Newton's second law for a body in equilibrium, together with the concept of a buoyant force. In part **(a)** the normal, gravitational, and buoyant force of water act on the cube. In part **(b)** there is an additional buoyant force of mercury, while the normal force goes to zero. Using $V_{\text{Hg}} = Ah$, solve for the height of mercury, $h$.

## SOLUTION

**(a)** Find the normal force on the cube when half-immersed in water.

$$V_{\text{Al}} = \frac{M_{\text{Al}}}{\rho_{\text{Al}}} = \frac{1.00 \times 10^3 \text{ kg}}{2.70 \times 10^3 \text{ kg/m}^3} = 0.370 \text{ m}^3$$

Calculate the volume $V$ of the cube and the length $d$ of one side, for future reference (both quantities will be needed for what follows):

$$d = V_{\text{Al}}^{1/3} = 0.718 \text{ m}$$

Write Newton's second law for the cube, and solve for the normal force. The buoyant force is equal to the weight of the displaced water (half the volume of the cube).

$$n - M_{Al}g + B_{water} = 0$$
$$n = M_{Al}g - B_{water} = M_{Al}g - \rho_{water}\,(V/2)g$$
$$= (1.00 \times 10^3 \text{ kg})(9.80 \text{ m/s}^2)$$
$$- (1.00 \times 10^3 \text{ kg/m}^3)(0.370 \text{ m}^3/2.00)(9.80 \text{ m/s}^2)$$
$$n = 9.80 \times 10^3 \text{ N} - 1.81 \times 10^3 \text{ N} = \boxed{7.99 \times 10^3 \text{ N}}$$

**(b)** Calculate the level $h$ of added mercury.

Apply Newton's second law to the cube:

$$n - M_{Al}g + B_{water} + B_{Hg} = 0$$

Set $n = 0$ and solve for the buoyant force of mercury:

$$B_{Hg} = (\rho_{Hg}\,Ah)g = M_{Al}g - B_{water} = 7.99 \times 10^3 \text{ N}$$

Solve for $h$, noting that $A = d^2$:

$$h = \frac{M_{Al}g - B_{water}}{\rho_{Hg}Ag} = \frac{7.99 \times 10^3 \text{ N}}{(13.6 \times 10^3 \text{ kg/m}^3)(0.718 \text{ m})^2(9.80 \text{ m/s}^2)}$$

$$h = \boxed{0.116 \text{ m}}$$

**REMARKS** Notice that the buoyant force of mercury calculated in part (b) is the same as the normal force in part (a). This is naturally the case, because enough mercury was added to exactly cancel out the normal force. We could have used this fact to take a shortcut, simply writing $B_{Hg} = 7.99 \times 10^3$ N immediately, solving for $h$, and avoiding a second use of Newton's law. Most of the time, however, we wouldn't be so lucky! Try calculating the normal force when the level of mercury is 4.00 cm.

**QUESTION 9.7** What would happen to the aluminum cube if more mercury were poured into the tank?

**EXERCISE 9.7** A cube of aluminum 1.00 m on a side is immersed one-third in water and two-thirds in glycerin. What is the normal force on the cube?

**ANSWER** $1.50 \times 10^4$ N

# 9.6 Fluids in Motion

When a fluid is in motion, its flow can be characterized in one of two ways. The flow is said to be **streamline,** or **laminar,** if every particle that passes a particular point moves along exactly the same smooth path followed by previous particles passing that point. This path is called a *streamline* (Fig. 9.25). Different streamlines can't cross each other under this steady-flow condition, and the streamline at any point coincides with the direction of the velocity of the fluid at that point.

In contrast, the flow of a fluid becomes irregular, or **turbulent,** above a certain velocity or under any conditions that can cause abrupt changes in velocity. Irregular motions of the fluid, called *eddy currents,* are characteristic in turbulent flow, as shown in Figure 9.26.

In discussions of fluid flow, the term **viscosity** is used for the degree of internal friction in the fluid. This internal friction is associated with the resistance between two adjacent layers of the fluid moving relative to each other. A fluid such as kerosene has a lower viscosity than does crude oil or molasses.

Zaichenko Olga/istockphoto.com

Andy Sacks/Stone/Getty Images

**Figure 9.25** An illustration of streamline flow around an automobile in a test wind tunnel. The streamlines in the airflow are made visible by smoke particles.

**Figure 9.26** Hot gases made visible by smoke particles. The smoke first moves in laminar flow at the bottom and then in turbulent flow above.

Many features of fluid motion can be understood by considering the behavior of an **ideal fluid,** which satisfies the following conditions:

1. **The fluid is nonviscous,** which means there is no internal friction force between adjacent layers.
2. **The fluid is incompressible,** which means its density is constant.
3. **The fluid motion is steady,** meaning that the velocity, density, and pressure at each point in the fluid don't change with time.
4. **The fluid moves without turbulence.** This implies that each element of the fluid has zero angular velocity about its center, so there can't be any eddy currents present in the moving fluid. A small wheel placed in the fluid would translate but not rotate.

## 9.6.1 Equation of Continuity

Figure 9.27a represents a fluid flowing through a pipe of nonuniform size. The particles in the fluid move along the streamlines in steady-state flow. In a small time interval $\Delta t$, the fluid entering the bottom end of the pipe moves a distance $\Delta x_1 = v_1 \Delta t$, where $v_1$ is the speed of the fluid at that location. If $A_1$ is the cross-sectional area in this region, then the mass contained in the bottom blue region is $\Delta M_1 = \rho_1 A_1 \Delta x_1 = \rho_1 A_1 v_1 \Delta t$, where $\rho_1$ is the density of the fluid at $A_1$. Similarly, the fluid that moves out of the upper end of the pipe in the same time interval $\Delta t$ has a mass of $\Delta M_2 = \rho_2 A_2 v_2 \Delta t$. However, **because mass is conserved and because the flow is steady,** the mass that flows into the bottom of the pipe through $A_1$ in the time $\Delta t$ must equal the mass that flows out through $A_2$ in the same interval. Therefore, $\Delta M_1 = \Delta M_2$, or

$$\rho_1 A_1 v_1 = \rho_2 A_2 v_2 \qquad [9.9]$$

For the case of an incompressible fluid, $\rho_1 = \rho_2$ and Equation 9.9 reduces to

◀ Equation of continuity

$$A_1 v_1 = A_2 v_2 \qquad [9.10]$$

This expression is called the **equation of continuity.** From this result, we see that **the product of the cross-sectional area of the pipe and the fluid speed at that cross section is a constant.** Therefore, the speed is high where the tube is constricted and low where the tube has a larger diameter. The product $Av$, which has dimensions of volume per unit time, is called the **flow rate. The condition $Av$ = constant is equivalent to the fact that the volume of fluid entering one end of the tube in a given time interval equals the volume of fluid leaving the tube in the same interval, assuming that the fluid is incompressible and there are no leaks.** Figure 9.27b is an example

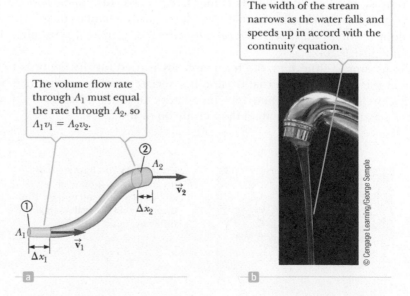

The width of the stream narrows as the water falls and speeds up in accord with the continuity equation.

The volume flow rate through $A_1$ must equal the rate through $A_2$, so $A_1 v_1 = A_2 v_2$.

**Figure 9.27** (a) A fluid moving with streamline flow through a pipe of varying cross-sectional area. The volume of fluid flowing through $A_1$ in a time interval $\Delta t$ must equal the volume flowing through $A_2$ in the same time interval. (b) Water flowing slowly out of a faucet.

© Cengage Learning/George Semple

of an application of the equation of continuity: As the stream of water flows continuously from a faucet, the width of the stream narrows as it falls and speeds up.

There are many instances in everyday experience that involve the equation of continuity. Reducing the cross-sectional area of a garden hose by putting a thumb over the open end makes the water spray out with greater speed; hence, the stream goes farther. Similar reasoning explains why smoke from a smoldering piece of wood first rises in a streamline pattern, getting thinner with height, eventually breaking up into a swirling, turbulent pattern. The smoke rises because it's less dense than air and the buoyant force of the air accelerates it upward. As the speed of the smoke stream increases, the cross-sectional area of the stream decreases, in accordance with the equation of continuity. The stream soon reaches a speed so great that streamline flow is not possible. We study the relationship between speed of fluid flow and turbulence in a later discussion on the Reynolds number.

> **Tip 9.3** Continuity Equations
>
> The rate of flow of fluid into a system equals the rate of flow out of the system. The incoming fluid occupies a certain volume and can enter the system only if an equal volume of fluid is expelled during the same time interval.

### EXAMPLE 9.8  NIAGARA FALLS

**GOAL** Apply the equation of continuity.

**PROBLEM** Each second, 5 525 $m^3$ of water flows over the 670-m-wide cliff of the Horseshoe Falls portion of Niagara Falls. The water is approximately 2 m deep as it reaches the cliff. Estimate its speed at that instant.

**STRATEGY** This is an estimate, so only one significant figure will be retained in the answer. The volume flow rate is given, and, according to the equation of continuity, is a constant equal to $Av$. Find the cross-sectional area, substitute, and solve for the speed.

· · · · · · · · · · · · · · · · · · · · · · · · · · · · · · · · · · · · · · · · · · · · · · · · · · · · · · · · · · · · · · · · · · · · · · · · · · · · · · ·

**SOLUTION**

Calculate the cross-sectional area of the water as it reaches the edge of the cliff:

$$A = (670 \text{ m})(2 \text{ m}) = 1\ 340 \text{ m}^2$$

Multiply this result by the speed and set it equal to the flow rate. Then solve for $v$:

$$Av = \text{volume flow rate}$$
$$(1\ 340 \text{ m}^2)v = 5\ 525 \text{ m}^3/\text{s} \quad \rightarrow \quad v \approx \boxed{4 \text{ m/s}}$$

· · · · · · · · · · · · · · · · · · · · · · · · · · · · · · · · · · · · · · · · · · · · · · · · · · · · · · · · · · · · · · · · · · · · · · · · · · · · · · ·

**QUESTION 9.8** What happens to the speed of blood in an artery when plaque starts to build up on the artery's sides?

**EXERCISE 9.8** The Garfield Thomas water tunnel at Pennsylvania State University has a circular cross section that constricts from a diameter of 3.6 m to the test section which has a diameter of 1.2 m. If the speed of flow is 3.0 m/s in the larger-diameter pipe, determine the speed of flow in the test section.

**ANSWER** 27 m/s

### EXAMPLE 9.9  WATERING A GARDEN

**GOAL** Combine the equation of continuity with concepts of flow rate and kinematics.

**PROBLEM** A water hose 2.50 cm in diameter is used by a gardener to fill a 30.0-liter bucket. (One liter = 1 000 $cm^3$.) The gardener notices that it takes 1.00 min to fill the bucket. A nozzle with an opening of cross-sectional area 0.500 $cm^2$ is then attached to the hose. The nozzle is held so that water is projected horizontally from a point 1.00 m above the ground. Over what horizontal distance can the water be projected?

**STRATEGY** We can find the volume flow rate through the hose by dividing the volume of the bucket by the time it takes to fill it. After finding the flow rate, apply the equation of continuity to find the speed at which the water shoots horizontally from the nozzle. The rest of the problem is an application of two-dimensional kinematics. The answer obtained is the same as would be found for a ball having the same initial velocity and height.

· · · · · · · · · · · · · · · · · · · · · · · · · · · · · · · · · · · · · · · · · · · · · · · · · · · · · · · · · · · · · · · · · · · · · · · · · · · · · · ·

**SOLUTION**

Calculate the volume flow rate into the bucket, and convert to $m^3/s$:

$$\text{volume flow rate} =$$
$$= \frac{30.0 \text{ L}}{1.00 \text{ min}} \left( \frac{1.00 \times 10^3 \text{ cm}^3}{1.00 \text{ L}} \right) \left( \frac{1.00 \text{ m}}{100.0 \text{ cm}} \right)^3 \left( \frac{1.00 \text{ min}}{60.0 \text{ s}} \right)$$
$$= 5.00 \times 10^{-4} \text{ m}^3/\text{s}$$

(Continued)

Solve the equation of continuity for $v_{0x}$, the $x$-component of the initial velocity of the stream exiting the hose:

$$A_1 v_1 = A_2 v_2 = A_2 v_{0x}$$

$$v_{0x} = \frac{A_1 v_1}{A_2} = \frac{5.00 \times 10^{-4} \text{ m}^3/\text{s}}{0.500 \times 10^{-4} \text{ m}^2} = 10.0 \text{ m/s}$$

Calculate the time for the stream to fall 1.00 m, using kinematics. Initially, the stream is horizontal, so $v_{0y}$ is zero:

$$\Delta y = v_{0y} t - \tfrac{1}{2} g t^2$$

Set $v_{0y} = 0$ in the preceding equation and solve for $t$, noting that $\Delta y = -1.00$ m:

$$t = \sqrt{\frac{-2\Delta y}{g}} = \sqrt{\frac{-2(-1.00 \text{ m})}{9.80 \text{ m/s}^2}} = 0.452 \text{ s}$$

Find the horizontal distance the stream travels:

$$x = v_{0x} t = (10.0 \text{ m/s})(0.452 \text{ s}) = \boxed{4.52 \text{ m}}$$

**REMARKS** It's interesting that the motion of fluids can be treated with the same kinematics equations as individual objects.

**QUESTION 9.9** By what factor would the range be changed if the flow rate were doubled?

**EXERCISE 9.9** The nozzle is replaced with a Y-shaped fitting that splits the flow in half. Garden hoses are connected to each end of the Y, with each hose having a 0.400 cm$^2$ nozzle. **(a)** How fast does the water come out of one of the nozzles? **(b)** How far would one of the nozzles squirt water if both were operated simultaneously and held horizontally 1.00 m off the ground? *Hint:* Find the volume flow rate through each 0.400-cm$^2$ nozzle, then follow the same steps as before.

**ANSWERS** **(a)** 6.25 m/s **(b)** 2.83 m

---

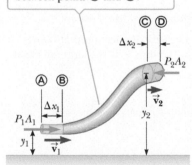

The tube of fluid between points Ⓐ and Ⓒ moves forward so it is between points Ⓑ and Ⓓ.

**Figure 9.28** By the work–energy theorem, the work done by the opposing pressures $P_1$ and $P_2$ equals the difference in mechanical energy between that of the fluid now between points Ⓒ and Ⓓ and the fluid that was formerly between Ⓐ and Ⓑ.

## 9.6.2 Bernoulli's Equation

As a fluid moves through a pipe of varying cross section and elevation, the pressure changes along the pipe. In 1738 the Swiss physicist Daniel Bernoulli (1700–1782) derived an expression that relates the pressure of a fluid to its speed and elevation. Bernoulli's equation is not a freestanding law of physics; rather, it's **a consequence of energy conservation as applied to an ideal fluid.**

In deriving Bernoulli's equation, we again assume the fluid is incompressible, nonviscous, and flows in a nonturbulent, steady-state manner. Consider the flow through a nonuniform pipe in the time $\Delta t$, as in Figure 9.28. The force on the lower end of the fluid is $P_1 A_1$, where $P_1$ is the pressure at the lower end. The work done on the lower end of the fluid by the fluid behind it is

$$W_1 = F_1 \Delta x_1 = P_1 A_1 \Delta x_1 = P_1 V$$

where $V$ is the volume of the lower blue region in the figure. In a similar manner, the work done on the fluid on the upper portion in the time $\Delta t$ is

$$W_2 = -P_2 A_2 \Delta x_2 = -P_2 V$$

The volume is the same because, by the equation of continuity, the volume of fluid that passes through $A_1$ in the time $\Delta t$ equals the volume that passes through $A_2$ in the same interval. The work $W_2$ is negative because the force on the fluid at the top is opposite its displacement. The net work done by these forces in the time $\Delta t$ is

$$W_{\text{fluid}} = P_1 V - P_2 V$$

Part of this work goes into changing the fluid's kinetic energy, and part goes into changing the gravitational potential energy of the fluid–Earth system. If $m$ is the mass of the fluid passing through the pipe in the time interval $\Delta t$, then the change in kinetic energy of the volume of fluid is

$$\Delta KE = \tfrac{1}{2} m v_2^2 - \tfrac{1}{2} m v_1^2$$

The change in the gravitational potential energy is

$$\Delta PE = m g y_2 - m g y_1$$

Because the net work done by the fluid on the segment of fluid shown in Figure 9.28 changes the kinetic energy and the potential energy of the nonisolated system, we have

$$W_{\text{fluid}} = \Delta KE + \Delta PE$$

The three terms in this equation are those we have just evaluated. Substituting expressions for each of the terms gives

$$P_1 V - P_2 V = \tfrac{1}{2}mv_2{}^2 - \tfrac{1}{2}mv_1{}^2 + mgy_2 - mgy_1$$

If we divide each term by $V$ and recall that $\rho = m/V$, this expression becomes

$$P_1 - P_2 = \tfrac{1}{2}\rho v_2{}^2 - \tfrac{1}{2}\rho v_1{}^2 + \rho g y_2 - \rho g y_1$$

Rearrange the terms as follows:

$$P_1 + \tfrac{1}{2}\rho v_1{}^2 + \rho g y_1 = P_2 + \tfrac{1}{2}\rho v_2{}^2 + \rho g y_2 \qquad [9.11]$$

This is **Bernoulli's equation,** often expressed as

$$P + \tfrac{1}{2}\rho v^2 + \rho g y = \text{constant} \qquad [9.12]$$

Bernoulli's equation states that the sum of the pressure $P$, the kinetic energy per unit volume, $\tfrac{1}{2}\rho v^2$, and the potential energy per unit volume, $\rho g y$, has the same value at all points along a streamline.

An important consequence of Bernoulli's equation can be demonstrated by considering Figure 9.29, which shows water flowing through a horizontal constricted pipe from a region of large cross-sectional area into a region of smaller cross-sectional area. This device, called a **Venturi tube,** can be used to measure the speed of fluid flow. Because the pipe is horizontal, $y_1 = y_2$, and Equation 9.11 applied to points 1 and 2 gives

$$P_1 + \tfrac{1}{2}\rho v_1{}^2 = P_2 + \tfrac{1}{2}\rho v_2{}^2 \qquad [9.13]$$

Because the water is not backing up in the pipe, its speed $v_2$ in the constricted region must be greater than its speed $v_1$ in the region of greater diameter. From Equation 9.13, we see that $P_2$ must be less than $P_1$ because $v_2 > v_1$. This result is often expressed by the statement that **swiftly moving fluids exert less pressure than do slowly moving fluids.** This important fact enables us to understand a wide range of everyday phenomena.

**DANIEL BERNOULLI**
**Swiss physicist and mathematician (1700–1782)**

In his most famous work, *Hydrodynamica,* Bernoulli showed that, as the velocity of fluid flow increases, its pressure decreases. In this same publication, Bernoulli also attempted the first explanation of the behavior of gases with changing pressure and temperature; this was the beginning of the kinetic theory of gases.

◀ Bernoulli's equation

**Tip 9.4** Bernoulli's Principle for Gases

Equation 9.11 isn't strictly true for gases because they aren't incompressible. The qualitative behavior is the same, however: As the speed of the gas increases, its pressure decreases.

The pressure $P_1$ is greater than the pressure $P_2$, because $v_1 < v_2$.

Pressure is lower at the narrow part of the tube, so the fluid level is higher.

a                b

**Figure 9.29** (a) This device can be used to measure the speed of fluid flow. (b) A Venturi tube, located at the top of the photograph. The higher level of fluid in the middle column shows that the pressure at the top of the column, which is in the constricted region of the Venturi tube, is lower than the pressure elsewhere in the column.

© Cengage Learning/Charles D. Winters

*Quick Quiz*

**9.7** You observe two helium balloons floating next to each other at the ends of strings secured to a table. The facing surfaces of the balloons are separated by 1–2 cm. You blow through the opening between the balloons. What happens to the balloons? (a) They move toward each other. (b) They move away from each other. (c) They are unaffected.

## EXAMPLE 9.10  SHOOT-OUT AT THE OLD WATER TANK

**GOAL** Apply Bernoulli's equation to find the speed of a fluid.

**PROBLEM** A nearsighted sheriff fires at a cattle rustler with his trusty six-shooter. Fortunately for the rustler, the bullet misses him and penetrates the town water tank, causing a leak (Fig. 9.30). **(a)** If the top of the tank is open to the atmosphere, determine the speed at which the water leaves the hole when the water level is 0.500 m above the hole. **(b)** Where does the stream hit the ground if the hole is 3.00 m above the ground?

**Figure 9.30** (Example 9.10) The water speed $v_1$ from the hole in the side of the container is given by $v_1 = \sqrt{2gh}$.

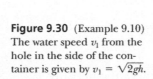

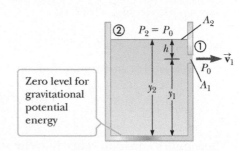

**STRATEGY** **(a)** Assume the tank's cross-sectional area is large compared to the hole's ($A_2 \gg A_1$), so the water level drops very slowly and $v_2 \approx 0$. Apply Bernoulli's equation to points ① and ② in Figure 9.30, noting that $P_1$ equals atmospheric pressure $P_0$ at the hole and is approximately the same at the top of the water tank. Part **(b)** can be solved with kinematics, just as if the water were a ball thrown horizontally.

......................................................

**SOLUTION**

**(a)** Find the speed of the water leaving the hole.

Substitute $P_1 = P_2 = P_0$ and $v_2 \approx 0$ into Bernoulli's equation, and solve for $v_1$:

$$P_0 + \tfrac{1}{2}\rho v_1^2 + \rho g y_1 = P_0 + \rho g y_2$$
$$v_1 = \sqrt{2g(y_2 - y_1)} = \sqrt{2gh}$$
$$v_1 = \sqrt{2(9.80 \text{ m/s}^2)(0.500 \text{ m})} = \boxed{3.13 \text{ m/s}}$$

**(b)** Find where the stream hits the ground.

Use the displacement equation to find the time of the fall, noting that the stream is initially horizontal, so $v_{0y} = 0$.

$$\Delta y = -\tfrac{1}{2}gt^2 + v_{0y}t$$
$$-3.00 \text{ m} = -(4.90 \text{ m/s}^2)t^2$$
$$t = 0.782 \text{ s}$$

Compute the horizontal distance the stream travels in this time:

$$x = v_{0x}t = (3.13 \text{ m/s})(0.782 \text{ s}) = \boxed{2.45 \text{ m}}$$

......................................................

**REMARKS** As the analysis of part (a) shows, the speed of the water emerging from the hole is equal to the speed acquired by an object falling freely through the vertical distance $h$. This is known as **Torricelli's law.**

**QUESTION 9.10** As time passes, what happens to the speed of the water leaving the hole?

**EXERCISE 9.10** Suppose, in a similar situation, the water hits the ground 4.20 m from the hole in the tank. If the hole is 2.00 m above the ground, how far above the hole is the water level?

**ANSWER** 2.21 m above the hole

## EXAMPLE 9.11 | FLUID FLOW IN A PIPE

**GOAL** Solve a problem combining Bernoulli's equation and the equation of continuity.

**PROBLEM** A large pipe with a cross-sectional area of $1.00 \text{ m}^2$ descends 5.00 m and narrows to $0.500 \text{ m}^2$, where it terminates in a valve at point ① (Fig. 9.31). If the pressure at point ② is atmospheric pressure, and the valve is opened wide and water allowed to flow freely, find the speed of the water leaving the pipe.

**STRATEGY** The equation of continuity, together with Bernoulli's equation, constitute two equations in two unknowns: the speeds $v_1$ and $v_2$. Eliminate $v_2$ from Bernoulli's equation with the equation of continuity, and solve for $v_1$.

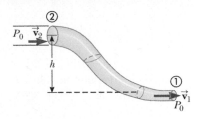

**Figure 9.31** (Example 9.11)

**SOLUTION**

Write Bernoulli's equation:

**(1)** $P_1 + \frac{1}{2}\rho v_1^2 + \rho g y_1 = P_2 + \frac{1}{2}\rho v_2^2 + \rho g y_2$

Solve the equation of continuity for $v_2$:

$A_2 v_2 = A_1 v_1$

**(2)** $v_2 = \dfrac{A_1}{A_2} v_1$

In Equation (1), set $P_1 = P_2 = P_0$, and substitute the expression for $v_2$. Then solve for $v_1$.

**(3)** $P_0 + \frac{1}{2}\rho v_1^2 + \rho g y_1 = P_0 + \frac{1}{2}\rho \left(\dfrac{A_1}{A_2} v_1\right)^2 + \rho g y_2$

$$v_1^2\left[1 - \left(\dfrac{A_1}{A_2}\right)^2\right] = 2g(y_2 - y_1) = 2gh$$

$$v_1 = \dfrac{\sqrt{2gh}}{\sqrt{1 - (A_1/A_2)^2}}$$

Substitute the given values:

$v_1 = \boxed{11.4 \text{ m/s}}$

**REMARKS** Calculating actual flow rates of real fluids through pipes is in fact much more complex than presented here, due to viscosity, the possibility of turbulence, and other factors.

**QUESTION 9.11** Find a symbolic expression for the limit of speed $v_1$ as the lower cross-sectional area $A_1$ opening becomes negligibly small compared to cross section $A_2$. What is this result called?

**EXERCISE 9.11** Water flowing in a horizontal pipe is at a pressure of $1.40 \times 10^5$ Pa at a point where its cross-sectional area is $1.00 \text{ m}^2$. When the pipe narrows to $0.400 \text{ m}^2$, the pressure drops to $1.16 \times 10^5$ Pa. Find the water's speed **(a)** in the wider pipe and **(b)** in the narrower pipe.

**ANSWERS** **(a)** 3.02 m/s    **(b)** 7.56 m/s

# 9.7 Other Applications of Fluid Dynamics

In this section we describe some common phenomena that can be explained, at least in part, by Bernoulli's equation.

In general, an object moving through a fluid is acted upon by a net upward force as the result of any effect that causes the fluid to change direction as it flows past the object. For example, a golf ball struck with a club is given a rapid backspin, as shown in Figure 9.32. The dimples on the ball help entrain the air along the curve of the ball's surface. The figure shows a thin layer of air wrapping partway around

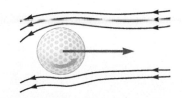

**Figure 9.32** A spinning golf ball is acted upon by a lifting force that allows it to travel much farther than it would if it were not spinning.

the ball and being deflected downward as a result. Because the ball pushes the air down, by Newton's third law the air must push up on the ball and cause it to rise. Without the dimples, the air isn't as well entrained, so the golf ball doesn't travel as far. A tennis ball's fuzz performs a similar function, though the desired result is accurate placement rather than greater distance.

Many devices operate in the manner illustrated in Figure 9.33. A stream of air passing over an open tube reduces the pressure above the tube, causing the liquid to rise into the airstream. The liquid is then dispersed into a fine spray of droplets. You might recognize that this so-called atomizer is used in perfume bottles and paint sprayers. The same principle is used in the carburetor of a gasoline engine. In that case, the low-pressure region in the carburetor is produced by air drawn in by the piston through the air filter. The gasoline vaporizes, mixes with the air, and enters the cylinder of the engine for combustion.

In a person with advanced arteriosclerosis, the Bernoulli effect produces a symptom called **vascular flutter.** In this condition, the artery is constricted as a result of accumulated plaque on its inner walls, as shown in Figure 9.34. To maintain a constant flow rate, the blood must travel faster than normal through the constriction. If the speed of the blood is sufficiently high in the constricted region, the blood pressure is low, and the artery may collapse under external pressure, causing a momentary interruption in blood flow. During the collapse there is no Bernoulli effect, so the vessel reopens under arterial pressure. As the blood rushes through the constricted artery, the internal pressure drops and the artery closes again. Such variations in blood flow can be heard with a stethoscope. If the plaque becomes dislodged and ends up in a smaller vessel that delivers blood to the heart, it can cause a heart attack.

An **aneurysm** is a weakened spot on an artery where the artery walls have ballooned outward. Blood flows more slowly though this region, as can be seen from the equation of continuity, resulting in an increase in pressure in the vicinity of the aneurysm relative to the pressure in other parts of the artery. This condition is dangerous because the excess pressure can cause the artery to rupture.

The lift on an aircraft wing can also be explained in part by the Bernoulli effect. Airplane wings are designed so that the air speed above the wing is greater than the speed below. As a result, the air pressure above the wing is less than the pressure below, and there is a net upward force on the wing, called the *lift*. (There is also a horizontal component called the *drag*.) Another factor influencing the lift on a wing, shown in Figure 9.35, is the slight upward tilt of the wing. This causes air molecules striking the bottom to be deflected downward, producing a reaction force upward by Newton's third law. Finally, turbulence also has an effect. If the wing is tilted too much, the flow of air across the upper surface becomes turbulent, and the pressure difference across the wing is not as great as that predicted by the Bernoulli effect. In an extreme case, this turbulence may cause the aircraft to stall.

**Figure 9.33** A stream of air passing over a tube dipped in a liquid causes the liquid to rise in the tube. This effect is used in perfume bottles and paint sprayers.

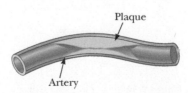

Plaque

Artery

**Figure 9.34** Blood must travel faster than normal through a constricted region of an artery.

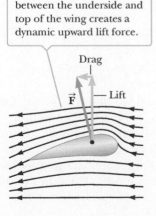

The difference in pressure between the underside and top of the wing creates a dynamic upward lift force.

Drag

$\vec{F}$ — Lift

**Figure 9.35** Streamline flow around an airplane wing. The pressure above is less than the pressure below, and there is a dynamic upward lift force.

## EXAMPLE 9.12 | LIFT ON AN AIRFOIL

**GOAL** Use Bernoulli's equation to calculate the lift on an airplane wing.

**PROBLEM** An airplane has wings, each with area 4.00 m$^2$, designed so that air flows over the top of the wing at 245 m/s and underneath the wing at 222 m/s. Find the mass of the airplane such that the lift on the plane will support its weight, assuming the force from the pressure difference across the wings is directed straight upward.

**STRATEGY** This problem can be solved by substituting values into Bernoulli's equation to find the pressure difference between the air under the wing and the air over the wing, followed by applying Newton's second law to find the mass the airplane can lift.

**SOLUTION**

Apply Bernoulli's equation to the air flowing under the wing (point 1) and over the wing (point 2). Gravitational potential energy terms are small compared with the other terms, and can be neglected.

$$P_1 + \tfrac{1}{2}\rho v_1^2 = P_2 + \tfrac{1}{2}\rho v_2^2$$

Solve this equation for the pressure difference:

$$\Delta P = P_1 - P_2 = \tfrac{1}{2}\rho v_2^2 - \tfrac{1}{2}\rho v_1^2 = \tfrac{1}{2}\rho\left(v_2^2 - v_1^2\right)$$

Substitute the given speeds and $\rho = 1.29$ kg/m$^3$, the density of air:

$$\Delta P = \tfrac{1}{2}(1.29 \text{ kg/m}^3)(245^2 \text{ m}^2/\text{s}^2 - 222^2 \text{ m}^2/\text{s}^2)$$
$$\Delta P = 6.93 \times 10^3 \text{ Pa}$$

Apply Newton's second law. To support the plane's weight, the sum of the lift and gravity forces must equal zero. Solve for the mass $m$ of the plane.

$$2A\,\Delta P - mg = 0 \quad \rightarrow \quad m = \boxed{5.66 \times 10^3 \text{ kg}}$$

**REMARKS** Note the factor of two in the last equation, needed because the airplane has two wings. The density of the atmosphere drops steadily with increasing height, reducing the lift. As a result, all aircraft have a maximum operating altitude.

**QUESTION 9.12** Why is the maximum lift affected by increasing altitude?

**EXERCISE 9.12** Approximately what size wings would an aircraft need on Mars if its engine generates the same differences in speed as in the example and the total mass of the craft is 400 kg? The density of air on the surface of Mars is approximately one percent Earth's density at sea level, and the acceleration of gravity on the surface of Mars is about 3.8 m/s$^2$.

**ANSWER** Rounding to one significant digit, each wing would have to have an area of about 10 m$^2$. There have been proposals for solar-powered robotic Mars aircraft, which would have to be gossamer-light with large wings.

## APPLYING PHYSICS 9.4 | SAILING UPWIND

How can a sailboat accomplish the seemingly impossible task of sailing into the wind?

**EXPLANATION** As shown in Figure 9.36, the wind blowing in the direction of the arrow causes the sail to billow out and take on a shape similar to that of an airplane wing. By Bernoulli's equation, just as for an airplane wing, there is a force, $\vec{\mathbf{F}}_{\text{wind}}$, on the sail in the direction shown. The component of that force perpendicular to the boat tends to make the boat move sideways in the water, but the force of the water on the keel, $\vec{\mathbf{F}}_{\text{water}}$, prevents this sideways motion. The component of the force in the forward direction, $\vec{\mathbf{F}}_R$, drives the boat almost against the wind. The word *almost* is used because a sailboat can move forward only when the wind direction is about 10° to 15° or more with respect to the forward direction. This means that to sail

directly against the wind, a boat must follow a zigzag path, a procedure called *tacking*, so that the wind is always at some angle with respect to the direction of travel. ■

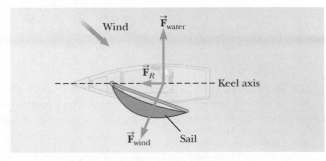

**Figure 9.36** (Applying Physics 9.4)

Consider the portion of a home plumbing system shown in Figure 9.37. The water trap in the pipe below the sink captures a plug of water that prevents sewer gas from finding its way from the sewer pipe, up the sink drain, and into the home. Suppose the dishwasher is draining and the water is moving to the left in the sewer pipe. What is the purpose of the vent, which is open to the air above the roof of the house? In which direction is air moving at the opening of the vent, upward or downward?

**EXPLANATION** Imagine that the vent isn't present so that the drainpipe for the sink is simply connected through the trap to the sewer pipe. As water from the dishwasher moves to the left in the sewer pipe, the pressure in the sewer pipe is reduced below atmospheric pressure, in accordance with Bernoulli's principle. The pressure at the drain in the sink is still at atmospheric pressure. This pressure difference can push the plug of water in the water trap of the sink down the drainpipe and into the sewer pipe, removing it as a barrier to sewer gas. With the addition of the vent to the roof, the reduced pressure from the dishwasher water will result in air entering the vent pipe at the roof. This inflow of air will keep the pressure in the vent pipe and the right-hand side of the sink drainpipe close to atmospheric pressure so that the plug of water in the water trap will remain in place. ■

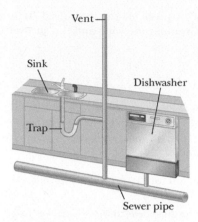

**Figure 9.37** (Applying Physics 9.5)

APPLICATION
Rocket Engines

The exhaust speed of a rocket engine can also be understood qualitatively with Bernoulli's equation, although, in actual practice, a large number of additional variables need to be taken into account. Rockets actually work better in vacuum than in the atmosphere, contrary to an early *New York Times* article criticizing rocket pioneer Robert Goddard, which held that they wouldn't work at all, having no air to push against. The pressure inside the combustion chamber is $P$, and the pressure just outside the nozzle is the ambient atmospheric pressure, $P_{atm}$. Differences in height between the combustion chamber and the end of the nozzle result in negligible contributions of gravitational potential energy. In addition, the gases inside the chamber flow at negligible speed compared to gases going through the nozzle. The exhaust speed can be found from Bernoulli's equation,

$$v_{ex} = \sqrt{\frac{2(P - P_{atm})}{\rho}}$$

This equation shows that the exhaust speed is reduced in the atmosphere, so rockets are actually more effective in the vacuum of space. Also of interest is the appearance of the density $\rho$ in the denominator. A lower density working fluid or gas will give a higher exhaust speed, which partly explains why liquid hydrogen, which has a very low density, is a fuel of choice.

# 9.8 Surface Tension, Capillary Action, and Viscous Fluid Flow

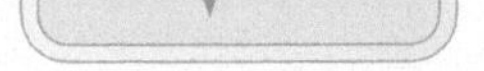

**Figure 9.38** The net force on a molecule at $A$ is zero because such a molecule is completely surrounded by other molecules. The net force on a surface molecule at $B$ is downward because it isn't completely surrounded by other molecules.

If you look closely at a dewdrop sparkling in the morning sunlight, you will find that the drop is spherical. The drop takes this shape because of a property of liquid surfaces called **surface tension.** To understand the origin of surface tension, consider a molecule at point $A$ in a container of water, as in Figure 9.38. Although nearby molecules exert forces on this molecule, the net force on it is zero because it's completely surrounded by other molecules and hence is attracted equally in all directions. The molecule at $B$, however, is not attracted equally in all directions. Because there are no molecules above it to exert upward forces, the molecule at $B$ is pulled toward the interior of the liquid. The contraction at the surface of the liquid ceases when the inward pull exerted on the surface molecules is balanced by the outward repulsive forces that arise from collisions with molecules in the interior of the liquid. **The net effect of this pull on all the surface molecules is to make the surface of the liquid contract and, consequently, to make the surface area of the liquid as small as possible.** Drops of water take on a spherical shape because a sphere has the smallest surface area for a given volume.

If you place a sewing needle very carefully on the surface of a bowl of water, you will find that the needle floats even though the density of steel is about eight times that of water. This phenomenon can also be explained by surface tension. A close examination of the needle shows that it actually rests in a depression in the liquid surface as shown in Figure 9.39. The water surface acts like an elastic membrane under tension. The weight of the needle produces a depression, increasing the surface area of the film. Molecular forces now act at all points along the depression, tending to restore the surface to its original horizontal position. The vertical components of these forces act to balance the force of gravity on the needle. The floating needle can be sunk by adding a little detergent to the water, which reduces the surface tension.

The **surface tension** $\gamma$ in a film of liquid is defined as the magnitude of the surface tension force $F$ divided by the length $L$ along which the force acts:

$$\gamma \equiv \frac{F}{L} \qquad [9.14]$$

The SI unit of surface tension is the newton per meter, and values for a few representative materials are given in Table 9.2.

Surface tension can be thought of as the energy content of the fluid at its surface per unit surface area. To see that this is reasonable, we can manipulate the units of surface tension $\gamma$ as follows:

$$\frac{N}{m} = \frac{N \cdot m}{m^2} = \frac{J}{m^2}$$

In general, in **any equilibrium configuration of an object, the energy is a minimum.** Consequently, a fluid will take on a shape such that its surface area is as small as possible. For a given volume, a spherical shape has the smallest surface area; therefore, a drop of water takes on a spherical shape.

An apparatus used to measure the surface tension of liquids is shown in Figure 9.40. A circular wire with a circumference $L$ is lifted from a body of liquid. The surface film clings to the inside and outside edges of the wire, holding back the wire and causing the spring to stretch. If the spring is calibrated, the force required to overcome the surface tension of the liquid can be measured. In this case, the surface tension is given by

$$\gamma = \frac{F}{2L}$$

We use $2L$ for the length because the surface film exerts forces on both the inside and outside of the ring.

The surface tension of liquids decreases with increasing temperature because the faster moving molecules of a hot liquid aren't bound together as strongly as are those in a cooler liquid. In addition, certain ingredients called surfactants decrease surface tension when added to liquids. For example, soap or detergent decreases the surface tension of water, making it easier for soapy water to penetrate the cracks and crevices of your clothes to clean them better than plain

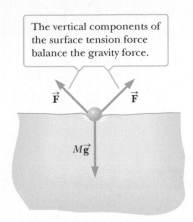

The vertical components of the surface tension force balance the gravity force.

**Figure 9.39** End view of a needle resting on the surface of water.

**Table 9.2** Surface Tensions for Various Liquids

| Liquid | $T\,(°C)$ | Surface Tension (N/m) |
|---|---|---|
| Ethyl alcohol | 20 | 0.022 |
| Mercury | 20 | 0.465 |
| Soapy water | 20 | 0.025 |
| Water | 20 | 0.073 |
| Water | 100 | 0.059 |

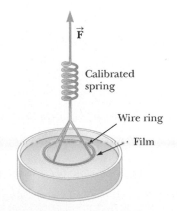

**Figure 9.40** An apparatus for measuring the surface tension of liquids. The force on the wire ring is measured just before the ring breaks free of the liquid.

water does. A similar effect occurs in the lungs. The surface tissue of the air sacs in the lungs contains a fluid that has a surface tension of about 0.050 N/m. A liquid with a surface tension this high would make it very difficult for the lungs to expand during inhalation. However, as the area of the lungs increases with inhalation, the body secretes into the tissue a substance that gradually reduces the surface tension of the liquid. At full expansion, the surface tension of the lung fluid can drop to as low as 0.005 N/m.

**BIO APPLICATION**
Air Sac Surface Tension

## EXAMPLE 9.13   WALKING ON WATER   BIO

**GOAL** Apply the surface tension equation.

**PROBLEM** Many insects can literally walk on water, using surface tension for their support. To show this is feasible, assume the insect's "foot" is spherical. When the insect steps onto the water with all six legs, a depression is formed in the water around each foot, as shown in Figure 9.41a. The surface tension of the water produces upward forces on the water that tend to restore the water surface to its normally flat shape. If the insect's mass is $2.0 \times 10^{-5}$ kg and the radius of each foot is $1.5 \times 10^{-4}$ m, find the angle $\theta$.

**STRATEGY** Find an expression for the magnitude of the net force $F$ directed tangentially to the depressed part of the water surface, and obtain the part that is acting vertically, in opposition to the downward force of gravity. Assume the radius of depression is the same as the radius of the insect's foot. Because the insect has six legs, one-sixth of the insect's weight must be supported by one of the legs, assuming the weight is distributed evenly. The length $L$ is just the distance around a circle. Using Newton's second law for a body in equilibrium (zero acceleration), solve for $\theta$.

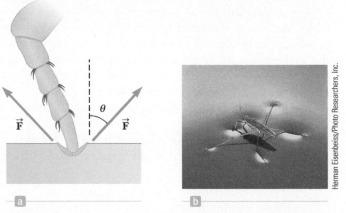

**Figure 9.41** (Example 9.13) (a) One foot of an insect resting on the surface of water. (b) This water strider resting on the surface of a lake remains on the surface, rather than sinking, because an upward surface tension force acts on each leg, balancing the force of gravity on the insect.

### SOLUTION

Start with the surface tension equation:

$$F = \gamma L$$

Focus on one circular foot, substituting $L = 2\pi r$. Multiply by $\cos \theta$ to get the vertical component $F_v$:

$$F_v = \gamma(2\pi r) \cos \theta$$

Write Newton's second law for the insect's one foot, which supports one-sixth of the insect's weight:

$$\sum F = F_v - F_{grav} = 0$$
$$\gamma(2\pi r) \cos \theta - \tfrac{1}{6}mg = 0$$

Solve for $\cos \theta$ and substitute:

$$(1) \quad \cos \theta = \frac{mg}{12\pi r \gamma}$$
$$= \frac{(2.0 \times 10^{-5} \text{ kg})(9.80 \text{ m/s}^2)}{12\pi(1.5 \times 10^{-4} \text{ m})(0.073 \text{ N/m})} = 0.47$$

Take the inverse cosine of both sides to find the angle $\theta$:

$$\theta = \cos^{-1}(0.47) = \boxed{62°}$$

**REMARKS** If the weight of the insect were great enough to make the right side of Equation (1) greater than 1, a solution for $\theta$ would be impossible because the cosine of an angle can never be greater than 1. In this circumstance, the insect would sink.

**QUESTION 9.13** True or False: Warm water gives more support to walking insects than cold water.

**EXERCISE 9.13** A typical sewing needle floats on water when its long dimension is parallel to the water's surface. Estimate the needle's maximum possible mass, assuming the needle is two inches long. *Hint:* The cosine of an angle is never larger than 1.

**ANSWER** 0.8 g

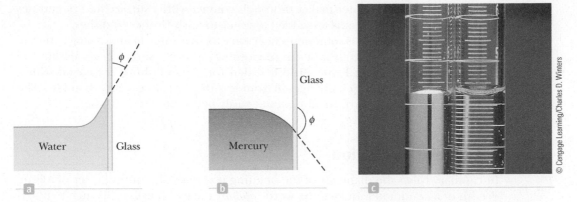

**Figure 9.42** A liquid in contact with a solid surface. (a) For water, the adhesive force is greater than the cohesive force. (b) For mercury, the adhesive force is less than the cohesive force. (c) The surface of mercury (*left*) curves downward in a glass container, whereas the surface of water (*right*) curves upward, as you move from the center to the edge.

## 9.8.1 The Surface of Liquid

If you have ever closely examined the surface of water in a glass container, you may have noticed that the surface of the liquid near the walls of the glass curves upward as you move from the center to the edge, as shown in Figure 9.42a. However, if mercury is placed in a glass container, the mercury surface curves downward, as in Figure 9.42b. These surface effects can be explained by considering the forces between molecules. In particular, we must consider the forces that the molecules of the liquid exert on one another and the forces that the molecules of the glass surface exert on those of the liquid. In general terms, forces between like molecules, such as the forces between water molecules, are called **cohesive forces,** and forces between unlike molecules, such as those exerted by glass on water, are called **adhesive forces.**

Water tends to cling to the walls of the glass because the adhesive forces between the molecules of water and the glass molecules are *greater* than the cohesive forces between the water molecules. In effect, the water molecules cling to the surface of the glass rather than fall back into the bulk of the liquid. When this condition prevails, the liquid is said to "wet" the glass surface. The surface of the mercury curves downward near the walls of the container because the cohesive forces between the mercury atoms are greater than the adhesive forces between mercury and glass. A mercury atom near the surface is pulled more strongly toward other mercury atoms than toward the glass surface, so mercury doesn't wet the glass surface.

The angle $\phi$ between the solid surface and a line drawn tangent to the liquid at the surface is called the **contact angle** (Fig. 9.43). The angle $\phi$ is less than 90° for any substance in which adhesive forces are stronger than cohesive forces and greater than 90° if cohesive forces predominate. For example, if a drop of water is placed on paraffin, the contact angle is approximately 107° (Fig. 9.43a). If certain chemicals, called wetting agents or detergents, are added to the water, the contact angle becomes less than 90°, as shown in Figure 9.43b. The addition of such substances to

**APPLICATION**

Detergents and Waterproofing Agents

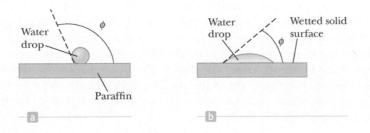

**Figure 9.43** (a) The contact angle between water and paraffin is about 107°. In this case, the cohesive force is greater than the adhesive force. (b) When a chemical called a wetting agent is added to the water, it wets the paraffin surface, and $\phi < 90°$. In this case, the adhesive force is greater than the cohesive force.

water ensures that the water makes thorough contact with a surface and penetrates it. For this reason, detergents are added to water to wash clothes or dishes.

On the other hand, it is sometimes necessary to *keep* water from making intimate contact with a surface, as in waterproof clothing, where a situation somewhat the reverse of that shown in Figure 9.43 is called for. The clothing is sprayed with a waterproofing agent, which changes $\phi$ from less than 90° to greater than 90°. The water beads up on the surface and doesn't easily penetrate the clothing.

### 9.8.2 Capillary Action

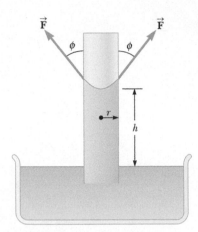

**Figure 9.44** A liquid rises in a narrow tube because of capillary action, a result of surface tension and adhesive forces.

In capillary tubes the diameter of the opening is very small, on the order of a hundredth of a centimeter. In fact, the word *capillary* means "hairlike." If such a tube is inserted into a fluid for which adhesive forces dominate over cohesive forces, the liquid rises into the tube, as shown in Figure 9.44. The rising of the liquid in the tube can be explained in terms of the shape of the liquid's surface and surface tension effects. At the point of contact between liquid and solid, the upward force of surface tension is directed as shown in the figure. From Equation 9.14, the magnitude of this force is

$$F = \gamma L = \gamma(2\pi r)$$

(We use $L = 2\pi r$ here because the liquid is in contact with the surface of the tube at all points around its circumference.) The vertical component of this force due to surface tension is

$$F_v = \gamma(2\pi r)(\cos \phi) \qquad [9.15]$$

For the liquid in the capillary tube to be in equilibrium, this upward force must be equal to the weight of the cylinder of water of height $h$ inside the capillary tube. The weight of this water is

$$w = Mg = \rho Vg = \rho g \pi r^2 h \qquad [9.16]$$

Equating $F_v$ in Equation 9.15 to $w$ in Equation 9.16 (applying Newton's second law for equilibrium), we have

$$\gamma(2\pi r)(\cos \phi) = \rho g \pi r^2 h$$

Solving for $h$ gives the height to which water is drawn into the tube:

$$h = \frac{2\gamma}{\rho g r} \cos \phi \qquad [9.17]$$

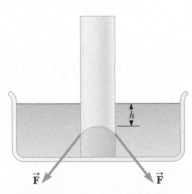

**Figure 9.45** When cohesive forces between molecules of a liquid exceed adhesive forces, the level of the liquid in the capillary tube is below the surface of the surrounding fluid.

**BIO APPLICATION**
Blood Samples with Capillary Tubes

**BIO APPLICATION**
Capillary Action in Plants

If a capillary tube is inserted into a liquid in which cohesive forces dominate over adhesive forces, the level of the liquid in the capillary tube will be below the surface of the surrounding fluid, as shown in Figure 9.45. An analysis similar to the above would show that the distance $h$ to the depressed surface is given by Equation 9.17.

Capillary tubes are often used to draw small samples of blood from a needle prick in the skin. Capillary action must also be considered in the construction of concrete-block buildings because water seepage through capillary pores in the blocks or the mortar may cause damage to the inside of the building. To prevent such damage, the blocks are usually coated with a waterproofing agent either outside or inside the building. Water seepage through a wall is an undesirable effect of capillary action, but there are many useful effects. Plants depend on capillary action to transport water and nutrients, and sponges and paper towels use capillary action to absorb spilled fluids.

## EXAMPLE 9.14 | RISING WATER

**GOAL** Apply surface tension to capillary action.

**PROBLEM** Find the height to which water would rise in a capillary tube with a radius equal to $5.0 \times 10^{-5}$ m. Assume the contact angle between the water and the material of the tube is small enough to be considered zero.

**STRATEGY** This problem requires substituting values into Equation 9.17.

**SOLUTION**

Substitute the known values into Equation 9.17:

$$h = \frac{2\gamma \cos 0°}{\rho g r}$$

$$= \frac{2(0.073 \text{ N/m})}{(1.00 \times 10^3 \text{ kg/m}^3)(9.80 \text{ m/s}^2)(5.0 \times 10^{-5} \text{ m})}$$

$$= \boxed{0.30 \text{ m}}$$

**QUESTION 9.14** Based on the result of this calculation, is capillary action likely to be the sole mechanism of water and nutrient transport in plants? Explain.

**EXERCISE 9.14** Suppose ethyl alcohol rises 0.250 m in a thin tube. Estimate the radius of the tube, assuming the contact angle is approximately zero.

**ANSWER** $2.2 \times 10^{-5}$ m

### 9.8.3 Viscous Fluid Flow

It is considerably easier to pour water out of a container than to pour honey. This is because honey has a higher viscosity than water. In a general sense, **viscosity refers to the internal friction of a fluid.** It's very difficult for layers of a viscous fluid to slide past one another. Likewise, it's difficult for one solid surface to slide past another if there is a highly viscous fluid, such as soft tar, between them.

When an ideal (nonviscous) fluid flows through a pipe, the fluid layers slide past one another with no resistance. If the pipe has a uniform cross section, each layer has the same velocity, as shown in Figure 9.46a. In contrast, the layers of a viscous fluid have different velocities, as Figure 9.46b indicates. The fluid has the greatest velocity at the center of the pipe, whereas the layer next to the wall doesn't move because of adhesive forces between molecules and the wall surface.

To better understand the concept of viscosity, consider a layer of liquid between two solid surfaces, as in Figure 9.47. The lower surface is fixed in position, and the top surface moves to the right with a velocity $\vec{v}$ under the action of an external force $\vec{F}$. Because of this motion, a portion of the liquid is distorted from its original shape, *ABCD*, to the shape *AEFD* a moment later. The force required to move the upper plate and distort the liquid is proportional to both the area $A$ in contact with the fluid and the speed $v$ of the fluid. Further, the force is inversely proportional

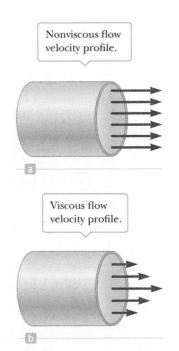

Nonviscous flow velocity profile.

a

Viscous flow velocity profile.

b

**Figure 9.46** (a) The particles in an ideal (nonviscous) fluid all move through the pipe with the same velocity. (b) In a viscous fluid, the velocity of the fluid particles is zero at the surface of the pipe and increases to a maximum value at the center of the pipe.

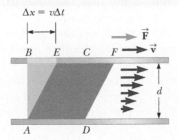

$\Delta x = v\Delta t$

$B$   $E$   $C$   $F$ $\vec{F}$

$\vec{v}$

$d$

$A$   $D$

**Figure 9.47** A layer of liquid between two solid surfaces in which the lower surface is fixed and the upper surface moves to the right with a velocity $\vec{v}$.

**Table 9.3** Viscosities of Various Fluids

| Fluid | $T$ (°C) | Viscosity $\eta$ (N · s/m²) |
|---|---|---|
| Water | 20 | $1.0 \times 10^{-3}$ |
| Water | 100 | $0.3 \times 10^{-3}$ |
| Whole blood | 37 | $2.7 \times 10^{-3}$ |
| Glycerin | 20 | $1\,500 \times 10^{-3}$ |
| 10-wt motor oil | 30 | $250 \times 10^{-3}$ |

to the distance $d$ between the two plates. We can express these proportionalities as $F \propto Av/d$. The force required to move the upper plate at a fixed speed $v$ is therefore

$$F = \eta \frac{Av}{d} \qquad [9.18]$$

where $\eta$ (the lowercase Greek letter *eta*) is the **coefficient of viscosity** of the fluid.

The SI units of viscosity are N · s/m². The units of viscosity in many reference sources are expressed in dyne · s/cm², called 1 **poise**, in honor of the French scientist J. L. Poiseuille (1799–1869). The relationship between the SI unit of viscosity and the poise is

$$1 \text{ poise} = 10^{-1} \text{ N} \cdot \text{s/m}^2 \qquad [9.19]$$

Small viscosities are often expressed in centipoise (cp), where 1 cp = $10^{-2}$ poise. The coefficients of viscosity for some common substances are listed in Table 9.3.

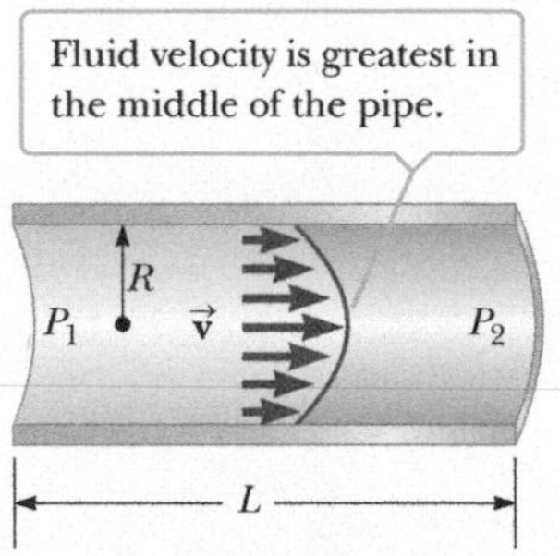

Fluid velocity is greatest in the middle of the pipe.

**Figure 9.48** Velocity profile of a fluid flowing through a uniform pipe of circular cross section. The rate of flow is given by Poiseuille's law.

### 9.8.4 Poiseuille's Law

Figure 9.48 shows a section of a tube of length $L$ and radius $R$ containing a fluid under a pressure $P_1$ at the left end and a pressure $P_2$ at the right. Because of this pressure difference, the fluid flows through the tube. The rate of flow (volume per unit time) depends on the pressure difference ($P_1 - P_2$), the dimensions of the tube, and the viscosity of the fluid. The result, known as **Poiseuille's law,** is

Poiseuille's law ▶
$$\text{Rate of flow} = \frac{\Delta V}{\Delta t} = \frac{\pi R^4 (P_1 - P_2)}{8\eta L} \qquad [9.20]$$

where $\eta$ is the coefficient of viscosity of the fluid. We won't attempt to derive this equation here because the methods of integral calculus are required. However, it is reasonable that the rate of flow should increase if the pressure difference across the tube or the tube radius increases. Likewise, the flow rate should decrease if the viscosity of the fluid or the length of the tube increases. So the presence of $R$ and the pressure difference in the numerator of Equation 9.20 and of $L$ and $\eta$ in the denominator make sense.

**BIO APPLICATION**
Poiseuille's Law and Blood Flow

From Poiseuille's law, we see that in order to maintain a constant flow rate, the pressure difference across the tube has to increase as the viscosity of the fluid increases. This fact is important in understanding the flow of blood through the circulatory system. The viscosity of blood increases as the number of red blood cells rises. Blood with a high concentration of red blood cells requires greater pumping pressure from the heart to keep it circulating than does blood of lower red blood cell concentration.

Note that the flow rate varies as the radius of the tube raised to the fourth power. Consequently, if a constriction occurs in a vein or artery, the heart will have to work considerably harder in order to produce a higher pressure drop and hence maintain the required flow rate.

## EXAMPLE 9.15    A BLOOD TRANSFUSION    BIO

**GOAL** Apply Poiseuille's law.

**PROBLEM** A patient receives a blood transfusion through a needle of radius 0.20 mm and length 2.0 cm. The density of blood is 1 050 kg/m³. The bottle supplying the blood is 0.500 m above the patient's arm. What is the rate of flow through the needle?

**STRATEGY** Find the pressure difference between the level of the blood and the patient's arm. Substitute into Poiseuille's law, using the value for the viscosity of whole blood in Table 9.3.

**SOLUTION**

Calculate the pressure difference:

$$P_1 - P_2 = \rho g h = (1\ 050\ \text{kg/m}^3)(9.80\ \text{m/s}^2)(0.500\ \text{m})$$
$$= 5.15 \times 10^3\ \text{Pa}$$

Substitute into Poiseuille's law:

$$\frac{\Delta V}{\Delta t} = \frac{\pi R^4 (P_1 - P_2)}{8\eta L}$$

$$= \frac{\pi (2.0 \times 10^{-4}\ \text{m})^4 (5.15 \times 10^3\ \text{Pa})}{8(2.7 \times 10^{-3}\ \text{N} \cdot \text{s/m}^2)(2.0 \times 10^{-2}\ \text{m})}$$

$$= \boxed{6.0 \times 10^{-8}\ \text{m}^3/\text{s}}$$

**REMARKS** Compare this to the volume flow rate in the absence of any viscosity. Using Bernoulli's equation, the calculated volume flow rate is approximately five times as great. As expected, viscosity greatly reduces flow rate.

**QUESTION 9.15** If the radius of a tube is doubled, by what factor will the flow rate change for a viscous fluid?

**EXERCISE 9.15** A pipe carrying water from a tank 20.0 m tall must cross $3.00 \times 10^2$ km of wilderness to reach a remote town. Find the radius of pipe so that the volume flow rate is at least 0.050 0 m³/s. (Use the viscosity of water at 20°C.)

**ANSWER** 0.118 m

## 9.8.5 Reynolds Number

At sufficiently high velocities, fluid flow changes from simple streamline flow to turbulent flow, characterized by a highly irregular motion of the fluid. Experimentally, the onset of turbulence in a tube is determined by a dimensionless factor called the **Reynolds number,** *RN*, given by

$$RN = \frac{\rho v d}{\eta} \qquad\qquad [9.21] \quad \blacktriangleleft \text{ Reynolds number}$$

where $\rho$ is the density of the fluid, $v$ is the average speed of the fluid along the direction of flow, $d$ is the diameter of the tube, and $\eta$ is the viscosity of the fluid. If *RN* is below about 2 000, the flow of fluid through a tube is streamline; turbulence occurs if *RN* is above 3 000. In the region between 2 000 and 3 000, the flow is unstable, meaning that the fluid can move in streamline flow, but any small disturbance will cause its motion to change to turbulent flow.

## EXAMPLE 9.16    TURBULENT FLOW OF BLOOD    BIO

**GOAL** Use the Reynolds number to determine a speed associated with the onset of turbulence.

**PROBLEM** Determine the speed at which blood flowing through an artery of diameter 0.20 cm will become turbulent.

**STRATEGY** The solution requires only the substitution of values into Equation 9.21 giving the Reynolds number and then solving it for the speed $v$.

*(Continued)*

## SOLUTION

Solve Equation 9.21 for $v$, and substitute the viscosity and density of blood from Example 9.15, the diameter $d$ of the artery, and a Reynolds number of $3.00 \times 10^3$:

$$v = \frac{\eta(RN)}{\rho d} = \frac{(2.7 \times 10^{-3}\ \text{N} \cdot \text{s/m}^2)(3.00 \times 10^3)}{(1.05 \times 10^3\ \text{kg/m}^3)(0.20 \times 10^{-2}\ \text{m})}$$

$$v = \boxed{3.9\ \text{m/s}}$$

**REMARKS** Exercise 9.16 shows that rapid ingestion of water through a straw may create a turbulent state.

**QUESTION 9.16** True or False: If the viscosity of a fluid flowing through a tube is increased, the speed associated with the onset of turbulence decreases.

**EXERCISE 9.16** Determine the speed $v$ at which water at 20°C sucked up a straw would become turbulent. The straw has a diameter of 0.006 0 m.

**ANSWER** $v = 0.50$ m/s

# 9.9 Transport Phenomena

When a fluid flows through a tube, the basic mechanism that produces the flow is a difference in pressure across the ends of the tube. This pressure difference is responsible for the transport of a mass of fluid from one location to another. The fluid may also move from place to place because of a second mechanism—one that depends on a difference in *concentration* between two points in the fluid, as opposed to a pressure difference. When the concentration (the number of molecules per unit volume) is higher at one location than at another, molecules will flow from the point where the concentration is high to the point where it is lower. The two fundamental processes involved in fluid transport resulting from concentration differences are called *diffusion* and *osmosis*.

## 9.9.1 Diffusion

**In a diffusion process, molecules move from a region where their concentration is high to a region where their concentration is lower.** To understand why diffusion occurs, consider Figure 9.49, which depicts a container in which a high concentration of molecules has been introduced into the left side. The dashed line in the figure represents an imaginary barrier separating the two regions. Because the molecules are moving with high speeds in random directions, many of them will cross the imaginary barrier moving from left to right. Very few molecules will pass through moving from right to left, simply because there are very few of them on the right side of the container at any instant. As a result, there will always be a *net* movement from the region with many molecules to the region with fewer molecules. For this reason, the concentration on the left side of the container will decrease, and that on the right side will increase with time. Once a concentration equilibrium has been reached, there will be no *net* movement across the cross-sectional area: The rate of movement of molecules from left to right will equal the rate from right to left.

The basic equation for diffusion is **Fick's law,**

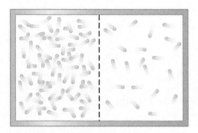

**Figure 9.49** When the concentration of gas molecules on the left side of the container exceeds the concentration on the right side, there will be a net motion (diffusion) of molecules from left to right.

**Table 9.4** Diffusion Coefficients of Various Substances at 20°C

| Substance | $D$ (m²/s) |
| --- | --- |
| Oxygen<br>through air | $6.4 \times 10^{-5}$ |
| Oxygen<br>through tissue | $1 \times 10^{-11}$ |
| Oxygen<br>through water | $1 \times 10^{-9}$ |
| Sucrose<br>through water | $5 \times 10^{-10}$ |
| Hemoglobin<br>through water | $76 \times 10^{-12}$ |

Fick's law ▶

$$\text{Diffusion rate} = \frac{\text{mass}}{\text{time}} = \frac{\Delta M}{\Delta t} = DA\left(\frac{C_2 - C_1}{L}\right) \qquad [9.22]$$

where $D$ is a constant of proportionality. The left side of this equation is called the *diffusion rate* and is a measure of the mass being transported per unit time. The equation says that the rate of diffusion is proportional to the cross-sectional area $A$ and to the change in concentration per unit distance, $(C_2 - C_1)/L$, which is called the *concentration gradient*. The concentrations $C_1$ and $C_2$ are measured in kilograms per cubic meter. The proportionality constant $D$ is called the **diffusion coefficient** and has units of square meters per second. Table 9.4 lists diffusion coefficients for a few substances.

## 9.9.2 The Size of Cells and Osmosis

Diffusion through cell membranes is vital in carrying oxygen to the cells of the body and in removing carbon dioxide and other waste products from them. Cells require oxygen for those metabolic processes in which substances are either synthesized or broken down. In such processes, the cell uses up oxygen and produces carbon dioxide as a by-product. A fresh supply of oxygen diffuses from the blood, where its concentration is high, into the cell, where its concentration is low. Likewise, carbon dioxide diffuses from the cell into the blood, where it is in lower concentration. Water, ions, and other nutrients also pass into and out of cells by diffusion.

A cell can function properly only if it can transport nutrients and waste products rapidly across the cell membrane. The surface area of the cell should be large enough so that the exposed membrane area can exchange materials effectively whereas the volume should be small enough so that materials can reach or leave particular locations rapidly. This requires a large surface-area-to-volume ratio.

Model a cell as a cube, each side with length $L$. The total surface area is $6L^2$ and the volume is $L^3$. The surface area to volume is then

$$\frac{\text{surface area}}{\text{volume}} = \frac{6L^2}{L^3} = \frac{6}{L}$$

Because $L$ is in the denominator, a smaller $L$ means a larger ratio. This shows that the smaller the size of a body, the more efficiently it can transport nutrients and waste products across the cell membrane. Cells range in size from a millionth of a meter to several millionths, so a good estimate of a typical cell's surface-to-volume ratio is $10^6$.

The diffusion of material through a membrane is partially determined by the size of the pores (holes) in the membrane wall. Small molecules, such as water, may pass through the pores easily, while larger molecules, such as sugar, may pass through only with difficulty or not at all. A membrane that allows passage of some molecules but not others is called a **selectively permeable** membrane.

**Osmosis is the diffusion of water across a selectively permeable membrane from a high water concentration to a low water concentration.** As in the case of diffusion, osmosis continues until the concentrations on the two sides of the membrane are equal.

To understand the effect of osmosis on living cells, consider a particular cell in the body with a sugar concentration of 1%. (A 1% solution is 1 g of sugar dissolved in enough water to make 100 mL of solution; "mL" is the abbreviation for milliliters, where $1\ \text{mL} = 10^{-3}\ \text{L} = 1\ \text{cm}^3$.) Assume this cell is immersed in a 5% sugar solution (5 g of sugar dissolved in enough water to make 100 mL). Compared to the 1% solution, there are five times as many sugar molecules per unit volume in the 5% sugar solution, so there must be fewer water molecules. Accordingly, water will diffuse from inside the cell, where its concentration is higher, across the cell membrane to the solution, where the concentration of water is lower. This loss of water from the cell would cause it to shrink and perhaps become damaged through dehydration. If the concentrations were reversed, water would diffuse *into* the cell, causing it to swell and perhaps burst. If solutions are introduced into the body intravenously, care must be taken to ensure that they don't disturb the osmotic balance of its cells, or damage can occur. For example, if a 9% saline solution surrounds a red blood cell, the cell will shrink. By contrast, if the solution is about 1%, the cell will eventually burst.

In the body, blood is cleansed of impurities by osmosis as it flows through the kidneys. (See Fig. 9.50a.) Arterial blood first passes through a bundle of capillaries known as a *glomerulus*, where most of the waste products and some essential salts and minerals are removed. From the glomerulus, a narrow tube emerges that is in intimate contact with other capillaries throughout its length. As blood passes through the tubules, most of the essential elements are returned to it; waste products are not allowed to reenter and are eventually removed in urine.

If the kidneys fail, an artificial kidney or a dialysis machine can filter the blood. Figure 9.50b shows how this is done. Blood from an artery in the arm is mixed with

**BIO APPLICATION**

Effect of Osmosis on Living Cells

**APPLICATION**

Kidney Function and Dialysis

**Figure 9.50** (a) Diagram of a single nephron in the human excretory system. (b) An artificial kidney.

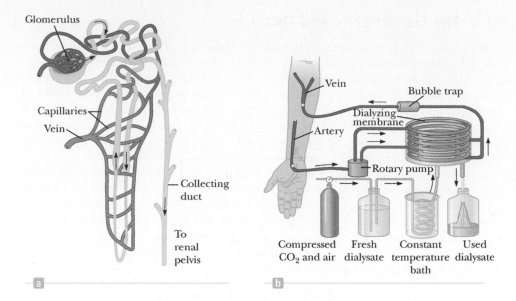

heparin, a blood thinner, and allowed to pass through a tube covered with a semi-permeable membrane. The tubing is immersed in a bath of a dialysate fluid with the same chemical composition as purified blood. Waste products from the blood enter the dialysate by diffusion through the membrane. The filtered blood is then returned to a vein.

### 9.9.3 Motion Through a Viscous Medium

When an object falls through air, its motion is impeded by the force of air resistance. In general, this force depends on the shape of the falling object and on its velocity. The force of air resistance acts on all falling objects, but the exact details of the motion can be calculated only for a few cases in which the object has a simple shape, such as a sphere. In this section we examine the motion of a tiny spherical object falling slowly through a viscous medium.

In 1845 a scientist named George Stokes found that the magnitude of the resistive force on a very small spherical object of radius $r$ falling slowly through a fluid of viscosity $\eta$ with speed $v$ is given by

$$F_r = 6\pi\eta rv \qquad [9.23]$$

This equation, called **Stokes' law,** has many important applications. For example, it describes the sedimentation of particulate matter in blood samples. It was used by Robert Millikan (1886–1953) to calculate the radius of charged oil droplets falling through air. From this, Millikan was ultimately able to determine the charge of the electron and was awarded the Nobel Prize in 1923 for his pioneering work on elemental charges.

As a sphere falls through a viscous medium, three forces act on it, as shown in Figure 9.51: $\vec{F}_r$, the force of friction; $\vec{B}$, the buoyant force of the fluid; and $\vec{w}$, the force of gravity acting on the sphere. The magnitude of $\vec{w}$ is given by

$$w = \rho g V = \rho g \left(\frac{4}{3}\pi r^3\right)$$

where $\rho$ is the density of the sphere and $\frac{4}{3}\pi r^3$ is its volume. According to Archimedes' principle, the magnitude of the buoyant force is equal to the weight of the fluid displaced by the sphere,

$$B = \rho_f g V = \rho_f g \left(\frac{4}{3}\pi r^3\right)$$

where $\rho_f$ is the density of the fluid.

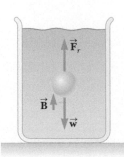

**Figure 9.51** A sphere falling through a viscous medium. The forces acting on the sphere are the resistive frictional force $\vec{F}_r$, the buoyant force $\vec{B}$, and the force of gravity $\vec{w}$ acting on the sphere.

At the instant the sphere begins to fall, the force of friction is zero because the speed of the sphere is zero. As the sphere accelerates, its speed increases and so does $\vec{F}_r$. Finally, **at a speed called the terminal speed $v_t$, the net force goes to zero.** This occurs when the net upward force balances the downward force of gravity. Therefore, the sphere reaches terminal speed when

$$F_r + B = w$$

or

$$6\pi\eta r v_t + \rho_f g\left(\frac{4}{3}\pi r^3\right) = \rho g\left(\frac{4}{3}\pi r^3\right)$$

When this equation is solved for $v_t$, we get

$$v_t = \frac{2r^2 g}{9\eta}(\rho - \rho_f) \qquad\qquad \text{[9.24]} \qquad \blacktriangleleft \text{ Terminal speed}$$

## 9.9.4 Sedimentation and Centrifugation

If an object isn't spherical, we can still use the basic approach just described to determine its terminal speed. The only difference is that we can't use Stokes' law for the resistive force. Instead, we assume that the resistive force has a magnitude given by $F_r = kv$, where $k$ is a coefficient that must be determined experimentally. As discussed previously, the object reaches its terminal speed when the downward force of gravity is balanced by the net upward force, or

$$w = B + F_r \qquad\qquad \text{[9.25]}$$

where $B = \rho_f g V$ is the buoyant force. The volume $V$ of the displaced fluid is related to the density $\rho$ of the falling object by $V = m/\rho$. Hence, we can express the buoyant force as

$$B = \frac{\rho_f}{\rho} mg$$

We substitute this expression for $B$ and $F_r = kv_t$ into Equation 9.25 (terminal speed condition):

$$mg = \frac{\rho_f}{\rho} mg + kv_t$$

or

$$v_t = \frac{mg}{k}\left(1 - \frac{\rho_f}{\rho}\right) \qquad\qquad \text{[9.26]}$$

The terminal speed for particles in biological samples is usually quite small. For example, the terminal speed for blood cells falling through plasma is about 5 cm/h in the gravitational field of the Earth. The terminal speeds for the molecules that make up a cell are many orders of magnitude smaller than this because of their much smaller mass. The speed at which materials fall through a fluid is called the **sedimentation rate** and is important in clinical analysis.

The sedimentation rate in a fluid can be increased by increasing the effective acceleration $g$ that appears in Equation 9.26. A fluid containing various biological molecules is placed in a centrifuge and whirled at very high angular speeds (Fig. 9.52). Under these conditions, the particles gain a large radial acceleration $a_c = v^2/r = \omega^2 r$ that is much greater than the free-fall acceleration, so we can replace $g$ in Equation 9.26 by $\omega^2 r$ and obtain

$$v_t = \frac{m\omega^2 r}{k}\left(1 - \frac{\rho_f}{\rho}\right) \qquad\qquad \text{[9.27]}$$

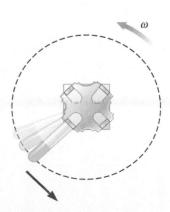

**Figure 9.52** Simplified diagram of a centrifuge (top view).

BIO APPLICATION
Separating Biological Molecules
with Centrifugation

This equation indicates that the sedimentation rate is enormously speeded up in a centrifuge ($\omega^2 r \gg g$) and that those particles with the greatest mass will have the largest terminal speed. Consequently the most massive particles will settle out on the bottom of a test tube first.

# 9.10 The Deformation of Solids

Although a solid may be thought of as having a definite shape and volume, it's possible to change its shape and volume by applying external forces. A sufficiently large force will permanently deform or break an object, but otherwise, when the external forces are removed, the object tends to return to its original shape and size. This is called *elastic behavior.*

The elastic properties of solids are discussed in terms of stress and strain. **Stress** is the force per unit area causing a deformation; **strain** is a measure of the amount of the deformation. For sufficiently small stresses, **stress is proportional to strain,** with the constant of proportionality depending on the material being deformed and on the nature of the deformation. We call this proportionality constant the **elastic modulus:**

$$\text{stress} = \text{elastic modulus} \times \text{strain} \qquad [9.28]$$

The elastic modulus is analogous to a spring constant. It can be taken as the stiffness of a material: A material having a large elastic modulus is very stiff and difficult to deform. There are three relationships having the form of Equation 9.28, corresponding to tensile, shear, and bulk deformation, and all of them satisfy an equation similar to Hooke's law for springs:

$$F = -k\,\Delta x \qquad [9.29]$$

where $F$ is the applied force, $k$ is the spring constant, and $\Delta x$ is essentially the amount by which the spring is stretched or compressed.

## 9.10.1 Young's Modulus: Elasticity in Length

Consider a long bar of cross-sectional area $A$ and length $L_0$, clamped at one end (Fig. 9.53). When an external force $\vec{\mathbf{F}}$ is applied along the bar, perpendicular to the cross section, internal forces in the bar resist the distortion ("stretching") that $\vec{\mathbf{F}}$ tends to produce. Nevertheless, the bar attains an equilibrium in which (1) its length is greater than $L_0$ and (2) the external force is balanced by internal forces. Under these circumstances, the bar is said to be *stressed.* We define the **tensile stress** as the ratio of the magnitude of the external force $F$ to the cross-sectional area $A$. The word "tensile" has the same root as the word "tension" and is used because the bar is under tension. The SI unit of stress is the newton per square meter (N/m$^2$), called the **pascal** (Pa), the same as the unit of pressure:

$$1\ \text{Pa} \equiv \text{N/m}^2$$

The **tensile strain** in this case is defined as the ratio of the change in length $\Delta L$ to the original length $L_0$ and is therefore a dimensionless quantity. Using Equation 9.28, we can write an equation relating tensile stress to tensile strain:

$$\frac{F}{A} = Y\frac{\Delta L}{L_0} \qquad [9.30]$$

In this equation, $Y$ is the constant of proportionality, called **Young's modulus.** Notice that Equation 9.30 could be solved for $F$ and put in the form $F = k\,\Delta L$, where $k = YA/L_0$, making it look just like Hooke's law, Equation 9.29.

A material having a large Young's modulus is difficult to stretch or compress. This quantity is typically used to characterize a rod or wire stressed under either

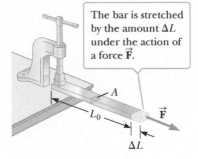

The bar is stretched by the amount $\Delta L$ under the action of a force $\vec{\mathbf{F}}$.

$A$

$L_0$

$\vec{\mathbf{F}}$

$\Delta L$

**Figure 9.53** A force is applied to a long bar clamped at one end.

The pascal ▶

**Table 9.5** Typical Values for the Elastic Modulus

| Substance | Young's Modulus (Pa) | Shear Modulus (Pa) | Bulk Modulus (Pa) |
|---|---|---|---|
| Aluminum | $7.0 \times 10^{10}$ | $2.5 \times 10^{10}$ | $7.0 \times 10^{10}$ |
| Bone | $1.8 \times 10^{10}$ | $8.0 \times 10^{10}$ | — |
| Brass | $9.1 \times 10^{10}$ | $3.5 \times 10^{10}$ | $6.1 \times 10^{10}$ |
| Copper | $11 \times 10^{10}$ | $4.2 \times 10^{10}$ | $14 \times 10^{10}$ |
| Steel | $20 \times 10^{10}$ | $8.4 \times 10^{10}$ | $16 \times 10^{10}$ |
| Tungsten | $35 \times 10^{10}$ | $14 \times 10^{10}$ | $20 \times 10^{10}$ |
| Glass | $6.5–7.8 \times 10^{10}$ | $2.6–3.2 \times 10^{10}$ | $5.0–5.5 \times 10^{10}$ |
| Quartz | $5.6 \times 10^{10}$ | $2.6 \times 10^{10}$ | $2.7 \times 10^{10}$ |
| Rib cartilage | $1.2 \times 10^{7}$ | — | — |
| Rubber | $0.1 \times 10^{7}$ | — | — |
| Tendon | $2 \times 10^{7}$ | — | — |
| Water | — | — | $0.21 \times 10^{10}$ |
| Mercury | — | — | $2.8 \times 10^{10}$ |

tension or compression. Because strain is a dimensionless quantity, $Y$ is in pascals. Typical values are given in Table 9.5. Experiments show that (1) the change in length for a fixed external force is proportional to the original length and (2) the force necessary to produce a given strain is proportional to the cross-sectional area. The value of Young's modulus for a given material depends on whether the material is stretched or compressed. A human femur, for example, is stronger under compression than tension. For many materials, such as metals, the moduli for compression and tension differ very little from each other.

It's possible to exceed the **elastic limit** of a substance by applying a sufficiently great stress (Fig. 9.54). At the elastic limit, the stress–strain curve departs from a straight line. A material subjected to a stress beyond this limit ordinarily doesn't return to its original length when the external force is removed. As the stress is increased further, it surpasses the **ultimate strength:** the greatest stress the substance can withstand without breaking. The **breaking point** for brittle materials is just beyond the ultimate strength. For ductile metals like copper and gold, after passing the point of ultimate strength, the metal thins and stretches at a lower stress level before breaking.

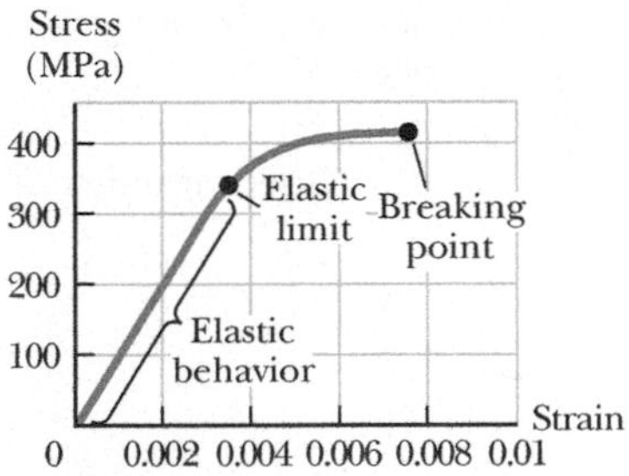

Stress (MPa)

**Figure 9.54** Stress-versus-strain curve for an elastic solid.

## 9.10.2 Shear Modulus: Elasticity of Shape

Another type of deformation occurs when an object is subjected to a force $\vec{\mathbf{F}}$ *parallel* to one of its faces while the opposite face is held fixed by a second force (Fig. 9.55a). If the object is originally a rectangular block, such a parallel force results in a shape with the cross section of a parallelogram. This kind of stress is called a **shear stress.** A book pushed sideways, as in Figure 9.55b, is being subjected to a shear stress. There is no change in volume with this kind of deformation. It's important to remember that in shear stress, the applied force is *parallel* to the cross-sectional

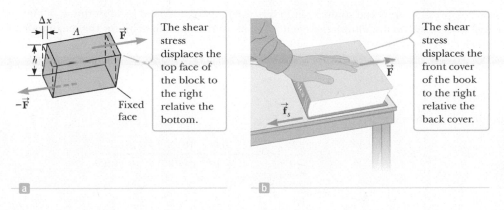

The shear stress displaces the top face of the block to the right relative the bottom.

The shear stress displaces the front cover of the book to the right relative the back cover.

**Figure 9.55** (a) A shear deformation in which a rectangular block is distorted by forces applied tangent to two of its faces. (b) A book under shear stress.

area, whereas in tensile stress the force is *perpendicular* to the cross-sectional area. We define **the shear stress as $F/A$, the ratio of the magnitude of the parallel force to the area $A$ of the face being sheared. The shear strain is the ratio $\Delta x/h$, where $\Delta x$ is the horizontal distance the sheared face moves and $h$ is the height of the object.** The shear stress is related to the shear strain according to

$$\frac{F}{A} = S\frac{\Delta x}{h}$$

[9.31]

where $S$ is the **shear modulus** of the material, with units of pascals (force per unit area). Once again, notice the similarity to Hooke's law.

A material having a large shear modulus is difficult to bend. Shear moduli for some representative materials are listed in Table 9.5.

### 9.10.3 Bulk Modulus: Volume Elasticity

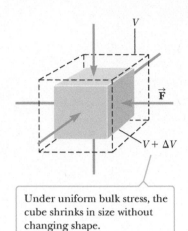

Under uniform bulk stress, the cube shrinks in size without changing shape.

**Figure 9.56** A solid cube is under uniform pressure and is therefore compressed on all sides by forces normal to its six faces. The arrowheads of force vectors on the sides of the cube that are not visible are hidden by the cube.

The bulk modulus characterizes the response of a substance to uniform squeezing. Suppose the external forces acting on an object are all perpendicular to the surface on which the force acts and are distributed uniformly over the surface of the object (Fig. 9.56). This occurs when an object is immersed in a fluid. An object subject to this type of deformation undergoes a change in volume but no change in shape. **The volume stress $\Delta P$ is defined as the ratio of the change in the magnitude of the applied force $\Delta F$ to the surface area $A$.** From the definition of pressure in Section 9.2, $\Delta P$ is also simply a change in pressure. The volume strain is equal to the change in volume $\Delta V$ divided by the original volume $V$. Again using Equation 9.28, we can relate a volume stress to a volume strain by the formula

Bulk modulus ▶

$$\Delta P = -B\frac{\Delta V}{V}$$

[9.32]

A material having a large bulk modulus doesn't compress easily. Note that a negative sign is included in this defining equation so that $B$ is always positive. An increase in pressure (positive $\Delta P$) causes a decrease in volume (negative $\Delta V$) and vice versa.

Table 9.5 lists bulk modulus values for some materials. If you look up such values in a different source, you may find that the reciprocal of the bulk modulus, called the **compressibility** of the material, is listed. Note from the table that both solids and liquids have bulk moduli. There is neither a Young's modulus nor shear modulus for liquids, however, because liquids simply flow when subjected to a tensile or shearing stress.

---

**EXAMPLE 9.17   BUILT TO LAST**

**GOAL** Calculate a compression due to tensile stress and maximum load.

**PROBLEM** A vertical steel beam in a building supports a load of $6.0 \times 10^4$ N. **(a)** If the length of the beam is 4.0 m and its cross-sectional area is $8.0 \times 10^{-3}$ m$^2$, find the distance the beam is compressed along its length. **(b)** What maximum load in newtons could the steel beam support before failing?

**STRATEGY** Equation 9.28 pertains to compressive stress and strain and can be solved for $\Delta L$, followed by substitution of known values. For part (b), set the compressive stress equal to the ultimate strength of steel from Table 9.6. Solve for the magnitude of the force, which is the total weight the structure can support.

. . . . . . . . . . . . . . . . . . . . . . . . . . . . . . . . . . . . . . . . . . . . . . . . . . . . . . . . . . . . . . . . . . . . . . . . . . . . . . . . . . . . . . .

**SOLUTION**

**(a)** Find the amount of compression in the beam.

Solve Equation 9.30 for $\Delta L$ and substitute, using the value of Young's modulus from Table 9.5:

$$\frac{F}{A} = Y\frac{\Delta L}{L_0}$$

$$\Delta L = \frac{FL_0}{YA} = \frac{(6.0 \times 10^4 \text{ N})(4.0 \text{ m})}{(2.0 \times 10^{11} \text{ Pa})(8.0 \times 10^{-3} \text{ m}^2)}$$

$$= 1.5 \times 10^{-4} \text{ m}$$

**(b)** Find the maximum load that the beam can support.

Set the compressive stress equal to the ultimate compressive strength from Table 9.6, and solve for $F$:

$$\frac{F}{A} = \frac{F}{8.0 \times 10^{-3} \text{ m}^2} = 5.0 \times 10^8 \text{ Pa}$$

$$F = \boxed{4.0 \times 10^6 \text{ N}}$$

**REMARKS** In designing load-bearing structures of any kind, it's always necessary to build in a safety factor. No one would drive a car over a bridge that had been designed to supply the minimum necessary strength to keep it from collapsing.

**QUESTION 9.17** Rank by the amount of fractional increase in length under increasing tensile stress, from smallest to largest: rubber, tungsten, steel, aluminum.

**EXERCISE 9.17** A cable used to lift heavy materials like steel I-beams must be strong enough to resist breaking even under a load of $1.0 \times 10^6$ N. For safety, the cable must support twice that load. **(a)** What cross-sectional area should the cable have if it's to be made of steel? **(b)** By how much will an 8.0-m length of this cable stretch when subject to the $1.0 \times 10^6$-N load?

**ANSWERS** **(a)** $4.0 \times 10^{-3}$ m² **(b)** $1.0 \times 10^{-2}$ m

**Table 9.6** Ultimate Strength of Materials

| Material | Tensile Strength (N/m²) | Compressive Strength (N/m²) |
|---|---|---|
| Iron | $1.7 \times 10^8$ | $5.5 \times 10^8$ |
| Steel | $5.0 \times 10^8$ | $5.0 \times 10^8$ |
| Aluminum | $2.0 \times 10^8$ | $2.0 \times 10^8$ |
| Bone | $1.2 \times 10^8$ | $1.5 \times 10^8$ |
| Marble | — | $8.0 \times 10^7$ |
| Brick | $1 \times 10^6$ | $3.5 \times 10^7$ |
| Concrete | $2 \times 10^6$ | $2 \times 10^7$ |

## EXAMPLE 9.18    FOOTBALL INJURIES    BIO

**GOAL** Obtain an estimate of shear stress.

**PROBLEM** A defensive lineman of mass $M = 125$ kg makes a flying tackle at $v_i = 4.00$ m/s on a stationary quarterback of mass $m = 85.0$ kg, and the lineman's helmet makes solid contact with the quarterback's femur. **(a)** What is the speed $v_f$ of the two athletes immediately after contact? Assume a linear perfectly inelastic collision. **(b)** If the collision lasts for 0.100 s, estimate the average force exerted on the quarterback's femur. **(c)** If the cross-sectional area of the quarterback's femur is equal to $5.00 \times 10^{-4}$ m², calculate the shear stress exerted on the bone in the collision.

**STRATEGY** The solution proceeds in three well-defined steps. In part **(a)**, use conservation of linear momentum to calculate the final speed of the system consisting of the quarterback and the lineman. Second, the speed found in part **(a)** can be used in the impulse-momentum theorem to obtain an estimate of the average force exerted on the femur. Third, dividing the average force by the cross-sectional area of the femur gives the desired estimate of the shear stress.

**SOLUTION**

**(a)** What is the speed of the system immediately after contact?

Apply momentum conservation to the system:

$$p_{\text{initial}} = p_{\text{final}}$$

Substitute expressions for the initial and final momenta:

$$Mv_i = (M + m)\, v_f$$

Solve for the final speed $v_f$:

$$v_f = \frac{Mv_i}{M + m} = \frac{(125 \text{ kg})(4.00 \text{ m/s})}{125 \text{ kg} + 85.0 \text{ kg}} = \boxed{2.38 \text{ m/s}}$$

*(Continued)*

**(b)** Obtain an estimate for the average force delivered to the quarterback's femur.

Apply the impulse-momentum theorem:

$$F_{av}\,\Delta t = \Delta p = Mv_f - Mv_i$$

Solve for the average force exerted on the quarterback's femur:

$$F_{av} = \frac{M(v_f - v_i)}{\Delta t}$$

$$= \frac{(125\ \text{kg})(4.00\ \text{m/s} - 2.38\ \text{m/s})}{0.100\ \text{s}} = \boxed{2.03 \times 10^3\ \text{N}}$$

**(c)** Obtain the average shear stress exerted on the quarterback's femur.

Divide the average force found in part (b) by the cross-sectional area of the femur:

$$\text{Shear stress} = \frac{F}{A} = \frac{2.03 \times 10^3\ \text{N}}{5.00 \times 10^{-4}\ \text{m}^2} = \boxed{4.06 \times 10^6\ \text{Pa}}$$

. . . . . . . . . . . . . . . . . . . . . . . . . . . . . . . . . . . . . . . . . . . . . . . . . . . . . . . . . . . . . . . . . . . . . . . . . . . . . . . .

**REMARKS** The ultimate shear strength of a femur is approximately $7 \times 10^7$ Pa, so this collision would not be expected to break the quarterback's leg.

**QUESTION 9.18** What kind of stress would be sustained by the lineman? What parts of his body would be affected?

**EXERCISE 9.18** Calculate the diameter of a horizontal steel bolt if it is expected to support a maximum load hung on it having a mass of $2.00 \times 10^3$ kg but for safety reasons must be designed to support three times that load. (Assume the ultimate shear strength of steel is $2.50 \times 10^8$ Pa.)

**ANSWER** 1.73 cm

---

## EXAMPLE 9.19 | LEAD BALLAST OVERBOARD

**GOAL** Apply the concepts of bulk stress and strain.

**PROBLEM** Ships and sailing vessels often carry lead ballast in various forms, such as bricks, to keep the ship properly oriented and upright in the water. Suppose a ship takes on cargo and the crew jettisons a total of 0.500 m³ of lead ballast into water 2.00 km deep. Calculate **(a)** the change in the pressure at that depth and **(b)** the change in volume of the lead upon reaching the bottom. Take the density of sea water to be $1.025 \times 10^3$ kg/m³, and take the bulk modulus of lead to be $4.2 \times 10^{10}$ Pa.

**STRATEGY** The pressure difference between the surface and a depth of 2.00 km is due to the weight of the water column. Calculate the weight of water in a column with cross section of 1.00 m². That number in newtons will be the same magnitude as the pressure difference in pascal. Substitute the pressure change into the bulk stress and strain equation to obtain the change in volume of the lead.

. . . . . . . . . . . . . . . . . . . . . . . . . . . . . . . . . . . . . . . . . . . . . . . . . . . . . . . . . . . . . . . . . . . . . . . . . . . . . . . .

### SOLUTION

**(a)** Calculate the pressure difference between the surface and at a depth of 2.00 km.

Use the density, volume, and acceleration of gravity $g$ to compute the weight of water in a column having cross-sectional area of 1.00 m²:

$$w = mg = (\rho V)g$$
$$= (1.025 \times 10^3\ \text{kg/m}^3)(2.00 \times 10^3\ \text{m}^3)(9.80\ \text{m/s}^2)$$
$$= 2.01 \times 10^7\ \text{N}$$

Divide by the area (in this case, 1.00 m²) to obtain the pressure difference due to the column of water:

$$\Delta P = \frac{F}{A} = \frac{2.01 \times 10^7\ \text{N}}{1.00\ \text{m}^2} = \boxed{2.01 \times 10^7\ \text{Pa}}$$

**(b)** Calculate the change in volume of the lead upon reaching the bottom.

Write the bulk stress and strain equation:

$$\Delta P = -B\frac{\Delta V}{V}$$

Solve for $\Delta V$:

$$\Delta V = -\frac{V\Delta P}{B} = -\frac{(0.500\ \text{m}^3)(2.01 \times 10^7\ \text{Pa})}{4.2 \times 10^{10}\ \text{Pa}} = \boxed{-2.4 \times 10^{-4}\ \text{m}^3}$$

. . . . . . . . . . . . . . . . . . . . . . . . . . . . . . . . . . . . . . . . . . . . . . . . . . . . . . . . . . . . . . . . . . . . . . . . . . . . . . . .

**REMARKS** The negative sign indicates a *decrease* in volume. The following exercise shows that even water can be compressed, although not by much.

**QUESTION 9.19** Rank the following substances in order of the fractional change in volume in response to increasing pressure, from smallest to largest: copper, steel, water, mercury.

**EXERCISE 9.19** (a) By what percentage does the volume of a ball of water shrink at that same depth? (b) What is the ratio of the new radius to the initial radius?

**ANSWERS** (a) 0.96%    (b) 0.997

## 9.10.4 Arches and the Ultimate Strength of Materials

As we have seen, the ultimate strength of a material is the maximum force per unit area the material can withstand before it breaks or fractures. Such values are of great importance, particularly in the construction of buildings, bridges, and roads. Table 9.6 gives the ultimate strength of a variety of materials under both tension and compression. Note that bone and a variety of building materials (concrete, brick, and marble) are stronger under compression than under tension. The greater ability of brick and stone to resist compression is the basis of the semicircular arch, developed and used extensively by the Romans in everything from memorial arches to expansive temples and aqueduct supports.

Before the development of the arch, the principal method of spanning a space was the simple post-and-beam construction (Fig. 9.57a), in which a horizontal beam is supported by two columns. This type of construction was used to build the great Greek temples. The columns of these temples were closely spaced because of the limited length of available stones and the low ultimate tensile strength of a sagging stone beam.

The semicircular arch (Fig. 9.57b) developed by the Romans was a great technological achievement in architectural design. It effectively allowed the heavy load of a wide roof span to be channeled into horizontal and vertical forces on narrow supporting columns. The stability of this arch depends on the compression between its wedge-shaped stones. The stones are forced to squeeze against each other by the uniform loading, as shown in the figure. This compression results in horizontal outward forces at the base of the arch where it starts curving away from the vertical. These forces must then be balanced by the stone walls shown on the sides of the arch. It's common to use very heavy walls (buttresses) on either side of the arch to provide horizontal stability. If the foundation of the arch should move, the compressive forces between the wedge-shaped stones may decrease to the extent that the arch collapses. The stone surfaces used in the Roman arches were cut to make very tight joints; mortar was usually not used. The resistance to slipping between stones was provided by the compression force and the friction between the stone faces.

Another important architectural innovation was the pointed Gothic arch, shown in Figure 9.57c. This type of structure was first used in Europe beginning in the

**APPLICATION**
Arch Structures in Buildings

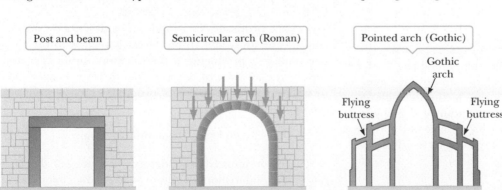

| Post and beam | Semicircular arch (Roman) | Pointed arch (Gothic) |

Gothic arch

Flying buttress    Flying buttress

a    b    c

**Figure 9.57** (a) A simple post-and-beam structure. (b) The semicircular arch developed by the Romans. (c) Gothic arch with flying buttresses to provide lateral support.

12th century, followed by the construction of several magnificent Gothic cathedrals in France in the 13th century. One of the most striking features of these cathedrals is their extreme height. For example, the cathedral at Chartres rises to 118 ft, and the one at Reims has a height of 137 ft. Such magnificent buildings evolved over a very short time, without the benefit of any mathematical theory of structures. However, Gothic arches required flying buttresses to prevent the spreading of the arch supported by the tall, narrow columns.

## SUMMARY

### 9.1 States of Matter

Matter is normally classified as being in one of three states: solid, liquid, or gaseous. The fourth state of matter is called a plasma, which consists of a neutral system of charged particles interacting electromagnetically.

### 9.2 Density and Pressure

The **density** $\rho$ of a substance of uniform composition is its mass per unit volume—kilograms per cubic meter (kg/m$^3$) in the SI system:

$$\rho \equiv \frac{M}{V} \qquad [9.1]$$

The **pressure** $P$ in a fluid, measured in pascals (Pa), is the force per unit area that the fluid exerts on an object immersed in it:

$$P \equiv \frac{F}{A} \qquad [9.2]$$

### 9.3 Variation of Pressure with Depth

The pressure in an incompressible fluid varies with depth $h$ according to the expression

$$P = P_0 + \rho g h \qquad [9.6]$$

where $P_0$ is atmospheric pressure ($1.013 \times 10^5$ Pa) and $\rho$ is the density of the fluid.

**Pascal's principle** states that when pressure is applied to an enclosed fluid, the pressure is transmitted undiminished to every point of the fluid and to the walls of the containing vessel.

### 9.5 Buoyant Forces and Archimedes' Principle

When an object is partially or fully submerged in a fluid, the fluid exerts an upward force, called the **buoyant force,** on the object. This force is, in fact, due to the net difference in pressure between the top and bottom of the object. It can be shown that the magnitude of the buoyant force $B$ is equal to the weight of the fluid displaced by the object, or

$$B = \rho_{\text{fluid}} V_{\text{fluid}} g \qquad [9.7b]$$

Equation 9.7b is known as **Archimedes' principle.**

Solving a buoyancy problem usually involves putting the buoyant force into Newton's second law and then proceeding as in Topic 4.

### 9.6 Fluids in Motion

Certain aspects of a fluid in motion can be understood by assuming the fluid is nonviscous and incompressible and that its motion is in a steady state with no turbulence:

1. The flow rate through the pipe is a constant, which is equivalent to stating that the product of the cross-sectional area $A$ and the speed $v$ at any point is constant. At any two points, therefore, we have

$$A_1 v_1 = A_2 v_2 \qquad [9.10]$$

This relation is referred to as the **equation of continuity.**

2. The sum of the pressure, the kinetic energy per unit volume, and the potential energy per unit volume is the same at any two points along a streamline:

$$P_1 + \tfrac{1}{2}\rho v_1^2 + \rho g y_1 = P_2 + \tfrac{1}{2}\rho v_2^2 + \rho g y_2 \qquad [9.11]$$

Equation 9.11 is known as **Bernoulli's equation.** Solving problems with Bernoulli's equation is similar to solving problems with the work–energy theorem, whereby two points are chosen, one point where a quantity is unknown and another where all quantities are known. Equation 9.11 is then solved for the unknown quantity.

### 9.10 The Deformation of Solids

The elastic properties of a solid can be described using the concepts of stress and strain. **Stress** is related to the force per unit area producing a deformation; **strain** is a measure of the amount of deformation. Stress is proportional to strain, and the constant of proportionality is the **elastic modulus:**

$$\text{stress} = \text{elastic modulus} \times \text{strain} \qquad [9.28]$$

Three common types of deformation are (1) the resistance of a solid to elongation or compression, characterized by **Young's modulus** $Y$; (2) the resistance to displacement of the faces of a solid sliding past each other, characterized by the shear modulus $S$; and (3) the resistance of a solid or liquid to a change in volume, characterized by the bulk modulus $B$.

All three types of deformation obey laws similar to Hooke's law for springs. Solving problems is usually a matter of identifying the given physical variables and solving for the unknown variable.

## CONCEPTUAL QUESTIONS

1. The three containers in Figure CQ9.1 are filled with water to the same level. Rank the pressures at the bottom of the containers (choose one): (a) $P_A > P_B > P_C$ (b) $P_A > P_B = P_C$ (c) $P_A = P_B > P_C$ (d) $P_A < P_B < P_C$ (e) $P_A = P_B = P_C$

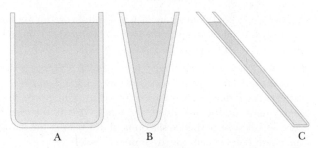

**Figure CQ9.1**

2. The density of air is 1.3 kg/m$^3$ at sea level. From your knowledge of air pressure at ground level, estimate the height of the atmosphere. As a simplifying assumption, take the atmosphere to be of uniform density up to some height, after which the density rapidly falls to zero. (In reality, the density of the atmosphere decreases as we go up.) (This question is courtesy of Edward F. Redish. For more questions of this type, see http://www.physics.umd.edu/perg/.)

3. Four solid, uniform objects are placed in a container of water (Fig. CQ9.3) Rank their densities from highest to lowest.

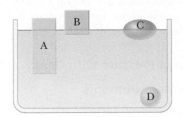

**Figure CQ9.3**

4. Figure CQ9.4 shows aerial views from directly above two dams. Both dams are equally long (the vertical dimension in the diagram) and equally deep (into the page in the diagram). The dam on the left holds back a very large lake, while the dam on the right holds back a narrow river. Which dam has to be built more strongly?

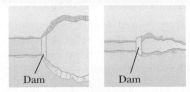

**Figure CQ9.4**

5. Equal volumes of two fluids are added to the U-shaped pipe as shown in Figure CQ9.5. The pipe is open at both ends and the fluids come to equilibrium without mixing. (a) Which fluid has the higher density, fluid A or fluid B? (b) What is the ratio $\rho_B/\rho_A$ of the fluid densities?

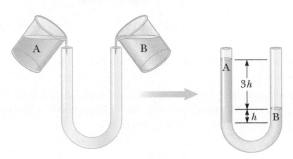

**Figure CQ9.5**

6. Many people believe that a vacuum created inside a vacuum cleaner causes particles of dirt to be drawn in. Actually, the dirt is pushed in. Explain.

7. Water flows along a streamline down a river of constant width (Fig. CQ9.7). Over a short distance the water slows from speed $v$ to $v/3$. Which of the following can you correctly conclude about the river's depth? (a) It became deeper by a factor of 3. (b) It became shallower by a factor of 3. (c) It became deeper by a factor of $3^2$. (d) It became shallower by a factor of $3^2$.

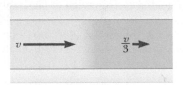

**Figure CQ9.7**

8. **BIO** During inhalation, the pressure in the lungs is slightly less than external pressure and the muscles controlling exhalation are relaxed. Under water, the body equalizes internal and external pressures. Discuss the condition of the muscles if a person under water is breathing through a snorkel. Would a snorkel work in deep water?

9. The water supply for a city is often provided from reservoirs built on high ground. Water flows from the reservoir, through pipes, and into your home when you turn the tap on your faucet. Why is the water flow more rapid out of a faucet on the first floor of a building than in an apartment on a higher floor?

**10.** An ice cube is placed in a glass of water. What happens to the level of the water as the ice melts?

**11.** Water flows along a streamline through the pipe shown in Figure CQ9.11. Point A is higher than points B and C, and the pipe has a constant radius until it expands between B and C. From highest to lowest, (a) rank the flow speeds at points A, B, and C and (b) rank the pressures at points A, B, and C.

**Figure CQ9.11**

**12.** Will an ice cube float higher in water or in an alcoholic beverage?

**13.** Tornadoes and hurricanes often lift the roofs of houses. Use the Bernoulli effect to explain why. Why should you keep your windows open under these conditions?

**14.** Once ski jumpers are airborne (Fig. CQ9.14), why do they bend their bodies forward and keep their hands at their sides?

**Figure CQ9.14**

**15.** A person in a boat floating in a small pond throws an anchor overboard. What happens to the level of the pond? (a) It rises. (b) It falls. (c) It remains the same.

**16.** One of the predicted problems due to global warming is that ice in the polar ice caps will melt and raise sea levels everywhere in the world. Is that more of a worry for ice (a) at the North Pole, where most of the ice floats on water; (b) at the South Pole, where most of the ice sits on land; (c) both at the North and South Poles equally; or (d) at neither pole?

# PROBLEMS

*See the Preface for an explanation of the icons used in this problems set.*

## 9.2 Density and Pressure

**1.** An 81.5-kg man stands on a horizontal surface. (a) What is the volume of the man's body if his average density is 985 kg/m³? (b) What average pressure from his weight is exerted on the horizontal surface if the man's two feet have a combined area of $4.50 \times 10^{-2}$ m²?

**2.** The British gold sovereign coin is an alloy of gold and copper having a total mass of 7.988 g, and is 22-karat gold. (a) Find the mass of gold in the sovereign in kilograms using the fact that the number of karats = $24 \times$ (mass of gold)/(total mass). (b) Calculate the volumes of gold and copper, respectively, used to manufacture the coin. (c) Calculate the density of the British sovereign coin.

**3.** The weight of Earth's atmosphere exerts an average pressure of $1.01 \times 10^5$ Pa on the ground at sea level. Use the definition of pressure to estimate the weight of Earth's atmosphere by approximating Earth as a sphere of radius $R_E = 6.38 \times 10^6$ m and surface area $A = 4\pi R_E^2$.

**4.** [T] Calculate the mass of a solid gold rectangular bar that has dimensions of 4.50 cm × 11.0 cm × 26.0 cm.

**5.** [Q|C] The nucleus of an atom can be modeled as several protons and neutrons closely packed together. Each particle has a mass of $1.67 \times 10^{-27}$ kg and radius on the order of $10^{-15}$ m. (a) Use this model and the data provided to estimate the density of the nucleus of an atom. (b) Compare your result with the density of a material such as iron. What do your result and comparison suggest about the structure of matter?

**6.** The four tires of an automobile are inflated to a gauge pressure of $2.0 \times 10^5$ Pa. Each tire has an area of 0.024 m² in contact with the ground. Determine the weight of the automobile.

**7.** Suppose a distant world with surface gravity of 7.44 m/s² has an atmospheric pressure of $8.04 \times 10^4$ Pa at the surface. (a) What force is exerted by the atmosphere on a disk-shaped region 2.00 m in radius at the surface of a methane ocean? (b) What is the weight of a 10.0-m deep cylindrical column of methane with radius 2.00 m? (c) Calculate the pressure at a depth of 10.0 m in the methane ocean. *Note:* The density of liquid methane is 415 kg/m³.

## 9.3 Variation of Pressure with Depth

## 9.4 Pressure Measurements

**8.** [BIO] A normal blood pressure reading is less than 120/80 where both numbers are gauge pressures measured in millimeters of mercury (mmHg). What are the (a) absolute and (b) gauge pressures in pascals at the base of a 0.120 m column of mercury?

**9.** (a) Calculate the absolute pressure at the bottom of a freshwater lake at a depth of 27.5 m. Assume the density of the water is $1.00 \times 10^3$ kg/m³ and the air above is at a pressure of 101.3 kPa. (b) What force is exerted by the water on the window of an underwater vehicle at this depth if the window is circular and has a diameter of 35.0 cm?

**10.** [V] Mercury is poured into a U-tube as shown in Figure P9.10a. The left arm of the tube has cross-sectional area $A_1$ of 10.0 cm², and the right arm has a cross-sectional area $A_2$ of 5.00 cm². One hundred grams of water are then poured into the right arm as shown in Figure P9.10b. (a) Determine the length of the water column in the right arm of the U-tube. (b) Given that the density of mercury is 13.6 g/cm³, what distance $h$ does the mercury rise in the left arm?

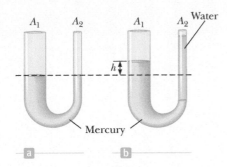

**Figure P9.10**

11. **BIO** A collapsible plastic bag (Fig. P9.11) contains a glucose solution. If the average gauge pressure in the vein is $1.33 \times 10^3$ Pa, what must be the minimum height $h$ of the bag to infuse glucose into the vein? Assume the specific gravity of the solution is 1.02.

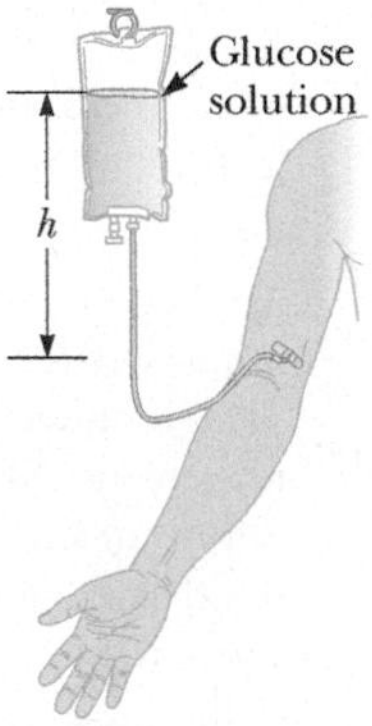

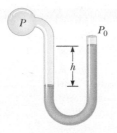

**Figure P9.11**

12. A hydraulic jack has an input piston of area 0.050 m² and an output piston of area 0.70 m². How much force on the input piston is required to lift a car weighing $1.2 \times 10^4$ N?

13. **V** A container is filled to a depth of 20.0 cm with water. On top of the water floats a 30.0-cm-thick layer of oil with specific gravity 0.700. What is the absolute pressure at the bottom of the container?

14. Blaise Pascal duplicated Torricelli's barometer using a red Bordeaux wine, of density 984 kg/m³ as the working liquid (Fig. P9.14). (a) What was the height $h$ of the wine column for normal atmospheric pressure? (b) Would you expect the vacuum above the column to be as good as for mercury?

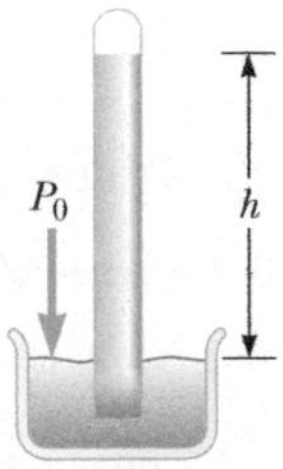

**Figure P9.14**

15. **BIO** A sphygmomanometer is a device used to measure blood pressure, typically consisting of an inflatable cuff and a manometer used to measure air pressure in the cuff. In a mercury sphygmomanometer, blood pressure is related to the difference in heights between two columns of mercury.

The mercury sphygmomanometer shown in Figure P9.15 contains air at the cuff pressure $P$. The difference in mercury heights between the left tube and the right tube is $h = 115$ mmHg $= 0.115$ m, a normal systolic reading. What is the gauge systolic blood pressure $P_{gauge}$ in pascals? The density of mercury is $\rho = 13.6 \times 10^3$ kg/m³ and the ambient pressure is $P_0 = 1.01 \times 10^5$ Pa.

**Figure P9.15**

16. **V** Piston ① in Figure P9.16 has a diameter of 0.25 in.; piston ② has a diameter of 1.5 in. In the absence of friction, determine the force $\vec{\mathbf{F}}$ necessary to support the 500-lb weight.

**Figure P9.16**

## 9.5 Buoyant Forces and Archimedes' Principle

17. A table-tennis ball has a diameter of 3.80 cm and average density of 0.084 0 g/cm³. What force is required to hold it completely submerged under water?

18. A 20.0-kg lead mass rests on the bottom of a pool. (a) What is the volume of the lead? (b) What buoyant force acts on the lead? (c) Find the lead's weight. (d) What is the normal force acting on the lead?

19. A small ferryboat is 4.00 m wide and 6.00 m long. When a loaded truck pulls onto it, the boat sinks an additional 4.00 cm into the river. What is the weight of the truck?

20. **GP** A 62.0-kg survivor of a cruise line disaster rests atop a block of Styrofoam insulation, using it as a raft. The Styrofoam has dimensions 2.00 m × 2.00 m × 0.090 0 m. The bottom 0.024 m of the raft is submerged. (a) Draw a force diagram of the system consisting of the survivor and raft. (b) Write Newton's second law for the system in one dimension, using $B$ for buoyancy, $w$ for the weight of the survivor, and $w_r$ for the weight of the raft. (Set $a = 0$.) (c) Calculate the numeric value for the buoyancy, $B$. (Seawater has density 1 025 kg/m³.) (d) Using the value of $B$ and the weight $w$ of the survivor, calculate the weight $w_r$ of the Styrofoam. (e) What is the density of the Styrofoam? (f) What is the maximum buoyant force, corresponding to the raft being submerged up to its top surface? (g) What total mass of survivors can the raft support?

21. A hot-air balloon consists of a basket hanging beneath a large envelope filled with hot air. A typical hot-air balloon has a total mass of 545 kg, including passengers in its basket, and holds $2.55 \times 10^3$ m³ of hot air in its envelope. If the ambient air density is 1.25 kg/m³, determine the density of hot air inside the envelope when the balloon is neutrally buoyant. Neglect the volume of air displaced by the basket and passengers.

22. **GP** A large balloon of mass 226 kg is filled with helium gas until its volume is 325 m³. Assume the density of air is 1.29 kg/m³ and the density of helium is 0.179 kg/m³. (a) Draw a force diagram for the balloon. (b) Calculate the buoyant force acting on the balloon. (c) Find the net force on the balloon and determine whether the balloon will rise or fall after it is released. (d) What maximum additional mass can the balloon support in equilibrium? (e) What happens to the

balloon if the mass of the load is *less* than the value calculated in part (d)? (f) What limits the height to which the balloon can rise?

23. QC A spherical weather balloon is filled with hydrogen until its radius is 3.00 m. Its total mass including the instruments it carries is 15.0 kg. (a) Find the buoyant force acting on the balloon, assuming the density of air is 1.29 kg/m³. (b) What is the net force acting on the balloon and its instruments after the balloon is released from the ground? (c) Why does the radius of the balloon tend to increase as it rises to higher altitude?

24. BIO QC The average human has a density of 945 kg/m³ after inhaling and 1 020 kg/m³ after exhaling. (a) Without making any swimming movements, what percentage of the human body would be above the surface in the Dead Sea (a body of water with a density of about 1 230 kg/m³) in each of these cases? (b) Given that bone and muscle are denser than fat, what physical characteristics differentiate "sinkers" (those who tend to sink in water) from "floaters" (those who readily float)?

25. QC On October 21, 2001, Ian Ashpole of the United Kingdom achieved a record altitude of 3.35 km (11 000 ft) powered by 600 toy balloons filled with helium. Each filled balloon had a radius of about 0.50 m and an estimated mass of 0.30 kg. (a) Estimate the total buoyant force on the 600 balloons. (b) Estimate the net upward force on all 600 balloons. (c) Ashpole parachuted to Earth after the balloons began to burst at the high altitude and the system lost buoyancy. Why did the balloons burst?

26. V The gravitational force exerted on a solid object is 5.00 N as measured when the object is suspended from a spring scale as in Figure P9.26a. When the suspended object is submerged in water, the scale reads 3.50 N (Fig. P9.26b). Find the density of the object.

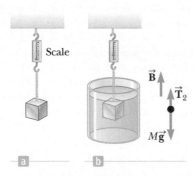

Figure P9.26

27. T A cube of wood having an edge dimension of 20.0 cm and a density of 650. kg/m³ floats on water. (a) What is the distance from the horizontal top surface of the cube to the water level? (b) What mass of lead should be placed on the cube so that the top of the cube will be just level with the water surface?

28. A light spring of force constant $k$ = 160 N/m rests vertically on the bottom of a large beaker of water (Fig. P9.28a). A 5.00-kg block of wood (density = 650 kg/m³) is connected to the spring, and the block–spring system is allowed to come to static equilibrium (Fig. P9.28b). What is the elongation $\Delta L$ of the spring?

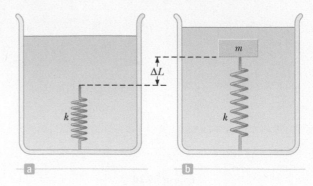

Figure P9.28

29. A sample of an unknown material appears to weigh 300. N in air and 200. N when immersed in alcohol of specific gravity 0.700. What are (a) the volume and (b) the density of the material?

30. An object weighing 300 N in air is immersed in water after being tied to a string connected to a balance. The scale now reads 265 N. Immersed in oil, the object appears to weigh 275 N. Find (a) the density of the object and (b) the density of the oil.

31. A 1.00-kg beaker containing 2.00 kg of oil (density = 916 kg/m³) rests on a scale. A 2.00-kg block of iron is suspended from a spring scale and is completely submerged in the oil (Fig. P9.31). Find the equilibrium readings of both scales.

Figure P9.31

## 9.6 Fluids in Motion

## 9.7 Other Applications of Fluid Dynamics

32. A horizontal pipe narrows from a radius of 0.250 m to 0.100 m. If the speed of the water in the pipe is 1.00 m/s in the larger-radius pipe, what is the speed in the smaller pipe?

33. A large water tank is 3.00 m high and filled to the brim, the top of the tank open to the air. A small pipe with a faucet is attached to the side of the tank, 0.800 m above the ground. If the valve is opened, at what speed will water come out of the pipe?

34. Water flowing through a garden hose of diameter 2.74 cm fills a 25.0-L bucket in 1.50 min. (a) What is the speed of the water leaving the end of the hose? (b) A nozzle is now attached to the end of the hose. If the nozzle diameter is one-third the diameter of the hose, what is the speed of the water leaving the nozzle?

35. BIO (a) Calculate the mass flow rate (in grams per second) of blood ($\rho$ = 1.0 g/cm³) in an aorta with a cross-sectional area of 2.0 cm² if the flow speed is 40. cm/s. (b) Assume that the aorta branches to form a large number of capillaries with a combined cross-sectional area of 3.0 × 10³ cm². What is the flow speed in the capillaries?

36. V A liquid ($\rho$ = 1.65 g/cm³) flows through a horizontal pipe of varying cross section as in Figure P9.36. In the first section, the cross-sectional area is 10.0 cm²,

Figure P9.36

the flow speed is 275 cm/s, and the pressure is $1.20 \times 10^5$ Pa. In the second section, the cross-sectional area is 2.50 cm². Calculate the smaller section's (a) flow speed and (b) pressure.

37. **BIO** A hypodermic syringe contains a medicine with the density of water (Fig. P9.37). The barrel of the syringe has a cross-sectional area of $2.50 \times 10^{-5}$ m². In the absence of a force on the plunger, the pressure everywhere is 1.00 atm. A force $\vec{F}$ of magnitude 2.00 N is exerted on the plunger, making medicine squirt from the needle. Determine the medicine's flow speed through the needle. Assume the pressure in the needle remains equal to 1.00 atm and that the syringe is horizontal.

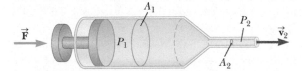

**Figure P9.37**

38. **BIO** When a person inhales, air moves down the bronchus (windpipe) at 15 cm/s. The average flow speed of the air doubles through a constriction in the bronchus. Assuming incompressible flow, determine the pressure drop in the constriction.

39. **Q|C** A jet airplane in level flight has a mass of $8.66 \times 10^4$ kg, and the two wings have an estimated total area of 90.0 m². (a) What is the pressure difference between the lower and upper surfaces of the wings? (b) If the speed of air under the wings is 225 m/s, what is the speed of the air over the wings? Assume air has a density of 1.29 kg/m³. (c) Explain why all aircraft have a "ceiling," a maximum operational altitude.

40. A man attaches a divider to an outdoor faucet so that water flows through a single pipe of radius 9.00 mm into two pipes, each with a radius of 6.00 mm. If water flows through the single pipe at 1.25 m/s, calculate the speed of the water in the narrower pipes.

41. **GP** In a water pistol, a piston drives water through a larger tube of radius 1.00 cm into a smaller tube of radius 1.00 mm as in Figure P9.41. (a) If the pistol is fired horizontally at a height of 1.50 m, use ballistics to determine the time it takes water to travel from the nozzle to the ground. (Neglect air resistance and assume atmospheric pressure is 1.00 atm.) (b) If the range of the stream is to be 8.00 m, with what speed must the stream leave the nozzle? (c) Given the areas of the nozzle and cylinder, use the equation of continuity to calculate the speed at which the plunger must be moved. (d) What is the pressure at the nozzle? (e) Use Bernoulli's equation to find the pressure needed in the larger cylinder. Can gravity terms be neglected? (f) Calculate the force that must be exerted on the trigger to achieve the desired range. (The force that must be exerted is due to pressure over and above atmospheric pressure.)

42. Water moves through a constricted pipe in steady, ideal flow. At the lower point shown in Figure P9.42, the pressure is $1.75 \times 10^5$ Pa and the pipe radius is 3.00 cm. At the higher point located at $y = 2.50$ m, the pressure is $1.20 \times 10^5$ Pa and the pipe radius is 1.50 cm. Find the

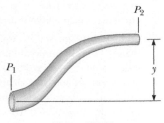

**Figure P9.42**

speed of flow (a) in the lower section and (b) in the upper section. (c) Find the volume flow rate through the pipe.

43. **T** A jet of water squirts out horizontally from a hole near the bottom of the tank shown in Figure P9.43. If the hole has a diameter of 3.50 mm, what is the height $h$ of the water level in the tank?

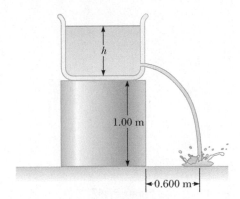

**Figure P9.43**

44. A large storage tank, open to the atmosphere at the top and filled with water, develops a small hole in its side at a point 16.0 m below the water level. If the rate of flow from the leak is $2.50 \times 10^{-3}$ m³/min, determine (a) the speed at which the water leaves the hole and (b) the diameter of the hole.

45. The inside diameters of the larger portions of the horizontal pipe depicted in Figure P9.45 are 2.50 cm. Water flows to the right at a rate of $1.80 \times 10^{-4}$ m³/s. Determine the inside diameter of the constriction.

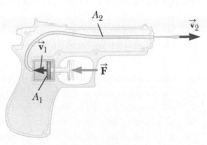

**Figure P9.41**

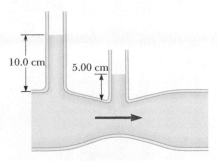

**Figure P9.45**

46. Water is pumped through a pipe of diameter 15.0 cm from the Colorado River up to Grand Canyon Village, on the rim of the canyon. The river is at 564 m elevation and the village is at 2 096 m. (a) At what minimum pressure must the water be pumped to arrive at the village? (b) If 4 500 m³ are pumped per day, what is the speed of the water in the pipe? (c) What additional pressure is necessary to deliver this flow? *Note:* You may assume the free-fall acceleration and the density of air are constant over the given range of elevations.

47. Old Faithful geyser in Yellowstone Park erupts at approximately 1-hour intervals, and the height of the fountain reaches 40.0 m (Fig. P9.47). (a) Consider the rising stream as a series of separate drops. Analyze the free-fall motion of one of the drops to determine the speed at which the water leaves the ground. (b) Treat the rising stream as an ideal fluid in streamline flow. Use Bernoulli's equation to determine the speed of the water as it leaves ground level. (c) What is the pressure (above atmospheric pressure) in the heated underground chamber 175 m below the vent? You may assume the chamber is large compared with the geyser vent.

**Figure P9.47**

48. The Venturi tube shown in Figure P9.48 may be used as a fluid flowmeter. Suppose the device is used at a service station to measure the flow rate of gasoline ($\rho = 7.00 \times 10^2$ kg/m³) through a hose having an outlet radius of 1.20 cm. If the difference in pressure is measured to be $P_1 - P_2 = 1.20$ kPa and the radius of the inlet tube to the meter is 2.40 cm, find (a) the speed of the gasoline as it leaves the hose and (b) the fluid flow rate in cubic meters per second.

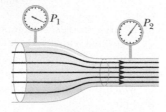

**Figure P9.48**

## 9.8 Surface Tension, Capillary Action, and Viscous Fluid Flow

49. A square metal sheet 3.0 cm on a side and of negligible thickness is attached to a balance and inserted into a container of fluid. The contact angle is found to be zero, as shown in Figure P9.49a, and the balance to which the metal sheet is attached reads

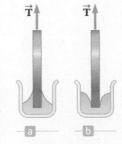

**Figure P9.49**

0.40 N. A thin veneer of oil is then spread over the sheet, and the contact angle becomes 180°, as shown in Figure P9.49b. The balance now reads 0.39 N. What is the surface tension of the fluid?

50. **BIO** To lift a wire ring of radius 1.75 cm from the surface of a container of blood plasma, a vertical force of $1.61 \times 10^{-2}$ N greater than the weight of the ring is required. Calculate the surface tension of blood plasma from this information.

51. A certain fluid has a density of 1 080 kg/m³ and is observed to rise to a height of 2.1 cm in a 1.0-mm-diameter tube. The contact angle between the wall and the fluid is zero. Calculate the surface tension of the fluid.

52. **BIO** Whole blood has a surface tension of 0.058 N/m and a density of 1 050 kg/m³. To what height can whole blood rise in a capillary blood vessel that has a radius of $2.0 \times 10^{-6}$ m if the contact angle is zero?

53. The block of ice (temperature 0°C) shown in Figure P9.53 is drawn over a level surface lubricated by a layer of water 0.10 mm thick. Determine the magnitude of the force $\vec{F}$ needed to pull the block with a constant speed of 0.50 m/s. At 0°C, the viscosity of water has the value $\eta = 1.79 \times 10^{-3}$ N · s/m².

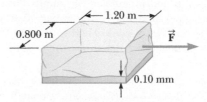

**Figure P9.53**

54. A thin 1.5-mm coating of glycerine has been placed between two microscope slides of width 1.0 cm and length 4.0 cm. Find the force required to pull one of the microscope slides at a constant speed of 0.30 m/s relative to the other slide.

55. A straight horizontal pipe with a diameter of 1.0 cm and a length of 50 m carries oil with a coefficient of viscosity of 0.12 N · s/m². At the output of the pipe, the flow rate is $8.6 \times 10^{-5}$ m³/s and the pressure is 1.0 atm. Find the *gauge* pressure at the pipe input.

56. **BIO** The pulmonary artery, which connects the heart to the lungs, has an inner radius of 2.6 mm and is 8.4 cm long. If the pressure drop between the heart and lungs is 400 Pa, what is the average speed of blood in the pulmonary artery?

57. Spherical particles of a protein of density 1.8 g/cm³ are shaken up in a solution of 20°C water. The solution is allowed to stand for 1.0 h. If the depth of water in the tube is 5.0 cm, find the radius of the largest particles that remain in solution at the end of the hour.

58. **BIO** A hypodermic needle is 3.0 cm in length and 0.30 mm in diameter. What pressure difference between the input and output of the needle is required so that the flow rate of water through it will be 1 g/s? (Use $1.0 \times 10^{-3}$ Pa · s as the viscosity of water.)

59. **BIO** What radius needle should be used to inject a volume of 500. cm³ of a solution into a patient in 30.0 min? Assume the length of the needle is 2.5 cm and the solution is elevated 1.0 m above the point of injection. Further, assume the viscosity and density of the solution are those of pure water, and that the pressure inside the vein is atmospheric.

60. **BIO** The aorta in humans has a diameter of about 2.0 cm, and at certain times the blood speed through it is about 55 cm/s. Is the blood flow turbulent? The density of whole blood is 1 050 kg/m³, and its coefficient of viscosity is $2.7 \times 10^{-3}$ N · s/m².

## 9.9 Transport Phenomena

61. **BIO** Sucrose is allowed to diffuse along a 10.-cm length of tubing filled with water. The tube is 6.0 cm² in cross-sectional area. The diffusion coefficient is equal to $5.0 \times 10^{-10}$ m²/s, and $8.0 \times 10^{-14}$ kg is transported along the tube in 15 s. What is the difference in the concentration levels of sucrose at the two ends of the tube?

62. **BIO** Glycerin in water diffuses along a horizontal column that has a cross-sectional area of 2.0 cm². The concentration gradient is $3.0 \times 10^{-2}$ kg/m⁴, and the diffusion rate is found to be $5.7 \times 10^{-15}$ kg/s. Determine the diffusion coefficient.

63. The viscous force on an oil drop is measured to be equal to $3.0 \times 10^{-13}$ N when the drop is falling through air with a speed of $4.5 \times 10^{-4}$ m/s. If the radius of the drop is $2.5 \times 10^{-6}$ m, what is the viscosity of air?

64. Small spheres of diameter 1.00 mm fall through 20°C water with a terminal speed of 1.10 cm/s. Calculate the density of the spheres.

## 9.10 The Deformation of Solids

65. **T** A 200.-kg load is hung on a wire of length 4.00 m, cross-sectional area $0.200 \times 10^{-4}$ m², and Young's modulus $8.00 \times 10^{10}$ N/m². What is its increase in length?

66. A 25.0-m long steel cable with a cross-sectional area of $2.03 \times 10^{-3}$ m² is used to suspend a $3.50 \times 10^{3}$-kg container. By how much will the cable stretch once bearing the load?

67. A plank 2.00 cm thick and 15.0 cm wide is firmly attached to the railing of a ship by clamps so that the rest of the board extends 2.00 m horizontally over the sea below. A man of mass 80.0 kg is forced to stand on the very end. If the end of the board drops by 5.00 cm because of the man's weight, find the shear modulus of the wood.

68. Artificial diamonds can be made using high-pressure, high-temperature presses. Suppose an artificial diamond of volume $1.00 \times 10^{-6}$ m³ is formed under a pressure of 5.00 GPa. Find the change in its volume when it is released from the press and brought to atmospheric pressure. Take the diamond's bulk modulus to be $B = 194$ GPa.

69. For safety in climbing, a mountaineer uses a nylon rope that is 50. m long and 1.0 cm in diameter. When supporting a 90.-kg climber, the rope elongates 1.6 m. Find its Young's modulus.

70. **T** Assume that if the shear stress in steel exceeds about $4.00 \times 10^{8}$ N/m², the steel ruptures. Determine the shearing force necessary to (a) shear a steel bolt 1.00 cm in diameter and (b) punch a 1.00-cm-diameter hole in a steel plate 0.500 cm thick.

71. **BIO** Bone has a Young's modulus of $18 \times 10^{9}$ Pa. Under compression, it can withstand a stress of about $160 \times 10^{6}$ Pa before breaking. Assume that a femur (thigh bone) is 0.50 m long, and calculate the amount of compression this bone can withstand before breaking.

72. **BIO** A stainless-steel orthodontic wire is applied to a tooth, as in Figure P9.72. The wire has an unstretched length of 3.1 cm and a radius of 0.11 mm. If the wire is stretched 0.10 mm, find the magnitude and direction of the force on the tooth. Disregard the width of the tooth and assume Young's modulus for stainless steel is $18 \times 10^{10}$ Pa.

**Figure P9.72**

73. A high-speed lifting mechanism supports an 800.-kg object with a steel cable that is 25.0 m long and 4.00 cm² in cross-sectional area. (a) Determine the elongation of the cable. (b) By what additional amount does the cable increase in length if the object is accelerated upward at a rate of 3.0 m/s²? (c) What is the greatest mass that can be accelerated upward at 3.0 m/s² if the stress in the cable is not to exceed the elastic limit of the cable, which is $2.2 \times 10^{8}$ Pa?

74. The deepest point in the ocean is in the Mariana Trench, about 11 km deep. The pressure at the ocean floor is huge, about $1.13 \times 10^{8}$ N/m². (a) Calculate the change in volume of 1.00 m³ of water carried from the surface to the bottom of the Pacific. (b) The density of water at the surface is $1.03 \times 10^{3}$ kg/m³. Find its density at the bottom.

75. Determine the elongation of the rod in Figure P9.75 if it is under a tension of $5.8 \times 10^{3}$ N.

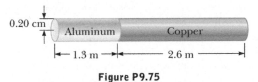

**Figure P9.75**

76. **BIO** The total cross-sectional area of the load-bearing calcified portion of the two forearm bones (radius and ulna) is approximately 2.4 cm². During a car crash, the forearm is slammed against the dashboard. The arm comes to rest from an initial speed of 80 km/h in 5.0 ms. If the arm has an effective mass of 3.0 kg and bone material can withstand a maximum compressive stress of $16 \times 10^{7}$ Pa, is the arm likely to withstand the crash?

## Additional Problems

77. An iron block of volume 0.20 m³ is suspended from a spring scale and immersed in a flask of water. Then the iron block is removed, and an aluminum block of the same volume replaces it. (a) In which case is the buoyant force the greatest, for the iron block or the aluminum block? (b) In which case does the spring scale read the largest value? (c) Use the known densities of these materials to calculate the quantities requested in parts (a) and (b). Are your calculations consistent with your previous answers to parts (a) and (b)?

78. **S** Suppose two worlds, each having mass $M$ and radius $R$, coalesce into a single world. Due to gravitational contraction, the combined world has a radius of only $\frac{3}{4}R$. What is the average density of the combined world as a multiple of $\rho_0$, the average density of the original two worlds?

79. **BIO** In most species of clingfish (family Gobiesocidae), pelvic and pectoral fins converge to form a suction cup edged

by hairy structures that allow a good seal even on rough surfaces. Experiments have shown that a clingfish's suction cup can support up to 230 times the fish's body weight. Suppose a 30.0-g northern clingfish has a suction cup disk area of 15.0 cm$^2$ and the ambient water pressure is $1.10 \times 10^5$ Pa. What ratio $P_{cup}/P_{ambient}$ of the pressure inside the suction cup to the ambient pressure allows the fish to support 230 times its body weight?

80. **BIO** Take the density of blood to be $\rho$ and the distance between the feet and the heart to be $h_H$. Ignore the flow of blood. (a) Show that the difference in blood pressure between the feet and the heart is given by $P_F - P_H = \rho g h_H$. (b) Take the density of blood to be $1.05 \times 10^3$ kg/m$^3$ and the distance between the heart and the feet to be 1.20 m. Find the difference in blood pressure between these two points. This problem indicates that pumping blood from the extremities is very difficult for the heart. The veins in the legs have valves in them that open when blood is pumped toward the heart and close when blood flows away from the heart. Also, pumping action produced by physical activities such as walking and breathing assists the heart.

81. **BIO** The approximate diameter of the aorta is 0.50 cm; that of a capillary is 10. $\mu$m. The approximate average blood flow speed is 1.0 m/s in the aorta and 1.0 cm/s in the capillaries. If all the blood in the aorta eventually flows through the capillaries, estimate the number of capillaries in the circulatory system.

82. Superman attempts to drink water through a very long vertical straw as in Figure P9.82. With his great strength, he achieves maximum possible suction. The walls of the straw don't collapse. (a) Find the maximum height through which he can lift the water. (b) Still thirsty, the Man of Steel repeats his attempt on the Moon, which has no atmosphere. Find the difference between the water levels inside and outside the straw.

83. **BIO** The human brain and spinal cord are immersed in the cerebrospinal fluid. The fluid is normally continuous between the cranial and spinal cavities and exerts a pressure of 100 to 200 mm of H$_2$O above the prevailing atmospheric pressure. In medical work, pressures are often measured in units of mm of H$_2$O because body fluids, including the cerebrospinal fluid, typically have nearly the same density as water. The pressure of the cerebrospinal fluid can be measured by means of a *spinal tap*. A hollow tube is inserted into the spinal column, and the height to which the fluid rises is observed, as shown in Figure P9.83. If the fluid rises to a height of 160. mm, we write its gauge pressure as 160. mm H$_2$O. (a) Express this pressure in pascals, in atmospheres, and in millimeters of mercury. (b) Sometimes it is necessary to determine whether an accident victim has suffered a crushed vertebra that is

**Figure P9.82**

**Figure P9.83**

blocking the flow of cerebrospinal fluid in the spinal column. In other cases, a physician may suspect that a tumor or other growth is blocking the spinal column and inhibiting the flow of cerebrospinal fluid. Such conditions can be investigated by means of the *Queckensted test*. In this procedure, the veins in the patient's neck are compressed to make the blood pressure rise in the brain. The increase in pressure in the blood vessels is transmitted to the cerebrospinal fluid. What should be the normal effect on the height of the fluid in the spinal tap? (c) Suppose compressing the veins had no effect on the level of the fluid. What might account for this phenomenon?

84. **BIO** **S** A *hydrometer* is an instrument used to determine liquid density. A simple one is sketched in Figure P9.84. The bulb of a syringe is squeezed and released to lift a sample of the liquid of interest into a tube containing a calibrated rod of known density. (Assume the rod is cylindrical.) The rod, of length $L$ and average density $\rho_0$, floats partially immersed in the liquid of density $\rho$. A length $h$ of the rod protrudes above the surface of the liquid. Show that the density of the liquid is given by

$$\rho = \frac{\rho_0 L}{L - h}$$

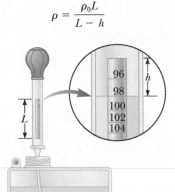

**Figure P9.84**

85. Figure P9.85 shows a water tank with a valve. If the valve is opened, what is the maximum height attained by the stream of water coming out of the right side of the tank? Assume $h = 10.0$ m, $L = 2.00$ m, and $\theta = 30.0°$, and that the cross-sectional area at $A$ is very large compared with that at $B$.

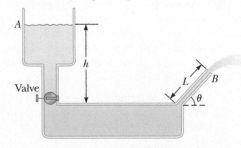

**Figure P9.85**

86. A helium-filled balloon, whose envelope has a mass of 0.25 kg, is tied to a 2.0-m-long, 0.050-kg string. The balloon is spherical with a radius of 0.40 m. When released, it lifts a length $h$ of the string and then remains in equilibrium, as in Figure P9.86. Determine the value of $h$. *Hint:* Only that part of the string above the floor contributes to the load being supported by the balloon.

**Figure P9.86**

87. A light spring of constant $k$ = 90.0 N/m is attached vertically to a table (Fig. P9.87a). A 2.00-g balloon is filled with helium (density = 0.179 kg/m³) to a volume of 5.00 m³ and is then connected to the spring, causing the spring to stretch as shown in Figure P9.87b. Determine the extension distance $L$ when the balloon is in equilibrium.

88. A U-tube open at both ends is partially filled with water (Fig. P9.88a). Oil ($\rho$ = 750 kg/m³) is then poured into the right arm and forms a column $L$ = 5.00 cm high (Fig. P9.88b). (a) Determine the difference $h$ in the heights of the two liquid surfaces. (b) The right arm is then shielded from any air motion while air is blown across the top of the left arm until the surfaces of the two liquids are at the same height (Fig. P9.88c). Determine the speed of the air being blown across the left arm. Assume the density of air is 1.29 kg/m³.

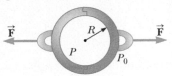
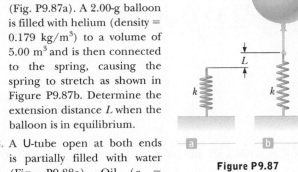

**Figure P9.87**

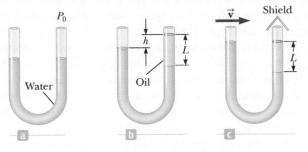

**Figure P9.88**

89. **S** In about 1657, Otto von Guericke, inventor of the air pump, evacuated a sphere made of two brass hemispheres (Fig. P9.89). Two teams of eight horses each could pull the hemispheres apart only on some trials and then "with greatest difficulty," with the resulting sound likened to a cannon firing. Find the force $F$ required to pull the thin-walled evacuated hemispheres apart in terms of $R$, the radius of the hemispheres, $P$ the pressure inside the hemispheres, and atmospheric pressure $P_0$.

**Figure P9.89**

90. Oil having a density of 930 kg/m³ floats on water. A rectangular block of wood 4.00 cm high and with a density of 960 kg/m³ floats partly in the oil and partly in the water. The oil completely covers the block. How far below the interface between the two liquids is the bottom of the block?

91. A water tank open to the atmosphere at the top has two small holes punched in its side, one above the other. The holes are 5.00 cm and 12.0 cm above the floor. How high does water stand in the tank if the two streams of water hit the floor at the same place?

HOW CAN TRAPPED WATER BLOW OFF the top of a volcano in a giant explosion? What causes a sidewalk or road to fracture and buckle spontaneously when the temperature changes? How can thermal energy be harnessed to do work, running the engines that make everything in modern living possible?

Answering these and related questions is the domain of **thermal physics,** the study of temperature, heat, and how they affect matter. Quantitative descriptions of thermal phenomena require careful definitions of the concepts of temperature, heat, and internal energy. Heat leads to changes in internal energy and thus to changes in temperature, which cause the expansion or contraction of matter. Such changes can damage roadways and buildings, create stress fractures in metal, and render flexible materials stiff and brittle, the latter resulting in compromised O-rings and the *Challenger* disaster. Changes in internal energy can also be harnessed for transportation, construction, and food preservation.

Gases are critical in the harnessing of thermal energy to do work. Within normal temperature ranges, a gas acts like a large collection of noninteracting point particles, called an ideal gas. Such gases can be studied on either a macroscopic or microscopic scale. On the macroscopic scale, the pressure, volume, temperature, and number of particles associated with a gas can be related in a single equation known as the ideal gas law. On the microscopic scale, a model called the kinetic theory of gases pictures the components of a gas as small particles. That model will enable us to understand how processes on the atomic scale affect macroscopic properties like pressure, temperature, and internal energy.

# 10.1 Temperature and the Zeroth Law of Thermodynamics

Temperature is commonly associated with how hot or cold an object feels when we touch it. While our senses provide us with qualitative indications of temperature, they are unreliable and often misleading. A metal ice tray feels colder to the hand, for example, than a package of frozen vegetables at the same temperature, because metals conduct thermal energy more rapidly than a cardboard package. What we need is a reliable and reproducible method of making quantitative measurements that establish the relative "hotness" or "coldness" of objects—a method related solely to temperature. Scientists have developed a variety of thermometers for making such measurements.

When placed in contact with each other, two objects at different initial temperatures will eventually reach a common intermediate temperature. If a cup of hot coffee is cooled with an ice cube, for example, the ice rises in temperature and eventually melts while the temperature of the coffee decreases.

Understanding the concept of temperature requires understanding *thermal contact* and *thermal equilibrium.* Two objects are in **thermal contact** if energy can be exchanged between them. Two objects are in **thermal equilibrium** if they are in thermal contact and there is no net exchange of energy.

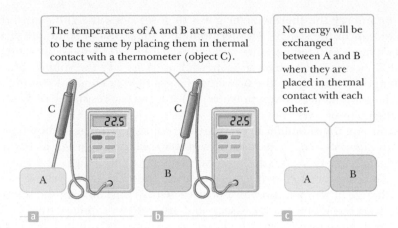

**Figure 10.1** The zeroth law of thermodynamics.

The temperatures of A and B are measured to be the same by placing them in thermal contact with a thermometer (object C).

No energy will be exchanged between A and B when they are placed in thermal contact with each other.

The exchange of energy between two objects due to differences in their temperatures is called **heat,** a concept examined in more detail in Topic 11.

Using these ideas, we can develop a formal definition of temperature. Consider two objects A and B that are not in thermal contact with each other, and a third object C that acts as a **thermometer**—a device calibrated to measure the temperature of an object. We wish to determine whether A and B would be in thermal equilibrium if they were placed in thermal contact. The thermometer (object C) is first placed in thermal contact with A until thermal equilibrium is reached, as in Figure 10.1a, whereupon the reading of the thermometer is recorded. The thermometer is then placed in thermal contact with B, and its reading is again recorded at equilibrium (Fig. 10.1b). If the two readings are the same, then A and B in thermal equilibrium with each other. If A and B are placed in thermal contact with each other, as in Figure 10.1c, there is no net transfer of energy between them.

We can summarize these results in a statement known as the **zeroth law of thermodynamics (the law of equilibrium):**

> If objects A and B are separately in thermal equilibrium with a third object C, then A and B are in thermal equilibrium with each other.

◀ Zeroth law of thermodynamics

This statement is important because it makes it possible to define **temperature.** We can think of temperature as the property that determines whether or not an object is in thermal equilibrium with other objects. **Two objects in thermal equilibrium with each other are at the same temperature.**

### Quick Quiz

**10.1** Two objects with different sizes, masses, and temperatures are placed in thermal contact. Choose the best answer: Energy travels (a) from the larger object to the smaller object (b) from the object with more mass to the one with less mass (c) from the object at higher temperature to the object at lower temperature.

# 10.2 Thermometers and Temperature Scales

Thermometers are devices used to measure the temperature of an object or a system. When a thermometer is in thermal contact with a system, energy is exchanged until the thermometer and the system are in thermal equilibrium with each other. For accurate readings, the thermometer must be much smaller than the system,

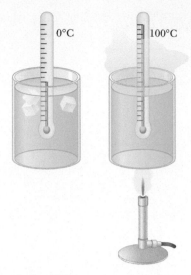

**Figure 10.2** Schematic diagram of a mercury thermometer. Because of thermal expansion, the level of the mercury rises as the temperature of the mercury changes from 0°C (the ice point) to 100°C (the steam point).

so that the energy the thermometer gains or loses doesn't significantly alter the energy content of the system. All thermometers make use of some physical property that changes with temperature and can be calibrated to make the temperature measurable. Some of the physical properties used are (1) the volume of a liquid, (2) the length of a solid, (3) the pressure of a gas held at constant volume, (4) the volume of a gas held at constant pressure, (5) the electric resistance of a conductor, and (6) the color of a very hot object.

One common thermometer in everyday use consists of a mass of liquid — usually mercury or alcohol — that expands into a glass capillary tube when its temperature rises (Fig. 10.2). In this case, the physical property that changes is the volume of a liquid. To serve as an effective thermometer, the change in volume of the liquid with change in temperature must be very nearly constant over the temperature ranges of interest. When the cross-sectional area of the capillary tube is constant as well, the change in volume of the liquid varies linearly with its length along the tube. We can then define a temperature in terms of the length of the liquid column. The thermometer can be calibrated by placing it in thermal contact with environments that remain at constant temperature. One such environment is a mixture of water and ice in thermal equilibrium at atmospheric pressure. Another commonly used system is a mixture of water and steam in thermal equilibrium at atmospheric pressure.

Once we have marked the ends of the liquid column for our chosen environment on our thermometer, we need to define a scale of numbers associated with various temperatures. An example of such a scale is the **Celsius temperature scale,** formerly called the centigrade scale. On the Celsius scale, the temperature of the ice–water mixture is defined to be zero degrees Celsius, written 0°C, and called the **ice point** or **freezing point** of water. The temperature of the water–steam mixture is defined as 100°C, called the **steam point** or **boiling point** of water. Once the ends of the liquid column in the thermometer have been marked at these two points, the distance between marks is divided into 100 equal segments, each corresponding to a change in temperature of one degree Celsius.

Thermometers calibrated in this way present problems when extremely accurate readings are needed. For example, an alcohol thermometer calibrated at the ice and steam points of water might agree with a mercury thermometer only at the calibration points. Because mercury and alcohol have different thermal expansion properties, when one indicates a temperature of 50°C, say, the other may indicate a slightly different temperature. The discrepancies between different types of thermometers are especially large when the temperatures to be measured are far from the calibration points.

## 10.2.1 The Constant-Volume Gas Thermometer and the Kelvin Scale

We can construct practical thermometers such as the mercury thermometer, but these types of thermometers don't define temperature in a fundamental way. One thermometer, however, *is* more fundamental, and offers a way to define temperature and relate it directly to internal energy: the **gas thermometer.** In a gas thermometer, the temperature readings are nearly independent of the substance used in the thermometer. One type of gas thermometer is the constant-volume unit shown in Figure 10.3. The behavior observed in this device is the variation of pressure with temperature of a fixed volume of gas. When the constant-volume gas thermometer was developed, it was calibrated using the ice and steam points of water as follows (a different calibration procedure, to be discussed shortly, is now used): The gas flask is inserted into an ice–water bath, and mercury reservoir B is raised or lowered until the volume of the confined gas is at some value, indicated by the zero point on the scale. The height *h*, the difference between the levels in the reservoir and column A, indicates the pressure in the flask at 0°C. The flask is inserted into

The volume of gas in the flask is kept constant by raising or lowering reservoir *B* to keep the mercury level in column *A* constant.

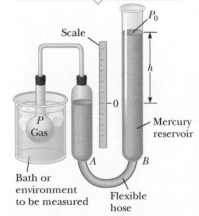

**Figure 10.3** A constant-volume gas thermometer measures the pressure of the gas contained in the flask immersed in the bath.

water at the steam point, and reservoir B is readjusted until the height in column A is again brought to zero on the scale, ensuring that the gas volume is the same as it had been in the ice bath (hence the designation "constant-volume"). A measure of the new value for $h$ gives a value for the pressure at 100°C. These pressure and temperature values are then plotted on a graph, as in Figure 10.4. The line connecting the two points serves as a calibration curve for measuring unknown temperatures. If we want to measure the temperature of a substance, we place the gas flask in thermal contact with the substance and adjust the column of mercury until the level in column A returns to zero. The height of the mercury column tells us the pressure of the gas, and we could then find the temperature of the substance from the calibration curve.

Now suppose that temperatures are measured with various gas thermometers containing different gases. Experiments show that the thermometer readings are nearly independent of the type of gas used, as long as the gas pressure is low and the temperature is well above the point at which the gas liquefies.

We can also perform the temperature measurements with the gas in the flask at different starting pressures at 0°C. As long as the pressure is low, we will generate straight-line calibration curves for each starting pressure, as shown for three experimental trials (solid lines) in Figure 10.5.

If the lines in Figure 10.5 are extended back toward negative temperatures, we find a startling result: In every case, regardless of the type of gas or the value of the low starting pressure, **the pressure extrapolates to zero when the temperature is −273.15°C.** This fact suggests that this particular temperature is universal in its importance, because it doesn't depend on the substance used in the thermometer. In addition, because the lowest possible pressure is $P = 0$, a perfect vacuum, the temperature −273.15°C must represent a lower bound for physical processes. We define this temperature as **absolute zero.**

Absolute zero is used as the basis for the **Kelvin temperature scale,** which sets −273.15°C as its zero point (0 K). The size of a "degree" on the Kelvin scale is chosen to be identical to the size of a degree on the Celsius scale. The relationship between these two temperature scales is

$$T_C = T - 273.15 \qquad\qquad [10.1]$$

where $T_C$ is the Celsius temperature and $T$ is the Kelvin temperature (sometimes called the **absolute temperature**).

Technically, Equation 10.1 should have units on the right-hand side so that it reads $T_C = T \,°C/K - 273.15°C$. The units are rather cumbersome in this context, so we will usually suppress them in such calculations except in the final answer. (This will also be the case when discussing the Celsius and Fahrenheit scales.)

Early gas thermometers made use of ice and steam points according to the procedure just described. These points are experimentally difficult to duplicate, however, because they are pressure-sensitive. Consequently, a procedure based on two new points was adopted in 1954 by the International Committee on Weights and Measures. The first point is absolute zero. The second point is **the triple point of water, which is the single temperature and pressure at which water, water vapor, and ice can coexist in equilibrium.** This point is a convenient and reproducible reference temperature for the Kelvin scale; it occurs at a temperature of 0.01°C and a pressure of 4.58 mm of mercury. The temperature at the triple point of water on the Kelvin scale occurs at 273.16 K. Therefore, **the SI unit of temperature, the kelvin, is defined as 1/273.16 of the temperature of the triple point of water.** Figure 10.6 shows the Kelvin temperatures for various physical processes and structures. Absolute zero has been closely approached but never achieved.

What would happen to a substance if its temperature could reach 0 K? As Figure 10.5 indicates, the substance would exert zero pressure on the walls of its container (assuming the gas doesn't liquefy or solidify on the way to absolute zero). In Section 10.5 we show that the pressure of a gas is proportional to the kinetic

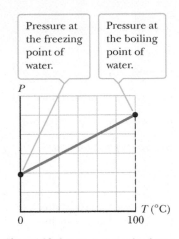

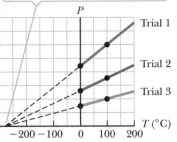

**Figure 10.4** A typical graph of pressure versus temperature taken with a constant-volume gas thermometer.

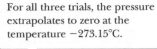

**Figure 10.5** Pressure versus temperature for experimental trials in which gases have different pressures in a constant-volume gas thermometer.

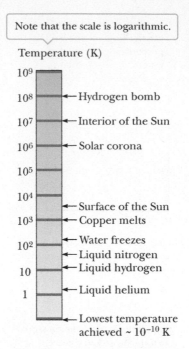

Note that the scale is logarithmic.

Temperature (K)

| | |
|---|---|
| $10^9$ | |
| $10^8$ | ← Hydrogen bomb |
| $10^7$ | ← Interior of the Sun |
| $10^6$ | ← Solar corona |
| $10^5$ | |
| $10^4$ | |
| | ← Surface of the Sun |
| $10^3$ | ← Copper melts |
| $10^2$ | ← Water freezes |
| | ← Liquid nitrogen |
| 10 | ← Liquid hydrogen |
| | ← Liquid helium |
| 1 | |
| | ← Lowest temperature achieved ~ $10^{-10}$ K |

**Figure 10.6** Absolute temperatures at which various selected physical processes take place.

energy of the molecules of that gas. According to classical physics, therefore, the kinetic energy of the gas would go to zero and there would be no motion at all of the individual components of the gas. According to quantum theory, however (to be discussed in Topic 27), the gas would always retain some residual energy, called the *zero-point energy*, at that low temperature.

## 10.2.2 The Celsius, Kelvin, and Fahrenheit Temperature Scales

Equation 10.1 shows that the Celsius temperature $T_C$ is shifted from the absolute (Kelvin) temperature $T$ by 273.15. Because the size of a Celsius degree is the same as a Kelvin, a temperature difference of 5°C is equal to a temperature difference of 5 K. The two scales differ only in the choice of zero point. The ice point (273.15 K) corresponds to 0.00°C, and the steam point (373.15 K) is equivalent to 100.00°C.

The most common temperature scale in use in the United States is the Fahrenheit scale. It sets the temperature of the ice point at 32°F and the temperature of the steam point at 212°F. The relationship between the Celsius and Fahrenheit temperature scales is

$$T_F = \tfrac{9}{5}T_C + 32 \qquad\qquad [10.2a]$$

For example, a temperature of 50.0°F corresponds to a Celsius temperature of 10.0°C and an absolute temperature of 283 K.

Equation 10.2a can be inverted to give Celsius temperatures in terms of Fahrenheit temperatures:

$$T_C = \tfrac{5}{9}(T_F - 32) \qquad\qquad [10.2b]$$

Equation 10.2 can also be used to find a relationship between changes in temperature on the Celsius and Fahrenheit scales. If the Celsius temperature changes by $\Delta T_C$, the Fahrenheit temperature changes by the amount

$$\Delta T_F = \tfrac{9}{5}\Delta T_C \qquad\qquad [10.3]$$

Figure 10.7 compares the Celsius, Fahrenheit, and Kelvin scales. Although less commonly used, other scales do exist, such as the Rankine scale. That scale has Fahrenheit degrees and a zero point at absolute zero.

**Figure 10.7** A comparison of the Celsius, Fahrenheit, and Kelvin temperature scales.

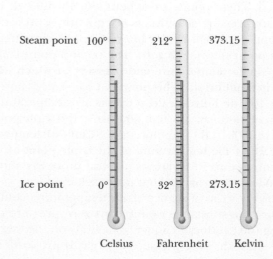

## EXAMPLE 10.1 SKIN TEMPERATURE BIO

**GOAL** Apply the temperature conversion formulas.

**PROBLEM** The temperature gradient between the skin and the air is regulated by cutaneous (skin) blood flow. If the cutaneous blood vessels are constricted, the skin temperature and the temperature of the environment will be about the same. When the vessels are dilated, more blood is brought to the surface. Suppose during dilation the skin warms from 72.0°F to 84.0°F. **(a)** Convert these temperatures to Celsius and find the difference. **(b)** Convert the temperatures to Kelvin, again finding the difference.

**STRATEGY** This is a matter of applying the conversion formulas, Equations 10.1 and 10.2. For part **(b)** it's easiest to use the answers for Celsius rather than develop another set of conversion equations.

**SOLUTION**

**(a)** Convert the temperatures from Fahrenheit to Celsius and find the difference.

Convert the lower temperature, using Equation 10.2b:

$$T_C = \tfrac{5}{9}(T_F - 32.0) = \tfrac{5}{9}(72.0 - 32.0) = \boxed{22.2°C}$$

Convert the upper temperature:

$$T_C = \tfrac{5}{9}(T_F - 32.0) = \tfrac{5}{9}(84.0 - 32.0) = \boxed{28.9°C}$$

Find the difference of the two temperatures:

$$\Delta T_C = 28.9°C - 22.2°C = \boxed{6.7°C}$$

**(b)** Convert the temperatures from Fahrenheit to Kelvin and find their difference.

Convert the lower temperature, using the answers for Celsius found in part **(a)**:

$$T_C = T - 273.15 \quad \rightarrow \quad T = T_C + 273.15$$
$$T = 22.2 + 273.15 = \boxed{295.4 \text{ K}}$$

Convert the upper temperature:

$$T = 28.9 + 273.15 = \boxed{302.1 \text{ K}}$$

Find the difference of the two temperatures:

$$\Delta T = 302.1 \text{ K} - 295.4 \text{ K} = \boxed{6.7 \text{ K}}$$

**REMARKS** The change in temperature in Kelvin and Celsius is the same, as it should be.

**QUESTION 10.1** Which represents a larger temperature change, a Celsius degree or a Fahrenheit degree?

**EXERCISE 10.1** Core body temperature can rise from 98.6°F to 107°F during extreme exercise, such as a marathon run. Such elevated temperatures can also be caused by viral or bacterial infections or tumors and are dangerous if sustained. **(a)** Convert the given temperatures to Celsius and find the difference. **(b)** Convert the temperatures to Kelvin, again finding the difference.

**ANSWERS** **(a)** 37.0°C, 41.7°C, 4.7°C **(b)** 310.2 K, 314.9 K, 4.7 K

## EXAMPLE 10.2 EXTRATERRESTRIAL TEMPERATURE SCALE

**GOAL** Understand how to relate different temperature scales.

**PROBLEM** An extraterrestrial scientist invents a temperature scale such that water freezes at −75°E and boils at 325°E, where E stands for an extraterrestrial scale. Find an equation that relates temperature in °E to temperature in °C.

**STRATEGY** Using the given data, find the ratio of the number of °E between the two temperatures to the number of °C. This ratio will be the same as a similar ratio for any other such process—say, from the freezing point to an unknown temperature—corresponding to $T_E$ and $T_C$. Setting the two ratios equal and solving for $T_E$ in terms of $T_C$ yields the desired relationship. For clarity, the rules of significant figures will not be applied here.

**SOLUTION**

Find the change in temperature in °E between the freezing and boiling points of water:

$$\Delta T_E = 325°E - (-75°E) = 400°E$$

Find the change in temperature in °C between the freezing and boiling points of water:

$$\Delta T_C = 100°C - 0°C = 100°C$$

*(Continued)*

Form the ratio of these two quantities.

$$\frac{\Delta T_E}{\Delta T_C} = \frac{400°E}{100°C} = 4 \frac{°E}{°C}$$

This ratio is the same between any other two temperatures—say, from the freezing point to an unknown final temperature. Set the two ratios equal to each other:

$$\frac{\Delta T_E}{\Delta T_C} = \frac{T_E - (-75°E)}{T_C - 0°C} = 4 \frac{°E}{°C}$$

Solve for $T_E$:

$$T_E - (-75°E) = 4(°E/°C)(T_C - 0°C)$$
$$T_E = \boxed{4T_C - 75}$$

**REMARKS** The relationship between any other two temperature scales can be derived in the same way.

**QUESTION 10.2** True or False: Finding the relationship between two temperature scales using knowledge of the freezing and boiling point of water in each system is equivalent to finding the equation of a straight line.

**EXERCISE 10.2** Find the equation converting °F to °E.

**ANSWER** $T_E = \frac{20}{9}T_F - 146$

---

Without these joints to separate sections of roadway on bridges, the surface would buckle due to thermal expansion on very hot days or crack due to contraction on very cold days.

© Cengage Learning/George Sample

a

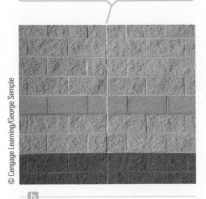

The long, vertical joint is filled with a soft material that allows the wall to expand and contract as the temperature of the bricks changes.

© Cengage Learning/George Sample

b

**Figure 10.8** Thermal expansion joints in (a) bridges and (b) walls.

# 10.3 Thermal Expansion of Solids and Liquids

Our discussion of the liquid thermometer made use of one of the best-known changes that occur in most substances: As the temperature of a substance increases, its volume increases. This phenomenon, known as **thermal expansion,** plays an important role in numerous applications. Thermal expansion joints, for example, must be included in buildings, concrete highways, and bridges to compensate for changes in dimensions with variations in temperature (Fig. 10.8).

The overall thermal expansion of an object is a consequence of the change in the average separation between its constituent atoms or molecules. To understand this idea, consider how the atoms in a solid substance behave. These atoms are located at fixed equilibrium positions; if an atom is pulled away from its position, a restoring force pulls it back. We can imagine that the atoms are particles connected by springs to their neighboring atoms. (See Fig. 9.1 in the previous topic.) If an atom is pulled away from its equilibrium position, the distortion of the springs provides a restoring force.

At ordinary temperatures, the atoms vibrate around their equilibrium positions with an amplitude (maximum distance from the center of vibration) of about $10^{-11}$ m, with an average spacing between the atoms of about $10^{-10}$ m. As the temperature of the solid increases, the atoms vibrate with greater amplitudes and the average separation between them increases. Consequently, the solid as a whole expands.

If the thermal expansion of an object is sufficiently small compared with the object's initial dimensions, then the change in any dimension is, to a good approximation, proportional to the first power of the temperature change. Suppose an object has an initial length $L_0$ along some direction at some temperature $T_0$. Then the length increases by $\Delta L$ for a change in temperature $\Delta T$. So for small changes in temperature,

$$\Delta L = \alpha L_0 \Delta T \qquad \text{[10.4]}$$

or

$$L - L_0 = \alpha L_0 (T - T_0)$$

where $L$ is the object's final length, $T$ is its final temperature, and the proportionality constant $\alpha$ is called the **coefficient of linear expansion** for a given material and has units of $(°C)^{-1}$.

Table 10.1 lists the coefficients of linear expansion for various materials. Note that for these materials $\alpha$ is positive, indicating an increase in length with increasing temperature.

Thermal expansion affects the choice of glassware used in kitchens and laboratories. If hot liquid is poured into a cold container made of ordinary glass, the container may well break due to thermal stress. The inside surface of the glass becomes hot and expands, while the outside surface is at room temperature, and ordinary glass may not withstand the difference in expansion without breaking. Pyrex glass has a coefficient of linear expansion of about one-third that of ordinary glass, so the thermal stresses are smaller. Kitchen measuring cups and laboratory beakers are often made of Pyrex so they can be used with hot liquids.

**APPLICATION**
Pyrex Glass

**Table 10.1** Average Coefficients of Expansion for Some Materials Near Room Temperature

| Material | Average Coefficient of Linear Expansion $[(°C)^{-1}]$ | Material | Average Coefficient of Volume Expansion $[(°C)^{-1}]$ |
|---|---|---|---|
| Aluminum | $24 \times 10^{-6}$ | Acetone | $1.5 \times 10^{-4}$ |
| Brass and bronze | $19 \times 10^{-6}$ | Benzene | $1.24 \times 10^{-4}$ |
| Concrete | $12 \times 10^{-6}$ | Ethyl alcohol | $1.12 \times 10^{-4}$ |
| Copper | $17 \times 10^{-6}$ | Gasoline | $9.6 \times 10^{-4}$ |
| Glass (ordinary) | $9 \times 10^{-6}$ | Glycerin | $4.85 \times 10^{-4}$ |
| Glass (Pyrex) | $3.2 \times 10^{-6}$ | Mercury | $1.82 \times 10^{-4}$ |
| Invar (Ni–Fe alloy) | $0.9 \times 10^{-6}$ | Turpentine | $9.0 \times 10^{-4}$ |
| Lead | $29 \times 10^{-6}$ | Air[a] at 0°C | $3.67 \times 10^{-3}$ |
| Steel | $11 \times 10^{-6}$ | Helium | $3.665 \times 10^{-3}$ |

[a]Gases do not have a specific value for the volume expansion coefficient because the amount of expansion depends on the type of process through which the gas is taken. The values given here assume the gas undergoes an expansion at constant pressure.

**Tip 10.1** Coefficients of Expansion Are Not Constants

The coefficients of expansion can vary somewhat with temperature, so the given coefficients are actually averages.

## EXAMPLE 10.3    EXPANSION OF A RAILROAD TRACK

**GOAL** Apply the concept of linear expansion and relate it to stress.

**PROBLEM** (a) A steel railroad track has a length of 30.000 m when the temperature is 0°C. What is its length on a hot day when the temperature is 40.0°C? (b) Suppose the track is nailed down so that it can't expand. What stress results in the track due to the temperature change (Fig. 10.9)?

**STRATEGY** (a) Apply the linear expansion equation, using Table 10.1 and Equation 10.4. (b) A track that cannot expand by $\Delta L$ due to external constraints is equivalent to compressing the track by $\Delta L$, creating a stress in the track. Using the equation relating tensile stress to tensile strain together with the linear expansion equation, the amount of (compressional) stress can be calculated using Equation 9.30.

**Figure 10.9** (Example 10.3) Thermal expansion: The extreme heat of a July day in Asbury Park, New Jersey, caused these railroad tracks to buckle.

AP Images/Wide World Photos

. . . . . . . . . . . . . . . . . . . . . . . . . . . . .

**SOLUTION**

**(a)** Find the length of the track at 40.0°C.

Substitute given quantities into Equation 10.4, finding the change in length:

$$\Delta L = \alpha L_0 \Delta T = [11 \times 10^{-6}(°C)^{-1}](30.000\ \text{m})(40.0°C)$$
$$= 0.013\ \text{m}$$

*(Continued)*

Add the change to the original length to find the final length:

$$L = L_0 + \Delta L = \boxed{30.013 \text{ m}}$$

**(b)** Find the stress if the track cannot expand.

Substitute into Equation 9.30 to find the stress:

$$\frac{F}{A} = Y\frac{\Delta L}{L_0} = (2.00 \times 10^{11} \text{ Pa})\left(\frac{0.013 \text{ m}}{30.0 \text{ m}}\right)$$

$$= \boxed{8.7 \times 10^7 \text{ Pa}}$$

**REMARKS** Repeated heating and cooling is an important part of the weathering process that gradually wears things out, weakening structures over time.

**QUESTION 10.3** What happens to the tension of wires in a piano when the temperature decreases?

**EXERCISE 10.3** What is the length of the same railroad track on a cold winter day when the temperature is 0°F?

**ANSWER** 29.994 m

---

## APPLYING PHYSICS 10.1    BIMETALLIC STRIPS AND THERMOSTATS

How can different coefficients of expansion for metals be used as a temperature gauge and control electronic devices such as air conditioners?

**EXPLANATION** When the temperatures of a brass rod and a steel rod of equal length are raised by the same amount from some common initial value, the brass rod expands more than the steel rod because brass has a larger coefficient of expansion than steel. A simple device that uses this principle is a **bimetallic strip.** Such strips can be found in the thermostats of certain home heating systems. The strip is made by securely bonding two different metals together. As the temperature of the strip increases, the two metals expand by different amounts and the strip bends, as in Figure 10.10. The change in shape can make or break an electrical connection. ■

**Figure 10.10** (Applying Physics 10.1) (a) A bimetallic strip bends as the temperature changes because the two metals have different coefficients of expansion. (b) A bimetallic strip used in a thermostat to break or make electrical contact. (c) The interior of a thermostat, showing the coiled bimetallic strip. Why do you suppose the strip is coiled?

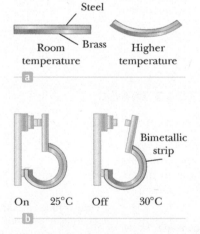

It may be helpful to picture a thermal expansion as a magnification or a photographic enlargement. For example, as the temperature of a metal washer increases (Fig. 10.11), all dimensions, including the radius of the hole, increase according to Equation 10.4.

One practical application of thermal expansion is the common technique of using hot water to loosen a metal lid stuck on a glass jar. This works because the circumference of the lid expands more than the rim of the jar.

Because the linear dimensions of an object change due to variations in temperature, it follows that surface area and volume of the object also change. Consider a square of material having an initial length $L_0$ on a side and therefore an initial area $A_0 = L_0^2$. As the temperature is increased, the length of each side increases to

$$L = L_0 + \alpha L_0 \Delta T$$

The new area $A$ is

$$A = L^2 = (L_0 + \alpha L_0 \Delta T)(L_0 + \alpha L_0 \Delta T) = L_0^2 + 2\alpha L_0^2 \Delta T + \alpha^2 L_0^2 (\Delta T)^2$$

The last term in this expression contains the quantity $\alpha \Delta T$ raised to the second power. Because $\alpha \Delta T$ is much less than one, squaring it makes it even smaller. Consequently, we can neglect this term to get a simpler expression:

$$A = L_0^2 + 2\alpha L_0^2 \Delta T$$

$$A = A_0 + 2\alpha A_0 \Delta T$$

so that

$$\Delta A = A - A_0 = \gamma A_0 \Delta T \qquad [10.5]$$

where $\gamma = 2\alpha$. The quantity $\gamma$ (Greek letter gamma) is called the **coefficient of area expansion.**

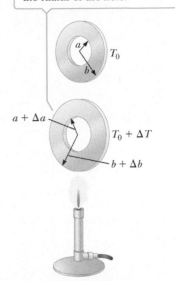

As the washer is heated, all dimensions increase, including the radius of the hole.

**Figure 10.11** Thermal expansion of a homogeneous metal washer. (Note that the expansion is exaggerated in this figure.)

## EXAMPLE 10.4   RINGS AND RODS

**GOAL**  Apply the equation of area expansion.

**PROBLEM**  (a) A circular copper ring at 20.0°C has a hole with an area of 9.980 cm². What minimum temperature must it have so that it can be slipped onto a steel metal rod having a cross-sectional area of 10.000 cm²? (b) Suppose the ring and the rod are heated simultaneously. What minimum change in temperature of both will allow the ring to be slipped onto the end of the rod? (Assume no significant change in the coefficients of linear expansion over this temperature range.)

**STRATEGY**  In part (a), finding the necessary temperature change is just a matter of substituting given values into Equation 10.5, the equation of area expansion. Remember that $\gamma = 2\alpha$. Part (b) is a little harder because now the rod is also expanding. If the ring is to slip onto the rod, however, the final cross-sectional areas of both ring and rod must be equal. Write this condition in mathematical terms, using Equation 10.5 on both sides of the equation, and solve for $\Delta T$.

. . . . . . . . . . . . . . . . . . . . . . . . . . . . . . . . . . . . . . . . . . . . . . . . . . . . . . . . . . . . . . . . . . . . . . . . . . . . . . . . . . . . .

**SOLUTION**

**(a)** Find the temperature of the ring that will allow it to slip onto the rod.

Write Equation 10.5 and substitute known values, leaving $\Delta T$ as the sole unknown:

$$\Delta A = \gamma A_0 \Delta T$$
$$0.020 \text{ cm}^2 = [34 \times 10^{-6} \text{ (°C)}^{-1}](9.980 \text{ cm}^2)(\Delta T)$$

Solve for $\Delta T$, then add this change to the initial temperature to get the final temperature:

$$\Delta T = 59°C$$
$$T = T_0 + \Delta T = 20.0°C + 59°C = \boxed{79°C}$$

**(b)** If both ring and rod are heated, find the minimum change in temperature that will allow the ring to be slipped onto the rod.

Set the final areas of the copper ring and steel rod equal to each other:

$$A_C + \Delta A_C = A_S + \Delta A_S$$

Substitute for each change in area, $\Delta A$:

$$A_C + \gamma_C A_C \Delta T = A_S + \gamma_S A_S \Delta T$$

*(Continued)*

Rearrange terms to get $\Delta T$ on one side only, factor it out, and solve:

$$\gamma_C A_C \Delta T - \gamma_S A_S \Delta T = A_S - A_C$$

$$(\gamma_C A_C - \gamma_S A_S)\Delta T = A_S - A_C$$

$$\Delta T = \frac{A_S - A_C}{\gamma_C A_C - \gamma_S A_S}$$

$$= \frac{10.000 \text{ cm}^2 - 9.980 \text{ cm}^2}{(34 \times 10^{-6}\,^\circ\text{C}^{-1})(9.980 \text{ cm}^2) - (22 \times 10^{-6}\,(^\circ\text{C})^{-1})(10.000 \text{ cm}^2)}$$

$$\Delta T = \boxed{170^\circ\text{C}}$$

**REMARKS** Warming and cooling strategies are sometimes useful for separating glass parts in a chemistry lab, such as the glass stopper in a bottle of reagent.

**QUESTION 10.4** If instead of heating the copper ring in part **(a)** the steel rod is cooled, would the magnitude of the required temperature change be larger, smaller, or the same? Why? (Don't calculate it!)

**EXERCISE 10.4** A steel ring with a hole having area of 3.990 cm² is to be placed on an aluminum rod with cross-sectional area of 4.000 cm². Both rod and ring are initially at a temperature of 35.0°C. At what common temperature can the steel ring be slipped onto one end of the aluminum rod?

**ANSWER** −61°C

---

We can also show that the *increase in volume* of an object accompanying a change in temperature is

$$\Delta V = \beta V_0 \Delta T \qquad [10.6]$$

where $\beta$, the **coefficient of volume expansion,** is equal to $3\alpha$. (Note that $\gamma = 2\alpha$ and $\beta = 3\alpha$ only if the coefficient of linear expansion of the object is the same in all directions.) The proof of Equation 10.6 is similar to the proof of Equation 10.5.

As Table 10.1 indicates, each substance has its own characteristic coefficients of expansion.

**APPLICATION**
Rising Sea Levels

The thermal expansion of water has a profound influence on rising ocean levels. At current rates of global warming, scientists predict that about one-half of the expected rise in sea level will be caused by thermal expansion; the remainder will be due to the melting of polar ice.

### Quick Quiz

**10.2** If you quickly plunge a room-temperature mercury thermometer into very hot water, the mercury level will (a) go up briefly before reaching a final reading, (b) go down briefly before reaching a final reading, or (c) not change.

**10.3** If you are asked to make a very sensitive glass thermometer, which of the following working fluids would you choose? (a) mercury (b) alcohol (c) gasoline (d) glycerin

**10.4** Two spheres are made of the same metal and have the same radius, but one is hollow and the other is solid. The spheres are taken through the same temperature increase. Which sphere expands more? (a) solid sphere, (b) hollow sphere, (c) they expand by the same amount, or (d) not enough information to say.

---

## EXAMPLE 10.5 GLOBAL WARMING AND COASTAL FLOODING BIO

**GOAL** Apply the volume expansion equation together with linear expansion.

**PROBLEM** **(a)** Estimate the fractional change in the volume of Earth's oceans due to an average temperature change of 1°C. **(b)** Use the fact that the average depth of the ocean is $4.00 \times 10^3$ m to estimate the change in depth. Note that $\beta_{water} = 2.07 \times 10^{-4}(^\circ\text{C})^{-1}$.

**STRATEGY** In part **(a)** solve the volume expansion expression, Equation 10.6, for $\Delta V/V$. For part **(b)** use linear expansion to estimate the increase in depth. Neglect the expansion of landmasses, which would reduce the rise in sea level only slightly.

**SOLUTION**

**(a)** Find the fractional change in volume.

Divide the volume expansion equation by $V_0$ and substitute:

$$\Delta V = \beta V_0 \Delta T$$

$$\frac{\Delta V}{V_0} = \beta \, \Delta T = (2.07 \times 10^{-4} (°\text{C})^{-1})(1°\text{C}) = \boxed{2 \times 10^{-4}}$$

**(b)** Find the approximate increase in depth.

Use the linear expansion equation. Divide the volume expansion coefficient of water by 3 to get the equivalent linear expansion coefficient:

$$\Delta L = \alpha L_0 \, \Delta T = \left(\frac{\beta}{3}\right) L_0 \, \Delta T$$

$$\Delta L = (6.90 \times 10^{-5} (°\text{C})^{-1})(4\,000 \text{ m})(1°\text{C}) \approx \boxed{0.3 \text{ m}}$$

**REMARKS** Three-tenths of a meter may not seem significant, but combined with increased melting of land-based polar ice, some coastal areas could experience flooding. An increase of several degrees increases the value of $\Delta L$ several times and could significantly reduce the value of waterfront property.

**QUESTION 10.5** Assuming all have the same initial volume, rank the following substances by the amount of volume expansion due to an increase in temperature, from least to most: glass, mercury, aluminum, ethyl alcohol.

**EXERCISE 10.5** A 1.00-liter aluminum cylinder at 5.00°C is filled to the brim with gasoline at the same temperature. If the aluminum and gasoline are warmed to 65.0°C, how much of the gasoline spills out? *Hint:* Be sure to account for the expansion of the container. Also, ignore the possibility of evaporation, and assume the volume coefficients are good to three digits.

**ANSWER** The volume spilled is 53.3 cm³. Forgetting to take into account the expansion of the cylinder results in a (wrong) answer of 57.6 cm³.

---

> ## Quick Quiz
>
> **10.5** Why doesn't the melting of ocean-based ice raise as much concern as the melting of land-based ice?

## 10.3.1 The Unusual Behavior of Water

Liquids generally increase in volume with increasing temperature and have volume expansion coefficients about ten times greater than those of solids. Over a small temperature range, water is an exception to this rule, as we can see from its density-versus-temperature curve in Figure 10.12. As the temperature increases from

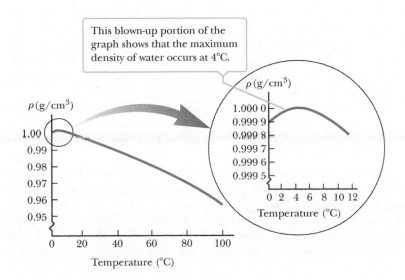

**Figure 10.12** The density of water as a function of temperature.

0°C to 4°C, water contracts, so its density increases. Above 4°C, water exhibits the expected expansion with increasing temperature. The density of water reaches its maximum value of 1 000 kg/m$^3$ at 4°C.

We can use this unusual thermal expansion behavior of water to explain why a pond freezes slowly from the top down. When the atmospheric temperature drops from 7°C to 6°C, say, the water at the surface of the pond also cools and consequently decreases in volume. This means the surface water is more dense than the water below it, which has not yet cooled nor decreased in volume. As a result, the surface water sinks and warmer water from below is forced to the surface to be cooled, a process called *upwelling*. When the atmospheric temperature is between 4°C and 0°C, however, the surface water expands as it cools, becoming less dense than the water below it. The sinking process stops, and eventually the surface water freezes. As the water freezes, the ice remains on the surface because ice is less dense than water. The ice continues to build up on the surface, and water near the bottom of the pond remains at 4°C. Further, the ice forms an insulating layer that slows heat loss from the underlying water, offering thermal protection for marine life.

**BIO APPLICATION**
The Expansion of Water on Freezing and Life on Earth

Without buoyancy and the expansion of water upon freezing, life on Earth may not have been possible. If ice had been more dense than water, it would have sunk to the bottom of the ocean and built up over time. This could have led to a freezing of the oceans, turning Earth into an icebound world similar to Hoth in the Star Wars epic *The Empire Strikes Back*.

**APPLICATION**
Bursting Water Pipes in Winter

The same peculiar thermal expansion properties of water sometimes cause pipes to burst in winter. As energy leaves the water through the pipe by heat and is transferred to the outside cold air, the outer layers of water in the pipe freeze first. The continuing energy transfer causes ice to form ever closer to the center of the pipe. As long as there is still an opening through the ice, the water can expand as its temperature approaches 0°C or as it freezes into more ice, pushing itself into another part of the pipe. Eventually, however, the ice will freeze to the center somewhere along the pipe's length, forming a plug of ice at that point. If there is still liquid water between this plug and some other obstruction, such as another ice plug or a spigot, then no additional volume is available for further expansion and freezing. The pressure in the pipe builds and can rupture the pipe.

# 10.4 The Ideal Gas Law

The properties of gases are important in a number of thermodynamic processes. Our weather is a good example of the types of processes that depend on the behavior of gases.

If we introduce a gas into a container, it expands to fill the container uniformly, with its pressure depending on the size of the container, the temperature, and the amount of gas. A larger container results in a lower pressure, whereas higher temperatures or larger amounts of gas result in a higher pressure. The pressure $P$, volume $V$, temperature $T$, and amount $n$ of gas in a container are related to each other by an *equation of state*.

The equation of state can be very complicated, but is found experimentally to be relatively simple if the gas is maintained at a low pressure (or a low density). Such a low-density gas approximates what is called an **ideal gas.** Most gases at room temperature and atmospheric pressure behave approximately as ideal gases. **An ideal gas is a collection of atoms or molecules that move randomly and exert no long-range forces on each other. Each particle of the ideal gas is individually pointlike, occupying a negligible volume.**

A gas usually consists of a very large number of particles, so it's convenient to express the amount of gas in a given volume in terms of the number of **moles,** $n$. A mole is a number. The same number of particles is found in a mole of helium as in a mole of iron or aluminum. This number is known as *Avogadro's number* and is given by

$$N_A = 6.02 \times 10^{23} \text{ particles/mole}$$

◀ Avogadro's number

Avogadro's number and the definition of a mole are fundamental to chemistry and related branches of physics. The number of moles of a substance is related to its mass $m$ by the expression

$$n = \frac{m}{\text{molar mass}}$$ [10.7]

where the molar mass of the substance is defined as the mass of one mole of that substance, usually expressed in grams per mole.

There are lots of atoms in the world, so it's natural and convenient to choose a very large number like Avogadro's number when describing collections of atoms. At the same time, Avogadro's number must be special in some way because otherwise why not just count things in terms of some large power of ten, like $10^{24}$?

It turns out that Avogadro's number was chosen so that the mass in grams of one Avogadro's number of an element is numerically the same as the mass of one atom of the element, expressed in atomic mass units (u).

This relationship is very convenient. Looking at the periodic table of the elements in the back of the book, we find that carbon has an atomic mass of 12 u, so 12 g of carbon consists of exactly $6.02 \times 10^{23}$ atoms of carbon. The atomic mass of oxygen is 16 u, so in 16 g of oxygen there are again $6.02 \times 10^{23}$ atoms of oxygen. The same holds true for molecules: The molecular mass of molecular hydrogen, $H_2$, is 2 u, and there is an Avogadro's number of molecules in 2 g of molecular hydrogen.

The technical definition of a mole is as follows: **One mole (mol) of any substance is that amount of the substance that contains as many particles (atoms, molecules, or other particles) as there are atoms in 12 g of the isotope carbon-12.**

Taking carbon-12 as a test case, let's find the mass of an Avogadro's number of carbon-12 atoms. A carbon-12 atom has an atomic mass of 12 u, or 12 atomic mass units. One atomic mass unit is equal to $1.66 \times 10^{-24}$ g, about the same as the mass of a neutron or proton—particles that make up atomic nuclei. The mass $m$ of an Avogadro's number of carbon-12 atoms is then given by

$$m = N_A(12 \text{ u}) = 6.02 \times 10^{23}(12 \text{ u})\left(\frac{1.66 \times 10^{-24} \text{ g}}{\text{u}}\right) = 12.0 \text{ g}$$

So we see that Avogadro's number is deliberately chosen to be the inverse of the number of grams in an atomic mass unit. In this way the atomic mass of an atom expressed in atomic mass units is numerically the same as the mass of an Avogadro's number of that kind of atom expressed in grams. Because there are $6.02 \times 10^{23}$ particles in one mole of *any* element, the mass per atom for a given element is

$$m_{\text{atom}} = \frac{\text{molar mass}}{N_A}$$

For example, the mass of a helium atom is

$$m_{\text{He}} = \frac{4.00 \text{ g/mol}}{6.02 \times 10^{23} \text{ atoms/mol}} = 6.64 \times 10^{-24} \text{ g/atom}$$

Now suppose an ideal gas is confined to a cylindrical container with a volume that can be changed by moving a piston, as in Figure 10.13. Assume that the cylinder doesn't leak, so the number of moles remains constant. Experiments yield the following observations: First, when the gas is kept at a constant temperature,

**Figure 10.13** A gas confined to a cylinder whose volume can be varied with a movable piston.

**Tip 10.2** Only Kelvin Works!

Temperatures used in the ideal gas law must always be in Kelvins.

its pressure is inversely proportional to its volume (Boyle's law). Second, when the pressure of the gas is kept constant, the volume of the gas is directly proportional to the temperature (Charles' law). Third, when the volume of the gas is held constant, the pressure is directly proportional to the temperature (Gay-Lussac's law). These observations can be summarized by the following equation of state, known as the **ideal gas law:**

Equation of state for ▶
an ideal gas

$$PV = nRT \qquad\qquad [10.8]$$

In this equation, $R$ is a constant for a specific gas that must be determined from experiments, whereas $T$ is the temperature in kelvins. Each point on a $P$ versus $V$ diagram would represent a different state of the system. Experiments on several gases show that, as the pressure approaches zero, the quantity $PV/nT$ approaches the same value of $R$ for all gases. For this reason, $R$ is called the **universal gas constant.** In SI units, where pressure is expressed in pascals and volume in cubic meters,

The universal gas constant ▶

$$R = 8.31 \text{ J/mol} \cdot \text{K} \qquad\qquad [10.9]$$

**Tip 10.3** Standard Temperature and Pressure

Chemists often define standard temperature and pressure (STP) to be 20°C and 1.0 atm. We choose STP to be 0°C and 1.0 atm.

If the pressure is expressed in atmospheres and the volume is given in liters (recall that $1 \text{ L} = 10^3 \text{ cm}^3 = 10^{-3} \text{ m}^3$), then

$$R = 0.082\,1 \text{ L} \cdot \text{atm/mol} \cdot \text{K}$$

Using this value of $R$ and Equation 10.8, the volume occupied by 1 mol of any ideal gas at atmospheric pressure and at 0°C (273 K) is 22.4 L.

---

### EXAMPLE 10.6    AN EXPANDING GAS

**GOAL**  Use the ideal gas law to analyze a system of gas.

**PROBLEM**  An ideal gas at 20.0°C and a pressure of $1.50 \times 10^5$ Pa is in a container having a volume of 1.00 L. **(a)** Determine the number of moles of gas in the container. **(b)** The gas pushes against a piston, expanding to twice its original volume, while the pressure falls to atmospheric pressure. Find the final temperature.

**STRATEGY**  In part **(a)** solve the ideal gas equation of state for the number of moles, $n$, and substitute the known quantities. Be sure to convert the temperature from Celsius to Kelvin! When comparing two states of a gas as in part **(b)** it's often most convenient to divide the ideal gas equation of the final state by the equation of the initial state. Then quantities that don't change can immediately be canceled, simplifying the algebra.

. . . . . . . . . . . . . . . . . . . . . . . . . . . . . . . . . . . . . . . . . . . . . . . . . . . . . . . . . . . . . . . . . . . . . . . . . . . . . . . . . . . . . . . . . . . . . . . . . . . . . . . . . . . .

**SOLUTION**

**(a)** Find the number of moles of gas.

Convert the temperature to kelvins:

$$T = T_C + 273 = 20.0 + 273 = 293 \text{ K}$$

Solve the ideal gas law for $n$ and substitute:

$$PV = nRT$$

$$n = \frac{PV}{RT} = \frac{(1.50 \times 10^5 \text{ Pa})(1.00 \times 10^{-3} \text{ m}^3)}{(8.31 \text{ J/mol} \cdot \text{K})(293 \text{ K})}$$

$$= \boxed{6.16 \times 10^{-2} \text{ mol}}$$

**(b)** Find the temperature after the gas expands to 2.00 L.

Divide the ideal gas law for the final state by the ideal gas law for the initial state:

$$\frac{P_f V_f}{P_i V_i} = \frac{nRT_f}{nRT_i}$$

Cancel the number of moles $n$ and the gas constant $R$, and solve for $T_f$:

$$\frac{P_f V_f}{P_i V_i} = \frac{T_f}{T_i}$$

$$T_f = \frac{P_f V_f}{P_i V_i} T_i = \frac{(1.01 \times 10^5 \, \text{Pa})(2.00 \, \text{L})}{(1.50 \times 10^5 \, \text{Pa})(1.00 \, \text{L})} (293 \, \text{K})$$

$$= \boxed{395 \, \text{K}}$$

**REMARKS** Remember the trick used in part (b); it's often useful in ideal gas problems. Notice that it wasn't necessary to convert units from liters to cubic meters because the units were going to cancel anyway.

**QUESTION 10.6** Assuming constant temperature, does a helium balloon expand, contract, or remain at constant volume as it rises through the air?

**EXERCISE 10.6** Suppose the temperature of 4.50 L of ideal gas drops from 375 K to 275 K. **(a)** If the volume remains constant and the initial pressure is atmospheric pressure, find the final pressure. **(b)** Find the number of moles of gas.

**ANSWERS** **(a)** $7.41 \times 10^4$ Pa   **(b)** 0.146 mol

---

## EXAMPLE 10.7   MESSAGE IN A BOTTLE

**GOAL** Apply the ideal gas law in tandem with Newton's second law.

**PROBLEM** A beachcomber finds a corked bottle containing a message. The air in the bottle is at atmospheric pressure and a temperature of 30.0°C. The cork has a cross-sectional area of 2.30 cm². The beachcomber places the bottle over a fire, figuring the increased pressure will push out the cork. At a temperature of 99°C the cork is ejected from the bottle. **(a)** What was the pressure in the bottle just before the cork left it? **(b)** What force of friction held the cork in place? Neglect any change in volume of the bottle.

**STRATEGY** In part **(a)** the number of moles of air in the bottle remains the same as it warms over the fire. Take the ideal gas equation for the final state and divide by the ideal gas equation for the initial state. Solve for the final pressure. In part **(b)** there are three forces acting on the cork: a friction force, the exterior force of the atmosphere pushing in, and the force of the air inside the bottle pushing out. Apply Newton's second law. Just before the cork begins to move, the three forces are in equilibrium and the static friction force has its maximum value.

### SOLUTION

**(a)** Find the final pressure.

Divide the ideal gas law at the final point by the ideal gas law at the initial point:

$$(1) \qquad \frac{P_f V_f}{P_i V_i} = \frac{nRT_f}{nRT_i}$$

Cancel $n$, $R$, and $V$, which don't change, and solve for $P_f$:

$$\frac{P_f}{P_i} = \frac{T_f}{T_i} \quad \rightarrow \quad P_f = P_i \frac{T_f}{T_i}$$

Substitute known values, obtaining the final pressure:

$$P_f = (1.01 \times 10^5 \, \text{Pa}) \frac{372 \, \text{K}}{303 \, \text{K}} = \boxed{1.24 \times 10^5 \, \text{Pa}}$$

**(b)** Find the magnitude of the friction force acting on the cork.

Apply Newton's second law to the cork just before it leaves the bottle. $P_{\text{in}}$ is the pressure inside the bottle, and $P_{\text{out}}$ is the pressure outside.

$$\sum F = 0 \quad \rightarrow \quad P_{\text{in}} A - P_{\text{out}} A - F_{\text{friction}} = 0$$

$$F_{\text{friction}} = P_{\text{in}} A - P_{\text{out}} A = (P_{\text{in}} - P_{\text{out}}) A$$

$$= (1.24 \times 10^5 \, \text{Pa} - 1.01 \times 10^5 \, \text{Pa})(2.30 \times 10^{-4} \, \text{m}^2)$$

$$F_{\text{friction}} = \boxed{5.29 \, \text{N}}$$

**REMARKS** Notice the use, once again, of the ideal gas law in Equation (1). Whenever comparing the state of a gas at two different points, this is the best way to do the math. One other point: Heating the gas blasted the cork out of the bottle, which meant the gas did work on the cork. The work done by an expanding gas—driving pistons and generators—is one of the foundations of modern technology and will be studied extensively in Topic 12.

**QUESTION 10.7** As the cork begins to move, what happens to the pressure inside the bottle?

*(Continued)*

**EXERCISE 10.7** A tire contains air and a gauge pressure of $5.00 \times 10^4$ Pa and a temperature of 30.0°C. After nightfall, the temperature drops to $-10.0$°C. Find the new gauge pressure in the tire. (Recall that gauge pressure is absolute pressure minus atmospheric pressure. Assume constant volume.)

**ANSWER** $3.01 \times 10^4$ Pa

---

## EXAMPLE 10.8  SUBMERGING A BALLOON

**GOAL** Combine the ideal gas law with the equation of hydrostatic equilibrium and buoyancy.

**PROBLEM** A sturdy balloon with volume 0.500 m³ is attached to a $2.50 \times 10^2$-kg iron weight and tossed overboard into a freshwater lake. The balloon is made of a light material of negligible mass and elasticity (although it can be compressed). The air in the balloon is initially at atmospheric pressure. The system fails to sink and there are no more weights, so a skin diver decides to drag it deep enough so that the balloon will remain submerged. **(a)** Find the volume of the balloon at the point where the system will remain submerged, in equilibrium. **(b)** What's the balloon's pressure at that point? **(c)** Assuming constant temperature, to what minimum depth must the balloon be dragged?

**STRATEGY** As the balloon and weight are dragged deeper into the lake, the air in the balloon is compressed and the volume is reduced along with the buoyancy. At some depth $h$ the total buoyant force acting on the balloon and weight, $B_{bal} + B_{Fe}$, will equal the total weight, $w_{bal} + w_{Fe}$, and the balloon will remain at that depth. Substitute these forces into Newton's second law and solve for the unknown volume of the balloon, answering part **(a)**. Then use the ideal gas law to find the pressure, and the equation of hydrostatic equilibrium to find the depth.

· · · · · · · · · · · · · · · · · · · · · · · · · · · · · · · · · · · · · · · · · · · · · · · · · · · · · · · · · · · · · · · · · · · · · · ·

**SOLUTION**

**(a)** Find the volume of the balloon at the equilibrium point.

Find the volume of the iron, $V_{Fe}$:

$$V_{Fe} = \frac{m_{Fe}}{\rho_{Fe}} = \frac{2.50 \times 10^2 \text{ kg}}{7.86 \times 10^3 \text{ kg/m}^3} = 0.031\ 8 \text{ m}^3$$

Find the mass of the balloon, which is equal to the mass of the air if we neglect the mass of the balloon's material:

$$m_{bal} = \rho_{air} V_{bal} = (1.29 \text{ kg/m}^3)(0.500 \text{ m}^3) = 0.645 \text{ kg}$$

Apply Newton's second law to the system when it's in equilibrium:

$$B_{Fe} - w_{Fe} + B_{bal} - w_{bal} = 0$$

Substitute the appropriate expression for each term:

$$\rho_{wat} V_{Fe}\, g - m_{Fe}\, g + \rho_{wat} V_{bal}\, g - m_{bal}\, g = 0$$

Cancel the $g$'s and solve for the volume of the balloon, $V_{bal}$:

$$V_{bal} = \frac{m_{bal} + m_{Fe} - \rho_{wat} V_{Fe}}{\rho_{wat}}$$

$$= \frac{0.645 \text{ kg} + 2.50 \times 10^2 \text{ kg} - (1.00 \times 10^3 \text{ kg/m}^3)(0.031\ 8 \text{ m}^3)}{1.00 \times 10^3 \text{kg/m}^3}$$

$$V_{bal} = \boxed{0.219 \text{ m}^3}$$

**(b)** What's the balloon's pressure at the equilibrium point?

Now use the ideal gas law to find the pressure, assuming constant temperature, so that $T_i = T_f$.

$$\frac{P_f V_f}{P_i V_i} = \frac{nRT_f}{nRT_i} = 1$$

$$P_f = \frac{V_i}{V_f} P_i = \frac{0.500 \text{ m}^3}{0.219 \text{ m}^3}(1.01 \times 10^5 \text{ Pa})$$

$$= \boxed{2.31 \times 10^5 \text{ Pa}}$$

**(c)** To what minimum depth must the balloon be dragged?

Use the equation of hydrostatic equilibrium to find the depth:

$$P_f = P_{atm} + \rho g h$$

$$h = \frac{P_f - P_{atm}}{\rho g} = \frac{2.31 \times 10^5 \text{ Pa} - 1.01 \times 10^5 \text{ Pa}}{(1.00 \times 10^3 \text{ kg/m}^3)(9.80 \text{ m/s}^2)}$$

$$= \boxed{13.3 \text{ m}}$$

**REMARKS** Once again, the ideal gas law was used to good effect. This problem shows how even answering a fairly simple question can require the application of several different physical concepts: density, buoyancy, the ideal gas law, and hydrostatic equilibrium.

**QUESTION 10.8** If a glass is turned upside down and then submerged in water, what happens to the volume of the trapped air as the glass is pushed deeper under water?

**EXERCISE 10.8** A boy takes a 30.0-cm³ balloon holding air at 1.00 atm at the surface of a freshwater lake down to a depth of 4.00 m. Find the volume of the balloon at this depth. Assume the balloon is made of light material of little elasticity (although it can be compressed) and the temperature of the trapped air remains constant.

**ANSWER** 21.6 cm³

---

As previously stated, the number of molecules contained in one mole of any gas is Avogadro's number, $N_A = 6.02 \times 10^{23}$ particles/mol, so

$$n = \frac{N}{N_A} \qquad [10.10]$$

where $n$ is the number of moles and $N$ is the number of molecules in the gas. With Equation 10.10, we can rewrite the ideal gas law in terms of the total number of molecules as

$$PV = nRT = \frac{N}{N_A} RT$$

or

$$PV = Nk_B T \qquad [10.11] \qquad \blacktriangleleft \text{Ideal gas law}$$

where

$$k_B = \frac{R}{N_A} = 1.38 \times 10^{-23} \, \text{J/K} \qquad [10.12] \qquad \blacktriangleleft \text{Boltzmann's constant}$$

is **Boltzmann's constant.** This reformulation of the ideal gas law is used in the next section to relate the temperature of a gas to the average kinetic energy of particles in the gas.

# 10.5 The Kinetic Theory of Gases

In Section 10.4 we discussed the macroscopic properties of an ideal gas, including pressure, volume, number of moles, and temperature. In this section we consider the ideal gas model from the microscopic point of view. We show that the macroscopic properties can be understood on the basis of what is happening on the atomic scale. In addition, we reexamine the ideal gas law in terms of the behavior of the individual molecules that make up the gas.

Using the model of an ideal gas, we describe the **kinetic theory of gases.** With this theory, we can interpret the pressure and temperature of an ideal gas in terms of microscopic variables. The kinetic theory of gases model makes the following assumptions:

1. **The number of molecules in the gas is large, and the average separation between them is large compared with their dimensions.** Because the number of molecules is large, we can analyze their behavior statistically. The large separation between molecules means that the molecules occupy a negligible volume in the container. This assumption is consistent with the ideal gas model, in which we imagine the molecules to be pointlike.

$\blacktriangleleft$ Assumptions of kinetic theory for an ideal gas

2. **The molecules obey Newton's laws of motion, but as a whole, they move randomly.** By "randomly," we mean that any molecule can move in any direction with equal probability, with a wide distribution of speeds.
3. **The molecules interact only through short-range forces during elastic collisions.** This assumption is consistent with the ideal gas model, in which the molecules exert no long-range forces on each other.
4. **The molecules make elastic collisions with the walls.**
5. **All molecules in the gas are identical.**

Although we often picture an ideal gas as consisting of single atoms, *molecular* gases exhibit ideal behavior at low pressures. On average, effects associated with molecular structure have no effect on the motions considered, so we can apply the results of the following development to both molecular gases and monatomic gases.

## 10.5.1 Molecular Model for the Pressure of an Ideal Gas

As a first application of kinetic theory, we derive an expression for the pressure of an ideal gas in a container in terms of microscopic quantities. The pressure of the gas is the result of collisions between the gas molecules and the walls of the container. During these collisions, the gas molecules undergo a change of momentum as a result of the force exerted on them by the walls.

We now derive an expression for the pressure of an ideal gas consisting of $N$ molecules in a container of volume $V$. In this section we use $m$ to represent the mass of one molecule. The container is a cube with edges of length $d$ (Fig. 10.14). Consider the collision of one molecule moving with a velocity $-v_x$ toward the left-hand face of the box (Fig. 10.15). After colliding elastically with the wall, the molecule moves in the positive $x$-direction with a velocity $+v_x$. Because the momentum of the molecule is $-mv_x$ before the collision and $+mv_x$ afterward, the change in its momentum is

$$\Delta p_x = mv_x - (-mv_x) = 2mv_x$$

If $F_1$ is the magnitude of the average force exerted by a molecule on the *wall* in the time $\Delta t$, then applying Newton's second law to the wall gives

$$F_1 = \frac{\Delta p_x}{\Delta t} = \frac{2mv_x}{\Delta t}$$

For the molecule to make two collisions with the same wall, it must travel a distance $2d$ along the $x$-direction in a time $\Delta t$. Therefore, the time interval between two collisions with the same wall is $\Delta t = 2d/v_x$, and the force imparted to the wall by a single molecule is

$$F_1 = \frac{2mv_x}{\Delta t} = \frac{2mv_x}{2d/v_x} = \frac{mv_x^2}{d}$$

The total force $F$ exerted by all the molecules on the wall is found by adding the forces exerted by the individual molecules:

$$F = \frac{m}{d}\left(v_{1x}^2 + v_{2x}^2 + \cdots\right)$$

In this equation $v_{1x}$ is the $x$-component of velocity of molecule 1, $v_{2x}$ is the $x$-component of velocity of molecule 2, and so on. The summation terminates when we reach $N$ molecules because there are $N$ molecules in the container.

Note that the average value of the square of the velocity in the $x$-direction for $N$ molecules is

$$\overline{v_x^2} = \frac{v_{1x}^2 + v_{2x}^2 + \cdots + v_{Nx}^2}{N}$$

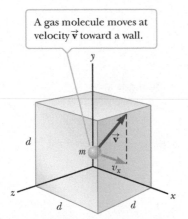

A gas molecule moves at velocity $\vec{v}$ toward a wall.

**Figure 10.14** A cubical box with sides of length $d$ containing an ideal gas.

Before collision

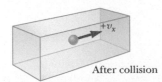

After collision

**Figure 10.15** A molecule moving along the $x$-axis in a container collides elastically with a wall, reversing its momentum and exerting a force on the wall.

where $\overline{v_x^2}$ is the average value of $v_x^2$. The total force on the wall can then be written

$$F = \frac{Nm}{d}\,\overline{v_x^2}$$

Now we focus on one molecule in the container traveling in some arbitrary direction with velocity $\vec{v}$ and having components $v_x$, $v_y$, and $v_z$. In this case we must express the total force on the wall in terms of the speed of the molecules rather than just a single component. The Pythagorean theorem relates the square of the speed to the square of these components according to the expression $v^2 = v_x^2 + v_y^2 + v_z^2$. Hence, the average value of $v^2$ for all the molecules in the container is related to the average values $\overline{v_x^2}$, $\overline{v_y^2}$, and $\overline{v_z^2}$ according to the expression $\overline{v^2} = \overline{v_x^2} + \overline{v_y^2} + \overline{v_z^2}$. Because the motion is completely random, the average values $\overline{v_x^2}$, $\overline{v_y^2}$, and $\overline{v_z^2}$ are equal to each other. Using this fact and the earlier equation for $\overline{v_x^2}$, we find that

$$\overline{v_x^2} = \tfrac{1}{3}\,\overline{v^2}$$

The total force on the wall, then, is

$$F = \frac{N}{3}\left(\frac{m\overline{v^2}}{d}\right)$$

This expression allows us to find the total pressure exerted on the wall by dividing the force by the area:

$$P = \frac{F}{A} = \frac{F}{d^2} = \frac{1}{3}\left(\frac{N}{d^3}\,m\overline{v^2}\right) = \frac{1}{3}\left(\frac{N}{V}\right)m\overline{v^2}$$

$$P = \frac{2}{3}\left(\frac{N}{V}\right)\left(\tfrac{1}{2}m\overline{v^2}\right) \qquad [10.13]$$

◀ Pressure of an ideal gas

**Figure 10.16** The glass vessel contains dry ice (solid carbon dioxide). Carbon dioxide gas is denser than air, hence, it falls when poured from the cylinder. The gas is colorless but is made visible by the formation of tiny ice crystals from water vapor.

Equation 10.13 says that **the pressure is proportional to the number of molecules per unit volume and to the average translational kinetic energy of a molecule,** $\tfrac{1}{2}m\overline{v^2}$. With this simplified model of an ideal gas, we have arrived at an important result that relates the large-scale quantity of pressure to an atomic quantity: the average value of the square of the molecular speed. This relationship provides a key link between the atomic world and the large-scale world (Fig. 10.16).

Equation 10.13 captures some familiar features of pressure. One way to increase the pressure inside a container is to increase the number of molecules per unit volume in the container. You do this when you add air to a tire. The pressure in the tire can also be increased by increasing the average translational kinetic energy of the molecules in the tire. As we will see shortly, this can be accomplished by increasing the temperature of the gas inside the tire. That's why the pressure inside a tire increases as the tire warms up during long trips. The continuous flexing of the tires as they move along the road transfers energy to the air inside them, increasing the air's temperature, which in turn raises the pressure.

## EXAMPLE 10.9  HIGH-ENERGY ELECTRON BEAM

**GOAL** Calculate the pressure of an electron particle beam.

**PROBLEM** A beam of electrons moving in the positive $x$-direction impacts a target in a vacuum chamber. **(a)** If $1.25 \times 10^{14}$ electrons traveling at a speed of $3.00 \times 10^7$ m/s strike the target during each brief pulse lasting $5.00 \times 10^{-8}$ s, what average force is exerted on the target during the pulse? Assume all the electrons penetrate the target and are absorbed. **(b)** What average pressure is exerted on the beam spot, which has radius 4.00 mm? *Note*: The beam spot is the region of the target struck by the beam.

**STRATEGY** The average force exerted by the target on an electron is the change in electron's momentum divided by the time required to bring the electron to rest. By the third law, an equal and opposite force is exerted on the target. During the pulse, $N$ such collisions take place in a total time $\Delta t$, so multiplying the negative of a single electron's change in momentum by $N$ and

*(Continued)*

dividing by the pulse duration $\Delta t$ gives the average force exerted on the target during the pulse. Dividing that force by the area of the beam spot yields the average pressure on the beam spot.

## SOLUTION

**(a)** The force on the target is equal to the negative of the change in momentum of each electron multiplied by the number $N$ of electrons and divided by the pulse duration:

$$F = -\frac{N\Delta p}{\Delta t}$$

Substitute the expression $\Delta p = mv_f - mv_i$ and note that $v_f = 0$ by assumption:

$$F = -\frac{N(mv_f - mv_i)}{\Delta t} = -\frac{Nm(0 - v_i)}{\Delta t}$$

Substitute values:

$$F = -\frac{(1.25 \times 10^{14})(9.11 \times 10^{-31}\text{ kg})(0 - 3.00 \times 10^7\text{ m/s})}{(5.00 \times 10^{-8}\text{ s})}$$

$$= \boxed{0.068\ 3\text{ N}}$$

**(b)** Calculate the pressure of the beam.

Use the definition of average pressure, the force divided by area:

$$P = \frac{F}{A} = \frac{F}{\pi r^2} = \frac{0.068\ 3\text{ N}}{\pi (0.004\ 00\text{ m})^2}$$

$$= \boxed{1.36 \times 10^3\text{ Pa}}$$

**REMARKS** High-energy electron beams can be used for welding and shock strengthening of materials. Relativistic effects (see Topic 26) were neglected in this calculation and would be relatively small in any case at a tenth the speed of light. This example illustrates how numerous collisions by atomic or, in this case, subatomic particles can result in macroscopic physical effects such as forces and pressures.

**QUESTION 10.9** If the same beam were directed at a material that reflected all the electrons, how would the final pressure be affected?

**EXERCISE 10.9** A beam of protons traveling at $2.00 \times 10^6$ m/s strikes a target during a brief pulse that lasts $7.40 \times 10^{-9}$ s. **(a)** If there are $4.00 \times 10^9$ protons in the beam and all are assumed to be reflected elastically, what force is exerted on the target? **(b)** What average pressure is exerted on the beam spot, which has radius of 2.00 mm?

**ANSWERS** **(a)** 0.003 61 N   **(b)** 287 Pa

## 10.5.2 Molecular Interpretation of Temperature

Having related the pressure of a gas to the average kinetic energy of the gas molecules, we now relate temperature to a microscopic description of the gas. We can obtain some insight into the meaning of temperature by multiplying Equation 10.13 by the volume:

$$PV = \tfrac{2}{3} N\left(\tfrac{1}{2}m\overline{v^2}\right)$$

Comparing this equation with the equation of state for an ideal gas in the form of Equation 10.11, $PV = Nk_\text{B}T$, we note that the left-hand sides of the two equations are identical. Equating the right-hand sides, we obtain

◀ Temperature is proportional to the average kinetic energy

$$T = \frac{2}{3k_\text{B}}\left(\tfrac{1}{2}m\overline{v^2}\right) \qquad\qquad [10.14]$$

This means that **the temperature of a gas is a direct measure of the average molecular kinetic energy of the gas.** As the temperature of a gas increases, the molecules move with higher average kinetic energy.

Rearranging Equation 10.14, we can relate the translational molecular kinetic energy to the temperature:

◀ Average kinetic energy per molecule

$$\tfrac{1}{2}m\overline{v^2} = \tfrac{3}{2}k_\text{B}T \qquad\qquad [10.15]$$

So the average translational kinetic energy per molecule is $\frac{3}{2}k_BT$. The total translational kinetic energy of $N$ molecules of gas is simply $N$ times the average energy per molecule,

$$KE_{total} = N\left(\tfrac{1}{2}m\overline{v^2}\right) = \tfrac{3}{2}Nk_BT = \tfrac{3}{2}nRT \qquad [10.16]$$

◄ Total kinetic energy of $N$ molecules

where we have used $k_B = R/N_A$ for Boltzmann's constant and $n = N/N_A$ for the number of moles of gas. From this result, we see that **the total translational kinetic energy of a system of molecules is proportional to the absolute temperature of the system.**

For a monatomic gas, translational kinetic energy is the only type of energy the molecules can have, so Equation 10.16 gives the **internal energy $U$ for a monatomic gas:**

$$U = \tfrac{3}{2}nRT \qquad \text{(monatomic gas)} \qquad [10.17]$$

For diatomic and polyatomic molecules, additional possibilities for energy storage are available in the vibration and rotation of the molecule.

The square root of $\overline{v^2}$ is called the **root-mean-square (rms) speed** of the molecules. From Equation 10.15, we get, for the rms speed,

$$v_{rms} = \sqrt{\overline{v^2}} = \sqrt{\frac{3k_BT}{m}} = \sqrt{\frac{3RT}{M}} \qquad [10.18]$$

◄ Root-mean-square speed

where $M$ is the molar mass in *kilograms per mole*, if $R$ is given in SI units. Equation 10.18 shows that, at a given temperature, lighter molecules tend to move faster than heavier molecules. For example, if gas in a vessel consists of a mixture of hydrogen and oxygen, the hydrogen ($H_2$) molecules, with a molar mass of $2.0 \times 10^{-3}$ kg/mol, move four times faster than the oxygen ($O_2$) molecules, with molar mass $32 \times 10^{-3}$ kg/mol. If we calculate the rms speed for hydrogen at room temperature ($\sim$300 K), we find

**Tip 10.4** Kilograms Per Mole, Not Grams Per Mole

In the equation for the rms speed, the units of molar mass $M$ must be consistent with the units of the gas constant $R$. In particular, if $R$ is in SI units, $M$ must be expressed in kilograms per mole, not grams per mole.

$$v_{rms} = \sqrt{\frac{3RT}{M}} = \sqrt{\frac{3(8.31 \text{ J/mol}\cdot\text{K})(300 \text{ K})}{2.0\times10^{-3}\text{kg/mol}}} = 1.9\times10^3 \text{ m/s}$$

This speed is about 17% of the escape speed for Earth, as calculated in Topic 7. Because it is an average speed, a large number of molecules have much higher speeds and can therefore escape from Earth's atmosphere. This is why Earth's atmosphere doesn't currently contain hydrogen: it has all bled off into space.

Table 10.2 lists the rms speeds for various molecules at 20°C. A system of gas at a given temperature will exhibit a variety of speeds. This distribution of speeds is known as the *Maxwell velocity distribution*. An example of such a distribution for nitrogen gas at two different temperatures is given in Figure 10.17. The horizontal axis is speed, and the vertical axis is the number of molecules per unit speed. Notice that three speeds are of special interest: the most probable speed, corresponding to the peak in the graph; the average speed, which is found by averaging over all the possible speeds; and the rms speed. For every gas, note that $v_{mp} < v_{av} < v_{rms}$. As the temperature rises, these three speeds shift to the right.

**Table 10.2** Some rms Speeds

| Gas | Molar Mass (kg/mol) | $v_{rms}$ at 20°C (m/s) |
|---|---|---|
| $H_2$ | $2.02\times10^{-3}$ | 1 902 |
| He | $4.0\times10^{-3}$ | 1 352 |
| $H_2O$ | $18\times10^{-3}$ | 637 |
| Ne | $20.2\times10^{-3}$ | 602 |
| $N_2$ and CO | $28.0\times10^{-3}$ | 511 |
| NO | $30.0\times10^{-3}$ | 494 |
| $O_2$ | $32.0\times10^{-3}$ | 478 |
| $CO_2$ | $44.0\times10^{-3}$ | 408 |
| $SO_2$ | $64.1\times10^{-3}$ | 338 |

*Quick Quiz*

**10.6** One container is filled with argon gas and another with helium gas. Both containers are at the same temperature. Which atoms have the higher rms speed? (a) argon, (b) helium, (c) they have the same speed, or (d) not enough information to say.

**Figure 10.17** The Maxwell speed distribution for $10^5$ nitrogen molecules at 300 K and 900 K.

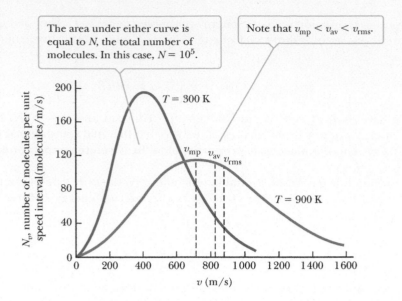

The area under either curve is equal to $N$, the total number of molecules. In this case, $N = 10^5$.

Note that $v_{mp} < v_{av} < v_{rms}$.

## APPLYING PHYSICS 10.2   EXPANSION AND TEMPERATURE

Imagine a gas in an insulated cylinder with a movable piston. The piston has been pushed inward, compressing the gas, and is now released. As the molecules of the gas strike the piston, they move it outward. Explain, from the point of view of the kinetic theory, how the expansion of this gas causes its temperature to drop.

**EXPLANATION** From the point of view of kinetic theory, a molecule colliding with the piston causes the piston to move

with some velocity. According to the conservation of momentum, the molecule must rebound with less speed than it had before the collision. As these collisions occur, the average speed of the collection of molecules is therefore reduced. Because temperature is related to the average speed of the molecules, the temperature of the gas drops. ■

## EXAMPLE 10.10   A CYLINDER OF HELIUM

**GOAL** Calculate the internal energy of a system and the average kinetic energy per molecule.

**PROBLEM** A cylinder contains 2.00 mol of helium gas at 20.0°C. Assume the helium behaves like an ideal gas. **(a)** Find the total internal energy of the system. **(b)** What is the average kinetic energy per molecule? **(c)** How much energy would have to be added to the system to double the rms speed? The molar mass of helium is equal to $4.00 \times 10^{-3}$ kg/mol.

**STRATEGY** This problem requires substitution of given information into the appropriate equations: Equation 10.17 for part **(a)** and Equation 10.15 for part **(b)**. In part **(c)** use the equations for the rms speed and internal energy together. A *change* in the internal energy must be computed.

. . . . . . . . . . . . . . . . . . . . . . . . . . . . . . . . . . . . . . . . . . . . . . . . . . . . . . . . . . . . . . . . . . . . . . . . . . . . . . . . . . . . . . . .

**SOLUTION**

**(a)** Find the total internal energy of the system.

Substitute values into Equation 10.17 with $n = 2.00$ and $T = 293$ K:

$$U = \tfrac{3}{2}(2.00 \text{ mol})(8.31 \text{ J/mol} \cdot \text{K})(293 \text{ K}) = \boxed{7.30 \times 10^3 \text{ J}}$$

**(b)** What is the average kinetic energy per molecule?

Substitute given values into Equation 10.15:

$$\tfrac{1}{2}m\overline{v^2} = \tfrac{3}{2}k_B T = \tfrac{3}{2}(1.38 \times 10^{-23} \text{ J/K})(293 \text{ K})$$
$$= \boxed{6.07 \times 10^{-21} \text{ J}}$$

**(c)** How much energy must be added to double the rms speed?

From Equation 10.18, doubling the rms speed requires quadrupling $T$. Calculate the required change of internal energy, which is the energy that must be put into the system:

$$\Delta U = U_f - U_i = \tfrac{3}{2}nRT_f - \tfrac{3}{2}nRT_i = \tfrac{3}{2}nR(T_f - T_i)$$
$$\Delta U = \tfrac{3}{2}(2.00 \text{ mol})(8.31 \text{ J/mol} \cdot \text{K})[(4.00 \times 293 \text{ K}) - 293 \text{ K}]$$
$$= \boxed{2.19 \times 10^4 \text{ J}}$$

**REMARKS** Computing changes in internal energy will be important in understanding engine cycles in Topic 12.

**QUESTION 10.10** True or False: At the same temperature, 1 mole of helium gas has the same internal energy as 1 mole of argon gas.

**EXERCISE 10.10** The temperature of 5.00 moles of argon gas is lowered from $3.00 \times 10^2$ K to $2.40 \times 10^2$ K. **(a)** Find the change in the internal energy, $\Delta U$, of the gas. **(b)** Find the change in the average kinetic energy per atom.

**ANSWERS** **(a)** $\Delta U = -3.74 \times 10^3$ J    **(b)** $-1.24 \times 10^{-21}$ J

---

# SUMMARY

## 10.1 Temperature and the Zeroth Law of Thermodynamics

Two systems are in **thermal contact** if energy can be exchanged between them, and in **thermal equilibrium** if they're in contact and there is no net exchange of energy. The exchange of energy between two objects because of differences in their temperatures is called **heat.**

The **zeroth law of thermodynamics** states that if two objects A and B are separately in thermal equilibrium with a third object, then A and B are in thermal equilibrium with each other. Equivalently, if the third object is a **thermometer,** then the **temperature** it measures for A and B will be the same. Two objects in thermal equilibrium are at the same temperature.

## 10.2 Thermometers and Temperature Scales

Thermometers measure temperature and are based on physical properties, such as the temperature-dependent expansion or contraction of a solid, liquid, or gas. These changes in volume are related to a linear scale, the most common being the **Fahrenheit, Celsius,** and **Kelvin scales.** The Kelvin temperature scale takes its zero point as **absolute zero** ($0$ K $= -273.15°C$), the point at which, by extrapolation, the pressure of all gases falls to zero.

The relationship between the Celsius temperature $T_C$ and the Kelvin (absolute) temperature $T$ is

$$T_C = T - 273.15 \qquad [10.1]$$

The relationship between the Fahrenheit and Celsius temperatures is

$$T_F = \tfrac{9}{5}T_C + 32 \qquad [10.2a]$$

## 10.3 Thermal Expansion of Solids and Liquids

Ordinarily a substance expands when heated. If an object has an initial length $L_0$ at some temperature and undergoes a change in temperature $\Delta T$, its linear dimension changes by the amount $\Delta L$, which is proportional to the object's initial length and the temperature change:

$$\Delta L = \alpha L_0 \, \Delta T \qquad [10.4]$$

The parameter $\alpha$ is called the **coefficient of linear expansion.** The change in area of a substance with change in temperature is given by

$$\Delta A = \gamma A_0 \, \Delta T \qquad [10.5]$$

where $\gamma = 2\alpha$ is the **coefficient of area expansion.** Similarly, the change in volume with temperature of most substances is proportional to the initial volume $V_0$ and the temperature change $\Delta T$:

$$\Delta V = \beta V_0 \, \Delta T \qquad [10.6]$$

where $\beta = 3\alpha$ is the **coefficient of volume expansion.**

The expansion and contraction of material due to changes in temperature create stresses and strains, sometimes sufficient to cause fracturing.

## 10.4 The Ideal Gas Law

Avogadro's number is $N_A = 6.02 \times 10^{23}$ particles/mol. A mole of anything, by definition, consists of an Avogadro's number of particles. The number is defined so that one mole of carbon-12 atoms has a mass of exactly 12 g. The mass of one mole of a pure substance in grams is the same, numerically, as that substance's atomic (or molecular) mass.

An **ideal gas** obeys the equation

$$PV = nRT \qquad [10.8]$$

where $P$ is the pressure of the gas, $V$ is its volume, $n$ is the number of moles of gas, $R$ is the universal gas constant ($8.31$ J/mol $\cdot$ K), and $T$ is the absolute temperature in kelvins. A real gas at very low pressures behaves approximately as an ideal gas.

Solving problems usually entails comparing two different states of the same system of gas, dividing the ideal gas equation for the final state by the ideal gas equation for the initial state, canceling factors that don't change, and solving for the unknown quantity.

## 10.5 The Kinetic Theory of Gases

The **pressure** of $N$ molecules of an ideal gas contained in a volume $V$ is given by

$$P = \frac{2}{3}\left(\frac{N}{V}\right)\left(\tfrac{1}{2}\,m\overline{v^2}\right) \qquad [10.13]$$

where $\tfrac{1}{2}m\overline{v^2}$ is the **average kinetic energy per molecule.**

The average kinetic energy of the molecules of a gas is directly proportional to the absolute temperature of the gas:

$$\tfrac{1}{2}m\overline{v^2} = \tfrac{3}{2}k_B T \qquad [10.15]$$

The quantity $k_B$ is **Boltzmann's constant** $(1.38 \times 10^{-23}\,\text{J/K})$. The internal energy of $n$ moles of a monatomic ideal gas is

$$U = \tfrac{3}{2}nRT \qquad\qquad [10.17]$$

The **root-mean-square (rms) speed** of the molecules of a gas is

$$v_{\text{rms}} = \sqrt{\frac{3k_B T}{m}} = \sqrt{\frac{3RT}{M}} \qquad [10.18]$$

## CONCEPTUAL QUESTIONS

1. (a) Why does an ordinary glass dish usually break when placed on a hot stove? (b) Dishes made of Pyrex glass don't break as easily. What characteristic of Pyrex prevents breakage?

2. A sealed container contains a fixed volume of a monatomic ideal gas. If the gas temperature is increased by a factor of two, what is the ratio of the final to the initial (a) pressure, (b) average molecular kinetic energy, (c) root-mean-square speed, and (d) internal energy.

3. Some thermometers are made of a mercury column in a glass tube. Based on the operation of these common thermometers, which has the larger coefficient of linear expansion, glass or mercury? (Don't answer this question by looking in a table.)

4. A rubber balloon is blown up and the end tied. Is the pressure inside the balloon greater than, less than, or equal to the ambient atmospheric pressure? Explain.

5. Objects deep beneath the surface of the ocean are subjected to extremely high pressures, as we saw in Topic 9. Some bacteria in these environments have adapted to pressures as much as a thousand times atmospheric pressure. How might such bacteria be affected if they were rapidly moved to the surface of the ocean?

6. A container filled with an ideal gas is connected to a reservoir of the same gas so that the number of moles in the container can change. If the pressure and volume of the container are each doubled while the temperature is held constant, what is the ratio of the final to the initial number of moles in the container?

7. Why do vapor bubbles in a pot of boiling water get larger as they approach the surface?

8. Markings to indicate length are placed on a steel tape in a room that is at a temperature of 22°C. Measurements are then made with the same tape on a day when the temperature is 27°C. Are the measurements too long, too short, or accurate?

9. Figure CQ10.9 shows Maxwell speed distributions for three different samples of oxygen ($O_2$) gas. (a) Is the temperature of sample B greater than, less than, or equal to the temperature of sample A? (b) Is the temperature of sample C greater than, less than, or equal to the temperature of sample A?

10. The air we breathe is largely composed of nitrogen ($N_2$) and oxygen ($O_2$) molecules. The mass of an $N_2$ molecule is less than the mass of an $O_2$ molecule. (a) For air at 300 K, is the average kinetic energy of an $N_2$ molecule greater than, less than, or equal to the average kinetic energy of an $O_2$ molecule? (b) Is the rms speed in air of an $N_2$ molecule greater than, less than, or equal to the rms speed in air of an $O_2$ molecule?

11. Metal lids on glass jars can often be loosened by running hot water over them. Why does that work?

12. Suppose the volume of an ideal gas is doubled while the pressure is reduced by half. Does the internal energy of the gas increase, decrease, or remain the same? Explain.

13. An automobile radiator is filled to the brim with water when the engine is cool. What happens to the water when the engine is running and the water has been raised to a high temperature?

14. Figure CQ10.14 shows a metal washer being heated by a Bunsen burner. The red arrows in options **a**, **b**, and **c** indicate the possible directions of expansion caused by the heating. Which option correctly illustrates the washer's expansion?

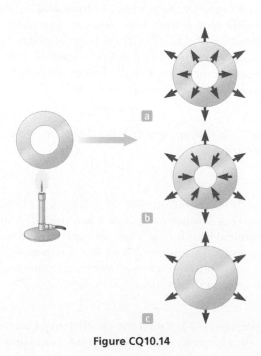

**Figure CQ10.14**

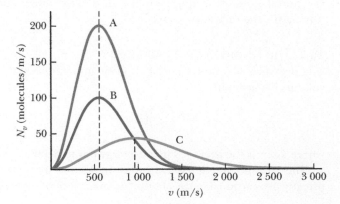

**Figure CQ10.9**

# PROBLEMS

*See the Preface for an explanation of the icons used in this problems set.*

## 10.1 Temperature and the Zeroth Law of Thermodynamics

## 10.2 Thermometers and Temperature Scales

1. For each of the following temperatures, find the equivalent temperature on the indicated scale: (a) $-273.15°C$ on the Fahrenheit scale, (b) $98.6°F$ on the Celsius scale, and (c) $1.00 \times 10^2$ K on the Fahrenheit scale.

2. The pressure in a constant-volume gas thermometer is 0.700 atm at $1.00 \times 10^{2}°C$ and 0.512 atm at $0°C$. (a) What is the temperature when the pressure is 0.040 0 atm? (b) What is the pressure at $450°C$?

3. The boiling point of liquid hydrogen is 20.3 K at atmospheric pressure. What is this temperature on (a) the Celsius scale and (b) the Fahrenheit scale?

4. **V** Death Valley holds the record for the highest recorded temperature in the United States. On July 10, 1913, at a place called Furnace Creek Ranch, the temperature rose to $134°F$. The lowest U.S. temperature ever recorded occurred at Prospect Creek Camp in Alaska on January 23, 1971, when the temperature plummeted to $-79.8°F$. (a) Convert these temperatures to the Celsius scale. (b) Convert the Celsius temperatures to Kelvin.

5. On January 22, 1943, in Spearfish, South Dakota, the temperature rose from $-4.00°F$ to $45.0°F$ over the course of two minutes (the current world record for the fastest recorded temperature change). By how much did the temperature change on the Kelvin scale?

6. **T** In a student experiment, a constant-volume gas thermometer is calibrated in dry ice ($-78.5°C$) and in boiling ethyl alcohol ($78.0°C$). The separate pressures are 0.900 atm and 1.635 atm. (a) What value of absolute zero in degrees Celsius does the calibration yield? What pressures would be found at (b) the freezing and (c) boiling points of water? *Hint*: Use the linear relationship $P = A + BT$, where $A$ and $B$ are constants.

7. **BIO** A person's body temperature is $101.6°F$, indicating a fever of $3.0°F$ above the normal average body temperature of $98.6°F$. How many degrees above normal is this body temperature on the Celsius scale?

8. The temperature difference between the inside and the outside of a home on a cold winter day is $57.0°F$. Express this difference on (a) the Celsius scale and (b) the Kelvin scale.

9. **Q|C** A nurse measures the temperature of a patient to be $41.5°C$. (a) What is this temperature on the Fahrenheit scale? (b) Do you think the patient is seriously ill? Explain.

10. **S** Temperature differences on the Rankine scale are identical to differences on the Fahrenheit scale, but absolute zero is given as $0°R$. (a) Find a relationship converting the temperatures $T_F$ of the Fahrenheit scale to the corresponding temperatures $T_R$ of the Rankine scale. (b) Find a second relationship converting temperatures $T_R$ of the Rankine scale to the temperatures $T_K$ of the Kelvin scale.

## 10.3 Thermal Expansion of Solids and Liquids

11. The New River Gorge bridge in West Virginia is a 518-m-long steel arch. How much will its length change between temperature extremes of $-20.0°C$ and $35.0°C$?

12. **Q|C** A grandfather clock is controlled by a swinging brass pendulum that is 1.3 m long at a temperature of $20.0°C$. (a) What is the length of the pendulum rod when the temperature drops to $0.0°C$? (b) If a pendulum's period is given by $T = 2\pi\sqrt{L/g}$, where $L$ is its length, does the change in length of the rod cause the clock to run fast or slow?

13. A pair of eyeglass frames are made of epoxy plastic (coefficient of linear expansion $= 1.30 \times 10^{-4}$ $(°C)^{-1}$). At room temperature ($20.0°C$), the frames have circular lens holes 2.20 cm in radius. To what temperature must the frames be heated if lenses 2.21 cm in radius are to be inserted into them?

14. A spherical steel ball bearing has a diameter of 2.540 cm at $25.00°C$. (a) What is its diameter when its temperature is raised to $100.0°C$? (b) What temperature change is required to increase its volume by 1.000%?

15. A brass ring of diameter 10.00 cm at $20.0°C$ is heated and slipped over an aluminum rod of diameter 10.01 cm at $20.0°C$. Assuming the average coefficients of linear expansion are constant, (a) to what temperature must the combination be cooled to separate the two metals? Is that temperature attainable? (b) What if the aluminum rod were 10.02 cm in diameter?

16. A wire is 25.0 m long at $2.00°C$ and is 1.19 cm longer at $30.0°C$. Find the wire's coefficient of linear expansion.

17. The density of lead is $1.13 \times 10^4$ kg/m$^3$ at $20.0°C$. Find its density at $105°C$.

18. **Q|C** The Golden Gate Bridge in San Francisco has a main span of length 1.28 km, one of the longest in the world. Imagine that a steel wire with this length and a cross-sectional area of $4.00 \times 10^{-6}$ m$^2$ is laid on the bridge deck with its ends attached to the towers of the bridge, on a summer day when the temperature of the wire is $35.0°C$. (a) When winter arrives, the towers stay the same distance apart and the bridge deck keeps the same shape as its expansion joints open. When the temperature drops to $-10.0°C$, what is the tension in the wire? Take Young's modulus for steel to be $20.0 \times 10^{10}$ N/m$^2$. (b) Permanent deformation occurs if the stress in the steel exceeds its elastic limit of $3.00 \times 10^8$ N/m$^2$. At what temperature would the wire reach its elastic limit? (c) Explain how your answers to (a) and (b) would change if the Golden Gate Bridge were twice as long.

19. **T** An underground gasoline tank can hold $1.00 \times 10^3$ gallons of gasoline at $52.0°F$. If the tank is being filled on a day when the outdoor temperature (and the gasoline in a tanker truck) is $95.0°F$, how many gallons from the truck can be poured into the tank? Assume the temperature of the gasoline quickly cools from $95.0°F$ to $52.0°F$ upon entering the tank.

20. **S** Show that the coefficient of volume expansion, $\beta$, is related to the coefficient of linear expansion, $\alpha$, through the expression $\beta = 3\alpha$.

21. A hollow aluminum cylinder 20.0 cm deep has an internal capacity of 2.000 L at 20.0°C. It is completely filled with turpentine at 20.0°C. The turpentine and the aluminum cylinder are then slowly warmed together to 80.0°C. (a) How much turpentine overflows? (b) What is the volume of the turpentine remaining in the cylinder at 80.0°C? (c) If the combination with this amount of turpentine is then cooled back to 20.0°C, how far below the cylinder's rim does the turpentine's surface recede?

22. A construction worker uses a steel tape to measure the length of an aluminum support column. If the measured length is 18.700 m when the temperature is 21.2°C, what is the measured length when the temperature rises to 29.4°C? *Note*: Don't neglect the expansion of the tape.

23. The band in Figure P10.23 is stainless steel (coefficient of linear expansion = $17.3 \times 10^{-6}$ (°C)$^{-1}$; Young's modulus = $18 \times 10^{10}$ N/m$^2$). It is essentially circular with an initial mean radius of 5.0 mm, a height of 4.0 mm, and a thickness of 0.50 mm. If the band just fits snugly over the tooth when heated to a temperature of 80.0°C, what is the tension in the band when it cools to a temperature of 37°C?

**Figure P10.23**

24. Q|C The Trans-Alaskan pipeline is 1 300 km long, reaching from Prudhoe Bay to the port of Valdez, and is subject to temperatures ranging from −73°C to +35°C. (a) How much does the steel pipeline expand due to the difference in temperature? (b) How can one compensate for this expansion?

25. The average coefficient of volume expansion for carbon tetrachloride is $5.81 \times 10^{-4}$ (°C)$^{-1}$. If a 50.0-gal steel container is filled completely with carbon tetrachloride when the temperature is 10.0°C, how much will spill over when the temperature rises to 30.0°C?

26. GP The density of gasoline is $7.30 \times 10^2$ kg/m$^3$ at 0°C. Its average coefficient of volume expansion is $9.60 \times 10^{-4}$(°C)$^{-1}$, and note that 1.00 gal = 0.003 80 m$^3$. (a) Calculate the mass of 10.0 gal of gas at 0°C. (b) If 1.000 m$^3$ of gasoline at 0°C is warmed by 20.0°C, calculate its new volume. (c) Using the answer to part (b), calculate the density of gasoline at 20.0°C. (d) Calculate the mass of 10.0 gal of gas at 20.0°C. (e) How many extra kilograms of gasoline would you get if you bought 10.0 gal of gasoline at 0°C rather than at 20.0°C from a pump that is not temperature compensated?

27. Figure P10.27 shows a circular steel casting with a gap. If the casting is heated, (a) does the width of the gap increase or decrease? (b) The gap width is 1.600 cm when the temperature is 30.0°C. Determine the gap width when the temperature is 190°C.

**Figure P10.27**

28. V The concrete sections of a certain superhighway are designed to have a length of 25.0 m. The sections are poured and cured at 10.0°C. What minimum spacing should the engineer leave between the sections to eliminate buckling if the concrete is to reach a temperature of 50.0°C?

## 10.4 The Ideal Gas Law

29. A sample of pure copper has a mass of 12.5 g. Calculate the number of (a) moles in the sample and (b) copper atoms in the sample.

30. Formaldehyde has the chemical formula $CH_2O$. Calculate the number of (a) moles, and (b) $CH_2O$ molecules in 275 g of formaldehyde.

31. One mole of oxygen gas is at a pressure of 6.00 atm and a temperature of 27.0°C. (a) If the gas is heated at constant volume until the pressure triples, what is the final temperature? (b) If the gas is heated so that both the pressure and volume are doubled, what is the final temperature?

32. A container holds 0.500 m$^3$ of oxygen at an absolute pressure of 4.00 atm. A valve is opened, allowing the gas to drive a piston, increasing the volume of the gas until the pressure drops to 1.00 atm. If the temperature remains constant, what new volume does the gas occupy?

33. (a) An ideal gas occupies a volume of 1.0 cm$^3$ at 20.°C and atmospheric pressure. Determine the number of molecules of gas in the container. (b) If the pressure of the 1.0-cm$^3$ volume is reduced to $1.0 \times 10^{-11}$ Pa (an extremely good vacuum) while the temperature remains constant, how many moles of gas remain in the container?

34. T An automobile tire is inflated with air originally at 10.0°C and normal atmospheric pressure. During the process, the air is compressed to 28.0% of its original volume and the temperature is increased to 40.0°C. (a) What is the tire pressure in pascals? (b) After the car is driven at high speed, the tire's air temperature rises to 85.0°C and the tire's interior volume increases by 2.00%. What is the new tire pressure (absolute) in pascals?

35. Gas is confined in a tank at a pressure of 11.0 atm and a temperature of 25.0°C. If two-thirds of the gas is withdrawn and the temperature is raised to 75.0°C, what is the new pressure of the gas remaining in the tank?

36. V Gas is contained in an 8.00-L vessel at a temperature of 20.0°C and a pressure of 9.00 atm. (a) Determine the number of moles of gas in the vessel. (b) How many molecules are in the vessel?

37. V A weather balloon is designed to expand to a maximum radius of 20 m at its working altitude, where the air pressure is 0.030 atm and the temperature is 200 K. If the balloon is filled at atmospheric pressure and 300 K, what is its radius at liftoff?

38. The density of helium gas at 0°C is $\rho_0 = 0.179$ kg/m$^3$. The temperature is then raised to $T = 100$°C, but the pressure is kept constant. Assuming the helium is an ideal gas, calculate the new density $\rho_f$ of the gas.

39. An air bubble has a volume of 1.50 cm$^3$ when it is released by a submarine $1.00 \times 10^2$ m below the surface of a lake. What is the volume of the bubble when it reaches the surface? Assume the temperature and the number of air molecules in the bubble remain constant during its ascent.

40. BIO During inhalation, a person's diaphragm and intercostal muscles contract, expanding the chest cavity and lowering the internal air pressure below ambient so that air flows in through the mouth and nose to the lungs. Suppose a person's lungs hold 1 250 mL of air at a pressure of 1.00 atm.

If the person expands the chest cavity by 525 mL while keeping the nose and mouth closed so that no air is inhaled, what will be the air pressure in the lungs in atm? Assume the air temperature remains constant.

## 10.5 The Kinetic Theory of Gases

41. [V] What is the average kinetic energy of a molecule of oxygen at a temperature of 300. K?

42. Calculate the root-mean-square (rms) speed of methane ($CH_4$) gas molecules at a temperature of 325 K.

43. Three moles of an argon gas are at a temperature of 275 K. Calculate (a) the kinetic energy per molecule, (b) the root-mean-square (rms) speed of an atom in the gas, and (c) the internal energy of the gas.

44. A sealed cubical container 20.0 cm on a side contains a gas with three times Avogadro's number of neon atoms at a temperature of 20.0°C. (a) Find the internal energy of the gas. (b) Find the total translational kinetic energy of the gas. (c) Calculate the average kinetic energy per atom. (d) Use Equation 10.13 to calculate the gas pressure. (e) Calculate the gas pressure using the ideal gas law (Eq. 10.8).

45. Use Avogadro's number to find the mass of a helium atom.

46. Two gases in a mixture pass through a filter at rates proportional to the gases' rms speeds. (a) Find the ratio of speeds for the two isotopes of chlorine, $^{35}Cl$ and $^{37}Cl$, as they pass through the air. (b) Which isotope moves faster?

47. At what temperature would the rms speed of helium atoms equal (a) the escape speed from Earth, $1.12 \times 10^4$ m/s and (b) the escape speed from the Moon, $2.37 \times 10^3$ m/s? (See Topic 7 for a discussion of escape speed.) *Note*: The mass of a helium atom is $6.64 \times 10^{-27}$ kg.

48. [Q|C] A 7.00-L vessel contains 3.50 moles of ideal gas at a pressure of $1.60 \times 10^6$ Pa. Find (a) the temperature of the gas and (b) the average kinetic energy of a gas molecule in the vessel. (c) What additional information would you need if you were asked to find the average speed of a gas molecule?

49. Superman leaps in front of Lois Lane to save her from a volley of bullets. In a 1-minute interval, an automatic weapon fires 150 bullets, each of mass 8.0 g, at $4.00 \times 10^2$ m/s. The bullets strike his mighty chest, which has an area of 0.75 m². Find the average force exerted on Superman's chest if the bullets bounce back after an elastic, head-on collision.

50. [T] In a period of 1.0 s, $5.0 \times 10^{23}$ nitrogen molecules strike a wall of area 8.0 cm². If the molecules move at $3.00 \times 10^2$ m/s and strike the wall head-on in a perfectly elastic collision, find the pressure exerted on the wall. (The mass of one $N_2$ molecule is $4.68 \times 10^{-26}$ kg.)

## Additional Problems

51. Inside the wall of a house, an L-shaped section of hot-water pipe consists of three parts: a straight horizontal piece $h = 28.0$ cm long, an elbow, and a straight, vertical piece $\ell = 134$ cm long (Fig. P10.51). A stud and a second-story floorboard hold the ends of this section of copper pipe stationary. Find the magnitude and direction of the

**Figure P10.51**

displacement of the pipe elbow when the water flow is turned on, raising the temperature of the pipe from 18.0°C to 46.5°C.

52. [T] The active element of a certain laser is made of a glass rod 30.0 cm long and 1.50 cm in diameter. Assume the average coefficient of linear expansion of the glass is $9.00 \times 10^{-6}$ (°C)$^{-1}$. If the temperature of the rod increases by 65.0°C, what is the increase in (a) its length, (b) its diameter, and (c) its volume?

53. A popular brand of cola contains 6.50 g of carbon dioxide dissolved in 1.00 L of soft drink. If the evaporating carbon dioxide is trapped in a cylinder at 1.00 atm and 20.0°C, what volume does the gas occupy?

54. [Q|C] Consider an object with any one of the shapes displayed in Table 8.1. What is the percentage increase in the moment of inertia of the object when it is warmed from 0°C to 100.°C if it is composed of (a) copper or (b) aluminum? Assume the average linear expansion coefficients shown in Table 10.1 do not vary between 0°C and 100.°C. (c) Why are the answers for parts (a) and (b) the same for all the shapes?

55. A steel beam being used in the construction of a skyscraper has a length of 35.000 m when delivered on a cold day at a temperature of 15.000°F. What is the length of the beam when it is being installed later on a warm day when the temperature is 90.000°F?

56. A 1.5-m-long glass tube that is closed at one end is weighted and lowered to the bottom of a freshwater lake. When the tube is recovered, an indicator mark shows that water rose to within 0.40 m of the closed end. Determine the depth of the lake. Assume constant temperature.

57. Long-term space missions require reclamation of the oxygen in the carbon dioxide exhaled by the crew. In one method of reclamation, 1.00 mol of carbon dioxide produces 1.00 mol of oxygen, with 1.00 mol of methane as a by-product. The methane is stored in a tank under pressure and is available to control the attitude of the spacecraft by controlled venting. A single astronaut exhales 1.09 kg of carbon dioxide each day. If the methane generated in the recycling of three astronauts' respiration during one week of flight is stored in an originally empty 150-L tank at −45.0°C, what is the final pressure in the tank?

58. [Q|C] [S] A vertical cylinder of cross-sectional area $A$ is fitted with a tight-fitting, frictionless piston of mass $m$ (Fig. P10.58). (a) If $n$ moles of an ideal gas are in the cylinder at a temperature of $T$, use Newton's second law for equilibrium to show that the height $h$ at which the piston is in equilibrium under its own weight is given by

$$h = \frac{nRT}{mg + P_0 A}$$

where $P_0$ is atmospheric pressure.
(b) Is the pressure inside the cylinder less than, equal to, or greater than atmospheric pressure? (c) If the gas in the cylinder is warmed, how would the answer for $h$ be affected?

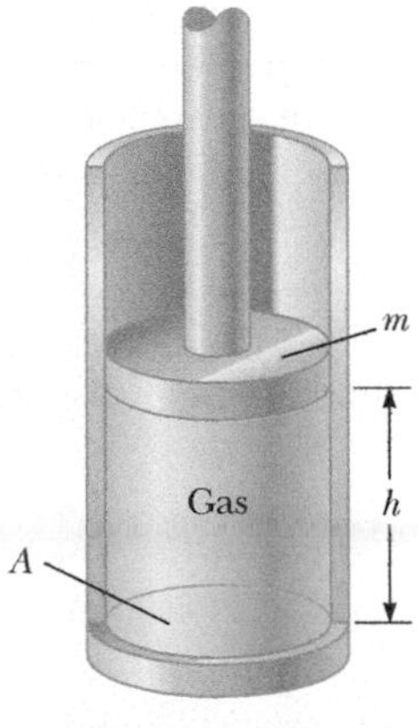

**Figure P10.58**

**59.** A flask made of Pyrex is calibrated at 20.0°C. It is filled to the 100-mL mark on the flask with 35.0°C acetone. (a) What is the volume of the acetone when both it and the flask cool to 20.0°C? (b) Would the temporary increase in the Pyrex flask's volume make an appreciable difference in the answer? Why or why not?

**60.** GP A 20.0-L tank of carbon dioxide gas ($CO_2$) is at a pressure of $9.50 \times 10^5$ Pa and temperature of 19.0°C. (a) Calculate the temperature of the gas in Kelvin. (b) Use the ideal gas law to calculate the number of moles of gas in the tank. (c) Use the periodic table to compute the molecular weight of carbon dioxide, expressing it in grams per mole. (d) Obtain the number of grams of carbon dioxide in the tank. (e) A fire breaks out, raising the ambient temperature by 224.0 K while 82.0 g of gas leak out of the tank. Calculate the new temperature and the number of moles of gas remaining in the tank. (f) Using a technique analogous to that in Example 10.6b, find a symbolic expression for the final pressure, neglecting the change in volume of the tank. (g) Calculate the final pressure in the tank as a result of the fire and leakage.

**61.** QC S A liquid with a coefficient of volume expansion of $\beta$ just fills a spherical flask of volume $V_0$ at temperature $T_i$ (Fig. P10.61). The flask is made of a material that has a coefficient of linear expansion of $\alpha$. The liquid is free to expand into a capillary of cross-sectional area $A$ at the top. (a) Show that if the temperature increases by $\Delta T$, the liquid rises in the capillary by the amount $\Delta h = (V_0/A)(\beta - 3\alpha)\Delta T$. (b) For a typical system, such as a mercury thermometer, why is it a good approximation to neglect the expansion of the flask?

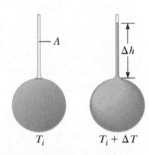

**Figure P10.61**

**62.** Before beginning a long trip on a hot day, a driver inflates an automobile tire to a gauge pressure of 1.80 atm at 300. K. At the end of the trip, the gauge pressure has increased to 2.20 atm. (a) Assuming the volume has remained constant, what is the temperature of the air inside the tire? (b) What percentage of the original mass of air in the tire should be released so the pressure returns to its original value? Assume the temperature remains at the value found in part (a) and the volume of the tire remains constant as air is released.

**63.** Two concrete spans of a 250-m-long bridge are placed end to end so that no room is allowed for expansion (Fig. P10.63a). If the temperature increases by 20.0°C, what is the height $y$ to which the spans rise when they buckle (Fig. P10.63b)?

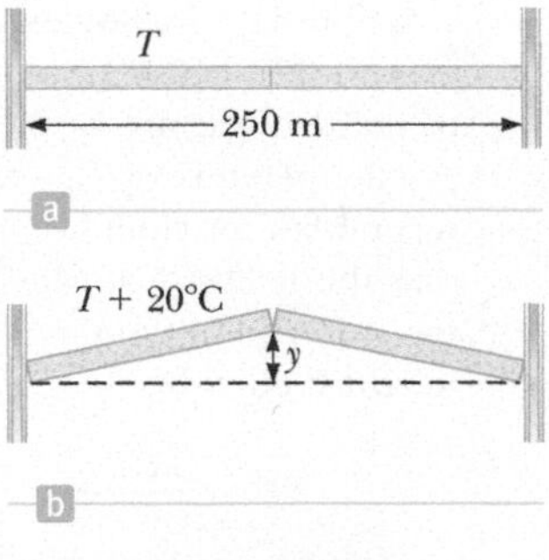

**Figure P10.63**

**64.** An expandable cylinder has its top connected to a spring with force constant $2.00 \times 10^3$ N/m (Fig. P10.64). The cylinder is filled with 5.00 L of gas with the spring relaxed at a pressure of 1.00 atm and a temperature of 20.0°C. (a) If the lid has a cross-sectional area of 0.010 0 m²

and negligible mass, how high will the lid rise when the temperature is raised to 250°C? (b) What is the pressure of the gas at 250°C?

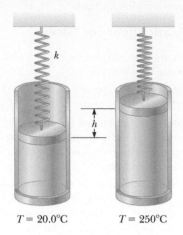

**Figure P10.64**

**65.** QC S A bimetallic strip of length $L$ is made of two ribbons of different metals bonded together. (a) First assume the strip is originally straight. As the strip is warmed, the metal with the greater average coefficient of expansion expands more than the other, forcing the strip into an arc, with the outer radius having a greater circumference (Fig. P10.65). Derive an expression for the angle of bending, $\theta$, as a function of the initial length of the strips, their average coefficients of linear expansion, the change in temperature, and the separation of the centers of the strips ($\Delta r = r_2 - r_1$). (b) Show that the angle of bending goes to zero when $\Delta T$ goes to zero and also when the two average coefficients of expansion become equal. (c) What happens if the strip is cooled?

**Figure P10.65**

**66.** A 250-m-long bridge is improperly designed so that it cannot expand with temperature. It is made of concrete with $\alpha = 12 \times 10^{-6}$ (°C)$^{-1}$. (a) Assuming the maximum change in temperature at the site is expected to be 20°C, find the change in length the span would undergo if it were free to expand. (b) Show that the stress on an object with Young's modulus $Y$ when raised by $\Delta T$ with its ends firmly fixed is given by $\alpha Y \Delta T$. (c) If the maximum stress the bridge can withstand without crumbling is $2.0 \times 10^7$ Pa, will it crumble because of this temperature increase? Young's modulus for concrete is about $2.0 \times 10^{10}$ Pa.

**67.** QC Following a collision in outer space, a copper disk at 850°C is rotating about its axis with an angular speed of 25.0 rad/s. As the disk radiates infrared light, its temperature falls to 20.0°C. No external torque acts on the disk. (a) Does the angular speed change as the disk cools? Explain how it changes or why it does not. (b) What is its angular speed at the lower temperature?

**68.** Two small containers, each with a volume of $1.00 \times 10^2$ cm³, contain helium gas at 0°C and 1.00 atm pressure. The two containers are joined by a small open tube of negligible volume, allowing gas to flow from one container to the other. What common pressure will exist in the two containers if the temperature of one container is raised to $1.00 \times 10^2$ °C while the other container is kept at 0°C?

# Energy in Thermal Processes

WHEN TWO OBJECTS WITH DIFFERENT TEMPERATURES are placed in thermal contact, the temperature of the warmer object decreases while the temperature of the cooler object increases. With time they reach a common equilibrium temperature somewhere in between their initial temperatures. During this process, we say that energy is transferred from the warmer object to the cooler one.

Until about 1850 the subjects of thermodynamics and mechanics were considered two distinct branches of science, and the principle of conservation of energy seemed to describe only certain kinds of mechanical systems. Experiments performed by English physicist James Joule (1818–1889) and others showed that the decrease in mechanical energy (kinetic plus potential) of an isolated system was equal to the increase in internal energy of the system. Today, internal energy is treated as a form of energy that can be transformed into mechanical energy and vice versa. Once the concept of energy was broadened to include internal energy, the law of conservation of energy emerged as a universal law of nature.

This topic focuses on some of the processes of energy transfer between a system and its surroundings.

## 11.1 Heat and Internal Energy

A major distinction must be made between heat and internal energy. These terms are not interchangeable: Heat involves a *transfer* of internal energy from one location to another. The following formal definitions will make the distinction precise.

**Internal energy** $U$ is the energy associated with the atoms and molecules of the system. The internal energy includes kinetic and potential energy associated with the random translational, rotational, and vibrational motion of the particles that make up the system, and any potential energy bonding the particles together.

◄ Internal energy

In Topic 10 we showed that the internal energy of a monatomic ideal gas is associated with the translational motion of its atoms. In this special case, the internal energy is the total translational kinetic energy of the atoms; the higher the temperature of the gas, the greater the kinetic energy of the atoms and the greater the internal energy of the gas. For more complicated diatomic and polyatomic gases, internal energy includes other forms of molecular energy, such as rotational kinetic energy and the kinetic and potential energy associated with molecular vibrations. Internal energy is also associated with the intermolecular potential energy ("bond energy") between molecules in a liquid or solid.

Heat was introduced in Topic 5 as one possible method of transferring energy between a system and its environment, and we provide a formal definition here:

**Heat** is the transfer of energy between a system and its environment due to a temperature difference between them.

**JAMES PRESCOTT JOULE**
**British physicist (1818–1889)**

Joule received some formal education in mathematics, philosophy, and chemistry from John Dalton, but was in large part self-educated. Joule's most active research period, from 1837 through 1847, led to the establishment of the principle of conservation of energy and the relationship between heat and other forms of energy transfer. His study of the quantitative relationship among electrical, mechanical, and chemical effects of heat culminated in his announcement in 1843 of the amount of work required to produce a unit of internal energy.

The symbol $Q$ is used to represent the amount of energy transferred by heat between a system and its environment. For brevity, we will often use the phrase "the energy $Q$ transferred to a system . . ." rather than "the energy $Q$ transferred by heat to a system . . ."

If a pan of water is heated on the burner of a stove, it's incorrect to say more heat is in the water. Heat is the *transfer* of thermal energy, just as work is the transfer of mechanical energy. When an object is pushed, it doesn't have more work; rather, it has more mechanical energy transferred *by* work. Similarly, the pan of water has more thermal energy transferred by heat.

### 11.1.1 Units of Heat

Early in the development of thermodynamics, before scientists realized the connection between thermodynamics and mechanics, heat was defined in terms of the temperature changes it produced in an object, and a separate unit of energy, the **calorie,** was used for heat. The calorie (cal) is defined as **the energy necessary to raise the temperature of 1 g of water from 14.5°C to 15.5°C.** (The "Calorie," with a capital "C," used in describing the energy content of foods, is actually a kilocalorie.) Likewise, the unit of heat in the U.S. customary system, the **British thermal unit** (Btu), was defined as **the energy required to raise the temperature of 1 lb of water from 63°F to 64°F.**

◀ Definition of the calorie

In 1948, scientists agreed that because heat (like work) is a measure of the transfer of energy, its SI unit should be the joule. The calorie is now defined to be exactly 4.186 J:

◀ The mechanical equivalent of heat

$$1 \text{ cal} \equiv 4.186 \text{ J} \qquad [11.1]$$

This definition makes no reference to raising the temperature of water. The calorie is a general energy unit, introduced here for historical reasons, although we will make little use of it. The definition in Equation 11.1 is known, from the historical background we have discussed, as the **mechanical equivalent of heat.**

---

## EXAMPLE 11.1   WORKING OFF BREAKFAST   BIO

**GOAL**  Relate caloric energy to mechanical energy.

**PROBLEM**  A student eats a breakfast consisting of a bowl of cereal and milk, containing a total of $3.20 \times 10^2$ Calories of energy. He wishes to do an equivalent amount of work in the gymnasium by performing curls with a 25.0-kg barbell (Fig. 11.1). How many times must he raise the weight to expend that much energy? Assume he raises it through a vertical displacement of 0.400 m each time, the distance from his lap to his upper chest.

**STRATEGY**  Convert the energy in Calories to joules, then equate that energy to the work necessary to do $n$ repetitions of the barbell exercise. The work he does lifting the barbell can be found from the work–energy theorem and the change in potential energy of the barbell. He does negative work on the barbell going down, to keep it from speeding up. The net work on the barbell during one repetition is zero, but his muscles expend the same energy both in raising and lowering.

**Figure 11.1**  (Example 11.1)

..................................................................................................................

**SOLUTION**

Convert his breakfast Calories, $E$, to joules:

$$E = (3.20 \times 10^2 \text{ Cal}) \left( \frac{1.00 \times 10^3 \text{ cal}}{1.00 \text{ Cal}} \right) \left( \frac{4.186 \text{ J}}{\text{cal}} \right)$$

$$= 1.34 \times 10^6 \text{ J}$$

Use the work–energy theorem to find the work necessary to lift the barbell up to its maximum height:

$$W = \Delta KE + \Delta PE = (0 - 0) + (mgh - 0) = mgh$$

The student must expend the same amount of energy lowering the barbell, making $2mgh$ per repetition. Multiply this amount by $n$ repetitions and set it equal to the food energy $E$:

$$n(2mgh) = E$$

Solve for $n$, substituting the food energy for $E$:

$$n = \frac{E}{2mgh} = \frac{1.34 \times 10^6 \, \text{J}}{2(25.0 \, \text{kg})(9.80 \, \text{m/s}^2)(0.400 \, \text{m})}$$

$$= 6.84 \times 10^3 \text{ times}$$

**REMARKS** If the student does one repetition every 5 seconds, it will take him 9.5 hours to work off his breakfast! In exercising, a large fraction of energy is lost through heat, however, due to the inefficiency of the body in doing work. The efficiency depends on the metabolic rate, which increases as activity becomes more vigorous. The transfer of energy dramatically reduces the exercise requirement by at least three-quarters, a little over two hours. Further, some small fraction of the energy content of the cereal may not actually be absorbed. All the same, it might be best to forgo a second bowl of cereal!

**QUESTION 11.1** From the point of view of physics, does the answer depend on how fast the repetitions are performed? How do faster repetitions affect human metabolism?

**EXERCISE 11.1** How many sprints from rest to a speed of 5.0 m/s would a 65-kg woman have to complete to burn off $5.0 \times 10^2$ Calories? (Assume 100% efficiency in converting food energy to mechanical energy.)

**ANSWER** $2.6 \times 10^3$ sprints

Getting proper exercise is an important part of staying healthy and keeping weight under control. As seen in the preceding example, the body expends energy when doing mechanical work, and these losses are augmented by the inefficiency of converting the body's internal stores of energy into useful work, with three-quarters or more leaving the body through heat. In addition, exercise tends to elevate the body's general metabolic rate, which persists even after the exercise is over. The increase in metabolic rate due to exercise, more so than the exercise itself, is helpful in weight reduction.

**BIO APPLICATION**
Physiology of Exercise

# 11.2 Specific Heat

The historical definition of the calorie is the amount of energy necessary to raise the temperature of one gram of a specific substance—water—by one degree. That amount is 4.186 J. Raising the temperature of one kilogram of water by 1°C requires 4 186 J of energy. The amount of energy required to raise the temperature of one kilogram of an arbitrary substance by 1°C varies with the substance. For example, the energy required to raise the temperature of one kilogram of copper by 1.0°C is 387 J. Every substance requires a unique amount of energy per unit mass to change the temperature of that substance by 1.0°C.

If a quantity of energy $Q$ is transferred to a substance of mass $m$, changing its temperature by $\Delta T = T_f - T_i$, the **specific heat** $c$ of the substance is defined by

$$c \equiv \frac{Q}{m \, \Delta T} \qquad [11.2]$$

**SI unit: Joule per kilogram-degree Celsius (J/kg · °C)**

Table 11.1 lists specific heats for several substances. From the definition of the calorie, the specific heat of water is 4 186 J/kg · °C. The values quoted are typical, but vary depending on the temperature and whether the matter is in a solid, liquid, or gaseous state.

**Table 11.1** Specific Heats of Some Materials at Atmospheric Pressure

| Substance | J/kg · °C | cal/g · °C |
|---|---|---|
| Aluminum | 900 | 0.215 |
| Beryllium | 1 820 | 0.436 |
| Cadmium | 230 | 0.055 |
| Copper | 387 | 0.092 4 |
| Ethyl alcohol | 2 430 | 0.581 |
| Germanium | 322 | 0.077 |
| Glass | 837 | 0.200 |
| Gold | 129 | 0.030 8 |
| Human tissue | 3 470 | 0.829 |
| Ice | 2 090 | 0.500 |
| Iron | 448 | 0.107 |
| Lead | 128 | 0.030 5 |
| Mercury | 138 | 0.033 |
| Silicon | 703 | 0.168 |
| Silver | 234 | 0.056 |
| Steam | 2 010 | 0.480 |
| Tin | 227 | 0.054 2 |
| Water | 4 186 | 1.00 |

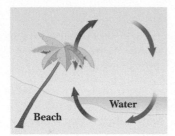

**Figure 11.2** Circulation of air at the beach. On a hot day, the air above the sand warms faster than the air above the cooler water. The warmer air floats upward due to Archimedes' principle, resulting in the movement of cooler air toward the beach.

From the definition of specific heat, we can express the energy $Q$ needed to raise the temperature of a system of mass $m$ by $\Delta T$ as

$$Q = mc\,\Delta T \qquad [11.3]$$

The energy required to raise the temperature of 0.500 kg of water by 3.00°C, for example, is $Q = (0.500 \text{ kg})(4\ 186 \text{ J/kg} \cdot °\text{C})(3.00°\text{C}) = 6.28 \times 10^3$ J. Note that when the temperature increases, $\Delta T$ and $Q$ are *positive,* corresponding to energy flowing *into* the system. When the temperature decreases, $\Delta T$ and $Q$ are *negative,* and energy flows *out* of the system.

Table 11.1 shows that water has the highest specific heat relative to most other temperatures found in regions near large bodies of water. As the temperature of a body of water decreases during winter, the water transfers energy to the air, which carries the energy landward when prevailing winds are toward the land. Off the western coast of the United States, the energy liberated by the Pacific Ocean is carried to the east, keeping coastal areas much warmer than they would be otherwise. Winters are generally colder in the eastern coastal states, because the prevailing winds tend to carry the energy away from land.

The fact that the specific heat of water is higher than the specific heat of sand is responsible for the pattern of airflow at a beach. During the day, the Sun adds roughly equal amounts of energy to the beach and the water, but the lower specific heat of sand causes the beach to reach a higher temperature than the water. As a result, the air above the land reaches a higher temperature than the air above the water. The denser cold air pushes the less dense hot air upward (due to Archimedes' principle), resulting in a breeze from ocean to land during the day. Because the hot air gradually cools as it rises, it subsequently sinks, setting up the circulation pattern shown in Figure 11.2.

A similar effect produces rising layers of air called *thermals* that can help eagles soar higher and hang gliders stay in flight longer. A thermal is created when a portion of the Earth reaches a higher temperature than neighboring regions. Thermals often occur in plowed fields, which are warmed by the Sun to higher temperatures than nearby fields shaded by vegetation. The cooler, denser air over the vegetation-covered fields pushes the expanding air over the plowed field upward, and a thermal is formed.

---

### Quick Quiz

**11.1** Suppose you have 1 kg each of iron, glass, and water, and all three samples are at 10°C. (a) Rank the samples from lowest to highest temperature after 100 J of energy is added to each by heat. (b) Rank them from least to greatest amount of energy transferred by heat if enough energy is transferred so that each increases in temperature by 20°C.

---

## EXAMPLE 11.2  STRESSING A STRUT

**GOAL** Use the energy transfer equation in the context of linear expansion and compressional stress.

**PROBLEM** A steel strut near a ship's furnace is 2.00 m long, with a mass of 1.57 kg and cross-sectional area of $1.00 \times 10^{-4}$ m². During operation of the furnace, the strut absorbs a net thermal energy of $2.50 \times 10^5$ J. (a) Find the change in temperature of the strut. (b) Find the increase in length of the strut. (c) If the strut is not allowed to expand because it's bolted at each end, find the compressional stress developed in the strut.

**STRATEGY** This problem can be solved by substituting given quantities into three different equations. In part (a),

the change in temperature can be computed by substituting into Equation 11.3, which relates temperature change to the energy transferred by heat. In part (b), substituting the result of part (a) into the linear expansion equation yields the change in length. If that change of length is thwarted by poor design, as in part (c), the result is compressional stress, found with the compressional stress–strain equation. *Note:* The specific heat of steel may be taken to be the same as that of iron.

## SOLUTION

**(a)** Find the change in temperature.

Solve Equation 11.3 for the change in temperature and substitute:

$$Q = m_s c_s \Delta T \quad \rightarrow \quad \Delta T = \frac{Q}{m_s c_s}$$

$$\Delta T = \frac{(2.50 \times 10^5 \text{ J})}{(1.57 \text{ kg})(448 \text{ J/kg} \cdot {}^\circ\text{C})} = \boxed{355^\circ\text{C}}$$

**(b)** Find the change in length of the strut if it's allowed to expand.

Substitute into the linear expansion equation:

$$\Delta L = \alpha L_0 \Delta T = (11 \times 10^{-6} \ ({}^\circ\text{C})^{-1})(2.00 \text{ m})(355^\circ\text{C})$$

$$= \boxed{7.8 \times 10^{-3} \text{ m}}$$

**(c)** Find the compressional stress in the strut if it is not allowed to expand.

Substitute into the compressional stress–strain equation:

$$\frac{F}{A} = Y\frac{\Delta L}{L} = (2.00 \times 10^{11} \text{ Pa})\frac{7.8 \times 10^{-3} \text{ m}}{2.01 \text{ m}}$$

$$= \boxed{7.8 \times 10^8 \text{ Pa}}$$

**REMARKS** Notice the use of 2.01 m in the denominator of the last calculation, rather than 2.00 m. This is because, in effect, the strut was compressed back to the original length from the length to which it would have expanded. (The difference is negligible, however.) The answer exceeds the ultimate compressive strength of steel and underscores the importance of allowing for thermal expansion. Of course, it's likely the strut would bend, relieving some of the stress (creating some shear stress in the process). Finally, if the strut is attached at both ends by bolts, thermal expansion and contraction would exert sheer stresses on the bolts, possibly weakening or loosening them over time.

**QUESTION 11.2** Which of the following combinations of properties will result in the smallest expansion of a substance due to the absorption of a given amount $Q$ of thermal energy? **(a)** small specific heat, large coefficient of expansion **(b)** small specific heat, small coefficient of expansion **(c)** large specific heat, small coefficient of expansion **(d)** large specific heat, large coefficient of expansion

**EXERCISE 11.2** Suppose a steel strut having a cross-sectional area of $5.00 \times 10^{-4} \text{ m}^2$ and length 2.50 m is bolted between two rigid bulkheads in the engine room of a submarine. Assume the density of the steel is the same as that of iron. **(a)** Calculate the change in temperature of the strut if it absorbs $3.00 \times 10^5 \text{ J}$ of thermal energy. **(b)** Calculate the compressional stress in the strut.

**ANSWERS** **(a)** 68.2°C   **(b)** $1.50 \times 10^8$ Pa

# 11.3 Calorimetry

One technique for measuring the specific heat of a solid or liquid is to raise the temperature of the substance to some value, place it into a vessel containing cold water of known mass and temperature, and measure the temperature of the combination after equilibrium is reached. Define the system as the substance and the water. If the vessel is assumed to be a good insulator, so that energy doesn't leave the system, then we can assume the system is isolated. Vessels having this property are called **calorimeters,** and analysis performed using such vessels is called **calorimetry.**

The principle of conservation of energy for this isolated system requires that the net result of all energy transfers is zero. If one part of the system loses energy, another part has to gain the energy because the system is isolated and the energy has nowhere else to go. When a warm object is placed in the cooler water of a calorimeter, the warm object becomes cooler while the water becomes warmer. This principle can be written

$$Q_{\text{cold}} = -Q_{\text{hot}} \tag{11.4}$$

$Q_{\text{cold}}$ is positive because energy is flowing into cooler objects, and $Q_{\text{hot}}$ is negative because energy is leaving the hot object. The negative sign on the right-hand side of Equation 11.4 ensures that the right-hand side is a positive number, consistent with the left-hand side. The equation is valid only when the system it describes is isolated.

Calorimetry problems involve solving Equation 11.4 for an unknown quantity, usually either a specific heat or a temperature.

## EXAMPLE 11.3   FINDING A SPECIFIC HEAT

**GOAL** Solve a calorimetry problem involving only two substances.

**PROBLEM** A 125-g block of an unknown substance with a temperature of 90.0°C is placed in a Styrofoam cup containing 0.326 kg of water at 20.0°C. The system reaches an equilibrium temperature of 22.4°C. What is the specific heat, $c_x$, of the unknown substance if the heat capacity of the cup is neglected?

**STRATEGY** The water gains thermal energy $Q_{cold}$ while the block loses thermal energy $Q_{hot}$. Using Equation 11.3, substitute expressions into Equation 11.4 and solve for the unknown specific heat, $c_x$.

**SOLUTION**

Let $T$ be the final temperature, and let $T_w$ and $T_x$ be the initial temperatures of the water and block, respectively. Apply Equations 11.3 and 11.4:

$$Q_{cold} = -Q_{hot}$$

$$m_w c_w (T - T_w) = -m_x c_x (T - T_x)$$

Solve for $c_x$ and substitute numerical values:

$$c_x = \frac{m_w c_w (T - T_w)}{m_x (T_x - T)}$$

$$= \frac{(0.326 \text{ kg})(4\,190 \text{ J/kg} \cdot {}^\circ\text{C})(22.4{}^\circ\text{C} - 20.0{}^\circ\text{C})}{(0.125 \text{ kg})(90.0{}^\circ\text{C} - 22.4{}^\circ\text{C})}$$

$$c_x = 388 \text{ J/kg} \cdot {}^\circ\text{C} \quad \rightarrow \quad \boxed{390 \text{ J/kg} \cdot {}^\circ\text{C}}$$

**REMARKS** Comparing our results to values given in Table 11.1, the unknown substance is probably copper. Note that because the factor $(22.4{}^\circ\text{C} - 20.0{}^\circ\text{C}) = 2.4{}^\circ\text{C}$ has only two significant figures, the final answer must similarly be rounded to two figures, as indicated.

**QUESTION 11.3** Objects $A$, $B$, and $C$ are at different temperatures, $A$ lowest and $C$ highest. The three objects are put in thermal contact with each other simultaneously. Without doing a calculation, is it possible to determine whether object $B$ will gain or lose thermal energy?

**EXERCISE 11.3** A 255-g block of gold at 85.0°C is immersed in 155 g of water at 25.0°C. Find the equilibrium temperature, assuming the system is isolated and the heat capacity of the cup can be neglected.

**ANSWER** 27.9°C

---

**Tip 11.2** Celsius Versus Kelvin

In equations in which $T$ appears, such as the ideal gas law, the Kelvin temperature must be used. In equations involving $\Delta T$, such as calorimetry equations, it's possible to use either Celsius or Kelvin temperatures because a change in temperature is the same on both scales. When in doubt, use Kelvin.

As long as there are no more than two substances involved, Equation 11.4 can be used to solve elementary calorimetry problems. Sometimes, however, there may be three (or more) substances exchanging thermal energy, each at a different temperature. If the problem requires finding the final temperature, it may not be clear whether the substance with the middle temperature gains or loses thermal energy. In such cases, Equation 11.4 can't be used reliably.

For example, suppose we want to calculate the final temperature of a system consisting initially of a glass beaker at 25°C, hot water at 40°C, and a block of aluminum at 37°C. We know that after the three are combined, the glass beaker warms up and the hot water cools, but we don't know for sure whether the aluminum block gains or loses energy because the final temperature is unknown.

Fortunately, we can still solve such a problem as long as it's set up correctly. With an unknown final temperature $T_f$, the expression $Q = mc(T_f - T_i)$ will be positive if $T_f > T_i$ and negative if $T_f < T_i$. Equation 11.4 can be written as

$$\sum Q_k = 0 \qquad \text{[11.5]}$$

where $Q_k$ is the energy change in the $k$th object. Equation 11.5 says that the sum of all the gains and losses of thermal energy must add up to zero, as required by the conservation of energy for an isolated system. Each term in Equation 11.5 will have the correct sign automatically. Applying Equation 11.5 to the water, aluminum, and glass problem, we get

$$Q_w + Q_{al} + Q_g = 0$$

There's no need to decide in advance whether a substance in the system is gaining or losing energy. This equation is similar in style to the conservation of mechanical energy equation, where the gains and losses of kinetic and potential energies sum to zero for an isolated system: $\Delta K + \Delta PE = 0$. As will be seen, changes in thermal energy can be included on the left-hand side of this equation.

When more than two substances exchange thermal energy, it's easy to make errors substituting numbers, so it's a good idea to construct a table to organize and assemble all the data. This strategy is illustrated in the next example.

## EXAMPLE 11.4  CALCULATE AN EQUILIBRIUM TEMPERATURE

**GOAL**  Solve a calorimetry problem involving three substances at three different temperatures.

**PROBLEM**  Suppose 0.400 kg of water initially at 40.0°C is poured into a 0.300-kg glass beaker having a temperature of 25.0°C. A 0.500-kg block of aluminum at 37.0°C is placed in the water and the system insulated. Calculate the final equilibrium temperature of the system.

**STRATEGY**  The energy transfer for the water, aluminum, and glass will be designated $Q_w$, $Q_{al}$, and $Q_g$, respectively. The sum of these transfers must equal zero, by conservation of energy. Construct a table, assemble the three terms from the given data, and solve for the final equilibrium temperature, $T$.

**SOLUTION**

Apply Equation 11.5 to the system:

**(1)**  $Q_w + Q_{al} + Q_g = 0$

**(2)**  $m_w c_w (T - T_w) + m_{al} c_{al} (T - T_{al}) + m_g c_g (T - T_g) = 0$

Construct a data table:

| Q (J) | m (kg) | c (J/kg · °C) | $T_f$ | $T_i$ |
|-------|--------|----------------|-------|-------|
| $Q_w$ | 0.400 | 4 190 | T | 40.0°C |
| $Q_{al}$ | 0.500 | $9.00 \times 10^2$ | T | 37.0°C |
| $Q_g$ | 0.300 | 837 | T | 25.0°C |

Using the table, substitute into Equation (2):

$$(1.68 \times 10^3 \, \text{J/°C})(T - 40.0°C)$$
$$+ (4.50 \times 10^2 \, \text{J/°C})(T - 37.0°C)$$
$$+ (2.51 \times 10^2 \, \text{J/°C})(T - 25.0°C) = 0$$
$$(1.68 \times 10^3 \, \text{J/°C} + 4.50 \times 10^2 \, \text{J/°C} + 2.51 \times 10^2 \, \text{J/°C})T$$
$$= 9.01 \times 10^4 \, \text{J}$$

$$T = \boxed{37.8°C}$$

**REMARKS**  The answer turned out to be very close to the aluminum's initial temperature, so it would have been impossible to guess in advance whether the aluminum would lose or gain energy. Notice the way the table was organized, mirroring the order of factors in the different terms. This kind of organization helps prevent substitution errors, which are common in these problems.

**QUESTION 11.4**  Suppose thermal energy $Q$ leaked from the system. How should the right side of Equation (1) be adjusted? **(a)** No change is needed.  **(b)** $+Q$  **(c)** $-Q$.

**EXERCISE 11.4**  A 20.0-kg gold bar at 35.0°C is placed in a large, insulated 0.800-kg glass container at 15.0°C and 2.00 kg of water at 25.0°C. Calculate the final equilibrium temperature.

**ANSWER**  26.6°C

# 11.4 Latent Heat and Phase Change

A substance usually undergoes a change in temperature when energy is transferred between the substance and its environment. In some cases, however, the transfer of energy doesn't result in a change in temperature. This can occur when the physical characteristics of the substance change from one form to another, commonly referred to as a **phase change.** Some common phase changes are solid to liquid (melting), liquid to gas (boiling), and a change in the crystalline structure of a solid. Any such phase change involves a change in the internal energy, but *no change in the temperature.*

Latent heat ▶

The energy $Q$ needed to change the phase of a given pure substance is

$$Q = \pm mL \qquad [11.6]$$

where $L$, called the **latent heat** of the substance, depends on the nature of the phase change as well as on the substance.

**Tip 11.3** Signs Are Critical

For phase changes, use the correct explicit sign in Equation 11.6, positive if you are adding energy to the substance, negative if you're taking it away.

The unit of latent heat is the joule per kilogram (J/kg). The word *latent* means "lying hidden within a person or thing." The positive sign in Equation 11.6 is chosen when energy is absorbed by a substance, as when ice is melting. The negative sign is chosen when energy is removed from a substance, as when steam condenses to water.

The **latent heat of fusion** $L_f$ is used when a phase change occurs during melting or freezing, whereas the **latent heat of vaporization** $L_v$ is used when a phase change occurs during boiling or condensing.[1] For example, at atmospheric pressure the latent heat of fusion for water is $3.33 \times 10^5$ J/kg and the latent heat of vaporization for water is $2.26 \times 10^6$ J/kg. The latent heats of different substances vary considerably, as can be seen in Table 11.2.

Another process, sublimation, is the passage from the solid to the gaseous phase without going through a liquid phase. The fuming of dry ice (frozen carbon dioxide) illustrates this process, which has its own latent heat associated with it, the heat of sublimation.

To better understand the physics of phase changes, consider the addition of energy to a 1.00-g cube of ice at $-30.0°C$ in a container held at constant pressure. Suppose this input of energy turns the ice to steam (water vapor) at $120.0°C$. Figure 11.3 (page 357) is a plot of the experimental measurement of temperature as energy is added to the system. We examine each portion of the curve separately.

**Table 11.2** Latent Heats of Fusion and Vaporization

| Substance | Melting Point (°C) | Latent Heat of Fusion (J/kg) | cal/g | Boiling Point (°C) | Latent Heat of Vaporization (J/kg) | cal/g |
|---|---|---|---|---|---|---|
| Helium | −269.65 | $5.23 \times 10^3$ | 1.25 | −268.93 | $2.09 \times 10^4$ | 4.99 |
| Nitrogen | −209.97 | $2.55 \times 10^4$ | 6.09 | −195.81 | $2.01 \times 10^5$ | 48.0 |
| Oxygen | −218.79 | $1.38 \times 10^4$ | 3.30 | −182.97 | $2.13 \times 10^5$ | 50.9 |
| Ethyl alcohol | −114 | $1.04 \times 10^5$ | 24.9 | 78 | $8.54 \times 10^5$ | 204 |
| Water | 0.00 | $3.33 \times 10^5$ | 79.7 | 100.00 | $2.26 \times 10^6$ | 540 |
| Sulfur | 119 | $3.81 \times 10^4$ | 9.10 | 444.60 | $3.26 \times 10^5$ | 77.9 |
| Lead | 327.3 | $2.45 \times 10^4$ | 5.85 | 1 750 | $8.70 \times 10^5$ | 208 |
| Aluminum | 660 | $3.97 \times 10^5$ | 94.8 | 2 450 | $1.14 \times 10^7$ | 2 720 |
| Silver | 960.80 | $8.82 \times 10^4$ | 21.1 | 2 193 | $2.33 \times 10^6$ | 558 |
| Gold | 1 063.00 | $6.44 \times 10^4$ | 15.4 | 2 660 | $1.58 \times 10^6$ | 377 |
| Copper | 1 083 | $1.34 \times 10^5$ | 32.0 | 1 187 | $5.06 \times 10^6$ | 1 210 |

[1]When a gas cools, it eventually returns to the liquid phase, or *condenses.* The energy per unit mass given up during the process is called the *heat of condensation,* and it equals the heat of vaporization. When a liquid cools, it eventually solidifies, and the *heat of solidification* equals the heat of fusion.

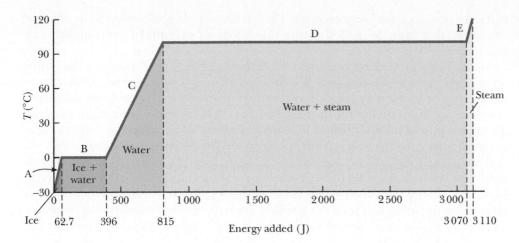

**Figure 11.3** A plot of temperature versus energy added when 1.00 g of ice, initially at $-30.0°C$, is converted to steam at 120°C.

**Part A** During this portion of the curve, the temperature of the ice changes from $-30.0°C$ to $0.0°C$. Because the specific heat of ice is 2 090 J/kg · °C, we can calculate the amount of energy added from Equation 11.3:

$$Q = mc_{ice}\,\Delta T = (1.00 \times 10^{-3}\,\text{kg})(2\,090\,\text{J/kg} \cdot °\text{C})(30.0°\text{C}) = 62.7\,\text{J}$$

**Part B** When the ice reaches 0°C, the ice–water mixture remains at that temperature—even though energy is being added—until all the ice melts to become water at 0°C. According to Equation 11.6, the energy required to melt 1.00 g of ice at 0°C is

$$Q = mL_f = (1.00 \times 10^{-3}\,\text{kg})(3.33 \times 10^5\,\text{J/kg}) = 333\,\text{J}$$

**Part C** Between 0°C and 100°C, no phase change occurs. The energy added to the water is used to increase its temperature, as in part A. The amount of energy necessary to increase the temperature from 0°C to 100°C is

$$Q = mc_{water}\,\Delta T = (1.00 \times 10^{-3}\,\text{kg})(4.19 \times 10^3\,\text{J/kg} \cdot °\text{C})(1.00 \times 10^2\,°\text{C})$$

$$Q = 4.19 \times 10^2\,\text{J}$$

**Part D** At 100°C, another phase change occurs as the water changes to steam at 100°C. As in Part B, the water–steam mixture remains at constant temperature, this time at 100°C—even though energy is being added—until all the liquid has been converted to steam. The energy required to convert 1.00 g of water at 100°C to steam at 100°C is

$$Q = mL_v = (1.00 \times 10^{-3}\,\text{kg})(2.26 \times 10^6\,\text{J/kg}) = 2.26 \times 10^3\,\text{J}$$

**Part E** During this portion of the curve, as in parts A and C, no phase change occurs, so all the added energy goes into increasing the temperature of the steam. The energy that must be added to raise the temperature of the steam to 120.0°C is

$$Q = mc_{steam}\,\Delta T = (1.00 \times 10^{-3}\,\text{kg})(2.01 \times 10^3\,\text{J/kg} \cdot °\text{C})(20.0°\text{C}) = 40.2\,\text{J}$$

The total amount of energy that must be added to change 1.00 g of ice at $-30.0°C$ to steam at 120.0°C is the sum of the results from all five parts of the curve, $3.11 \times 10^3$ J. Conversely, to cool 1.00 g of steam at 120.0°C to the point at which it becomes ice cooled to $-30.0°C$, $3.11 \times 10^3$ J of energy must be removed.

Phase changes can be described in terms of rearrangements of molecules when energy is added to or removed from a substance. Consider first the liquid-to-gas phase change. The molecules in a liquid are close together, and the forces between them are stronger than the forces between the more widely separated molecules of a gas. Work must therefore be done on the liquid against these attractive molecular forces so as to separate the molecules. The latent heat of vaporization is the amount of energy that must be added to the one kilogram of liquid to accomplish this separation.

Similarly, at the melting point of a solid, the amplitude of vibration of the atoms about their equilibrium positions becomes great enough to allow the atoms to pass the barriers of adjacent atoms and move to their new positions. On average, these new positions are less symmetrical than the old ones and therefore have higher energy. The latent heat of fusion is equal to the work required at the molecular level to transform the mass from the ordered solid phase to the disordered liquid phase.

The average distance between atoms is much greater in the gas phase than in either the liquid or the solid phase. Each atom or molecule is removed from its neighbors, overcoming the attractive forces of nearby neighbors. Therefore, more work is required at the molecular level to vaporize a given mass of a substance than to melt it, so in general, the latent heat of vaporization is much greater than the latent heat of fusion (see Table 11.2).

### Quick Quiz

**11.2** Calculate the slopes for the A, C, and E portions of Figure 11.3. Rank the slopes from least to greatest and explain what your ranking means. (a) A, C, E (b) C, A, E (c) E, A, C (d) E, C, A

### PROBLEM-SOLVING STRATEGY

### Calorimetry with Phase Changes

1. **Make a table for all data.** Include separate rows for different phases and for any transition between phases. Include columns for each quantity used and a final column for the combination of the quantities. Transfers of thermal energy in this last column are given by $Q = mc\Delta T$, whereas phase changes are given by $Q = \pm mL_f$ for changes between liquid and solid and by $Q = \pm mL_v$ for changes between liquid and gas.
2. **Apply conservation of energy.** If the system is isolated, use $\Sigma Q_k = 0$ (Eq. 11.5). For a nonisolated system, the net energy change should replace the zero on the right-hand side of that equation. Here, $\Sigma Q_k$ is just the sum of all the terms in the last column of the table.
3. **Solve** for the unknown quantity.

---

### EXAMPLE 11.5   ICE WATER

**GOAL**  Solve a problem involving heat transfer and a phase change from solid to liquid.

**PROBLEM**  At a party, 6.00 kg of ice at $-5.00°C$ is added to a cooler holding 30.0 liters of water at $20.0°C$. What is the temperature of the water when it comes to equilibrium?

**STRATEGY**  In this problem, it's best to make a table. With the addition of thermal energy $Q_{ice}$ the ice will warm to $0°C$, then melt at $0°C$ with the addition of energy $Q_{melt}$. Next, the melted ice will warm to some final temperature $T$ by absorbing energy $Q_{ice-water}$, obtained from the energy change of the original liquid water, $Q_{water}$. By conservation of energy, these quantities must sum to zero.

..........................................................................................

### SOLUTION

Calculate the mass of liquid water:

$$m_{water} = \rho_{water}V$$

$$=(1.00 \times 10^3 \text{ kg/m}^3)(30.0 \text{ L}) \frac{1.00 \text{ m}^3}{1.00 \times 10^3 \text{ L}}$$

$$= 30.0 \text{ kg}$$

Write the equation of thermal equilibrium:        **(1)**   $Q_{ice} + Q_{melt} + Q_{ice-water} + Q_{water} = 0$

Construct a comprehensive table:

| $Q$ | $m$ (kg) | $c$ (J/kg · °C) | $L$ (J/kg) | $T_f$ (°C) | $T_i$ (°C) | Expression |
|---|---|---|---|---|---|---|
| $Q_{ice}$ | 6.00 | 2 090 | | 0 | −5.00 | $m_{ice}c_{ice}(T_f - T_i)$ |
| $Q_{melt}$ | 6.00 | | $3.33 \times 10^5$ | 0 | 0 | $m_{ice}L_f$ |
| $Q_{ice-water}$ | 6.00 | 4 190 | | $T$ | 0 | $m_{ice}c_{water}(T_f - T_i)$ |
| $Q_{water}$ | 30.0 | 4 190 | | $T$ | 20.0 | $m_{water}c_{water}(T_f - T_i)$ |

Substitute all quantities in the second through sixth columns into the last column and sum, which is the evaluation of Equation (1), and solve for $T$:

$$6.27 \times 10^4 \, J + 2.00 \times 10^6 \, J$$
$$+ (2.51 \times 10^4 \, J/°C)(T - 0°C)$$
$$+ (1.26 \times 10^5 \, J/°C)(T - 20.0°C) = 0$$

$$T = \boxed{3.03°C}$$

**REMARKS** Making a table is optional. However, simple substitution errors are extremely common, and the table makes such errors less likely.

**QUESTION 11.5** Can a closed system containing different substances at different initial temperatures reach an equilibrium temperature that is lower than all the initial temperatures?

**EXERCISE 11.5** What mass of ice at −10.0°C is needed to cool a whale's water tank, holding $1.20 \times 10^3$ m³ of water, from 20.0°C to a more comfortable 10.0°C?

**ANSWER** $1.27 \times 10^5$ kg

## EXAMPLE 11.6 PARTIAL MELTING

**GOAL** Understand how to handle an incomplete phase change.

**PROBLEM** A 5.00-kg block of ice at 0°C is added to an insulated container partially filled with 10.0 kg of water at 15.0°C. **(a)** Find the final temperature, neglecting the heat capacity of the container. **(b)** Find the mass of the ice that was melted.

**STRATEGY** Part **(a)** is tricky because the ice does not entirely melt in this example. When there is any doubt concerning whether there will be a complete phase change, some preliminary calculations are necessary. First, find the total energy required to melt the ice, $Q_{melt}$, and then find $Q_{water}$, the maximum energy that can be delivered by the water above 0°C. If the energy delivered by the water is high enough, all the ice melts. If not, there will usually be a final mixture of ice and water at 0°C, unless the ice starts at a temperature far below 0°C, in which case all the liquid water freezes.

**SOLUTION**
**(a)** Find the equilibrium temperature.

First, compute the amount of energy necessary to completely melt the ice:

$$Q_{melt} = m_{ice}L_f = (5.00 \text{ kg})(3.33 \times 10^5 \, J/kg)$$
$$= 1.67 \times 10^6 \, J$$

Next, calculate the maximum energy that can be lost by the initial mass of liquid water without freezing it:

$$Q_{water} = m_{water}c\Delta T$$
$$= (10.0 \text{ kg})(4 \, 190 \, J/kg \cdot °C)(0°C - 15.0°C)$$
$$= -6.29 \times 10^5 \, J$$

This result is less than half the energy necessary to melt all the ice, so the final state of the system is a mixture of water and ice at the freezing point:

$$T = \boxed{0°C}$$

**(b)** Compute the mass of ice melted.

Set the total available energy equal to the heat of fusion of $m$ grams of ice, $mL_f$, and solve for $m$:

$$6.29 \times 10^5 \, J = mL_f = m(3.33 \times 10^5 \, J/kg)$$
$$m = \boxed{1.89 \text{ kg}}$$

**REMARKS** If this problem is solved assuming (wrongly) that all the ice melts, a final temperature of $T = -16.5°C$ is obtained. The only way that could happen is if the system were not isolated, contrary to the statement of the problem. In Exercise 11.6, you must also compute the thermal energy needed to warm the ice to its melting point.

(Continued)

**QUESTION 11.6** What effect would doubling the initial amount of liquid water have on the amount of ice melted?

**EXERCISE 11.6** If 8.00 kg of ice at $-5.00°C$ is added to 12.0 kg of water at 20.0°C, compute the final temperature. How much ice remains, if any?

**ANSWER** $T = 0°C$, 5.23 kg

Sometimes problems involve changes in mechanical energy. During a collision, for example, some kinetic energy can be transformed to the internal energy of the colliding objects. This kind of transformation is illustrated in Example 11.7, which involves a possible impact of a comet on Earth. In this example, a number of liberties will be taken in order to estimate the magnitude of the destructive power of such a catastrophic event. The specific heats depend on temperature and pressure, for example, but that will be ignored. Also, the ideal gas law doesn't apply at the temperatures and pressures attained, and the result of the collision wouldn't be superheated steam, but a plasma of charged particles. Despite all these simplifications, the example yields good order-of-magnitude results.

## EXAMPLE 11.7    ARMAGEDDON!

**GOAL** Link mechanical energy to thermal energy, phase changes, and the ideal gas law to create an estimate.

**PROBLEM** A comet half a kilometer in radius consisting of ice at 273 K hits Earth at a speed of $4.00 \times 10^4$ m/s. For simplicity, assume all the kinetic energy converts to thermal energy on impact and that all the thermal energy goes into warming the comet. **(a)** Calculate the volume and mass of the ice. **(b)** Use conservation of energy to find the final temperature of the comet material. Assume, contrary to fact, that the result is superheated steam and that the usual specific heats are valid, although in fact, they depend on both temperature and pressure. **(c)** Assuming the steam retains a spherical shape and has the same initial volume as the comet, calculate the pressure of the steam using the ideal gas law. This law actually doesn't apply to a system at such high pressure and temperature, but can be used to get an estimate.

**STRATEGY** Part **(a)** requires the volume formula for a sphere and the definition of density. In part **(b)** conservation of energy can be applied. There are four processes involved: (1) melting the ice, (2) warming the ice water to the boiling point, (3) converting the boiling water to steam, and (4) warming the steam. The energy needed for these processes will be designated $Q_{melt}$, $Q_{water}$, $Q_{vapor}$, and $Q_{steam}$, respectively. These quantities plus the change in kinetic energy $\Delta K$ sum to zero because they are assumed to be internal to the system. In this case, the first three $Q$'s can be neglected compared to the (extremely large) kinetic energy term. Solve for the unknown temperature and substitute it into the ideal gas law in part **(c)**.

· · · · · · · · · · · · · · · · · · · · · · · · · · · · · · · · · · · · · · · · · · · · · · · · · · · · · · · · · · · · · · · · ·

**SOLUTION**

**(a)** Find the volume and mass of the ice.

Apply the volume formula for a sphere:

$$V = \frac{4}{3}\pi r^3 = \frac{4}{3}(3.14)(5.00 \times 10^2 \text{ m})^3$$

$$= \boxed{5.23 \times 10^8 \text{ m}^3}$$

Apply the density formula to find the mass of the ice:

$$m = \rho V = (917 \text{ kg/m}^3)(5.23 \times 10^8 \text{ m}^3)$$

$$= \boxed{4.80 \times 10^{11} \text{ kg}}$$

**(b)** Find the final temperature of the cometary material.

Use conservation of energy:

**(1)** $\quad Q_{melt} + Q_{water} + Q_{vapor} + Q_{steam} + \Delta K = 0$

**(2)** $\quad mL_f + mc_{water}\Delta T_{water} + mL_v + mc_{steam}\Delta T_{steam}$

$$+ (0 - \tfrac{1}{2}mv^2) = 0$$

The first three terms are negligible compared to the kinetic energy. The steam term involves the unknown final temperature, so retain only it and the kinetic energy, canceling the mass and solving for $T$:

$$mc_{steam}(T - 373 \text{ K}) - \tfrac{1}{2}mv^2 = 0$$

$$T = \frac{\tfrac{1}{2}v^2}{c_{steam}} + 373 \text{ K} = \frac{\tfrac{1}{2}(4.00 \times 10^4 \text{ m/s})^2}{2\,010 \text{ J/kg} \cdot \text{K}} + 373 \text{ K}$$

$$T = \boxed{3.98 \times 10^5 \text{ K}}$$

**(c)** Estimate the pressure of the gas, using the ideal gas law.

First, compute the number of moles of steam:

$$n = (4.80 \times 10^{11}\text{kg})\left(\frac{1 \text{ mol}}{0.018 \text{ kg}}\right) = 2.67 \times 10^{13} \text{ mol}$$

Solve for the pressure, using $PV = nRT$:

$$P = \frac{nRT}{V}$$

$$= \frac{(2.67 \times 10^{13} \text{ mol})(8.31 \text{ J/mol} \cdot \text{K})(3.98 \times 10^{5} \text{ K})}{5.23 \times 10^{8} \text{ m}^3}$$

$$P = \boxed{1.69 \times 10^{11} \text{ Pa}}$$

**REMARKS** The estimated pressure is several hundred times greater than the ultimate shear stress of steel! This high-pressure region would expand rapidly, destroying everything within a very large radius. Fires would ignite across a continent-sized region, and tidal waves would wrap around the world, wiping out coastal regions everywhere. The Sun would be obscured for at least a decade, and numerous species, possibly including *Homo sapiens,* would become extinct. Such extinction events are rare, but in the long run represent a significant threat to life on Earth.

**QUESTION 11.7** Why would a nickel–iron asteroid be more dangerous than an asteroid of the same size made mainly of ice?

**EXERCISE 11.7** Suppose a lead bullet with mass 5.00 g and an initial temperature of 65.0°C hits a wall and completely liquefies. What minimum speed did it have before impact? (*Hint:* The minimum speed corresponds to the case where all the kinetic energy becomes internal energy of the lead and the final temperature of the lead is at its melting point. Don't neglect any terms here!)

**ANSWER** 341 m/s

# 11.5 Energy Transfer

For some applications it's necessary to know the rate at which energy is transferred between a system and its surroundings and the mechanisms responsible for the transfer. This information is particularly important when weatherproofing buildings or in medical applications, such as approximating human survival time when exposed to the elements.

Earlier in this topic we defined heat as a transfer of energy between a system and its surroundings due to a temperature difference between them. In this section we take a closer look at heat as a means of energy transfer and consider the processes of thermal conduction, convection, and radiation.

**Figure 11.4** Conduction makes the metal handle of a cooking pan hot.

© Cengage Learning/George Semple

## 11.5.1 Thermal Conduction

The energy transfer process most closely associated with a temperature difference is called **thermal conduction** or simply **conduction.** In this process the transfer can be viewed on an atomic scale as an exchange of kinetic energy between microscopic particles—molecules, atoms, and electrons—with less energetic particles gaining energy as they collide with more energetic particles. An inexpensive pot, as in Figure 11.4, may have a metal handle with no surrounding insulation. As the pot is warmed, the temperature of the metal handle increases, and the cook must hold it with a cloth potholder to avoid being burned.

The way the handle warms up can be understood by looking at what happens to the microscopic particles in the metal. Before the pot is placed on the stove, the particles are vibrating about their equilibrium positions. As the stove coil warms up, those particles in contact with it begin to vibrate with larger amplitudes. These particles collide with their neighbors and transfer some of their energy in the collisions. Metal atoms and electrons farther and farther from the coil gradually increase the amplitude of their vibrations, until eventually those in the handle

**Tip 11.4** Blankets and Coats in Cold Weather

When you sleep under a blanket in the winter or wear a warm coat outside, the blanket or coat serves as a layer of material with low thermal conductivity that reduces the transfer of energy away from your body by heat. The primary insulating medium is the air trapped in small pockets within the material.

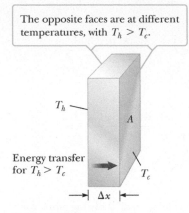

The opposite faces are at different temperatures, with $T_h > T_c$.

$T_h$

$A$

Energy transfer for $T_h > T_c$

$T_c$

$\Delta x$

**Figure 11.5** Energy transfer through a conducting slab of cross-sectional area $A$ and thickness $\Delta x$.

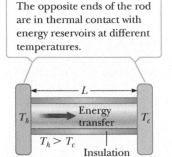

The opposite ends of the rod are in thermal contact with energy reservoirs at different temperatures.

$L$

$T_h$ — Energy transfer — $T_c$

$T_h > T_c$

Insulation

**Figure 11.6** Conduction of energy through a uniform, insulated rod of length $L$.

**Table 11.3** Thermal Conductivities

| Substance | Thermal Conductivity $(J/s \cdot m \cdot °C)$ |
|---|---|
| **Metals (at 25°C)** | |
| Aluminum | 238 |
| Copper | 397 |
| Gold | 314 |
| Iron | 79.5 |
| Lead | 34.7 |
| Silver | 427 |
| **Gases (at 20°C)** | |
| Air | 0.023 4 |
| Helium | 0.138 |
| Hydrogen | 0.172 |
| Nitrogen | 0.023 4 |
| Oxygen | 0.023 8 |
| **Nonmetals** | |
| **(approximate values)** | |
| Asbestos | 0.08 |
| Concrete | 0.8 |
| Glass | 0.8 |
| Ice | 2 |
| Rubber | 0.2 |
| Water | 0.6 |
| Wood | 0.08 |

are affected. This increased vibration represents an increase in temperature of the metal (and possibly a burned hand!).

Although the transfer of energy through a substance can be partly explained by atomic vibrations, the rate of conduction depends on the properties of the substance. For example, it's possible to hold a piece of asbestos in a flame indefinitely, which implies that very little energy is conducted through the asbestos. In general, metals are good thermal conductors because they contain large numbers of electrons that are relatively free to move through the metal and can transport energy from one region to another. In a good conductor such as copper, conduction takes place via the vibration of atoms and the motion of free electrons. Materials such as asbestos, cork, paper, and fiberglass are poor thermal conductors. Gases are also poor thermal conductors because of the large distance between their molecules.

Conduction occurs only if there is a difference in temperature between two parts of the conducting medium. The temperature difference drives the flow of energy. Consider a slab of material of thickness $\Delta x$ and cross-sectional area $A$ with its opposite faces at different temperatures $T_c$ and $T_h$, where $T_h > T_c$ (Fig. 11.5). The slab allows energy to transfer from the region of higher temperature to the region of lower temperature by thermal conduction. The rate of energy transfer, $P = Q/\Delta t$, is proportional to the cross-sectional area of the slab and the temperature difference and is inversely proportional to the thickness of the slab:

$$P = \frac{Q}{\Delta t} \propto A \frac{\Delta T}{\Delta x}$$

Note that $P$ has units of watts when $Q$ is in joules and $\Delta t$ is in seconds.

Suppose a substance is in the shape of a long, uniform rod of length $L$, as in Figure 11.6. We assume the rod is insulated, so thermal energy can't escape by conduction from its surface except at the ends. One end is in thermal contact with an energy reservoir at temperature $T_c$ and the other end is in thermal contact with a reservoir at temperature $T_h > T_c$. When a steady state is reached, the temperature at each point along the rod is constant in time. In this case $\Delta T = T_h - T_c$ and $\Delta x = L$, so

$$\frac{\Delta T}{\Delta x} = \frac{T_h - T_c}{L}$$

The rate of energy transfer by conduction through the rod is given by

$$P = kA \frac{(T_h - T_c)}{L} \qquad [11.7]$$

where $k$, a proportionality constant that depends on the material, is called the **thermal conductivity.** Substances that are good conductors have large thermal conductivities, whereas good insulators have low thermal conductivities. Table 11.3 lists the thermal conductivities for various substances.

### Quick Quiz

**11.3** Will an ice cube wrapped in a wool blanket remain frozen for (a) less time, (b) the same length of time, or (c) a longer time than an identical ice cube exposed to air at room temperature?

**11.4** Two rods of the same length and diameter are made from different materials. The rods are to connect two regions of different temperature so that energy will transfer through the rods by heat. They can be connected in series, as in Figure 11.7a (page 363), or in parallel, as in Figure 11.7b. In which case is the rate of energy transfer by heat larger? (a) When the rods are in series (b) When the rods are in parallel (c) The rate is the same in both cases.

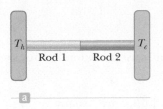
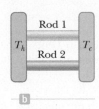

**Figure 11.7** (Quick Quiz 11.4) In which case is the rate of energy transfer larger?

---

## EXAMPLE 11.8  CONDUCTIVE LOSSES FROM THE HUMAN BODY  BIO

**GOAL** Apply the conduction equation to a human being.

**PROBLEM** In a human being, a layer of fat and muscle lies under the skin having various thicknesses depending on location. In response to a cold environment, capillaries near the surface of the body constrict, reducing blood flow and thereby reducing the conductivity of the tissues. These tissues form a shell up to an inch thick having a thermal conductivity of about 0.21 W/m · K, the same as skin or fat. **(a)** Estimate the rate of loss of thermal energy due to conduction from the human core region to the skin surface, assuming a shell thickness of 2.0 cm and a skin temperature of 33.0°C. (Skin temperature varies, depending on external conditions.)

**(b)** Calculate the thermal energy lost due to conduction in 1.0 h. **(c)** Estimate the change in body temperature in 1.0 h if the energy is not replenished. Assume a body mass of 75 kg and a skin surface area of 1.73 m².

**STRATEGY** The solution to part **(a)** requires applying Equation 11.7 for the rate of energy transfer due to conduction. Multiplying the power found in part **(a)** by the elapsed time yields the total thermal energy transfer in the given time. In part **(c)**, an estimate for the change in temperature if the energy is not replenished can be developed using Equation 11.3, $Q = mc\Delta T$.

**SOLUTION**

**(a)** Estimate the rate of loss of thermal energy due to conduction.

Write the thermal conductivity equation:

$$P = \frac{kA(T_h - T_c)}{L}$$

Substitute values:

$$P = \frac{(0.21 \text{ W/m} \cdot \text{K})(1.73 \text{ m}^2)(37.0°\text{C} - 33.0°\text{C})}{2.0 \times 10^{-2} \text{ m}} = \boxed{73 \text{ W}}$$

**(b)** Calculate the thermal energy lost due to conduction in 1.0 h.

Multiply the power $P$ by the time $\Delta t$:

$$Q = P\Delta t = (73 \text{ W})(3\,600 \text{ s}) = \boxed{2.6 \times 10^5 \text{ J}}$$

**(c)** Estimate the change in body temperature in 1.0 h if the energy is not replenished.

Write Equation 11.3 and solve it for $\Delta T$:

$$Q = mc\Delta T$$

$$\Delta T = \frac{Q}{mc} = \frac{2.6 \times 10^5 \text{ J}}{(75 \text{ kg})(3\,470 \text{ J/kg} \cdot \text{K})} = \boxed{1.0°\text{C}}$$

**REMARKS** The calculation doesn't take into account the thermal gradient, which further reduces the rate of conduction through the shell. Whereas thermal energy transfers through the shell by conduction, other mechanisms remove that energy from the body's surface because air is a poor conductor of thermal energy. Convection, radiation, and evaporation of sweat are the primary mechanisms that remove thermal energy from the skin. The calculation shows that even under mild conditions the body must constantly replenish its internal energy. It's possible to die of exposure even in temperatures well above freezing.

**QUESTION 11.8** Why does a long distance runner require very little in the way of warm clothing when running in cold weather, but puts on a sweater after finishing the run?

**EXERCISE 11.8** BIO A female minke whale has a core body temperature of 35°C and a core/blubber interface temperature of 29°C, with an average blubber thickness of 4.0 cm and thermal conductivity of 0.25 W/m · K. **(a)** At what rate is energy lost from the whale's core by conduction from the core/blubber interface through the blubber to the skin? Assume a skin temperature of 12°C and a total body area of 22 m². **(b)** What percent of the daily energy budget is this number? (The average female minke whale requires $8.0 \times 10^8$ J of energy per day—that's a lot of plankton and krill.)

**ANSWERS** (a) $2.3 \times 10^3$ W    (b) 25%

**Figure 11.8** A worker installs fiberglass insulation in a home. The mask protects the worker against the inhalation of microscopic fibers, which could be hazardous to his health.

## 11.5.2 Home Insulation

To determine whether to add insulation (Fig. 11.8) to a ceiling or some other part of a building, the preceding discussion of conduction must be extended for two reasons:

1. The insulating properties of materials used in buildings are usually expressed in engineering (U.S. customary) rather than SI units. Measurements stamped on a package of fiberglass insulating board will be in units such as British thermal units, feet, and degrees Fahrenheit.
2. In dealing with the insulation of a building, conduction through a compound slab must be considered, with each portion of the slab having a certain thickness and a specific thermal conductivity. A typical wall in a house consists of an array of materials, such as wood paneling, drywall, insulation, sheathing, and wood siding.

The rate of energy transfer by conduction through a compound slab is

$$\frac{Q}{\Delta t} = \frac{A(T_h - T_c)}{\sum_i L_i / k_i} \qquad [11.8]$$

where $T_h$ and $T_c$ are the temperatures of the *outer extremities* of the slab and the summation is over all portions of the slab. This formula can be derived algebraically, using the facts that the temperature at the interface between two insulating materials must be the same and that the rate of energy transfer through one insulator must be the same as through all the other insulators. If the slab consists of three different materials, the denominator is the sum of three terms. In engineering practice, the term $L/k$ for a particular substance is referred to as the $R$-value of the material, so Equation 11.8 reduces to

$$\frac{Q}{\Delta t} = \frac{A(T_h - T_c)}{\sum_i R_i} \qquad [11.9]$$

The $R$-values for a few common building materials are listed in Table 11.4. Note the unit of $R$ and the fact that the $R$-values are defined for specific thicknesses.

**Table 11.4** $R$-Values for Some Common Building Materials

| Material | $R$ value[a] $(\text{ft}^2 \cdot {}^\circ\text{F} \cdot \text{h/Btu})$ |
|---|---|
| Hardwood siding (1.0 in. thick) | 0.91 |
| Wood shingles (lapped) | 0.87 |
| Brick (4.0 in. thick) | 4.00 |
| Concrete block (filled cores) | 1.93 |
| Styrofoam (1.0 in. thick) | 5.0 |
| Fiberglass batting (3.5 in. thick) | 10.90 |
| Fiberglass batting (6.0 in. thick) | 18.80 |
| Fiberglass board (1.0 in. thick) | 4.35 |
| Cellulose fiber (1.0 in. thick) | 3.70 |
| Flat glass (0.125 in. thick) | 0.89 |
| Insulating glass (0.25-in. space) | 1.54 |
| Vertical air space (3.5 in. thick) | 1.01 |
| Stagnant layer of air | 0.17 |
| Drywall (0.50 in. thick) | 0.45 |
| Sheathing (0.50 in. thick) | 1.32 |

[a]The values in this table can be converted to SI units by multiplying the values by 0.176 1.

Next to any vertical outside surface is a very thin, stagnant layer of air that must be considered when the total $R$-value for a wall is computed. The thickness of this stagnant layer depends on the speed of the wind. As a result, energy loss by conduction from a house on a day when the wind is blowing is greater than energy loss on a day when the wind speed is zero. A representative $R$-value for a stagnant air layer is given in Table 11.4. The values are typically given in British units, but they can be converted to the equivalent metric units by multiplying the values in the table by 0.176 1.

### EXAMPLE 11.9 | CONSTRUCTION AND THERMAL INSULATION

**GOAL** Calculate the $R$-value of several layers of insulating material and its effect on thermal energy transfer.

**PROBLEM** (a) Find the energy transferred in 1.00 h by conduction through a concrete wall 2.0 m high, 3.65 m long, and 0.20 m thick if one side of the wall is held at 5.00°C and the other side is at 20.0°C (Fig. 11.9). Assume the concrete has a thermal conductivity of 0.80 J/s·m·°C. (b) The owner of the home decides to increase the insulation, so he installs 0.50 in. of thick sheathing, 3.5 in. of fiberglass batting, and a drywall 0.50-in. thick. Calculate the $R$-factor. (c) Calculate the energy transferred in 1.00 h by conduction. (d) What is the temperature between the concrete wall and the sheathing? Assume there is an air layer on the exterior of the concrete wall but not between the concrete and the sheathing.

**STRATEGY** The $R$-value of the concrete wall is given by $L/k$. Add this to the $R$-value of two air layers and then substitute into Equation 11.8, multiplying by the seconds in an

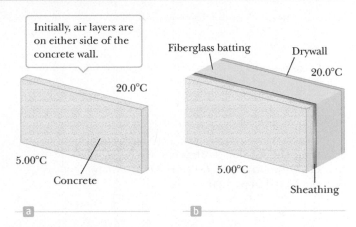

**Figure 11.9** (Example 11.9) A cross-sectional view of (a) a concrete wall with two air spaces and (b) the same wall with sheathing, fiberglass batting, drywall, and two air layers.

hour to get the total energy transferred through the wall in an hour. Repeat this process, with different materials, for parts (b) and (c). Part (d) requires finding the $R$-value for an air layer and the concrete wall and then substituting into the thermal conductivity equation. In this problem metric units are used, so be sure to convert the $R$-values in the table. (Converting to SI requires multiplication of the British units by 0.176 1.)

### SOLUTION

(a) Find the energy transferred in 1.00 h by conduction through a concrete wall.

Calculate the $R$-value of concrete plus two air layers:

$$\sum R = \frac{L}{k} + 2R_{\text{air layer}} = \frac{0.20 \text{ m}}{0.80 \text{ J/s} \cdot \text{m} \cdot °\text{C}} + 2\left(0.030 \frac{\text{m}^2}{\text{J/s} \cdot °\text{C}}\right)$$

$$= 0.31 \frac{\text{m}^2}{\text{J/s} \cdot °\text{C}}$$

Write the thermal conduction equation:

$$P = \frac{A(T_h - T_c)}{\sum R}$$

Substitute values:

$$P = \frac{(7.3 \text{ m}^2)(20.0°\text{C} - 5.00°\text{C})}{0.31 \text{ m}^2 \cdot \text{s} \cdot °\text{C/J}} = 353 \text{ W} \quad \rightarrow \quad 350 \text{ W}$$

Multiply the power in watts times the seconds in an hour:

$$Q = P\Delta t = (350 \text{ W})(3\ 600 \text{ s}) = \boxed{1.3 \times 10^6 \text{ J}}$$

(b) Calculate the $R$-factor of the newly insulated wall.

Refer to Table 11.4 and sum the appropriate quantities after converting them to SI units:

$$R_{\text{total}} = R_{\text{outside air layer}} + R_{\text{concrete}} + R_{\text{sheath}}$$

$$+ R_{\text{fiberglass}} + R_{\text{drywall}} + R_{\text{inside air layer}}$$

$$= (0.030 + 0.25 + 0.232 + 1.92 + 0.079 + 0.030)$$

$$= \boxed{2.5 \text{ m}^2 \cdot °\text{C} \cdot \text{s/J}}$$

*(Continued)*

**(c)** Calculate the energy transferred in 1.00 h by conduction.

Write the thermal conduction equation:

$$P = \frac{A(T_h - T_c)}{\sum R}$$

Substitute values:

$$P = \frac{(7.3 \text{ m}^2)(20.0°C - 5.00°C)}{2.5 \text{ m}^2 \cdot \text{s} \cdot °C/J} = 44 \text{ W}$$

Multiply the power in watts times the seconds in an hour:

$$Q = P\Delta t = (44 \text{ W})(3\,600 \text{ s}) = \boxed{1.6 \times 10^5 \text{ J}}$$

**(d)** Calculate the temperature between the concrete and the sheathing.

Write the thermal conduction equation:

$$P = \frac{A(T_h - T_c)}{\sum R}$$

Solve algebraically for $T_h$ by multiplying both sides by $\sum R$ and dividing both sides by area $A$:

$$P\sum R = A(T_h - T_c) \quad \rightarrow \quad (T_h - T_c) = \frac{P\sum R}{A}$$

Add $T_c$ to both sides:

$$T_h = \frac{P\sum R}{A} + T_c$$

Substitute the $R$-value for the concrete wall from part **(a)**, but subtract the $R$-value of one air layer from that calculated in part **(a)**:

$$T_h = \frac{(44 \text{ W})(0.31 \text{ m}^2 \cdot \text{s} \cdot °C/J - 0.03 \text{ m}^2 \cdot \text{s} \cdot °C/J)}{7.3 \text{ m}^2} + 5.00°C$$

$$= \boxed{6.7°C}$$

• • • • • • • • • • • • • • • • • • • • • • • • • • • • • • • • • • • • • • • • • • • • • • • • • • • • • • • • • • • • • • • •

**REMARKS** Notice the enormous energy savings that can be realized with good insulation!

**QUESTION 11.9** Which of the following choices results in the best possible $R$-value? **(a)** Use material with a small thermal conductivity and large thickness. **(b)** Use thin material with a large thermal conductivity. **(c)** Use material with a small thermal conductivity and small thickness.

**EXERCISE 11.9** Instead of the layers of insulation, the owner installs a brick wall on the exterior of the concrete wall. **(a)** Calculate the $R$-factor, including the two stagnant air layers on the inside and outside of the wall. **(b)** Calculate the energy transferred in 1.00 h by conduction, under the same conditions as in the example. **(c)** What is the temperature between the concrete and the brick?

**ANSWERS** **(a)** $1.01 \text{ m}^2 \cdot °C \cdot \text{s/J}$ **(b)** $3.9 \times 10^5 \text{ J}$ **(c)** $16°C$

---

## 11.5.3 Convection

**Figure 11.10** Warming a hand by convection.

When you warm your hands over an open flame, as illustrated in Figure 11.10, the air directly above the flame, being warmed, expands. As a result, the density of this air decreases and the air rises, warming your hands as it flows by. **The transfer of energy by the movement of a substance is called convection.** When the movement results from differences in density, as with air around a fire, it's referred to as *natural convection*. Airflow at a beach is an example of natural convection, as is the mixing that occurs as surface water in a lake cools and sinks. When the substance is forced to move by a fan or pump, as in some hot air and hot water heating systems, the process is called *forced convection*.

Convection currents assist in the boiling of water (Fig. 11.11). In a teakettle on a hot stovetop, the lower layers of water are warmed first. The warmed water has a lower density and rises to the top, while the denser, cool water at the surface sinks to the bottom of the kettle and is warmed.

The same process occurs when a radiator raises the temperature of a room. The hot radiator warms the air in the lower regions of the room. The warm air expands and, because of its lower density, rises to the ceiling. The denser, cooler air from above sinks, setting up the continuous air current pattern shown in Figure 11.12.

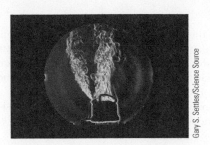

**Figure 11.11** Photograph of a teakettle, showing steam and turbulent convection air currents.

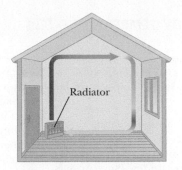

**Figure 11.12** Convection currents are set up in a room warmed by a radiator.

An automobile engine is maintained at a safe operating temperature by a combination of conduction and forced convection. Water (actually, a mixture of water and antifreeze) circulates in the interior of the engine. As the metal of the engine block increases in temperature, energy passes from the hot metal to the cooler water by thermal conduction. The water pump forces water out of the engine and into the radiator, carrying energy along with it (by forced convection). In the radiator the hot water passes through metal pipes that are in contact with the cooler outside air, and energy passes into the air by conduction. The cooled water is then returned to the engine by the water pump to absorb more energy. The process of air being pulled past the radiator by the fan is also forced convection.

The algal blooms often seen in temperate lakes and ponds during the spring or fall are caused by convection currents in the water. To understand this process, consider Figure 11.13. During the summer, bodies of water develop temperature gradients, with a warm upper layer of water separated from a cold lower layer by a buffer zone called a thermocline. In the spring and fall, temperature changes in the water break down this thermocline, setting up convection currents that mix the water. The mixing process transports nutrients from the bottom to the surface. The nutrient-rich water forming at the surface can cause a rapid, temporary increase in the algae population.

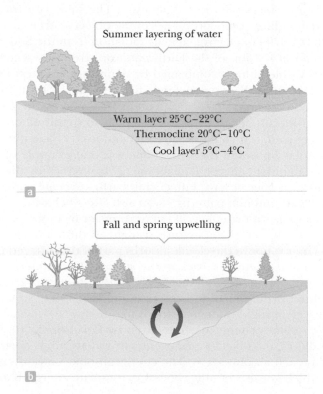

Summer layering of water

Warm layer 25°C–22°C
Thermocline 20°C–10°C
Cool layer 5°C–4°C

a

Fall and spring upwelling

b

**Figure 11.13** (a) During the summer, a warm upper layer of water is separated from a cooler lower layer by a thermocline. (b) Convection currents during the spring and fall mix the water and can cause algal blooms.

**APPLYING PHYSICS 11.1** BODY TEMPERATURE BIO

The body temperature of mammals ranges from about 35°C to 38°C, whereas that of birds ranges from about 40°C to 43°C. How can these narrow ranges of body temperature be maintained in cold weather?

**EXPLANATION** A natural method of maintaining body temperature is via layers of fat beneath the skin. Fat protects against both conduction and convection because of its low thermal conductivity and because there are few blood vessels in fat to carry blood to the surface, where energy losses by convection can occur. Birds ruffle their feathers in cold weather to trap a layer of air with a low thermal conductivity between the feathers and the skin. Bristling the fur produces the same effect in fur-bearing animals.

Humans keep warm with wool sweaters and down jackets that trap the warmer air in regions close to their bodies, reducing energy loss by convection and conduction. ∎

### 11.5.4 Radiation

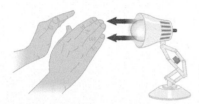

**Figure 11.14** Warming hands by radiation.

Another process of transferring energy is through **radiation.** Figure 11.14 shows how your hands can be warmed by a lamp through radiation. Because your hands aren't in physical contact with the lamp and the conductivity of air is very low, conduction can't account for the energy transfer. Nor can convection be responsible for any transfer of energy because your hands aren't above the lamp in the path of convection currents. The warmth felt in your hands must therefore come from the transfer of energy by radiation.

All objects radiate energy continuously in the form of electromagnetic waves due to thermal vibrations of their molecules. These vibrations create the orange glow of an electric stove burner, an electric space heater, and the coils of a toaster.

The rate at which an object radiates energy is proportional to the fourth power of its absolute temperature. This is known as **Stefan's law,** expressed in equation form as

Stefan's law ▶ $$P = \sigma A e T^4 \qquad\qquad [11.10]$$

where $P$ is the power in watts (or joules per second) radiated by the object, $\sigma$ is the Stefan–Boltzmann constant, equal to $5.669\ 6 \times 10^{-8}\ \text{W/m}^2 \cdot \text{K}^4$, $A$ is the surface area of the object in square meters, $e$ is a constant called the **emissivity** of the object, and $T$ is the object's Kelvin temperature. The value of $e$ can vary between zero and one, depending on the properties of the object's surface.

Approximately 1 370 J of electromagnetic radiation from the Sun passes through each square meter at the top of the Earth's atmosphere every second. This radiation is primarily visible light, accompanied by significant amounts of infrared and ultraviolet light. We study these types of radiation in detail in Topic 21. Some of this energy is reflected back into space, and some is absorbed by the atmosphere, but enough arrives at the surface of the Earth each day to supply all our energy needs hundreds of times over, if it could be captured and used efficiently. The growth in the number of solar houses in the United States is one example of an attempt to make use of this abundant energy. Radiant energy from the Sun affects our day-to-day existence in a number of ways, influencing Earth's average temperature, ocean currents, agriculture, and rain patterns. It can also affect behavior.

As another example of the effects of energy transfer by radiation, consider what happens to the atmospheric temperature at night. If there is a cloud cover above the Earth, the water vapor in the clouds absorbs part of the infrared radiation emitted by the Earth and re-emits it back to the surface. Consequently, the temperature at the surface remains at moderate levels. In the absence of cloud cover, there is nothing to prevent the radiation from escaping into space, so the temperature drops more on a clear night than on a cloudy night.

As an object radiates energy at a rate given by Equation 11.10, it also absorbs radiation. If it didn't, the object would eventually radiate all its energy and its temperature would reach absolute zero. The energy an object absorbs comes from its environment, which consists of other bodies that radiate energy. If an object is at a temperature $T$ and its surroundings are at a temperature $T_0$, the net energy gained or lost each second by the object as a result of radiation is

$$P_{net} = \sigma A e (T^4 - T_0^4)$$   [11.11]

When an object is in equilibrium with its surroundings, it radiates and absorbs energy at the same rate, so its temperature remains constant. When an object is hotter than its surroundings, it radiates more energy than it absorbs and therefore cools.

An **ideal absorber** is an object that absorbs all the light radiation incident on it, including invisible infrared and ultraviolet light. Such an object is called a **black body** because a room-temperature black body would look black. Because a black body doesn't reflect radiation at any wavelength, any light coming from it is due to atomic and molecular vibrations alone. A perfect black body has emissivity $e = 1$. An ideal absorber is also an ideal radiator of energy. The Sun, for example, is nearly a perfect black body. This statement may seem contradictory because the Sun is bright, not dark; the light that comes from the Sun, however, is emitted, not reflected. Black bodies are perfect absorbers that look black at room temperature because they don't reflect any light. All black bodies, except those at absolute zero, emit light that has a characteristic spectrum, discussed in Topic 27. In contrast to black bodies, an object for which $e = 0$ absorbs none of the energy incident on it, reflecting it all. Such a body is an **ideal reflector.**

White clothing is more comfortable to wear in the summer than black clothing. Black fabric acts as a good absorber of incoming sunlight and as a good emitter of this absorbed energy. About half of the emitted energy, however, travels toward the body, causing the person wearing the garment to feel uncomfortably warm. White or light-colored clothing reflects away much of the incoming energy.

The amount of energy radiated by an object can be measured with temperature-sensitive recording equipment via a technique called **thermography.** An image of the pattern formed by varying radiation levels, called a **thermogram,** is brightest in the warmest areas. Figure 11.15 reproduces a thermogram of a house. More energy escapes in the lighter regions, such as the door and windows. The owners of this house could conserve energy and reduce their heating costs by adding insulation to the attic area and by installing thermal draperies over the windows. Thermograms have also been used to image injured or diseased tissue in medicine (Fig. 11.16),

**APPLICATION**
Light-Colored Summer Clothing

**BIO APPLICATION**
Thermography

**BIO APPLICATION**
Radiation Thermometers for Measuring Body Temperature

Blue and purple indicate areas of least energy loss.

White and yellow indicate areas of greatest energy loss.

**Figure 11.15** Thermogram of a house, made during cold weather.

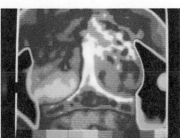

**Figure 11.16** Thermogram of a woman's breasts. Her left breast is diseased (red and orange) and her right breast (blue) is healthy.

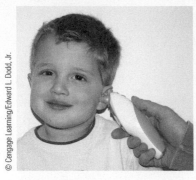

**Figure 11.17** A radiation thermometer measures a patient's temperature by monitoring the intensity of infrared radiation leaving the ear.

because such areas are often at a different temperature than surrounding healthy tissue, although many radiologists consider thermograms inadequate as a diagnostic tool.

Figure 11.17 shows a recently developed radiation thermometer that has removed most of the risk of taking the temperature of young children or the aged with a rectal thermometer, such as bowel perforation or bacterial contamination. The instrument measures the intensity of the infrared radiation leaving the eardrum and surrounding tissues and converts this information to a standard numerical reading. The eardrum is a particularly good location to measure body temperature because it's near the hypothalamus, the body's temperature control center.

### Quick Quiz

**11.5** Stars A and B have the same temperature, but star A has twice the radius of star B. (a) What is the ratio of star A's power output to star B's output due to electromagnetic radiation? The emissivity of both stars can be assumed to be 1. (b) Repeat the question if the stars have the same radius, but star A has twice the absolute temperature of star B. (c) What's the ratio if star A has both twice the radius and twice the absolute temperature of star B?

## APPLYING PHYSICS 11.2  THERMAL RADIATION AND NIGHT VISION

How can thermal radiation be used to see objects in near total darkness?

**EXPLANATION** There are two methods of night vision, one enhancing a combination of very faint visible light and infrared light, and another using infrared light only. The latter is valuable for creating images in absolute darkness. Because all objects above absolute zero emit thermal radiation due to the vibrations of their atoms, the infrared (invisible) light can be focused by a special lens and scanned by an array of infrared detector elements. These elements create a thermogram. The information from thousands of separate points in the field of view is converted to electrical impulses and translated by a microchip into a form suitable for display. Different temperature areas are assigned different colors, which can then be easily discerned on the display. ■

## EXAMPLE 11.10  POLAR BEAR CLUB  BIO

**GOAL** Apply Stefan's law.

**PROBLEM** A member of the Polar Bear Club, dressed only in bathing trunks of negligible size, prepares to plunge into the Gulf of Finland from the beach in St. Petersburg, Russia. The air is calm, with a temperature of 5°C. If the swimmer's surface body temperature is 25°C, compute the net rate of energy loss from his skin due to radiation. How much energy is lost in 10.0 min? Assume his emissivity is 0.900 and his surface area is 1.50 m².

**STRATEGY** Use Equation 11.11, the thermal radiation equation, substituting the given information. Remember to convert temperatures to Kelvin by adding 273 to each value in degrees Celsius!

- - - - - - - - - - - - - - - - - - - - - - - - - - - - - - - - - - - - - - - - - - - - - - - - - - - - - - -

**SOLUTION**

Convert temperatures from Celsius to Kelvin:

$$T_{5°C} = T_C + 273 = 5 + 273 = 278 \text{ K}$$
$$T_{25°C} = T_C + 273 = 25 + 273 = 298 \text{ K}$$

Compute the net rate of energy loss, using Equation 11.11:

$$P_{net} = \sigma A e(T^4 - T_0^4)$$
$$= (5.67 \times 10^{-8} \text{ W/m}^2 \cdot \text{K}^4)(1.50 \text{ m}^2)$$
$$\times (0.900)[(298 \text{ K})^4 - (278 \text{ K})^4]$$
$$P_{net} = \boxed{146 \text{ W}}$$

Multiply the preceding result by the time, 10 minutes, to get the energy lost in that time due to radiation:

$$Q = P_{net} \times \Delta t = (146 \text{ J/s})(6.00 \times 10^2 \text{ s}) = \boxed{8.76 \times 10^4 \text{ J}}$$

- - - - - - - - - - - - - - - - - - - - - - - - - - - - - - - - - - - - - - - - - - - - - - - - - - - - - - -

**REMARKS** Energy is also lost from the body through convection and conduction. Clothing traps layers of air next to the skin, which are warmed by radiation and conduction. In still air these warm layers are more readily retained. Even a Polar Bear Club member enjoys some benefit from the still air, better retaining a stagnant air layer next to the surface of his skin.

**QUESTION 11.10** Suppose that at a given temperature the rate of an object's energy loss due to radiation is equal to its loss by conduction. When the object's temperature is raised, is the energy loss due to radiation **(a)** greater than, **(b)** equal to, or **(c)** less than the rate of energy loss due to conduction? (Assume the temperature of the environment is constant.)

**EXERCISE 11.10** Repeat the calculation when the man is standing in his bedroom, with an ambient temperature of $20.0°C$. Assume his body surface temperature is $27.0°C$, with emissivity of $0.900$.

**ANSWER** $55.9$ W, $3.35 \times 10^4$ J

---

## EXAMPLE 11.11  PLANET OF ALPHA CENTAURI B

**GOAL** Apply Stefan's law to stars and their planets.

**PROBLEM** The star Alpha Centauri B is one member of the triple star system, Alpha Centauri AB-C, the closest star system to Earth. **(a)** Calculate the power output $P$ of Alpha Centauri B, given its surface temperature of $5\,790$ K and radius $R = 6.02 \times 10^8$ m. **(b)** Calculate the power $P_I$ intercepted by a possible Earth-sized planet, Alpha Centauri Bb, with radius $r = 6.64 \times 10^6$ m, orbiting its star at a distance of $r_O = 6.00 \times 10^9$ m. **(c)** Estimate the temperature of the planet using Stefan's equation. Assume all worlds are black bodies, with $e = 1$.

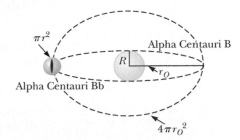

**STRATEGY** Calculating the power output in part **(a)** is a matter of substitution. To solve part **(b)**, it's necessary to find the fraction of the star's power intercepted by the planet. The star's energy crosses a sphere of area $A_O = 4\pi r_O^2$, where radius $r_O$ is the planet's distance from Alpha Centauri B. The cross-sectional area of the planet's disk, $A_{pd} = \pi r^2$, intercepts a fraction of this energy given by $A_{pd}/A_O$. (See Fig. 11.18.) Multiplying the star's power output by the fraction gives the amount of power the planet must both absorb and emit if in equilibrium, which is the answer to part **(b)**. Substitute it into Stefan's equation and solve for the planet's temperature, the answer for part **(c)**.

**Figure 11.18** (Example 11.11) The power emitted by Alpha Centauri B travels radially outward, crossing a sphere with the same radius, $r_O$, as Alpha Centauri Bb's orbital radius. The cross-sectional area of the planet intercepts a small part of that radiation. (*Note:* Figure not drawn to scale.)

......................................................

### SOLUTION

**(a)** Calculate the power output of Alpha Centauri B.

Compute the surface area of Alpha Centauri B:

$A = 4\pi R^2 = 4\pi (6.02 \times 10^8 \text{ m})^2 = 4.55 \times 10^{18} \text{ m}^2$

Write Stefan's equation and substitute values:

$P = \sigma A e T^4 = (5.67 \times 10^{-8} \text{ W/m}^2 \cdot \text{K}^4)(4.55 \times 10^{18} \text{ m}^2)$
$(1.00)(5\,790 \text{ K})^4 = \boxed{2.90 \times 10^{26} \text{ W}}$

**(b)** Calculate the power $P_I$ intercepted by a possible Earth-sized planet, Alpha Centauri Bb.

Calculate the area of the planet's disk, $A_{pd}$, and the area of a sphere, $A_O$, with the same radius as the planet's orbital radius:

$A_{pd} = \pi r^2 = \pi (6.64 \times 10^6 \text{ m})^2 = 1.39 \times 10^{14} \text{ m}^2$
$A_O = 4\pi r_O^2 = 4\pi (6.00 \times 10^9 \text{ m})^2 = 4.52 \times 10^{20} \text{ m}^2$

Find the fraction of the star's power intercepted by the planet:

$P_I = \left( \dfrac{A_{pd}}{A_O} \right) P = \left( \dfrac{1.39 \times 10^{14} \text{ m}^2}{4.52 \times 10^{20} \text{ m}^2} \right)(2.90 \times 10^{26} \text{ W})$

$= \boxed{8.92 \times 10^{19} \text{ W}}$

**(c)** Estimate the temperature of the planet using Stefan's equation.

Write Stefan's equation, set it equal to the intercepted power, $P_I$, and solve for the temperature. Note that the full planetary area, $4\pi r^2$, not just the disk area, must be used:

$P_I = \sigma A e T^4 = (5.67 \times 10^{-8} \text{ W/m}^2 \cdot \text{K}^4)(5.54 \times 10^{14} \text{ m}^2)(1.00)T^4$
$= (3.15 \times 10^7 \text{ W/K}^4)T^4 = 8.92 \times 10^{19} \text{ W}$

$T = \boxed{1.30 \times 10^3 \text{ K}}$

......................................................

**REMARKS** This calculation is only an estimate because the planet may not be a perfect black body, and the effects of an atmosphere—unlikely in this case—can greatly affect the typical average temperature on a given world.

*(Continued)*

**QUESTION 11.11** An implicit premise of Example 11.11 is that the planet will radiate away all the energy that it intercepts. Why is this a reasonable assumption?

**EXERCISE 11.11** (a) Calculate how much power Earth emits, using Stefan's equation and the Earth's average temperature of about 15.0°C. (b) Assuming a planet with identical characteristics to Earth orbits Alpha Centauri B and intercepts the power calculated in part (a) with its disk, estimate how far it must be from Alpha Centauri B. (The answer is a little greater than the distance from the Sun to Venus.)

**ANSWERS** (a) $2.00 \times 10^{17}$ W    (b) $1.21 \times 10^{11}$ m

### 11.5.5 The Dewar Flask

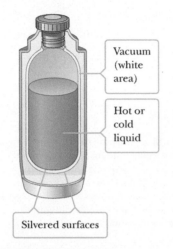

Vacuum (white area)

Hot or cold liquid

Silvered surfaces

**Figure 11.19** A cross-sectional view of a Thermos bottle designed to store hot or cold liquids.

The Thermos bottle, also called a **Dewar flask** (after its inventor), is designed to minimize energy transfer by conduction, convection, and radiation. The insulated bottle can store either cold or hot liquids for long periods. The standard vessel (Fig. 11.19) is a double-walled Pyrex glass with silvered walls. The space between the walls is evacuated to minimize energy transfer by conduction and convection. The silvered surface minimizes energy transfer by radiation because silver is a very good reflector and has very low emissivity. A further reduction in energy loss is achieved by reducing the size of the neck. Dewar flasks are commonly used to store liquid nitrogen (boiling point 77 K) and liquid oxygen (boiling point 90 K).

To confine liquid helium (boiling point 4.2 K), which has a very low heat of vaporization, it's often necessary to use a double Dewar system in which the Dewar flask containing the liquid is surrounded by a second Dewar flask. The space between the two flasks is filled with liquid nitrogen.

Some of the principles of the Thermos bottle are used in the protection of sensitive electronic instruments in orbiting space satellites. In half of its orbit around the Earth, a satellite is exposed to intense radiation from the Sun, and in the other half, it lies in the Earth's cold shadow. Without protection, its interior would be subjected to tremendous extremes of temperature. The interior of the satellite is wrapped with blankets of highly reflective aluminum foil. The foil's shiny surface reflects away much of the Sun's radiation while the satellite is in the unshaded part of the orbit and helps retain interior energy while the satellite is in the Earth's shadow.

**APPLICATION**
Thermos Bottles

# 11.6 Climate Change and Greenhouse Gases BIO

Many of the principles of energy transfer, and opposition to it, can be understood by studying the operation of a glass greenhouse. During the day, sunlight passes into the greenhouse and is absorbed by the walls, soil, plants, and so on. This absorbed visible light is subsequently reradiated as infrared radiation, causing the temperature of the interior to rise.

In addition, convection currents are inhibited in a greenhouse. As a result, warmed air can't rapidly pass over the surfaces of the greenhouse that are exposed to the outside air and thereby cause an energy loss by conduction through those surfaces. Most experts now consider this restriction to be a more important warming effect than the trapping of infrared radiation. In fact, experiments have shown that when the glass over a greenhouse is replaced by a special glass known to transmit infrared light, the temperature inside is lowered only slightly. On the basis of this evidence, the primary mechanism that raises the temperature of a greenhouse is not the trapping of infrared radiation, but the inhibition of airflow that occurs under any roof (in an attic, for example).

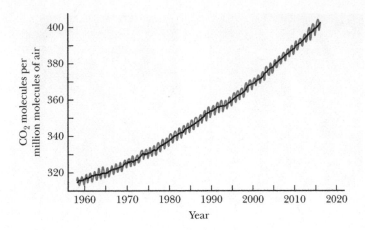

**Figure 11.20** The concentration of atmospheric carbon dioxide in parts per million (ppm) of dry air as a function of time during the latter part of the 20th century. These data were recorded at Mauna Loa Observatory in Hawaii. The yearly variations (rust-colored curve) coincide with growing seasons because vegetation absorbs carbon dioxide from the air. The steady increase (black curve) is of concern to scientists.

A phenomenon commonly known as the **greenhouse effect** can also play a major role in determining the Earth's temperature. First, note that the Earth's atmosphere is a good transmitter (and hence a poor absorber) of visible radiation and a good absorber of infrared radiation. The visible light that reaches the Earth's surface is absorbed and reradiated as infrared light, which in turn is absorbed (trapped) by the Earth's atmosphere. An extreme case is the warmest planet, Venus, which has a carbon dioxide ($CO_2$) atmosphere and temperatures approaching 850°F.

As fossil fuels (coal, oil, and natural gas) are burned, large amounts of carbon dioxide are released into the atmosphere, causing it to retain more energy. These emissions are of great concern to scientists and governments throughout the world. Many scientists are convinced that the 10% increase in the amount of atmospheric carbon dioxide since 1970 could lead to drastic changes in world climate. The increase in concentration of atmospheric carbon dioxide in the latter part of the 20th century and the first years of the 21st century is shown in Figure 11.20. According to one estimate, doubling the carbon dioxide content in the atmosphere will cause temperatures to increase by 2°C. In temperate regions such as Europe and the United States, a 2°C temperature rise would save billions of dollars per year in fuel costs. Unfortunately, it would also melt a large amount of land-based ice from Greenland and Antarctica, raising the level of the oceans and destroying many coastal regions. A 2°C rise would also increase the frequency of droughts and consequently decrease already low crop yields in tropical and sub-tropical countries. Even slightly higher average temperatures might make it impossible for certain plants and animals to survive in their customary ranges.

At present, about $3.5 \times 10^{10}$ tons of $CO_2$ are released into the atmosphere each year. Most of this gas results from human activities such as the burning of fossil fuels, the cutting of forests, and manufacturing processes. Another greenhouse gas is methane ($CH_4$), which is released in the digestive process of cows and other ruminants. This gas originates from that part of the animal's stomach called the *rumen*, where cellulose is digested. Termites are also major producers of this gas. Greenhouse gases such as nitrous oxide ($N_2O$) and sulfur dioxide ($SO_2$) are increasing due to automobile and industrial pollution.

The Montreal Protocol on Substances that Deplete the Ozone Layer is widely viewed as the most successful international environmental agreement, resulting in a 97% reduction in ozone-depleting substances. Many of these substances were replaced with hydrofluorocarbons (HFCs); it is now understood that HFCs can be up to 10 000 times more potent than $CO_2$ in contributing to climate change. The international community is working together on a 2016 Amendment to the Montreal Protocol to reduce the production and consumption of HFCs.

Whether the increasing greenhouse gases are responsible or not, there is convincing evidence that climate change is under way. Evidence comes from the

**Figure 11.21** *Death of an ice shelf.* The image in (a), taken on January 9, 1995, in the near-visible part of the spectrum, shows James Ross Island (spidery-shaped, just off center) before the iceberg calved, but after the disintegration of the ice shelf between James Ross Island and the Antarctic peninsula. In the image in part (b), taken on February 12, 1995, the iceberg has calved and begun moving away from land. The iceberg is about 78 km by 27 km and 200 m thick. A century ago James Ross Island was completely surrounded in ice that joined it to Antarctica.

 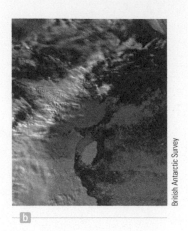

British Antarctic Survey

melting of ice in Antarctica and the retreat of glaciers at widely scattered sites throughout the world (see Fig. 11.21). For example, satellite images of Antarctica show James Ross Island completely surrounded by water for the first time since maps were made, about 100 years ago. Previously, the island was connected to the mainland by an ice bridge. In addition, at various places across the continent, ice shelves are retreating, some at a rapid rate.

Perhaps at no place in the world are glaciers monitored with greater interest than in Switzerland. There, it is found that the Alps have lost about 50% of their glacial ice compared to 130 years ago. The retreat of glaciers on high-altitude peaks in the tropics is even more severe than in Switzerland. The Lewis glacier on Mount Kenya and the snows of Kilimanjaro are two examples. In certain regions of the planet where glaciers are near large bodies of water and are fed by large and frequent snows, however, glaciers continue to advance, so the overall picture of a catastrophic global warming scenario may be premature. In about 50 years, though, the amount of carbon dioxide in the atmosphere is expected to be about twice what it was in the preindustrial era. Because of the possible catastrophic consequences, most scientists voice the concern that reductions in greenhouse gas emissions need to be made now.

# SUMMARY

## 11.1 Heat and Internal Energy

**Internal energy** is associated with a system's microscopic components. Internal energy includes the kinetic energy of translation, rotation, and vibration of molecules, as well as potential energy.

**Heat** is the transfer of energy across the boundary of a system resulting from a temperature difference between the system and its surroundings. The symbol $Q$ represents the amount of energy transferred.

The **calorie** is the amount of energy necessary to raise the temperature of 1 g of water from 14.5°C to 15.5°C. The **mechanical equivalent of heat** is 4.186 J/cal.

## 11.2 Specific Heat

## 11.3 Calorimetry

The energy required to change the temperature of a substance of mass $m$ by an amount $\Delta T$ is

$$Q = mc\Delta T \qquad [11.3]$$

where $c$ is the **specific heat** of the substance. In calorimetry problems the specific heat of a substance can be determined by placing it in water of known temperature, isolating the system, and measuring the temperature at equilibrium. The sum of all energy gains and losses for all the objects in an isolated system is given by

$$\sum Q_k = 0 \qquad [11.5]$$

where $Q_k$ is the energy change in the $k$th object in the system. This equation can be solved for the unknown specific heat or used to determine an equilibrium temperature.

## 11.4 Latent Heat and Phase Change

The energy required to change the phase of a pure substance of mass $m$ is

$$Q = \pm mL \qquad [11.6]$$

where $L$ is the **latent heat** of the substance. The latent heat of fusion, $L_f$, describes an energy transfer during a change from a solid phase to a liquid phase (or vice versa), while the latent heat of vaporization, $L_v$, describes an energy transfer during a change from a liquid phase to a gaseous phase (or vice versa). Calorimetry problems involving phase changes are handled with Equation 11.5, with latent heat terms added to the specific heat terms.

## 11.5 Energy Transfer

Energy can be transferred by several different processes, including work, discussed in Topic 5, and by conduction, convection, and radiation. **Conduction** can be viewed as an exchange of kinetic energy between colliding molecules or electrons. The rate at which energy transfers by conduction through a slab of area $A$ and thickness $L$ is

$$P = kA \frac{(T_h - T_c)}{L} \qquad [11.7]$$

where $k$ is the **thermal conductivity** of the material making up the slab (Fig. 11.22).

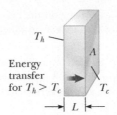

**Figure 11.22** Energy transfer through a slab is proportional to the cross-sectional area and temperature difference, and inversely proportional to the thickness.

Energy is transferred by **convection** as a substance moves from one place to another.

All objects emit **radiation** from their surfaces in the form of electromagnetic waves at a net rate of

$$P_{net} = \sigma Ae(T^4 - T_0^4) \qquad [11.11]$$

where $T$ is the temperature of the object and $T_0$ is the temperature of the surroundings. An object that is hotter than its surroundings radiates more energy than it absorbs, whereas a body that is cooler than its surroundings absorbs more energy than it radiates.

## CONCEPTUAL QUESTIONS

1. Rub the palm of your hand on a metal surface for 30 to 45 seconds. Place the palm of your other hand on an unrubbed portion of the surface and then on the rubbed portion. The rubbed portion will feel warmer. Now repeat this process on a wooden surface. Why does the temperature difference between the rubbed and unrubbed portions of the wood surface seem larger than for the metal surface?

2. On a clear, cold night, why does frost tend to form on the tops, rather than on the sides, of mailboxes and cars?

3. Substance A has twice the specific heat of substance B. Equal masses of the two substances, at different temperatures, are placed in thermal contact and allowed to come to equilibrium. (a) What is the ratio $Q_B/Q_A$ of the energy transferred to (or from) the samples? (*Hint:* Remember to include the correct signs.) (b) What is the ratio $\Delta T_B/\Delta T_A$ of their temperature changes?

4. Equal masses of substance A at 10.0°C and substance B at 90.0°C are placed in a well-insulated container of negligible mass and allowed to come to equilibrium. If the equilibrium temperature is 75.0°C, which substance has the larger specific heat? (a) substance A (b) substance B (c) The specific heats are identical. (d) The answer depends on the exact initial temperatures. (e) More information is required.

5. Identical amounts of thermal energy are added to each of three isolated, equal-mass samples A, B, and C of unknown substances. If their temperature changes are ordered as $\Delta T_B > \Delta T_C > \Delta T_A$, which sample has the largest specific heat?

6. Different amounts of thermal energy are added to each of three isolated samples A, B, and C of lead. If the energy transfers are ordered as $Q_B > Q_C > Q_A$ and each sample undergoes the same temperature change, which sample has the largest mass?

7. Cups of water for coffee or tea can be warmed with a coil that is immersed in the water and raised to a high temperature by means of electricity. (a) Why do the instructions warn users not to operate the coils in the absence of water? (b) Can the immersion coil be used to warm up a cup of stew?

8. The U.S. penny is now made of copper-coated zinc. Can a calorimetric experiment be devised to test for the metal content in a collection of pennies? If so, describe the procedure.

9. A tile floor may feel uncomfortably cold to your bare feet, but a carpeted floor in an adjoining room at the same temperature feels warm. Why?

10. In a calorimetry experiment, three samples A, B, and C with $T_A > T_B > T_C$ are placed in thermal contact. When the samples have reached thermal equilibrium at a common temperature $T$, which one of the following *must* be true? (a) $Q_A > Q_B > Q_C$ (b) $Q_A < 0$, $Q_B < 0$, and $Q_C > 0$ (c) $T > T_B$ (d) $T < T_B$ (e) $T_A > T > T_C$.

11. Figure CQ11.11 shows a composite bar made of three different materials that connects a hot reservoir at 100°C to a cold reservoir at 0°C. If the sections A, B, and C all have the same dimensions and the temperatures shown in the figure are constant, rank the thermal conductivities from largest to smallest.

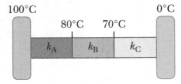

**Figure CQ11.11**

12. Objects A and B have the same size and shape with emissivities $e_A$ and $e_B$ and temperatures $T_A$ and $T_B$, respectively. (a) If $e_A = e_B$ and $T_B = 4T_A$, what is the ratio $P_B/P_A$ of their radiated powers? (b) If, instead, they radiate the same power and $e_A = 4e_B$, what is the ratio $T_B/T_A$ of their Kelvin temperatures?

13. A poker is a stiff, nonflammable rod used to push burning logs around in a fireplace. Suppose it is to be made of a single material. For best functionality and safety, should the poker be made from a material with (a) high specific heat and high thermal conductivity, (b) low specific heat and low thermal conductivity, (c) low specific heat and high thermal conductivity, (d) high specific heat and low thermal conductivity, or (e) low specific heat and low density?

14. On a very hot day, it's possible to cook an egg on the hood of a car. Would you select a black car or a white car on which to cook your egg? Why?

15. A person shakes a sealed, insulated bottle containing coffee for a few minutes. What is the change in the temperature of the coffee? (a) a large decrease (b) a slight decrease (c) no change (d) a slight increase (e) a large increase

16. Star A has twice the radius and twice the absolute temperature of star B. What is the ratio of the power output of star A to that of star B? The emissivity of both stars can be assumed to be 1. (a) 4 (b) 8 (c) 16 (d) 32 (e) 64

## PROBLEMS

*See the Preface for an explanation of the icons used in this problems set.*

### 11.1 Heat and Internal Energy

1. Convert $3.50 \times 10^3$ cal to the equivalent number of (a) kilocalories (also known as Calories, used to describe the energy content of food) and (b) joules.

2. A medium-sized banana provides about 105 Calories of energy. (a) Convert 105 Cal to joules. (b) Suppose that amount of energy is transformed into kinetic energy of a 1.00-kg object initially at rest. Calculate the final speed of the object. (c) If that same amount of energy is added to 3.79 kg (about 1 gal) of water at 20.0°C, what is the water's final temperature?

### 11.2 Specific Heat

3. **BIO** **Q|C** A 75-kg sprinter accelerates from rest to a speed of 11.0 m/s in 5.0 s. (a) Calculate the mechanical work done by the sprinter during this time. (b) Calculate the average power the sprinter must generate. (c) If the sprinter converts food energy to mechanical energy with an efficiency of 25%, at what average rate is he burning Calories? (d) What happens to the other 75% of the food energy being used?

4. **BIO** A 55-kg student eats a 540-Calorie (540 kcal) jelly doughnut for breakfast. (a) How many joules of energy are the equivalent of one jelly doughnut? (b) How many stairs must the student climb to perform an amount of mechanical work equivalent to the food energy in one jelly doughnut? Assume the height of a single stair is 15 cm. (c) If the human body is only 25% efficient in converting chemical energy to mechanical energy, how many stairs must the woman climb to work off her breakfast?

5. **BIO** A person's basal metabolic rate (BMR) is the rate at which energy is expended while resting in a neutrally temperate environment. A typical BMR is $7.00 \times 10^6$ J/day. Convert this BMR to units of (a) watts and (b) kilocalories (or Calories) per hour. (c) Suppose a 1.00-kg object's gravitation potential energy is increased at a rate equal to this typical BMR. Find the rate of change of the object's height in m/s.

6. **T** The temperature of a silver bar rises by 10.0°C when it absorbs 1.23 kJ of energy by heat. The mass of the bar is 525 g. Determine the specific heat of silver from these data.

7. The highest recorded waterfall in the world is found at Angel Falls in Venezuela. Its longest single waterfall has a height of 807 m. If water at the top of the falls is at 15.0°C, what is the maximum temperature of the water at the bottom of the falls? Assume all the kinetic energy of the water as it reaches the bottom goes into raising the water's temperature.

8. An aluminum rod is 20.0 cm long at 20.0°C and has a mass of 0.350 kg. If $1.00 \times 10^4$ J of energy is added to the rod by heat, what is the change in length of the rod?

9. Lake Erie contains roughly $4.00 \times 10^{11}$ m$^3$ of water. (a) How much energy is required to raise the temperature of that volume of water from 11.0°C to 12.0°C? (b) How many years would it take to supply this amount of energy by using the $1.00 \times 10^3$ MW exhaust energy of an electric power plant?

10. **Q|C** A 3.00-g copper coin at 25.0°C drops 50.0 m to the ground. (a) Assuming 60.0% of the change in gravitational potential energy of the coin–Earth system goes into increasing the internal energy of the coin, determine the coin's final temperature. (b) Does the result depend on the mass of the coin? Explain.

11. **V** A 5.00-g lead bullet traveling at $3.00 \times 10^2$ m/s is stopped by a large tree. If half the kinetic energy of the bullet is transformed into internal energy and remains with the bullet while the other half is transmitted to the tree, what is the increase in temperature of the bullet?

12. The apparatus shown in Figure P11.12 was used by Joule to measure the mechanical equivalent of heat. Work is done on the water by a rotating paddle wheel, which is driven by two blocks falling at a constant speed. The temperature of the stirred water increases due to the friction between the water and the paddles. If the energy lost in the bearings and through the walls is neglected, then the loss in potential energy associated with the blocks equals the work done by the paddle wheel on the water. If each block has a mass of 1.50 kg and the insulated tank is filled with 0.200 kg of water, what is the increase in temperature of the water after the blocks fall through a distance of 3.00 m?

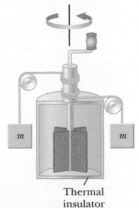

**Figure P11.12** The falling weights rotate the paddles, causing the temperature of the water to increase.

13. A 0.200-kg aluminum cup contains 800. g of water in thermal equilibrium with the cup at 80.°C. The combination of cup and water is cooled uniformly so that the temperature decreases by 1.5°C per minute. At what rate is energy being removed? Express your answer in watts.

14. A 1.5-kg copper block is given an initial speed of 3.0 m/s on a rough horizontal surface. Because of friction, the block finally

comes to rest. (a) If the block absorbs 85% of its initial kinetic energy as internal energy, calculate its increase in temperature. (b) What happens to the remaining energy?

15. A swimming pool filled with water has dimensions of 5.00 m × 10.0 m × 1.78 m. (a) Find the mass of water in the pool. (b) Find the thermal energy required to heat the pool water from 15.5°C to 26.5°C. (c) Calculate the cost of heating the pool from 15.5°C to 26.5°C if electrical energy costs $0.100 per kilowatt-hour.

16. **GP** In the summer of 1958 in St. Petersburg, Florida, a new sidewalk was poured near the childhood home of one of the authors. No expansion joints were supplied, and by mid-July, the sidewalk had been completely destroyed by thermal expansion and had to be replaced, this time with the important addition of expansion joints! This event is modeled here.

A slab of concrete 4.00 cm thick, 1.00 m long, and 1.00 m wide is poured for a sidewalk at an ambient temperature of 25.0°C and allowed to set. The slab is exposed to direct sunlight and placed in a series of such slabs without proper expansion joints, so linear expansion is prevented. (a) Using the linear expansion equation (Eq. 10.4), eliminate $\Delta L$ from the equation for compressive stress and strain (Eq. 9.3). (b) Use the expression found in part (a) to eliminate $\Delta T$ from Equation 11.3, obtaining a symbolic equation for thermal energy transfer $Q$. (c) Compute the mass of the concrete slab given that its density is $2.40 \times 10^3$ kg/m³. (d) Concrete has an ultimate compressive strength of $2.00 \times 10^7$ Pa, specific heat of 880 J/kg · °C, and Young's modulus of $2.1 \times 10^{10}$ Pa. How much thermal energy must be transferred to the slab to reach this compressive stress? (e) What temperature change is required? (f) If the Sun delivers $1.00 \times 10^3$ W of power to the top surface of the slab and if half the energy, on the average, is absorbed and retained, how long does it take the slab to reach the point at which it is in danger of cracking due to compressive stress?

## 11.3 Calorimetry

17. What mass of water at 25.0°C must be allowed to come to thermal equilibrium with a 1.85-kg cube of aluminum initially at $1.50 \times 10^2$°C to lower the temperature of the aluminum to 65.0°C? Assume any water turned to steam subsequently recondenses.

18. Lead pellets, each of mass 1.00 g, are heated to 200.°C. How many pellets must be added to 0.500 kg of water that is initially at 20.0°C to make the equilibrium temperature 25.0°C? Neglect any energy transfer to or from the container.

19. **V** An aluminum cup contains 225 g of water and a 40-g copper stirrer, all at 27°C. A 400-g sample of silver at an initial temperature of 87°C is placed in the water. The stirrer is used to stir the mixture until it reaches its final equilibrium temperature of 32°C. Calculate the mass of the aluminum cup.

20. A large room in a house holds 975 kg of dry air at 30.0°C. A woman opens a window briefly and a cool breeze brings in an additional 50.0 kg of dry air at 18.0°C. At what temperature will the two air masses come into thermal equilibrium, assuming they form a closed system? (The specific heat of dry air is 1 006 J/kg · °C, although that value will cancel out of the calorimetry equation.)

21. **V QC** An aluminum calorimeter with a mass of 0.100 kg contains 0.250 kg of water. The calorimeter and water are in thermal equilibrium at 10.0°C. Two metallic blocks are placed into the water. One is a 50.0-g piece of copper at 80.0°C. The other has a mass of 70.0 g and is originally at a temperature of 100.°C. The entire system stabilizes at a final temperature of 20.0°C. (a) Determine the specific heat of the unknown sample. (b) Using the data in Table 11.1, can you make a positive identification of the unknown material? Can you identify a possible material? (c) Explain your answers for part (b).

22. **T** A 1.50-kg iron horseshoe initially at 600°C is dropped into a bucket containing 20.0 kg of water at 25.0°C. What is the final temperature of the water–horseshoe system? Ignore the heat capacity of the container and assume a negligible amount of water boils away.

23. A student drops two metallic objects into a 120-g steel container holding 150 g of water at 25°C. One object is a 200-g cube of copper that is initially at 85°C, and the other is a chunk of aluminum that is initially at 5.0°C. To the surprise of the student, the water reaches a final temperature of 25°C, precisely where it started. What is the mass of the aluminum chunk?

24. When a driver brakes an automobile, the friction between the brake drums and the brake shoes converts the car's kinetic energy to thermal energy. If a 1 500-kg automobile traveling at 30 m/s comes to a halt, how much does the temperature rise in each of the four 8.0-kg iron brake drums? (The specific heat of iron is 448 J/kg · °C.)

25. A Styrofoam cup holds 0.275 kg of water at 25.0°C. Find the final equilibrium temperature after a 0.100-kg block of copper at 90.0°C is placed in the water. Neglect any thermal energy transfer with the Styrofoam cup.

26. An unknown substance has a mass of 0.125 kg and an initial temperature of 95.0°C. The substance is then dropped into a calorimeter made of aluminum containing 0.285 kg of water initially at 25.0°C. The mass of the aluminum container is 0.150 kg, and the temperature of the calorimeter increases to a final equilibrium temperature of 32.0°C. Assuming no thermal energy is transferred to the environment, calculate the specific heat of the unknown substance.

## 11.4 Latent Heat and Phase Change

27. Suppose $9.30 \times 10^5$ J of energy are transferred to 2.00 kg of ice at 0°C. (a) Calculate the energy required to melt all the ice into liquid water. (b) How much energy remains to raise the temperature of the liquid water? (c) Determine the final temperature of the liquid water in Celsius.

28. How much thermal energy is required to boil 2.00 kg of water at 100.0°C into steam at 125°C? The latent heat of vaporization of water is $2.26 \times 10^6$ J/kg and the specific heat of steam is 2 010 J/(kg · °C).

29. A 75-g ice cube at 0°C is placed in 825 g of water at 25°C. What is the final temperature of the mixture?

30. A 50.-kg ice cube at 0°C is heated until 45 g has become water at 100.°C and 5.0 g has become steam at 100.°C. How much energy was added to accomplish the transformation?

31. **V** A 100.-g cube of ice at 0°C is dropped into 1.0 kg of water that was originally at 80.°C. What is the final temperature of the water after the ice has melted?

32. How much energy is required to change a 40.-g ice cube from ice at −10.°C to steam at 110.°C?

33. **T** A 75-kg cross-country skier glides over snow as in Figure P11.33. The coefficient of friction between skis and snow is 0.20. Assume all the snow beneath her skis is at 0°C and that all the internal energy generated by friction is added to snow, which sticks to her skis until it melts. How far would she have to ski to melt 1.0 kg of snow?

**Figure P11.33**

34. **GP** Into a 0.500-kg aluminum container at 20.0°C is placed 6.00 kg of ethyl alcohol at 30.0°C and 1.00 kg ice at −10.0°C. Assume the system is insulated from its environment. (a) Identify all five thermal energy transfers that occur as the system goes to a final equilibrium temperature *T*. Use the form "substance at *X*°C to substance at *Y*°C." (b) Construct a table similar to the table in Example 11.5. (c) Sum all terms in the right-most column of the table and set the sum equal to zero. (d) Substitute information from the table into the equation found in part (c) and solve for the final equilibrium temperature, *T*.

35. A 40.-g block of ice is cooled to −78°C and is then added to 560 g of water in an 80.-g copper calorimeter at a temperature of 25°C. Determine the final temperature of the system consisting of the ice, water, and calorimeter. (If not all the ice melts, determine how much ice is left.) Remember that the ice must first warm to 0°C, melt, and then continue warming as water. (The specific heat of ice is 0.500 cal/g · °C = 2 090 J/kg · °C.)

36. **BIO** When you jog, most of the food energy you burn above your basal metabolic rate (BMR) ends up as internal energy that would raise your body temperature if it were not eliminated. The evaporation of perspiration is the primary mechanism for eliminating this energy. Determine the amount of water you lose to evaporation when running for 30. minutes at a rate that uses $4.00 \times 10^2$ kcal/h above your BMR. (That amount is often considered to be the "maximum fat-burning" energy output.) The metabolism of 1.0 grams of fat generates approximately 9.0 kcal of energy and produces approximately 1.0 grams of water. (The hydrogen atoms in the fat molecule are transferred to oxygen to form water.) What fraction of your need for water will be provided by fat metabolism? (The latent heat of vaporization of water at room temperature is $2.5 \times 10^6$ J/kg.)

37. A high-end gas stove usually has at least one burner rated at 14 000 Btu/h. (a) If you place a 0.25-kg aluminum pot containing 2.0 liters of water at 20.°C on this burner, how long will it take to bring the water to a boil, assuming all the heat from the burner goes into the pot? (b) Once boiling begins, how much time is required to boil all the water out of the pot?

38. **BIO** A 60.0-kg runner expends $3.00 \times 10^2$ W of power while running a marathon. Assuming 10.0% of the energy is delivered to the muscle tissue and that the excess energy is removed from the body primarily by sweating, determine the volume of bodily fluid (assume it is water) lost per hour. (At 37.0°C, the latent heat of vaporization of water is $2.41 \times 10^6$ J/kg.)

39. Steam at 100.°C is added to ice at 0°C. (a) Find the amount of ice melted and the final temperature when the mass of steam is 10. g and the mass of ice is 50. g. (b) Repeat with steam of mass 1.0 g and ice of mass 50. g.

40. **BIO** The excess internal energy of metabolism is exhausted through a variety of channels, such as through radiation and evaporation of perspiration. Consider another pathway for energy loss: moisture in exhaled breath. Suppose you breathe out 22.0 breaths per minute, each with a volume of 0.600 L. Suppose also that you inhale dry air and exhale air at 37°C containing water vapor with a vapor pressure of 3.20 kPa. The vapor comes from the evaporation of liquid water in your body. Model the water vapor as an ideal gas. Assume its latent heat of evaporation at 37°C is the same as its heat of vaporization at 100.°C. Calculate the rate at which you lose energy by exhaling humid air.

41. **QIC** A 3.00-g lead bullet at 30.0°C is fired at a speed of 2.40 $\times 10^2$ m/s into a large, fixed block of ice at 0°C, in which it becomes embedded. (a) Describe the energy transformations that occur as the bullet is cooled. What is the final temperature of the bullet? (b) What quantity of ice melts?

## 11.5 Energy Transfer

42. A glass windowpane in a home is 0.62 cm thick and has dimensions of 1.0 m × 2.0 m. On a certain day, the indoor temperature is 25°C and the outdoor temperature is 0°C. (a) What is the rate at which energy is transferred by heat through the glass? (b) How much energy is lost through the window in one day, assuming the temperatures inside and outside remain constant?

43. A pond with a flat bottom has a surface area of 820 m² and a depth of 2.0 m. On a warm day, the surface water is at a temperature of 25°C, while the bottom of the pond is at 12°C. Find the rate at which energy is transferred by conduction from the surface to the bottom of the pond.

44. **BIO** The thermal conductivities of human tissues vary greatly. Fat and skin have conductivities of about 0.20 W/m·K and 0.020 W/m·K, respectively, while other tissues inside the body have conductivities of about 0.50 W/m·K. Assume that between the core region of the body and the skin surface lies a skin layer of 1.0 mm, fat layer of 0.50 cm, and 3.2 cm of other tissues. (a) Find the *R*-factor for each of these layers, and the equivalent *R*-factor for all layers taken together, retaining two digits. (b) Find the rate of energy loss when the core temperature is 37°C and the exterior temperature is 0°C. Assume that both a protective layer of clothing and an insulating layer of unmoving air are absent, and a body area of 2.0 m².

45. A steam pipe is covered with 1.50-cm-thick insulating material of thermal conductivity 0.200 cal/cm · °C · s. How much energy is lost every second when the steam is at 200.°C and the

surrounding air is at 20.0°C? The pipe has a circumference of 800. cm and a length of 50.0 m. Neglect losses through the ends of the pipe.

46. The average thermal conductivity of the walls (including windows) and roof of a house in Figure P11.46 is $4.8 \times 10^{-4}$ kW/m·°C, and their average thickness is 21.0 cm. The house is heated with natural gas, with a heat of combustion (energy released per cubic meter of gas burned) of 9 300 kcal/m³. How many cubic meters of gas must be burned each day to maintain an inside temperature of 25.0°C if the outside temperature is 0.0°C? Disregard surface air layers, radiation, and energy loss by heat through the ground.

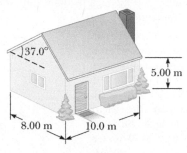

37.0°

5.00 m

8.00 m    10.0 m

**Figure P11.46**

47. Consider two cooking pots of the same dimensions, each containing the same amount of water at the same initial temperature. The bottom of the first pot is made of copper, while the bottom of the second pot is made of aluminum. Both pots are placed on a hot surface having a temperature of 145°C. The water in the copper-bottomed pot boils away completely in 425 s. How long does it take the water in the aluminum-bottomed pot to boil away completely?

48. A thermopane window consists of two glass panes, each 0.50 cm thick, with a 1.0-cm-thick sealed layer of air in between. (a) If the inside surface temperature is 23°C and the outside surface temperature is 0.0°C, determine the rate of energy transfer through 1.0 m² of the window. (b) Compare your answer to (a) with the rate of energy transfer through 1.0 m² of a single 1.0-cm-thick pane of glass. Disregard surface air layers.

49. **T** A copper rod and an aluminum rod of equal diameter are joined end to end in good thermal contact. The temperature of the free end of the copper rod is held constant at 100.°C and that of the far end of the aluminum rod is held at 0°C. If the copper rod is 0.15 m long, what must be the length of the aluminum rod so that the temperature at the junction is 50.°C?

50. A Styrofoam box has a surface area of 0.80 m² and a wall thickness of 2.0 cm. The temperature of the inner surface is 5.0°C, and the outside temperature is 25°C. If it takes 8.0 h for 5.0 kg of ice to melt in the container, determine the thermal conductivity of the Styrofoam.

51. **V** A rectangular glass window pane on a house has a width of 1.0 m, a height of 2.0 m, and a thickness of 0.40 cm. Find the energy transferred through the window by conduction in 12 hours on a day when the inside temperature of the house is 22°C and the outside temperature is 2.0°C. Take surface air layers into consideration.

52. A granite ball of radius 2.00 m and emissivity 0.450 is heated to 135°C. (a) Convert the given temperature to Kelvin. (b) What is the surface area of the ball? (c) If the ambient temperature is 25.0°C, what net power does the ball radiate?

53. Measurements on two stars indicate that Star X has a surface temperature of 5 727°C and Star Y has a surface temperature of 11 727°C. If both stars have the same radius, what is the ratio of the luminosity (total power output) of Star Y to the luminosity of Star X? Both stars can be considered to have an emissivity of 1.0.

54. The filament of a 75-W light bulb is at a temperature of 3 300 K. Assuming the filament has an emissivity $e = 1.0$, find its surface area.

## Additional Problems

55. The bottom of a copper kettle has a 10.0-cm radius and is 2.00 mm thick. The temperature of the outside surface is 102°C, and the water inside the kettle is boiling at 1 atm of pressure. Find the rate at which energy is being transferred through the bottom of the kettle.

56. A family comes home from a long vacation with laundry to do and showers to take. The water heater has been turned off during the vacation. If the heater has a capacity of 50.0 gallons and a 4 800-W heating element, how much time is required to raise the temperature of the water from 20.0°C to 60.0°C? Assume the heater is well insulated and no water is withdrawn from the tank during that time.

57. **T** A 0.040.-kg ice cube floats in 0.200 kg of water in a 0.100-kg copper cup; all are at a temperature of 0°C. A piece of lead at 98°C is dropped into the cup, and the final equilibrium temperature is 12°C. What is the mass of the lead?

58. **BIO** The surface area of an unclothed person is 1.50 m², and his skin temperature is 33.0°C. The person is located in a dark room with a temperature of 20.0°C, and the emissivity of the skin is $e = 0.95$. (a) At what rate is energy radiated by the body? (b) What is the significance of the sign of your answer?

59. **QC** A student measures the following data in a calorimetry experiment designed to determine the specific heat of aluminum:

| | |
|---|---|
| Initial temperature of water and calorimeter: | 70.0°C |
| Mass of water: | 0.400 kg |
| Mass of calorimeter: | 0.040 kg |
| Specific heat of calorimeter: | 0.63 kJ/kg·°C |
| Initial temperature of aluminum: | 27.0°C |
| Mass of aluminum: | 0.200 kg |
| Final temperature of mixture: | 66.3°C |

Use these data to determine the specific heat of aluminum. Explain whether your result is within 15% of the value listed in Table 11.1.

60. **BIO** Overall, 80% of the energy used by the body must be eliminated as excess thermal energy and needs to be dissipated. The mechanisms of elimination are radiation, evaporation of sweat (2 430 kJ/kg), evaporation from the lungs (38 kJ/h), conduction, and convection.

A person working out in a gym has a metabolic rate of 2 500 kJ/h. His body temperature is 37°C, and the outside temperature 24°C. Assume the skin has an area of 2.0 m² and emissivity of 0.97. (a) At what rate is his excess thermal energy dissipated by radiation? (b) If he eliminates 0.40 kg of perspiration during that hour, at what rate is thermal energy dissipated by evaporation of sweat? (c) At what rate is energy eliminated by evaporation from the lungs? (d) At what rate must the remaining excess energy be eliminated through conduction and convection?

61. Liquid helium has a very low boiling point, 4.2 K, as well as a very low latent heat of vaporization, $2.00 \times 10^4$ J/kg. If energy is transferred to a container of liquid helium at the boiling point from an immersed electric heater at a rate of 10.0 W, how long does it take to boil away 2.00 kg of the liquid?

62. A class of 10 students taking an exam has a power output per student of about 200 W. Assume the initial temperature of the room is 20°C and that its dimensions are 6.0 m by 15.0 m by 3.0 m. What is the temperature of the room at the end of 1.0 h if all the energy remains in the air in the room and none is added by an outside source? The specific heat of air is 837 J/kg · °C, and its density is about $1.3 \times 10^{-3}$ g/cm$^3$.

63. A bar of gold (Au) is in thermal contact with a bar of silver (Ag) of the same length and area (Fig. P11.63). One end of the compound bar is maintained at 80.0°C, and the opposite end is at 30.0°C. Find the temperature at the junction when the energy flow reaches a steady state.

```
          ┌─────────┐
          │ 80.0°C  │
          └─────────┘
              Au
Insulation ──┤
              Ag
          ┌─────────┐
          │ 30.0°C  │
          └─────────┘
```

**Figure P11.63**

64. An iron plate is held against an iron wheel so that a sliding frictional force of 50. N acts between the two pieces of metal. The relative speed at which the two surfaces slide over each other is 40. m/s. (a) Calculate the rate at which mechanical energy is converted to internal energy. (b) The plate and the wheel have masses of 5.0 kg each, and each receives 50% of the internal energy. If the system is run as described for 10. s and each object is then allowed to reach a uniform internal temperature, what is the resultant temperature increase?

65. An automobile has a mass of 1 500 kg, and its aluminum brakes have an overall mass of 6.00 kg. (a) Assuming all the internal energy transformed by friction when the car stops is deposited in the brakes and neglecting energy transfer, how many times could the car be braked to rest starting from 25.0 m/s before the brakes would begin to melt? (Assume an initial temperature of 20.0°C.) (b) Identify some effects that are neglected in part (a) but are likely to be important in a more realistic assessment of the temperature increase of the brakes.

66. Three liquids are at temperatures of 10°C, 20°C, and 30°C, respectively. Equal masses of the first two liquids are mixed, and the equilibrium temperature is 17°C. Equal masses of the second and third are then mixed, and the equilibrium temperature is 28°C. Find the equilibrium temperature when equal masses of the first and third are mixed.

67. Earth's surface absorbs an average of about 960. W/m$^2$ from the Sun's irradiance. The power absorbed is $P_{abs} = (960.$ W/m$^2)$ $(A_{disc})$, where $A_{disc} = \pi R_E^2$ is Earth's projected area. An equal amount of power is radiated so that Earth remains in thermal equilibrium with its environment at nearly 0 K. Estimate Earth's surface temperature by setting the radiated power from Stefan's law equal to the absorbed power and solving for the temperature in Kelvin. In Stefan's law, assume $e = 1$ and take the area to be $A = 4\pi R_E^2$, the surface area of a spherical Earth. (*Note*: Earth's atmosphere acts like a blanket and warms

the planet to a global average about 30 K above the value calculated here.)

68. A wood stove is used to heat a single room. The stove is cylindrical in shape, with a diameter of 40.0 cm and a length of 50.0 cm, and operates at a temperature of 400.°F. (a) If the temperature of the room is 70.0°F, determine the amount of radiant energy delivered to the room by the stove each second if the emissivity is 0.920. (b) If the room is a square with walls that are 8.00 ft high and 25.0 ft wide, determine the *R*-value needed in the walls and ceiling to maintain the inside temperature at 70.0°F if the outside temperature is 32.0°F. Note that we are ignoring any heat conveyed by the stove via convection and any energy lost through the walls (and windows!) via convection or radiation.

69. A "solar cooker" consists of a curved reflecting mirror that focuses sunlight onto the object to be heated (Fig. P11.69). The solar power per unit area reaching the Earth at the location of a 0.50-m-diameter solar cooker is 600. W/m$^2$. Assuming 50% of the incident energy is converted to thermal energy, how long would it take to boil away 1.0 L of water initially at 20.°C? (Neglect the specific heat of the container.)

**Figure P11.69**

70. **BIO** For bacteriological testing of water supplies and in medical clinics, samples must routinely be incubated for 24 h at 37°C. A standard constant-temperature bath with electric heating and thermostatic control is not suitable in developing nations without continuously operating electric power lines. Peace Corps volunteer and MIT engineer Amy Smith invented a low-cost, low-maintenance incubator to fill the need. The device consists of a foam-insulated box containing several packets of a waxy material that melts at 37.0°C, interspersed among tubes, dishes, or bottles containing the test samples and growth medium (food for bacteria). Outside the box, the waxy material is first melted by a stove or solar energy collector. Then it is put into the box to keep the test samples warm as it solidifies. The heat of fusion of the phase-change material is 205 kJ/kg. Model the insulation as a panel with surface area 0.490 m$^2$, thickness 9.50 cm, and conductivity 0.012 0 W/m · °C. Assume the exterior temperature is 23.0°C for 12.0 h and 16.0°C for 12.0 h. (a) What mass of the waxy material is required to conduct the bacteriological test? (b) Explain why your calculation can be done without knowing the mass of the test samples or of the insulation.

71. The surface of the Sun has a temperature of about 5 800 K. The radius of the Sun is $6.96 \times 10^8$ m. Calculate the total energy radiated by the Sun each second. Assume the emissivity of the Sun is 0.986.

72. **BIO** The evaporation of perspiration is the primary mechanism for cooling the human body. Estimate the amount of water you will lose when you bake in the sun on the beach for an hour. Use a value of 1 000 W/m$^2$ for the intensity of sunlight and note that the energy required to evaporate a liquid at a particular temperature is approximately equal to the sum of the energy required to raise its temperature to the boiling

point and the latent heat of vaporization (determined at the boiling point).

73. At time $t = 0$, a vessel contains a mixture of 10. kg of water and an unknown mass of ice in equilibrium at 0°C. The temperature of the mixture is measured over a period of an hour, with the following results: During the first 50. min, the mixture remains at 0°C; from 50. min to 60. min, the temperature increases steadily from 0°C to 2.0°C. Neglecting the heat capacity of the vessel, determine the mass of ice that was initially placed in it. Assume a constant power input to the container.

74. **Q|C** An ice-cube tray is filled with 75.0 g of water. After the filled tray reaches an equilibrium temperature 20.0°C, it is placed in a freezer set at −8.00°C to make ice cubes. (a) Describe the processes that occur as energy is being removed from the water to make ice. (b) Calculate the energy that must be removed from the water to make ice cubes at −8.00°C.

75. An aluminum rod and an iron rod are joined end to end in good thermal contact. The two rods have equal lengths and radii. The free end of the aluminum rod is maintained at a temperature of 100.°C, and the free end of the iron rod is maintained at 0°C. (a) Determine the temperature of the interface between the two rods. (b) If each rod is 15 cm long and each has a cross-sectional area of 5.0 cm², what quantity of energy is conducted across the combination in 30. min?

# 12 The Laws of Thermodynamics

**ACCORDING TO THE FIRST LAW OF THERMODYNAMICS**, the internal energy of a system can be increased either by adding energy to the system or by doing work on it. That means the internal energy of a system, which is just the sum of the molecular kinetic and potential energies, can change as a result of two separate types of energy transfer across the boundary of the system. Although the first law imposes conservation of energy for both energy added by heat and work done on a system, it doesn't predict which of several possible energy-conserving processes actually occur in nature.

The second law of thermodynamics constrains the first law by establishing which processes allowed by the first law actually occur. For example, the second law tells us that energy never flows by heat spontaneously from a cold object to a hot object. One important application of this law is in the study of heat engines (such as the internal combustion engine or the woman in Fig. 12.1) and the principles that limit their efficiency.

## 12.1 Work in Thermodynamic Processes

**Figure 12.1** A cyclist is an engine: she requires fuel and oxygen to burn it, and the result is work that drives her forward as her excess waste energy is expelled in her evaporating sweat.

Energy can be transferred to a system by heat and by work done on the system. In most cases of interest treated here, the system is a volume of gas, which is important in understanding engines. All such systems of gas will be assumed to be in thermodynamic equilibrium, so that every part of the gas is at the same temperature and pressure. If that were not the case, the ideal gas law wouldn't apply and most of the results presented here wouldn't be valid. Consider a gas contained by a cylinder fitted with a movable piston (Fig. 12.2a) and in equilibrium. The gas occupies a volume $V$ and exerts a uniform pressure $P$ on the cylinder walls and the piston. The gas is compressed slowly enough so the system remains essentially in thermodynamic equilibrium at all times. As the piston is pushed downward by an external force $F$ through a displacement $\Delta y$, the work done on the gas is

$$W = -F\Delta y = -PA\Delta y$$

where we have set the magnitude $F$ of the external force equal to $PA$, possible because the pressure is the same everywhere in the system (by the assumption of equilibrium). Note that if the piston is pushed downward, $\Delta y = y_f - y_i$ is negative, so we need an explicit negative sign in the expression for $W$ to make the work positive. The change in volume of the gas is $\Delta V = A\Delta y$, which leads to the following definition:

The **work $W$ done on a gas** at constant pressure is given by

$$W = -P\,\Delta V \qquad [12.1]$$

where $P$ is the pressure throughout the gas and $\Delta V$ is the change in volume of the gas during the process.

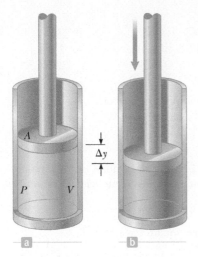

**Figure 12.2** (a) A gas in a cylinder occupying a volume $V$ at a pressure $P$. (b) Pushing the piston down compresses the gas.

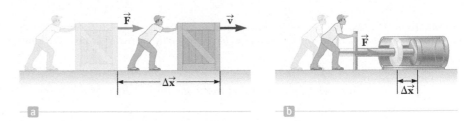

**Figure 12.3** (a) When a force is exerted on a crate, the work done by that force increases the crate's mechanical energy. (b) When a piston is pushed, the gas in the container is compressed, increasing the thermal energy of the gas.

If the gas is compressed as in Figure 12.2b, $\Delta V$ is negative and the work done on the gas is positive. If the gas expands, $\Delta V$ is positive and the work done on the gas is negative. The work done by the gas on its environment, $W_{env}$, is simply the negative of the work done on the gas. In the absence of a change in volume, the work is zero.

The definition of work $W$ in Equation 12.1 specifies **work done on** a gas. In many texts, work $W$ is defined as **work done by** a gas. In this text, work done by a gas is denoted by $W_{env}$. In every case, $W = -W_{env}$, so the two definitions differ by a minus sign. The reason it's important to define work $W$ as work done *on* a gas is to make the concept of work in thermodynamics consistent with the concept of work in mechanics. In mechanics, the system is some object, and when positive work is done on that object, its energy increases. When work $W$ done on a gas as defined in Equation 12.1 is positive, the internal energy of the gas increases, which is consistent with the mechanics definition.

In Figure 12.3a, the man pushes a crate, doing positive work on it, so the crate's speed and therefore its kinetic energy both increase. In Figure 12.3b, a man pushes a piston to the right, compressing the gas in the container and doing positive work on the gas. The average speed of the molecules of gas increases, so the temperature and therefore the internal energy of the gas increase. Consequently, just as doing work on a crate increases its kinetic energy, doing work on a system of gas increases its internal energy.

**Tip 12.1** Work Done *on* Versus Work Done *by*

Work done *on* the gas is labeled $W$. That definition focuses on the internal energy of the system. Work done *by* the gas, say on a piston, is labeled $W_{env}$, where the focus is on harnessing a system's internal energy to do work on something external to the gas. $W$ and $W_{env}$ are two different ways of looking at the same thing. It's always true that $W = -W_{env}$.

---

## EXAMPLE 12.1   WORK DONE BY AN EXPANDING GAS

**GOAL** Apply the definition of work at constant pressure.

**PROBLEM** In a system similar to that shown in Figure 12.2, the gas in the cylinder is at a pressure equal to $1.01 \times 10^5$ Pa and the piston has an area of $0.100$ m$^2$. As energy is slowly added to the gas by heat, the piston is pushed up a distance of 4.00 cm. Calculate the work done by the expanding gas on the surroundings, $W_{env}$, assuming the pressure remains constant.

**STRATEGY** The work done on the environment is the negative of the work done on the gas given in Equation 12.1. Compute the change in volume and multiply by the pressure.

*(Continued)*

## SOLUTION

Find the change in volume of the gas, $\Delta V$, which is the cross-sectional area times the displacement:

$$\Delta V = A\,\Delta y = (0.100 \text{ m}^2)(4.00 \times 10^{-2} \text{ m})$$
$$= 4.00 \times 10^{-3} \text{ m}^3$$

Multiply this result by the pressure, getting the work the gas does on the environment, $W_{env}$:

$$W_{env} = P\,\Delta V = (1.01 \times 10^5 \text{ Pa})(4.00 \times 10^{-3} \text{ m}^3)$$
$$= \boxed{404\,\text{J}}$$

**REMARKS** The volume of the gas increases, so the work done on the environment is positive. The work done on the system during this process is $W = -404\,\text{J}$. The energy required to perform positive work on the environment must come from the energy of the gas.

**QUESTION 12.1** If no energy were added to the gas during the expansion, could the pressure remain constant?

**EXERCISE 12.1** Gas in a cylinder similar to Figure 12.2 moves a piston with area $0.200 \text{ m}^2$ as energy is slowly added to the system. If $2.00 \times 10^3$ J of work is done on the environment and the pressure of the gas in the cylinder remains constant at $1.01 \times 10^5$ Pa, find the displacement of the piston.

**ANSWER** $9.90 \times 10^{-2}$ m

---

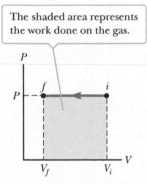

The shaded area represents the work done on the gas.

**Figure 12.4** The *PV* diagram for a gas being compressed at constant pressure.

Equation 12.1 can be used to calculate the work done on the system *only* when the pressure of the gas remains constant during the expansion or compression. A process in which the pressure remains constant is called an **isobaric process**. The pressure versus volume graph, or **PV diagram,** of an isobaric process is shown in Figure 12.4. The curve on such a graph is called the *path* taken between the initial and final states, with the arrow indicating the direction the process is going, in this case from larger to smaller volume. The area under the graph is

$$\text{Area} = P(V_f - V_i) = P\,\Delta V$$

**The area under the graph in a PV diagram is equal in magnitude to the work done on the gas.**

That statement is true in general, whether or not the process proceeds at constant pressure. Just draw the *PV* diagram of the process, find the area underneath the graph (and above the horizontal axis), and that area will be the equal to the magnitude of the work done on the gas. If the arrow on the graph points toward larger volumes, the work done on the gas is negative. If the arrow on the graph points toward smaller volumes, the work done on the gas is positive.

Whenever negative work is done on a system, positive work is done by the system on its environment. The negative work done on the system represents a loss of energy from the system—the cost of doing positive work on the environment.

---

**Quick Quiz**

**12.1** By visual inspection, order the *PV* diagrams shown in Figure 12.5 from the most negative work done on the system to the most positive work done on the system. (a) a, b, c, d (b) a, c, b, d (c) d, b, c, a (d) d, a, c, b

Notice that the graphs in Figure 12.5 all have the same endpoints, but the areas beneath the curves are different. The work done on a system depends on the path taken in the *PV* diagram.

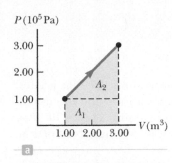

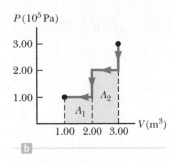

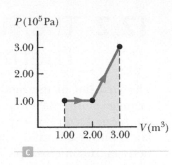

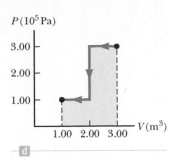

**Figure 12.5** (Quick Quiz 12.1 and Example 12.2)

## EXAMPLE 12.2    WORK AND *PV* DIAGRAMS

**GOAL** Calculate work from a *PV* diagram.

**PROBLEM** Find the numeric value of the work done on the gas in **(a)** Figure 12.5a and **(b)** Figure 12.5b.

**STRATEGY** The regions in question are composed of rectangles and triangles. Use basic geometric formulas to find the area underneath each curve. Check the direction of the arrow to determine signs.

. . . . . . . . . . . . . . . . . . . . . . . . . . . . . . . . . . . . . . . . . . . . . . . . . . . . . . . . . . . . . . . . . . . . . . . . . . . . . . . . . . . .

### SOLUTION

**(a)** Find the work done on the gas in Figure 12.5a.

Compute the areas $A_1$ and $A_2$ in Figure 12.5a. $A_1$ is a rectangle and $A_2$ is a triangle.

$$A_1 = \text{height} \times \text{width} = (1.00 \times 10^5 \text{ Pa})(2.00 \text{ m}^3)$$
$$= 2.00 \times 10^5 \text{ J}$$
$$A_2 = \tfrac{1}{2} \text{ base} \times \text{height}$$
$$= \tfrac{1}{2}(2.00 \text{ m}^3)(2.00 \times 10^5 \text{ Pa}) = 2.00 \times 10^5 \text{ J}$$

Sum the areas (the arrows point to increasing volume, so the work done on the gas is negative):

$$\text{Area} = A_1 + A_2 = 4.00 \times 10^5 \text{ J}$$
$$W = \boxed{-4.00 \times 10^5 \text{ J}}$$

**(b)** Find the work done on the gas in Figure 12.5b.

Compute the areas of the two rectangular regions:

$$A_1 = \text{height} \times \text{width} = (1.00 \times 10^5 \text{ Pa})(1.00 \text{ m}^3)$$
$$= 1.00 \times 10^5 \text{ J}$$
$$A_2 = \text{height} \times \text{width} = (2.00 \times 10^5 \text{ Pa})(1.00 \text{ m}^3)$$
$$= 2.00 \times 10^5 \text{ J}$$

Sum the areas (the arrows point to decreasing volume, so the work done on the gas is positive):

$$\text{Area} = A_1 + A_2 = 3.00 \times 10^5 \text{ J}$$
$$W = \boxed{+3.00 \times 10^5 \text{ J}}$$

. . . . . . . . . . . . . . . . . . . . . . . . . . . . . . . . . . . . . . . . . . . . . . . . . . . . . . . . . . . . . . . . . . . . . . . . . . . . . . . . . . . .

**REMARKS** Notice that in both cases the paths in the *PV* diagrams start and end at the same points, but the answers are different.

**QUESTION 12.2** Is work done on a system during a process in which its volume remains constant?

**EXERCISE 12.2** Compute the work done on the system in Figures 12.5c and 12.5d.

**ANSWERS** $-3.00 \times 10^5 \text{ J}$, $+4.00 \times 10^5 \text{ J}$

# 12.2 The First Law of Thermodynamics

The **first law of thermodynamics** is another energy conservation law that relates changes in internal energy—the energy associated with the position and motion of all the molecules of a system—to energy transfers due to heat and work. The first law is universally valid, applicable to all kinds of processes, providing a connection between the microscopic and macroscopic worlds.

There are two ways energy can be transferred between a system and its surrounding environment: by doing work, which requires a macroscopic displacement of an object through the application of a force, and by a direct exchange of energy across the system boundary, often by heat. Heat is the transfer of energy between a system and its environment due to a temperature difference and usually occurs through one or more of the mechanisms of radiation, conduction, and convection. For example, in Figure 12.6 hot gases and radiation impinge on the cylinder, raising its temperature, and energy $Q$ is transferred by conduction to the gas, where it is distributed mainly through convection. Other processes for transferring energy into a system are possible, such as a chemical reaction or an electrical discharge. Any energy $Q$ exchanged between the system and the environment and *any work done through the expansion or compression of the system results in a change in the internal energy, ΔU, of the system*. A change in internal energy results in measurable changes in the macroscopic variables of the system such as the pressure, temperature, and volume. The relationship between the change in internal energy, $\Delta U$, energy $Q$, and the work $W$ done on the system is given by the **first law of thermodynamics:**

First law of ▶
thermodynamics

If a system undergoes a change from an initial state to a final state, then the change in the internal energy $\Delta U$ is given by

$$\Delta U = U_f - U_i = Q + W \qquad [12.2]$$

where $Q$ is the energy exchanged between the system and the environment, and $W$ is the work done on the system.

The quantity $Q$ is positive when energy is transferred into the system and negative when energy is removed from the system.

Figure 12.6 illustrates the first law for a cylinder of gas and how the system interacts with the environment. The gas cylinder contains a frictionless piston, and the block is initially at rest. Energy $Q$ is introduced into the gas as the gas expands against the piston with constant pressure $P$. Until the piston hits the stops, it exerts a force on the block, which accelerates on a frictionless surface. Negative work $W$ is

**Figure 12.6** Thermal energy $Q$ is transferred to the gas, increasing its internal energy. The gas presses against the piston, displacing it and performing mechanical work on the environment or, equivalently, doing negative work on the gas, reducing the internal energy.

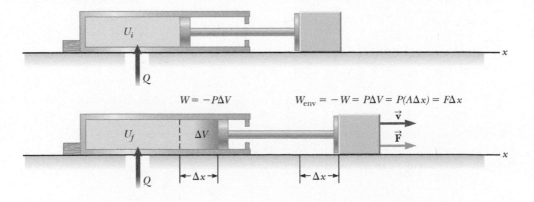

done on the gas, and at the same time positive work $W_{env} = -W$ is done by the gas on the block. Adding the work done on the environment, $W_{env}$, to the work done on the gas, $W$, gives zero net work, as it must because energy must be conserved.

From Equation 12.2, we also see that the internal energy of any isolated system must remain constant, so that $\Delta U = 0$. Even when a system isn't isolated, the change in internal energy will be zero if the system goes through a cyclic process in which all the thermodynamic variables—pressure, volume, temperature, and moles of gas—return to their original values.

It's important to remember that the quantities in Equation 12.2 concern a *system*, not the effect on the system's environment through work. If the system is hot gas expanding against a piston, as in Figure 12.6, the system work $W$ is *negative* because the piston can only expand at the expense of the internal energy of the gas. The work $W_{env}$ done by the hot gas on the *environment*—in this case, moving a piston which moves the block—is positive, but that's not the work $W$ in Equation 12.2. This way of defining work in the first law makes it consistent with the concept of work defined in Topic 5. In both the mechanical and thermal cases, the effect on the system is the same: positive work increases the system's energy, and negative work decreases the system's energy.

Some textbooks identify $W$ as the work done by the gas on its environment. This is an equivalent formulation, but it means that $W$ must carry a minus sign in the first law. That convention isn't consistent with previous discussions of the energy of a system, because when $W$ is positive the system *loses* energy, whereas in Topic 5 positive $W$ means that the system *gains* energy. For that reason, the old convention is not used in this book.

**Tip 12.2** Dual Sign Conventions

Many physics and engineering textbooks present the first law as $\Delta U = Q - W$, with a minus sign between the transferred energy and the work. The reason is that work is defined in these treatments as the work done *by* the system rather than *on* the system, as in our treatment. Using our notation, this equivalent first law would read $\Delta U = Q - W_{env}$.

---

**EXAMPLE 12.3   HEATING A GAS**

**GOAL** Combine the first law of thermodynamics with work done during a constant pressure process.

**PROBLEM** An ideal gas absorbs $5.00 \times 10^3$ J of energy while doing $2.00 \times 10^3$ J of work on the environment during a constant pressure process. (a) Compute the change in the internal energy of the gas. (b) If the internal energy now drops by $4.50 \times 10^3$ J and $7.50 \times 10^3$ J is expelled from the system, find the change in volume, assuming a constant pressure process at $1.01 \times 10^5$ Pa.

**STRATEGY** Part (a) requires substitution of the given information into the first law, Equation 12.2. Notice, however, that the given work is done on the *environment*. The negative of this amount is the work done on the *system*, representing a loss of internal energy. Part (b) is a matter of substituting the equation for work at constant pressure into the first law and solving for the change in volume.

**SOLUTION**

(a) Compute the change in internal energy of the gas.

Substitute values into the first law, noting that the work done on the gas is negative:

$\Delta U = Q + W = 5.00 \times 10^3$ J $- 2.00 \times 10^3$ J
$= \boxed{3.00 \times 10^3 \text{ J}}$

(b) Find the change in volume, noting that $\Delta U$ and $Q$ are both negative in this case.

Substitute the equation for work done at constant pressure into the first law:

$\Delta U = Q + W = Q - P\Delta V$
$-4.50 \times 10^3$ J $= -7.50 \times 10^3$ J $- (1.01 \times 10^5 \text{ Pa})\Delta V$

Solve for the change in volume, $\Delta V$:

$\Delta V = \boxed{-2.97 \times 10^{-2} \text{ m}^3}$

**REMARKS** The change in volume is negative, so the system contracts, doing negative work on the environment, whereas the work $W$ on the system is positive.

*(Continued)*

**QUESTION 12.3** True or False: When an ideal gas expands at constant pressure, the change in the internal energy must be positive.

**EXERCISE 12.3** Suppose the internal energy of an ideal gas rises by $3.00 \times 10^3$ J at a constant pressure of $1.00 \times 10^5$ Pa, while the system gains $4.20 \times 10^3$ J of energy by heat. Find the change in volume of the system.

**ANSWER** $1.20 \times 10^{-2}$ m$^3$

---

Recall that an expression for the internal energy of an ideal gas is

$$U = \tfrac{3}{2}nRT \qquad \text{[12.3a]}$$

This expression is valid only for a *monatomic* ideal gas, which means the particles of the gas consist of single atoms. The change in the internal energy, $\Delta U$, for such a gas is given by

$$\Delta U = \tfrac{3}{2}nR\,\Delta T \qquad \text{[12.3b]}$$

The **molar specific heat at constant volume** of a monatomic ideal gas, $C_v$, is defined by

$$C_v \equiv \tfrac{3}{2}R \qquad \text{[12.4]}$$

The change in internal energy of an ideal gas can then be written

$$\Delta U = nC_v\,\Delta T \qquad \text{[12.5]}$$

For ideal gases, this expression is always valid, even when the volume isn't constant. The value of the molar specific heat, however, depends on the gas and can vary under different conditions of temperature and pressure.

A gas with a larger molar specific heat requires more energy to realize a given temperature change. The size of the molar specific heat depends on the structure of the gas molecule and how many different ways it can store energy. A monatomic gas such as helium can store energy as motion in three different directions. A gas such as hydrogen, on the other hand, is diatomic in normal temperature ranges, and aside from moving in three directions, it can also tumble, rotating in two different directions. So hydrogen molecules can store energy in the form of translational motion and in addition can store energy through tumbling. Further, molecules can also store energy in the vibrations of their constituent atoms. A gas composed of molecules with more ways to store energy will have a larger molar specific heat.

Each different way a gas molecule can store energy is called a *degree of freedom*. Each degree of freedom contributes $\tfrac{1}{2}R$ to the molar specific heat. Because an atomic ideal gas can move in three directions, it has a molar specific heat capacity $C_v = 3(\tfrac{1}{2}R) = \tfrac{3}{2}R$. A diatomic gas like molecular oxygen, $O_2$, can also tumble in two different directions. This adds $2 \times \tfrac{1}{2}R = R$ to the molar specific heat, so $C_v = \tfrac{5}{2}R$ for diatomic gases. The spinning about the long axis connecting the two atoms is generally negligible. Vibration of the atoms in a molecule can also contribute to the heat capacity. A full analysis of a given system is often complex, so in general, molar specific heats must be determined by experiment. Some representative values of $C_v$ can be found in Table 12.1.

**Table 12.1** Molar Specific Heats of Various Gases

| Gas | Molar Specific Heat (J/mol · K)[a] | | | |
|---|---|---|---|---|
| | $C_p$ | $C_v$ | $C_p - C_v$ | $\gamma = C_p/C_v$ |
| **Monatomic Gases** | | | | |
| He | 20.8 | 12.5 | 8.33 | 1.67 |
| Ar | 20.8 | 12.5 | 8.33 | 1.67 |
| Ne | 20.8 | 12.7 | 8.12 | 1.64 |
| Kr | 20.8 | 12.3 | 8.49 | 1.69 |
| **Diatomic Gases** | | | | |
| $H_2$ | 28.8 | 20.4 | 8.33 | 1.41 |
| $N_2$ | 29.1 | 20.8 | 8.33 | 1.40 |
| $O_2$ | 29.4 | 21.1 | 8.33 | 1.40 |
| CO | 29.3 | 21.0 | 8.33 | 1.40 |
| $Cl_2$ | 34.7 | 25.7 | 8.96 | 1.35 |
| **Polyatomic Gases** | | | | |
| $CO_2$ | 37.0 | 28.5 | 8.50 | 1.30 |
| $SO_2$ | 40.4 | 31.4 | 9.00 | 1.29 |
| $H_2O$ | 35.4 | 27.0 | 8.37 | 1.30 |
| $CH_4$ | 35.5 | 27.1 | 8.41 | 1.31 |

[a]All values except that for water were obtained at 300 K.

# 12.3 Thermal Processes in Gases

Thermal processes can be complex. Fortunately, they can often be broken down into a series of simple processes. In this section, the four most common processes will be studied and illustrated by their effect on an ideal gas. Each process corresponds to making one of the variables in the ideal gas law a constant or assuming one of the three quantities in the first law of thermodynamics is zero. The four processes are called isobaric (constant pressure), adiabatic (no thermal energy transfer, or $Q = 0$), isovolumetric (constant volume, corresponding to $W = 0$), and isothermal (constant temperature, corresponding to $\Delta U = 0$). Naturally, many other processes don't fall into one of these four categories, so they will be covered in a fifth category, called a general process. What is essential in each case is to be able to calculate the three thermodynamic quantities from the first law: the work $W$, the thermal energy transfer $Q$, and the change in the internal energy $\Delta U$.

## 12.3.1 Isobaric Processes

Recall from Section 12.1 that in an isobaric process the pressure remains constant as the gas expands or is compressed. An expanding gas does work on its environment, given by $W_{env} = P \Delta V$. The $PV$ diagram of an isobaric expansion is given in Figure 12.4. As previously discussed, the magnitude of the work done on the gas is just the area under the path in its $PV$ diagram: height times length, or $P \Delta V$. The negative of this quantity, $W = -P \Delta V$, is the energy lost by the gas because the gas does work as it expands. This is the quantity that should be substituted into the first law.

The work done by the gas on its environment must come at the expense of the change in its internal energy, $\Delta U$. Because the change in the internal energy of an ideal gas is given by $\Delta U = nC_v \Delta T$, the temperature of an expanding gas must decrease as the internal energy decreases. Expanding volume and decreasing temperature means the pressure must also decrease, in conformity with the ideal gas law, $PV = nRT$. Consequently, the only way such a process can remain at constant pressure is if thermal energy $Q$ is transferred into the gas by heat. Rearranging the first law, we obtain

$$Q = \Delta U - W = \Delta U + P\Delta V$$

Now we can substitute the expression in Equation 12.3b for $\Delta U$ and use the ideal gas law to substitute $P\Delta V = nR\Delta T$:

$$Q = \tfrac{3}{2}nR\Delta T + nR\Delta T = \tfrac{5}{2}nR\Delta T$$

Another way to express this transfer by heat is

$$Q = nC_p\Delta T \qquad [12.6]$$

where $C_p = \tfrac{5}{2}R$. For ideal gases, the molar heat capacity at constant pressure, $C_p$, is the sum of the molar heat capacity at constant volume, $C_v$, and the gas constant $R$:

$$C_p = C_v + R \qquad [12.7]$$

This can be seen in the fourth column of Table 12.1, where $C_p - C_v$ is calculated for a number of different gases. The difference works out to be approximately $R$ in virtually every case.

## EXAMPLE 12.4 | EXPANDING GAS

**GOAL** Use molar specific heats and the first law in a constant pressure process.

**PROBLEM** Suppose a system of monatomic ideal gas at $2.00 \times 10^5$ Pa and an initial temperature of 293 K slowly expands at constant pressure from a volume of 1.00 L to 2.50 L. **(a)** Find the work done on the environment. **(b)** Find the change in internal energy of the gas. **(c)** Use the first law of thermodynamics to obtain the thermal energy absorbed by the gas during the process. **(d)** Use the molar heat capacity at constant pressure to find the thermal energy absorbed. **(e)** How would the answers change for a diatomic ideal gas?

**STRATEGY** This problem mainly involves substituting values into the appropriate equations. Substitute into the equation for work at constant pressure to obtain the answer to part **(a)**. In part **(b)** use the ideal gas law twice: to find the temperature when $V = 2.00$ L and to find the number of moles of the gas. These quantities can then be used to obtain the change in internal energy, $\Delta U$. Part **(c)** can then be solved by substituting into the first law, yielding $Q$, the answer checked in part **(d)** with Equation 12.6. Repeat these steps for part **(e)** after increasing the molar specific heats by $R$ because of the extra two degrees of freedom associated with a diatomic gas.

**SOLUTION**

**(a)** Find the work done on the environment.

Apply the definition of work at constant pressure:

$$W_{env} = P\Delta V = (2.00 \times 10^5 \text{ Pa})(2.50 \times 10^{-3} \text{ m}^3$$
$$- 1.00 \times 10^{-3} \text{ m}^3)$$
$$W_{env} = \boxed{3.00 \times 10^2 \text{ J}}$$

**(b)** Find the change in the internal energy of the gas.

First, obtain the final temperature, using the ideal gas law, noting that $P_i = P_f$:

$$\frac{P_f V_f}{P_i V_i} = \frac{T_f}{T_i} \rightarrow T_f = T_i \frac{V_f}{V_i} = (293 \text{ K})\frac{(2.50 \times 10^{-3} \text{ m}^3)}{(1.00 \times 10^{-3} \text{ m}^3)}$$
$$T_f = 733 \text{ K}$$

Again using the ideal gas law, obtain the number of moles of gas:

$$n = \frac{P_i V_i}{RT_i} = \frac{(2.00 \times 10^5 \text{ Pa})(1.00 \times 10^{-3} \text{ m}^3)}{(8.31 \text{ J/K} \cdot \text{mol})(293 \text{ K})}$$
$$= 8.21 \times 10^{-2} \text{ mol}$$

Use these results and given quantities to calculate the change in internal energy, $\Delta U$:

$$\Delta U = nC_v\Delta T = \tfrac{3}{2}nR\Delta T$$
$$= \tfrac{3}{2}(8.21 \times 10^{-2} \text{ mol})(8.31 \text{ J/K} \cdot \text{mol})(733 \text{ K} - 293 \text{ K})$$
$$\Delta U = \boxed{4.50 \times 10^2 \text{ J}}$$

**(c)** Use the first law to obtain the energy transferred by heat.

Solve the first law for $Q$, and substitute $\Delta U$ and $W = -W_{env} = -3.00 \times 10^2$ J:

$$\Delta U = Q + W \rightarrow Q = \Delta U - W$$
$$Q = 4.50 \times 10^2 \text{ J} - (-3.00 \times 10^2 \text{ J}) = \boxed{7.50 \times 10^2 \text{ J}}$$

**(d)** Use the molar heat capacity at constant pressure to obtain $Q$.

Substitute values into Equation 12.6:

$$Q = nC_p\Delta T = \tfrac{5}{2}nR\Delta T$$
$$= \tfrac{5}{2}(8.21 \times 10^{-2}\,\text{mol})(8.31\,\text{J/K}\cdot\text{mol})(733\,\text{K} - 293\,\text{K})$$
$$= \boxed{7.50 \times 10^2\,\text{J}}$$

**(e)** How would the answers change for a diatomic gas?

Obtain the new change in internal energy, $\Delta U$, noting that $C_v = \tfrac{5}{2}R$ for a diatomic gas:

$$\Delta U = nC_v\Delta T = \left(\tfrac{3}{2} + 1\right)nR\,\Delta T$$
$$= \tfrac{5}{2}(8.21 \times 10^{-2}\,\text{mol})(8.31\,\text{J/K}\cdot\text{mol})(733\,\text{K} - 293\,\text{K})$$
$$\Delta U = \boxed{7.50 \times 10^2\,\text{J}}$$

Obtain the new energy transferred by heat, $Q$:

$$Q = nC_p\Delta T = \left(\tfrac{5}{2} + 1\right)nR\Delta T$$
$$= \tfrac{7}{2}(8.21 \times 10^{-2}\,\text{mol})(8.31\,\text{J/K}\cdot\text{mol})(733\,\text{K} - 293\,\text{K})$$
$$Q = \boxed{1.05 \times 10^3\,\text{J}}$$

· · · · · · · · · · · · · · · · · · · · · · · · · · · · · · · · · · · · · · · · · · · · · · · · · · · · · · · · · · · · · · · · · · · ·

**REMARKS** Part **(b)** could also be solved with fewer steps by using the ideal gas equation $PV = nRT$ once the work is known. The pressure and number of moles are constant, and the gas is ideal, so $P\Delta V = nR\Delta T$. Given that $C_v = \tfrac{3}{2}R$, the change in the internal energy $\Delta U$ can then be calculated in terms of the expression for work:

$$\Delta U = nC_v\Delta T = \tfrac{3}{2}nR\Delta T = \tfrac{3}{2}P\Delta V = \tfrac{3}{2}W$$

Similar methods can be used in other processes.

**QUESTION 12.4** True or False: During a constant pressure compression, the temperature of an ideal gas must always decrease, and the gas must always exhaust thermal energy ($Q < 0$).

**EXERCISE 12.4** Suppose an ideal monatomic gas at an initial temperature of 475 K is compressed from 3.00 L to 2.00 L while its pressure remains constant at $1.00 \times 10^5$ Pa. Find **(a)** the work done on the gas, **(b)** the change in internal energy, and **(c)** the energy transferred by heat, $Q$.

**ANSWERS** **(a)** $1.00 \times 10^2\,\text{J}$  **(b)** $-1.50 \times 10^2\,\text{J}$  **(c)** $-2.50 \times 10^2\,\text{J}$

## 12.3.2 Adiabatic Processes

In an adiabatic process, no energy enters or leaves the system by heat. Such a system is insulated, thermally isolated from its environment. In general, however, the system isn't mechanically isolated, so it can still do work. A sufficiently rapid process may be considered approximately adiabatic because there isn't time for any significant transfer of energy by heat.

For adiabatic processes $Q = 0$, so the first law becomes

$$\Delta U = W \quad \text{(adiabatic processes)}$$

The work done during an adiabatic process can be calculated by finding the change in the internal energy. Alternately, the work can be computed from a $PV$ diagram. For an ideal gas undergoing an adiabatic process, it can be shown that

$$PV^\gamma = \text{constant} \qquad\qquad \text{[12.8a]}$$

where

$$\gamma = \frac{C_p}{C_v} \qquad\qquad \text{[12.8b]}$$

is called the *adiabatic index* of the gas. Values of the adiabatic index for several different gases are given in Table 12.1. After computing the constant on the right-hand side of Equation 12.8a and solving for the pressure $P$, the area under the curve in the $PV$ diagram can be found by counting boxes, yielding the work.

If a hot gas is allowed to expand so quickly that there is no time for energy to enter or leave the system by heat, the work done on the gas is negative and the internal energy decreases. This decrease occurs because kinetic energy is transferred from the gas molecules to the moving piston. Such an adiabatic expansion is of practical importance and is nearly realized in an internal combustion engine when a gasoline–air mixture is ignited and expands rapidly against a piston. The following example illustrates this process.

## EXAMPLE 12.5    WORK AND AN ENGINE CYLINDER

**GOAL**  Use the first law to find the work done in an adiabatic expansion.

**PROBLEM**  In a car engine operating at a frequency of $1.80 \times 10^3$ rev/min, the expansion of hot, high-pressure gas against a piston occurs in about 10 ms. Because energy transfer by heat typically takes a time on the order of minutes or hours, it's safe to assume little energy leaves the hot gas during the expansion. Find the work done by the gas on the piston during this adiabatic expansion by assuming the engine cylinder contains 0.100 moles of an ideal monatomic gas that goes from $1.20 \times 10^3$ K to $4.00 \times 10^2$ K, typical engine temperatures, during the expansion.

**STRATEGY**  Find the change in internal energy using the given temperatures. For an adiabatic process, this equals the work done on the gas, which is the negative of the work done on the environment—in this case, the piston.

### SOLUTION

Start with the first law, taking $Q = 0$:

$$W = \Delta U - Q = \Delta U - 0 = \Delta U$$

Find $\Delta U$ from the expression for the internal energy of an ideal monatomic gas:

$$\Delta U = U_f - U_i = \tfrac{3}{2}nR(T_f - T_i)$$
$$= \tfrac{3}{2}(0.100 \text{ mol})(8.31 \text{ J/mol} \cdot \text{K})(4.00 \times 10^2 \text{ K} - 1.20 \times 10^3 \text{ K})$$
$$\Delta U = -9.97 \times 10^2 \text{ J}$$

The change in internal energy equals the work done on the system, which is the negative of the work done on the piston:

$$W_{\text{piston}} = -W = -\Delta U = \boxed{9.97 \times 10^2 \text{ J}}$$

**REMARKS**  The work done on the piston comes at the expense of the internal energy of the gas. In an ideal adiabatic expansion, the loss of internal energy is completely converted into useful work. In a real engine, there are always losses.

**QUESTION 12.5**  In an adiabatic expansion of an ideal gas, why must the change in temperature always be negative?

**EXERCISE 12.5**  A monatomic ideal gas with volume 0.200 L is rapidly compressed, so the process can be considered adiabatic. If the gas is initially at $1.01 \times 10^5$ Pa and $3.00 \times 10^2$ K and the final temperature is 477 K, find the work done by the gas on the environment, $W_{\text{env}}$.

**ANSWER**  $-17.9$ J

## EXAMPLE 12.6    AN ADIABATIC EXPANSION

**GOAL**  Use the adiabatic pressure vs. volume relation to find a change in pressure and the work done on a gas.

**PROBLEM**  A monatomic ideal gas at an initial pressure of $1.01 \times 10^5$ Pa expands adiabatically from an initial volume of 1.50 m³, doubling its volume (Fig. 12.7). **(a)** Find the new pressure. **(b)** Sketch the $PV$ diagram and estimate the work done on the gas.

**STRATEGY**  There isn't enough information to solve this problem with the ideal gas law. Instead, use Equation 12.8a,b and the given information to find the adiabatic index and the constant $C$ for the process. For part **(b)**, sketch the $PV$ diagram and count boxes to estimate the area under the graph, which gives the work.

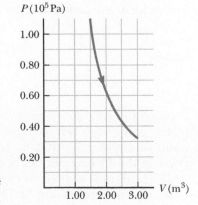

**Figure 12.7**  (Example 12.6) The $PV$ diagram of an adiabatic expansion: the graph of $P = CV^{-\gamma}$, where $C$ is a constant and $\gamma = C_p/C_v$.

## SOLUTION

**(a)** Find the new pressure.

First, calculate the adiabatic index:

$$\gamma = \frac{C_p}{C_v} = \frac{\frac{5}{2}R}{\frac{3}{2}R} = \frac{5}{3}$$

Use Equation 12.8a to find the constant $C$:

$$C = P_1 V_1^{\gamma} = (1.01 \times 10^5 \text{ Pa})(1.50 \text{ m}^3)^{5/3}$$
$$= 1.99 \times 10^5 \text{ Pa} \cdot \text{m}^5$$

The constant $C$ is fixed for the entire process and can be used to find $P_2$:

$$C = P_2 V_2^{\gamma} = P_2 (3.00 \text{ m}^3)^{5/3}$$
$$1.99 \times 10^5 \text{ Pa} \cdot \text{m}^5 = P_2 (6.24 \text{ m}^5)$$
$$P_2 = \boxed{3.19 \times 10^4 \text{ Pa}}$$

**(b)** Estimate the work done on the gas from a $PV$ diagram.

Count the boxes between $V_1 = 1.50 \text{ m}^3$ and $V_2 = 3.00 \text{ m}^3$ in the graph of $P = (1.99 \times 10^5 \text{ Pa} \cdot \text{m}^5) V^{-5/3}$ in the $PV$ diagram shown in Figure 12.7:

Number of boxes $\approx 17$

Each box has an 'area' of $5.00 \times 10^3$ J.

$$W \approx -17 \cdot 5.00 \times 10^3 \text{ J} = \boxed{-8.5 \times 10^4 \text{ J}}$$

**REMARKS** The exact answer, obtained with calculus, is $-8.43 \times 10^4$ J, so our result is a very good estimate. The answer is negative because the gas is expanding, doing positive work on the environment, thereby reducing its own internal energy.

**QUESTION 12.6** For an adiabatic expansion between two given volumes and an initial pressure, which gas does more work, a monatomic gas or a diatomic gas?

**EXERCISE 12.6** Repeat the preceding calculations for an ideal diatomic gas expanding adiabatically from an initial volume of 0.500 m³ to a final volume of 1.25 m³, starting at a pressure of $P_1 = 1.01 \times 10^5$ Pa. Use the same techniques as in the example.

**ANSWERS** $P_2 = 2.80 \times 10^4$ Pa, $W \approx -4 \times 10^4$ J

## 12.3.3 Isovolumetric Processes

An **isovolumetric process,** sometimes called an *isochoric* process (which is harder to remember), proceeds at constant volume, corresponding to vertical lines in a $PV$ diagram. If the volume doesn't change, no work is done on or by the system, so $W = 0$ and the first law of thermodynamics reads

$$\Delta U = Q \quad \text{(isovolumetric process)}$$

This result tells us that **in an isovolumetric process, the change in internal energy of a system equals the energy transferred to the system by heat.** From Equation 12.5, the energy transferred by heat in constant volume processes is given by

$$Q = nC_v \Delta T \qquad [12.9]$$

## EXAMPLE 12.7    AN ISOVOLUMETRIC PROCESS

**GOAL** Apply the first law to a constant-volume process.

**PROBLEM** A monatomic ideal gas has a temperature $T = 3.00 \times 10^2$ K and a constant volume of 1.50 L. If there are 5.00 moles of gas, **(a)** how much thermal energy must be added in order to raise the temperature of the gas to $3.80 \times 10^2$ K? **(b)** Calculate the change in pressure of the gas, $\Delta P$. **(c)** How much thermal energy would be required if the gas were ideal and diatomic? **(d)** Calculate the change in the pressure for the diatomic gas.

*(Continued)*

**SOLUTION**

**(a)** How much thermal energy must be added in order to raise the temperature of the gas to $3.80 \times 10^2$ K?

Apply Equation 12.9, using the fact that $C_v = 3R/2$ for an ideal monatomic gas:

$$(1) \quad Q = \Delta U = nC_v\Delta T = \tfrac{3}{2}nR\Delta T$$
$$= \tfrac{3}{2}(5.00 \text{ mol})(8.31 \text{ J/K} \cdot \text{mol})(80.0 \text{ K})$$
$$Q = \boxed{4.99 \times 10^3 \text{ J}}$$

**(b)** Calculate the change in pressure, $\Delta P$.

Use the ideal gas equation $PV = nRT$ and Equation (1) to relate $\Delta P$ to $Q$:

$$\Delta(PV) = (\Delta P)V = nR\Delta T = \tfrac{2}{3}Q$$

Solve for $\Delta P$:

$$\Delta P = \frac{2}{3}\frac{Q}{V} = \frac{2}{3}\frac{4.99 \times 10^3 \text{ J}}{1.50 \times 10^{-3} \text{ m}^3}$$
$$= \boxed{2.22 \times 10^6 \text{ Pa}}$$

**(c)** How much thermal energy would be required if the gas were ideal and diatomic?

Repeat the calculation with $C_v = 5R/2$:

$$Q = \Delta U = nC_v\Delta T = \tfrac{5}{2}nR\Delta T = \boxed{8.31 \times 10^3 \text{ J}}$$

**(d)** Calculate the change in the pressure for the diatomic gas.

Use the result of part **(c)** and repeat the calculation of part **(b)**, with 2/3 replaced by 2/5 because the gas is diatomic:

$$\Delta P = \frac{2}{5}\frac{Q}{V} = \frac{2}{5}\frac{8.31 \times 10^3 \text{ J}}{1.50 \times 10^{-3} \text{ m}^3}$$
$$= \boxed{2.22 \times 10^6 \text{ Pa}}$$

**REMARKS** The constant volume diatomic gas, under the same conditions, requires more thermal energy per degree of temperature change because there are more ways for the diatomic molecules to store energy. Despite the extra energy added, the diatomic gas reaches the same final pressure as the monatomic gas.

**QUESTION 12.7** If the same amount of energy as found in part **(a)** were transferred to 5.00 moles of carbon dioxide at the same initial temperature, would the final temperature be lower, higher, or unchanged?

**EXERCISE 12.7** **(a)** Find the change in temperature $\Delta T$ of 22.0 mol of a monatomic ideal gas if it absorbs 9 750 J at a constant volume of 2.40 L. **(b)** What is the change in pressure, $\Delta P$? **(c)** If the system is an ideal diatomic gas, find the change in its temperature. **(d)** Find the change in pressure of the diatomic gas.

**ANSWERS** **(a)** 35.6 K  **(b)** $2.71 \times 10^6$ Pa  **(c)** 21.3 K  **(d)** $1.63 \times 10^6$ Pa

## 12.3.4 Isothermal Processes

During an isothermal process, the temperature of a system doesn't change. In an ideal gas, the internal energy $U$ depends only on the temperature, so it follows that $\Delta U = 0$ because $\Delta T = 0$. In this case, the first law of thermodynamics gives

$$W = -Q \quad \text{(isothermal process)}$$

We see that if the system is an ideal gas undergoing an isothermal process, the work done on the system is equal to the negative of the thermal energy transferred to the system. Such a process can be visualized in Figure 12.8. A cylinder filled with gas is in contact with a large energy reservoir that can exchange energy with the gas without changing its temperature. For a constant temperature ideal gas,

$$P = \frac{nRT}{V}$$

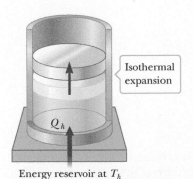

Isothermal expansion

$Q_h$

Energy reservoir at $T_h$

**Figure 12.8** The gas in the cylinder expands isothermally while in contact with a reservoir at temperature $T_h$.

where the numerator on the right-hand side is constant. The $PV$ diagram of a typical isothermal process is graphed in Figure 12.9, contrasted with an adiabatic process. The pressure falls off more rapidly for an adiabatic expansion because thermal energy can't be transferred into the system. In an isothermal expansion, the system loses energy by doing work on the environment but regains an equal amount of energy across the boundary.

Using methods of calculus, it can be shown that the work done on the environment during an isothermal process is given by

$$W_{env} = nRT \ln\left(\frac{V_f}{V_i}\right)$$    [12.10]

The symbol "ln" in Equation 12.10 is an abbreviation for the natural logarithm, discussed in Appendix A. The work $W$ done on the gas is just the negative of $W_{env}$.

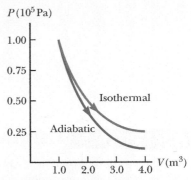

**Figure 12.9** The $PV$ diagram of an isothermal expansion, graph of $P = CV^{-1}$, where $C$ is a constant, compared to an adiabatic expansion, $P = C_A V^{-\gamma}$. $C_A$ is a constant equal in magnitude to $C$ in this case but carrying different units.

---

### EXAMPLE 12.8    AN ISOTHERMALLY EXPANDING BALLOON

**GOAL** Find the work done during an isothermal expansion.

**PROBLEM** A balloon contains 5.00 moles of a monatomic ideal gas. As energy is added to the system by heat (say, by absorption from the Sun), the volume increases by 25% at a constant temperature of 27.0°C. Find the work $W_{env}$ done by the gas in expanding the balloon, the thermal energy $Q$ transferred to the gas, and the work $W$ done on the gas.

**STRATEGY** Be sure to convert temperatures to kelvins. Use Equation 12.10 for isothermal work $W_{env}$ done on the environment to find the work $W$ done on the balloon, which satisfies $W = -W_{env}$. Further, for an isothermal process, the thermal energy $Q$ transferred to the system equals the work $W_{env}$ done by the system on the environment.

..................................................................................................................

**SOLUTION**

Substitute into Equation 12.10, finding the work done during the isothermal expansion. Note that $T = 27.0°C = 3.00 \times 10^2$ K.

$$W_{env} = nRT \ln\left(\frac{V_f}{V_i}\right)$$

$$= (5.00 \text{ mol})(8.31 \text{ J/K} \cdot \text{mol})(3.00 \times 10^2 \text{ K})$$

$$\times \ln\left(\frac{1.25 V_0}{V_0}\right)$$

$$W_{env} = \boxed{2.78 \times 10^3 \text{ J}}$$

$$Q = W_{env} = \boxed{2.78 \times 10^3 \text{ J}}$$

The negative of this amount is the work done on the gas:

$$W = -W_{env} = \boxed{-2.78 \times 10^3 \text{ J}}$$

..................................................................................................................

**REMARKS** Notice the relationship between the work done on the gas, the work done on the environment, and the energy transferred. These relationships are true of all isothermal processes.

**QUESTION 12.8** True or False: In an isothermal process, no thermal energy transfer takes place.

**EXERCISE 12.8** Suppose that subsequent to this heating, $1.50 \times 10^4$ J of thermal energy is removed from the gas isothermally. Find the final volume in terms of the initial volume of the example, $V_0$. (*Hint:* Follow the same steps as in the example, but in reverse. Also note that the initial volume in this exercise is $1.25 V_0$.)

**ANSWER** $0.375 V_0$

---

## 12.3.5 General Case

When a process follows none of the four given models, it's still possible to use the first law to get information about it. The work can be computed from the area under the curve of the $PV$ diagram, and if the temperatures at the endpoints can be found, $\Delta U$ follows from Equation 12.5, as illustrated in the following example.

### EXAMPLE 12.9   A GENERAL PROCESS

**GOAL** Find thermodynamic quantities for a process that doesn't fall into any of the four previously discussed categories.

**PROBLEM** A quantity of 4.00 moles of a monatomic ideal gas expands from an initial volume of 0.100 m³ to a final volume of 0.300 m³ and pressure of $2.5 \times 10^5$ Pa (Fig. 12.10a). Compute **(a)** the work done on the gas, **(b)** the change in internal energy of the gas, and **(c)** the thermal energy transferred to the gas.

**STRATEGY** The work done on the gas is equal to the negative of the area under the curve in the $PV$ diagram. Use the ideal gas law to get the temperature change and, subsequently, the change in internal energy. Finally, the first law gives the thermal energy transferred by heat.

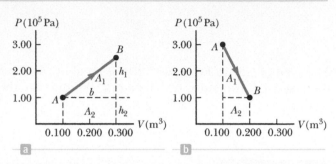

**Figure 12.10** (a) (Example 12.9) (b) (Exercise 12.9)

. . . . . . . . . . . . . . . . . . . . . . . . . . . . . . . . . . . . . . . . . . . . . . . . . . . . . . . . . . . . . . . . . . . . . . . . . . . . . . . . . . . . . . . . . . . . . . . . . .

### SOLUTION

**(a)** Find the work done on the gas by computing the area under the curve in Figure 12.10a.

Find $A_1$, the area of the triangle:

$$A_1 = \tfrac{1}{2}bh_1 = \tfrac{1}{2}(0.200 \text{ m}^3)(1.50 \times 10^5 \text{ Pa}) = 1.50 \times 10^4 \text{ J}$$

Find $A_2$, the area of the rectangle:

$$A_2 = bh_2 = (0.200 \text{ m}^3)(1.00 \times 10^5 \text{ Pa}) = 2.00 \times 10^4 \text{ J}$$

Sum the two areas (the gas is expanding, so the work done on the gas is negative and a minus sign must be supplied):

$$W = -(A_1 + A_2) = \boxed{-3.50 \times 10^4 \text{ J}}$$

**(b)** Find the change in the internal energy during the process.

Compute the temperature at points $A$ and $B$ with the ideal gas law:

$$T_A = \frac{P_A V_A}{nR} = \frac{(1.00 \times 10^5 \text{ Pa})(0.100 \text{ m}^3)}{(4.00 \text{ mol})(8.31 \text{ J/K} \cdot \text{mol})} = 301 \text{ K}$$

$$T_B = \frac{P_B V_B}{nR} = \frac{(2.50 \times 10^5 \text{ Pa})(0.300 \text{ m}^3)}{(4.00 \text{ mol})(8.31 \text{ J/K} \cdot \text{mol})} = 2.26 \times 10^3 \text{ K}$$

Compute the change in internal energy:

$$\Delta U = \tfrac{3}{2}nR \, \Delta T$$
$$= \tfrac{3}{2}(4.00 \text{ mol})(8.31 \text{ J/K} \cdot \text{mol})(2.26 \times 10^3 \text{ K} - 301 \text{ K})$$
$$\Delta U = \boxed{9.77 \times 10^4 \text{ J}}$$

**(c)** Compute $Q$ with the first law:

$$Q = \Delta U - W = 9.77 \times 10^4 \text{ J} - (-3.50 \times 10^4 \text{ J})$$
$$= \boxed{1.33 \times 10^5 \text{ J}}$$

. . . . . . . . . . . . . . . . . . . . . . . . . . . . . . . . . . . . . . . . . . . . . . . . . . . . . . . . . . . . . . . . . . . . . . . . . . . . . . . . . . . . . . . . . . . . . . . . . .

**REMARKS** As long as it's possible to compute the work, cycles involving these more exotic processes can be completely analyzed. Usually, however, it's necessary to use calculus. Note that the solution to part **(b)** could have been facilitated by yet another application of $PV = nRT$:

$$\Delta U = \tfrac{3}{2}nR\Delta T = \tfrac{3}{2}\Delta(PV) = \tfrac{3}{2}(P_B V_B - P_A V_A)$$

This result means that even in the absence of information about the number of moles or the temperatures, the problem could be solved knowing the initial and final pressures and volumes.

**QUESTION 12.9** For a curve with lower pressures but the same endpoints as in Figure 12.10a, would the thermal energy transferred be (a) smaller than, (b) equal to, or (c) greater than the thermal energy transfer of the straight-line path?

**EXERCISE 12.9** Figure 12.10b represents a process involving 3.00 moles of a monatomic ideal gas expanding from 0.100 m³ to 0.200 m³. Find the work done on the system, the change in the internal energy of the system, and the thermal energy transferred in the process.

**ANSWERS** $W = -2.00 \times 10^4$ J, $\Delta U = -1.50 \times 10^4$ J, $Q = 5.00 \times 10^3$ J

---

Given all the different processes and formulas, it's easy to become confused when approaching one of these ideal gas problems, although most of the time only substitution into the correct formula is required. The essential facts and formulas are compiled in Table 12.2, both for easy reference and also to display the similarities and differences between the processes.

**Table 12.2** The First Law and Thermodynamic Processes (Ideal Gases)

| Process | $\Delta U$ | $Q$ | $W$ |
|---|---|---|---|
| Isobaric | $nC_v \Delta T$ | $nC_p \Delta T$ | $-P\Delta V$ |
| Adiabatic | $nC_v \Delta T$ | 0 | $\Delta U$ |
| Isovolumetric | $nC_v \Delta T$ | $\Delta U$ | 0 |
| Isothermal | 0 | $-W$ | $-nRT \ln\left(\dfrac{V_f}{V_i}\right)$ |
| General | $nC_v \Delta T$ | $\Delta U - W$ | $(PV \text{Area})$ |

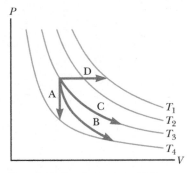

**Figure 12.11** (Quick Quiz 12.2) Identify the nature of paths A, B, C, and D.

# 12.4 Heat Engines and the Second Law of Thermodynamics

A **heat engine** takes in energy by heat and partially converts it to other forms, such as electrical and mechanical energy. In a typical process for producing electricity in a power plant, for instance, coal or some other fuel is burned, and the resulting internal energy is used to convert water to steam. The steam is then directed at the blades of a turbine, setting it rotating. Finally, the mechanical energy associated with this rotation is used to drive an electric generator. In another heat engine—the internal combustion engine in an automobile—energy enters the engine as fuel is injected into the cylinder and combusted, and a fraction of this energy is converted to mechanical energy.

In general, a heat engine carries some working substance through a **cyclic process**[1] during which (1) energy is transferred by heat from a source at a high temperature, (2) work is done by the engine, and (3) energy is expelled from the engine by heat to a source at lower temperature. As an example,

◀ Cyclic process

---

[1]Strictly speaking, the internal combustion engine is not a heat engine according to the description of the cyclic process, because the air–fuel mixture undergoes only one cycle and is then expelled through the exhaust system.

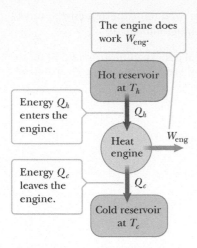

The engine does work $W_{eng}$.

Energy $Q_h$ enters the engine.

Hot reservoir at $T_h$

$Q_h$

Heat engine

$W_{eng}$

Energy $Q_c$ leaves the engine.

$Q_c$

Cold reservoir at $T_c$

**Figure 12.12** In this schematic representation of a heat engine, part of the thermal energy from the hot reservoir is turned into work while the rest is expelled to the cold reservoir.

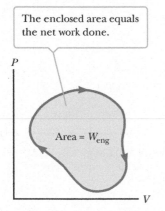

The enclosed area equals the net work done.

$P$

Area = $W_{eng}$

$V$

**Figure 12.13** The $PV$ diagram for an arbitrary cyclic process.

consider the operation of a steam engine in which the working substance is water. The water in the engine is carried through a cycle in which it first evaporates into steam in a boiler and then expands against a piston. After the steam is condensed with cooling water, it returns to the boiler, and the process is repeated.

It's useful to draw a heat engine schematically, as in Figure 12.12. The engine absorbs energy $Q_h$ from the hot reservoir, does work $W_{eng}$, then gives up energy $Q_c$ to the cold reservoir. (Note that *negative* work is done *on* the engine, so that $W = -W_{eng}$.) Because the working substance goes through a cycle, always returning to its initial thermodynamic state, its initial and final internal energies are equal, so $\Delta U = 0$. From the first law of thermodynamics, therefore,

$$\Delta U = 0 = Q + W \quad \rightarrow \quad Q_{net} = -W = W_{eng}$$

The last equation shows that **the work $W_{eng}$ done by a heat engine equals the net energy absorbed by the engine.** As we can see from Figure 12.12, $Q_{net} = |Q_h| - |Q_c|$. Therefore,

$$W_{eng} = |Q_h| - |Q_c| \qquad [12.11]$$

Ordinarily, a transfer of thermal energy $Q$ can be either positive or negative, so the use of absolute value signs makes the signs of $Q_h$ and $Q_c$ explicit.

If the working substance is a gas, then **the work done by the engine for a cyclic process is the area enclosed by the curve representing the process on a $PV$ diagram.** This area is shown for an arbitrary cyclic process in Figure 12.13.

The **thermal efficiency** $e$ of a heat engine is defined as the work done by the engine, $W_{eng}$, divided by the energy absorbed during one cycle:

$$e \equiv \frac{W_{eng}}{|Q_h|} = \frac{|Q_h| - |Q_c|}{|Q_h|} = 1 - \frac{|Q_c|}{|Q_h|} \qquad [12.12]$$

We can think of thermal efficiency as the ratio of the benefit received (work) to the cost incurred (energy transfer at the higher temperature). Equation 12.12 shows that a heat engine has 100% efficiency ($e = 1$) only if $Q_c = 0$, meaning no energy is expelled to the cold reservoir. In other words, a heat engine with perfect efficiency would have to use all the input energy for doing mechanical work. That isn't possible, as will be seen in Section 12.5.

---

### EXAMPLE 12.10 THE EFFICIENCY OF AN ENGINE

**GOAL** Apply the efficiency formula to a heat engine.

**PROBLEM** During one cycle, an engine extracts $2.00 \times 10^3$ J of energy from a hot reservoir and transfers $1.50 \times 10^3$ J to a cold reservoir. **(a)** Find the thermal efficiency of the engine. **(b)** How much work does this engine do in one cycle? **(c)** What average power does the engine generate if it goes through four cycles in 2.50 s?

**STRATEGY** Apply Equation 12.12 to obtain the thermal efficiency, then use the first law, adapted to engines (Eq. 12.11), to find the work done in one cycle. To obtain the power generated, divide the work done in four cycles by the time it takes to run those cycles.

## SOLUTION

**(a)** Find the engine's thermal efficiency.

Substitute $Q_c$ and $Q_h$ into Equation 12.12:

$$e = 1 - \frac{|Q_c|}{|Q_h|} = 1 - \frac{1.50 \times 10^3\,\text{J}}{2.00 \times 10^3\,\text{J}} = \boxed{0.250, \text{ or } 25.0\%}$$

**(b)** How much work does this engine do in one cycle?

Apply the first law in the form of Equation 12.11 to find the work done by the engine:

$$W_{\text{eng}} = |Q_h| - |Q_c| = 2.00 \times 10^3\,\text{J} - 1.50 \times 10^3\,\text{J}$$
$$= \boxed{5.00 \times 10^2\,\text{J}}$$

**(c)** Find the average power output of the engine.

Multiply the answer of part **(b)** by four and divide by time:

$$P = \frac{W}{\Delta t} = \frac{4.00 \times (5.00 \times 10^2\,\text{J})}{2.50\,\text{s}} = \boxed{8.00 \times 10^2\,\text{W}}$$

**REMARKS** Problems like this usually reduce to solving two equations and two unknowns, as here, where the two equations are the efficiency equation and the first law and the unknowns are the efficiency and the work done by the engine.

**QUESTION 12.10** Can the efficiency of an engine always be improved by increasing the thermal energy put into the system during a cycle? Explain.

**EXERCISE 12.10** The energy absorbed by an engine is three times as great as the work it performs. **(a)** What is its thermal efficiency? **(b)** What fraction of the energy absorbed is expelled to the cold reservoir? **(c)** What is the average power output of the engine if the energy input is 1 650 J each cycle and it goes through two cycles every 3 seconds?

**ANSWERS** **(a)** $\frac{1}{3}$   **(b)** $\frac{2}{3}$   **(c)** 367 W

---

## EXAMPLE 12.11   ANALYZING AN ENGINE CYCLE

**GOAL** Combine several concepts to analyze an engine cycle.

**PROBLEM** A heat engine contains an ideal monatomic gas confined to a cylinder by a movable piston. The gas starts at $A$, where $T = 3.00 \times 10^2$ K. (See Fig. 12.14a.) The process $B \to C$ is an isothermal expansion. **(a)** Find the number $n$ of moles of gas and the temperature at $B$. **(b)** Find $\Delta U$, $Q$, and $W$ for the isovolumetric process $A \to B$. **(c)** Repeat for the isothermal process $B \to C$. **(d)** Repeat for the isobaric process $C \to A$. **(e)** Find the net change in the internal energy for the complete cycle. **(f)** Find the thermal energy $Q_h$ transferred into the system, the thermal energy rejected, $Q_c$, the thermal efficiency, and net work on the environment performed by the engine.

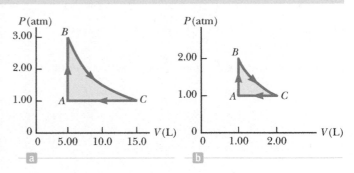

**Figure 12.14** (a) (Example 12.11) (b) (Exercise 12.11)

**STRATEGY** In part **(a)** $n$ and $T$ can be found from the ideal gas law, which connects the equilibrium values of $P$, $V$, and $T$. Once the temperature $T$ is known at the points $A$, $B$, and $C$, the change in internal energy, $\Delta U$, can be computed from the formula in Table 12.2 for each process. $Q$ and $W$ can be similarly computed, or deduced from the first law, using the techniques applied in the single-process examples.

---

## SOLUTION

**(a)** Find $n$ and $T_B$ with the ideal gas law:

$$n = \frac{P_A V_A}{R T_A} = \frac{(1.00\,\text{atm})(5.00\,\text{L})}{(0.082\,1\,\text{L}\cdot\text{atm}/\text{mol}\cdot\text{K})(3.00 \times 10^2\,\text{K})}$$
$$= \boxed{0.203\,\text{mol}}$$

$$T_B = \frac{P_B V_B}{nR} = \frac{(3.00\,\text{atm})(5.00\,\text{L})}{(0.203\,\text{mol})(0.082\,1\,\text{L}\cdot\text{atm}/\text{mol}\cdot\text{K})}$$
$$= \boxed{9.00 \times 10^2\,\text{K}}$$

*(Continued)*

**(b)** Find $\Delta U_{AB}$, $Q_{AB}$, and $W_{AB}$ for the constant volume process $A \rightarrow B$.

Compute $\Delta U_{AB}$, noting that $C_v = \frac{3}{2}R = 12.5 \text{ J/mol} \cdot \text{K}$:

$$\Delta U_{AB} = nC_v \Delta T = (0.203 \text{ mol})(12.5 \text{ J/mol} \cdot \text{K})$$
$$\times (9.00 \times 10^2 \text{ K} - 3.00 \times 10^2 \text{ K})$$
$$\Delta U_{AB} = \boxed{1.52 \times 10^3 \text{ J}}$$

$\Delta V = 0$ for isovolumetric processes, so no work is done:

$$W_{AB} = \boxed{0}$$

We can find $Q_{AB}$ from the first law:

$$Q_{AB} = \Delta U_{AB} = \boxed{1.52 \times 10^3 \text{ J}}$$

**(c)** Find $\Delta U_{BC}$, $Q_{BC}$, and $W_{BC}$ for the isothermal process $B \rightarrow C$.

This process is isothermal, so the temperature doesn't change, and the change in internal energy is zero:

$$\Delta U_{BC} = nC_v \Delta T = \boxed{0}$$

Compute the work done on the system, using the negative of Equation 12.10:

$$W_{BC} = -nRT \ln\left(\frac{V_C}{V_B}\right)$$
$$= -(0.203 \text{ mol})(8.31 \text{ J/mol} \cdot \text{K})(9.00 \times 10^2 \text{ K})$$
$$\times \ln\left(\frac{1.50 \times 10^{-2} \text{ m}^3}{5.00 \times 10^{-3} \text{ m}^3}\right)$$
$$W_{BC} = \boxed{-1.67 \times 10^3 \text{ J}}$$

Compute $Q_{BC}$ from the first law:

$$0 = Q_{BC} + W_{BC} \quad \rightarrow \quad Q_{BC} = -W_{BC} = \boxed{1.67 \times 10^3 \text{ J}}$$

**(d)** Find $\Delta U_{CA}$, $Q_{CA}$, and $W_{CA}$ for the isobaric process $C \rightarrow A$.

Compute the work on the system, with pressure constant:

$$W_{CA} = -P\Delta V = -(1.01 \times 10^5 \text{ Pa})(5.00 \times 10^{-3} \text{ m}^3$$
$$- 1.50 \times 10^{-2} \text{ m}^3)$$
$$W_{CA} = \boxed{1.01 \times 10^3 \text{ J}}$$

Find the change in internal energy, $\Delta U_{CA}$:

$$\Delta U_{CA} = \frac{3}{2}nR\Delta T = \frac{3}{2}(0.203 \text{ mol})(8.31 \text{ J/K} \cdot \text{mol})$$
$$\times (3.00 \times 10^2 \text{ K} - 9.00 \times 10^2 \text{ K})$$
$$\Delta U_{CA} = \boxed{-1.52 \times 10^3 \text{ J}}$$

Compute the thermal energy, $Q_{CA}$, from the first law:

$$Q_{CA} = \Delta U_{CA} - W_{CA} = -1.52 \times 10^3 \text{ J} - 1.01 \times 10^3 \text{ J}$$
$$= \boxed{-2.53 \times 10^3 \text{ J}}$$

**(e)** Find the net change in internal energy, $\Delta U_{net}$, for the cycle:

$$\Delta U_{net} = \Delta U_{AB} + \Delta U_{BC} + \Delta U_{CA}$$
$$= 1.52 \times 10^3 \text{ J} + 0 - 1.52 \times 10^3 \text{ J} = \boxed{0}$$

**(f)** Find the energy input, $Q_h$; the energy rejected, $Q_c$; the thermal efficiency; and the net work performed by the engine:

Sum all the positive contributions to find $Q_h$:

$$Q_h = Q_{AB} + Q_{BC} = 1.52 \times 10^3 \text{ J} + 1.67 \times 10^3 \text{ J}$$
$$= \boxed{3.19 \times 10^3 \text{ J}}$$

Sum any negative contributions (in this case, there is only one):

$$Q_c = \boxed{-2.53 \times 10^3 \text{ J}}$$

Find the engine efficiency and the net work done by the engine:

$$e = 1 - \frac{|Q_c|}{|Q_h|} = 1 - \frac{2.53 \times 10^3 \, \text{J}}{3.19 \times 10^3 \, \text{J}} = \boxed{0.207}$$

$$W_{\text{eng}} = -(W_{AB} + W_{BC} + W_{CA})$$
$$= -(0 - 1.67 \times 10^3 \, \text{J} + 1.01 \times 10^3 \, \text{J})$$
$$= \boxed{6.60 \times 10^2 \, \text{J}}$$

**REMARKS** Cyclic problems are rather lengthy, but the individual steps are often short substitutions. Notice that the change in internal energy for the cycle is zero and that the net work done on the environment is identical to the net thermal energy transferred, both as they should be.

**QUESTION 12.11** If $BC$ were a straight-line path, would the work done by the cycle be affected? How?

**EXERCISE 12.11** $4.05 \times 10^{-2}$ mol of monatomic ideal gas goes through the process shown in Figure 12.14b. The temperature at point $A$ is $3.00 \times 10^2$ K and is $6.00 \times 10^2$ K during the isothermal process $B \to C$. **(a)** Find $Q$, $\Delta U$, and $W$ for the constant volume process $A \to B$. **(b)** Do the same for the isothermal process $B \to C$. **(c)** Repeat, for the constant pressure process $C \to A$. **(d)** Find $Q_h$, $Q_c$, and the efficiency. **(e)** Find $W_{\text{eng}}$.

**ANSWERS** **(a)** $Q_{AB} = \Delta U_{AB} = 151$ J, $W_{AB} = 0$ **(b)** $\Delta U_{BC} = 0$, $Q_{BC} = -W_{BC} = 1.40 \times 10^2$ J **(c)** $Q_{CA} = -252$ J, $\Delta U_{CA} = -151$ J, $W_{CA} = 101$ J **(d)** $Q_h = 291$ J, $Q_c = -252$ J, $e = 0.134$ **(e)** $W_{\text{eng}} = 39$ J

## 12.4.1 Refrigerators and Heat Pumps

Heat engines can operate in reverse. In this case, energy is injected into the engine, modeled as work $W$ in Figure 12.15, resulting in energy being extracted from the cold reservoir and transferred to the hot reservoir. The system now operates as a heat pump, a common example being a refrigerator (Fig. 12.16, page 402). Energy $Q_c$ is extracted from the interior of the refrigerator and delivered as energy $Q_h$ to the warmer air in the kitchen. The work is done in the compressor unit of the refrigerator, compressing a refrigerant such as freon, causing its temperature to increase.

A household air conditioner is another example of a heat pump. Some homes are both heated and cooled by heat pumps. In winter, the heat pump extracts energy $Q_c$ from the cool outside air and delivers energy $Q_h$ to the warmer air inside. In summer, energy $Q_c$ is removed from the cool inside air, while energy $Q_h$ is ejected to the warm air outside.

For a refrigerator or an air conditioner—a heat pump operating in cooling mode—work $W$ is what you pay for, in terms of electrical energy running the compressor, whereas $Q_c$ is the desired benefit. The most efficient refrigerator or air conditioner is one that removes the greatest amount of energy from the cold reservoir in exchange for the least amount of work.

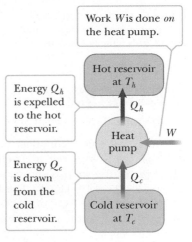

Work $W$ is done *on* the heat pump.

Energy $Q_h$ is expelled to the hot reservoir.

Energy $Q_c$ is drawn from the cold reservoir.

**Figure 12.15** In this schematic representation of a heat pump, thermal energy is extracted from the cold reservoir and "pumped" to the hot reservoir.

The coefficient of performance (COP) for a refrigerator or an air conditioner is the magnitude of the energy extracted from the cold reservoir, $|Q_c|$, divided by the work $W$ performed by the device:

$$\text{COP(cooling mode)} = \frac{|Q_c|}{W} \qquad \text{[12.13]}$$

**SI unit: dimensionless**

The larger this ratio, the better the performance, because more energy is being removed for a given amount of work. A good refrigerator or air conditioner will have a COP of 5 or 6.

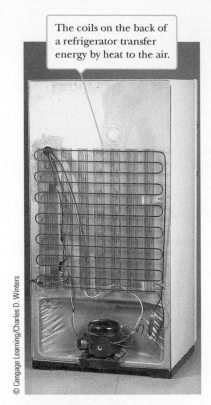

The coils on the back of a refrigerator transfer energy by heat to the air.

**Figure 12.16** The back of a household refrigerator. The air surrounding the coils is the hot reservoir.

A heat pump operating in heating mode warms the inside of a house in winter by extracting energy from the colder outdoor air. This statement may seem paradoxical, but recall that this process is equivalent to a refrigerator removing energy from its interior and ejecting it into the kitchen.

The coefficient of performance of a heat pump operating in the heating mode is the magnitude of the energy rejected to the hot reservoir, $|Q_h|$, divided by the work $W$ done by the pump:

$$\text{COP(heating mode)} = \frac{|Q_h|}{W} \qquad \text{[12.14]}$$

**SI unit: dimensionless**

In effect, the COP of a heat pump in the heating mode is the ratio of what you gain (energy delivered to the interior of your home) to what you give (work input). Typical values for this COP are greater than 1, because $|Q_h|$ is usually greater than $W$.

In a groundwater heat pump, energy is extracted in the winter from water deep in the ground rather than from the outside air, while energy is delivered to that water in the summer. This strategy increases the year-round efficiency of the heating and cooling unit because the groundwater is at a higher temperature than the air in winter and at a cooler temperature than the air in summer.

---

**EXAMPLE 12.12   COOLING THE LEFTOVERS**

**GOAL** Apply the coefficient of performance of a refrigerator.

**PROBLEM** A 2.00-L container of leftover soup at a temperature of 323 K is placed in a refrigerator. Assume the specific heat of the soup is the same as that of water and the density is $1.25 \times 10^3$ kg/m$^3$. The refrigerator cools the soup to 283 K. **(a)** If the COP of the refrigerator is 5.00, find the energy needed, in the form of work, to cool the soup. **(b)** If the compressor has a power rating of 0.250 hp, for what minimum length of time must it operate to cool the soup to 283 K? (The minimum time assumes the soup cools at the same rate that the heat pump ejects thermal energy from the refrigerator.)

**STRATEGY** The solution to this problem requires three steps. First, find the total mass $m$ of the soup. Second, using $Q = mc\,\Delta T$, where $Q = Q_c$, find the energy transfer required to cool the soup. Third, substitute $Q_c$ and the COP into Equation 12.13, solving for $W$. Divide the work by the power to get an estimate of the time required to cool the soup.

. . . . . . . . . . . . . . . . . . . . . . . . . . . . . . . . . . . . . . . . . . . . . . . . . . . . . . . . . . . . . . . . . . . . . . . . . . . . . . . . . . . . . . .

**SOLUTION**

**(a)** Find the work needed to cool the soup.

Calculate the mass of the soup:

$$m = \rho V = (1.25 \times 10^3 \text{ kg/m}^3)(2.00 \times 10^{-3} \text{ m}^3) = 2.50 \text{ kg}$$

Find the energy transfer required to cool the soup:

$$Q_c = Q = mc\,\Delta T$$
$$= (2.50 \text{ kg})(4\,190 \text{ J/kg} \cdot \text{K})(283 \text{ K} - 323 \text{ K})$$
$$= -4.19 \times 10^5 \text{ J}$$

Substitute $Q_c$ and the COP into Equation 12.13:

$$\text{COP} = \frac{|Q_c|}{W} = \frac{4.19 \times 10^5 \text{ J}}{W} = 5.00$$
$$W = 8.38 \times 10^4 \text{ J}$$

**(b)** Find the time needed to cool the soup.

Convert horsepower to watts:

$$P = (0.250 \text{ hp})(746 \text{ W/1 hp}) = 187 \text{ W}$$

Divide the work by the power to find the elapsed time:

$$\Delta t = \frac{W}{P} = \frac{8.38 \times 10^4 \, \text{J}}{187 \, \text{W}} = \boxed{448 \, \text{s}}$$

**REMARKS** This example illustrates how cooling different substances requires differing amounts of work due to differences in specific heats. The problem doesn't take into account the insulating properties of the soup container and of the soup itself, which retard the cooling process.

**QUESTION 12.12** If the refrigerator door is left open, does the kitchen become cooler? Why or why not?

**EXERCISE 12.12** (a) How much work must a heat pump with a COP of 2.50 do to extract 1.00 MJ of thermal energy from the outdoors (the cold reservoir)? (b) If the unit operates at 0.500 hp, how long will the process take? (Be sure to use the correct COP!)

**ANSWERS** (a) $6.67 \times 10^5 \, \text{J}$   (b) $1.79 \times 10^3 \, \text{s}$

## 12.4.2 The Second Law of Thermodynamics

There are limits to the efficiency of heat engines. The ideal engine would convert all input energy into useful work, but it turns out that such an engine is impossible to construct. The Kelvin-Planck formulation of the **second law of thermodynamics** can be stated as follows:

> No heat engine operating in a cycle can absorb energy from a reservoir and use it entirely for the performance of an equal amount of work.

This form of the second law means that the efficiency $e = W_{\text{eng}}/|Q_h|$ of engines must always be less than 1. Some energy $|Q_c|$ must always be lost to the environment. In other words, it's theoretically impossible to construct a heat engine with an efficiency of 100%.

To summarize, the first law says **we can't get a greater amount of energy out of a cyclic process than we put in,** and the second law says **we can't break even.** No matter what engine is used, some energy must be transferred by heat to the cold reservoir. In Equation 12.11, the second law simply means $|Q_c|$ is always greater than zero.

There is another equivalent statement of the second law:

> If two systems are in thermal contact, net thermal energy transfers spontaneously by heat from the hotter system to the colder system.

Here, spontaneous means the energy transfer occurs naturally, with no work being done. Thermal energy naturally transfers from hotter systems to colder systems. Work must be done to transfer thermal energy from a colder system to a hotter system, however. An example is the refrigerator, which transfers thermal energy from inside the refrigerator to the warmer kitchen.

## 12.4.3 Reversible and Irreversible Processes

No engine can operate with 100% efficiency, but different designs yield different efficiencies, and it turns out that one design in particular delivers the maximum possible efficiency. This design is the Carnot cycle, discussed in the next subsection. Understanding it requires the concepts of reversible and irreversible processes. In a **reversible** process, every state along the path is an equilibrium state, so the system can return to its initial conditions by going along the same path in the reverse direction. A process that doesn't satisfy this requirement is **irreversible.**

Most natural processes are known to be irreversible; the reversible process is an idealization. Although real processes are always irreversible, some are *almost* reversible. If a real process occurs so slowly that the system is virtually always in equilibrium, the process can be considered reversible. Imagine compressing a gas very slowly by

**LORD KELVIN**
**British Physicist and Mathematician (1824–1907)**

Born William Thomson in Belfast, Kelvin was the first to propose the use of an absolute scale of temperature. His study of Carnot's theory led to the idea that energy cannot pass spontaneously from a colder object to a hotter object; this principle is known as the second law of thermodynamics.

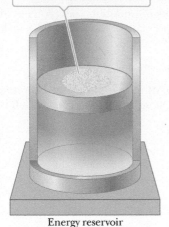

Individual grains of sand drop onto the piston, slowly compressing the gas.

Energy reservoir

**Figure 12.17** A method for compressing a gas in a reversible isothermal process.

**SADI CARNOT**
**French Engineer (1796–1832)**

Carnot is considered to be the founder of the science of thermodynamics. Some of his notes found after his death indicate that he was the first to recognize the relationship between work and heat.

**Tip 12.3** Don't Shop for a Carnot Engine

The Carnot engine is only an idealization. If a Carnot engine were developed in an effort to maximize efficiency, it would have zero power output because for all of the processes to be reversible, the engine would have to run infinitely slowly.

dropping grains of sand onto a frictionless piston, as in Figure 12.17. The temperature can be kept constant by placing the gas in thermal contact with an energy reservoir. The pressure, volume, and temperature of the gas are well defined during this isothermal compression. Each added grain of sand represents a change to a new equilibrium state. The process can be reversed by slowly removing grains of sand from the piston.

### 12.4.4 The Carnot Engine

In 1824, in an effort to understand the efficiency of real engines, a French engineer named Sadi Carnot (1796–1832) described a theoretical engine now called a *Carnot engine* that is of great importance from both a practical and a theoretical viewpoint. He showed that a heat engine operating in an ideal, reversible cycle—now called a **Carnot cycle**—between two energy reservoirs is the most efficient engine possible. Such an engine establishes an upper limit on the efficiencies of all real engines. **Carnot's theorem** can be stated as follows:

No real engine operating between two energy reservoirs can be more efficient than a Carnot engine operating between the same two reservoirs.

In a Carnot cycle, an ideal gas is contained in a cylinder with a movable piston at one end. The temperature of the gas varies between $T_c$ and $T_h$. The cylinder walls and the piston are thermally nonconducting. Figure 12.18 shows the four stages of the Carnot cycle, and Figure 12.19 is the $PV$ diagram for the cycle. The cycle consists of two adiabatic and two isothermal processes, all reversible:

1. The process $A \rightarrow B$ is an isothermal expansion at temperature $T_h$ in which the gas is placed in thermal contact with a hot reservoir (a large oven, for example) at temperature $T_h$ (Fig. 12.18a). During the process, the gas absorbs energy $Q_h$ from the reservoir and does work $W_{AB}$ in raising the piston.
2. In the process $B \rightarrow C$, the base of the cylinder is replaced by a thermally nonconducting wall and the gas expands adiabatically, so no energy enters or leaves the system by heat (Fig. 12.18b). During the process, the temperature falls from $T_h$ to $T_c$ and the gas does work $W_{BC}$ in raising the piston.
3. In the process $C \rightarrow D$, the gas is placed in thermal contact with a cold reservoir at temperature $T_c$ (Fig. 12.18c) and is compressed isothermally at temperature $T_c$. During this time, the gas expels energy $Q_c$ to the reservoir and the work done on the gas is $W_{CD}$.
4. In the final process, $D \rightarrow A$, the base of the cylinder is again replaced by a thermally nonconducting wall (Fig. 12.18d), and the gas is compressed adiabatically. The temperature of the gas increases to $T_h$, and the work done on the gas is $W_{DA}$.

For a Carnot engine, the following relationship between the thermal energy transfers and the absolute temperatures can be derived:

$$\frac{|Q_c|}{|Q_h|} = \frac{T_c}{T_h} \qquad [12.15]$$

Substituting this expression into Equation 12.12, we find that the thermal efficiency of a Carnot engine is

$$e_C = 1 - \frac{T_c}{T_h} \qquad [12.16]$$

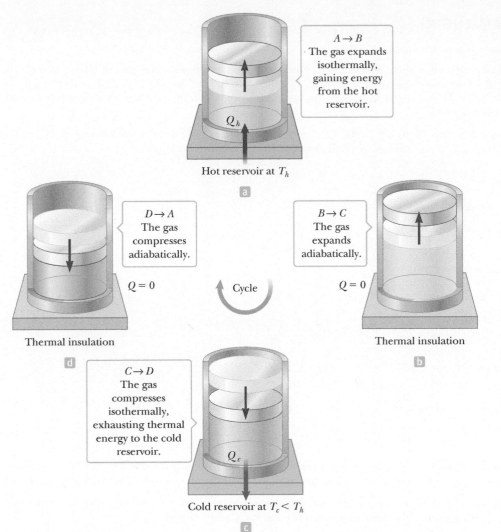

Figure 12.18 The Carnot cycle. The letters A, B, C, and D refer to the states of the gas shown in Figure 12.19. The arrows on the piston indicate the direction of its motion during each process.

$A \rightarrow B$
The gas expands isothermally, gaining energy from the hot reservoir.

$Q_h$

Hot reservoir at $T_h$

a

$D \rightarrow A$
The gas compresses adiabatically.

$Q = 0$

$B \rightarrow C$
The gas expands adiabatically.

$Q = 0$

Cycle

Thermal insulation

d

Thermal insulation

b

$C \rightarrow D$
The gas compresses isothermally, exhausting thermal energy to the cold reservoir.

$Q_c$

Cold reservoir at $T_c < T_h$

c

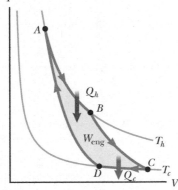

Figure 12.19 The PV diagram for the Carnot cycle. The net work done, $W_{eng}$, equals the net energy transferred into the Carnot engine in one cycle, $|Q_h| - |Q_c|$.

where *T* must be in kelvins. From this result, we see that **all Carnot engines operating reversibly between the same two temperatures have the same efficiency.**

Equation 12.16 can be applied to any working substance operating in a Carnot cycle between two energy reservoirs. According to that equation, the efficiency is zero if $T_c = T_h$. The efficiency increases as $T_c$ is lowered and as $T_h$ is increased. The efficiency can be one (100%), however, only if $T_c = 0$ K. According to the **third law of thermodynamics,** it's impossible to lower the temperature of a system to absolute zero in a finite number of steps, so such reservoirs are not available and the maximum efficiency is always less than 1. In most practical cases, the cold reservoir is near room temperature, about 300 K, so increasing the efficiency requires raising the temperature of the hot reservoir. **All real engines operate irreversibly, due to friction and the brevity of their cycles, and are therefore *less* efficient than the Carnot engine.**

◀ Third law of thermodynamics

*Quick Quiz*

**12.3** Three engines operate between reservoirs separated in temperature by 300 K. The reservoir temperatures are as follows:

Engine A: $T_h = 1\ 000$ K, $T_c = 700$ K

Engine B: $T_h = 800$ K, $T_c = 500$ K

Engine C: $T_h = 600$ K, $T_c = 300$ K

Rank the engines in order of their theoretically possible efficiency, from highest to lowest. (a) A, B, C (b) B, C, A (c) C, B, A (d) C, A, B

## EXAMPLE 12.13 | THE STEAM ENGINE

**GOAL** Apply the equations of an ideal (Carnot) engine.

**PROBLEM** A steam engine has a boiler that operates at $5.00 \times 10^2$ K. The energy from the boiler changes water to steam, which drives the piston. The temperature of the exhaust is that of the outside air, $3.00 \times 10^2$ K. **(a)** What is the engine's efficiency if it's an ideal engine? **(b)** If the $3.50 \times 10^3$ J of energy is supplied from the boiler, find the energy transferred to the cold reservoir and the work done by the engine on its environment.

**STRATEGY** This problem requires substitution into Equations 12.15 and 12.16, both applicable to a Carnot engine. The first equation relates the ratio $Q_c/Q_h$ to the ratio $T_c/T_h$, and the second gives the Carnot engine efficiency.

**SOLUTION**

**(a)** Find the engine's efficiency, assuming it's ideal.

Substitute into Equation 12.16, the equation for the efficiency of a Carnot engine:

$$e_C = 1 - \frac{T_c}{T_h} = 1 - \frac{3.00 \times 10^2 \text{ K}}{5.00 \times 10^2 \text{ K}} = \boxed{0.400, \text{ or } 40.0\%}$$

**(b)** Find the energy transferred to the cold reservoir and the work done on the environment if $3.50 \times 10^3$ J is delivered to the engine during one cycle.

Equation 12.15 shows that the ratio of energies equals the ratio of temperatures:

$$\frac{|Q_c|}{|Q_h|} = \frac{T_c}{T_h} \rightarrow |Q_c| = |Q_h| \frac{T_c}{T_h}$$

Substitute, finding the energy transferred to the cold reservoir:

$$|Q_c| = (3.50 \times 10^3 \text{ J})\left(\frac{3.00 \times 10^2 \text{ K}}{5.00 \times 10^2 \text{ K}}\right) = \boxed{2.10 \times 10^3 \text{ J}}$$

Use Equation 12.11 to find the work done by the engine:

$$W_{\text{eng}} = |Q_h| - |Q_c| = 3.50 \times 10^3 \text{ J} - 2.10 \times 10^3 \text{ J}$$
$$= \boxed{1.40 \times 10^3 \text{ J}}$$

**REMARKS** This problem differs from the earlier examples on work and efficiency because we used the special Carnot relationships, Equations 12.15 and 12.16. Remember that these equations can only be used when the cycle is identified as ideal or a Carnot.

**QUESTION 12.13** True or False: A nonideal engine operating between the same temperature extremes as a Carnot engine and having the same input thermal energy will perform the same amount of work as the Carnot engine.

**EXERCISE 12.13** The highest theoretical efficiency of a gasoline engine based on the Carnot cycle is 0.300, or 30.0%. **(a)** If this engine expels its gases into the atmosphere, which has a temperature of $3.00 \times 10^2$ K, what is the temperature in the cylinder immediately after combustion? **(b)** If the heat engine absorbs 837 J of energy from the hot reservoir during each cycle, how much work can it perform in each cycle?

**ANSWERS** **(a)** 429 K **(b)** 251 J

# 12.5 Entropy

Temperature and internal energy, associated with the zeroth and first laws of thermodynamics, respectively, are both state variables, meaning they can be used to describe the thermodynamic state of a system. A state variable called the **entropy** $S$ is related to the second law of thermodynamics. We define entropy on a macroscopic scale as German physicist Rudolf Clausius (1822–1888) first expressed it in 1865:

> Let $Q_r$ be the energy absorbed or expelled during a reversible, constant temperature process between two equilibrium states. Then the change in entropy during any constant temperature process connecting the two equilibrium states is defined as

$$\Delta S \equiv \frac{Q_r}{T}$$  [12.17]

**SI unit: joules/kelvin (J/K)**

A similar formula holds when the temperature isn't constant, but its derivation entails calculus and won't be considered here. Calculating the change in entropy, $\Delta S$, during a transition between two equilibrium states requires finding a reversible path that connects the states. The entropy change calculated on that reversible path is taken to be $\Delta S$ for the *actual* path. This approach is necessary because quantities such as the temperature of a system can be defined only for systems in equilibrium, and a reversible path consists of a sequence of equilibrium states. The subscript $r$ on the term $Q_r$ emphasizes that the path chosen for the calculation must be reversible. The change in entropy $\Delta S$, like changes in internal energy $\Delta U$ and changes in potential energy, depends only on the endpoints, and not on the path connecting them.

The concept of entropy gained wide acceptance in part because it provided another variable to describe the state of a system, along with pressure, volume, and temperature. Its significance was enhanced when it was found that **the entropy of the Universe increases in all natural processes.** This is yet another way of stating the second law of thermodynamics.

Although the entropy of the *Universe* increases in all natural processes, the entropy of a *system* can decrease. For example, if system A transfers energy $Q$ to system B by heat, the entropy of system A decreases. This transfer, however, can only occur if the temperature of system B is less than the temperature of system A. Because temperature appears in the denominator in the definition of entropy, system B's increase in entropy will be greater than system A's decrease, so taken together, the entropy of the Universe increases.

For centuries, individuals have attempted to build perpetual motion machines that operate continuously without any input of energy or increase in entropy. The laws of thermodynamics preclude the invention of any such machines.

The concept of entropy is satisfying because it enables us to present the second law of thermodynamics in the form of a mathematical statement. In the next section, we find that entropy can also be interpreted in terms of probabilities, a relationship that has profound implications.

**RUDOLF CLAUSIUS**
**German Physicist (1822–1888)**

Born with the name Rudolf Gottlieb, he adopted the classical name of Clausius, which was a popular thing to do in his time. "I propose . . . to call $S$ the entropy of a body, after the Greek word 'transformation.' I have designedly coined the word 'entropy' to be similar to energy, for these two quantities are so analogous in their physical significance, that an analogy of denominations seems to be helpful."

**Tip 12.4** Entropy ≠ Energy

Don't confuse energy and entropy. Although the names sound similar the concepts are different.

---

*Quick Quiz*

**12.4** Which of the following is true for the entropy change of a system that undergoes a reversible, adiabatic process? (a) $\Delta S < 0$ (b) $\Delta S = 0$ (c) $\Delta S > 0$

---

**EXAMPLE 12.14** **MELTING A PIECE OF LEAD**

**GOAL** Calculate the change in entropy due to a phase change.

**PROBLEM** (a) Find the change in entropy of $3.00 \times 10^2$ g of lead when it melts at 327°C. Lead has a latent heat of fusion of $2.45 \times 10^4$ J/kg. (b) Suppose the same amount of energy is used to melt part of a piece of silver, which is already at its melting point of 961°C. Find the change in the entropy of the silver.

**STRATEGY** This problem can be solved by substitution into Equation 12.17. Be sure to use the Kelvin temperature scale.

· · · · · · · · · · · · · · · · · · · · · · · · · · · · · · · · · · · · · · · · · · · · · · · · · · · · · · · · · · · · · · · · · · · · · · · · · · · · · · · · · · ·

**SOLUTION**

**(a)** Find the entropy change of the lead.

Find the energy necessary to melt the lead:   $Q = mL_f = (0.300 \text{ kg})(2.45 \times 10^4 \text{ J/kg}) = 7.35 \times 10^3 \text{ J}$

Convert the temperature in degrees Celsius to Kelvins:   $T = T_C + 273 = 327 + 273 = 6.00 \times 10^2 \text{ K}$

Substitute the quantities found into the entropy equation:   $\Delta S = \dfrac{Q}{T} = \dfrac{7.35 \times 10^3 \text{ J}}{6.00 \times 10^2 \text{ K}} = \boxed{12.3 \text{ J/K}}$

*(Continued)*

**(b)** Find the entropy change of the silver.

The added energy is the same as in part **(a)**, by supposition. Substitute into the entropy equation, after first converting the melting point of silver to kelvins:

$$T = T_C + 273 = 961 + 273 = 1.234 \times 10^3 \text{ K}$$

$$\Delta S = \frac{Q}{T} = \frac{7.35 \times 10^3 \text{ J}}{1.234 \times 10^3 \text{ K}} = \boxed{5.96 \text{ J/K}}$$

**REMARKS** This example shows that adding a given amount of energy to a system increases its entropy, but adding the same amount of energy to another system at higher temperature results in a smaller increase in entropy. This is because the change in entropy is inversely proportional to the temperature.

**QUESTION 12.14** If the same amount of energy were used to melt ice at 0°C to water at 0°C, rank the entropy changes for ice, silver, and lead, from smallest to largest.

**EXERCISE 12.14** Find the change in entropy of a 2.00-kg block of gold at 1 063°C when it melts to become liquid gold at 1 063°C. (The latent heat of fusion for gold is $6.44 \times 10^4$ J/kg.)

**ANSWER** 96.4 J/K

---

**EXAMPLE 12.15** **ICE, STEAM, AND THE ENTROPY OF THE UNIVERSE**

**GOAL** Calculate the change in entropy for a system and its environment.

**PROBLEM** A block of ice at 273 K is put in thermal contact with a container of steam at 373 K, converting 25.0 g of ice to water at 273 K while condensing some of the steam to water at 373 K. Find **(a)** the change in entropy of the ice, **(b)** the change in entropy of the steam, and **(c)** the change in entropy of the Universe.

**STRATEGY** First, calculate the energy transfer necessary to melt the ice. The amount of energy gained by the ice is lost by the steam. Compute the entropy change for each process and sum to get the entropy change of the Universe.

**SOLUTION**

**(a)** Find the change in entropy of the ice.

Use the latent heat of fusion, $L_f$, to compute the thermal energy needed to melt 25.0 g of ice:

$$Q_{\text{ice}} = mL_f = (0.025 \text{ kg})(3.33 \times 10^5 \text{ J}) = 8.33 \times 10^3 \text{ J}$$

Calculate the change in entropy of the ice:

$$\Delta S_{\text{ice}} = \frac{Q_{\text{ice}}}{T_{\text{ice}}} = \frac{8.33 \times 10^3 \text{ J}}{273 \text{ K}} = \boxed{30.5 \text{ J/K}}$$

**(b)** Find the change in entropy of the steam.

By supposition, the thermal energy lost by the steam is equal to the thermal energy gained by the ice:

$$\Delta S_{\text{steam}} = \frac{Q_{\text{steam}}}{T_{\text{steam}}} = \frac{-8.33 \times 10^3 \text{ J}}{373 \text{ K}} = \boxed{-22.3 \text{ J/K}}$$

**(c)** Find the change in entropy of the Universe.

Sum the two changes in entropy:

$$\Delta S_{\text{universe}} = \Delta S_{\text{ice}} + \Delta S_{\text{steam}} = 30.5 \text{ J/k} - 22.3 \text{ J/K}$$

$$= \boxed{+8.2 \text{ J/K}}$$

**REMARKS** Notice that the entropy of the Universe increases, as it must in all natural processes.

**QUESTION 12.15** True or False: For a given magnitude of thermal energy transfer, the change in entropy is smaller for processes that proceed at lower temperature.

**EXERCISE 12.15** A 4.00-kg block of ice at 273 K encased in a thin plastic shell of negligible mass melts in a large lake at 293 K. At the instant the ice has completely melted in the shell and is still at 273 K, calculate the change in entropy of **(a)** the ice, **(b)** the lake (which essentially remains at 293 K), and **(c)** the Universe.

**ANSWERS** **(a)** $4.88 \times 10^3$ J/K **(b)** $-4.55 \times 10^3$ J/K **(c)** $+3.3 \times 10^2$ J/K

## EXAMPLE 12.16 | A FALLING BOULDER

**GOAL** Combine mechanical energy and entropy.

**PROBLEM** A chunk of rock of mass $1.00 \times 10^3$ kg at 293 K falls from a cliff of height 125 m into a large lake, also at 293 K. Find the change in entropy of the lake, assuming all the rock's kinetic energy upon entering the lake converts to thermal energy absorbed by the lake.

**STRATEGY** Gravitational potential energy when the rock is at the top of the cliff converts to kinetic energy of the rock before it enters the lake, then is transferred to the lake as thermal energy. The change in the lake's temperature is negligible (due to its mass). Divide the mechanical energy of the rock by the temperature of the lake to estimate the lake's change in entropy.

................................................................................................................................

**SOLUTION**

Calculate the gravitational potential energy associated with the rock at the top of the cliff:

$$PE = mgh = (1.00 \times 10^3 \text{ kg})(9.80 \text{ m/s}^2)(125 \text{ m})$$
$$= 1.23 \times 10^6 \text{ J}$$

This energy is transferred to the lake as thermal energy, resulting in an entropy increase of the lake:

$$\Delta S = \frac{Q}{T} = \frac{1.23 \times 10^6 \text{ J}}{293 \text{ K}} = \boxed{4.20 \times 10^3 \text{ J/K}}$$

................................................................................................................................

**REMARKS** This example shows how even simple mechanical processes can bring about increases in the Universe's entropy.

**QUESTION 12.16** If you carefully remove your physics book from a shelf and place it on the ground, what happens to the entropy of the Universe? Does it increase, decrease, or remain the same? Explain.

**EXERCISE 12.16** Estimate the change in entropy of a tree trunk at 15.0°C when a bullet of mass 5.00 g traveling at $1.00 \times 10^3$ m/s embeds itself in it. (Assume the kinetic energy of the bullet transforms to thermal energy, all of which is absorbed by the tree.)

**ANSWER** 8.68 J/K

---

## 12.5.1 Entropy and Disorder

A large element of chance is inherent in natural processes. The spacing between trees in a natural forest, for example, is random; if you discovered a forest where all the trees were equally spaced, you would conclude that it had been planted. Likewise, leaves fall to the ground with random arrangements. It would be highly unlikely to find the leaves laid out in perfectly straight rows. We can express the results of such observations by saying that **a disorderly arrangement is much more probable than an orderly one if the laws of nature are allowed to act without interference** (Fig. 12.20).

Entropy originally found its place in thermodynamics, but its importance grew tremendously as the field of statistical mechanics developed. This analytical approach employs an alternate interpretation of entropy. In statistical mechanics, the behavior of a substance is described by the statistical behavior of the atoms and molecules contained in it. One of the main conclusions of the statistical mechanical

**Figure 12.20** (a) A royal flush is a highly ordered poker hand with a low probability of occurrence. (b) A disordered and worthless poker hand. The probability of this *particular* hand occurring is the same as that of the royal flush. There are so many worthless hands, however, that the probability of being dealt a worthless hand is much higher than that of being dealt a royal flush. Can you calculate the probability of being dealt a full house (a pair and three of a kind) from a standard deck of 52 cards?

**Tip 12.5** Don't Confuse the *W*'s

The symbol *W* used here is a *probability*, not to be confused with the same symbol used for work.

approach is that **isolated systems tend toward greater disorder, and entropy is a measure of that disorder.**

In light of this new view of entropy, Boltzmann found another method for calculating entropy through use of the relation

$$S = k_B \ln W \qquad [12.18]$$

where $k_B = 1.38 \times 10^{-23}$ J/K is Boltzmann's constant and $W$ is a number proportional to the probability that the system has a particular configuration. The symbol "ln" again stands for natural logarithm, discussed in Appendix A.

Equation 12.18 could be applied to a bag of marbles. Imagine that you have 100 marbles—50 red and 50 green—stored in a bag. You are allowed to draw four marbles from the bag according to the following rules: Draw one marble, record its color, return it to the bag, and draw again. Continue this process until four marbles have been drawn. Note that because each marble is returned to the bag before the next one is drawn, the probability of drawing a red marble is always the same as the probability of drawing a green one.

The results of all possible drawing sequences are shown in Table 12.3. For example, the result RRGR means that a red marble was drawn first, a red one second, a green one third, and a red one fourth. The table indicates that there is only one possible way to draw four red marbles. There are four possible sequences that produce one green and three red marbles, six sequences that produce two green and two red, four sequences that produce three green and one red, and one sequence that produces all green. From Equation 12.18, we see that the state with the greatest disorder (two red and two green marbles) has the highest entropy because it is most probable. In contrast, the most ordered states (all red marbles and all green marbles) are least likely to occur and are states of lowest entropy.

The outcome of the draw can range between these highly ordered (lowest-entropy) and highly disordered (highest-entropy) states. Entropy can be regarded as an index of how far a system has progressed from an ordered to a disordered state.

**The second law of thermodynamics is really a statement of what is most probable rather than of what must be.** Imagine placing an ice cube in contact with a hot piece of pizza. There is nothing in nature that absolutely forbids the transfer of energy by heat from the ice to the much warmer pizza. Statistically, it's possible for a slow-moving molecule in the ice to collide with a faster-moving molecule in the pizza so that the slow one transfers some of its energy to the faster one. When the great number of molecules present in the ice and pizza are considered, however, the odds are overwhelmingly in favor of the transfer of energy from the faster-moving molecules to the slower-moving molecules. Furthermore, this example demonstrates that a system naturally tends to move from a state of order to a state of disorder. The initial state, in which all the pizza molecules have high kinetic energy and all the ice molecules have lower kinetic energy, is much more ordered than the final state after energy transfer has taken place and the ice has melted.

APPLICATION
The Direction of Time

Even more generally, the second law of thermodynamics defines the direction of time for all events as the direction in which the entropy of the universe increases. Although conservation of energy isn't violated if energy flows spontaneously from a cold object (the ice cube) to a hot object (the pizza slice), that event violates the

**Table 12.3** Possible Results of Drawing Four Marbles from a Bag

| End Result | Possible Draws | Total Number of Same Results |
|---|---|---|
| All R | RRRR | 1 |
| 1G, 3R | RRRG, RRGR, RGRR, GRRR | 4 |
| 2G, 2R | RRGG, RGRG, GRRG, RGGR, GRGR, GGRR | 6 |
| 3G, 1R | GGGR, GGRG, GRGG, RGGG | 4 |
| All G | GGGG | 1 |

second law because it represents a spontaneous increase in order. Of course, such an event also violates everyday experience. If the melting ice cube is filmed and the film speeded up, the difference between running the film in forward and reverse directions would be obvious to an audience. The same would be true of filming any event involving a large number of particles, such as a dish dropping to the floor and shattering.

As another example, suppose you were able to measure the velocities of all the air molecules in a room at some instant. It's very unlikely that you would find all molecules moving in the same direction with the same speed; that would be a highly ordered state, indeed. The most probable situation is a system of molecules moving haphazardly in all directions with a wide distribution of speeds, a highly disordered state. This physical situation can be compared to the drawing of marbles from a bag: If a container held $10^{23}$ molecules of a gas, the probability of finding all the molecules moving in the same direction with the same speed at some instant would be similar to that of drawing a marble from the bag $10^{23}$ times and getting a red marble on every draw, clearly an unlikely set of events.

The tendency of nature to move toward a state of disorder affects the ability of a system to do work. Consider a ball thrown toward a wall. The ball has kinetic energy, and its state is an ordered one, which means that all the atoms and molecules of the ball move in unison at the same speed and in the same direction (apart from their random internal motions). When the ball hits the wall, however, part of the ball's kinetic energy is transformed into the random, disordered, internal motion of the molecules in the ball and the wall, and the temperatures of the ball and the wall both increase slightly. Before the collision, the ball was capable of doing work. It could drive a nail into the wall, for example. With the transformation of part of the ordered energy into disordered internal energy, this capability of doing work is reduced. The ball rebounds with less kinetic energy than it originally had, because the collision is inelastic.

Various forms of energy can be converted to internal energy, as in the collision between the ball and the wall, but the reverse transformation is never complete. In general, given two kinds of energy, A and B, if A can be completely converted to B and vice versa, we say that A and B are of the *same grade*. However, if A can be completely converted to B and the reverse is never complete, then A is of a *higher grade* of energy than B. In the case of a ball hitting a wall, the kinetic energy of the ball is of a higher grade than the internal energy contained in the ball and the wall after the collision. When high-grade energy is converted to internal energy, it can never be fully recovered as high-grade energy.

This conversion of high-grade energy to internal energy is referred to as the **degradation of energy.** The energy is said to be degraded because it takes on a form that is less useful for doing work. In other words, **in all real processes, the energy available for doing work decreases.**

Finally, note once again that the statement that entropy must increase in all natural processes is true only for isolated systems. There are instances in which the entropy of some system decreases, but with a corresponding net increase in entropy for some other system. When all systems are taken together to form the Universe, **the entropy of the Universe always increases.**

Ultimately, the entropy of the Universe should reach a maximum. When it does, the Universe will be in a state of uniform temperature and density. All physical, chemical, and biological processes will cease, because a state of perfect disorder implies no available energy for doing work. This gloomy state of affairs is sometimes referred to as the ultimate "heat death" of the Universe.

### Quick Quiz

**12.5** Suppose you are throwing two dice in a friendly game of craps. For any given throw, the two numbers that are faceup can have a sum of 2, 3, 4, 5, 6, 7, 8, 9, 10, 11, or 12. Which outcome is most probable? Which is least probable?

# 12.6 Human Metabolism BIO

Animals do work and give off energy by heat, leading us to believe the first law of thermodynamics can be applied to living organisms to describe them in a general way. The internal energy stored in humans goes into other forms needed for maintaining and repairing the major body organs and is transferred out of the body by work as a person walks or lifts a heavy object, and by heat when the body is warmer than its surroundings. Because the rates of change of internal energy, energy loss by heat, and energy loss by work vary widely with the intensity and duration of human activity, it's best to measure the time rates of change of $\Delta U$, $Q$, and $W$. Rewriting the first law, these time rates of change are related by

$$\frac{\Delta U}{\Delta t} = \frac{Q}{\Delta t} + \frac{W}{\Delta t}$$ [12.19]

On average, energy $Q$ flows *out* of the body, and work is done *by* the body on its surroundings, so both $Q/\Delta t$ and $W/\Delta t$ are negative. This means that $\Delta U/\Delta t$ would be negative and the internal energy and body temperature would decrease with time if a human were a closed system with no way of ingesting matter or replenishing internal energy stores. Because all animals are actually open systems, they acquire internal energy (chemical potential energy) by eating and breathing, so their internal energy and temperature are kept constant. Overall, the energy from the oxidation of food ultimately supplies the work done by the body and energy lost from the body by heat, and this is the interpretation we give Equation 12.19. That is, $\Delta U/\Delta t$ is the rate at which internal energy is added to our bodies by food, and this term just balances the rate of energy loss by heat, $Q/\Delta t$, and by work, $W/\Delta t$. Finally, if we have a way of measuring $\Delta U/\Delta t$ and $W/\Delta t$ for a human, we can calculate $Q/\Delta t$ from Equation 12.19 and gain useful information on the efficiency of the body as a machine.

## 12.6.1 Measuring the Metabolic Rate $\Delta U/\Delta t$

The value of $W/\Delta t$, the work done by a person per unit time, can easily be determined by measuring the power output supplied by the person (in pedaling a bike, for example). **The metabolic rate $\Delta U/\Delta t$ is the rate at which chemical potential energy in food and oxygen are transformed into internal energy to just balance the body losses of internal energy by work and heat.** Although the mechanisms of food oxidation and energy release in the body are complicated, involving many intermediate reactions and enzymes (organic compounds that speed up the chemical reactions taking place at "low" body temperatures), an amazingly simple rule summarizes these processes: **The metabolic rate is directly proportional to the rate of oxygen consumption by volume.** It is found that for an average diet, the consumption of one liter of oxygen releases 4.8 kcal, or 20 kJ, of energy. We may write this important summary rule as

Metabolic rate equation ▶

$$\frac{\Delta U}{\Delta t} = 4.8 \frac{\Delta V_{O_2}}{\Delta t}$$ [12.20]

where the metabolic rate $\Delta U/\Delta t$ is measured in kcal/s and $\Delta V_{O_2}/\Delta t$, the volume rate of oxygen consumption, is in L/s. Measuring the rate of oxygen consumption during various activities ranging from sleep to intense bicycle racing effectively measures the variation of metabolic rate or the variation in the total power the body generates. A simultaneous measurement of the work per unit time done by a person along with the metabolic rate allows the efficiency of the body as a machine to be determined. Figure 12.21 shows a person monitored for oxygen consumption while riding a bike attached to a dynamometer, a device for measuring power output.

## 12.6.2 Metabolic Rate, Activity, and Weight Gain

Table 12.4 shows the measured rate of oxygen consumption in milliliters per minute per kilogram of body mass and the calculated metabolic rate for a 65-kg

**Figure 12.21** This bike rider is being monitored for oxygen consumption.

**Table 12.4** Oxygen Consumption and Metabolic Rates for Various Activities for a 65-kg Male[a]

| Activity | O₂ Use Rate (mL/min · kg) | Metabolic Rate (kcal/h) | Metabolic Rate (W) |
|---|---|---|---|
| Sleeping | 3.5 | 70 | 80 |
| Light activity (dressing, walking slowly, desk work) | 10 | 200 | 230 |
| Moderate activity (walking briskly) | 20 | 400 | 465 |
| Heavy activity (basketball, swimming a fast breaststroke) | 30 | 600 | 700 |
| Extreme activity (bicycle racing) | 70 | 1 400 | 1 600 |

[a]*Source: A Companion to Medical Studies,* 2/e, R. Passmore, Philadelphia, F. A. Davis, 1968.

male engaged in various activities. A sleeping person uses about 80 W of power, the **basal metabolic rate,** just to maintain and run different body organs such as the heart, lungs, liver, kidneys, brain, and skeletal muscles. More intense activity increases the metabolic rate to a maximum of about 1 600 W for a superb racing cyclist, although such a high rate can only be maintained for periods of a few seconds. When we sit watching a riveting film, we give off about as much energy by heat as a bright (250-W) lightbulb.

Regardless of level of activity, the daily food intake should just balance the loss in internal energy if a person is not to gain weight. Further, exercise is a poor substitute for dieting as a method of losing weight, although it has other benefits. For example, the loss of 1 pound of body fat requires the muscles to expend 4 100 kcal of energy. If the goal is to lose 1 pound of fat in 35 days, a jogger could run an extra mile a day, because a 65-kg jogger uses about 120 kcal to jog 1 mile (35 days × 120 kcal/day = 4 200 kcal). An easier way to lose the pound of fat would be to diet and eat two fewer slices of bread per day for 35 days, because bread has a calorie content of 60 kcal/slice (35 days × 2 slices/day × 60 kcal/slice = 4 200 kcal).

## EXAMPLE 12.17 | FIGHTING FAT

**GOAL** Estimate human energy usage during a typical day.

**PROBLEM** In the course of 24 hours, a 65-kg person spends 8 h at a desk, 2 h puttering around the house, 1 h jogging 5 miles, 5 h in moderate activity, and 8 h sleeping. What is the change in his internal energy during this period?

**STRATEGY** The time rate of energy usage—or power—multiplied by time gives the amount of energy used during a given activity. Use Table 12.4 to find the power $P_i$ needed for each activity, multiply each by the time, and sum them all up.

**SOLUTION**

$$\Delta U = -\sum P_i \Delta t_i = -(P_1 \Delta t_1 + P_2 \Delta t_2 + \cdots + P_n \Delta t_n)$$

$$= -(200 \text{ kcal/h})(10 \text{ h}) - (5 \text{ mi/h})(120 \text{ kcal/mi})(1 \text{ h}) - (400 \text{ kcal/h})(5 \text{ h}) - (70 \text{ kcal/h})(8 \text{ h})$$

$$\Delta U = \boxed{-5\,000 \text{ kcal}}$$

**REMARKS** If this is a typical day in the man's life, he will have to consume less than 5 000 kilocalories on a daily basis in order to lose weight. A complication lies in the fact that human metabolism tends to drop when food intake is reduced.

**QUESTION 12.17** How could completely skipping meals lead to weight gain?

**EXERCISE 12.17** If a 60.0-kg man ingests 3 000 kcal a day and spends 6 h sleeping, 4 h walking briskly, 8 h sitting at a desk job, 1 h swimming a fast breaststroke, and 5 h watching action movies on TV, about how much weight will the man gain or lose every day? (*Note:* Recall that using about 4 100 kcal of energy will burn off a pound of fat.)

**ANSWER** He'll lose a little more than one-half a pound of fat a day.

**Table 12.5** Physical Fitness and Maximum Oxygen Consumption Rate[a]

**Table 12.5** Physical Fitness and Maximum Oxygen Consumption Rate[a]

| Fitness Level | Maximum Oxygen Consumption Rate (mL/min · kg) |
|---|---|
| Very poor | 28 |
| Poor | 34 |
| Fair | 42 |
| Good | 52 |
| Excellent | 70 |

[a]*Source: Aerobics*, K. H. Cooper, Bantam Books, New York, 1968.

## 12.6.3 Physical Fitness and Efficiency of the Human Body as a Machine

One measure of a person's physical fitness is his or her maximum capacity to use or consume oxygen. This "aerobic" fitness can be increased and maintained with regular exercise, but falls when training stops. Typical maximum rates of oxygen consumption and corresponding fitness levels are shown in Table 12.5; we see that the maximum oxygen consumption rate varies from 28 mL/min · kg of body mass for poorly conditioned subjects to 70 mL/min · kg for superb athletes.

We have already pointed out that the first law of thermodynamics can be rewritten to relate the metabolic rate $\Delta U/\Delta t$ to the rate at which energy leaves the body by work and by heat:

$$\frac{\Delta U}{\Delta t} = \frac{Q}{\Delta t} + \frac{W}{\Delta t}$$

Now consider the body as a machine capable of supplying mechanical power to the outside world and ask for its efficiency. The body's efficiency $e$ is defined as the ratio of the mechanical power supplied by a human to the metabolic rate or the total power input to the body:

$$e = \text{body's efficiency} = \frac{\left|\dfrac{W}{\Delta t}\right|}{\left|\dfrac{\Delta U}{\Delta t}\right|} \qquad [12.21]$$

In this definition, absolute value signs are used to show that $e$ is a positive number and to avoid explicitly using minus signs required by our definitions of $W$ and $Q$ in the first law. Table 12.6 shows the efficiency of workers engaged in different activities for several hours. These values were obtained by measuring the power output and simultaneous oxygen consumption of mine workers and calculating the metabolic rate from their oxygen consumption. The table shows that a person can steadily supply mechanical power for several hours at about 100 W with an efficiency of about 17%. It also shows the dependence of efficiency on activity, and that $e$ can drop to values as low as 3% for highly inefficient activities like shoveling, which involves many starts and stops. Finally, it is interesting in comparison to the average results of Table 12.6 that a superbly conditioned athlete, efficiently coupled to a mechanical device for extracting power (a bike!), can supply a power of around 300 W for about 30 minutes at a peak efficiency of 22%.

**Table 12.6** Metabolic Rate, Power Output, and Efficiency for Different Activities[a]

| Activity | Metabolic Rate $\dfrac{\Delta U}{\Delta t}$ (watts) | Power Output $\dfrac{W}{\Delta t}$ (watts) | Efficiency $e$ |
|---|---|---|---|
| Cycling | 505 | 96 | 0.19 |
| Pushing loaded coal cars in a mine | 525 | 90 | 0.17 |
| Shoveling | 570 | 17.5 | 0.03 |

[a]*Source:* "Inter- and Intra-Individual Differences in Energy Expenditure and Mechanical Efficiency," C. H. Wyndham et al., *Ergonomics* **9**, 17 (1966).

# SUMMARY

## 12.1 Work in Thermodynamic Processes

The work done on a gas at a constant pressure is

$$W = -P\Delta V \qquad [12.1]$$

**Figure 12.22** Positive work is done on a gas by compressing it.

The work done on the gas is positive if the gas is compressed ($\Delta V$ is negative), as in Figure 12.22, and negative if the gas expands ($\Delta V$ is positive). In general, the work done on a gas that takes it from some initial state to some final state is the negative of the area under the curve on a $PV$ diagram.

## 12.2 The First Law of Thermodynamics

According to the first law of thermodynamics (Fig. 12.23), when a system undergoes a change from one state to another, the **change in its internal energy** $\Delta U$ is

$$\Delta U = U_f - U_i = Q + W \qquad [12.2]$$

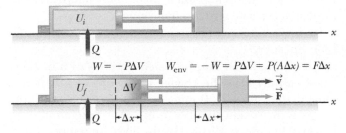

**Figure 12.23** Illustration of the first law of thermodynamics.

where $Q$ is the energy exchanged across the boundary between the system and the environment and $W$ is the work done on the system. The quantity $Q$ is positive when energy is transferred into the system by heating and negative when energy is removed from the system by cooling. $W$ is positive when work is done on the system (e.g., by compression) and negative when the system does positive work on its environment.

The change of the internal energy, $\Delta U$, of an ideal gas is given by

$$\Delta U = nC_v\Delta T \qquad [12.5]$$

where $C_v$ is the molar specific heat at constant volume.

## 12.3 Thermal Processes in Gases

An **isobaric process** (Fig. 12.24) is one that occurs at constant pressure. The work done on the system in such a process is $-P\Delta V$, whereas the thermal energy transferred by heat is given by

$$Q = nC_p\Delta T \qquad [12.6]$$

with the molar heat capacity at constant pressure given by $C_p = C_v + R$.

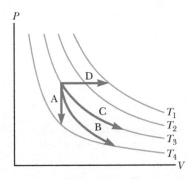

**Figure 12.24** Four gas processes: A is an isovolumetric process (constant volume); B is an adiabatic expansion (no thermal energy transfer); C is an isothermal process (constant temperature); D is an isobaric process (constant pressure).

In an **adiabatic process** (Fig. 12.24), no energy is transferred by heat between the system and its surroundings ($Q = 0$). In this case, the first law gives $\Delta U = W$, which means the internal energy changes solely as a consequence of work being done on the system. The pressure and volume in adiabatic processes are related by

$$PV^\gamma = \text{constant} \qquad [12.8a]$$

where $\gamma = C_p/C_v$ is the adiabatic index.

In an **isovolumetric process** (Fig. 12.24), the volume doesn't change and no work is done. For such processes, the first law gives $\Delta U = Q$.

An **isothermal process** (Fig. 12.24) occurs at constant temperature. The work done by an ideal gas on the environment is

$$W_{\text{env}} = nRT\ln\left(\frac{V_f}{V_i}\right) \qquad [12.10]$$

## 12.4 Heat Engines and the Second Law of Thermodynamics

In a **cyclic process** (in which the system returns to its initial state), $\Delta U = 0$ and therefore $Q = W_{\text{eng}}$, meaning the energy transferred into the system by heat equals the work done on the system during the cycle.

A **heat engine** (Fig. 12.25, page 416) takes in energy by heat and partially converts it to other forms of energy, such as mechanical and electrical energy. The work $W_{\text{eng}}$ done by a heat engine in carrying a working substance through a cyclic process ($\Delta U = 0$) is

$$W_{\text{eng}} = |Q_h| - |Q_c| \qquad [12.11]$$

where $Q_h$ is the energy absorbed from a hot reservoir and $Q_c$ is the energy expelled to a cold reservoir.

The **thermal efficiency** of a heat engine is defined as the ratio of the work done by the engine to the energy transferred into the engine per cycle:

$$e \equiv \frac{W_{eng}}{|Q_h|} = 1 - \frac{|Q_c|}{|Q_h|} \qquad [12.12]$$

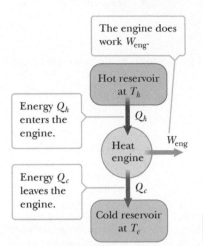

The engine does work $W_{eng}$.

Energy $Q_h$ enters the engine.

Energy $Q_c$ leaves the engine.

**Figure 12.25** Schematic diagram of a heat engine.

**Heat pumps** (Fig. 12.26) are heat engines in reverse. In a refrigerator the heat pump removes thermal energy from inside the refrigerator. Heat pumps operating in cooling mode have coefficient of performance given by

$$COP(\text{cooling mode}) = \frac{|Q_c|}{W} \qquad [12.13]$$

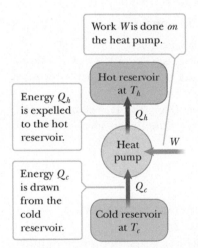

Work $W$ is done *on* the heat pump.

Energy $Q_h$ is expelled to the hot reservoir.

Energy $Q_c$ is drawn from the cold reservoir.

**Figure 12.26** Schematic diagram of a heat pump.

A heat pump in heating mode has coefficient of performance

$$COP(\text{heating mode}) = \frac{|Q_h|}{W} \qquad [12.14]$$

Real processes proceed in an order governed by the **second law of thermodynamics,** which can be stated in two ways:

1. Energy will not flow spontaneously by heat from a cold object to a hot object.
2. No heat engine operating in a cycle can absorb energy from a reservoir and perform an equal amount of work.

No real heat engine operating between the Kelvin temperatures $T_h$ and $T_c$ can exceed the efficiency of an engine operating between the same two temperatures in a **Carnot cycle** (Fig. 12.27), given by

$$e_C = 1 - \frac{T_c}{T_h} \qquad [12.16]$$

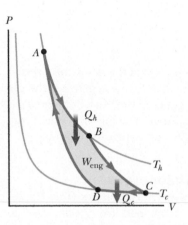

**Figure 12.27** *PV* diagram of a Carnot cycle.

Perfect efficiency of a Carnot engine requires a cold reservoir of 0 K, absolute zero. According to the **third law of thermodynamics,** however, it is impossible to lower the temperature of a system to absolute zero in a finite number of steps.

## 12.5 Entropy

The second law can also be stated in terms of a quantity called **entropy** (*S*). The **change in entropy** of a system is equal to the energy $Q_r$ flowing by heat into (or out of) the system as the system changes from one state to another by a reversible process, divided by the absolute temperature:

$$\Delta S \equiv \frac{Q_r}{T} \qquad [12.17]$$

One of the primary findings of statistical mechanics is that systems tend toward disorder, and entropy is a measure of that disorder. An alternate statement of the second law is that the entropy of the Universe increases in all natural processes.

---

## CONCEPTUAL QUESTIONS

1. Two identical containers each hold 1 mole of an ideal gas at 1 atm. Container A holds a monatomic gas and container B holds a diatomic gas. The gas in each container is compressed at constant pressure to half its original volume. (a) What is the ratio $W_A/W_B$ of the work done on gas A to the work done on gas B? (b) What is the ratio $\Delta U_A/\Delta U_B$ of the change in internal energy for gases A and B? (c) What is the ratio $Q_A/Q_B$ of the energy transferred to gases A and B?

2. Which one of the following statements is true? (a) The path on a *PV* diagram always goes from the smaller volume to the larger volume. (b) The path on a *PV* diagram always goes

from the smaller pressure to the larger pressure. (c) The area under the path on a *PV* diagram is always equal to the work done on a gas. (d) The area under the path on a *PV* diagram is always equal in magnitude to the work done on a gas.

3. **BIO** Consider the human body performing a strenuous exercise, such as lifting weights or riding a bicycle. Work is being done by the body, and energy is leaving by conduction from the skin into the surrounding air. According to the first law of thermodynamics, the temperature of the body should be steadily decreasing during the exercise. That isn't what happens, however. Is the first law invalid for this situation? Explain.

4. Clearly distinguish among temperature, heat, and internal energy.

5. For an ideal gas in an isothermal process, there is no change in internal energy. Suppose the gas does work *W* during such a process. How much energy is transferred by heat?

6. An ideal gas undergoes an adiabatic process so that no energy enters or leaves the gas by heat. Which one of the following statements is true? (a) Because no energy is added by heat, the temperature cannot change. (b) The temperature increases if the gas volume increases. (c) The temperature increases if the gas pressure increases. (d) The temperature decreases if the gas pressure increases. (e) The temperature decreases if the gas volume decreases.

7. Is it possible to construct a heat engine that creates no thermal pollution?

8. A heat engine does work $W_{eng}$ while absorbing energy $Q_h$ from the hot reservoir and expelling energy $Q_c$ to the cold reservoir. Which one of the following is impossible? (a) $|Q_h| > |Q_c| > W_{eng}$ (b) $|Q_h| > W_{eng} > |Q_c|$ (c) $|Q_h| > W_{eng} = |Q_c|$ (d) $W_{eng} > |Q_h|$ (e) $W_{eng} > |Q_c|$

9. When a sealed Thermos bottle full of hot coffee is shaken, what changes, if any, take place in (a) the temperature of the coffee and (b) its internal energy?

10. The first law of thermodynamics is $\Delta U = Q + W$. For each of the following cases, state whether the internal energy of an ideal gas increases, decreases, or remains constant: (a) No energy is transferred to the gas as it expands to twice its original volume. (b) The gas volume is held constant while energy $Q$ is removed. (c) The gas volume is held constant and no energy is transferred to or from the gas. (d) The gas temperature increases.

11. The first law of thermodynamics says we can't get more out of a process than we put in, but the second law says that we can't break even. Explain this statement.

12. Objects A and B with $T_A > T_B$ are placed in thermal contact and come to equilibrium. (a) For which object does the entropy increase? (b) For which object does the entropy decrease? (c) Which object has the greater magnitude of entropy change?

13. Using the first law of thermodynamics, explain why the *total* energy of an isolated system is always constant.

14. What is wrong with the following statement: "Given any two bodies, the one with the higher temperature contains more heat."

15. An ideal gas is compressed to half its initial volume by means of several possible processes. Which of the following processes results in the most work done on the gas? (a) isothermal (b) adiabatic (c) isobaric (d) The work done is independent of the process.

16. A thermodynamic process occurs in which the entropy of a system changes by $-6$ J/K. According to the second law of thermodynamics, what can you conclude about the entropy change of the environment? (a) It must be $+6$ J/K or less. (b) It must be equal to 6 J/K. (c) It must be between $+6$ J/K and 0. (d) It must be 0. (e) It must be $+6$ J/K or more.

17. A window air conditioner is placed on a table inside a well-insulated apartment, plugged in and turned on. What happens to the average temperature of the apartment? (a) It increases. (b) It decreases. (c) It remains constant. (d) It increases until the unit warms up and then decreases. (e) The answer depends on the initial temperature of the apartment.

# PROBLEMS

*See the Preface for an explanation of the icons used in this problems set.*

## 12.1 Work in Thermodynamic Processes

1. **Q|C** An ideal gas is enclosed in a cylinder with a movable piston on top of it. The piston has a mass of $8.00 \times 10^3$ g and an area of 5.00 cm$^2$ and is free to slide up and down, keeping the pressure of the gas constant. (a) How much work is done on the gas as the temperature of 0.200 mol of the gas is raised from 20.0°C to $3.00 \times 10^2$°C? (b) What does the sign of your answer to part (a) indicate?

2. Sketch a *PV* diagram and find the work done by the gas during the following stages. (a) A gas is expanded from a volume of 1.0 L to 3.0 L at a constant pressure of 3.0 atm. (b) The gas is then cooled at constant volume until the pressure falls to 2.0 atm. (c) The gas is then compressed at a constant pressure of 2.0 atm from a volume of 3.0 L to 1.0 L. *Note:* Be careful of signs. (d) The gas is heated until its pressure increases from 2.0 atm to 3.0 atm at a constant volume. (e) Find the net work done during the complete cycle.

3. **T** Gas in a container is at a pressure of 1.5 atm and a volume of 4.0 m$^3$. What is the work done *on* the gas (a) if it expands at constant pressure to twice its initial volume, and (b) if it is compressed at constant pressure to one-quarter its initial volume?

4. Find the numeric value of the work done on the gas in (a) Figure P12.4a and (b) Figure P12.4b.

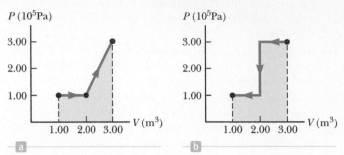

**Figure P12.4**

5. A gas expands from $I$ to $F$ along the three paths indicated in Figure P12.5. Calculate the work done *on* the gas along paths (a) *IAF*, (b) *IF*, and (c) *IBF*.

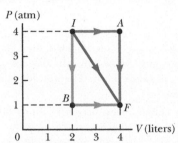

**Figure P12.5** Problems 5 and 15.

6. A gas follows the *PV* diagram in Figure P12.6. Find the work done on the gas along the paths (a) *AB*, (b) *BC*, (c) *CD*, (d) *DA*, and (e) *ABCDA*.

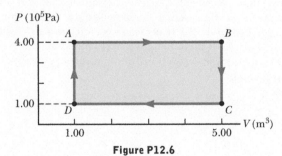

**Figure P12.6**

7. A sample of helium behaves as an ideal gas as it is heated at constant pressure from 273 K to 373 K. If 20.0 J of work is done by the gas during this process, what is the mass of helium present?

8. (a) Find the work done *by* an ideal gas as it expands from point $A$ to point $B$ along the path shown in Figure P12.8. (b) How much work is done by the gas if it compressed from $B$ to $A$ along the same path?

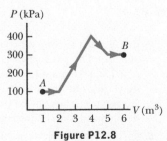

**Figure P12.8**

9. **V** One mole of an ideal gas initially at a temperature of $T_i = 0°C$ undergoes an expansion at a constant pressure of 1.00 atm to four times its original volume. (a) Calculate the new temperature $T_f$ of the gas. (b) Calculate the work done *on* the gas during the expansion.

10. (a) Determine the work done *on* a fluid that expands from $i$ to $f$ as indicated in Figure P12.10. (b) How much work is done *on* the fluid if it is compressed from $f$ to $i$ along the same path?

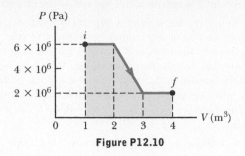

**Figure P12.10**

## 12.2 The First Law of Thermodynamics

## 12.3 Thermal Processes in Gases

11. A balloon holding 5.00 moles of helium gas absorbs 925 J of thermal energy while doing 102 J of work expanding to a larger volume. (a) Find the change in the balloon's internal energy. (b) Calculate the change in temperature of the gas.

12. A chemical reaction transfers 1 250 J of thermal energy into an ideal gas while the system expands by $2.00 \times 10^{-2}$ m³ at a constant pressure of $1.50 \times 10^5$ Pa. Find the change in the internal energy.

13. **S** The only form of energy possessed by molecules of a monatomic ideal gas is translational kinetic energy. Using the results from the discussion of kinetic theory in Section 10.5, show that the internal energy of a monatomic ideal gas at pressure $P$ and occupying volume $V$ may be written as $U = \frac{3}{2}PV$.

14. **GP** A cylinder of volume 0.300 m³ contains 10.0 mol of neon gas at 20.0°C. Assume neon behaves as an ideal gas. (a) What is the pressure of the gas? (b) Find the internal energy of the gas. (c) Suppose the gas expands at constant pressure to a volume of 1.000 m³. How much work is done on the gas? (d) What is the temperature of the gas at the new volume? (e) Find the internal energy of the gas when its volume is 1.000 m³. (f) Compute the change in the internal energy during the expansion. (g) Compute $\Delta U - W$. (h) Must thermal energy be transferred to the gas during the constant pressure expansion or be taken away? (i) Compute $Q$, the thermal energy transfer. (j) What symbolic relationship between $Q$, $\Delta U$, and $W$ is suggested by the values obtained?

15. **T** A gas expands from $I$ to $F$ in Figure P12.5. The energy added to the gas by heat is 418 J when the gas goes from $I$ to $F$ along the diagonal path. (a) What is the change in internal energy of the gas? (b) How much energy must be added to the gas by heat for the indirect path *IAF* to give the same change in internal energy?

16. **QC** In a running event, a sprinter does $4.8 \times 10^5$ J of work and her internal energy decreases by $7.5 \times 10^5$ J. (a) Determine the heat transferred between her body and surroundings during this event. (b) What does the sign of your answer to part (a) indicate?

17. **V** A gas is compressed at a constant pressure of 0.800 atm from 9.00 L to 2.00 L. In the process, 400. J of energy leaves the gas by heat. (a) What is the work done *on* the gas? (b) What is the change in its internal energy?

18. A quantity of a monatomic ideal gas undergoes a process in which both its pressure and volume are doubled as shown in Figure P12.18. What is the energy absorbed by heat into the gas during this process? *Hint:* The internal energy of a monatomic ideal gas at pressure $P$ and occupying volume $V$ is given by $U = \frac{3}{2}PV$.

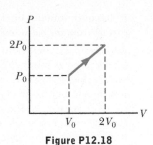

**Figure P12.18**

19. A gas is enclosed in a container fitted with a piston of cross-sectional area 0.150 m². The pressure of the gas is maintained at $6.00 \times 10^3$ Pa as the piston moves inward 20.0 cm. (a) Calculate the work done by the gas. (b) If the internal energy of the gas decreases by 8.00 J, find the amount of energy removed from the system by heat during the compression.

20. A monatomic ideal gas undergoes the thermodynamic process shown in the $PV$ diagram of Figure P12.20. Determine whether each of the values $\Delta U$, $Q$, and $W$ for the gas is positive, negative, or zero. *Hint:* The internal energy of a monatomic ideal gas at pressure $P$ and occupying volume $V$ is given by $U = \frac{3}{2}PV$.

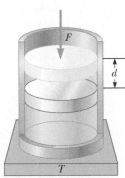

**Figure P12.20**

21. **Q|C** An ideal gas is compressed from a volume of $V_i = 5.00$ L to a volume of $V_f = 3.00$ L while in thermal contact with a heat reservoir at $T = 295$ K as in Figure P12.21. During the compression process, the piston moves down a distance of $d = 0.130$ m under the action of an average external force of $F = 25.0$ kN. Find (a) the work done on the gas, (b) the change in internal energy of the gas, and (c) the thermal energy exchanged

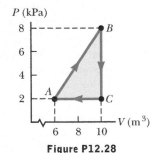

**Figure P12.21**

between the gas and the reservoir. (d) If the gas is thermally insulated so no thermal energy could be exchanged, what would happen to the temperature of the gas during the compression?

22. A system consisting of 0.025 6 moles of a diatomic ideal gas is taken from state $A$ to state $C$ along the path in Figure P12.22. (a) How much work is done *on* the gas during this process? (b) What is the lowest temperature of the gas during this process, and where does it occur? (c) Find the change in internal energy of the gas and (d) the energy delivered to the gas in going from $A$ to $C$. *Hint:* For part (c), adapt the equation in the remarks of Example 12.9 to a diatomic ideal gas.

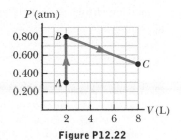

**Figure P12.22**

23. An ideal monatomic gas expands isothermally from 0.500 m³ to 1.25 m³ at a constant temperature of 675 K. If the initial pressure is $1.00 \times 10^5$ Pa, find (a) the work done on the gas, (b) the thermal energy transfer $Q$, and (c) the change in the internal energy.

24. **S** An ideal gas expands at constant pressure. (a) Show that $P\Delta V = nR\Delta T$. (b) If the gas is monatomic, start from the definition of internal energy and show that $\Delta U = \frac{3}{2}W_{env}$, where $W_{env}$ is the work done by the gas on its environment. (c) For the same monatomic ideal gas, show with the first law that $Q = \frac{5}{2}W_{env}$. (d) Is it possible for an ideal gas to expand at constant pressure while exhausting thermal energy? Explain.

25. An ideal monatomic gas contracts in an isobaric process from 1.25 m³ to 0.500 m³ at a constant pressure of $1.50 \times 10^5$ Pa. If the initial temperature is 425 K, find (a) the work done on the gas, (b) the change in internal energy, (c) the energy transfer $Q$, and (d) the final temperature.

26. An ideal diatomic gas expands adiabatically from 0.750 m³ to 1.50 m³. If the initial pressure and temperature are $1.50 \times 10^5$ Pa and 325 K, respectively, find (a) the number of moles in the gas, (b) the final gas pressure, (c) the final gas temperature, and (d) the work done on the gas.

27. **GP** An ideal monatomic gas is contained in a vessel of *constant* volume 0.200 m³. The initial temperature and pressure of the gas are 300. K and 5.00 atm, respectively. The goal of this problem is to find the temperature and pressure of the gas after 16.0 kJ of thermal energy is supplied to the gas. (a) Use the ideal gas law and initial conditions to calculate the number of moles of gas in the vessel. (b) Find the specific heat of the gas. (c) What is the work done by the gas during this process? (d) Use the first law of thermodynamics to find the change in internal energy of the gas. (e) Find the change in temperature of the gas. (f) Calculate the final temperature of the gas. (g) Use the ideal gas expression to find the final pressure of the gas.

28. Consider the cyclic process described by Figure P12.28. If $Q$ is negative for the process $BC$ and $\Delta U$ is negative for the process $CA$, determine the signs of $Q$, $W$, and $\Delta U$ associated with each process.

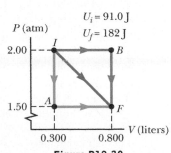

**Figure P12.28**

29. A 5.0-kg block of aluminum is heated from 20°C to 90°C at atmospheric pressure. Find (a) the work done by the aluminum, (b) the amount of energy transferred to it by heat, and (c) the increase in its internal energy.

30. One mole of gas initially at a pressure of 2.00 atm and a volume of 0.300 L has an internal energy equal to 91.0 J. In its final state, the gas is at a pressure of 1.50 atm and a volume of 0.800 L, and its internal energy equals 182 J. For the paths $IAF$, $IBF$, and $IF$ in Figure P12.30, calculate (a) the work done *on* the gas and (b) the net energy transferred to the gas by heat in the process.

## 12.4 Heat Engines and the Second Law of Thermodynamics

31. A gas increases in pressure from 2.00 atm to 6.00 atm at a constant volume of 1.00 m³ and then expands at constant pressure to a volume of 3.00 m³ before returning to its initial state as shown in Figure P12.31. How much work is done in one cycle?

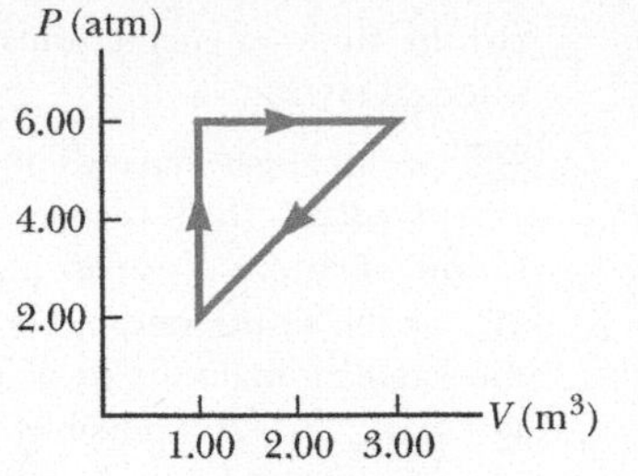

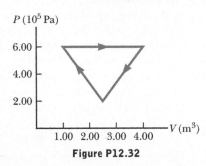

**Figure P12.31**

32. An ideal gas expands at a constant pressure of $6.00 \times 10^5$ Pa from a volume of 1.00 m³ to a volume of 4.00 m³ and then is compressed to one-third that pressure and a volume of 2.50 m³ as shown in Figure P12.32 before returning to its initial state. How much work is done in taking a gas through one cycle of the process shown in the figure?

**Figure P12.32**

33. A heat engine operates between a reservoir at 25°C and one at 375°C. What is the maximum efficiency possible for this engine?

34. **Q|C** A heat engine is being designed to have a Carnot efficiency of 65% when operating between two heat reservoirs. (a) If the temperature of the cold reservoir is 20°C, what must be the temperature of the hot reservoir? (b) Can the actual efficiency of the engine be equal to 65%? Explain.

35. The work done by an engine equals one-fourth the energy it absorbs from a reservoir. (a) What is its thermal efficiency? (b) What fraction of the energy absorbed is expelled to the cold reservoir?

36. In each cycle of its operation, a heat engine expels 2 400 J of energy and performs 1 800 J of mechanical work. (a) How much thermal energy must be added to the engine in each cycle? (b) Find the thermal efficiency of the engine.

37. One of the most efficient engines ever built is a coal-fired steam turbine engine in the Ohio River valley, driving an electric generator as it operates between 1 870°C and 430°C. (a) What is its maximum theoretical efficiency? (b) Its actual efficiency is 42.0%. How much mechanical power does the engine deliver if it absorbs $1.40 \times 10^5$ J of energy each second from the hot reservoir.

38. A lawnmower engine ejects $1.00 \times 10^4$ J each second while running with an efficiency of 0.200. Find the engine's horsepower rating, using the conversion factor 1 hp = 746 W.

39. An engine absorbs 1.70 kJ from a hot reservoir at 277°C and expels 1.20 kJ to a cold reservoir at 27°C in each cycle. (a) What is the engine's efficiency? (b) How much work is done by the engine in each cycle? (c) What is the power output of the engine if each cycle lasts 0.300 s?

40. A heat pump has a coefficient of performance of 3.80 and operates with a power consumption of $7.03 \times 10^3$ W. (a) How much energy does the heat pump deliver into a home during 8.00 h of continuous operation? (b) How much energy does it extract from the outside air in 8.00 h?

41. A freezer has a coefficient of performance of 6.30. The freezer is advertised as using 457 kW-h/y. (a) On average, how much energy does the freezer use in a single day? (b) On average, how much thermal energy is removed from the freezer each day? (c) What maximum mass of water at 20.0°C could the freezer freeze in a single day? *Note:* One kilowatt-hour (kW-h) is an amount of energy equal to operating a 1-kW appliance for one hour.

42. Two heat engines are operated in series so that part of the energy expelled from engine A is absorbed by engine B with $|Q_{hB}| = 0.750|Q_{cA}|$. Engines A and B have efficiencies $e_A = e_B = 0.250$ and engine A performs work $W_A = 275$ J. Find the overall efficiency of the two-engine combination, given by $e = \frac{W_A + W_B}{|Q_{hA}|}$.

43. In one cycle a heat engine absorbs 500 J from a high-temperature reservoir and expels 300 J to a low-temperature reservoir. If the efficiency of this engine is 60% of the efficiency of a Carnot engine, what is the ratio of the low temperature to the high temperature in the Carnot engine?

44. **Q|C** A power plant has been proposed that would make use of the temperature gradient in the ocean. The system is to operate between 20.0°C (surface water temperature) and 5.00°C (water temperature at a depth of about 1 km). (a) What is the maximum efficiency of such a system? (b) If the useful power output of the plant is 75.0 MW, how much energy is absorbed per hour? (c) In view of your answer to part (a), do you think such a system is worthwhile (considering that there is no charge for fuel)?

45. A certain nuclear power plant has an electrical power output of 435 MW. The rate at which energy must be supplied to the plant is 1 420 MW. (a) What is the thermal efficiency of the power plant? (b) At what rate is thermal energy expelled by the plant?

46. **T** A heat engine operates in a Carnot cycle between 80.0°C and 350°C. It absorbs 21 000 J of energy per cycle from the hot reservoir. The duration of each cycle is 1.00 s. (a) What is the mechanical power output of this engine? (b) How much energy does it expel in each cycle by heat?

## 12.5 Entropy

47. A Styrofoam cup holding 125 g of hot water at $1.00 \times 10^2$°C cools to room temperature, 20.0°C. What is the change in entropy of the room? (Neglect the specific heat of the cup and any change in temperature of the room.)

48. A 65-g ice cube is initially at 0.0°C. (a) Find the change in entropy of the cube after it melts completely at 0.0°C. (b) What is the change in entropy of the environment in this process? *Hint:* The latent heat of fusion for water is $3.33 \times 10^5$ J/kg.

49. A freezer is used to freeze 1.0 L of water completely into ice. The water and the freezer remain at a constant temperature of $T = 0$°C. Determine (a) the change in the entropy of the water and (b) the change in the entropy of the freezer.

50. **V** What is the change in entropy of 1.00 kg of liquid water at 100.°C as it changes to steam at 100.°C?

51. A 70.0-kg log falls from a height of 25.0 m into a lake. If the log, the lake, and the air are all at 300. K, find the change in entropy of the Universe during this process.

52. A sealed container holding 0.500 kg of liquid nitrogen at its boiling point of 77.3 K is placed in a large room at 21.0°C. Energy is transferred from the room to the nitrogen as the liquid nitrogen boils into a gas and then warms to the room's temperature. (a) Assuming the room's temperature remains essentially unchanged at 21.0°C, calculate the energy transferred from the room to the nitrogen. (b) Estimate the change in entropy of the room. Liquid nitrogen has a latent heat of vaporization of $2.01 \times 10^5$ J/kg. The specific heat of $N_2$ gas at constant pressure is $c_{N_2} = 1.04 \times 10^3$ J/kg · K.

53. **T** The surface of the Sun is approximately at $5.70 \times 10^3$ K, and the temperature of the Earth's surface is approximately 290. K. What entropy change occurs when $1.00 \times 10^3$ J of energy is transferred by heat from the Sun to the Earth?

54. **Q|C** When an aluminum bar is temporarily connected between a hot reservoir at 725 K and a cold reservoir at 310 K, 2.50 kJ of energy is transferred by heat from the hot reservoir to the cold reservoir. In this irreversible process, calculate the change in entropy of (a) the hot reservoir, (b) the cold reservoir, and (c) the Universe, neglecting any change in entropy of the aluminum rod. (d) Mathematically, why did the result for the Universe in part (c) have to be positive?

55. Prepare a table like Table 12.3 for the following occurrence: You toss four coins into the air simultaneously and record all the possible results of the toss in terms of the numbers of heads and tails that can result. (For example, HHTH and HTHH are two possible ways in which three heads and one tail can be achieved.) (a) On the basis of your table, what is the most probable result of a toss? In terms of entropy, (b) what is the most ordered state, and (c) what is the most disordered?

56. **S** *This is a symbolic version of Problem 54.* When a metal bar is temporarily connected between a hot reservoir at $T_h$ and a cold reservoir at $T_c$, the energy transferred by heat from the hot reservoir to the cold reservoir is $Q_h$. In this irreversible process, find expressions for the change in entropy of (a) the hot reservoir, (b) the cold reservoir, and (c) the Universe.

## 12.6 Human Metabolism

57. On a typical day, a 65-kg man sleeps for 8.0 h, does light chores for 3.0 h, walks slowly for 1.0 h, and jogs at moderate pace for 0.5 h. What is the change in his internal energy for all these activities?

58. **BIO** **Q|C** A weightlifter has a basal metabolic rate of 80.0 W. As he is working out, his metabolic rate increases by about 650 W. (a) How many hours does it take him to work off a 450-Calorie bagel if he stays in bed all day? (b) How long does it take him if he's working out? (c) Calculate the amount of mechanical work necessary to lift a 120-kg barbell 2.00 m. (d) He drops the barbell to the floor and lifts it repeatedly. How many times per minute must he repeat this process to do an amount of mechanical work equivalent to his metabolic rate increase of 650 W during exercise? (e) Could he actually do repetitions at the rate found in part (d) at the given metabolic level? Explain.

59. **BIO** Sweating is one of the main mechanisms with which the body dissipates heat. Sweat evaporates with a latent heat of 2 430 kJ/kg at body temperature, and the body can produce as much as 1.5 kg of sweat per hour. If sweating were the only

heat dissipation mechanism, what would be the maximum sustainable metabolic rate, in watts, if 80% of the energy used by the body goes into waste heat?

60. **BIO** A woman jogging has a metabolic rate of 625 W. (a) Calculate her volume rate of oxygen consumption in L/s. (b) Estimate her required respiratory rate in breaths/min if her lungs inhale 0.600 L of air in each breath and air is 20.9% oxygen.

61. **BIO** Suppose a highly trained athlete consumes oxygen at a rate of 70.0 mL/(min · kg) during a 30.0-min workout. If the athlete's mass is 78.0 kg and their body functions as a heat engine with a 20.0% efficiency, calculate (a) their metabolic rate in kcal/min and (b) the thermal energy in kcal released during the workout.

## Additional Problems

62. A Carnot engine operates between the temperatures $T_h = 1.00 \times 10^2$°C and $T_c = 20.0$°C. By what factor does the theoretical efficiency increase if the temperature of the hot reservoir is increased to $5.50 \times 10^2$°C?

63. A 1 500-kW heat engine operates at 25% efficiency. The heat energy expelled at the low temperature is absorbed by a stream of water that enters the cooling coils at 20.°C. If 60. L flows across the coils per second, determine the increase in temperature of the water.

64. **V** A Carnot engine operates between 100°C and 20°C. How much ice can the engine melt from its exhaust after it has done $5.0 \times 10^4$ J of work?

65. A substance undergoes the cyclic process shown in Figure P12.65. Work output occurs along path $AB$ while work input is required along path $BC$, and no work is involved in the constant volume process $CA$. Energy transfers by heat occur during each process involved in the cycle. (a) What is the work output during process $AB$? (b) How much work input is required during process $BC$? (c) What is the net energy input $Q$ during this cycle?

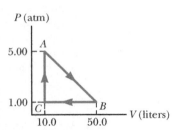

**Figure P12.65**

66. When a gas follows path 123 on the $PV$ diagram in Figure P12.66, 418 J of energy flows into the system by heat and −167 J of work is done *on* the gas. (a) What is the change in the internal energy of the system? (b) How much energy $Q$ flows into the system if the gas follows path 143? The work done on the gas

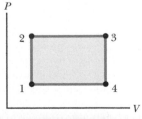

**Figure P12.66**

along this path is −63.0 J. What net work would be done on or by the system if the system followed (c) path 12341 and (d) path 14321? (e) What is the change in internal energy of the system in the processes described in parts (c) and (d)?

67. A $1.0 \times 10^2$-kg steel support rod in a building has a length of 2.0 m at a temperature of 20.0°C. The rod supports a hanging load of $6.0 \times 10^3$ kg. Find (a) the work done *on* the

rod as the temperature increases to 40.0°C, (b) the energy $Q$ added to the rod (assume the specific heat of steel is the same as that for iron), and (c) the change in internal energy of the rod.

68. **S** An ideal gas initially at pressure $P_0$, volume $V_0$, and temperature $T_0$ is taken through the cycle described in Figure P12.68. (a) Find the net work done *by* the gas per cycle in terms of $P_0$ and $V_0$. (b) What is the net energy $Q$ added to the system per cycle? (c) Obtain a numerical value for the net work done per cycle for 1.00 mol of gas initially at 0°C. *Hint:* Recall that the work done by the system equals the area under a *PV* curve.

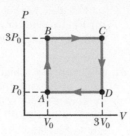

**Figure P12.68**

69. **T** One mole of neon gas is heated from 300. K to 420. K at constant pressure. Calculate (a) the energy $Q$ transferred to the gas, (b) the change in the internal energy of the gas, and (c) the work done *on* the gas. Note that neon has a molar specific heat of $c = 20.79$ J/mol·K for a constant-pressure process.

70. Every second at Niagara Falls, approximately $5.00 \times 10^3$ m³ of water falls a distance of 50.0 m. What is the increase in entropy per second due to the falling water? Assume the mass of the surroundings is so great that its temperature and that of the water stay nearly constant at 20.0°C. Also assume a negligible amount of water evaporates.

71. A cylinder containing 10.0 moles of a monatomic ideal gas expands from Ⓐ to Ⓑ along the path shown in Figure P12.71. (a) Find the temperature of the gas at point A and the temperature at point Ⓑ. (b) How much work is done by the gas during this expansion? (c) What is the change in internal energy of the gas? (d) Find the energy transferred to the gas by heat in this process.

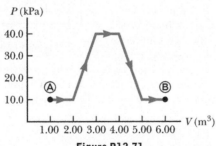

**Figure P12.71**

72. **GP** Two moles of molecular hydrogen ($H_2$) react with 1 mole of molecular oxygen ($O_2$) to produce 2 moles of water ($H_2O$) together with an energy release of 241.8 kJ/mole of water. Suppose a spherical vessel of radius 0.500 m contains 14.4 moles of $H_2$ and 7.2 moles of $O_2$ at 20.0°C. (a) What is the initial pressure in the vessel? (b) What is the initial internal energy of the gas? (c) Suppose a spark ignites the mixture and the gases burn completely into water vapor. How much

energy is produced? (d) Find the temperature and pressure of the steam, assuming it's an ideal gas. (e) Find the mass of steam and then calculate the steam's density. (f) If a small hole were put in the sphere, what would be the initial exhaust velocity of the exhausted steam if spewed out into a vacuum? (Use Bernoulli's equation.)

73. **BIO** Suppose you spend 30.0 minutes on a stair-climbing machine, climbing at a rate of 90.0 steps per minute, with each step 8.00 inches high. If you weigh 150. lb and the machine reports that 600. kcal have been burned at the end of the workout, what efficiency is the machine using in obtaining this result? If your actual efficiency is 0.18, how many kcal did you actually burn?

74. **BIO** Hydrothermal vents deep on the ocean floor spout water at temperatures as high as 570°C. This temperature is below the boiling point of water because of the immense pressure at that depth. Because the surrounding ocean temperature is at 4.0°C, an organism could use the temperature gradient as a source of energy. (a) Assuming the specific heat of water under these conditions is 1.0 cal/g · °C, how much energy is released when 1.0 L of water is cooled from 570°C to 4.0°C? (b) What is the maximum usable energy an organism can extract from this energy source? (Assume the organism has some internal type of heat engine acting between the two temperature extremes.) (c) Water from these vents contains hydrogen sulfide ($H_2S$) at a concentration of 0.90 mmole/L. Oxidation of 1.0 mole of $H_2S$ produces 310 kJ of energy. How much energy is available through $H_2S$ oxidation of 1.0 L of water?

75. An electrical power plant has an overall efficiency of 15%. The plant is to deliver 150 MW of electrical power to a city, and its turbines use coal as fuel. The burning coal produces steam at 190°C, which drives the turbines. The steam is condensed into water at 25°C by passing through coils that are in contact with river water. (a) How many metric tons of coal does the plant consume each day (1 metric ton = $1 \times 10^3$ kg)? (b) What is the total cost of the fuel per year if the delivery price is $8 per metric ton? (c) If the river water is delivered at 20°C, at what minimum rate must it flow over the cooling coils so that its temperature doesn't exceed 25°C? *Note:* The heat of combustion of coal is $7.8 \times 10^3$ cal/g.

76. A diatomic ideal gas expands from a volume of $V_A = 1.00$ m³ to 3.00 m³ along the path shown in Figure P12.76. If the initial pressure is $P_A = 2.00 \times 10^5$ Pa and there are 87.5 mol of gas, calculate (a) the work done on the gas during this process, (b) the change in temperature of the gas, and (c) the change in internal energy of the gas. (d) How much thermal energy is transferred to the system?

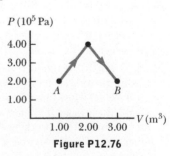

**Figure P12.76**

# Vibrations and Waves

PERIODIC MOTION, from masses on springs to vibrations of atoms, is one of the most important kinds of physical behavior. In this topic, we take a more detailed look at Hooke's law, where the force is proportional to the displacement, tending to restore objects to some equilibrium position. A large number of physical systems can be successfully modeled with this simple idea, including the vibrations of strings, the swinging of a pendulum, and the propagation of waves of all kinds. All these physical phenomena involve periodic motion.

Periodic vibrations can cause disturbances that move through a medium in the form of waves. Many kinds of waves occur in nature, such as sound waves, water waves, seismic waves, and electromagnetic waves. These very different physical phenomena are described by common terms and concepts introduced here.

## 13.1  Hooke's Law

One of the simplest types of vibrational motion is that of an object attached to a spring, previously discussed in the context of energy in Topic 5. We assume the object moves on a frictionless horizontal surface. If the spring is stretched or compressed a small distance $x$ from its unstretched or equilibrium position and then released, it exerts a force on the object as shown in Figure 13.1 (page 424). From experiment, the spring force $F_s$ is found to obey the equation

$$F_s = -kx \qquad\qquad [13.1]$$

◄ Hooke's law

where $x$ is the displacement of the object from its equilibrium position ($x = 0$) and $k$ is a positive constant called the **spring constant.** This force law for springs was discovered by Robert Hooke in 1678 and is known as **Hooke's law.** The value of $k$ is a measure of the stiffness of the spring. Stiff springs have large $k$ values, and soft springs have small $k$ values.

The negative sign in Equation 13.1 means that the force exerted by the spring is always directed *opposite* the displacement of the object. When the object is to the right of the equilibrium position, as in Figure 13.1a, $x$ is positive and $F_s$ is negative. This means that the force is in the negative direction, to the left. When the object is to the left of the equilibrium position, as in Figure 13.1c, $x$ is negative and $F_s$ is positive, indicating that the direction of the force is to the right. Of course, when $x = 0$, as in Figure 13.1b, the spring is unstretched and $F_s = 0$. Because the spring force always acts toward the equilibrium position, it is sometimes called a restoring force. **A restoring force always pushes or pulls the object toward the equilibrium position.**

Suppose the object is initially pulled a distance $A$ to the right and released from rest. The force exerted by the spring on the object pulls it back toward the equilibrium position. As the object moves toward $x = 0$, the magnitude of the force decreases (because $x$ decreases) and reaches zero at $x = 0$. The object gains speed as it moves toward the equilibrium position, however, reaching its

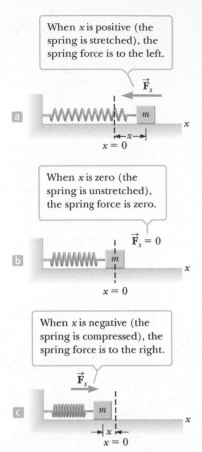

When $x$ is positive (the spring is stretched), the spring force is to the left.

$\vec{F}_s$

**a** $m$

$x$

$x = 0$

When $x$ is zero (the spring is unstretched), the spring force is zero.

$\vec{F}_s = 0$

**b** $m$

$x$

$x = 0$

When $x$ is negative (the spring is compressed), the spring force is to the right.

$\vec{F}_s$

**c** $m$

$x$

$x = 0$

**Figure 13.1** The force exerted by a spring on an object varies with the displacement of the object from the equilibrium position, $x = 0$.

▷ Acceleration in simple harmonic motion

**Tip 13.1** Constant-Acceleration Equations Don't Apply

The acceleration $a$ of a particle in simple harmonic motion is *not* constant; it changes, varying with $x$, so we can't apply the constant acceleration kinematic equations of Topic 2.

maximum speed when $x = 0$. The momentum gained by the object causes it to overshoot the equilibrium position and compress the spring. As the object moves to the left of the equilibrium position (negative $x$-values), the spring force acts on it to the right, steadily increasing in strength, and the speed of the object decreases. The object finally comes briefly to rest at $x = -A$ before accelerating back towards $x = 0$ and ultimately returning to the original position at $x = A$. The process is then repeated, and the object continues to oscillate back and forth over the same path. This type of motion is called **simple harmonic motion. Simple harmonic motion occurs when the net force along the direction of motion obeys Hooke's law—when the net force is proportional to the displacement from the equilibrium point and is always directed toward the equilibrium point.**

Not all periodic motions over the same path can be classified as simple harmonic motion. A ball being tossed back and forth between a parent and a child moves repetitively, but the motion isn't simple harmonic motion because the force acting on the ball doesn't take the form of Hooke's law, Equation 13.1.

The motion of an object suspended from a vertical spring is also simple harmonic. In this case the force of gravity acting on the attached object stretches the spring until equilibrium is reached and the object is suspended at rest. By definition, the equilibrium position of the object is $x = 0$. When the object is moved away from equilibrium by a distance $x$ and released, a net force acts toward the equilibrium position. Because the net force is proportional to $x$, the motion is simple harmonic.

The following three concepts are important in discussing any kind of periodic motion:

- The **amplitude** $A$ is the maximum distance of the object from its equilibrium position. In the absence of friction, an object in simple harmonic motion oscillates between the positions $x = -A$ and $x = +A$.
- The **period** $T$ is the time it takes the object to move through one complete cycle of motion, from $x = A$ to $x = -A$ and back to $x = A$.
- The **frequency** $f$ is the number of complete cycles or vibrations per unit of time, and is the reciprocal of the period ($f = 1/T$).

The acceleration of an object moving with simple harmonic motion can be found by using Hooke's law in the equation for Newton's second law, $F = ma$. This gives

$$ma = F = -kx$$

$$a = -\frac{k}{m}x \qquad [13.2]$$

Equation 13.2, an example of a *harmonic oscillator equation*, gives the acceleration as a function of position. Because the maximum value of $x$ is defined to be the amplitude $A$, the acceleration ranges over the values $-kA/m$ to $+kA/m$. In the next section we find equations for velocity as a function of position and for position as a function of time. Springs satisfying Hooke's law are also called *ideal* springs. In real springs, spring mass, internal friction, and varying elasticity affect the force law and motion.

*Quick Quiz*

**13.1** A block on the end of a horizontal spring is pulled from equilibrium at $x = 0$ to $x = A$ and released. Through what total distance does it travel in one full cycle of its motion? (a) $A/2$ (b) $A$ (c) $2A$ (d) $4A$

**13.2** For a simple harmonic oscillator, which of the following pairs of vector quantities can't both point in the same direction? (The position vector is the displacement from equilibrium.) (a) position and velocity (b) velocity and acceleration (c) position and acceleration

## EXAMPLE 13.1 | SIMPLE HARMONIC MOTION ON A FRICTIONLESS SURFACE

**GOAL** Calculate forces and accelerations for a horizontal spring system.

**PROBLEM** A 0.350-kg object attached to a spring of force constant $1.30 \times 10^2$ N/m is free to move on a frictionless horizontal surface, as in Figure 13.1. If the object is released from rest at $x = 0.100$ m, find the force on it and its acceleration at $x = 0.100$ m, $x = 0.050\ 0$ m, $x = 0$ m, $x = -0.050\ 0$ m, and $x = -0.100$ m.

**STRATEGY** Substitute given quantities into Hooke's law to find the forces, then calculate the accelerations with Newton's second law. The amplitude $A$ is the same as the point of release from rest, $x = 0.100$ m.

### SOLUTION

Write Hooke's force law:

$$F_s = -kx$$

Substitute the value for $k$ and take $x = A = 0.100$ m, finding the spring force at that point:

$$F_{max} = -kA = -(1.30 \times 10^2\ \text{N/m})(0.100\ \text{m})$$
$$= \boxed{-13.0\ \text{N}}$$

Solve Newton's second law for $a$ and substitute to find the acceleration at $x = A$:

$$ma = F_{max}$$
$$a = \frac{F_{max}}{m} = \frac{-13.0\ \text{N}}{0.350\ \text{kg}} = \boxed{-37.1\ \text{m/s}^2}$$

Repeat the same process for the other four points, assembling a table:

| Position (m) | Force (N) | Acceleration (m/s$^2$) |
|---|---|---|
| 0.100 | −13.0 | −37.1 |
| 0.050 | −6.50 | −18.6 |
| 0 | 0 | 0 |
| −0.050 | +6.50 | +18.6 |
| −0.100 | +13.0 | +37.1 |

**REMARKS** The table above shows that when the initial position is halved, the force and acceleration are also halved. Further, positive values of $x$ give negative values of the force and acceleration, whereas negative values of $x$ give positive values of the force and acceleration. As the object moves to the left and passes the equilibrium point, the spring force becomes positive (for negative values of $x$), slowing the object down.

**QUESTION 13.1** Will doubling a given displacement always result in doubling the magnitude of the spring force? Explain.

**EXERCISE 13.1** For the same spring and mass system, find the force exerted by the spring and the position $x$ when the object's acceleration is +9.00 m/s$^2$.

**ANSWERS** 3.15 N, −2.42 cm

## EXAMPLE 13.2 | MASS ON A VERTICAL SPRING

**GOAL** Apply Newton's second law together with the force of gravity and Hooke's law.

**PROBLEM** A spring is hung vertically (Fig. 13.2a), and an object of mass $m$ attached to the lower end is then slowly lowered a distance $d$ to the equilibrium point (Fig. 13.2b). **(a)** Find the value of the spring constant if the magnitude of the displacement $d$ is 2.0 cm and the mass is 0.55 kg. **(b)** If a second identical spring is attached to the object in parallel with the first spring (Fig. 13.2d), where is the new equilibrium point of the system? **(c)** What is the effective spring constant of the two springs acting as one?

**STRATEGY** This example is an application of Newton's second law. The spring force is upward, balancing the downward force of gravity $mg$ when the system is in equilibrium. (See Fig. 13.2c.) Because the suspended object is in equilibrium, the forces on the object sum to zero, and it's possible to solve for the spring constant $k$. Part **(b)** is solved the same way, but has two spring forces balancing the force of gravity. The spring constants are known, so the second law for equilibrium can be solved for the displacement of the spring.

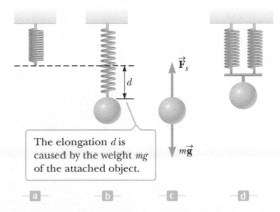

The elongation $d$ is caused by the weight $mg$ of the attached object.

**Figure 13.2** (Example 13.2) (a)–(c) Determining the spring constant. Because the upward spring force balances the weight when the system is in equilibrium, it follows that $k = mg/d$. (d) A system involving two springs in parallel.

*(Continued)*

Part (c) involves using the displacement found in part (b). Treating the two springs as one equivalent spring, the second law then leads to the effective spring constant of the two-spring system.

. . . . . . . . . . . . . . . . . . . . . . . . . . . . . . . . . . . . . . . . . . . . . . . . . . . . . . . . . . . . . . . . . . . . . . . . . . . . . . . . . . . . . . . . . . .

**SOLUTION**

(a) Find the value of the spring constant if the spring is displaced by 2.0 cm and the mass of the object is 0.55 kg.

Apply Newton's second law to the object (with $a = 0$) and solve for the spring constant $k$:

$$\sum F = F_g + F_s = -mg + kd = 0$$

$$k = \frac{mg}{d} = \frac{(0.55 \text{ kg})(9.80 \text{ m/s}^2)}{2.0 \times 10^{-2} \text{ m}} = \boxed{2.7 \times 10^2 \text{ N/m}}$$

(b) If a second identical spring is attached to the object in parallel with the first spring (Fig. 13.2d), find the new equilibrium point of the system.

Apply Newton's second law, but with two springs acting on the object:

$$\sum F = F_g + F_{s1} + F_{s2} = -mg + kd_2 + kd_2 = 0$$

Solve for $d_2$:

$$d_2 = \frac{mg}{2k} = \frac{(0.55 \text{ kg})(9.80 \text{ m/s}^2)}{2(2.7 \times 10^2 \text{ N/m})} = \boxed{1.0 \times 10^{-2} \text{ m}}$$

(c) What is the effective spring constant of the two springs acting as one?

Write the second law for the system, with an effective spring constant $k_{eff}$:

$$\sum F = F_g + F_s = -mg + k_{eff}d_2 = 0$$

Solve for $k_{eff}$:

$$k_{eff} = \frac{mg}{d_2} = \frac{(0.55 \text{ kg})(9.80 \text{ m/s}^2)}{1.0 \times 10^{-2} \text{ m}} = \boxed{5.4 \times 10^2 \text{ N/m}}$$

. . . . . . . . . . . . . . . . . . . . . . . . . . . . . . . . . . . . . . . . . . . . . . . . . . . . . . . . . . . . . . . . . . . . . . . . . . . . . . . . . . . . . . . . . . .

**REMARKS** In this example, the spring force is positive because it's directed upward. If the object were displaced from the equilibrium position and released, it would oscillate around that point, just like a horizontal spring. Notice that attaching an extra identical spring in parallel is equivalent to having a single spring with twice the force constant. With springs attached end to end in series, however, the exercise illustrates that, all other things being equal, longer springs have smaller force constants than shorter springs.

**QUESTION 13.2** Generalize: When two springs with force constants $k_1$ and $k_2$ act in parallel on an object, what is the spring constant $k_{eff}$ of the single spring that would be equivalent to the two springs, in terms of $k_1$ and $k_2$?

**EXERCISE 13.2** When a 75.0-kg man slowly adds his weight to a vertical spring attached to the ceiling, he reaches equilibrium when the spring is stretched by 6.50 cm. (a) Find the spring constant. (b) If a second, identical spring is hung on the first in series, and the man again adds his weight to the system, by how much does the system of springs stretch? (c) What would be the spring constant of a single, equivalent spring?

**ANSWERS** (a) $1.13 \times 10^4 \text{ N/m}$   (b) 13.0 cm   (c) $5.65 \times 10^3 \text{ N/m}$

# 13.2 Elastic Potential Energy

In this section, we review the material covered in Section 5 of Topic 5.

A system of interacting objects has potential energy associated with the configuration of the system. A compressed spring has potential energy that, when allowed to expand, can do work on an object, transforming spring potential energy into the object's kinetic energy. As an example, Figure 13.3 shows a ball being projected from a spring-loaded toy gun, where the spring is compressed a distance $x$. As the gun is fired, the compressed spring does work on the ball and imparts kinetic energy to it.

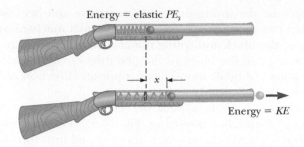

**Figure 13.3** A ball projected from a spring-loaded gun. The elastic potential energy stored in the spring is transformed into the kinetic energy of the ball.

Recall that the energy stored in a stretched or compressed spring or some other elastic material is called elastic potential energy, $PE_s$, given by

$$PE_s \equiv \tfrac{1}{2}kx^2 \qquad [13.3]$$

◄ Elastic potential energy

Recall also that the law of conservation of energy, including both gravitational and spring potential energy, is given by

$$(KE + PE_g + PE_s)_i = (KE + PE_g + PE_s)_f \qquad [13.4]$$

If nonconservative forces such as friction are present, then the change in mechanical energy must equal the work done by the nonconservative forces:

$$W_{nc} = (KE + PE_g + PE_s)_f - (KE + PE_g + PE_s)_i \qquad [13.5]$$

Rotational kinetic energy must be included in both Equation 13.4 and Equation 13.5 for systems involving torques.

As an example of the energy conversions that take place when a spring is included in a system, consider Figure 13.4. A block of mass $m$ slides on a frictionless horizontal surface with constant velocity $\vec{v}_i$ and collides with a coiled spring. The description that follows is greatly simplified by assuming the spring is very light (an ideal spring) and therefore has negligible kinetic energy. As the spring is compressed, it exerts a force to the left on the block. At maximum compression, the block comes to rest for just an instant (Fig. 13.4c). The initial total energy in the system (block plus spring) before the collision is the kinetic energy of the block. After the block collides with the spring and the spring is partially compressed, as in Figure 13.4b, the block has kinetic energy $\tfrac{1}{2}mv^2$ (where $v < v_i$) and the spring has potential energy $\tfrac{1}{2}kx^2$. When the block stops for an instant at the point of maximum compression, the kinetic

**APPLICATION**
Archery

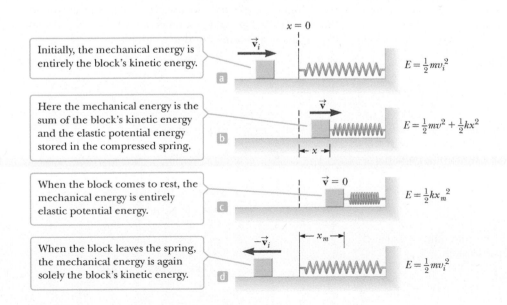

**Figure 13.4** A block sliding on a frictionless horizontal surface collides with a light spring. In the absence of friction, the mechanical energy in this process remains constant.

Initially, the mechanical energy is entirely the block's kinetic energy.

a    $E = \tfrac{1}{2}mv_i^2$

Here the mechanical energy is the sum of the block's kinetic energy and the elastic potential energy stored in the compressed spring.

b    $E = \tfrac{1}{2}mv^2 + \tfrac{1}{2}kx^2$

When the block comes to rest, the mechanical energy is entirely elastic potential energy.

c    $E = \tfrac{1}{2}kx_m^2$

When the block leaves the spring, the mechanical energy is again solely the block's kinetic energy.

d    $E = \tfrac{1}{2}mv_i^2$

energy is zero. Because the spring force is conservative and because there are no external forces that can do work on the system, **the total mechanical energy of the system consisting of the block and spring remains constant.** Energy is transformed from the kinetic energy of the block to the potential energy stored in the spring. As the spring expands, the block moves in the opposite direction and regains all its initial kinetic energy, as in Figure 13.4d.

When an archer pulls back on a bowstring, elastic potential energy is stored in both the bent bow and stretched bowstring (Fig. 13.5). When the arrow is released, the potential energy stored in the system is transformed into the kinetic energy of the arrow. Devices such as crossbows and slingshots work the same way.

**Figure 13.5** Elastic potential energy is stored in this drawn bow.

### Quick Quiz

**13.3** When an object moving in simple harmonic motion is at its maximum displacement from equilibrium, which of the following is at a maximum? (a) velocity, (b) acceleration, or (c) kinetic energy

---

### EXAMPLE 13.3    STOP THAT CAR!

**GOAL** Apply conservation of energy and the work–energy theorem with spring and gravitational potential energy.

**PROBLEM** A 13 000-N car starts at rest and rolls down a hill from a height of 10.0 m (Fig. 13.6). It then moves across a level surface and collides with a light spring-loaded guardrail. **(a)** Neglecting any losses due to friction, and ignoring the rotational kinetic energy of the wheels, find the maximum distance the spring is compressed. Assume a spring constant of $1.0 \times 10^6$ N/m. **(b)** Calculate the magnitude of the car's maximum acceleration after contact with the spring, assuming no frictional losses. **(c)** If the spring is compressed by only 0.30 m, find the change in the mechanical energy due to friction.

**Figure 13.6** (Example 13.3) A car starts from rest on a hill at the position shown. When the car reaches the bottom of the hill, it collides with a spring-loaded guardrail.

**STRATEGY** Because friction losses are neglected, use conservation of energy in the form of Equation 13.4 to solve for the spring displacement in part **(a)**. The initial and final values of the car's kinetic energy are zero, so the initial potential energy of the car–spring–Earth system is completely converted to elastic potential energy in the spring at the end of the ride. In part **(b)** apply Newton's second law, substituting the answer to part **(a)** for $x$ because the maximum compression will give the maximum acceleration. In part **(c)** friction is no longer neglected, so use the work–energy theorem, Equation 13.5. The change in mechanical energy must equal the mechanical energy lost due to friction.

................................................................

### SOLUTION

**(a)** Find the maximum spring compression, assuming no energy losses due to friction.

Apply conservation of mechanical energy. Initially, there is only gravitational potential energy, and at maximum compression of the guardrail, there is only spring potential energy.

$$(KE + PE_g + PE_s)_i = (KE + PE_g + PE_s)_f$$
$$0 + mgh + 0 = 0 + 0 + \tfrac{1}{2}kx^2$$

Solve for $x$:

$$x = \sqrt{\frac{2mgh}{k}} = \sqrt{\frac{2(13\,000\ \text{N})(10.0\ \text{m})}{1.0 \times 10^6\ \text{N/m}}} = \boxed{0.51\ \text{m}}$$

**(b)** Calculate the magnitude of the car's maximum acceleration by the spring, neglecting friction.

Apply Newton's second law to the car:

$$ma = -kx \quad \rightarrow \quad a = -\frac{kx}{m} = -\frac{kxg}{mg} = -\frac{kxg}{w}$$

Substitute values:

$$a = -\frac{(1.0 \times 10^6 \text{ N/m})(0.51 \text{ m})(9.8 \text{ m/s}^2)}{13\ 000 \text{ N}}$$

$$= -380 \text{ m/s}^2 \rightarrow |a| = \boxed{380 \text{ m/s}^2}$$

**(c)** If the compression of the guardrail is only 0.30 m, find the change in the mechanical energy due to friction.

Use the work–energy theorem:

$$W_{nc} = (KE + PE_g + PE_s)_f - (KE + PE_g + PE_s)_i$$

$$= (0 + 0 + \tfrac{1}{2}kx^2) - (0 + mgh + 0)$$

$$= \tfrac{1}{2}(1.0 \times 10^6 \text{ N/m})(0.30)^2 - (13\ 000 \text{ N})(10.0 \text{ m})$$

$$W_{nc} = \boxed{-8.5 \times 10^4 \text{ J}}$$

. . . . . . . . . . . . . . . . . . . . . . . . . . . . . . . . . . . . . . . . . . . . . . . . . . . . . . . . . . . . . . . . . . . . . . . . . . . . . . . . . . . . . . . . . . . .

**REMARKS** The answer to part (b) is about 40 times greater than the acceleration of gravity, so we'd better be wearing our seat belts. Note that the solution didn't require calculation of the velocity of the car.

**QUESTION 13.3** True or False: In the absence of energy losses due to friction, doubling the height of the hill doubles the maximum acceleration delivered by the spring.

**EXERCISE 13.3** A spring-loaded gun fires a 0.100-kg puck along a tabletop. The puck slides up a curved ramp and flies straight up into the air. **(a)** If the spring is displaced 12.0 cm from equilibrium and the spring constant is 875 N/m, how high does the puck rise, neglecting friction? **(b)** If instead it only rises to a height of 5.00 m because of friction, what is the change in mechanical energy?

**ANSWERS** **(a)** 6.43 m    **(b)** −1.40 J

In addition to studying the preceding example, it's a good idea to review those given in Section 5.5.

## 13.2.1 Velocity as a Function of Position

Conservation of energy provides a simple method of deriving an expression for the velocity of an object undergoing periodic motion as a function of position. The object in question is initially at its maximum extension $A$ (Fig. 13.7a) and is then released from rest. The initial energy of the system is entirely elastic potential energy stored in the spring, $\tfrac{1}{2}kA^2$. As the object moves toward the origin to some new position $x$ (Fig. 13.7b), part of this energy is transformed into kinetic energy, and the potential energy stored in the spring is reduced to $\tfrac{1}{2}kx^2$. Because the total energy of the system is equal to $\tfrac{1}{2}kA^2$ (the initial energy stored in the spring), we can equate this quantity to the sum of the kinetic and potential energies at the position $x$:

$$\tfrac{1}{2}kA^2 = \tfrac{1}{2}mv^2 + \tfrac{1}{2}kx^2$$

Solving for $v$, we get

$$v = \pm\sqrt{\frac{k}{m}(A^2 - x^2)} \qquad [13.6]$$

This expression shows that the object's speed is a maximum at $x = 0$ and is zero at the extreme positions $x = \pm A$.

The right side of Equation 13.6 is preceded by the $\pm$ sign because the square root of a number can be either positive or negative. If the object in Figure 13.7 is moving to the right, $v$ is positive; if the object is moving to the left, $v$ is negative.

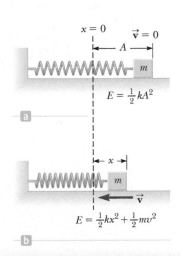

**Figure 13.7** (a) An object attached to a spring on a frictionless surface is released from rest with the spring extended a distance $A$. Just before the object is released, the total energy is the elastic potential energy $\tfrac{1}{2}kA^2$. (b) When the object reaches position $x$, it has kinetic energy $\tfrac{1}{2}mv^2$ and the elastic potential energy has decreased to $\tfrac{1}{2}kx^2$.

## EXAMPLE 13.4 THE OBJECT–SPRING SYSTEM REVISITED

**GOAL** Apply the time-independent velocity expression, Equation 13.6, to an object–spring system.

**PROBLEM** A 0.500-kg object connected to a light spring with a spring constant of 20.0 N/m oscillates on a frictionless horizontal surface. **(a)** Calculate the total energy of the system and the maximum speed of the object if the amplitude of the motion is 3.00 cm. **(b)** What is the velocity of the object when the displacement is 2.00 cm? **(c)** Compute the kinetic and potential energies of the system when the displacement is 2.00 cm.

**STRATEGY** The total energy of the system can be found most easily at $x = A$, where the kinetic energy is zero. There, the potential energy alone is equal to the total energy. Conservation of energy then yields the speed at $x = 0$. For part **(b)**, obtain the velocity by substituting the given value of $x$ into the time-independent velocity equation. Using this result, the kinetic energy asked for in part **(c)** can be found by substitution, and the potential energy can be found by substitution into Equation 13.3.

· · · · · · · · · · · · · · · · · · · · · · · · · · · · · · · · · · · · · · · · · · · · · · · · · · · · · · · · · · · · · · · · · · ·

### SOLUTION

**(a)** Calculate the total energy and maximum speed if the amplitude is 3.00 cm.

Substitute $x = A = 3.00$ cm and $k = 20.0$ N/m into the equation for the total mechanical energy $E$:

$$E = KE + PE_g + PE_s$$
$$= 0 + 0 + \tfrac{1}{2}kA^2 = \tfrac{1}{2}(20.0 \text{ N/m})(3.00 \times 10^{-2} \text{ m})^2$$
$$= \boxed{9.00 \times 10^{-3} \text{ J}}$$

Use conservation of energy with $x_i = A$ and $x_f = 0$ to compute the speed of the object at the origin:

$$(KE + PE_g + PE_s)_i = (KE + PE_g + PE_s)_f$$
$$0 + 0 + \tfrac{1}{2}kA^2 = \tfrac{1}{2}mv_{max}^2 + 0 + 0$$
$$\tfrac{1}{2}mv_{max}^2 = 9.00 \times 10^{-3} \text{ J}$$
$$v_{max} = \sqrt{\frac{18.0 \times 10^{-3} \text{ J}}{0.500 \text{ kg}}} = \boxed{0.190 \text{ m/s}}$$

**(b)** Compute the velocity of the object when the displacement is 2.00 cm.

Substitute known values directly into Equation 13.6:

$$v = \pm\sqrt{\frac{k}{m}(A^2 - x^2)}$$
$$= \pm\sqrt{\frac{20.0 \text{ N/m}}{0.500 \text{ kg}}[(0.030\ 0 \text{ m})^2 - (0.020\ 0 \text{ m})^2]}$$
$$= \boxed{\pm 0.141 \text{ m/s}}$$

**(c)** Compute the kinetic and potential energies when the displacement is 2.00 cm.

Substitute into the equation for kinetic energy:

$$KE = \tfrac{1}{2}mv^2 = \tfrac{1}{2}(0.500 \text{ kg})(0.141 \text{ m/s})^2 = \boxed{4.97 \times 10^{-3} \text{ J}}$$

Substitute into the equation for spring potential energy:

$$PE_s = \tfrac{1}{2}kx^2 = \tfrac{1}{2}(20.0 \text{ N/m})(2.00 \times 10^{-2} \text{ m})^2$$
$$= \boxed{4.00 \times 10^{-3} \text{ J}}$$

· · · · · · · · · · · · · · · · · · · · · · · · · · · · · · · · · · · · · · · · · · · · · · · · · · · · · · · · · · · · · · · · · · ·

**REMARKS** With the given information, it is impossible to choose between the positive and negative solutions in part **(b)**. Notice that the sum $KE + PE_s$ in part **(c)** equals the total energy $E$ found in part **(a)**, as it should (except for a small discrepancy due to rounding).

**QUESTION 13.4** True or False: Doubling the initial displacement doubles the speed of the object at the equilibrium point.

**EXERCISE 13.4** For what values of $x$ is the speed of the object 0.10 m/s?

**ANSWER** $\pm 2.55$ cm

# 13.3 Concepts of Oscillation Rates in Simple Harmonic Motion

The most essential physical concepts in simple harmonic motion are those pertaining to the rate of oscillation. Those concepts are the period, frequency, and angular frequency. Understanding simple harmonic oscillations and measures of oscillation rates can be facilitated by a comparison of simple harmonic motion with uniform circular motion.

## 13.3.1 Comparing Simple Harmonic Motion with Uniform Circular Motion

We can better understand and visualize many aspects of simple harmonic motion along a straight line by looking at its relationship to uniform circular motion. Figure 13.8 is a top view of an experimental arrangement that is useful for this purpose. A ball is attached to the rim of a turntable of radius $A$, illuminated from the side by a lamp. We find that **as the turntable rotates with constant angular speed, the shadow of the ball moves back and forth with simple harmonic motion.**

This fact can be understood from Equation 13.6, which says that the velocity of an object moving with simple harmonic motion is related to the displacement by

$$v = C\sqrt{A^2 - x^2}$$

where $C$ is a constant. To see that the shadow also obeys this relation, consider Figure 13.9, which shows the ball moving with a constant speed $v_0$ in a direction tangent to the circular path. At this instant, the velocity of the ball in the $x$-direction is given by $v = v_0 \sin \theta$, or

$$\sin \theta = \frac{v}{v_0}$$

From the larger triangle in Figure 13.9, we can obtain a second expression for $\sin \theta$:

$$\sin \theta = \frac{\sqrt{A^2 - x^2}}{A}$$

Equating the right-hand sides of the two expressions for $\sin \theta$, we find the following relationship between the velocity $v$ and the displacement $x$:

$$\frac{v}{v_0} = \frac{\sqrt{A^2 - x^2}}{A}$$

or

$$v = \frac{v_0}{A}\sqrt{A^2 - x^2} = C\sqrt{A^2 - x^2}$$

The velocity of the ball in the $x$-direction is related to the displacement $x$ in exactly the same way as the velocity of an object undergoing simple harmonic motion. The shadow therefore moves with simple harmonic motion.

A valuable example of the relationship between simple harmonic motion and circular motion can be seen in vehicles and machines that use the back-and-forth motion of a piston to create rotational motion in a wheel. Consider the drive wheel of a locomotive. In Figure 13.10, the rods are connected to a piston that moves back and forth in simple harmonic motion. The rods transform the back-and-forth motion of the piston into rotational motion of the wheels. A similar mechanism in an automobile engine transforms the back-and-forth motion of the pistons to rotational motion of the crankshaft.

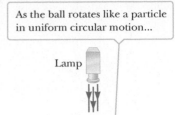

As the ball rotates like a particle in uniform circular motion...

...the ball's shadow on the screen moves back and forth with simple harmonic motion.

**Figure 13.8** An experimental setup for demonstrating the connection between simple harmonic motion and uniform circular motion.

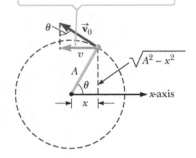

The $x$-component of the ball's velocity equals the projection of $\vec{v}_0$ on the $x$-axis.

**Figure 13.9** The ball rotates with constant speed $v_0$.

**APPLICATION**
Pistons and Drive Wheels

**Figure 13.10** The drive wheel mechanism of an old locomotive.

Gabriel GS./Shutterstock.com

### 13.3.2 Period, Frequency, and Angular Frequency

The period $T$ of the shadow in Figure 13.8, which represents the time required for one complete trip back and forth, is also the time it takes the ball to make one complete circular trip on the turntable. Because the ball moves through the distance $2\pi A$ (the circumference of the circle) in the time $T$, the speed $v_0$ of the ball around the circular path is

$$v_0 = \frac{2\pi A}{T}$$

and the period is

$$T = \frac{2\pi A}{v_0} \qquad [13.7]$$

Imagine that the ball moves from $P$ to $Q$, a quarter of a revolution, in Figure 13.8. The motion of the shadow is equivalent to the horizontal motion of an object on the end of a spring. For this reason, the radius $A$ of the circular motion is the same as the amplitude $A$ of the simple harmonic motion of the shadow. During the quarter of a cycle shown, the shadow moves from a point where the energy of the system (ball and spring) is solely elastic potential energy to a point where the energy is solely kinetic energy. By conservation of energy, we have

$$\tfrac{1}{2}kA^2 = \tfrac{1}{2}mv_0^2$$

which can be solved for $A/v_0$:

$$\frac{A}{v_0} = \sqrt{\frac{m}{k}}$$

Substituting this expression for $A/v_0$ in Equation 13.7, we find that the period is

**The period of an object–spring system moving with simple harmonic motion** ▶

$$T = 2\pi\sqrt{\frac{m}{k}} \qquad [13.8]$$

Equation 13.8 represents the time required for an object of mass $m$ attached to a spring with spring constant $k$ to complete one cycle of its motion. The square root of the mass is in the numerator, so a large mass will mean a large period, in agreement with intuition. The square root of the spring constant $k$ is in the denominator, so a large spring constant will yield a small period, again agreeing with intuition. It's also interesting that the period doesn't depend on the amplitude $A$.

The inverse of the period is the frequency of the motion:

$$f = \frac{1}{T} \qquad [13.9]$$

Therefore, the frequency of the periodic motion of a mass on a spring is

**Frequency of an object–spring system** ▶

$$f = \frac{1}{2\pi}\sqrt{\frac{k}{m}} \qquad [13.10]$$

The units of frequency are cycles per second ($s^{-1}$), or **hertz** (Hz). The **angular frequency** $\omega$ is

**Angular frequency of an object–spring system** ▶

$$\omega = 2\pi f = \sqrt{\frac{k}{m}} \qquad [13.11]$$

The frequency and angular frequency are actually closely related concepts. The unit of frequency is cycles per second, where a cycle may be thought of as a unit of angular measure corresponding to $2\pi$ radians, or $360°$. Viewed in this way, angular frequency is just a unit conversion of frequency. Radian measure is used for angles mainly because it provides a convenient and natural link between linear and angular quantities.

Although an ideal mass–spring system has a period proportional to the square root of the object's mass $m$, experiments show that a graph of $T^2$ versus $m$ doesn't pass through the origin. This is because the spring itself has a mass. The coils of the spring oscillate just like the object, except the amplitudes are smaller for all coils but the last. For a cylindrical spring, energy arguments can be used to show that the *effective* additional mass of a light spring is one-third the mass of the spring. The square of the period is proportional to the total oscillating mass, so a graph of $T^2$ versus total mass (the mass hung on the spring plus the effective oscillating mass of the spring) would pass through the origin.

> **Tip 13.2** Twin Frequencies
>
> The *frequency* gives the number of cycles per second, whereas the *angular frequency* gives the number of radians per second. These two physical concepts are nearly identical and are linked by the conversion factor $2\pi$ rad/cycle.

## Quick Quiz

**13.4** An object of mass $m$ is attached to a horizontal spring, stretched to a displacement $A$ from equilibrium, and released, undergoing harmonic oscillations on a frictionless surface with period $T_0$. The experiment is then repeated with a mass of $4m$. What's the new period of oscillation? (a) $2T_0$  (b) $T_0$  (c) $T_0/2$  (d) $T_0/4$

**13.5** Consider the situation in Quick Quiz 13.4. Is the subsequent total mechanical energy of the object with mass $4m$ (a) greater than, (b) less than, or (c) equal to the original total mechanical energy?

---

### APPLYING PHYSICS 13.1    BUNGEE JUMPING

A bungee cord can be roughly modeled as a spring. If you go bungee jumping, you will bounce up and down at the end of the elastic cord after your dive off a bridge (Fig. 13.11). Suppose you perform a dive and measure the frequency of your bouncing. You then move to another bridge, but find that the bungee cord is too long for dives off this bridge. What possible solutions might be applied? In terms of the original frequency, what is the frequency of vibration associated with the solution?

**EXPLANATION** There are two possible solutions: Make the bungee cord smaller or fold it in half. The latter would be the safer of the two choices, as we'll see. The force exerted by the bungee cord, modeled as a spring, is proportional to the separation of the coils as the spring is extended. First, we extend the spring by a given distance and measure the distance between coils. We then cut the spring in half. If one of the half-springs is now extended by the same distance, the coils will be twice as far apart as they were for the complete spring. Therefore, it takes twice as much force to stretch the half-spring through the same displacement, so the half-spring has a spring constant twice that of the complete spring. The folded bungee cord can then be modeled as two half-springs in parallel. Each half has a spring constant that is twice the original spring constant of the bungee cord. In addition, an object hanging on the folded bungee cord will experience two forces, one from each half-spring. As a result, the required force for a given extension will be four times as much as for the original bungee cord. The effective spring constant of the folded bungee cord is therefore four times as large as the original spring constant. Because the frequency of oscillation is proportional to the square root of the spring constant, your bouncing frequency on the folded cord will be twice what it was on the original cord.

This discussion neglects the fact that the coils of a spring have an initial separation. It's also important to remember that a shorter coil may lose elasticity more readily, possibly even going beyond the elastic limit for the material, with disastrous results. Bungee jumping is dangerous; discretion is advised! ■

**Figure 13.11** (Applying Physics 13.1) A bungee jumper relies on elastic forces to pull him up short of a deadly impact.

iStockphoto.com/Visual Communications

## EXAMPLE 13.5 THAT CAR NEEDS SHOCK ABSORBERS!

**GOAL** Understand the relationships between period, frequency, and angular frequency.

**PROBLEM** A $1.30 \times 10^3$-kg car is constructed on a frame supported by four springs. Each spring has a spring constant of $2.00 \times 10^4$ N/m. If two people riding in the car have a combined mass of $1.60 \times 10^2$ kg, find the frequency of vibration of the car when it is driven over a pothole in the road. Find also the period and the angular frequency. Assume the weight is evenly distributed.

**STRATEGY** Because the weight is evenly distributed, each spring supports one-fourth of the mass. Substitute this value and the spring constant into Equation 13.10 to get the frequency. The reciprocal is the period, and multiplying the frequency by $2\pi$ gives the angular frequency.

**SOLUTION**

Compute one-quarter of the total mass:

$$m = \tfrac{1}{4}(m_{car} + m_{pass}) = \tfrac{1}{4}(1.30 \times 10^3 \text{ kg} + 1.60 \times 10^2 \text{ kg})$$
$$= 365 \text{ kg}$$

Substitute into Equation 13.10 to find the frequency:

$$f = \frac{1}{2\pi}\sqrt{\frac{k}{m}} = \frac{1}{2\pi}\sqrt{\frac{2.00 \times 10^4 \text{ N/m}}{365 \text{ kg}}} = \boxed{1.18 \text{ Hz}}$$

Invert the frequency to get the period:

$$T = \frac{1}{f} = \frac{1}{1.18 \text{ Hz}} = \boxed{0.847 \text{ s}}$$

Multiply the frequency by $2\pi$ to get the angular frequency:

$$\omega = 2\pi f = 2\pi(1.18 \text{ Hz}) = \boxed{7.41 \text{ rad/s}}$$

**REMARKS** Solving this problem didn't require any knowledge of the size of the pothole because the frequency doesn't depend on the amplitude of the motion.

**QUESTION 13.5** True or False: The frequency of vibration of a heavy vehicle is greater than that of a lighter vehicle, assuming the two vehicles are supported by the same set of springs.

**EXERCISE 13.5** A 45.0-kg boy jumps on a 5.00-kg pogo stick with spring constant 3 650 N/m. Find **(a)** the angular frequency, **(b)** the frequency, and **(c)** the period of the boy's motion.

**ANSWERS** **(a)** 8.54 rad/s **(b)** 1.36 Hz **(c)** 0.735 s

# 13.4 Position, Velocity, and Acceleration as Functions of Time

We can obtain an expression for the position of an object moving with simple harmonic motion as a function of time by returning to the relationship between simple harmonic motion and uniform circular motion. Again, consider a ball on the rim of a rotating turntable of radius $A$, as in Figure 13.12. We refer to the circle made by the ball as the *reference circle* for the motion. We assume the turntable revolves at a *constant* angular speed $\omega$. As the ball rotates on the reference circle, the angle $\theta$ made by the line $OP$ with the $x$-axis changes with time. Meanwhile, the projection of $P$ on the $x$-axis, labeled point $Q$, moves back and forth along the axis with simple harmonic motion.

From the right triangle $OPQ$, we see that $\cos\theta = x/A$. Therefore, the $x$-coordinate of the ball is

$$x = A \cos\theta$$

Because the ball rotates with constant angular speed, it follows that $\theta = \omega t$ (see Topic 7), so we have

$$x = A\cos(\omega t) \qquad\qquad [13.12]$$

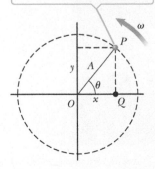

As the ball at $P$ rotates in a circle with uniform angular speed, its projection $Q$ along the $x$-axis moves with simple harmonic motion.

**Figure 13.12** A reference circle.

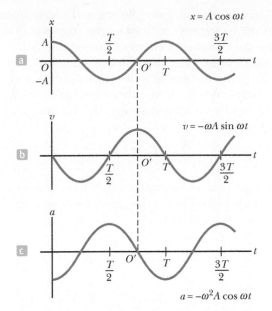

**Figure 13.13** (a) Displacement, (b) velocity, and (c) acceleration versus time for an object moving with simple harmonic motion under the initial conditions $x_0 = A$ and $v_0 = 0$ at $t = 0$.

In one complete revolution, the ball rotates through an angle of $2\pi$ rad in a time equal to the period $T$. In other words, the motion repeats itself every $T$ seconds. Therefore,

$$\omega = \frac{\Delta\theta}{\Delta t} = \frac{2\pi}{T} = 2\pi f \qquad\qquad \text{[13.13]}$$

where $f$ is the frequency of the motion. The angular speed of the ball as it moves around the reference circle is the same as the angular frequency of the projected simple harmonic motion. Consequently, Equation 13.12 can be written

$$x = A\cos(2\pi f t) \qquad\qquad \text{[13.14a]}$$

This cosine function represents the position of an object moving with simple harmonic motion as a function of time, and is graphed in Figure 13.13a. Because the cosine function varies between 1 and $-1$, $x$ varies between $A$ and $-A$. The shape of the graph is called *sinusoidal*.

Figures 13.13b and 13.13c represent curves for velocity and acceleration as a function of time. To find the equation for the velocity, use Equations 13.6 and 13.14a together with the identity $\cos^2\theta + \sin^2\theta = 1$, obtaining

$$v = -A\omega\sin(2\pi f t) \qquad\qquad \text{[13.14b]}$$

where we have used the fact that $\omega = \sqrt{k/m}$. The $\pm$ sign is no longer needed, because sine can take both positive and negative values. Deriving an expression for the acceleration involves substituting Equation 13.14a into Equation 13.2, Newton's second law for springs:

$$a = -A\omega^2\cos(2\pi f t) \qquad\qquad \text{[13.14c]}$$

The detailed steps of these derivations are left as an exercise for the student. Notice that when the displacement $x$ is at a maximum, at $x = A$ or $x = -A$, the velocity is zero, and when $x$ is zero, the magnitude of the velocity is a maximum. Further, when $x = +A$, its most positive value, the acceleration is a maximum but in the negative $x$-direction, and when $x$ is at its most negative position, $x = -A$, the acceleration has its maximum value in the positive $x$-direction. These facts are consistent with our earlier discussion of the points at which $v$ and $a$ reach their maximum, minimum, and zero values.

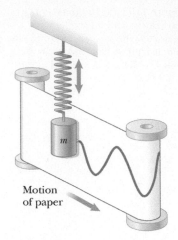

Motion of paper

**Figure 13.14** An experimental apparatus for demonstrating simple harmonic motion. A pen attached to the oscillating object traces out a sinusoidal wave on the moving chart paper.

The maximum values of the position, velocity, and acceleration are always equal to the magnitude of the expression in front of the trigonometric function in each equation because the largest value of either cosine or sine is 1.

Figure 13.14 illustrates one experimental arrangement that demonstrates the sinusoidal nature of simple harmonic motion. An object connected to a spring has a marking pen attached to it. While the object vibrates vertically, a sheet of paper is moved horizontally with constant speed. The pen traces out a sinusoidal pattern.

> ### Quick Quiz
>
> **13.6** If the amplitude of a system moving in simple harmonic motion is doubled, which of the following quantities *doesn't* change? (a) total energy (b) maximum speed (c) maximum acceleration (d) period

---

## EXAMPLE 13.6  THE VIBRATING OBJECT–SPRING SYSTEM

**GOAL** Identify the physical parameters of a harmonic oscillator from its mathematical description.

**PROBLEM** (a) Find the amplitude, frequency, and period of motion for an object vibrating at the end of a horizontal spring if the equation for its position as a function of time is

$$x = (0.250 \text{ m}) \cos\left(\frac{\pi}{8.00} t\right)$$

**(b)** Find the maximum magnitude of the velocity and acceleration. **(c)** What are the position, velocity, and acceleration of the object after 1.00 s has elapsed?

**STRATEGY** In part **(a)** the amplitude and frequency can be found by comparing the given equation with the standard form in Equation 13.14a, matching up the numerical values with the corresponding terms in the standard form. In part **(b)** the maximum speed will occur when the sine function in Equation 13.14b equals 1 or −1, the extreme values of the sine function (and similarly for the acceleration and the cosine function). In each case, find the magnitude of the expression in front of the trigonometric function. Part **(c)** is just a matter of substituting values into Equations 13.14a–13.14c.

......................................................................................

### SOLUTION

**(a)** Find the amplitude, frequency, and period.

Write the standard form given by Equation 13.14a and underneath it write the given equation:

$$(1) \quad x = A \cos(2\pi ft)$$

$$(2) \quad x = (0.250 \text{ m}) \cos\left(\frac{\pi}{8.00} t\right)$$

Equate the factors in front of the cosine functions to find the amplitude:

$$A = \boxed{0.250 \text{ m}}$$

The angular frequency $\omega$ is the factor in front of $t$ in Equations (1) and (2). Equate these factors:

$$\omega = 2\pi f = \frac{\pi}{8.00} \text{ rad/s} = 0.393 \text{ rad/s}$$

Divide $\omega$ by $2\pi$ to get the frequency $f$:

$$f = \frac{\omega}{2\pi} = \boxed{0.062\ 5 \text{ Hz}}$$

The period $T$ is the reciprocal of the frequency:

$$T = \frac{1}{f} = \boxed{16.0 \text{ s}}$$

**(b)** Find the maximum magnitudes of the velocity and the acceleration.

Calculate the maximum speed from the factor in front of the sine function in Equation 13.14b:

$$v_{max} = A\omega = (0.250 \text{ m})(0.393 \text{ rad/s}) = \boxed{0.098\ 3 \text{ m/s}}$$

Calculate the maximum acceleration from the factor in front of the cosine function in Equation 13.14c:

$$a_{max} = A\omega^2 = (0.250 \text{ m})(0.393 \text{ rad/s})^2 = \boxed{0.038\ 6 \text{ m/s}^2}$$

**(c)** Find the position, velocity, and acceleration of the object after 1.00 s.

Substitute $t = 1.00$ s in the given equation:

$$x = (0.250 \text{ m}) \cos (0.393 \text{ rad}) = \boxed{0.231 \text{ m}}$$

Substitute values into the velocity equation:

$$v = -A\omega \sin(\omega t)$$
$$= -(0.250 \text{ m})(0.393 \text{ rad/s}) \sin (0.393 \text{ rad/s} \cdot 1.00 \text{ s})$$
$$v = \boxed{-0.037\ 6 \text{ m/s}}$$

Substitute values into the acceleration equation:

$$a = -A\omega^2 \cos(\omega t)$$
$$= -(0.250 \text{ m})(0.393 \text{ rad/s}^2)^2 \cos (0.393 \text{ rad/s} \cdot 1.00 \text{ s})$$
$$a = \boxed{-0.035\ 7 \text{ m/s}^2}$$

**REMARKS** In evaluating the sine or cosine function, the angle is in radians, so you should either set your calculator to evaluate trigonometric functions based on radian measure or convert from radians to degrees.

**QUESTION 13.6** If the mass is doubled, is the magnitude of the acceleration of the system at any position **(a)** doubled, **(b)** halved, or **(c)** unchanged?

**EXERCISE 13.6** If the object–spring system is described by $x = (0.330 \text{ m}) \cos (1.50t)$, find **(a)** the amplitude, the angular frequency, the frequency, and the period; **(b)** the maximum magnitudes of the velocity and acceleration; and **(c)** the position, velocity, and acceleration when $t = 0.250$ s.

**ANSWERS** **(a)** $A = 0.330$ m, $\omega = 1.50$ rad/s, $f = 0.239$ Hz, $T = 4.18$ s   **(b)** $v_{max} = 0.495$ m/s, $a_{max} = 0.743$ m/s$^2$   **(c)** $x = 0.307$ m, $v = -0.181$ m/s, $a = -0.691$ m/s$^2$

# 13.5 Motion of a Pendulum

A simple pendulum is another mechanical system that exhibits periodic motion. It consists of a small bob of mass $m$ suspended by a light string of length $L$ fixed at its upper end, as in Figure 13.15. (By a light string, we mean that the string's mass is assumed to be very small compared with the mass of the bob and hence can be ignored.) When released, the bob swings to and fro over the same path, but is its motion simple harmonic?

Answering this question requires examining the restoring force—the force of gravity—that acts on the pendulum. The pendulum bob moves along a circular arc, rather than back and forth in a straight line. When the oscillations are small, however, the motion of the bob is nearly straight, so Hooke's law may apply approximately.

In Figure 13.15, $s$ is the displacement of the bob from equilibrium along the arc. Hooke's law is $F = -kx$, so we are looking for a similar expression involving $s$, $F_t = -ks$, where $F_t$ is the force acting in a direction tangent to the circular arc. From the figure, the restoring force is

$$F_t = -mg \sin \theta$$

Since $s = L\theta$, the equation for $F_t$ can be written as

$$F_t = -mg \sin \left(\frac{s}{L}\right)$$

This expression isn't of the form $F_t = -ks$, so in general, the motion of a pendulum is *not* simple harmonic. For small angles less than about 15 degrees, however, the angle $\theta$ measured in radians and the sine of the angle are approximately equal.

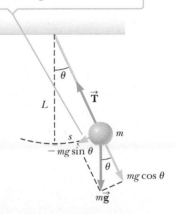

The restoring force causing the pendulum to oscillate harmonically is the tangential component of the gravity force $-mg \sin \theta$.

**Figure 13.15** A simple pendulum consists of a bob of mass $m$ suspended by a light string of length $L$. ($L$ is the distance from the pivot to the center of mass of the bob.)

For example, $\theta = 10.0° = 0.175$ rad, and sin $(10.0°) = 0.174$. Therefore, if we restrict the motion to *small* angles, the approximation sin $\theta \approx \theta$ is valid, and the restoring force can be written

$$F_t = -mg \sin \theta \approx -mg\theta$$

Substituting $\theta = s/L$, we obtain

$$F_t = -\left(\frac{mg}{L}\right) s$$

This equation follows the general form of Hooke's force law $F_t = -ks$, with $k = mg/L$. We are justified in saying that a pendulum undergoes simple harmonic motion only when it swings back and forth at small amplitudes (or, in this case, small values of $\theta$, so that sin $\theta \cong \theta$).

Recall that for the object–spring system, the angular frequency is given by Equation 13.11:

$$\omega = 2\pi f = \sqrt{\frac{k}{m}}$$

Substituting the expression of $k$ for a pendulum, we obtain

$$\omega = \sqrt{\frac{mg/L}{m}} = \sqrt{\frac{g}{L}}$$

This angular frequency can be substituted into Equation 13.12, which then mathematically describes the motion of a pendulum. The frequency is just the angular frequency divided by $2\pi$, while the period is the reciprocal of the frequency, or

$$T = 2\pi \sqrt{\frac{L}{g}} \qquad\qquad [13.15]$$

The period of a simple ▶ pendulum depends only on *L* and *g*

This equation reveals the somewhat surprising result that the period of a simple pendulum doesn't depend on the mass, but only on the pendulum's length and on the free-fall acceleration. Further, the amplitude of the motion isn't a factor as long as it's relatively small. The analogy between the motion of a simple pendulum and the object–spring system is illustrated in Figure 13.16.

Galileo first noted that the period of a pendulum was independent of its amplitude. He supposedly observed this while attending church services at the cathedral in Pisa. The pendulum he studied was a swinging chandelier that was set in motion when someone bumped it while lighting candles. Galileo was able to measure its period by timing the swings with his pulse.

APPLICATION
Pendulum Clocks

The dependence of the period of a pendulum on its length and on the free-fall acceleration allows us to use a pendulum as a timekeeper for a clock. A number of clock designs employ a pendulum, with the length adjusted so that its period serves as the basis for the rate at which the clock's hands turn. Of course, these clocks are used at different locations on the Earth, so there will be some variation of the free-fall acceleration. To compensate for this variation, the pendulum of a clock should have some movable mass so that the effective length can be adjusted.

APPLICATION
Use of Pendulum in Prospecting

Geologists often make use of the simple pendulum and Equation 13.15 when prospecting for oil or minerals. Deposits beneath the Earth's surface can produce irregularities in the free-fall acceleration over the region being studied. A specially designed pendulum of known length is used to measure the period, which in turn is used to calculate $g$. Although such a measurement in itself is inconclusive, it's an important tool for geological surveys.

**Tip 13.3** Pendulum Motion Is Not Harmonic

Remember that the pendulum *does not* exhibit true simple harmonic motion for *any* angle. If the angle is less than about 15°, the motion can be *modeled* as approximately simple harmonic.

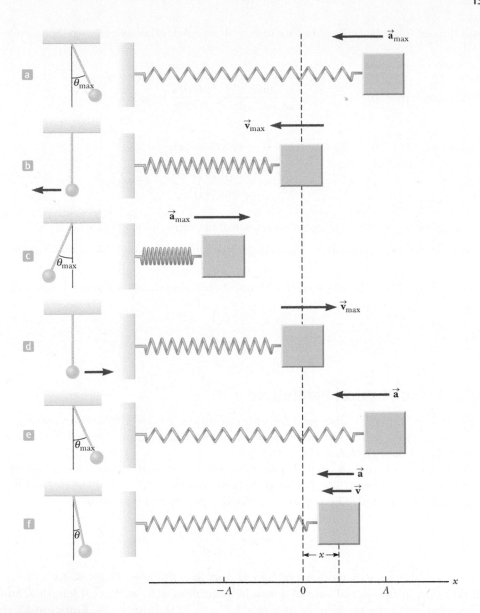

**Figure 13.16** Simple harmonic motion for an object–spring system, and its analogy, the motion of a simple pendulum.

## Quick Quiz

**13.7** A simple pendulum is suspended from the ceiling of a stationary elevator, and the period is measured. If the elevator moves with constant velocity, does the period (a) increase, (b) decrease, or (c) remain the same? If the elevator accelerates upward, does the period (a) increase, (b) decrease, or (c) remain the same?

**13.8** A pendulum clock depends on the period of a pendulum to keep correct time. Suppose a pendulum clock is keeping correct time and then Dennis the Menace slides the bob of the pendulum downward on the oscillating rod. Does the clock run (a) slow, (b) fast, or (c) correctly?

**13.9** The period of a simple pendulum is measured to be $T$ on the Earth. If the same pendulum were set in motion on the Moon, would its period be (a) less than $T$, (b) greater than $T$, or (c) equal to $T$?

## EXAMPLE 13.7    MEASURING THE VALUE OF *g*

**GOAL** Determine $g$ from pendulum motion.

**PROBLEM** Using a small pendulum of length 0.171 m, a geophysicist counts 72.0 complete swings in a time of 60.0 s. What is the value of $g$ in this location?

*(Continued)*

**STRATEGY** First calculate the period of the pendulum by dividing the total time by the number of complete swings. Solve Equation 13.15 for $g$ and substitute values.

**SOLUTION**

Calculate the period by dividing the total elapsed time by the number of complete oscillations:

$$T = \frac{\text{time}}{\text{\# of oscillations}} = \frac{60.0 \text{ s}}{72.0} = 0.833 \text{ s}$$

Solve Equation 13.15 for $g$ and substitute values:

$$T = 2\pi \sqrt{\frac{L}{g}} \quad \rightarrow \quad T^2 = 4\pi^2 \frac{L}{g}$$

$$g = \frac{4\pi^2 L}{T^2} = \frac{(39.5)(0.171 \text{ m})}{(0.833 \text{ s})^2} = \boxed{9.73 \text{ m/s}^2}$$

**REMARKS** Measuring such a vibration is a good way of determining the local value of the acceleration of gravity.

**QUESTION 13.7** True or False: A simple pendulum of length 0.50 m has a larger frequency of vibration than a simple pendulum of length 1.0 m.

**EXERCISE 13.7** What would be the period of the 0.171-m pendulum on the Moon, where the acceleration of gravity is 1.62 m/s²?

**ANSWER** 2.04 s

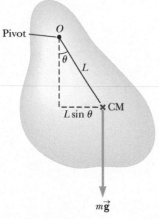

**Figure 13.17** A physical pendulum pivoted at *O*.

### 13.5.1 The Physical Pendulum

The simple pendulum discussed thus far consists of a mass attached to a string. A pendulum, however, can be made from an object of any shape. The general case is called the *physical pendulum*.

In Figure 13.17 a rigid object is pivoted at point *O*, which is a distance *L* from the object's center of mass. The center of mass oscillates along a circular arc, just like the simple pendulum. The period of a physical pendulum is given by

$$T = 2\pi \sqrt{\frac{I}{mgL}} \qquad [13.16]$$

where $I$ is the object's moment of inertia and $m$ is the object's mass. As a check, notice that in the special case of a simple pendulum with an arm of length $L$ and negligible mass, the moment of inertia is $I = mL^2$. Substituting into Equation 13.16 results in

$$T = 2\pi \sqrt{\frac{mL^2}{mgL}} = 2\pi \sqrt{\frac{L}{g}}$$

which is the correct period for a simple pendulum.

## 13.6 Damped Oscillations

The vibrating motions we have discussed so far have taken place in ideal systems that *oscillate indefinitely* under the action of a linear restoring force. In all real mechanical systems, forces of friction retard the motion, so the systems don't oscillate indefinitely. The friction reduces the mechanical energy of the system as time passes, and the motion is said to be **damped.**

APPLICATION

Shock Absorbers

Shock absorbers in automobiles (Fig. 13.18) are one practical application of damped motion. A shock absorber consists of a piston moving through a liquid such as oil. The upper part of the shock absorber is firmly attached to the body of the car. When the car travels over a bump in the road, holes in the piston allow it to move up and down through the fluid in a damped fashion.

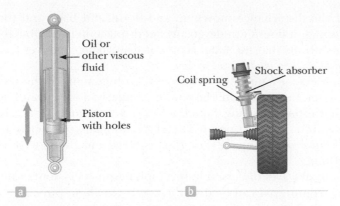

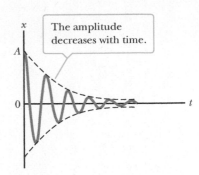

Figure 13.18 (a) A shock absorber consists of a piston oscillating in a chamber filled with oil. As the piston oscillates, the oil is squeezed through holes between the piston and the chamber, causing a damping of the piston's oscillations. (b) One type of automotive suspension system, in which a shock absorber is placed inside a coil spring at each wheel.

Figure 13.19 A graph of displacement versus time for an underdamped oscillator.

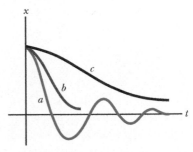

Figure 13.20 Plots of displacement versus time for (a) an underdamped oscillator, (b) a critically damped oscillator, and (c) an overdamped oscillator.

Damped motion varies with the fluid used. For example, if the fluid has a relatively low viscosity, the vibrating motion is preserved but the amplitude of vibration decreases in time and the motion ultimately ceases. This process is known as *underdamped* oscillation. The position versus time curve for an object undergoing such oscillation appears in Figure 13.19. Figure 13.20 compares three types of damped motion, with curve (a) representing underdamped oscillation. If the fluid viscosity is increased, the object returns rapidly to equilibrium after it's released and doesn't oscillate. In this case, the system is said to be *critically damped,* and is shown as curve (b) in Figure 13.20. The piston returns to the equilibrium position in the shortest time possible without once overshooting the equilibrium position. If the viscosity is made greater still, the system is said to be *overdamped.* In this case, the piston returns to equilibrium without ever passing through the equilibrium point, but the time required to reach equilibrium is greater than in critical damping, as illustrated by curve (c) in Figure 13.20.

To make automobiles more comfortable to ride in, shock absorbers are designed to be slightly underdamped. This can be demonstrated by a sharp downward push on the hood of a car. After the applied force is removed, the body of the car oscillates a few times about the equilibrium position before returning to its fixed position.

# 13.7 Waves

The world is full of waves: sound waves, waves on a string, seismic waves, and electromagnetic waves, such as visible light, radio waves, television signals, and x-rays. All these waves have as their source a vibrating object, so we can apply the concepts of simple harmonic motion in describing them.

In the case of sound waves, the vibrations that produce waves arise from sources such as a person's vocal chords or a plucked guitar string. The vibrations of electrons in an antenna produce radio or television waves, and the simple up-and-down motion of a hand can produce a wave on a string. Certain concepts are common to all waves, regardless of their nature. In the remainder of this topic, we focus our attention on the general properties of waves. In later topics, we study specific types of waves, such as sound waves and electromagnetic waves.

## 13.7.1 What Is a Wave?

When you drop a pebble into a pool of water, the disturbance produces water waves, which move away from the point where the pebble entered the water. A leaf floating near the disturbance moves up and down and back and forth about its original position, but doesn't undergo any net displacement attributable to the disturbance. This means that the water wave (or disturbance) moves from one place to another, *but the water isn't carried with it.*

When we observe a water wave, we see a rearrangement of the water's surface. Without the water, there wouldn't be a wave. Similarly, a wave traveling on a string wouldn't exist without the string. Sound waves travel through air as a result of pressure variations from point to point. Therefore, we can consider a wave to be *the motion of a disturbance.* In Topic 21, we discuss electromagnetic waves, which don't require a medium.

The mechanical waves discussed in this topic require (1) some source of disturbance, (2) a medium that can be disturbed, and (3) some physical connection or mechanism through which adjacent portions of the medium can influence each other. All waves carry energy and momentum. The amount of energy transmitted through a medium and the mechanism responsible for the transport of energy differ from case to case. The energy carried by ocean waves during a storm, for example, is much greater than the energy carried by a sound wave generated by a single human voice.

---

### APPLYING PHYSICS 13.2    BURYING BOND

At one point in *On Her Majesty's Secret Service,* a James Bond film from the 1960s, Bond was escaping on skis. He had a good lead and was a hard-to-hit moving target. There was no point in wasting bullets shooting at him, so why did the bad guys open fire?

**EXPLANATION** These misguided gentlemen had a good understanding of the physics of waves. An impulsive sound,

like a gunshot, can cause an acoustical disturbance that propagates through the air. If it impacts a ledge of snow that is ready to break free, an avalanche can result. Such a disaster occurred in 1916 during World War I when Austrian soldiers in the Alps were smothered by an avalanche caused by cannon fire. So the bad guys, who have never been able to hit Bond with a bullet, decided to use the sound of gunfire to start an avalanche. ■

---

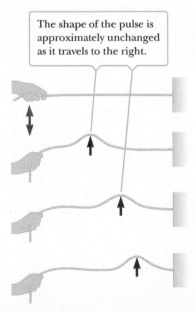

The shape of the pulse is approximately unchanged as it travels to the right.

**Figure 13.21** A hand moves the end of a stretched string up and down once (red arrow), causing a pulse to travel along the string.

### 13.7.2 Types of Waves

One of the simplest ways to demonstrate wave motion is to flip one end of a long string that is under tension and has its opposite end fixed, as in Figure 13.21. The bump (called a pulse) travels to the right with a definite speed. A disturbance of this type is called a **traveling wave.** The figure shows the shape of the string at three closely spaced times.

As such a wave pulse travels along the string, **each segment of the string that is disturbed moves in a direction perpendicular to the wave motion.** Figure 13.22 illustrates this point for a particular tiny segment *P.* The string never moves in the direction of the wave. A traveling wave in which the particles of the disturbed medium move in a direction perpendicular to the wave velocity is called a **transverse wave.** Figure 13.23a illustrates the formation of transverse waves on a long spring.

In another class of waves, called **longitudinal waves, the elements of the medium undergo displacements parallel to the direction of wave motion.** Sound waves in air are longitudinal. Their disturbance corresponds to a series of high- and low-pressure regions that may travel through air or through any material medium with a certain speed. A longitudinal pulse can easily be produced in a stretched spring, as in Figure 13.23b. The free end is pumped back and forth along the length of the spring. This action produces compressed and stretched regions of the coil that travel along the spring, parallel to the wave motion.

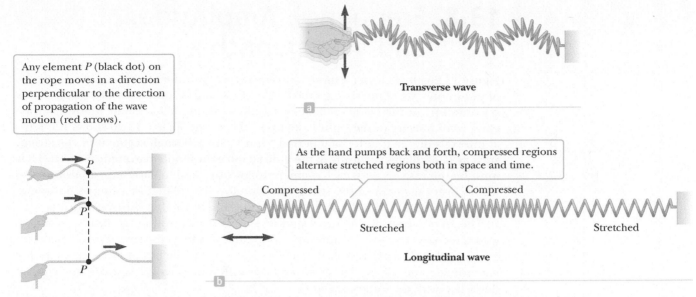

Any element *P* (black dot) on the rope moves in a direction perpendicular to the direction of propagation of the wave motion (red arrows).

**Transverse wave**

a

As the hand pumps back and forth, compressed regions alternate stretched regions both in space and time.

Compressed     Compressed

Stretched     Stretched

**Longitudinal wave**

b

**Figure 13.22** A pulse traveling on a stretched string is a transverse wave.

**Figure 13.23** (a) A transverse wave is set up in a spring by moving one end of the spring perpendicular to its length. (b) A longitudinal wave along a stretched spring.

Waves need not be purely transverse or purely longitudinal: ocean waves exhibit a superposition of both types. When an ocean wave encounters a cork, the cork executes a circular motion, going up and down while going forward and back.

Another type of wave, called a **soliton,** consists of a solitary wave front that propagates in isolation. Ordinary water waves generally spread out and dissipate, but solitons tend to maintain their form. The study of solitons began in 1849, when Scottish engineer John Scott Russell noticed a solitary wave leaving the turbulence in front of a barge and propagating forward all on its own. The wave maintained its shape and traveled down a canal at about 10 mi/h. Russell chased the wave two miles on horseback before losing it. Only in the 1960s did scientists take solitons seriously; they are now widely used to model physical phenomena, from elementary particles to the Giant Red Spot of Jupiter.

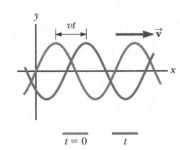

**Figure 13.24** A one-dimensional sinusoidal wave traveling to the right with a speed *v*. The brown curve is a snapshot of the wave at $t = 0$, and the blue curve is another snapshot at some later time *t*.

### 13.7.3 Picture of a Wave

Figure 13.24 shows the curved shape of a vibrating string. This pattern is a sinusoidal curve, the same as in simple harmonic motion. The brown curve can be thought of as a snapshot of a traveling wave taken at some instant of time, say, $t = 0$; the blue curve is a snapshot of the same traveling wave at a later time. This picture can also be used to represent a wave on water. In such a case, a high point would correspond to the *crest* of the wave and a low point to the *trough* of the wave.

The same waveform can be used to describe a longitudinal wave, even though no up-and-down motion is taking place. Consider a longitudinal wave traveling on a spring. Figure 13.25a is a snapshot of this wave at some instant, and Figure 13.25b shows the sinusoidal curve that represents the wave. Points where the coils of the spring are compressed correspond to the crests of the waveform, and stretched regions correspond to troughs.

The type of wave represented by the curve in Figure 13.25b is often called a *density wave* or *pressure wave,* because the crests, where the spring coils are compressed, are regions of high density, and the troughs, where the coils are stretched, are regions of low density. Sound waves are longitudinal waves, propagating as a series of high- and low-density regions.

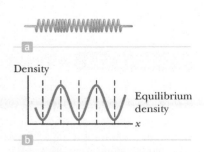

**Figure 13.25** (a) A longitudinal wave on a spring. (b) The crests of the waveform correspond to compressed regions of the spring, and the troughs correspond to stretched regions of the spring.

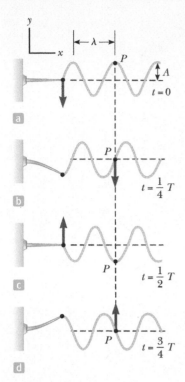

**Figure 13.26** One method for producing traveling waves on a continuous string. The left end of the string is connected to a blade that is set vibrating. Every part of the string, such as point $P$, oscillates vertically with simple harmonic motion.

# 13.8 Frequency, Amplitude, and Wavelength

Figure 13.26 illustrates a method of producing a continuous wave or a steady stream of pulses on a very long string. One end of the string is connected to a blade that is set vibrating. As the blade oscillates vertically with simple harmonic motion, a traveling wave moving to the right is set up in the string. Figure 13.26 shows the wave at intervals of one-quarter of a period. Note that **each small segment of the string, such as $P$, oscillates vertically in the $y$-direction with simple harmonic motion.** That must be the case because each segment follows the simple harmonic motion of the blade. Every segment of the string can therefore be treated as a simple harmonic oscillator vibrating with the same frequency as the blade that drives the string.

The frequencies of the waves studied in this course will range from rather low values for waves on strings and waves on water, to values for sound waves between 20 Hz and 20 000 Hz (recall that 1 Hz = 1 s$^{-1}$), to much higher frequencies for electromagnetic waves. These waves have different physical sources but can be described with the same concepts.

The horizontal dashed line in Figure 13.26 represents the position of the string when no wave is present. The maximum distance the string moves above or below this equilibrium value is called the **amplitude** $A$ of the wave. For the waves we work with, the amplitudes at the crest and the trough will be identical.

Figure 13.26a illustrates another characteristic of a wave. The horizontal arrows show the distance between two successive points that behave identically. This distance is called the **wavelength** $\lambda$ (the Greek letter lambda).

We can use these definitions to derive an expression for the speed of a wave. We start with the defining equation for the **wave speed** $v$:

$$v = \frac{\Delta x}{\Delta t}$$

The wave speed is the speed at which a particular part of the wave—say, a crest—moves through the medium.

A wave advances a distance of one wavelength in a time interval equal to one period of the vibration. Taking $\Delta x = \lambda$ and $\Delta t = T$, we see that

$$v = \frac{\lambda}{T}$$

Because the frequency is the reciprocal of the period, we have

Wave speed ▶ $\boxed{v = f\lambda}$ [13.17]

This important general equation applies to many different types of waves, such as sound waves and electromagnetic waves.

---

**EXAMPLE 13.8** **A TRAVELING WAVE**

**GOAL** Obtain information about a wave directly from its graph.

**PROBLEM** A wave traveling in the positive $x$-direction is pictured in Figure 13.27a. Find the amplitude, wavelength, speed, and period of the wave if it has a frequency of 8.00 Hz. In Figure 13.27a, $\Delta x = 40.0$ cm and $\Delta y = 15.0$ cm.

**STRATEGY** The amplitude and wavelength can be read directly from the figure: The maximum vertical displacement is the amplitude, and the distance from one crest to the next is the wavelength. Multiplying the wavelength by the frequency gives the speed, whereas the period is the reciprocal of the frequency.

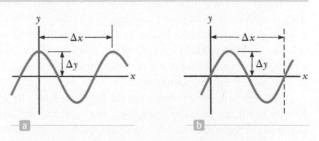

**Figure 13.27** (a) (Example 13.8) (b) (Exercise 13.8)

## SOLUTION

The maximum wave displacement is the amplitude $A$:

$A = \Delta y = 15.0 \text{ cm} = \boxed{0.150 \text{ m}}$

The distance from crest to crest is the wavelength:

$\lambda = \Delta x = 40.0 \text{ cm} = \boxed{0.400 \text{ m}}$

Multiply the wavelength by the frequency to get the speed:

$v = f\lambda = (8.00 \text{ Hz})(0.400 \text{ m}) = \boxed{3.20 \text{ m/s}}$

Take the reciprocal of the frequency to get the period:

$T = \dfrac{1}{f} = \dfrac{1}{8.00 \text{ Hz}} = \boxed{0.125 \text{ s}}$

**REMARKS** It's important not to confuse the wave with the medium it travels in. A wave is energy transmitted through a medium; some waves, such as light waves, don't require a medium.

**QUESTION 13.8** Is the frequency of a wave affected by the wave's amplitude?

**EXERCISE 13.8** A wave traveling in the positive $x$-direction is pictured in Figure 13.27b. Find the amplitude, wavelength, speed, and period of the wave if it has a frequency of 15.0 Hz. In the figure, $\Delta x = 72.0$ cm and $\Delta y = 25.0$ cm.

**ANSWERS** $A = 0.250$ m, $\lambda = 0.720$ m, $v = 10.8$ m/s, $T = 0.006\ 7$ s

---

### EXAMPLE 13.9 | SOUND AND LIGHT

**GOAL** Perform elementary calculations using speed, wavelength, and frequency.

**PROBLEM** A wave has a wavelength of 3.00 m. Calculate the frequency of the wave if it is (a) a sound wave and (b) a light wave. Take the speed of sound as 343 m/s and the speed of light as $3.00 \times 10^8$ m/s.

## SOLUTION

(a) Find the frequency of a sound wave with $\lambda = 3.00$ m.

Solve Equation 3.17 for the frequency and substitute:

$\textbf{(1)}\quad f = \dfrac{v}{\lambda} = \dfrac{343 \text{ m/s}}{3.00 \text{ m}} = \boxed{114 \text{ Hz}}$

(b) Find the frequency of a light wave with $\lambda = 3.00$ m.

Substitute into Equation (1), using the speed of light for $c$:

$f = \dfrac{c}{\lambda} = \dfrac{3.00 \times 10^8 \text{ m/s}}{3.00 \text{ m}} = \boxed{1.00 \times 10^8 \text{ Hz}}$

**REMARKS** The same equation can be used to find the frequency in each case, despite the great difference between the physical phenomena. Notice how much larger frequencies of light waves are than frequencies of sound waves.

**QUESTION 13.9** A wave in one medium encounters a new medium and enters it. Which of the following wave properties will be affected in this process? (a) wavelength (b) frequency (c) speed

**EXERCISE 13.9** (a) Find the wavelength of an electromagnetic wave with frequency 9.00 GHz $= 9.00 \times 10^9$ Hz (G $=$ giga $= 10^9$), which is in the microwave range. (b) Find the speed of a sound wave in an unknown fluid medium if a frequency of 567 Hz has a wavelength of 2.50 m.

**ANSWERS** (a) 0.033 3 m   (b) $1.42 \times 10^3$ m/s

---

# 13.9 The Speed of Waves on Strings

In this section, we focus our attention on the speed of a transverse wave on a stretched string.

For a vibrating string, there are two speeds to consider. One is the speed of the physical string that vibrates up and down, transverse to the string, in the $y$-direction. The other is the *wave speed*, which is the rate at which the disturbance propagates along the length of the string in the $x$-direction. We wish to find an expression for the wave speed.

If a horizontal string under tension is pulled vertically and released, it starts at its maximum displacement, $y = A$, and takes a certain amount of time to go to $y = -A$ and back to $A$ again. This amount of time is the period of the wave, and is the same as the time needed for the wave to advance *horizontally* by one wavelength. Dividing the wavelength by the period of one transverse oscillation gives the wave speed.

For a fixed wavelength, a string under greater tension $F$ has a greater wave speed because the period of vibration is shorter, and the wave advances one wavelength during one period. It also makes sense that a string with greater mass per unit length, $\mu$, vibrates more slowly, leading to a longer period and a slower wave speed. The wave speed is given by

$$v = \sqrt{\frac{F}{\mu}}$$ [13.18]

where $F$ is the tension in the string and $\mu$ is the mass of the string per unit length, called the *linear density*. From Equation 13.18, it's clear that a larger tension $F$ results in a larger wave speed, whereas a larger linear density $\mu$ gives a slower wave speed, as expected.

According to Equation 13.18, the propagation speed of a mechanical wave, such as a wave on a string, depends only on the properties of the string through which the disturbance travels. It doesn't depend on the amplitude of the vibration. This turns out to be generally true of waves in various media.

A proof of Equation 13.18 requires calculus, but dimensional analysis can easily verify that the expression is dimensionally correct. Note that the dimensions of $F$ are $ML/T^2$, and the dimensions of $\mu$ are $M/L$. The dimensions of $F/\mu$ are therefore $L^2/T^2$, so those of $\sqrt{F/\mu}$ are $L/T$, giving the dimensions of speed. No other combination of $F$ and $\mu$ is dimensionally correct, so in the case in which the tension and mass density are the only relevant physical factors, we have verified Equation 13.18 up to an overall constant.

APPLICATION
Bass Guitar Strings

According to Equation 13.18, we can increase the speed of a wave on a stretched string by increasing the tension in the string. Increasing the mass per unit length, on the other hand, decreases the wave speed. These physical facts lie behind the metallic windings on the bass strings of pianos and guitars. The windings increase the mass per unit length, $\mu$, decreasing the wave speed and hence the frequency, resulting in a lower tone. Tuning a string to a desired frequency is a simple matter of changing the tension in the string.

---

## EXAMPLE 13.10    A PULSE TRAVELING ON A STRING

**GOAL** Calculate the speed of a wave on a string.

**PROBLEM** A uniform string has a mass $M$ of 0.030 0 kg and a length $L$ of 6.00 m. Tension is maintained in the string by suspending a block of mass $m = 2.00$ kg from one end (Fig. 13.28). **(a)** Find the speed of a transverse wave pulse on this string. **(b)** Find the time it takes the pulse to travel from the wall to the pulley. Neglect the mass of the hanging part of the string.

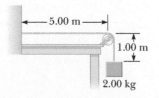

**Figure 13.28** (Example 13.10) The tension $F$ in the string is maintained by the suspended block. The wave speed is given by the expression $v = \sqrt{F/\mu}$.

**STRATEGY** The tension $F$ can be obtained from Newton's second law for equilibrium applied to the block, and the mass per unit length of the string is $\mu = M/L$. With these quantities, the speed of the transverse pulse can be found by substitution into Equation 13.18. Part (b) requires the formula $d = vt$.

......................................................................................................................

**SOLUTION**

**(a)** Find the speed of the wave pulse.

Apply the second law to the block: the tension $F$ is equal and opposite to the force of gravity.           $\sum F = F - mg = 0 \quad \rightarrow \quad F = mg$

Substitute expressions for $F$ and $\mu$ into Equation 13.18:

$$v = \sqrt{\frac{F}{\mu}} = \sqrt{\frac{mg}{M/L}}$$

$$= \sqrt{\frac{(2.00 \text{ kg})(9.80 \text{ m/s}^2)}{(0.030\ 0 \text{ kg})/(6.00 \text{ m})}} = \sqrt{\frac{19.6 \text{ N}}{0.005\ 00 \text{ kg/m}}}$$

$$= \boxed{62.6 \text{ m/s}}$$

**(b)** Find the time it takes the pulse to travel from the wall to the pulley.

Solve the distance formula for time:

$$t = \frac{d}{v} = \frac{5.00 \text{ m}}{62.6 \text{ m/s}} = \boxed{0.079\ 9 \text{ s}}$$

- - - - - - - - - - - - - - - - - - - - - - - - - - - - - - - - - - - - - - - - - - - - - - - - - - - - - - - - - - - - - - - - - - - -

**REMARKS** Don't confuse the speed of the wave on the string with the speed of the sound wave produced by the vibrating string. (See Topic 14.)

**QUESTION 13.10** If the mass of the block is quadrupled, what happens to the speed of the wave?

**EXERCISE 13.10** To what tension must a string with mass 0.010 0 kg and length 2.50 m be tightened so that waves will travel on it at a speed of 125 m/s?

**ANSWER** 62.5 N

# 13.10 Interference of Waves

Many interesting wave phenomena in nature require two or more waves passing through the same region of space at the same time. **Two traveling waves can meet and pass through each other without being destroyed or even altered.** For instance, when two pebbles are thrown into a pond, the expanding circular waves don't destroy each other. In fact, the ripples pass through each other. Likewise, when sound waves from two sources move through air, they pass through each other. In the region of overlap, the resultant wave is found by adding the displacements of the individual waves. For such analyses, the **superposition principle** applies:

> When two or more traveling waves encounter each other while moving through a medium, the resultant wave is found by adding together the displacements of the individual waves point by point.

Experiments show that the superposition principle is valid only when the individual waves have small amplitudes of displacement, which is an assumption we make in all our examples.

Figures 13.29a and 13.29b show two waves of the same amplitude and frequency. If at some instant of time these two waves were traveling through the same region of space, the resultant wave at that instant would have a shape like that shown in Figure 13.29c. For example, suppose the waves are water waves of amplitude 1 m. At the instant they overlap so that crest meets crest and trough meets trough, the resultant wave has an amplitude of 2 m. Waves coming together like that are said to be *in phase* and to exhibit **constructive interference.**

Figures 13.30a and 13.30b (page 448) show two similar waves. In this case, however, the crest of one coincides with the trough of the other, so one wave is *inverted* relative to the other. The resultant wave, shown in Figure 13.30c, is seen to be a state of complete cancellation. If these were water waves coming together, one of the waves would exert an upward force on an individual drop of water at the same instant the other wave was exerting a downward force. The result would be no motion of the water at all. In such a situation the two waves are said to be 180° out of phase and to exhibit **destructive interference.** Figure 13.31 (page 448) illustrates the interference of water waves produced by drops of water falling into a pond.

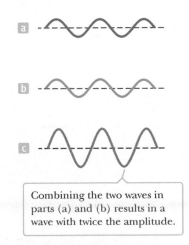

Combining the two waves in parts (a) and (b) results in a wave with twice the amplitude.

**Figure 13.29** Constructive interference. If two waves having the same frequency and amplitude are in phase, as in (a) and (b), the resultant wave when they combine (c) has the same frequency as the individual waves, but twice their amplitude.

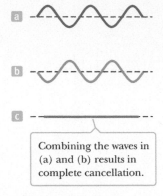

Combining the waves in (a) and (b) results in complete cancellation.

**Figure 13.30** Destructive interference. The two waves in (a) and (b) have the same frequency and amplitude but are 180° out of phase.

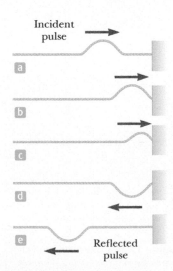

Martin Dohrn/SPL/Science Source

**Figure 13.31** Interference patterns produced by outward-spreading waves from many drops of liquid falling into a body of water.

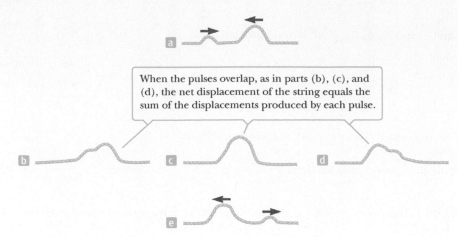

When the pulses overlap, as in parts (b), (c), and (d), the net displacement of the string equals the sum of the displacements produced by each pulse.

**Figure 13.32** Two wave pulses traveling on a stretched string in opposite directions pass through each other.

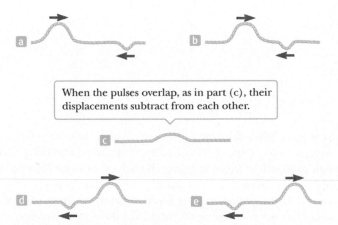

When the pulses overlap, as in part (c), their displacements subtract from each other.

**Figure 13.33** Two wave pulses traveling in opposite directions with displacements that are inverted relative to each other.

Figure 13.32 shows constructive interference in two pulses moving toward each other along a stretched string; Figure 13.33 shows destructive interference in two pulses. Notice in both figures that when the two pulses separate, their shapes are unchanged, as if they had never met!

# 13.11 Reflection of Waves

In our discussion so far, we have assumed waves could travel indefinitely without striking anything. Such conditions are not often realized in practice. Whenever a traveling wave reaches a boundary, part or all of the wave is reflected. For example, consider a pulse traveling on a string that is fixed at one end (Fig. 13.34). When the pulse reaches the wall, it is reflected.

Note that the reflected pulse is inverted. This can be explained as follows: When the pulse meets the wall, the string exerts an upward force on the wall. According to Newton's third law, the wall must exert an equal and opposite (downward) reaction force on the string. This downward force causes the pulse to invert on reflection.

Now suppose the pulse arrives at the string's end, and the end is attached to a ring of negligible mass that is free to slide along the post without friction (Fig. 13.35). Again the pulse is reflected, but this time it is not inverted. On reaching the post, the pulse exerts a force on the ring, causing it to accelerate upward. The ring is

Incident pulse

Reflected pulse

**Figure 13.34** The reflection of a traveling wave at the fixed end of a stretched string. Note that the reflected pulse is inverted, but its shape remains the same.

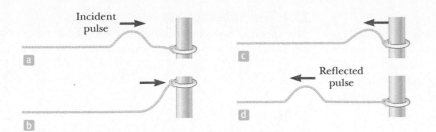

**Figure 13.35** The reflection of a traveling wave at the free end of a stretched string. In this case, the reflected pulse is not inverted.

then returned to its original position by the downward component of the tension in the string.

An alternate method of showing that a pulse is reflected without inversion when it strikes a free end of a string is to send the pulse down a string hanging vertically. When the pulse hits the free end, it's reflected without inversion, like the pulse in Figure 13.35.

Finally, when a pulse reaches a boundary, it's partly reflected and partly transmitted past the boundary into the new medium. This effect is easy to observe in the case of two ropes of different density joined at some boundary.

## SUMMARY

### 13.1 Hooke's Law

**Simple harmonic motion** occurs when the net force on an object along the direction of motion is proportional to the object's displacement and in the opposite direction:

$$F_s = -kx \qquad [13.1]$$

This equation is called Hooke's law. The time required for one complete vibration is called the **period** of the motion. The reciprocal of the period is the **frequency** of the motion, which is the number of oscillations per second.

When an object moves with simple harmonic motion, its **acceleration** as a function of position is

$$a = -\frac{k}{m}x \qquad [13.2]$$

### 13.2 Elastic Potential Energy

The energy stored in a stretched or compressed spring or in some other elastic material is called **elastic potential energy:**

$$PE_s \equiv \tfrac{1}{2}kx^2 \qquad [13.3]$$

The **velocity** of an object as a function of position, when the object is moving with simple harmonic motion, is

$$v = \pm\sqrt{\frac{k}{m}(A^2 - x^2)} \qquad [13.6]$$

### 13.3 Concepts of Oscillation Rates in Simple Harmonic Motion

The **period** of an object of mass $m$ moving with simple harmonic motion while attached to a spring of spring constant $k$ is

$$T = 2\pi\sqrt{\frac{m}{k}} \qquad [13.8]$$

where $T$ is independent of the amplitude $A$.

The **frequency** of an object–spring system is $f = 1/T$. The **angular frequency** $\omega$ of the system in rad/s is

$$\omega = 2\pi f = \sqrt{\frac{k}{m}} \qquad [13.11]$$

### 13.4 Position, Velocity, and Acceleration as Functions of Time

When an object is moving with simple harmonic motion, the **position, velocity,** and **acceleration** of the object as a function of time (Fig. 13.36) are given by

$$x = A\cos(2\pi ft) \qquad [13.14a]$$

$$v = -A\omega\sin(2\pi ft) \qquad [13.14b]$$

$$a = -A\omega^2\cos(2\pi ft) \qquad [13.14c]$$

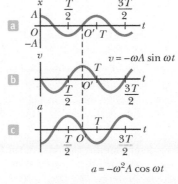

**Figure 13.36** (a) Displacement, (b) velocity, and (c) acceleration versus time for an object moving with simple harmonic motion under the initial conditions $x_0 = A$ and $v_0 = 0$ at $t = 0$.

### 13.5 Motion of a Pendulum

A **simple pendulum** of length $L$ moves with simple harmonic motion for small angular displacements from the vertical, with a period of

$$T = 2\pi\sqrt{\frac{L}{g}} \qquad [13.15]$$

## 13.7 Waves

In a **transverse wave**, the elements of the medium move in a direction perpendicular to the direction of the wave. An example is a wave on a stretched string.

In a **longitudinal wave**, the elements of the medium move parallel to the direction of the wave velocity. An example is a sound wave.

## 13.8 Frequency, Amplitude, and Wavelength

The relationship of the speed, wavelength, and frequency of a wave is

$$v = f\lambda \qquad [13.17]$$

This relationship holds for a wide variety of different waves.

## 13.9 The Speed of Waves on Strings

The speed of a wave traveling on a stretched string of mass per unit length $\mu$ and under tension $F$ is

$$v = \sqrt{\frac{F}{\mu}} \qquad [13.18]$$

## 13.10 Interference of Waves

The **superposition principle** states that if two or more traveling waves are moving through a medium, the resultant wave is found by adding the individual waves together point by point. When waves meet crest to crest and trough to trough, they undergo **constructive interference.** When crest meets trough, the waves undergo **destructive interference.**

## 13.11 Reflection of Waves

When a wave pulse reflects from a rigid boundary, the pulse is inverted. When the boundary is free, the reflected pulse is not inverted.

# CONCEPTUAL QUESTIONS

1. An object–spring system undergoes simple harmonic motion with an amplitude $A$. Does the total energy change if the mass is doubled but the amplitude isn't changed? Are the kinetic and potential energies at a given point in its motion affected by the change in mass? Explain.

2. If an object–spring system is hung vertically and set into oscillation, why does the motion eventually stop?

3. The spring in Figure CQ13.3 is stretched from its equilibrium position at $x = 0$ to a positive coordinate $x_0$. The force on the spring is $F_0$ and it stores elastic potential energy $PE_{s0}$. If the spring displacement is doubled to $2x_0$, determine (a) the ratio of the new force to the original force, $F_n/F_0$, and (b) the ratio of the new to the original elastic potential energy, $PE_{sn}/PE_{s0}$.

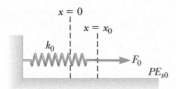

**Figure CQ13.3** Conceptual Questions 3 through 5.

4. If the spring constant shown in Figure CQ13.3 is doubled to $2k_0$, determine (a) the ratio of the new force to the original force, $F_n/F_0$, and (b) the ratio of the new to the original elastic potential energy, $PE_{sn}/PE_{s0}$.

5. If the spring shown in Figure CQ13.3 is compressed rather than stretched so that $x = -x_0$, determine (a) the ratio of the new force to the original force, $F_n/F_0$, and (b) the ratio of the new to the original elastic potential energy, $PE_{sn}/PE_{s0}$. (*Hint*: Force is a vector.)

6. If a spring is cut in half, what happens to its spring constant?

7. A pendulum bob is made from a sphere filled with water. What would happen to the frequency of vibration of this pendulum if the sphere had a hole in it that allowed the water to leak out slowly?

8. A block connected to a horizontal spring is in simple harmonic motion on a level, frictionless surface, oscillating with amplitude $A$ around $x = 0$. Identify whether each of the following statements is true or false: (a) If $x = \pm A$ then $|v| = |v_{max}|$ and $|a| = |a_{max}|$. (b) If $x = 0$ then $|v| = |v_{max}|$ and $|a| = 0$. (c) If $v > 0$ then $a < 0$. (d) If $x > 0$ then $a < 0$. (e) If $x > 0$ then $v > 0$.

9. (a) Is a bouncing ball an example of simple harmonic motion? (b) Is the daily movement of a student from home to school and back simple harmonic motion?

10. If a grandfather clock were running slow, how could we adjust the length of the pendulum to correct the time?

11. What happens to the speed of a wave on a string when the frequency is doubled? Assume the tension in the string remains the same.

12. If you stretch a rubber hose and pluck it, you can observe a pulse traveling up and down the hose. What happens to the speed of the pulse if you stretch the hose more tightly? What happens to the speed if you fill the hose with water?

13. Waves are traveling on a uniform string under tension. For each of the following changes, is the wave speed increased, decreased, or unchanged? Indicate your answers with "I" for increased, "D" for decreased, or "U" for unchanged. (a) The string tension is doubled while its mass and length are constant. (b) The string is folded in half and the tension is constant. (c) The string is cut to half its original length and the tension is constant. (d) The string is cut to half its original length and the tension is doubled.

14. Identify each of the following waves as either transverse or longitudinal: (a) The waves on a plucked guitar string. (b) The sound waves produced by a vibrating guitar string. (c) The waves on a spring with its end pumped back and forth along the spring's length.

# PROBLEMS

*See the Preface for an explanation of the icons used in this problems set.*

## 13.1 Hooke's Law

1. A block of mass $m = 0.60$ kg attached to a spring with force constant 130 N/m is free to move on a friction-less, horizontal surface as in Figure P13.1. The block is released from rest after the spring is stretched a distance $A = 0.13$ m. At that instant, find (a) the force on the block and (b) its acceleration.

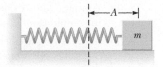

**Figure P13.1**

2. A spring oriented vertically is attached to a hard horizontal surface as in Figure P13.2. The spring has a force constant of 1.46 kN/m. How much is the spring compressed when a object of mass $m = 2.30$ kg is placed on top of the spring and the system is at rest?

**Figure P13.2**

3. The force constant of a spring is 137 N/m. Find the magnitude of the force required to (a) compress the spring by 4.80 cm from its unstretched length and (b) stretch the spring by 7.36 cm from its unstretched length.

4. A spring is hung from a ceiling, and an object attached to its lower end stretches the spring by a distance $d = 5.00$ cm from its unstretched position when the system is in equilibrium as in Figure P13.4. If the spring constant is 47.5 N/m, determine the mass of the object.

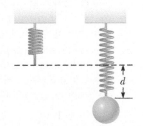

**Figure P13.4**

5. A biologist hangs a sample of mass 0.725 kg on a pair of identical, vertical springs in parallel and slowly lowers the sample to equilibrium, stretching the springs by 0.200 m. Calculate the value of the spring constant of one of the springs.

6. **V** An archer must exert a force of 375 N on the bowstring shown in Figure P13.6a such that the string makes an angle of $\theta = 35.0°$ with the vertical. (a) Determine the tension in the bowstring. (b) If the applied force is replaced by a stretched spring as in Figure P13.6b and the spring is stretched 30.0 cm from its unstretched length, what is the spring constant?

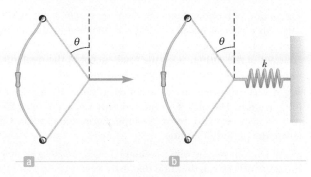

**Figure P13.6**

7. **Q/C** A spring 1.50 m long with force constant 475 N/m is hung from the ceiling of an elevator, and a block of mass 10.0 kg is attached to the bottom of the spring. (a) By how much is the spring stretched when the block is slowly lowered to its equilibrium point? (b) If the elevator subsequently accelerates upward at 2.00 m/s², what is the position of the block, taking the equilibrium position found in part (a) as $y = 0$ and upwards as the positive $y$-direction. (c) If the elevator cable snaps during the acceleration, describe the subsequent motion of the block relative to the freely falling elevator. What is the amplitude of its motion?

## 13.2 Elastic Potential Energy

8. **V** A block of mass $m = 2.00$ kg is attached to a spring of force constant $k = 5.00 \times 10^2$ N/m that lies on a horizontal fric-tionless surface as shown in Figure P13.8. The block is pulled to a position $x_i = 5.00$ cm to the right of equilibrium and released from rest. Find (a) the work required to stretch the spring and (b) the speed the block has as it passes through equilibrium.

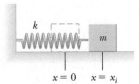

**Figure P13.8**

9. A slingshot consists of a light leather cup containing a stone. The cup is pulled back against two parallel rubber bands. It takes a force of 15.0 N to stretch either one of these bands 1.00 cm. (a) What is the potential energy stored in the two bands together when a 50.0-g stone is placed in the cup and pulled back 0.200 m from the equilibrium position? (b) With what speed does the stone leave the slingshot?

10. An archer pulls her bowstring back 0.400 m by exert-ing a force that increases uniformly from zero to 230 N. (a) What is the equivalent spring constant of the bow? (b) How much work is done in pulling the bow?

11. A student pushes the 1.50-kg block in Figure P13.11 against a horizontal spring, compressing it by 0.125 m. When released, the block travels across a horizontal surface and up an incline. Neglecting friction, find the block's maximum height if the spring constant is $k = 575$ N/m.

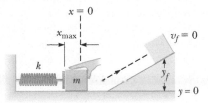

**Figure P13.11**

12. **V** An automobile having a mass of $1.00 \times 10^3$ kg is driven into a brick wall in a safety test. The bumper behaves like a spring with constant $5.00 \times 10^6$ N/m and is compressed 3.16 cm as the car is brought to rest. What was the speed of the car before impact, assuming no energy is lost in the collision with the wall?

13. A 10.0-g bullet is fired into, and embeds itself in, a 2.00-kg block attached to a spring with a force constant of 19.6 N/m and having negligible mass. How far is the spring compressed if the bullet has a speed of 300. m/s just before it strikes the block and the block slides on a frictionless surface? *Note:* You must use conservation of momentum in this problem because of the inelastic collision between the bullet and block.

14. **S** An object–spring system moving with simple harmonic motion has an amplitude $A$. (a) What is the total energy of the system in terms of $k$ and $A$ only? (b) Suppose at a certain instant the kinetic energy is twice the elastic potential energy. Write an equation describing this situation, using only the variables for the mass $m$, velocity $v$, spring constant $k$, and position $x$. (c) Using the results of parts (a) and (b) and the conservation of energy equation, find the positions $x$ of the object when its kinetic energy equals twice the potential energy stored in the spring. (The answer should in terms of $A$ only.)

15. **GP** A horizontal block–spring system with the block on a frictionless surface has total mechanical energy $E = 47.0$ J and a maximum displacement from equilibrium of 0.240 m. (a) What is the spring constant? (b) What is the kinetic energy of the system at the equilibrium point? (c) If the maximum speed of the block is 3.45 m/s, what is its mass? (d) What is the speed of the block when its displacement is 0.160 m? (e) Find the kinetic energy of the block at $x = 0.160$ m. (f) Find the potential energy stored in the spring when $x = 0.160$ m. (g) Suppose the same system is released from rest at $x = 0.240$ m on a rough surface so that it loses 14.0 J by the time it reaches its first turning point (after passing equilibrium at $x = 0$). What is its position at that instant?

16. A 0.250-kg block attached to a light spring oscillates on a frictionless, horizontal table. The oscillation amplitude is $A = 0.125$ m and the block moves at 3.00 m/s as it passes through equilibrium at $x = 0$. (a) Find the spring constant, $k$. (b) Calculate the total energy of the block–spring system. (c) Find the block's speed when $x = A/2$.

## 13.3 Concepts of Oscillation Rates in Simple Harmonic Motion

## 13.4 Position, Velocity, and Acceleration as Functions of Time

17. A block–spring system consists of a spring with constant $k = 425$ N/m attached to a 2.00-kg block on a frictionless surface. The block is pulled 8.00 cm from equilibrium and released from rest. For the resulting oscillation, find the (a) amplitude, (b) angular frequency, (c) frequency, and (d) period. What is the maximum value of the block's (e) velocity and (f) acceleration?

18. A 0.40-kg object connected to a light spring with a force constant of 19.6 N/m oscillates on a frictionless horizontal surface. If the spring is compressed 4.0 cm and released from rest, determine (a) the maximum speed of the object, (b) the speed of the object when the spring is compressed 1.5 cm, and (c) the speed of the object as it passes the point 1.5 cm from the equilibrium position. (d) For what value of $x$ does the speed equal one-half the maximum speed?

19. **T** At an outdoor market, a bunch of bananas attached to the bottom of a vertical spring of force constant 16.0 N/m is set into oscillatory motion with an amplitude of 20.0 cm. It is observed that the maximum speed of the bunch of bananas is 40.0 cm/s. What is the weight of the bananas in newtons?

20. A student stretches a spring, attaches a 1.00-kg mass to it, and releases the mass from rest on a frictionless surface. The resulting oscillation has a period of 0.500 s and an amplitude of 25.0 cm. Determine (a) the oscillation frequency, (b) the spring constant, and (c) the speed of the mass when it is halfway to the equilibrium position.

21. A horizontal spring attached to a wall has a force constant of $k = 8.50 \times 10^2$ N/m. A block of mass $m = 1.00$ kg is attached to the spring and rests on a frictionless, horizontal surface as in Figure P13.21. (a) The block is pulled to a position $x_i = 6.00$ cm from equilibrium and released. Find the potential energy stored in the spring when the block is 6.00 cm from equilibrium. (b) Find the speed of the block as it passes through the equilibrium position. (c) What is the speed of the block when it is at a position $x_i/2 = 3.00$ cm?

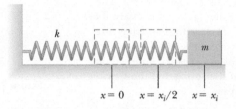

$x = 0$    $x = x_i/2$    $x = x_i$

**Figure P13.21**

22. An object moves uniformly around a circular path of radius 20.0 cm, making one complete revolution every 2.00 s. What are (a) the translational speed of the object, (b) the frequency of motion in hertz, and (c) the angular speed of the object?

23. The wheel in the simplified engine of Figure P13.23 has radius $A = 0.250$ m and rotates with angular frequency $\omega = 12.0$ rad/s. At $t = 0$, the piston is located at $x = A$. Calculate the piston's (a) position, (b) velocity, and (c) acceleration at $t = 1.15$ s.

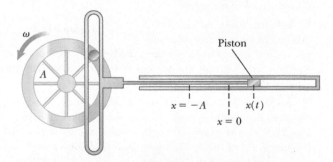

**Figure P13.23**

24. The period of motion of an object–spring system is $T = 0.528$ s when an object of mass $m = 238$ g is attached to the spring. Find (a) the frequency of motion in hertz and (b) the force constant of the spring. (c) If the total energy of the oscillating motion is 0.234 J, find the amplitude of the oscillations.

25. **T** A vertical spring stretches 3.9 cm when a 10.-g object is hung from it. The object is replaced with a block of mass 25 g that oscillates up and down in simple harmonic motion. Calculate the period of motion.

26. **V** When four people with a combined mass of 320 kg sit down in a $2.0 \times 10^3$-kg car, they find that their weight compresses the springs an additional 0.80 cm. (a) What is the effective force

constant of the springs? (b) The four people get out of the car and bounce it up and down. What is the frequency of the car's vibration?

27. The position of an object connected to a spring varies with time according to the expression $x = (5.2 \text{ cm}) \sin (8.0\pi t)$. Find (a) the period of this motion, (b) the frequency of the motion, (c) the amplitude of the motion, and (d) the first time after $t = 0$ that the object reaches the position $x = 2.6$ cm.

28. A harmonic oscillator is described by the function $x(t) = (0.200 \text{ m}) \cos (0.350t)$. Find the oscillator's (a) maximum velocity and (b) maximum acceleration. Find the oscillator's (c) position, (d) velocity, and (e) acceleration when $t = 2.00$ s.

29. **V** A 326-g object is attached to a spring and executes simple harmonic motion with a period of 0.250 s. If the total energy of the system is 5.83 J, find (a) the maximum speed of the object, (b) the force constant of the spring, and (c) the amplitude of the motion.

30. **S** An object executes simple harmonic motion with an amplitude $A$. (a) At what values of its position does its speed equal half its maximum speed? (b) At what values of its position does its potential energy equal half the total energy?

31. A 2.00-kg object on a frictionless horizontal track is attached to the end of a horizontal spring whose force constant is 5.00 N/m. The object is displaced 3.00 m to the right from its equilibrium position and then released, initiating simple harmonic motion. (a) What is the force (magnitude and direction) acting on the object 3.50 s after it is released? (b) How many times does the object oscillate in 3.50 s?

32. **GP** A spring of negligible mass stretches 3.00 cm from its relaxed length when a force of 7.50 N is applied. A 0.500-kg particle rests on a frictionless horizontal surface and is attached to the free end of the spring. The particle is displaced from the origin to $x = 5.00$ cm and released from rest at $t = 0$. (a) What is the force constant of the spring? (b) What are the angular frequency $\omega$, the frequency, and the period of the motion? (c) What is the total energy of the system? (d) What is the amplitude of the motion? (e) What are the maximum velocity and the maximum acceleration of the particle? (f) Determine the displacement $x$ of the particle from the equilibrium position at $t = 0.500$ s. (g) Determine the velocity and acceleration of the particle when $t = 0.500$ s.

33. Given that $x = A \cos (\omega t)$ is a sinusoidal function of time, show that $v$ (velocity) and $a$ (acceleration) are also sinusoidal functions of time. *Hint:* Use Equations 13.6 and 13.2.

## 13.5 Motion of a Pendulum

34. **V** A man enters a tall tower, needing to know its height. He notes that a long pendulum extends from the ceiling almost to the floor and that its period is 15.5 s. (a) How tall is the tower? (b) If this pendulum is taken to the Moon, where the free-fall acceleration is 1.67 m/s², what is the period there?

35. A simple pendulum has a length of 52.0 cm and makes 82.0 complete oscillations in 2.00 min. Find (a) the period of the pendulum and (b) the value of $g$ at the location of the pendulum.

36. A "seconds" pendulum is one that moves through its equilibrium position once each second. (The period of the pendulum is 2.000 s.) The length of a seconds pendulum is 0.992 7 m at Tokyo and 0.994 2 m at Cambridge, England. What is the ratio of the free-fall accelerations at these two locations?

37. A clock is constructed so that it keeps perfect time when its simple pendulum has a period of 1.000 s at locations where $g = 9.800$ m/s². The pendulum bob has length $L = 0.248\,2$ m, and instead of keeping perfect time, the clock runs slow by 1.500 minutes per day. (a) What is the free-fall acceleration at the clock's location? (b) What length of pendulum bob is required for the clock to keep perfect time?

38. A coat hanger of mass $m = 0.238$ kg oscillates on a peg as a physical pendulum as shown in Figure P13.38. The distance from the pivot to the center of mass of the coat hanger is $d = 18.0$ cm and the period of the motion is $T = 1.25$ s. Find the moment of inertia of the coat hanger about the pivot.

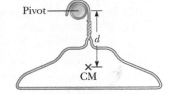

**Figure P13.38**

39. **T** The free-fall acceleration on Mars is 3.7 m/s². (a) What length of pendulum has a period of 1.0 s on Earth? (b) What length of pendulum would have a 1.0-s period on Mars? An object is suspended from a spring with force constant 10.0 N/m. Find the mass suspended from this spring that would result in a period of 1.0 s (c) on Earth and (d) on Mars.

40. A simple pendulum is 5.00 m long. (a) What is the period of simple harmonic motion for this pendulum if it is located in an elevator accelerating upward at 5.00 m/s²? (b) What is its period if the elevator is accelerating downward at 5.00 m/s²? (c) What is the period of simple harmonic motion for the pendulum if it is placed in a truck that is accelerating horizontally at 5.00 m/s²?

## 13.8 Frequency, Amplitude, and Wavelength

41. The sinusoidal wave shown in Figure P13.41 is traveling in the positive $x$-direction and has a frequency of 18.0 Hz. Find the (a) amplitude, (b) wavelength, (c) period, and (d) speed of the wave.

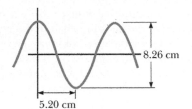

**Figure P13.41**

42. An object attached to a spring vibrates with simple harmonic motion as described by Figure P13.42. For this motion, find (a) the amplitude, (b) the period, (c) the angular frequency, (d) the maximum speed, (e) the maximum acceleration, and (f) an equation for its position $x$ in terms of a sine function.

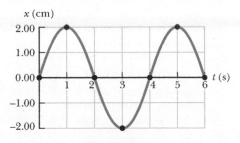

**Figure P13.42**

43. Light waves are electromagnetic waves that travel at $3.00 \times 10^8$ m/s. The eye is most sensitive to light having a wavelength of $5.50 \times 10^{-7}$ m. Find (a) the frequency of this light wave and (b) its period.

44. The distance between two successive minima of a transverse wave is 2.76 m. Five crests of the wave pass a given point along the direction of travel every 14.0 s. Find (a) the frequency of the wave and (b) the wave speed.

45. **V** A harmonic wave is traveling along a rope. It is observed that the oscillator that generates the wave completes 40.0 vibrations in 30.0 s. Also, a given maximum travels 425 cm along the rope in 10.0 s. What is the wavelength?

46. **BIO** A bat can detect small objects, such as an insect, whose size is approximately equal to one wavelength of the sound the bat makes. If bats emit a chirp at a frequency of $60.0 \times 10^3$ Hz and the speed of sound in air is 343 m/s, what is the smallest insect a bat can detect?

47. Orchestra instruments are commonly tuned to match an A-note played by the principal oboe. The Baltimore Symphony Orchestra tunes to an A-note at 440 Hz while the Boston Symphony Orchestra tunes to 442 Hz. If the speed of sound is constant at 343 m/s, find the magnitude of difference between the wavelengths of these two different A-notes.

48. Ocean waves are traveling to the east at 4.0 m/s with a distance of 20.0 m between crests. With what frequency do the waves hit the front of a boat (a) when the boat is at anchor and (b) when the boat is moving westward at 1.0 m/s?

## 13.9  The Speed of Waves on Strings

49. **V** An ethernet cable is 4.00 m long and has a mass of 0.200 kg. A transverse wave pulse is produced by plucking one end of the taut cable. The pulse makes four trips down and back along the cable in 0.800 s. What is the tension in the cable?

50. Workers attach a 25.0-kg mass to one end of a 20.0-m long cable and secure the other end to the top of a stationary crane, suspending the mass in midair. If the cable has a mass of 12.0 kg, determine the speed of transverse waves at (a) the middle and (b) the bottom end of the cable. (*Hint:* Don't neglect the cable's mass. Because of it, the tension increases from a minimum value at the bottom of the cable to a maximum value at the top.)

51. **V** A piano string of mass per unit length $5.00 \times 10^{-3}$ kg/m is under a tension of 1 350 N. Find the speed with which a wave travels on this string.

52. **QC** **S** A student taking a quiz finds on a reference sheet the two equations

$$f = \frac{1}{T} \quad \text{and} \quad v = \sqrt{\frac{T}{\mu}}$$

She has forgotten what $T$ represents in each equation. (a) Use dimensional analysis to determine the units required for $T$ in each equation. (b) Explain how you can identify the physical quantity each $T$ represents from the units.

53. Transverse waves with a speed of 50.0 m/s are to be produced on a stretched string. A 5.00-m length of string with a total mass of 0.060 0 kg is used. (a) What is the required tension in the string? (b) Calculate the wave speed in the string if the tension is 8.00 N.

54. **V** An astronaut on the Moon wishes to measure the local value of $g$ by timing pulses traveling down a wire that has a large object suspended from it. Assume a wire of mass 4.00 g is 1.60 m long and has a 3.00-kg object suspended from it. A pulse requires 36.1 ms to traverse the length of the wire. Calculate $g_{\text{Moon}}$ from these data. (You may neglect the mass of the wire when calculating the tension in it.)

55. A simple pendulum consists of a ball of mass 5.00 kg hanging from a uniform string of mass 0.060 0 kg and length $L$. If the period of oscillation of the pendulum is 2.00 s, determine the speed of a transverse wave in the string when the pendulum hangs vertically.

56. A string is 50.0 cm long and has a mass of 3.00 g. A wave travels at 5.00 m/s along this string. A second string has the same length, but half the mass of the first. If the two strings are under the same tension, what is the speed of a wave along the second string?

57. **T** Tension is maintained in a string as in Figure P13.57. The observed wave speed is $v = 24.0$ m/s when the suspended mass is $m = 3.00$ kg. (a) What is the mass per unit length of the string? (b) What is the wave speed when the suspended mass is $m = 2.00$ kg?

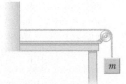

**Figure P13.57**

58. The elastic limit of a piece of steel wire is $2.70 \times 10^9$ Pa. What is the maximum speed at which transverse wave pulses can propagate along the wire without exceeding its elastic limit? (The density of steel is $7.86 \times 10^3$ kg/m$^3$.)

59. **QC** A 2.65-kg power line running between two towers has a length of 38.0 m and is under a tension of 12.5 N. (a) What is the speed of a transverse pulse set up on the line? (b) If the tension in the line was unknown, describe a procedure a worker on the ground might use to estimate the tension.

60. **S** A taut clothesline has length $L$ and a mass $M$. A transverse pulse is produced by plucking one end of the clothesline. If the pulse makes $n$ round trips along the clothesline in $t$ seconds, find expressions for (a) the speed of the pulse in terms of $n$, $L$, and $t$ and (b) the tension $F$ in the clothesline in terms of the same variables and mass $M$.

## 13.10  Interference of Waves

## 13.11  Reflection of Waves

61. A wave of amplitude 0.30 m interferes with a second wave of amplitude 0.20 m traveling in the same direction. What are (a) the largest and (b) the smallest resultant amplitudes that can occur, and under what conditions will these maxima and minima arise?

## Additional Problems

62. The position of a 0.30-kg object attached to a spring is described by

$$x = (0.25 \text{ m}) \cos (0.4\pi t)$$

Find (a) the amplitude of the motion, (b) the spring constant, (c) the position of the object at $t = 0.30$ s, and (d) the object's speed at $t = 0.30$ s.

63. An object of mass 2.00 kg is oscillating freely on a vertical spring with a period of 0.600 s. Another object of unknown mass on the same spring oscillates with a period of 1.05 s. Find (a) the spring constant $k$ and (b) the unknown mass.

64. **V** A certain tuning fork vibrates at a frequency of 196 Hz while each tip of its two prongs has an amplitude of 0.850 mm. (a) What is the period of this motion? (b) Find the wavelength of the sound produced by the vibrating fork, taking the speed of sound in air to be 343 m/s.

65. A simple pendulum has mass 1.20 kg and length 0.700 m. (a) What is the period of the pendulum near the surface of Earth? (b) If the same mass is attached to a spring, what spring constant would result in the period of motion found in part (a)?

66. A 0.500-kg block is released from rest and slides down a frictionless track that begins 2.00 m above the horizontal, as shown in Figure P13.66. At the bottom of the track, where the surface is horizontal, the block strikes and sticks to a light spring with a spring constant of 20.0 N/m. Find the maximum distance the spring is compressed.

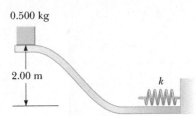

**Figure P13.66**

67. A 3.00-kg object is fastened to a light spring, with the intervening cord passing over a pulley (Fig. P13.67). The pulley is frictionless, and its inertia may be neglected. The object is released from rest when the spring is unstretched. If the object drops 10.0 cm before stopping, find (a) the spring constant of the spring and (b) the speed of the object when it is 5.00 cm below its starting point.

**Figure P13.67**

68. A 5.00-g bullet moving with an initial speed of 400. m/s is fired into and passes through a 1.00-kg block, as in Figure P13.68. The block, initially at rest on a

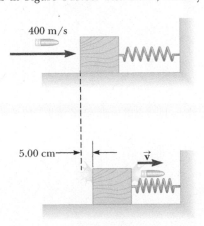

**Figure P13.68**

frictionless horizontal surface, is connected to a spring with a spring constant of 900. N/m. If the block moves 5.00 cm to the right after impact, find (a) the speed at which the bullet emerges from the block and (b) the mechanical energy lost in the collision.

69. **T** A large block $P$ executes horizontal simple harmonic motion as it slides across a frictionless surface with a frequency $f = 1.50$ Hz. Block $B$ rests on it, as shown in Figure P13.69, and the coefficient of static friction between the two is $\mu_s = 0.600$. What maximum amplitude of oscillation can the system have if block $B$ is not to slip?

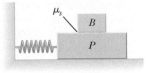

**Figure P13.69**

70. A spring in a toy gun has a spring constant of 9.80 N/m and can be compressed 20.0 cm beyond the equilibrium position. A 1.00-g pellet resting against the spring is propelled forward when the spring is released. (a) Find the muzzle speed of the pellet. (b) If the pellet is fired horizontally from a height of 1.00 m above the floor, what is its range?

71. **Q|C** A light balloon filled with helium of density 0.179 kg/m³ is tied to a light string of length $L = 3.00$ m. The string is tied to the ground, forming an "inverted" simple pendulum (Fig. P13.71a). If the balloon is displaced slightly from equilibrium, as in Figure P13.71b, (a) show that the motion is simple harmonic and (b) determine the period of the motion. Take the density of air to be 1.29 kg/m³. *Hint:* Use an analogy with the simple pendulum discussed in the text, and see Topic 9.

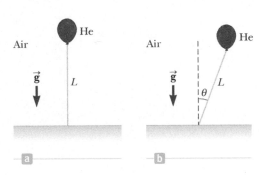

**Figure P13.71**

72. **S** An object of mass $m$ is connected to two rubber bands of length $L$, each under tension $F$, as in Figure P13.72. The object is displaced vertically by a small distance $y$. Assuming the tension does not change, show that (a) the restoring force is $-(2F/L)y$ and (b) the system exhibits simple harmonic motion with an angular frequency $\omega = \sqrt{2F/mL}$.

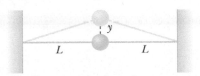

**Figure P13.72**

73. Assume a hole is drilled through the center of the Earth. It can be shown that an object of mass $m$ at a distance $r$ from the center of the Earth is pulled toward the center only by

the material in the shaded portion of Figure P13.73. Assume Earth has a uniform density $\rho$. Write down Newton's law of gravitation for an object at a distance $r$ from the center of the Earth and show that the force on it is of the form of Hooke's law, $F = -kr$, with an effective force constant of $k = (\frac{4}{3})\pi\rho Gm$, where $G$ is the gravitational constant.

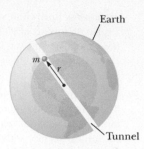

**Figure P13.73**

74. **BIO** Figure P13.74 shows a crude model of an insect wing. The mass $m$ represents the entire mass of the wing, which pivots about the fulcrum $F$. The spring represents the surrounding connective tissue. Motion of the wing corresponds to vibration of the spring. Suppose the mass of the wing is 0.30 g and the effective spring constant of the tissue is $4.7 \times 10^{-4}$ N/m. If the mass $m$ moves up and down a distance of 2.0 mm from its position of equilibrium, what is the maximum speed of the outer tip of the wing?

75. A 2.00-kg block hangs without vibrating at the end of a spring ($k = 500.$ N/m) that is attached to the ceiling of an elevator car. The car is rising with an upward acceleration of $g/3$ when the acceleration suddenly ceases (at $t = 0$). (a) What is the angular frequency of oscillation of the block after the acceleration ceases? (b) By what amount is the spring stretched during the time that the elevator car is accelerating?

76. **Q|C** **S** A system consists of a vertical spring with force constant $k = 1\,250$ N/m, length $L = 1.50$ m, and object of mass $m = 5.00$ kg attached to the end (Fig. P13.76). The object is placed at the level of the point of attachment with the spring unstretched, at position $y_i = L$, and then it is released so that it swings like a pendulum. (a) Write Newton's second law symbolically for the system as the object passes through its lowest point. (Note that at the lowest point, $r = L - y_f$.) (b) Write the conservation of energy equation symbolically, equating the total mechanical energies at the initial point and lowest point. (c) Find the coordinate position of the lowest point. (d) Will this pendulum's period be greater or less than the period of a simple pendulum with the same mass $m$ and length $L$? Explain.

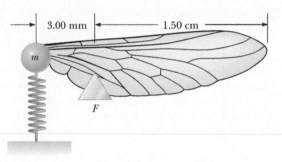

**Figure P13.74**

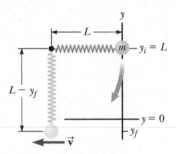

**Figure P13.76**

# Sound

SOUND WAVES ARE THE MOST IMPORTANT example of longitudinal waves. In this topic, we discuss the characteristics of sound waves: how they are produced, what they are, and how they travel through matter. We then investigate what happens when sound waves interfere with each other. The insights gained in this topic will help you understand how we hear.

## 14.1 Producing a Sound Wave

Whether it conveys the shrill whine of a jet engine or the soft melodies of a crooner, any sound wave has its source in a vibrating object. Musical instruments produce sounds in a variety of ways. The sound of a clarinet is produced by a vibrating reed, the sound of a drum by the vibration of the taut drumhead, the sound of a piano by vibrating strings, and the sound from a singer by vibrating vocal cords.

Sound waves are longitudinal waves traveling through a medium, such as air. In order to investigate how sound waves are produced, we focus our attention on the tuning fork, a common device for producing pure musical notes. A tuning fork consists of two metal prongs, or tines, that vibrate when struck. Their vibration disturbs the air near them, as shown in Figure 14.1. (The amplitude of vibration of the tine shown in the figure has been greatly exaggerated for clarity.) When a tine swings to the right, as in Figure 14.1a, the molecules in an element of air in front of its movement are forced closer together than normal. Such a region of high molecular density and high air pressure is called a **compression.** This compression moves away from the fork like a ripple on a pond. When the tine swings to the left, as in Figure 14.1b, the molecules in an element of air to the right of the tine spread apart, and the density and air pressure in this region are then lower than normal. Such a region of reduced density is called a **rarefaction** (pronounced "rare a fak' shun"). Molecules to the right of the rarefaction in the figure move to the left. The rarefaction itself therefore moves to the right, following the previously produced compression.

As the tuning fork continues to vibrate, a succession of compressions and rarefactions forms and spreads out from it. The resultant pattern in the air is somewhat like that pictured in Figure 14.2a. We can use a sinusoidal curve to represent a sound

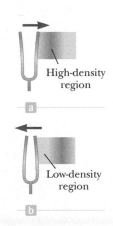

**Figure 14.1** A vibrating tuning fork. (a) As the right tine of the fork moves to the right, a high-density region (compression) of air is formed in front of its movement. (b) As the right tine moves to the left, a low-density region (rarefaction) of air is formed behind it.

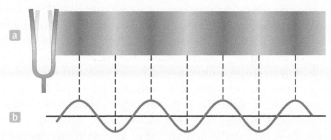

**Figure 14.2** (a) As the tuning fork vibrates, a series of compressions and rarefactions moves outward, away from the fork. (b) The crests of the wave correspond to compressions, the troughs to rarefactions.

457

wave, as in Figure 14.2b. Notice that there are crests in the sinusoidal wave at the points where the sound wave has compressions and troughs where the sound wave has rarefactions. The compressions and rarefactions of the sound waves are superposed on the random thermal motion of the atoms and molecules of the air (Eq. 10.18, page 341), so sound waves in gases travel at about the molecular rms speed.

# 14.2 Characteristics of Sound Waves

As already noted, the general motion of elements of air near a vibrating object is back and forth between regions of compression and rarefaction. This back-and-forth motion of elements of the medium in the direction of the disturbance is characteristic of a longitudinal wave. **The motion of the elements of the medium in a longitudinal sound wave is back and forth along the direction in which the wave travels.** By contrast, **in a transverse wave, the vibrations of the elements of the medium are at right angles to the direction of travel of the wave.**

## 14.2.1 Categories of Sound Waves

Sound waves fall into three categories covering different ranges of frequencies. **Audible waves** are longitudinal waves that lie within the range of sensitivity of the human ear, approximately 20 to 20 000 Hz. **Infrasonic waves** are longitudinal waves with frequencies below the audible range. Earthquake waves are an example. **Ultrasonic waves** are longitudinal waves with frequencies above the audible range for humans and are produced by certain types of whistles. Animals such as dogs can hear the waves emitted by these whistles.

## 14.2.2 Applications of Ultrasound

Ultrasonic waves are sound waves with frequencies greater than 20 kHz. Because of their high frequency and corresponding short wavelengths, ultrasonic waves can be used to produce images of small objects and are currently in wide use in medical applications, both as a diagnostic tool and in certain treatments. Internal organs can be examined via the images produced by the reflection and absorption of ultrasonic waves. Although ultrasonic waves are far safer than x-rays, their images don't always have as much detail. Certain organs, however, such as the liver and the spleen, are invisible to x-rays but can be imaged with ultrasonic waves.

Medical workers can measure the speed of the blood flow in the body with a device called an ultrasonic flow meter, which makes use of the Doppler effect (discussed in Section 14.6). The flow speed is found by comparing the frequency of the waves scattered by the flowing blood with the incident frequency.

Figure 14.3 illustrates the technique that produces ultrasonic waves for clinical use. Electrical contacts are made to the opposite faces of a crystal, such as quartz or strontium titanate. If an alternating voltage of high frequency is applied to these contacts, the crystal vibrates at the same frequency as the applied voltage, emitting a beam of ultrasonic waves. At one time, a technique like this was used to produce sound in nearly all headphones. This method of transforming electrical energy into mechanical energy, called the **piezoelectric effect**, is reversible: If some external source causes the crystal to vibrate, an alternating voltage is produced across it. A single crystal can therefore be used to both generate and receive ultrasonic waves.

The primary physical principle that makes ultrasound imaging possible is the fact that a sound wave is partially reflected whenever it is incident on a boundary between two materials having different densities. If a sound wave is traveling in a material of density $\rho_i$ and strikes a material of density $\rho_t$, the percentage of the incident sound wave intensity reflected, $PR$, is given by

$$PR = \left(\frac{\rho_i - \rho_t}{\rho_i + \rho_t}\right)^2 \times 100$$

**BIO APPLICATION**
Medical Uses of Ultrasound

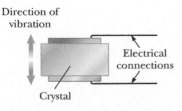

Direction of
vibration

Electrical
connections

Crystal

**Figure 14.3** An alternating voltage applied to the faces of a piezoelectric crystal causes the crystal to vibrate.

This equation assumes that the direction of the incident sound wave is perpendicular to the boundary and that the speed of sound is approximately the same in the two materials. The latter assumption holds very well for the human body because the speed of sound doesn't vary much in the organs of the body.

Physicians commonly use ultrasonic waves to observe fetuses (Fig. 14.4). This technique presents far less risk than do x-rays, which deposit more energy in cells and can produce birth defects. First, the abdomen of the mother is coated with a liquid, such as mineral oil. If that were not done, most of the incident ultrasonic waves from the piezoelectric source would be reflected at the boundary between the air and the mother's skin. Mineral oil has a density similar to that of skin, and a very small fraction of the incident ultrasonic wave is reflected when $\rho_i \approx \rho_t$. The ultrasound energy is emitted in pulses rather than as a continuous wave, so the same crystal can be used as a detector as well as a transmitter. An image of the fetus is obtained by using an array of transducers placed on the abdomen. The reflected sound waves picked up by the transducers are converted to an electric signal, which is used to form an image on a fluorescent screen. Difficulties such as the likelihood of spontaneous abortion or of breech birth are easily detected with this technique. Fetal abnormalities such as spina bifida and water on the brain are also readily observed.

An important medical application of ultrasonics is the *cavitron ultrasonic surgical aspirator* (CUSA). This device has made it possible to surgically remove brain tumors that were previously inoperable. The probe of the CUSA emits ultrasonic waves (at about 23 kHz) at its tip. When the tip touches a tumor, the part of the tumor near the probe is shattered and the residue can be sucked up (aspirated) through the hollow probe. Using this technique, neurosurgeons are able to remove brain tumors without causing serious damage to healthy surrounding tissue.

Ultrasound has been used not only for imaging purposes but also in surgery to destroy uterine fibroids and tumors of the prostate gland. An ultrasound device developed in 2009 allows neurosurgeons to perform brain surgery without opening the skull or cutting the skin. High-intensity focused ultrasound (HIFU) is created with an array of a thousand ultrasound transducers placed on the patient's skull. Each transducer can be individually focused on a selected region of the brain. The ultrasound heats the brain tissue in a small area and destroys it. Patients are conscious during the procedure and report momentary tingling or dizziness, sometimes a mild headache. A cooling system is required to keep the patient's skull from overheating. The device can eliminate tumors and malfunctioning neural tissue, and may have application in the treatment of Parkinson's disease and strokes. It may also be possible to use HIFU to target the delivery of therapeutic drugs in specific brain locations.

Ultrasound is also used to break up kidney stones that are otherwise too large to pass. Previously, invasive surgery was often required.

Another interesting application of ultrasound is the ultrasonic ranging unit used in some cameras to provide an almost instantaneous measurement of the distance between the camera and the object to be photographed. The principal component of this device is a crystal that acts as both a loudspeaker and a microphone. A pulse of ultrasonic waves is transmitted from the transducer to the object, which then reflects part of the signal, producing an echo that is detected by the device. The time interval between the outgoing pulse and the detected echo is electronically converted to a distance, because the speed of sound is a known quantity.

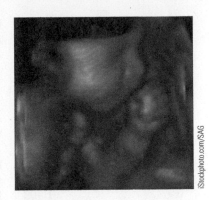

**Figure 14.4** An ultrasound image of a human fetus in the womb.

iStockphoto.com/SAG

**BIO** APPLICATION
Cavitron Ultrasonic Surgical Aspirator

**BIO** APPLICATION
High-intensity Focused Ultrasound (HIFU)

APPLICATION
Ultrasonic Ranging Unit for Cameras

# 14.3   The Speed of Sound

The speed of a sound wave in a fluid depends on the fluid's compressibility and inertia. If the fluid has a bulk modulus $B$ and an equilibrium density $\rho$, the speed of sound in it is

$$v = \sqrt{\frac{B}{\rho}}$$

[14.1]   ◀ Speed of sound in a fluid

**Table 14.1** Speeds of Sound in Various Media

| Medium | $v$ (m/s) |
|---|---|
| **Gases** | |
| Air (0°C) | 331 |
| Air (100°C) | 386 |
| Hydrogen (0°C) | 1 286 |
| Oxygen (0°C) | 317 |
| Helium (0°C) | 972 |
| **Liquids at 25°C** | |
| Water | 1 493 |
| Methyl alcohol | 1 143 |
| Sea water | 1 533 |
| **Solids**[a] | |
| Aluminum | 6 420 |
| Copper (rolled) | 5 010 |
| Steel | 5 950 |
| Lead (rolled) | 1 960 |
| Synthetic rubber | 1 600 |

[a]Values given are for propagation of longitudinal waves in bulk media. Speeds for longitudinal waves in thin rods are smaller, and speeds of transverse waves in bulk are smaller yet.

Equation 14.1 also holds true for a gas. Recall from Topic 9 that the bulk modulus is defined as the ratio of the change in pressure, $\Delta P$, to the resulting fractional change in volume, $\Delta V/V$:

$$B \equiv -\frac{\Delta P}{\Delta V/V} \qquad [14.2]$$

$B$ is always positive because an increase in pressure (positive $\Delta P$) results in a decrease in volume. Hence, the ratio $\Delta P/\Delta V$ is always negative.

It's interesting to compare Equation 14.1 with Equation 13.18 for the speed of transverse waves on a string, $v = \sqrt{F/\mu}$, discussed in Topic 13. In both cases, the wave speed depends on an elastic property of the medium ($B$ or $F$) and on an inertial property of the medium ($\rho$ or $\mu$). In fact, the speed of all mechanical waves follows an expression of the general form

$$v = \sqrt{\frac{\text{elastic property}}{\text{inertial property}}}$$

Another example of this general form is the **speed of a longitudinal wave in a solid rod**, which is

$$v = \sqrt{\frac{Y}{\rho}} \qquad [14.3]$$

where $Y$ is the Young's modulus of the solid (see Eq. 9.30) and $\rho$ is its density. This expression is valid only for a thin, solid rod.

Table 14.1 lists the speeds of sound in various media. The speed of sound is much higher in solids than in gases because the molecules in a solid interact more strongly with each other than do molecules in a gas. Striking a long steel rail with a hammer, for example, produces two sound waves, one moving through the rail and a slower wave moving through the air. A person with an ear pressed against the rail first hears the faster sound moving through the rail, then the sound moving through air. In general, sound travels faster through solids than liquids and faster through liquids than gases, although there are exceptions.

The speed of sound also depends on the temperature of the medium. For sound traveling through air, the relationship between the speed of sound and temperature is

$$v = (331 \text{ m/s}) \sqrt{\frac{T}{273 \text{ K}}} \qquad [14.4]$$

where 331 m/s is the speed of sound in air at 0°C and $T$ is the absolute (Kelvin) temperature. Using this equation, the speed of sound in air at 293 K (a typical room temperature) is approximately 343 m/s.

> **Quick Quiz**
>
> **14.1** Which of the following actions will increase the speed of sound in air?
> (a) decreasing the air temperature (b) increasing the frequency of the sound
> (c) increasing the air temperature (d) increasing the amplitude of the sound wave
> (e) reducing the pressure of the air

---

**APPLYING PHYSICS 14.1** **THE SOUNDS HEARD DURING A STORM**

How does lightning produce thunder, and what causes the extended rumble?

**EXPLANATION** Assume you're at ground level, and neglect ground reflections. When lightning strikes, a channel of ionized air carries a large electric current from a cloud to the ground. This results in a rapid temperature increase of the air in the channel as the current moves through it, causing a similarly rapid expansion of the air. The expansion is so sudden and so intense that a tremendous disturbance—thunder—is produced in the air. The entire length of the channel produces the sound at essentially the same instant of time. Sound produced at the bottom of the channel reaches you first because that's the point closest to you. Sounds from progressively higher portions of the channel reach you at later times, resulting in an extended roar. If the lightning channel were a perfectly straight line, the roar might be steady, but the zigzag shape of the path results in the rumbling variation in loudness, with different quantities of sound energy from different segments arriving at any given instant. ■

## EXAMPLE 14.1 | EXPLOSION OVER AN ICE SHEET

**GOAL** Calculate time of travel for sound through various media.

**PROBLEM** An explosion occurs 275 m above an 867-m-thick ice sheet that lies over ocean water. If the air temperature is $-7.00°C$, how long does it take the sound to reach a research vessel 1 250 m below the ice? Neglect any changes in the bulk modulus and density with temperature and depth. (Use $B_{ice} = 9.2 \times 10^9$ Pa.)

**STRATEGY** Calculate the speed of sound in air with Equation 14.4, and use $d = vt$ to find the time needed for the sound to reach the surface of the ice. Use Equation 14.1 to compute the speed of sound in ice, again finding the time with the distance equation. Finally, use the speed of sound in salt water to find the time needed to traverse the water and then sum the three times.

### SOLUTION

Calculate the speed of sound in air at $-7.00°C$, which is equivalent to 266 K:

$$v_{air} = (331 \text{ m/s})\sqrt{\frac{T}{273 \text{ K}}} = (331 \text{ m/s})\sqrt{\frac{266 \text{ K}}{273 \text{ K}}} = 327 \text{ m/s}$$

Calculate the travel time through the air:

$$t_{air} = \frac{d}{v_{air}} = \frac{275 \text{ m}}{327 \text{ m/s}} = 0.841 \text{ s}$$

Compute the speed of sound in ice, using the bulk modulus and density of ice:

$$v_{ice} = \sqrt{\frac{B}{\rho}} = \sqrt{\frac{9.2 \times 10^9 \text{ Pa}}{917 \text{ kg/m}^3}} = 3.2 \times 10^3 \text{ m/s}$$

Compute the travel time through the ice:

$$t_{ice} = \frac{d}{v_{ice}} = \frac{867 \text{ m}}{3 \, 200 \text{ m/s}} = 0.27 \text{ s}$$

Compute the travel time through the ocean water:

$$t_{water} = \frac{d}{v_{water}} = \frac{1 \, 250 \text{ m}}{1 \, 533 \text{ m/s}} = 0.815 \text{ s}$$

Sum the three times to obtain the total time of propagation:

$$t_{tot} = t_{air} + t_{ice} + t_{water} = 0.841 \text{ s} + 0.27 \text{ s} + 0.815 \text{ s}$$

$$= \boxed{1.93 \text{ s}}$$

**REMARKS** Notice that the speed of sound is highest in solid ice, second highest in liquid water, and slowest in air. The speed of sound depends on temperature, so the answer would have to be modified if the actual temperatures of ice and the sea water were known. At 0°C, for example, the speed of sound in sea water falls to 1 449 m/s.

**QUESTION 14.1** Is the speed of sound in rubber higher or lower than the speed of sound in aluminum? Explain.

**EXERCISE 14.1** Compute the speed of sound in the following substances at 273 K: **(a)** a thin lead rod ($Y = 1.6 \times 10^{10}$ Pa), **(b)** mercury ($B = 2.8 \times 10^{10}$ Pa), and **(c)** air at $-15.0°C$.

**ANSWERS** **(a)** $1.2 \times 10^3$ m/s   **(b)** $1.4 \times 10^3$ m/s   **(c)** 322 m/s

# 14.4 Energy and Intensity of Sound Waves

As the tines of a tuning fork move back and forth through the air, they exert a force on a layer of air and cause it to move. In other words, the tines do work on the layer of air. That the fork pours sound energy into the air is one reason the vibration of the fork slowly dies out. (Other factors, such as the energy lost to friction as the tines bend, are also responsible for the lessening of movement.)

The average **intensity** $I$ of a wave on a given surface is defined as the rate at which energy flows through the surface, $\Delta E/\Delta t$, divided by the surface area $A$:

$$I \equiv \frac{1}{A}\frac{\Delta E}{\Delta t} \qquad [14.5]$$

where the direction of energy flow is perpendicular to the surface at every point.

**SI unit: watt per meter squared (W/m$^2$)**

A rate of energy transfer is power, so Equation 14.5 can be written in the alternate form

◀ Intensity of a wave

$$I \equiv \frac{\text{power}}{\text{area}} = \frac{P}{A} \qquad [14.6]$$

where $P$ is the sound power passing through the surface, measured in watts, and the intensity again has units of watts per square meter.

The faintest sounds the human ear can detect at a frequency of 1 000 Hz have an intensity of about $1 \times 10^{-12}$ W/m$^2$. This intensity is called the **threshold of hearing.** The loudest sounds the ear can tolerate have an intensity of about 1 W/m$^2$ (the **threshold of pain**). At the threshold of hearing, the increase in pressure in the ear is approximately $3 \times 10^{-5}$ Pa over normal atmospheric pressure. Because atmospheric pressure is about $1 \times 10^5$ Pa, this means the ear can detect pressure fluctuations as small as about 3 parts in $10^{10}$! The maximum displacement of an air molecule at the threshold of hearing is about $1 \times 10^{-11}$ m, a remarkably small number! If we compare this displacement with the diameter of a molecule (about $10^{-10}$ m), we see that the ear is an extremely sensitive detector of sound waves.

The loudest sounds the human ear can tolerate at 1 kHz correspond to a pressure variation of about 29 Pa away from normal atmospheric pressure, with a maximum displacement of air molecules of $1 \times 10^{-5}$ m.

## 14.4.1 Intensity Level in Decibels

The loudest tolerable sounds have intensities about $1.0 \times 10^{12}$ times greater than the faintest detectable sounds. The most intense sound, however, isn't perceived as being $1.0 \times 10^{12}$ times louder than the faintest sound because the sensation of loudness is approximately logarithmic in the human ear. (For a review of logarithms, see Section A.3, heading G, in Appendix A.) The relative intensity of a sound is called the **intensity level** or **decibel level,** defined by

◀ Intensity level

$$\beta \equiv 10 \log \left(\frac{I}{I_0}\right) \qquad [14.7]$$

**Tip 14.1** Intensity Versus Intensity Level

Don't confuse intensity with intensity level. Intensity is a physical quantity with units of watts per meter squared; intensity level, or decibel level, is a convenient mathematical transformation of intensity to a logarithmic scale.

The constant $I_0 = 1.0 \times 10^{-12}$ W/m$^2$ is the reference intensity, the sound intensity at the threshold of hearing, $I$ is the intensity, and $\beta$ is the corresponding intensity level measured in decibels (dB). (The word *decibel*, which is one-tenth of a *bel*, comes from the name of the inventor of the telephone, Alexander Graham Bell [1847–1922].)

To get a feel for various decibel levels, we can substitute a few representative numbers into Equation 14.7, starting with $I = 1.0 \times 10^{-12}$ W/m$^2$:

$$\beta = 10 \log \left(\frac{1.0 \times 10^{-12}\ \text{W/m}^2}{1.0 \times 10^{-12}\ \text{W/m}^2}\right) = 10 \log (1) = 0\ \text{dB}$$

From this result, we see that the lower threshold of human hearing has been chosen to be zero on the decibel scale. Progressing upward by powers of ten yields

$$\beta = 10 \log \left( \frac{1.0 \times 10^{-11} \text{ W/m}^2}{1.0 \times 10^{-12} \text{ W/m}^2} \right) = 10 \log (10) = 10 \text{ dB}$$

$$\beta = 10 \log \left( \frac{1.0 \times 10^{-10} \text{ W/m}^2}{1.0 \times 10^{-12} \text{ W/m}^2} \right) = 10 \log (100) = 20 \text{ dB}$$

Notice the pattern: *Multiplying* a given intensity by ten *adds* 10 db to the intensity level. This pattern holds throughout the decibel scale. For example, a 50-dB sound is 10 times as intense as a 40-dB sound, whereas a 60-dB sound is 100 times as intense as a 40-dB sound.

On this scale, the threshold of pain ($I = 1.0$ W/m$^2$) corresponds to an intensity level of $\beta = 10 \log (1/1 \times 10^{-12}) = 10 \log (10^{12}) = 120$ dB. Nearby jet airplanes can create intensity levels of 150 dB, and subways and riveting machines have levels of 90–100 dB. The electronically amplified sound heard at rock concerts can attain levels of up to 120 dB, the threshold of pain. Exposure to such high intensity levels can seriously damage the ear. Earplugs are recommended whenever prolonged intensity levels exceed 90 dB. Recent evidence suggests that noise pollution, which is common in most large cities and in some industrial environments, may be a contributing factor to high blood pressure, anxiety, and nervousness. Table 14.2 gives the approximate intensity levels of various sounds.

**Table 14.2** Intensity Levels in Decibels for Different Sources

| Source of Sound | $\beta$ (dB) |
|---|---|
| Nearby jet airplane | 150 |
| Jackhammer, machine gun | 130 |
| Siren, rock concert | 120 |
| Subway, power mower | 100 |
| Busy traffic | 80 |
| Vacuum cleaner | 70 |
| Normal conversation | 50 |
| Mosquito buzzing | 40 |
| Whisper | 30 |
| Rustling leaves | 10 |
| Threshold of hearing | 0 |

## EXAMPLE 14.2    A NOISY GRINDING MACHINE

**GOAL** Working with watts and decibels.

**PROBLEM** A noisy grinding machine in a factory produces a sound intensity of $1.00 \times 10^{-5}$ W/m$^2$ at a certain location. Calculate (a) the decibel level of this machine at that point and (b) the new intensity level when a second, identical machine is added to the factory. (c) A certain number of additional such machines are put into operation alongside these two machines. When all the machines are running at the same time the decibel level is 77.0 dB. Find the sound intensity.

(Assume, in each part, that the sound intensity is measured at the same point, equidistant from all the machines.)

**STRATEGY** Parts (a) and (b) require substituting into the decibel formula, Equation 14.7, with the intensity in part (b) twice the intensity in part (a). In part (c), the intensity level in decibels is given, and it's necessary to work backwards, using the inverse of the logarithm function, to get the intensity in watts per meter squared.

. . . . . . . . . . . . . . . . . . . . . . . . . . . . . . . . . . . . . . . . . . . . . . . . . . . . . . . . . . . . . . . . . . . . .

**SOLUTION**

(a) Calculate the intensity level of the single grinder.

Substitute the intensity into the decibel formula:

$$\beta = 10 \log \left( \frac{1.00 \times 10^{-5} \text{ W/m}^2}{1.00 \times 10^{-12} \text{ W/m}^2} \right) = 10 \log (10^7)$$

$$= \boxed{70.0 \text{ dB}}$$

(b) Calculate the new intensity level when an additional machine is added.

Substitute twice the intensity of part (a) into the decibel formula:

$$\beta = 10 \log \left( \frac{2.00 \times 10^{-5} \text{ W/m}^2}{1.00 \times 10^{-12} \text{ W/m}^2} \right) = \boxed{73.0 \text{ dB}}$$

(c) Find the intensity corresponding to an intensity level of 77.0 dB.

Substitute 77.0 dB into the decibel formula and divide both sides by 10:

$$\beta = 77.0 \text{ dB} = 10 \log \left( \frac{I}{I_0} \right)$$

$$7.70 = \log \left( \frac{I}{10^{-12} \text{ W/m}^2} \right)$$

Make each side the exponent of 10. On the right-hand side, $10^{\log u} = u$, by definition of base 10 logarithms.

$$10^{7.70} = 5.01 \times 10^7 = \frac{I}{1.00 \times 10^{-12} \text{ W/m}^2}$$

$$I = \boxed{5.01 \times 10^{-5} \text{ W/m}^2}$$

*(Continued)*

**REMARKS** The answer is five times the intensity of the single grinder, so in part (c) there are five such machines operating simultaneously. Because of the logarithmic definition of intensity level, large changes in intensity correspond to small changes in intensity level.

**QUESTION 14.2** By how many decibels is the sound intensity level raised when the sound intensity is doubled?

**EXERCISE 14.2** Suppose a manufacturing plant has an average sound intensity level of 97.0 dB created by 25 identical machines. (a) Find the total intensity created by all the machines. (b) Find the sound intensity created by one such machine. (c) What's the sound intensity level if five such machines are running?

**ANSWERS** (a) $5.01 \times 10^{-3}\,\mathrm{W/m^2}$   (b) $2.00 \times 10^{-4}\,\mathrm{W/m^2}$   (c) 90.0 dB

---

**BIO APPLICATION**
OSHA Noise-Level Regulations

Federal OSHA regulations now demand that no office or factory worker be exposed to noise levels that average more than 85 dB over an 8-h day. From a management point of view, here's the good news: one machine in the factory may produce a noise level of 70 dB, but a second machine, though doubling the total intensity, increases the noise level by only 3 dB. Because of the logarithmic nature of intensity levels, doubling the intensity doesn't double the intensity level; in fact, it alters it by a surprisingly small amount. This means that equipment can be added to the factory without appreciably altering the intensity level of the environment.

Now here's the bad news: as you remove noisy machinery, the intensity level isn't lowered appreciably. In Exercise 14.2, reducing the intensity level by 7 dB would require the removal of 20 of the 25 machines! To lower the level another 7 dB would require removing 80% of the remaining machines, in which case only one machine would remain.

# 14.5 Spherical and Plane Waves

If a small spherical object oscillates so that its radius changes periodically with time, a spherical sound wave is produced (Fig. 14.5). The wave moves outward from the source at a constant speed.

Because all points on the vibrating sphere behave in the same way, we conclude that the energy in a spherical wave propagates equally in all directions. This means that no one direction is preferred over any other. If $P_{av}$ is the average power emitted by the source, then at any distance $r$ from the source, this power must be distributed over a spherical surface of area $4\pi r^2$, assuming no absorption in the medium. (Recall that $4\pi r^2$ is the surface area of a sphere.) Hence, the **intensity** of the sound at a distance $r$ from the source is

$$I = \frac{\text{average power}}{\text{area}} = \frac{P_{av}}{A} = \frac{P_{av}}{4\pi r^2} \qquad [14.8]$$

This equation shows that the intensity of a wave decreases with increasing distance from its source, as you might expect. The fact that $I$ varies as $1/r^2$ is a result of the assumption that the small source (sometimes called a **point source**) emits a spherical wave. (In fact, light waves also obey this so-called inverse-square relationship.) Because the average power is the same through any spherical surface centered at the source, we see that the intensities at distances $r_1$ and $r_2$ (Fig. 14.5) from the center of the source are

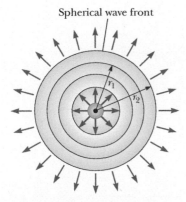

Spherical wave front

**Figure 14.5** A spherical wave propagating radially outward from an oscillating sphere. The intensity of the wave varies as $1/r^2$.

$$I_1 = \frac{P_{av}}{4\pi r_1^2} \quad \text{and} \quad I_2 = \frac{P_{av}}{4\pi r_2^2}$$

The ratio of the intensities at these two spherical surfaces is

$$\frac{I_1}{I_2} = \frac{r_2{}^2}{r_1{}^2} \qquad [14.9]$$

It's useful to represent spherical waves graphically with a series of circular arcs (lines of maximum intensity) concentric with the source representing part of a spherical surface, as in Figure 14.6. We call such an arc a **wave front.** The distance between adjacent wave fronts equals the wavelength λ. The radial lines pointing outward from the source and perpendicular to the arcs are called **rays.**

Now consider a small portion of a wave front that is at a *great* distance (relative to λ) from the source, as in Figure 14.7. In this case, the rays are nearly parallel to each other and the wave fronts are very close to being planes. At distances from the source that are great relative to the wavelength, therefore, we can approximate the wave front with parallel planes, called **plane waves.** Any small portion of a spherical wave that is far from the source can be considered a plane wave. Figure 14.8 illustrates a plane wave propagating along the x-axis. If the positive x-direction is taken to be the direction of the wave motion (or ray) in this figure, then the wave fronts are parallel to the plane containing the y- and z-axes.

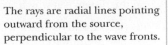

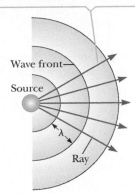

The rays are radial lines pointing outward from the source, perpendicular to the wave fronts.

**Figure 14.6** Spherical waves emitted by a point source. The circular arcs represent the spherical wave fronts concentric with the source.

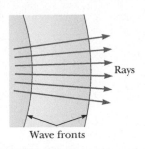

**Figure 14.7** Far away from a point source, the wave fronts are nearly parallel planes and the rays are nearly parallel lines perpendicular to the planes. A small segment of a spherical wave front is approximately a plane wave.

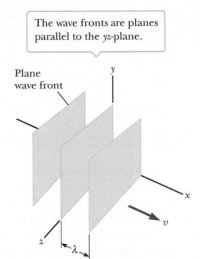

The wave fronts are planes parallel to the yz-plane.

**Figure 14.8** A representation of a plane wave moving in the positive x-direction with a speed v.

---

**EXAMPLE 14.3    INTENSITY VARIATIONS OF A POINT SOURCE**

**GOAL**  Relate sound intensities and their distances from a point source.

**PROBLEM**  A small source emits sound waves with a power output of 80.0 W. **(a)** Find the intensity 3.00 m from the source. **(b)** At what distance would the intensity be one-fourth as much as it is at r = 3.00 m? **(c)** Find the distance at which the sound intensity level is 40.0 dB.

**STRATEGY**  The source is small, so the emitted waves are spherical and the intensity in part **(a)** can be found by substituting values into Equation 14.8. Part **(b)** involves solving for r in Equation 14.8 followed by substitution (although Eq. 14.9 can be used instead). In part **(c)**, convert from the sound intensity level to the intensity in W/m², using Equation 14.7. Then substitute into Equation 14.9 (although Eq. 14.8 could be used instead) and solve for $r_2$.

................................................................

**SOLUTION**

**(a)** Find the intensity 3.00 m from the source.

Substitute $P_{av}$ = 80.0 W and r = 3.00 m into Equation 14.8:

$$I = \frac{P_{av}}{4\pi r^2} = \frac{80.0\ \text{W}}{4\pi(3.00\ \text{m})^2} = \boxed{0.707\ \text{W/m}^2}$$

**(b)** At what distance would the intensity be one-fourth as much as it is at r = 3.00 m?

Take I = (0.707 W/m²)/4, and solve for r in Equation 14.8:

$$r = \left(\frac{P_{av}}{4\pi I}\right)^{1/2} = \left[\frac{80.0\ \text{W}}{4\pi(0.707\ \text{W/m}^2)/4.0}\right]^{1/2} = \boxed{6.00\ \text{m}}$$

*(Continued)*

(c) Find the distance at which the intensity level is 40.0 dB.

Convert the intensity level of 40.0 dB to an intensity in W/m$^2$ by solving Equation 14.7 for $I$:

$$40.0 = 10 \log\left(\frac{I}{I_0}\right) \quad \rightarrow \quad 4.00 = \log\left(\frac{I}{I_0}\right)$$

$$10^{4.00} = \frac{I}{I_0} \quad \rightarrow \quad I = 10^{4.00} I_0 = 1.00 \times 10^{-8} \text{ W/m}^2$$

Solve Equation 14.9 for $r_2{}^2$, substitute the intensity and the result of part (a), and take the square root:

$$\frac{I_1}{I_2} = \frac{r_2{}^2}{r_1{}^2} \quad \rightarrow \quad r_2{}^2 = r_1{}^2 \frac{I_1}{I_2}$$

$$r_2{}^2 = (3.00 \text{ m})^2 \left(\frac{0.707 \text{ W/m}^2}{1.00 \times 10^{-8} \text{ W/m}^2}\right)$$

$$r_2 = \boxed{2.52 \times 10^4 \text{ m}}$$

- - - - - - - - - - - - - - - - - - - - - - - - - - - - - - - - - - - - - - - - - - - - - - - - - - - - - - - - - - - - - - - - - - - - - - - - - - - - -

**REMARKS** Once the intensity is known at one position a certain distance away from the source, it's easier to use Equation 14.9 rather than Equation 14.8 to find the intensity at any other location. This is particularly true for part (b), where, using Equation 14.9, we can see right away that doubling the distance reduces the intensity to one- fourth its previous value.

**QUESTION 14.3** The power output of a sound system is increased by a factor of 25. By what factor should you adjust your distance from the speakers so the sound intensity is the same?

**EXERCISE 14.3** Suppose a certain jet plane creates an intensity level of 125 dB at a distance of 5.00 m. What intensity level does it create on the ground directly underneath it when flying at an altitude of 2.00 km?

**ANSWER** 73.0 dB

# 14.6 The Doppler Effect

If a car or truck is moving while its horn is blowing, the frequency of the sound you hear is higher as the vehicle approaches you and lower as it moves away from you. This phenomenon is one example of the *Doppler effect*, named for Austrian physicist Christian Doppler (1803–1853), who discovered it. The same effect is heard if you're on a motorcycle and the horn is stationary: the frequency is higher as you approach the source and lower as you move away.

Although the Doppler effect is most often associated with sound, it's common to all waves, including light.

In deriving the Doppler effect, we assume the air is stationary and that all speed measurements are made relative to this stationary medium. In the general case, the speed of the observer $v_O$, the speed of the source $v_S$, and the speed of sound $v$ are all measured relative to the medium in which the sound is propagated.

## 14.6.1 Case 1: The Observer Is Moving Relative to a Stationary Source

In Figure 14.9, an observer is moving with a speed of $v_O$ toward the source (considered a point source), which is at rest ($v_S = 0$).

We take the frequency of the source to be $f_S$, the wavelength of the source to be $\lambda_S$, and the speed of sound in air to be $v$. If both observer and source are stationary, the observer detects $f_S$ wave fronts per second. (That is, when $v_O = 0$ and $v_S = 0$, the observed frequency $f_O$ equals the source frequency $f_S$.) When moving toward the source, the observer moves a distance of $v_O t$ in $t$ seconds. During this interval, **the observer detects an additional number of wave fronts.** The number of extra wave fronts is equal to the distance traveled, $v_O t$, divided by the wavelength $\lambda_S$:

$$\text{Additional wave fronts detected} = \frac{v_O t}{\lambda_S}$$

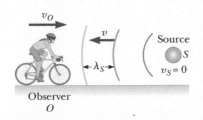

**Figure 14.9** An observer moving with a speed $v_O$ *toward* a stationary point source ($S$) hears a frequency $f_O$ that is *greater* than the source frequency $f_S$.

Divide this equation by the time $t$ to get the number of additional wave fronts detected *per second*, $v_O/\lambda_S$. Hence, the frequency heard by the observer is *increased* to

$$f_O = f_S + \frac{v_O}{\lambda_S}$$

Substituting $\lambda_S = v/f_S$ into this expression for $f_O$, we obtain

$$f_O = f_S\left(\frac{v + v_O}{v}\right) \qquad [14.10]$$

When the observer is *moving away* from a stationary source (Fig. 14.10), the observed frequency decreases. A derivation yields the same result as Equation 14.10, but with $v - v_O$ in the numerator. Therefore, when the observer is moving away from the source, substitute $-v_O$ for $v_O$ in Equation 14.10.

## 14.6.2 Case 2: The Source Is Moving Relative to a Stationary Observer

Now consider a source moving toward an observer at rest, as in Figure 14.11. Here, the wave fronts passing observer $A$ are closer together because the source is moving in the direction of the outgoing wave. As a result, the wavelength $\lambda_O$ measured by observer $A$ is shorter than the wavelength $\lambda_S$ of the source at rest. During each vibration, which lasts for an interval $T$ (the period), the source moves a distance $v_S T = v_S/f_S$ and **the wavelength is shortened by that amount.** The observed wavelength is therefore given by

$$\lambda_O = \lambda_S - \frac{v_S}{f_S}$$

Because $\lambda_S = v/f_S$, the frequency observed by $A$ is

$$f_O = \frac{v}{\lambda_O} = \frac{v}{\lambda_S - \dfrac{v_S}{f_S}} = \frac{v}{\dfrac{v}{f_S} - \dfrac{v_S}{f_S}}$$

or

$$f_O = f_S\left(\frac{v}{v - v_S}\right) \qquad [14.11]$$

As expected, **the observed frequency increases when the source is moving toward the observer.** When the source is *moving away* from an observer at rest, the minus sign in the denominator must be replaced with a plus sign, so the factor becomes $(v + v_S)$.

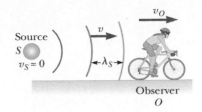

**Figure 14.10** An observer moving with a speed of $v_O$ *away* from a stationary source hears a frequency $f_O$ that is *lower* than the source frequency $f_S$.

> **Tip 14.2** Doppler Effect Doesn't Depend on Distance
>
> The sound from a source approaching at constant speed will increase in intensity, but the observed (elevated) frequency will remain unchanged. The Doppler effect doesn't depend on distance.

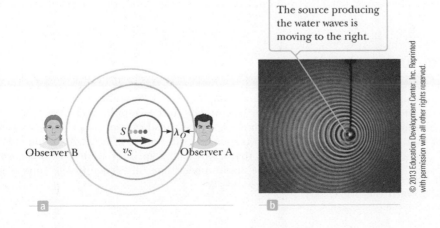

The source producing the water waves is moving to the right.

**Figure 14.11** (a) A source $S$ moving with speed $v_S$ toward stationary observer $A$ and away from stationary observer $B$. Observer $A$ hears an *increased* frequency, and observer $B$ hears a *decreased* frequency. (b) The Doppler effect in water, observed in a ripple tank.

### 14.6.3 General Case

When both the source and the observer are in motion relative to Earth, Equations 14.10 and 14.11 can be combined to give

Doppler shift: observer and ▶
source in motion

$$f_O = f_S \left( \frac{v + v_O}{v - v_S} \right)$$  [14.12]

In this expression, the signs for the values substituted for $v_O$ and $v_S$ depend on the direction of the velocity. When the observer moves *toward* the source, a *positive* speed is substituted for $v_O$; when the observer moves *away from* the source, a *negative* speed is substituted for $v_O$. Similarly, a *positive* speed is substituted for $v_S$ when the source moves *toward* the observer, a *negative* speed when the source moves away from the observer.

Choosing incorrect signs is the most common mistake made in working a Doppler effect problem. The following rules may be helpful: The word *toward* is associated with an *increase* in the observed frequency; the words *away from* are associated with a *decrease* in the observed frequency.

These two rules derive from the physical insight that when the observer is moving toward the source (or the source toward the observer), there is a smaller observed period between wave crests, hence a larger frequency, with the reverse holding—a smaller observed frequency—when the observer is moving away from the source (or the source away from the observer). Keep the physical insight in mind whenever you're in doubt about the signs in Equation 14.12: Adjust them as necessary to get the correct physical result.

The second most common mistake made in applying Equation 14.12 is to accidentally reverse numerator and denominator. Some find it helpful to remember the equation in the following form:

$$\frac{f_O}{v + v_O} = \frac{f_S}{v - v_S}$$

The advantage of this form is its symmetry: both sides are very nearly the same, with $O$'s on the left and $S$'s on the right. Forgetting which side has the plus sign and which has the minus sign is not a serious problem as long as physical insight is used to check the answer and make adjustments as necessary.

### Quick Quiz

**14.2** Suppose you're on a hot air balloon ride, carrying a buzzer that emits a sound of frequency $f$. If you accidentally drop the buzzer over the side while the balloon is rising at constant speed, what can you conclude about the sound you hear as the buzzer falls toward the ground? (a) The frequency and intensity increase. (b) The frequency decreases and the intensity increases. (c) The frequency decreases and the intensity decreases. (d) The frequency remains the same, but the intensity decreases.

---

### APPLYING PHYSICS 14.2    OUT-OF-TUNE SPEAKERS

Suppose you place your stereo speakers far apart and run past them from right to left or left to right. If you run rapidly enough and have excellent pitch discrimination, you may notice that the music playing seems to be out of tune when you're between the speakers. Why?

**EXPLANATION** When you are between the speakers, you are running away from one of them and toward the other, so there is a Doppler shift downward for the sound from the speaker behind you and a Doppler shift upward for the sound from the speaker ahead of you. As a result, the sound from the two speakers will not be in tune. A simple calculation shows that a world-class sprinter could run fast enough to generate about a semitone difference in the sound from the two speakers. ∎

## EXAMPLE 14.4  LISTEN, BUT DON'T STAND ON THE TRACK

**GOAL**  Solve a Doppler shift problem when only the source is moving.

**PROBLEM**  A train moving at a speed of 40.0 m/s sounds its whistle, which has a frequency of $5.00 \times 10^2$ Hz. Determine the frequency heard by a stationary observer as the train *approaches* the observer. The ambient temperature is 24.0°C.

**STRATEGY**  Use Equation 14.4 to get the speed of sound at the ambient temperature, then substitute values into Equation 14.12 for the Doppler shift. Because the train approaches the observer, the observed frequency will be larger. Choose the sign of $v_S$ to reflect this fact.

### SOLUTION

Use Equation 14.4 to calculate the speed of sound in air at $T = 24.0°C$:

$$v = (331 \text{ m/s}) \sqrt{\frac{T}{273 \text{ K}}}$$

$$= (331 \text{ m/s}) \sqrt{\frac{(273 + 24.0) \text{ K}}{273 \text{ K}}} = \boxed{345 \text{ m/s}}$$

The observer is stationary, so $v_O = 0$. The train is moving *toward* the observer, so $v_S = +40.0$ m/s (*positive*). Substitute these values and the speed of sound into the Doppler shift equation:

$$f_O = f_S \left( \frac{v + v_O}{v - v_S} \right)$$

$$= (5.00 \times 10^2 \text{ Hz}) \left( \frac{345 \text{ m/s}}{345 \text{ m/s} - 40.0 \text{ m/s}} \right)$$

$$= \boxed{566 \text{ Hz}}$$

**REMARKS**  If the train were going away from the observer, $v_S = -40.0$ m/s would have been chosen instead.

**QUESTION 14.4**  Does the Doppler shift change due to temperature variations? If so, why? For typical daily variations in temperature in a moderate climate, would any change in the Doppler shift be best characterized as (a) nonexistent, (b) small, or (c) large?

**EXERCISE 14.4**  Determine the frequency heard by the stationary observer as the train *recedes* from the observer.

**ANSWER**  448 Hz

## EXAMPLE 14.5  THE NOISY SIREN

**GOAL**  Solve a Doppler shift problem when both the source and observer are moving.

**PROBLEM**  An ambulance travels down a highway at a speed of 75.0 mi/h, its siren emitting sound at a frequency of $4.00 \times 10^2$ Hz. What frequency is heard by a passenger in a car traveling at 55.0 mi/h in the opposite direction as the car and ambulance (a) *approach* each other and (b) pass and *move away* from each other? Take the speed of sound in air to be $v = 345$ m/s.

**STRATEGY**  Aside from converting mi/h to m/s, this problem only requires substitution into the Doppler formula, but two signs must be chosen correctly in each part. In part (a) the observer moves toward the source and the source moves toward the observer, so both $v_O$ and $v_S$ should be chosen to be positive. Switch signs after they pass each other.

### SOLUTION

Convert the speeds from mi/h to m/s:

$$v_S = (75.0 \text{ mi/h}) \left( \frac{0.447 \text{ m/s}}{1.00 \text{ mi/h}} \right) = 33.5 \text{ m/s}$$

$$v_O = (55.0 \text{ mi/h}) \left( \frac{0.447 \text{ m/s}}{1.00 \text{ mi/h}} \right) = 24.6 \text{ m/s}$$

(a) Compute the observed frequency as the ambulance and car approach each other.

Each vehicle goes toward the other, so substitute $v_O = +24.6$ m/s and $v_S = +33.5$ m/s into the Doppler shift formula:

$$f_O = f_S \left( \frac{v + v_O}{v - v_S} \right)$$

$$= (4.00 \times 10^2 \text{ Hz}) \left( \frac{345 \text{ m/s} + 24.6 \text{ m/s}}{345 \text{ m/s} - 33.5 \text{ m/s}} \right) = \boxed{475 \text{ Hz}}$$

*(Continued)*

**(b)** Compute the observed frequency as the ambulance and car recede from each other.

Each vehicle goes away from the other, so substitute $v_O = -24.6$ m/s and $v_S = -33.5$ m/s into the Doppler shift formula:

$$f_O = f_S \left( \frac{v + v_O}{v - v_S} \right)$$

$$= (4.00 \times 10^2 \text{ Hz}) \left( \frac{345 \text{ m/s} + (-24.6 \text{ m/s})}{345 \text{ m/s} - (-33.5 \text{ m/s})} \right)$$

$$= \boxed{339 \text{ Hz}}$$

**REMARKS** Notice how the signs were handled. In part **(b)** the negative signs were required on the speeds because both observer and source were moving away from each other. Sometimes, of course, one of the speeds is negative and the other is positive.

**QUESTION 14.5** Is the Doppler shift affected by sound intensity level?

**EXERCISE 14.5** Repeat this problem, but assume the ambulance and car are going the same direction, with the ambulance initially behind the car. The speeds and the frequency of the siren are the same as in the example. Find the frequency heard by the observer in the car **(a)** before and **(b)** after the ambulance passes the car. *Note:* The highway patrol subsequently gives the driver of the car a ticket for not pulling over for an emergency vehicle!

**ANSWERS** **(a)** 411 Hz    **(b)** 391 Hz

### 14.6.4 Shock Waves

What happens when the source speed $v_S$ *exceeds* the wave velocity $v$? Figure 14.12 describes this situation graphically. The circles represent spherical wave fronts emitted by the source at various times during its motion. At $t = 0$, the source is at point $S_0$, and at some later time $t$, the source is at point $S_n$. In the interval $t$, the wave front centered at $S_0$ reaches a radius of $vt$. In this same interval, the source travels to $S_n$, a distance of $v_S t$. At the instant the source is at $S_n$, the waves just beginning to be generated at this point have wave fronts of zero radius. The line drawn from $S_n$ to the wave front centered on $S_0$ is tangent to all other wave fronts generated at intermediate times. All such tangent lines lie on the surface of a cone. The angle $\theta$ between one of these tangent lines and the direction of travel is given by

$$\sin \theta = \frac{v}{v_s}$$

The ratio $v_S / v$ is called the **Mach number.** The conical wave front produced when $v_S > v$ (supersonic speeds) is known as a **shock wave.** An interesting example of a shock wave is the V-shaped wave front produced by a boat (the bow wave) when the boat's speed exceeds the speed of the water waves (Fig. 14.13).

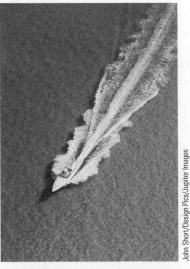

**Figure 14.13** The V-shaped bow wave is formed because the boat travels at a speed greater than the speed of the water waves. A bow wave is analogous to a shock wave formed by a jet traveling faster than sound.

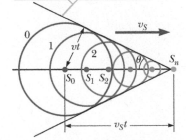

The envelope of the wave fronts forms a cone with half-angle of $\sin \theta = v/v_S$.

**Figure 14.12** A representation of a shock wave, produced when a source moves from $S_0$ to $S_n$ with a speed $v_S$ that is *greater* than the wave speed $v$ in that medium.

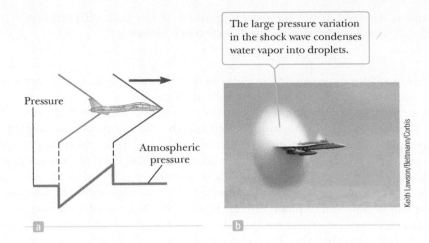

The large pressure variation in the shock wave condenses water vapor into droplets.

Pressure

Atmospheric pressure

Keith Lawson/Bettmann/Corbis

a

b

**Figure 14.14** (a) The two shock waves produced by the nose and tail of a jet airplane traveling at super-sonic speed. (b) A shock wave due to a jet traveling at the speed of sound is made visible as a fog of water vapor.

Jet aircraft and space shuttles traveling at supersonic speeds produce shock waves that are responsible for the loud explosion, or sonic boom, heard on the ground. A shock wave carries a great deal of energy concentrated on the surface of the cone, with correspondingly great pressure variations. Shock waves are unpleasant to hear and can damage buildings when aircraft fly supersonically at low altitudes. In fact, an airplane flying at supersonic speeds produces a double boom because two shock waves are formed: one from the nose of the plane and one from the tail (Fig. 14.14).

### Quick Quiz

**14.3** As an airplane flying with constant velocity moves from a cold air mass into a warm air mass, does the Mach number (a) increase, (b) decrease, or (c) remain the same?

# 14.7 Interference of Sound Waves

Sound waves can be made to interfere with each other, a phenomenon that can be demonstrated with the device shown in Figure 14.15. Sound from a loud-speaker at S is sent into a tube at $P$, where there is a T-shaped junction. The sound splits and follows two separate pathways, indicated by the red arrows. Half of the sound travels upward, half downward. Finally, the two sounds merge at an opening where a listener places her ear. If the two paths $r_1$ and $r_2$ have the same length, waves that enter the junction will separate into two halves, travel the two paths, and then combine again at the ear. This reuniting of the two waves produces *constructive interference*, and the listener hears a loud sound. If the upper path is adjusted to be one full wavelength longer than the lower path, constructive interference of the two waves occurs again, and a loud sound is detected at the receiver. We have the following result: **If the path difference $r_2 - r_1$ is zero or some integer multiple of wavelengths, then constructive inter-ference occurs and**

$$r_2 - r_1 = n\lambda \qquad (n = 0, 1, 2, \ldots) \qquad [14.13]$$

Suppose, however, that one of the path lengths, $r_2$, is adjusted so that the upper path is half a wavelength *longer* than the lower path $r_1$. In this case, an entering sound wave splits and travels the two paths as before, but now the wave along the upper path must travel a distance equivalent to half a wavelength farther than the wave traveling along the lower path. As a result, the crest of one wave meets the trough of the other when they merge at the receiver, causing the two waves to cancel each other. This phenomenon is called *totally destructive interference*, and

A sound wave from the speaker (S) enters the tube and splits into two parts at point $P$.

Path length $r_2$

S

P

R

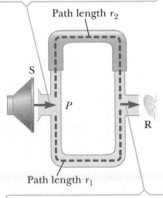

Path length $r_1$

The two waves combine at the opposite side and are detected at the receiver (R).

**Figure 14.15** An acoustical system for demonstrating interference of sound waves. The upper path length is varied by the sliding section.

no sound is detected at the receiver. In general, **if the path difference $r_2 - r_1$ is $\frac{1}{2}, 1\frac{1}{2}, 2\frac{1}{2} \ldots$ wavelengths, destructive interference occurs** and

◀ Condition for destructive interference

$$r_2 - r_1 = (n + \tfrac{1}{2})\lambda \qquad (n = 0, 1, 2, \ldots) \qquad \text{[14.14]}$$

Nature provides many other examples of interference phenomena, most notably in connection with light waves, described in Topic 24.

**APPLICATION**
Connecting Your Stereo Speakers

In connecting the wires between your stereo system and loudspeakers, you may notice that the wires are usually color coded and that the speakers have positive and negative signs on the connections. The reason for this is that the speakers need to be connected with the same "polarity." If they aren't, then the same electrical signal fed to both speakers will result in one speaker cone moving outward at the same time that the other speaker cone is moving inward. In this case, the sound leaving the two speakers will be 180° out of phase with each other. If you are sitting midway between the speakers, the sounds from both speakers travel the same distance and preserve the phase difference they had when they left. In an ideal situation, for a 180° phase difference, you would get complete destructive interference and no sound! In reality, the cancellation is not complete and is much more significant for bass notes (which have a long wavelength) than for the shorter wavelength treble notes. Nevertheless, to avoid a significant reduction in the intensity of bass notes, the color-coded wires and the signs on the speaker connections should be carefully noted.

**Tip 14.3** Do Waves Really Interfere?

In popular usage, to *interfere* means "to come into conflict with" or "to intervene to affect an outcome." This differs from its use in physics, where waves pass through each other and interfere, but don't affect each other in any way.

---

**EXAMPLE 14.6**    **TWO SPEAKERS DRIVEN BY THE SAME SOURCE**

**GOAL**  Use the concept of interference to compute a frequency.

**PROBLEM**  Two speakers placed 3.00 m apart are driven by the same oscillator (Fig. 14.16). A listener is originally at point $O$, which is located 8.00 m from the center of the line connecting the two speakers. The listener then walks to point $P$, which is a perpendicular distance 0.350 m from $O$, before reaching the *first minimum* in sound intensity. What is the frequency of the oscillator? Take the speed of sound in air to be $v_s = 343$ m/s.

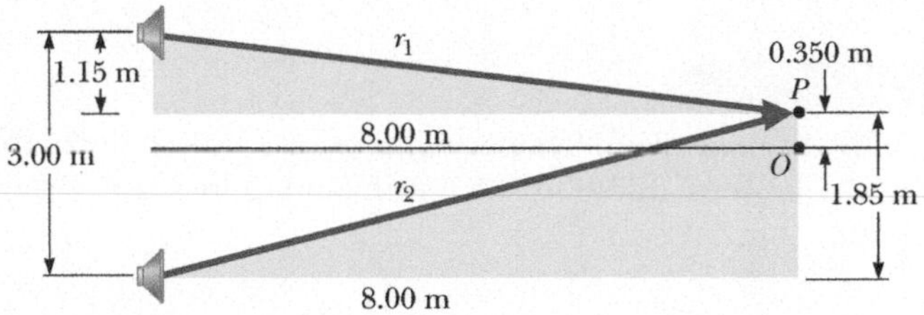

**Figure 14.16**  (Example 14.6) Two loudspeakers driven by the same source can produce interference.

**STRATEGY**  The position of the first minimum in sound intensity is given, which is a point of destructive interference. We can find the path lengths $r_1$ and $r_2$ with the Pythagorean theorem and then use Equation 14.14 for destructive interference to find the wavelength $\lambda$. Using $v = f\lambda$ then yields the frequency.

---

**SOLUTION**

Use the Pythagorean theorem to find the path lengths $r_1$ and $r_2$:

$$r_1 = \sqrt{(8.00 \text{ m})^2 + (1.15 \text{ m})^2} = 8.08 \text{ m}$$

$$r_2 = \sqrt{(8.00 \text{ m})^2 + (1.85 \text{ m})^2} = 8.21 \text{ m}$$

Substitute these values and $n = 0$ into Equation 14.14, solving for the wavelength:

$$r_2 - r_1 = (n + \tfrac{1}{2})\lambda$$

$$8.21 \text{ m} - 8.08 \text{ m} = 0.13 \text{ m} = \lambda/2 \quad \rightarrow \quad \lambda = 0.26 \text{ m}$$

Solve $v = \lambda f$ for the frequency $f$ and substitute the speed of sound and the wavelength:

$$f = \frac{v}{\lambda} = \frac{343 \text{ m/s}}{0.26 \text{ m}} = \boxed{1.3 \text{ kHz}}$$

---

**REMARKS**  For problems involving constructive interference, the only difference is that Equation 14.13, $r_2 - r_1 = n\lambda$, would be used instead of Equation 14.14.

**QUESTION 14.6**  True or False: In the same context, smaller wavelengths of sound would create more interference maxima and minima than longer wavelengths.

**EXERCISE 14.6**  If the oscillator frequency is adjusted so that the location of the first minimum is at a distance of 0.750 m from $O$, what is the new frequency?

**ANSWER**  0.62 kHz

# 14.8 Standing Waves

Standing waves can be set up in a stretched string by connecting one end of the string to a stationary clamp and connecting the other end to a vibrating object, such as the end of a tuning fork, or by shaking the hand holding the string up and down at a steady rate (Fig. 14.17). Traveling waves then reflect from the ends and move in both directions on the string. The incident and reflected waves combine according to the **superposition principle.** (See Section 13.10.) If the string vibrates at exactly the right frequency, the wave appears to stand still, hence its name, **standing wave.** A **node** occurs where the two traveling waves always have the same magnitude of displacement but the opposite sign, so the net displacement is zero at that point. There is no motion in the string at the nodes, but midway between two adjacent nodes, at an **antinode,** the string vibrates with the largest amplitude.

Figure 14.18 shows snapshots of the oscillation of a standing wave during half of a cycle. The red arrows indicate the direction of motion of different parts of the string. Notice that **all points on the string oscillate together vertically with the same frequency, but different points have different amplitudes of motion.** The points of attachment to the wall and all other stationary points on the string are called nodes, labeled N in Figure 14.18a. From the figure, observe that the distance between adjacent nodes is one-half the wavelength of the wave:

$$d_{NN} = \tfrac{1}{2}\lambda$$

Consider a string of length $L$ that is fixed at both ends, as in Figure 14.19. For a string, we can set up standing-wave patterns at many frequencies—the more loops, the higher the frequency. Figure 14.20 (page 474) is a multiflash photograph of a standing wave on a string.

First, **the ends of the string must be nodes, because these points are fixed.** If the string is displaced at its midpoint and released, the vibration shown in Figure 14.19b can be produced, in which case the center of the string is an antinode, labeled A. Note that from end to end, the pattern is N–A–N. The distance from a node to its adjacent antinode, N–A, is always equal to a quarter wavelength, $\lambda_1/4$.

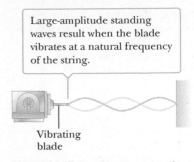

Large-amplitude standing waves result when the blade vibrates at a natural frequency of the string.

Vibrating blade

**Figure 14.17** Standing waves can be set up in a stretched string by connecting one end of the string to a vibrating blade.

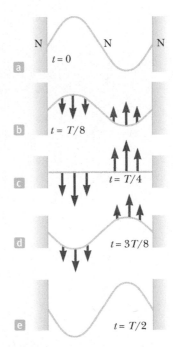

**Figure 14.18** A standing-wave pattern in a stretched string, shown by snapshots of the string during one-half of a cycle. In part (a) N denotes a node.

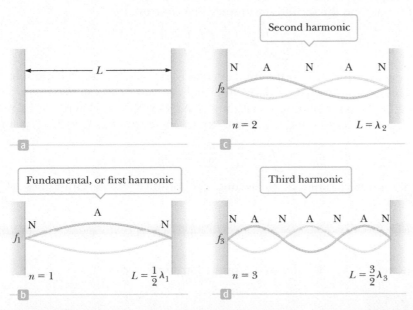

**Figure 14.19** (a) Standing waves in a stretched string of length $L$ fixed at both ends. The characteristic frequencies of vibration form a harmonic series: (b) the fundamental frequency, or first harmonic; (c) the second harmonic; and (d) the third harmonic. Note that N denotes a node, A an antinode.

**Figure 14.20** Multiflash photograph of a standing-wave two-loop pattern in a second harmonic ($n = 2$), using a cord driven by a vibrator at the left end.

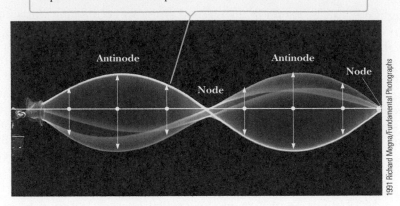

The amplitude of the vertical oscillation of any element of the string depends on the horizontal position of the element.

Antinode    Antinode

Node        Node

There are two such segments, N–A and A–N, so $L = 2(\lambda_1/4) = \lambda_1/2$, and $\lambda_1 = 2L$. The frequency of this vibration is therefore

$$f_1 = \frac{v}{\lambda_1} = \frac{v}{2L} \qquad [14.15]$$

Recall that the speed of a wave on a string is $v = \sqrt{F/\mu}$, where $F$ is the tension in the string and $\mu$ is its mass per unit length (Topic 13). Substituting into Equation 14.15, we obtain

$$f_1 = \frac{1}{2L}\sqrt{\frac{F}{\mu}} \qquad [14.16]$$

This lowest frequency of vibration is called the **fundamental frequency** of the vibrating string, or the **first harmonic.**

The first harmonic has nodes only at the ends: the points of attachment, with node–antinode pattern of N–A–N. The next harmonic, called the **second harmonic** (also called the **first overtone**), can be constructed by inserting an additional node–antinode segment between the endpoints. This makes the pattern N–A–N–A–N, as in Figure 14.19c. We count the node–antinode pairs: N–A, A–N, N–A, and A–N, four segments in all, each representing a quarter wavelength. We then have $L = 4(\lambda_2/4) = \lambda_2$, and the second harmonic (first overtone) is

$$f_2 = \frac{v}{\lambda_2} = \frac{v}{L} = 2\left(\frac{v}{2L}\right) = 2f_1$$

This frequency is equal to *twice* the fundamental frequency. The **third harmonic (second overtone)** is constructed similarly. Inserting one more N–A segment, we obtain Figure 14.19d, the pattern of nodes reading N–A–N–A–N–A–N. There are six node–antinode segments, so $L = 6(\lambda_3/4) = 3(\lambda_3/2)$, which means that $\lambda_3 = 2L/3$, giving

$$f_3 = \frac{v}{\lambda_3} = \frac{3v}{2L} = 3f_1$$

All the higher harmonics, it turns out, are positive integer multiples of the fundamental:

Natural frequencies of a ▶ string fixed at both ends

$$f_n = nf_1 = \frac{n}{2L}\sqrt{\frac{F}{\mu}} \qquad n = 1, 2, 3, \ldots \qquad [14.17]$$

The frequencies $f_1$, $2f_1$, $3f_1$, and so on form a **harmonic series.**

### Quick Quiz

**14.4** Which of the following frequencies are higher harmonics of a string with fundamental frequency of 150 Hz? (a) 200 Hz (b) 300 Hz (c) 400 Hz (d) 500 Hz (e) 600 Hz

When a stretched string is distorted to a shape that corresponds to any one of its harmonics, after being released it vibrates only at the frequency of that harmonic. If the string is struck or bowed, however, the resulting vibration includes different amounts of various harmonics, including the fundamental frequency. Waves not in the harmonic series are quickly damped out on a string fixed at both ends. In effect, when disturbed, the string "selects" the standing-wave frequencies. As we'll see later, the presence of several harmonics on a string gives stringed instruments their characteristic sound, which enables us to distinguish one from another even when they are producing identical fundamental frequencies.

The frequency of a string on a musical instrument can be changed by varying either the tension or the length. The tension in guitar and violin strings is varied by turning pegs on the neck of the instrument. As the tension is increased, the frequency of the harmonic series increases according to Equation 14.17. Once the instrument is tuned, the musician varies the frequency by pressing the strings against the neck at a variety of positions, thereby changing the effective lengths of the vibrating portions of the strings. As the length is reduced, the frequency again increases, as follows from Equation 14.17.

Finally, Equation 14.17 shows that a string of fixed length can be made to vibrate at a lower fundamental frequency by increasing its mass per unit length. This increase is achieved in the bass strings of guitars and pianos by wrapping the strings with metal windings.

**APPLICATION**
Tuning a Musical Instrument

---

### EXAMPLE 14.7 GUITAR FUNDAMENTALS

**GOAL** Apply standing-wave concepts to a stringed instrument.

**PROBLEM** The high E string on a certain guitar measures 64.0 cm in length and has a fundamental frequency of 329 Hz. When a guitarist presses down so that the string is in contact with the first fret (Fig. 14.21a), the string is shortened so that it plays an F note that has a frequency of 349 Hz. **(a)** How far is the fret from the nut? **(b)** Overtones can be produced on a guitar string by gently placing the index finger in the location of a node of a higher harmonic. The string should be touched, but not depressed against a fret. (Given the width of a finger, pressing too hard will damp out higher harmonics as well.) The fundamental frequency is thereby suppressed, making it possible to hear overtones. Where on the guitar string relative to the nut should the finger be lightly placed so as to hear the second harmonic of the high E string? The fourth harmonic? (This is equivalent to finding the location of the nodes in each case.)

**STRATEGY** For part **(a)** use Equation 14.15, corresponding to the fundamental frequency, to find the speed of waves on the string. Shortening the string by playing a higher note doesn't affect the wave speed, which depends only on the tension and linear density of the string (which are unchanged). Solve Equation 14.15 for the new length $L$, using the new fundamental frequency, and subtract this length from the original length to find the distance from the nut to the first fret. In part **(b)** remember that the distance from node to node is half a wavelength. Calculate the wavelength, divide it in two, and locate the nodes, which are integral numbers of half-wavelengths from the nut. *Note:* The nut is a small piece of wood or ebony at the top of the fret board. The distance from the nut to the bridge (below the sound hole) is the length of the string. (See Fig. 14.21b.)

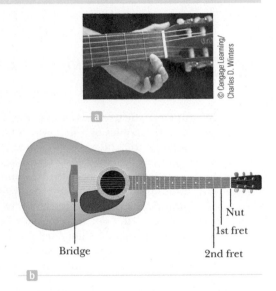

**Figure 14.21** (Example 14.7) (a) Playing an F note on a guitar. (b) Some parts of a guitar.

Bridge | Nut | 1st fret | 2nd fret

---

#### SOLUTION

**(a)** Find the distance from the nut to the first fret.

Substitute $L_0 = 0.640$ m and $f_1 = 329$ Hz into Equation 14.15, finding the wave speed on the string:

$$f_1 = \frac{v}{2L_0}$$

$$v = 2L_0 f_1 = 2(0.640 \text{ m})(329 \text{ Hz}) = 421 \text{ m/s}$$

Solve Equation 14.15 for the length $L$, and substitute the wave speed and the frequency of an F note.

$$L = \frac{v}{2f} = \frac{421 \text{ m/s}}{2(349 \text{ Hz})} = 0.603 \text{ m} = 60.3 \text{ cm}$$

*(Continued)*

Subtract this length from the original length $L_0$ to find the distance from the nut to the first fret:

$$\Delta x = L_0 - L = 64.0 \text{ cm} - 60.3 \text{ cm} = \boxed{3.7 \text{ cm}}$$

**(b)** Find the locations of nodes for the second and fourth harmonics.

The second harmonic has a wavelength $\lambda_2 = L_0 = 64.0$ cm. The distance from nut to node corresponds to half a wavelength.

$$\Delta x = \tfrac{1}{2}\lambda_2 = \tfrac{1}{2}L_0 = 32.0 \text{ cm}$$

The fourth harmonic, of wavelength $\lambda_4 = \tfrac{1}{2}L_O = 32.0$ cm, has three nodes between the endpoints:

$$\Delta x = \tfrac{1}{2}\lambda_4 = \boxed{16.0 \text{ cm}}, \Delta x = 2(\lambda_4/2) = \boxed{32.0 \text{ cm}},$$

$$\Delta x = 3(\lambda_4/2) = \boxed{48.0 \text{ cm}}$$

• • • • • • • • • • • • • • • • • • • • • • • • • • • • • • • • • • • • • • • • • • • • • • • • • • • • • • • • • • • • • • • • • • • • • • • • • • • • • • • • • • • • • • • •

**REMARKS** Placing a finger at the position $\Delta x = 32.0$ cm damps out the fundamental and odd harmonics, but not all the higher even harmonics. The second harmonic dominates, however, because the rest of the string is free to vibrate. Placing the finger at $\Delta x = 16.0$ cm or 48.0 cm damps out the first through third harmonics, allowing the fourth harmonic to be heard.

**QUESTION 14.7** True or False: If a guitar string has length $L$, gently placing a thin object at the position $(\tfrac{1}{2})^n L$ will always result in the sounding of a higher harmonic, where $n$ is a positive integer.

**EXERCISE 14.7** Pressing the E string down on the fret board just above the second fret pinches the string firmly against the fret, giving an F-sharp, which has frequency $3.70 \times 10^2$ Hz. **(a)** Where should the second fret be located? **(b)** Find two locations where you could touch the open E string and hear the third harmonic.

**ANSWERS** **(a)** 7.1 cm from the nut and 3.4 cm from the first fret. Note that the distance from the first to the second fret isn't the same as from the nut to the first fret. **(b)** 21.3 cm and 42.7 cm from the nut

---

## EXAMPLE 14.8 | HARMONICS OF A STRETCHED WIRE

**GOAL** Calculate string harmonics, relate them to sound, and combine them with tensile stress.

**PROBLEM** **(a)** Find the frequencies of the fundamental, second, and third harmonics of a steel wire 1.00 m long with a mass per unit length of $2.00 \times 10^{-3}$ kg/m and under a tension of 80.0 N. **(b)** Find the wavelengths of the sound waves created by the vibrating wire for all three modes. Assume the speed of sound in air is 345 m/s. **(c)** Suppose the wire is carbon steel with a density of $7.80 \times 10^3$ kg/m³, a cross-sectional area $A = 2.56 \times 10^{-7}$ m², and an elastic limit of $2.80 \times 10^8$ Pa. Find the fundamental frequency if the wire is tightened to the elastic limit. Neglect any stretching of the wire (which would slightly reduce the mass per unit length).

**STRATEGY** **(a)** It's easiest to find the speed of waves on the wire then substitute into Equation 14.15 to find the first harmonic. The next two are multiples of the first, given by Equation 14.17. **(b)** The frequencies of the sound waves are the same as the frequencies of the vibrating wire, but the wavelengths are different. Use $v_s = f\lambda$, where $v_s$ is the speed of sound in air, to find the wavelengths in air. **(c)** Find the force corresponding to the elastic limit and substitute it into Equation 14.16.

• • • • • • • • • • • • • • • • • • • • • • • • • • • • • • • • • • • • • • • • • • • • • • • • • • • • • • • • • • • • • • • • • • • • • • • • • • • • • • • • • • • • • • • •

**SOLUTION**

**(a)** Find the first three harmonics at the given tension.

Use Equation 13.18 to calculate the speed of the wave on the wire:

$$v = \sqrt{\frac{F}{\mu}} = \sqrt{\frac{80.0 \text{ N}}{2.00 \times 10^{-3} \text{ kg/m}}} = 2.00 \times 10^2 \text{ m/s}$$

Find the wire's fundamental frequency from Equation 14.15:

$$f_1 = \frac{v}{2L} = \frac{2.00 \times 10^2 \text{ m/s}}{2(1.00 \text{ m})} = \boxed{1.00 \times 10^2 \text{ Hz}}$$

Find the next two harmonics by multiplication:

$$f_2 = 2f_1 = \boxed{2.00 \times 10^2 \text{ Hz}}, f_3 = 3f_1 = \boxed{3.00 \times 10^2 \text{ Hz}}$$

**(b)** Find the wavelength of the sound waves produced.

Solve $v_s = f\lambda$ for the wavelength and substitute the frequencies:

$$\lambda_1 = v_s/f_1 = (345 \text{ m/s})/(1.00 \times 10^2 \text{ Hz}) = \boxed{3.45 \text{ m}}$$

$$\lambda_2 = v_s/f_2 = (345 \text{ m/s})/(2.00 \times 10^2 \text{ Hz}) = \boxed{1.73 \text{ m}}$$

$$\lambda_3 = v_s/f_3 = (345 \text{ m/s})/(3.00 \times 10^2 \text{ Hz}) = \boxed{1.15 \text{ m}}$$

**(c)** Find the fundamental frequency corresponding to the elastic limit.

Calculate the tension in the wire from the elastic limit:

$$\frac{F}{A} = \text{elastic limit} \quad \rightarrow \quad F = (\text{elastic limit})A$$

$$F = (2.80 \times 10^8 \text{ Pa})(2.56 \times 10^{-7} \text{ m}^2) = 71.7 \text{ N}$$

Substitute the values of $F$, $\mu$, and $L$ into Equation 14.16:

$$f_1 = \frac{1}{2L}\sqrt{\frac{F}{\mu}}$$

$$f_1 = \frac{1}{2(1.00 \text{ m})}\sqrt{\frac{71.7 \text{ N}}{2.00 \times 10^{-3} \text{ kg/m}}} = \boxed{94.7 \text{ Hz}}$$

**REMARKS** From the answer to part **(c)**, it appears we need to choose a thicker wire or use a better grade of steel with a higher elastic limit. The frequency corresponding to the elastic limit is smaller than the fundamental!

**QUESTION 14.8** A string on a guitar is replaced with one of lower linear density. To obtain the same frequency sound as previously, must the tension of the new string be **(a)** greater than, **(b)** less than, or **(c)** equal to the tension in the old string?

**EXERCISE 14.8** **(a)** Find the fundamental frequency and second harmonic if the tension in the wire is increased to 115 N. (Assume the wire doesn't stretch or break.) **(b)** Using a sound speed of 345 m/s, find the wavelengths of the sound waves produced.

**ANSWERS** **(a)** $1.20 \times 10^2$ Hz, $2.40 \times 10^2$ Hz    **(b)** 2.88 m, 1.44 m

# 14.9 Forced Vibrations and Resonance

In Topic 13, we learned that the energy of a damped oscillator decreases over time because of friction. It's possible to compensate for this energy loss by applying an external force that does positive work on the system.

For example, suppose an object–spring system having some natural frequency of vibration $f_0$ is pushed back and forth by a periodic force with frequency $f$. The system vibrates at the frequency $f$ of the driving force. This type of motion is referred to as a **forced vibration.** Its amplitude reaches a maximum when the frequency of the driving force equals the natural frequency of the system $f_0$, called the **resonant frequency** of the system. Under this condition, the system is said to be in **resonance.**

In Section 14.8, we learned that a stretched string can vibrate in one or more of its natural modes. Here again, if a periodic force is applied to the string, the amplitude of vibration increases as the frequency of the applied force approaches one of the string's natural frequencies of vibration.

Resonance vibrations occur in a wide variety of circumstances. Figure 14.22 illustrates one experiment that demonstrates a resonance condition. Several pendulums of different lengths are suspended from a flexible beam. If one of them, such as $A$, is set in motion, the others begin to oscillate because of vibrations in the flexible beam. Pendulum $C$, the same length as $A$, oscillates with the greatest amplitude because its natural frequency matches that of pendulum $A$ (the driving force).

Another simple example of resonance is a child being pushed on a swing, which is essentially a pendulum with a natural frequency that depends on its length. The swing is kept in motion by a series of appropriately timed pushes. For its amplitude to increase, the swing must be pushed each time it returns to the person's hands. This corresponds to a frequency equal to the natural frequency of the swing. If the energy put into the system per cycle of motion equals the energy lost due to friction, the amplitude remains constant.

Opera singers have been known to make audible vibrations in crystal goblets with their powerful voices. This is yet another example of resonance: The sound

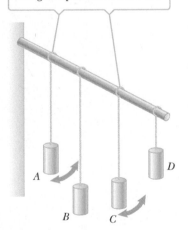

If pendulum $A$ is set in oscillation, only pendulum $C$, with a length matching that of $A$, will eventually oscillate with a large amplitude, or resonate.

**Figure 14.22** A demonstration of resonance.

**APPLICATION**

Shattering Goblets with the Voice

**Figure 14.23** (a) In 1940, turbulent winds set up torsional vibrations in the Tacoma Narrows Bridge, causing it to oscillate at a frequency near one of the natural frequencies of the bridge structure. (b) Once established, this resonance condition led to the bridge's collapse. A number of scientists, however, have challenged the resonance interpretation.

waves emitted by the singer can set up large-amplitude vibrations in the glass. If a highly amplified sound wave has the right frequency, the amplitude of forced vibrations in the glass increases to the point where the glass becomes heavily strained and shatters.

APPLICATION
Structural Integrity and Resonance

The classic example of structural resonance occurred in 1940, when the Tacoma Narrows Bridge in the state of Washington was put into oscillation by the wind (Fig. 14.23). The amplitude of the oscillations increased rapidly and reached a high value until the bridge ultimately collapsed (probably because of metal fatigue). In recent years, however, a number of researchers have called this explanation into question. Gusts of wind, in general, don't provide the periodic force necessary for a sustained resonance condition, and the bridge exhibited large twisting oscillations, rather than the simple up-and-down oscillations expected of resonance.

A more recent example of destruction by structural resonance occurred during the Loma Prieta earthquake near Oakland, California, in 1989. In a mile-long section of the double-decker Nimitz Freeway, the upper deck collapsed onto the lower deck, killing several people. The collapse occurred because that particular section was built on mud fill, whereas other parts were built on bedrock. As seismic waves pass through mud fill or other loose soil, their speed decreases and their amplitude increases. The section of the freeway that collapsed oscillated at the same frequency as other sections, but at a much larger amplitude.

# 14.10  Standing Waves in Air Columns

Standing longitudinal waves can be set up in a tube of air, such as an organ pipe, as the result of interference between sound waves traveling in opposite directions. The relationship between the incident wave and the reflected wave depends on whether the reflecting end of the tube is open or closed. A portion of the sound wave is reflected back into the tube even at an open end. **If one end is closed, a node must exist at that end because the movement of air is restricted. If the end is open, the elements of air have complete freedom of motion, and an antinode exists.**

Figure 14.24a shows the first three modes of vibration of a pipe open at both ends. When air is directed against an edge at the left, longitudinal standing waves are formed and the pipe vibrates at its natural frequencies. Note that, from end to end, the pattern is A–N–A, the same pattern as in the vibrating string, except node and antinode have exchanged positions. As before, an antinode and its adjacent node, A–N, represent a quarter-wavelength, and there are two, A–N and N–A, so $L = 2(\lambda_1/4) = \lambda_1/2$ and $\lambda_1 = 2L$. The fundamental frequency of the pipe open at both ends is then $f_1 = v/\lambda_1 = v/2L$. The next harmonic has an additional node and antinode between the ends, creating the pattern A–N–A–N–A. We count the pairs: A–N, N–A, A–N, and N–A, making four segments, each with length $\lambda_2/4$. We have

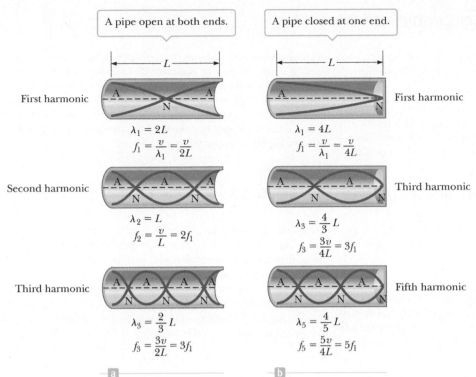

Figure 14.24 (a) Standing longitudinal waves in an organ pipe open at both ends. The natural frequencies $f_1$, $2f_1$, $3f_1$ ... form a harmonic series. (b) Standing longitudinal waves in an organ pipe closed at one end. Only *odd* harmonics are present, and the natural frequencies are $f_1$, $3f_1$, $5f_1$, and so on.

**Tip 14.4** Sound Waves Are Not Transverse

The standing longitudinal waves in Figure 14.23 are drawn as transverse waves only because it's difficult to draw longitudinal displacements: they're in the same direction as the wave propagation. In the figure, the vertical axis represents either pressure or horizontal displacement of the elements of the medium.

$L = 4(\lambda_2/4) = \lambda_2$, and the second harmonic (first overtone) is $f_2 = v/\lambda_2 = v/L = 2(v/2L) = 2f_1$. All higher harmonics, it turns out, are positive integer multiples of the fundamental:

$$f_n = n\frac{v}{2L} = nf_1 \qquad n = 1, 2, 3, \ldots \qquad [14.18]$$

◀ Pipe open at both ends; all harmonics are present

where $v$ is the speed of sound in air. Notice the similarity to Equation 14.17, which also involves multiples of the fundamental.

If a pipe is open at one end and closed at the other, the open end is an antinode and the closed end is a node (Fig. 14.24b). In such a pipe, the fundamental frequency consists of a single antinode–node pair, A–N, so $L = \lambda_1/4$ and $\lambda_1 = 4L$. The fundamental harmonic for a pipe closed at one end is then $f_1 = v/\lambda_1 = v/4L$. The first overtone has another node and antinode between the open end and closed end, making the pattern A–N–A–N. There are three antinode–node segments in this pattern (A–N, N–A, and A–N), so $L = 3(\lambda_3/4)$ and $\lambda_3 = 4L/3$. The first overtone therefore has frequency $f_3 = v/\lambda_3 = 3v/4L = 3f_1$. Similarly, $f_5 = 5f_1$. In contrast to the pipe open at both ends, **there are no even multiples of the fundamental harmonic.** The odd harmonics for a pipe open at one end only are given by

$$f_n = n\frac{v}{4L} = nf_1 \qquad n = 1, 3, 5, \ldots \qquad [14.19]$$

◀ Pipe closed at one end; only odd harmonics are present

### Quick Quiz

**14.5** A pipe open at both ends resonates at a fundamental frequency $f_{open}$. When one end is covered and the pipe is again made to resonate, the fundamental frequency is $f_{closed}$. Which of the following expressions describes how these two resonant frequencies compare? (a) $f_{closed} = f_{open}$ (b) $f_{closed} = \frac{3}{2} f_{open}$ (c) $f_{closed} = 2 f_{open}$ (d) $f_{closed} = \frac{1}{2} f_{open}$ (e) none of these

**14.6** Balboa Park in San Diego has an outdoor organ. When the air temperature increases, the fundamental frequency of one of the organ pipes (a) increases, (b) decreases, (c) stays the same, or (d) is impossible to determine. (The thermal expansion of the pipe is negligible.)

## APPLYING PHYSICS 14.3   OSCILLATIONS IN A HARBOR

Why do passing ocean waves sometimes cause the water in a harbor to undergo very large oscillations, called a *seiche* (pronounced *saysh*)?

**EXPLANATION** Water in a harbor is enclosed and possesses a natural frequency based on the size of the harbor. This is similar to the natural frequency of the enclosed air in a bottle, which can be excited by blowing across the edge of the opening.

Ocean waves pass by the opening of the harbor at a certain frequency. If this frequency matches that of the enclosed harbor, then a large standing wave can be set up in the water by resonance. This situation can be simulated by carrying a fish tank filled with water. If your walking frequency matches the natural frequency of the water as it sloshes back and forth, a large standing wave develops in the fish tank. ■

## APPLYING PHYSICS 14.4   WHY ARE INSTRUMENTS WARMED UP?

Why do the strings go flat and the wind instruments go sharp during a performance if an orchestra doesn't warm up beforehand?

**EXPLANATION** Without warming up, all the instruments will be at room temperature at the beginning of the concert. As the wind instruments are played, they fill with warm air from the player's exhalation. The increase in temperature of the air in the instruments causes an increase in the speed of

sound, which raises the resonance frequencies of the air columns. As a result, the instruments go sharp. The strings on the stringed instruments also increase in temperature due to the friction of rubbing with the bow. This results in thermal expansion, which causes a decrease in tension in the strings. With the decrease in tension, the wave speed on the strings drops and the fundamental frequencies decrease, so the stringed instruments go flat. ■

## APPLYING PHYSICS 14.5   HOW DO BUGLES WORK?

A bugle has no valves, keys, slides, or finger holes. How can it be used to play a song?

**EXPLANATION** Songs for the bugle are limited to harmonics of the fundamental frequency because there is no control over frequencies without valves, keys, slides, or finger holes. The player obtains different notes by changing the tension

in the lips as the bugle is played, exciting different harmonics. The normal playing range of a bugle is among the third, fourth, fifth, and sixth harmonics of the fundamental. "Reveille," for example, is played with just the three notes G, C, and F, and "Taps" is played with these three notes and the G one octave above the lower G. ■

## EXAMPLE 14.9   HARMONICS OF A PIPE

**GOAL** Find frequencies of open and closed pipes.

**PROBLEM** A pipe is 2.46 m long. **(a)** Determine the frequencies of the first three harmonics if the pipe is open at both ends. Take 343 m/s as the speed of sound in air. **(b)** How many harmonic frequencies of this pipe lie in the audible range, from 20 Hz to 20 000 Hz? **(c)** What are the three lowest possible frequencies if the pipe is closed at one end and open at the other?

**STRATEGY** Substitute into Equation 14.18 for part **(a)** and Equation 14.19 for part **(c)**. All harmonics, $n = 1, 2, 3 \ldots$ are available for the pipe open at both ends, but only the harmonics with $n = 1, 3, 5, \ldots$ for the pipe closed at one end. For part **(b)**, set the frequency in Equation 14.18 equal to $2.00 \times 10^4$ Hz.

. . . . . . . . . . . . . . . . . . . . . . . . . . . . . . . . . . . . . . . . . . . . . . . . . . . . . . . . . . . . . . . . . . . . . . . . . . . . . . . . . . . . . . . . . . . . . . . . . . . . . . .

**SOLUTION**

**(a)** Find the frequencies if the pipe is open at both ends.

Substitute into Equation 14.18, with $n = 1$:

$$f_1 = \frac{v}{2L} = \frac{343 \text{ m/s}}{2(2.46 \text{ m})} = \boxed{69.7 \text{ Hz}}$$

Multiply to find the second and third harmonics:

$$f_2 = 2f_1 = \boxed{139 \text{ Hz}} \qquad f_3 = 3f_1 = \boxed{209 \text{ Hz}}$$

**(b)** How many harmonics lie between 20 Hz and 20 000 Hz for this pipe?

Set the frequency in Equation 14.18 equal to $2.00 \times 10^4$ Hz and solve for $n$:

$$f_n = n\frac{v}{2L} = n\frac{343 \text{ m/s}}{2(2.46 \text{ m})} = 2.00 \times 10^4 \text{ Hz}$$

This works out to $n = 286.88$, which must be truncated down ($n = 287$ gives a frequency over $2.00 \times 10^4$ Hz):

$$n = \boxed{286}$$

(c) Find the frequencies for the pipe closed at one end.

Apply Equation 14.19 with $n = 1$:

$$f_1 = \frac{v}{4L} = \frac{343 \text{ m/s}}{4(2.46 \text{ m})} = \boxed{34.9 \text{ Hz}}$$

The next two harmonics are odd multiples of the first:

$$f_3 = 3f_1 = \boxed{105 \text{ Hz}} \qquad f_5 = 5f_1 = \boxed{175 \text{ Hz}}$$

**REMARKS** For a pipe closed at one end, referring to the "second harmonic" can be confusing, because that corresponds to $f_3$. Calling it the first overtone, however, is unambiguous.

**QUESTION 14.9** True or False: The fundamental wavelength of a longer pipe is greater than the fundamental wavelength of a shorter pipe.

**EXERCISE 14.9** (a) What length pipe open at both ends has a fundamental frequency of $3.70 \times 10^2$ Hz? Find the first overtone. (b) If the one end of this pipe is now closed, what is the new fundamental frequency? Find the first overtone. (c) If the pipe is open at one end only, how many harmonics are possible in the normal hearing range from 20 to 20 000 Hz?

**ANSWERS** (a) 0.464 m, $7.40 \times 10^2$ Hz   (b) 185 Hz, 555 Hz   (c) 54

---

## EXAMPLE 14.10   RESONANCE IN A TUBE OF VARIABLE LENGTH

**GOAL** Understand resonance in tubes and perform elementary calculations.

**PROBLEM** Figure 14.25a shows a simple apparatus for demonstrating resonance in a tube. A long tube open at both ends is partially submerged in a beaker of water, and a vibrating tuning fork of unknown frequency is placed near the top of the tube. The length of the air column, $L$, is adjusted by moving the tube vertically. The sound waves generated by the fork are reinforced when the length of the air column corresponds to one of the resonant frequencies of the tube. Suppose the smallest value of $L$ for which a peak occurs in the sound intensity is 9.00 cm. (a) With this measurement, determine the frequency of the tuning fork. (b) Find the wavelength and the next two air-column lengths giving resonance. Take the speed of sound to be 343 m/s.

**STRATEGY** Once the tube is in the water, the setup is the same as a pipe closed at one end. For part (a), substitute values for $v$ and $L$ into Equation 14.19 with $n = 1$, and find the frequency of the tuning fork. (b) The next resonance maximum occurs when the water level is low enough in the straw to allow a second node (see Fig. 14.25b), which is another half-wavelength in distance. The third resonance occurs when the third node is reached, requiring yet another half-wavelength of distance. The frequency in each case is the same because it's generated by the tuning fork.

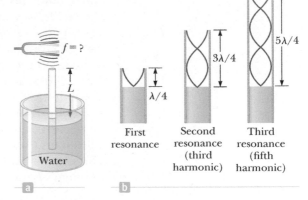

**Figure 14.25** (Example 14.10) (a) Apparatus for demonstrating the resonance of sound waves in a tube closed at one end. The length $L$ of the air column is varied by moving the tube vertically while it is partially submerged in water. (b) The first three resonances of the system.

---

**SOLUTION**

**(a)** Find the frequency of the tuning fork.

Substitute $n = 1$, $v = 343$ m/s, and $L_1 = 9.00 \times 10^{-2}$ m into Equation 14.19:

$$f_1 = \frac{v}{4L_1} = \frac{343 \text{ m/s}}{4(9.00 \times 10^{-2} \text{ m})} = \boxed{953 \text{ Hz}}$$

**(b)** Find the wavelength and the next two water levels giving resonance.

Calculate the wavelength, using the fact that, for a tube open at one end, $\lambda = 4L$ for the fundamental.

$$\lambda = 4L_1 = 4(9.00 \times 10^{-2} \text{ m}) = \boxed{0.360 \text{ m}}$$

Add a half-wavelength of distance to $L_1$ to get the next resonance position:

$$L_2 = L_1 + \lambda/2 = 0.090\,0 \text{ m} + 0.180 \text{ m} = \boxed{0.270 \text{ m}}$$

Add another half-wavelength to $L_2$ to obtain the third resonance position:

$$L_3 = L_2 + \lambda/2 = 0.270 \text{ m} + 0.180 \text{ m} = \boxed{0.450 \text{ m}}$$

*(Continued)*

**REMARKS** This experimental arrangement is often used to measure the speed of sound, in which case the frequency of the tuning fork must be known in advance.

**QUESTION 14.10** True or False: The resonant frequency of an air column depends on the length of the column and the speed of sound.

**EXERCISE 14.10** An unknown gas is introduced into the aforementioned apparatus using the same tuning fork, and the first resonance occurs when the air column is 5.84 cm long. Find the speed of sound in the gas.

**ANSWER** 223 m/s

# 14.11 Beats

The interference phenomena we have been discussing so far have involved the superposition of two or more waves with the same frequency, traveling in opposite directions. Another type of interference effect results from the superposition of two waves with slightly different frequencies. In such a situation, the waves at some fixed point are periodically in and out of phase, corresponding to an alternation in time between constructive and destructive interference. To understand this phenomenon, consider Figure 14.26. The two waves shown in Figure 14.26a were emitted by two tuning forks having slightly different frequencies; Figure 14.26b shows the superposition of these waves. At some time $t_a$ the waves are in phase and constructive interference occurs, as demonstrated by the resultant curve in Figure 14.26b. At some later time, however, the vibrations of the two forks move out of step with each other. At time $t_b$, one fork emits a compression while the other emits a rarefaction, and destructive interference occurs, as demonstrated by the curve shown. As time passes, the vibrations of the two forks move out of phase, then into phase again, and so on. As a consequence, a listener at some fixed point hears an alternation in loudness, known as **beats.** The number of beats per second, or the *beat frequency*, equals the difference in frequency between the two sources:

Beat frequency ▶

$$f_b = |f_2 - f_1|$$  [14.20]

where $f_b$ is the beat frequency and $f_1$ and $f_2$ are the two frequencies. The absolute value is used because the beat frequency is a positive quantity and will occur regardless of the order of subtraction.

APPLICATION
Using Beats to Tune a Musical
Instrument

A stringed instrument such as a piano can be tuned by beating a note on the instrument against a note of known frequency. The string can then be tuned to the desired frequency by adjusting the tension until no beats are heard.

**Figure 14.26** Beats are formed by the combination of two waves of slightly different frequencies traveling in the same direction. (a) The individual waves heard by an observer at a fixed point in space. (b) The combined wave has an amplitude (dashed line) that oscillates in time.

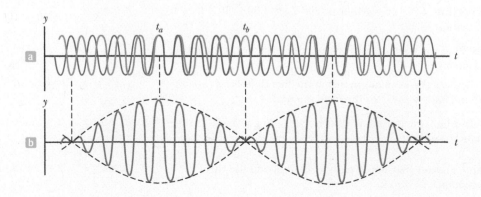

**14.7** You are tuning a guitar by comparing the sound of the string with that of a standard tuning fork. You notice a beat frequency of 5 Hz when both sounds are present. As you tighten the guitar string, the beat frequency rises steadily to 8 Hz. To tune the string exactly to the tuning fork, you should (a) continue to tighten the string, (b) loosen the string, or (c) impossible to determine from the given information.

---

**EXAMPLE 14.11**    **SOUR NOTES**

**GOAL** Apply the beat frequency concept.

**PROBLEM** A certain piano string is supposed to vibrate at a frequency of $4.40 \times 10^2$ Hz. To check its frequency, a tuning fork known to vibrate at a frequency of $4.40 \times 10^2$ Hz is sounded at the same time the piano key is struck, and a beat frequency of 4 beats per second is heard. (a) Find the two possible frequencies at which the string could be vibrating. (b) Suppose the piano tuner runs toward the piano, holding the vibrating tuning fork while his assistant plays the note, which is at 436 Hz. At his maximum speed, the piano tuner notices the beat frequency drops from 4 Hz to 2 Hz (without going through a beat frequency of zero). How fast is he moving? Use a sound speed of 343 m/s. (c) While the piano tuner is running, what beat frequency is observed by the assistant? *Note:* Assume all numbers are accurate to two decimal places, necessary for this last calculation.

**STRATEGY** (a) The beat frequency is equal to the absolute value of the difference in frequency between the two sources of sound and occurs if the piano string is tuned either too high or too low. Solve Equation 14.20 for these two possible frequencies. (b) Moving toward the piano raises the observed piano string frequency. Solve the Doppler shift formula, Equation 14.12, for the speed of the observer. (c) The assistant observes a Doppler shift for the tuning fork. Apply Equation 14.12.

. . . . . . . . . . . . . . . . . . . . . . . . . . . . . . . . . . . . . . . . . . . . . . . . . . . . . . . . . . . . . . . . . . . . . . . . . . . . . . . . . . . .

**SOLUTION**

**(a)** Find the two possible frequencies.

Case 1: $f_2 - f_1$ is already positive, so just drop the absolute-value signs:

$$f_b = f_2 - f_1 \;\rightarrow\; 4\ \text{Hz} = f_2 - 4.40 \times 10^2\ \text{Hz}$$

$$f_2 = \boxed{444\ \text{Hz}}$$

Case 2: $f_2 - f_1$ is negative, so drop the absolute-value signs, but apply an overall negative sign:

$$f_b = -(f_2 - f_1) \;\rightarrow\; 4\ \text{Hz} = -(f_2 - 4.40 \times 10^2\ \text{Hz})$$

$$f_2 = \boxed{436\ \text{Hz}}$$

**(b)** Find the speed of the observer if running toward the piano results in a beat frequency of 2 Hz.

Apply the Doppler shift to the case where frequency of the piano string heard by the running observer is $f_O = 438$ Hz:

$$f_O = f_S \left( \frac{v + v_O}{v - v_S} \right)$$

$$438\ \text{Hz} = (436\ \text{Hz}) \left( \frac{343\ \text{m/s} + v_O}{343\ \text{m/s}} \right)$$

$$v_O = \left( \frac{438\ \text{Hz} - 436\ \text{Hz}}{436\ \text{Hz}} \right)(343\ \text{m/s}) = \boxed{1.57\ \text{m/s}}$$

**(c)** What beat frequency does the assistant observe?

Apply Equation 14.12. Now the source is the tuning fork, so $f_S = 4.40 \times 10^2$ Hz.

$$f_O = f_S \left( \frac{v + v_O}{v - v_S} \right)$$

$$= (4.40 \times 10^2\ \text{Hz}) \left( \frac{343\ \text{m/s}}{343\ \text{m/s} - 1.57\ \text{m/s}} \right) = 442\ \text{Hz}$$

Compute the beat frequency:

$$f_b = f_2 - f_1 = 442\ \text{Hz} - 436\ \text{Hz} = \boxed{6\ \text{Hz}}$$

*(Continued)*

**REMARKS** The assistant on the piano bench and the tuner running with the fork observe different beat frequencies. Many physical observations depend on the state of motion of the observer, a subject discussed more fully in Topic 26, on relativity.

**QUESTION 14.11** Why aren't beats heard when two different notes are played on the piano?

**EXERCISE 14.11** The assistant adjusts the tension in the same piano string, and a beat frequency of 2.00 Hz is heard when the note and the tuning fork are struck at the same time. **(a)** Find the two possible frequencies of the string. **(b)** Assume the actual string frequency is the higher frequency. If the piano tuner runs away from the piano at 4.00 m/s while holding the vibrating tuning fork, what beat frequency does he hear? **(c)** What beat frequency does the assistant on the bench hear? Use 343 m/s for the speed of sound.

**ANSWERS** **(a)** 438 Hz, 442 Hz **(b)** 3 Hz **(c)** 7 Hz

# 14.12 Quality of Sound

**Tip 14.5** Pitch Is Not the Same as Frequency

Although pitch is related mostly (but not completely) to frequency, the two terms are not the same. A phrase such as "the pitch of the sound" is incorrect because pitch is not a physical property of the sound. Frequency is the physical measurement of the number of oscillations per second of the sound. Pitch is a psychological reaction to sound that enables a human being to place the sound on a scale from high to low or from treble to bass. Frequency is the stimulus and pitch is the response.

The sound-wave patterns produced by most musical instruments are complex. Figure 14.27 shows characteristic waveforms (pressure is plotted on the vertical axis, time on the horizontal axis) produced by a tuning fork, a flute, and a clarinet, each playing the same steady note. Although each instrument has its own characteristic pattern, the figure reveals that each of the waveforms is periodic. Note that the tuning fork produces only one harmonic (the fundamental frequency), but the two instruments emit mixtures of harmonics. Figure 14.28 graphs the harmonics of the waveforms of Figure 14.27. When the note is played on the flute (Fig. 14.27b), part of the sound consists of a vibration at the fundamental frequency, an even higher intensity is contributed by the second harmonic, the fourth harmonic produces about the same intensity as the fundamental, and so on. These sounds add together according to the principle of superposition to give the complex waveform shown. The clarinet emits a certain intensity at a frequency of the first harmonic, about half as much intensity at the frequency of the second harmonic, and so forth. The resultant superposition of these frequencies produces the pattern shown in Figure 14.27c. The tuning fork (Figs. 14.27a and 14.28a) emits sound only at the frequency of the first harmonic.

In music, the characteristic sound of any instrument is referred to as the *quality*, or *timbre*, of the sound. The quality depends on the mixture of harmonics in the sound. We say that the note C on a flute differs in quality from the same C on a clarinet. Instruments such as the bugle, trumpet, violin, and tuba are rich in harmonics. A musician playing a wind instrument can emphasize one or another of these harmonics by changing the configuration of the lips, thereby playing different musical notes with the same valve openings.

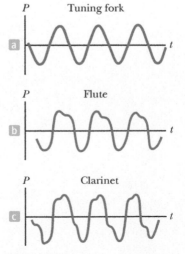

**Figure 14.27** Sound wave patterns produced by various instruments.

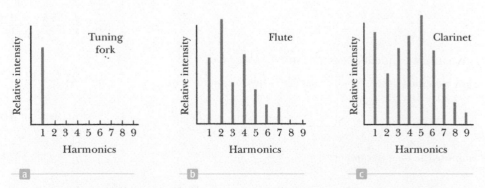

**Figure 14.28** Harmonics of the waveforms in Figure 14.27. Note their variation in intensity.

**APPLYING PHYSICS 14.6    WHY DOES THE PROFESSOR SOUND LIKE DONALD DUCK?**

A professor performs a demonstration in which he breathes helium and then speaks with a comical voice. One student explains, "The velocity of sound in helium is higher than in air, so the fundamental frequency of the standing waves in the mouth is increased." Another student says, "No, the fundamental frequency is determined by the vocal folds and cannot be changed. Only the quality of the voice has changed." Which student is correct?

**EXPLANATION** The second student is correct. The fundamental frequency of the complex tone from the voice is determined by the vibration of the vocal folds and is not changed by substituting a different gas in the mouth. The introduction of the helium into the mouth results in harmonics of higher frequencies being excited more than in the normal voice, but the fundamental frequency of the voice is the same, only the quality has changed. The unusual inclusion of the higher frequency harmonics results in a common description of this effect as a "high-pitched" voice, but that description is incorrect. (It is really a "quacky" timbre.) ∎

# 14.13 The Ear BIO

The human ear is divided into three regions: the outer ear, the middle ear, and the inner ear (Fig. 14.29). The *outer ear* consists of the ear canal (which is open to the atmosphere), terminating at the eardrum (tympanum). Sound waves travel down the ear canal to the eardrum, which vibrates in and out in phase with the pushes and pulls caused by the alternating high and low pressures of the waves. Behind the eardrum are three small bones of the *middle ear,* called the hammer, the anvil, and the stirrup because of their shapes. These bones transmit the vibration to the *inner ear,* which contains the cochlea, a snail-shaped tube about 2 cm long. The cochlea makes contact with the stirrup at the oval window and is divided along its length by the basilar membrane, which consists of small hairs (cilia) and nerve fibers. This membrane varies in mass per unit length and in tension along its length, and different portions of it resonate at different frequencies. (Recall that the natural frequency of a string depends on its mass per unit length and on the tension in it.) Along the basilar membrane are numerous nerve endings, which sense the vibration of the membrane and in turn transmit impulses to the brain. The brain interprets the impulses as sounds of varying frequency, depending on the locations along the basilar membrane of the impulse-transmitting nerves and on the rates at which the impulses are transmitted.

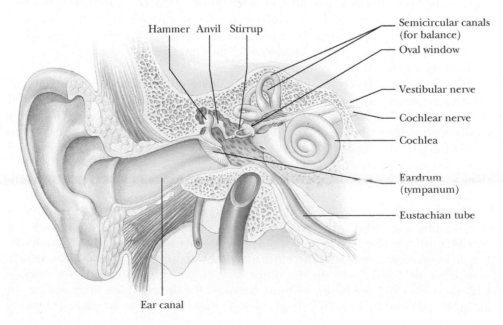

**Figure 14.29** The structure of the human ear. The three tiny bones (ossicles) that connect the eardrum to the window of the cochlea act as a double-lever system to decrease the amplitude of vibration and hence increase the pressure on the fluid in the cochlea.

Hammer  Anvil  Stirrup

Semicircular canals (for balance)

Oval window

Vestibular nerve

Cochlear nerve

Cochlea

Eardrum (tympanum)

Eustachian tube

Ear canal

**Figure 14.30** Curves of intensity level versus frequency for sounds that are perceived to be of equal loudness. Note that the ear is most sensitive at a frequency of about 3 300 Hz. The lowest curve corresponds to the threshold of hearing for only about 1% of the population.

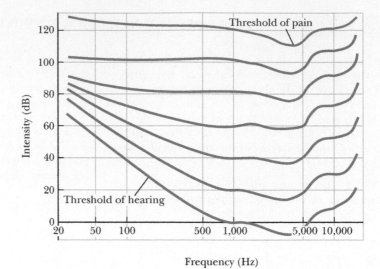

Figure 14.30 shows the frequency response curves of an average human ear for sounds of equal loudness, ranging from 0 to 120 dB. To interpret this series of graphs, take the bottom curve as the threshold of hearing. Compare the intensity level on the vertical axis for the two frequencies 100 Hz and 1 000 Hz. The vertical axis shows that the 100-Hz sound must be about 38 dB greater than the 1 000-Hz sound to be at the threshold of hearing, which means that the threshold of hearing is very strongly dependent on frequency. The easiest frequencies to hear are around 3 300 Hz; those above 12 000 Hz or below about 50 Hz must be relatively intense to be heard.

Now consider the curve labeled 80. This curve uses a 1 000-Hz tone at an intensity level of 80 dB as its reference. The curve shows that a tone of frequency 100 Hz would have to be about 4 dB louder than the 80-dB, 1 000-Hz tone in order to sound as loud. Notice that the curves flatten out as the intensity levels of the sounds increase, so when sounds are loud, all frequencies can be heard equally well.

The small bones in the middle ear represent an intricate lever system that increases the force on the oval window. The pressure is greatly magnified because the surface area of the eardrum is about 20 times that of the oval window (in analogy with a hydraulic press). The middle ear, together with the eardrum and oval window, in effect acts as a matching network between the air in the outer ear and the liquid in the inner ear. The overall energy transfer between the outer ear and the inner ear is highly efficient, with pressure amplification factors of several thousand. In other words, pressure variations in the inner ear are much greater than those in the outer ear.

The ear has its own built-in protection against loud sounds. The muscles connecting the three middle-ear bones to the walls control the volume of the sound by changing the tension on the bones as sound builds up, thus hindering their ability to transmit vibrations. In addition, the eardrum becomes stiffer as the sound intensity increases. These two events make the ear less sensitive to loud incoming sounds. There is a time delay between the onset of a loud sound and the ear's protective reaction, however, so a very sudden loud sound can still damage the ear.

The complex structure of the human ear is believed to be related to the fact that mammals evolved from seagoing creatures. In comparison, insect ears are considerably simpler in design because insects have always been land residents. A typical insect ear consists of an eardrum exposed directly to the air on one side and to an air-filled cavity on the other side. Nerve cells communicate directly with the cavity and the brain, without the need for the complex intermediary of an inner and middle ear. This simple design allows the ear to be placed virtually anywhere on the body. For example, a grasshopper has its ears on its legs. One advantage of the simple insect ear is that the distance and orientation of the ears can be varied so that it is easier to locate sources of sound, such as other insects.

One of the most amazing medical advances in recent decades is the cochlear implant, allowing the deaf to hear. Deafness can occur when the hairlike sensors (cilia) in the cochlea break off over a lifetime or sometimes because of prolonged exposure to loud sounds. Because the cilia don't grow back, the ear loses sensitivity to certain frequencies of sound. The cochlear implant stimulates the nerves in the ear electronically to restore hearing loss that is due to damaged or absent cilia.

**BIO** APPLICATION
Cochlear Implants

## SUMMARY

### 14.2 Characteristics of Sound Waves

**Sound waves** are longitudinal waves. **Audible waves** are sound waves with frequencies between 20 and 20 000 Hz. **Infrasonic waves** have frequencies below the audible range, and **ultrasonic waves** have frequencies above the audible range.

### 14.3 The Speed of Sound

The speed of sound in a medium of bulk modulus $B$ and density $\rho$ is

$$v = \sqrt{\frac{B}{\rho}} \qquad [14.1]$$

The speed of a longitudinal wave in a solid rod is

$$v = \sqrt{\frac{Y}{\rho}} \qquad [14.3]$$

where $Y$ is Young's modulus of the solid and $\rho$ is its density. Equation 14.3 is only valid for a thin, solid rod.

The speed of sound also depends on the temperature of the medium. The relationship between temperature and the speed of sound in air is

$$v = (331 \text{ m/s}) \sqrt{\frac{T}{273 \text{ K}}} \qquad [14.4]$$

where $T$ is the absolute (Kelvin) temperature and 331 m/s is the speed of sound in air at 0°C.

### 14.4 Energy and Intensity of Sound Waves

The **average intensity** of sound incident on a surface is defined by

$$I \equiv \frac{\text{power}}{\text{area}} = \frac{P}{A} \qquad [14.6]$$

where the power $P$ is the energy per unit time flowing through the surface, which has area $A$. The **intensity level** of a sound wave is given by

$$\beta \equiv 10 \log \left( \frac{I}{I_0} \right) \qquad [14.7]$$

The constant $I_0 = 1.0 \times 10^{-12}$ W/m$^2$ is a reference intensity, usually taken to be at the threshold of hearing, and $I$ is the intensity at level $\beta$, with $\beta$ measured in **decibels** (dB).

### 14.5 Spherical and Plane Waves

The **intensity** of a *spherical wave* produced by a point source is proportional to the average power emitted and inversely proportional to the square of the distance from the source:

$$I = \frac{P_{av}}{4\pi r^2} \qquad [14.8]$$

### 14.6 The Doppler Effect

The change in frequency heard by an observer whenever there is relative motion between a source of sound and the observer is called the **Doppler effect.** If the observer is moving with speed $v_O$ and the source is moving with speed $v_S$, the observed frequency is

$$f_O = f_S \left( \frac{v + v_O}{v - v_S} \right) \qquad [14.12]$$

where $v$ is the speed of sound. A positive speed is substituted for $v_O$ when the observer moves toward the source, a negative speed when the observer moves away from the source. Similarly, a positive speed is substituted for $v_S$ when the sources moves toward the observer, a negative speed when the source moves away. Speeds are measured relative to the medium in which the sound is propagated.

### 14.7 Interference of Sound Waves

When waves interfere, the resultant wave is found by adding the individual waves together point by point. When crest meets crest and trough meets trough, the waves undergo **constructive interference,** with path length difference

$$r_2 - r_1 = n\lambda \qquad n = 0, 1, 2, \ldots \qquad [14.13]$$

When crest meets trough, **destructive interference** occurs, with path length difference

$$r_2 - r_1 = (n + \tfrac{1}{2})\lambda \qquad n = 0, 1, 2, \ldots \qquad [14.14]$$

### 14.8 Standing Waves

**Standing waves** are formed when two waves having the same frequency, amplitude, and wavelength travel in opposite directions through a medium. The natural frequencies of vibration of a stretched string of length $L$, fixed at both ends, are

$$f_n = nf_1 = \frac{n}{2L}\sqrt{\frac{F}{\mu}} \qquad n = 1, 2, 3, \ldots \quad [14.17]$$

where $F$ is the tension in the string and $\mu$ is its mass per unit length.

## 14.9 Forced Vibrations and Resonance

A system capable of oscillating is said to be in **resonance** with some driving force whenever the frequency of the driving force matches one of the natural frequencies of the system. When the system is resonating, it oscillates with maximum amplitude.

## 14.10 Standing Waves in Air Columns

Standing waves can be produced in a tube of air. If the reflecting end of the tube is *open*, all harmonics are present and the natural frequencies of vibration are

$$f_n = n\frac{v}{2L} = nf_1 \qquad n = 1, 2, 3, \ldots \quad [14.18]$$

If the tube is *closed* at the reflecting end, only the *odd* harmonics are present and the natural frequencies of vibration are

$$f_n = n\frac{v}{4L} = nf_1 \qquad n = 1, 3, 5, \ldots \quad [14.19]$$

## 14.11 Beats

The phenomenon of **beats** is an interference effect that occurs when two waves with slightly different frequencies combine at a fixed point in space. For sound waves, the intensity of the resultant sound changes periodically with time. The *beat frequency* is

$$f_b = |f_2 - f_1| \qquad [14.20]$$

where $f_2$ and $f_1$ are the two source frequencies.

# CONCEPTUAL QUESTIONS

1. (a) You are driving down the highway in your car when a police car sounding its siren overtakes you and passes you. If its frequency at rest is $f_0$, is the frequency you hear while the car is catching up to you higher or lower than $f_0$? (b) What about the frequency you hear after the car has passed you?

2. When dealing with sound intensities and decibel levels, a convenient approximation (accurate to 2 significant figures) is: For every doubling of the intensity, the decibel level increases by 3.0. Suppose the sound level at some location is 85 dB. Find the decibel levels if the sound intensity is increased by factors of (a) 2.0, (b) 4.0, (c) 8.0, and (d) 16.

3. Fill in the blanks with the correct values (to two significant figures), assuming sound propagates as a spherical wave. If the intensity of a sound is $I_0$ at a given location, and then the distance to the sound source is doubled, the intensity decreases to (a) _____ $I_0$ and the decibel level decreases by (b) _____ dB.

4. Explain how the distance to a lightning bolt (Fig. CQ14.4) can be determined by counting the seconds between the flash and the sound of thunder.

**Figure CQ14.4**

5. Two cars are on the same straight road. Car A moves east at 55 mph and sounds its horn at 625 Hz. Rank from highest to lowest the frequency observed by Car B's driver in each of these cases: Car B is moving east at 70 mph and is (a) behind and (b) in front of Car A. Car B is moving west at 70 mph and is (c) behind and (d) in front of Car A.

6. Why does a vibrating guitar string sound louder when placed on the instrument than it would if allowed to vibrate in the air while off the instrument?

7. You are driving toward the base of a cliff and you honk your horn. (a) Is there a Doppler shift of the sound when you hear the echo? If so, is it like a moving source or moving observer? (b) What if the reflection occurs not from a cliff, but from the forward edge of a huge alien spacecraft moving toward you as you drive?

8. Each of the following statements is related to standing waves on a string. Choose the words that make each statement correct. (i) The harmonic number is equal to the number of [(a) nodes; (b) antinodes]. (ii) The distance from a node to its adjacent antinode is always equal to a [(c) quarter; (d) half] wavelength. (iii) The fundamental frequency has harmonic number [(e) zero; (f) one].

9. Each of the following questions is related to standing waves in air columns. Choose the words that make each statement correct. (i) For a pipe open at both ends, the harmonic number is equal to the number of [(a) nodes; (b) antinodes]. (ii) The distance from a node to its adjacent antinode is always equal to a [(c) quarter; (d) half] wavelength. (iii) For a pipe closed at one end, the wavelength of the fundamental harmonic equals [(e) two; (f) four] times the pipe length.

10. A soft drink bottle resonates as air is blown across its top. What happens to the resonant frequency as the level of fluid in the bottle decreases?

11. An airplane mechanic notices that the sound from a twin-engine aircraft varies rapidly in loudness when both engines are running. What could be causing this variation from loud to soft?

# PROBLEMS

*See the Preface for an explanation of the icons used in this problems set.*

## 14.2 Characteristics of Sound Waves

## 14.3 The Speed of Sound

*Note:* Unless otherwise specified, assume the speed of sound in air is 343 m/s, its value at an air temperature of 20.0°C. At any other Celsius temperature $T_C$, the speed of sound in air is described by Equation 14.4:

$$v = 331\sqrt{1 + \frac{T_C}{273}}$$

where $v$ is in m/s and $T$ is in °C. Use Table 14.1 to find speeds of sound in other media.

1. **V** **Q|C** Suppose you hear a clap of thunder 16.2 s after seeing the associated lightning stroke. The speed of light in air is $3.00 \times 10^8$ m/s. (a) How far are you from the lightning stroke? (b) Do you need to know the value of the speed of light to answer? Explain.

2. Earthquakes at fault lines in Earth's crust create seismic waves, which are longitudinal (P-waves) or transverse (S-waves). The P-waves have a speed of about 7 km/s. Estimate the average bulk modulus of Earth's crust given that the density of rock is about 2 500 kg/m$^3$.

3. On a hot summer day, the temperature of air in Arizona reaches 114°F. What is the speed of sound in air at this temperature?

4. **BIO** A dolphin located in seawater at a temperature of 25°C emits a sound directed toward the bottom of the ocean 150 m below. How much time passes before it hears an echo?

5. A group of hikers hears an echo 3.00 s after shouting. How far away is the mountain that reflected the sound wave?

6. **BIO** The range of human hearing extends from approximately 20 Hz to 20 000 Hz. Find the wavelengths of these extremes at a temperature of 27°C.

7. **BIO** Calculate the reflected percentage of an ultrasound wave passing from human muscle into bone. Muscle has a typical density of $1.06 \times 10^3$ kg/m$^3$ and bone has a typical density of $1.90 \times 10^3$ kg/m$^3$.

8. A stone is dropped from rest into a well. The sound of the splash is heard exactly 2.00 s later. Find the depth of the well if the air temperature is 10.0°C.

9. **Q|C** A hammer strikes one end of a thick steel rail of length 8.50 m. A microphone located at the opposite end of the rail detects two pulses of sound, one that travels through the air and a longitudinal wave that travels through the rail. (a) Which pulse reaches the microphone first? (b) Find the separation in time between the arrivals of the two pulses.

## 14.4 Energy and Intensity of Sound Waves

## 14.5 Spherical and Plane Waves

10. A person standing 1.00 m from a portable speaker hears its sound at an intensity of $7.50 \times 10^{-3}$ W/m$^2$. (a) Find the corresponding decibel level. (b) Find the sound intensity at a distance of 35.0 m, assuming the sound propagates as a spherical wave. (c) Find the decibel level at a distance of 35.0 m.

11. **BIO** The mating call of a male cicada is among the loudest noises in the insect world, reaching decibel levels of 105 dB at a distance of 1.00 m from the insect. (a) Calculate the corresponding sound intensity. (b) Calculate the sound intensity at a distance of 20.0 m from the insect, assuming the sound propagates as a spherical wave. (c) Calculate the decibel level at a distance of 20.0 m from 100 male cicadas each producing the same sound intensity.

12. **Q|C** The intensity level produced by a jet airplane at a certain location is 150 dB. (a) Calculate the intensity of the sound wave generated by the jet at the given location. (b) Compare the answer to part (a) to the threshold of pain and explain why employees directing jet airplanes at airports must wear hearing protection equipment.

13. **V** One of the loudest sounds in recent history was that made by the explosion of Krakatoa on August 26–27, 1883. According to barometric measurements, the sound had a decibel level of 180 dB at a distance of 161 km. Assuming the intensity falls off as the inverse of the distance squared, what was the decibel level on Rodriguez Island, 4 800 km away?

14. A sound wave from a siren has an intensity of 100.0 W/m$^2$ at a certain point, and a second sound wave from a nearby ambulance has an intensity level 10 dB greater than the siren's sound wave at the same point. What is the intensity level of the sound wave due to the ambulance?

15. **BIO** A person wears a hearing aid that uniformly increases the intensity level of all audible frequencies of sound by 30.0 dB. The hearing aid picks up sound having a frequency of 250 Hz at an intensity of $3.0 \times 10^{-11}$ W/m$^2$. What is the intensity delivered to the eardrum?

16. **BIO** The area of a typical eardrum is about $5.0 \times 10^{-5}$ m$^2$. Calculate the sound power (the energy per second) incident on an eardrum at (a) the threshold of hearing and (b) the threshold of pain.

17. **BIO** The toadfish makes use of resonance in a closed tube to produce very loud sounds. The tube is its swim bladder, used as an amplifier. The sound level of this creature has been measured as high as 100. dB. (a) Calculate the intensity of the sound wave emitted. (b) What is the intensity level if three of these toadfish try to make a sound at the same time?

18. **GP** A trumpet creates a sound intensity level of $1.15 \times 10^2$ dB at a distance of 1.00 m. (a) What is the sound intensity of a trumpet at this distance? (b) What is the sound intensity of five trumpets at this distance? (c) Find the sound intensity of five trumpets at the location of the first row of an audience, 8.00 m away, assuming, for simplicity, the sound energy propagates uniformly in all directions. (d) Calculate the decibel level of the five trumpets in the first row. (e) If the trumpets are being played in an outdoor auditorium, how far away, in theory, can their combined sound be heard? (f) In practice such a sound could not be heard once the listener was 2–3 km away. Why can't the sound be heard at the distance found in part (e)? *Hint:* In a very quiet room the ambient sound intensity level is about 30 dB.

19. There is evidence that elephants communicate via in-frasound, generating rumbling vocalizations as low as 14 Hz that can travel up to 10. km. The intensity level of these sounds can reach 103 dB, measured a distance of 5.0 m from the source. Determine the intensity level of the infrasound 10. km from the source, assuming the sound energy radiates uniformly in all directions.

20. A family ice show is held at an enclosed arena. The skaters perform to music playing at a level of 80.0 dB. This intensity level is too loud for your baby, who yells at 75.0 dB. (a) What total sound intensity engulfs you? (b) What is the combined sound level?

21. **T** A train sounds its horn as it approaches an intersection. The horn can just be heard at a level of 50. dB by an observer 10 km away. (a) What is the average power generated by the horn? (b) What intensity level of the horn's sound is observed by someone waiting at an intersection 50. m from the train? Treat the horn as a point source and neglect any absorption of sound by the air.

22. An outside loudspeaker (considered a small source) emits sound waves with a power output of 100 W. (a) Find the intensity 10.0 m from the source. (b) Find the intensity level in decibels at that distance. (c) At what distance would you experience the sound at the threshold of pain, 120 dB?

23. **S** Show that the difference in decibel levels $\beta_1$ and $\beta_2$ of a sound source is related to the ratio of its distances $r_1$ and $r_2$ from the receivers by the formula

$$\beta_2 - \beta_1 = 20 \log\left(\frac{r_1}{r_2}\right)$$

24. A skyrocket explodes 100 m above the ground (Fig. P14.24). Three observers are spaced 100 m apart, with the first (A) directly under the explosion. (a) What is the ratio of the sound intensity heard by observer A to that heard by observer B? (b) What is the ratio of the intensity heard by observer A to that heard by observer C?

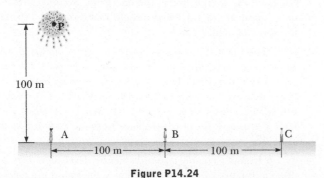

**Figure P14.24**

## 14.6 The Doppler Effect

25. A baseball hits a car, breaking its window and triggering its alarm which sounds at a frequency of 1 250 Hz. What frequency is heard by a boy on a bicycle riding away from the car at 6.50 m/s?

26. A train is moving past a crossing where cars are waiting for it to pass. While waiting, the driver of the lead car becomes sleepy and rests his head on the steering wheel, unintentionally activating the car's horn. A passenger in the back of the train hears the horn's sound at a frequency of 428 Hz and a passenger in the front hears it at 402 Hz. Find (a) the train's speed and (b) the horn's frequency, assuming the sound travels along the tracks.

27. A commuter train passes a passenger platform at a constant speed of 40.0 m/s. The train horn is sounded at its characteristic frequency of 320. Hz. (a) What overall change in frequency is detected by a person on the platform as the train moves from approaching to receding? (b) What wavelength is detected by a person on the platform as the train approaches?

28. An airplane traveling at half the speed of sound emits a sound of frequency 5.00 kHz. At what frequency does a stationary listener hear the sound (a) as the plane approaches? (b) After it passes?

29. **V** Two trains on separate tracks move toward each other. Train 1 has a speed of $1.30 \times 10^2$ km/h; train 2, a speed of 90.0 km/h. Train 2 blows its horn, emitting a frequency of $5.00 \times 10^2$ Hz. What is the frequency heard by the engineer on train 1?

30. At rest, a car's horn sounds the note A (440 Hz). The horn is sounded while the car is moving down the street. A bicyclist moving in the same direction with one-third the car's speed hears a frequency of 415 Hz. (a) Is the cyclist ahead of or behind the car? (b) What is the speed of the car?

31. An alert physics student stands beside the tracks as a train rolls slowly past. He notes that the frequency of the train whistle is 465 Hz when the train is approaching him and 441 Hz when the train is receding from him. Using these frequencies, he calculates the speed of the train. What value does he find?

32. **BIO** **Q|C** A bat flying at 5.00 m/s is chasing an insect flying in the same direction. If the bat emits a 40.0-kHz chirp and receives back an echo at 40.4 kHz, (a) what is the speed of the insect? (b) Will the bat be able to catch the insect? Explain.

33. A tuning fork vibrating at 512 Hz falls from rest and accelerates at 9.80 m/s². How far below the point of release is the tuning fork when waves of frequency 485 Hz reach the release point?

34. **BIO** Expectant parents are thrilled to hear their unborn baby's heartbeat, revealed by an ultrasonic motion detector. Suppose the fetus's ventricular wall moves in simple harmonic motion with amplitude 1.80 mm and frequency 115 beats per minute. The motion detector in contact with the mother's abdomen produces sound at precisely 2 MHz, which travels through tissue at 1.50 km/s. (a) Find the maximum linear speed of the heart wall. (b) Find the maximum frequency at which sound arrives at the wall of the baby's heart. (c) Find the maximum frequency at which reflected sound is received by the motion detector. (By electronically "listening" for echoes at a frequency different from the broadcast frequency, the motion detector can produce beeps of audible sound in synchrony with the fetal heartbeat.)

35. **T** A supersonic jet traveling at Mach 3.00 at an altitude of $h = 2.00 \times 10^4$ m is directly over a person at time $t = 0$ as shown in Figure P14.35. Assume the average speed of sound in air is 335 m/s over the path of the sound. (a) At what time will the person encounter the shock wave due to the sound emitted at $t = 0$? (b) Where will the plane be when this shock wave is heard?

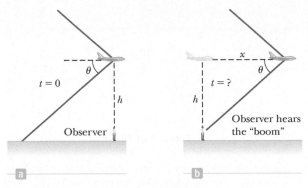

**Figure P14.35**

36. **GP** A yellow submarine traveling horizontally at 11.0 m/s uses sonar with a frequency of $5.27 \times 10^3$ Hz. A red submarine is in front of the yellow submarine and moving 3.00 m/s relative to the water in the same direction. A crewman in the red submarine observes sound waves ("pings") from the yellow submarine. Take the speed of sound in seawater as 1 533 m/s. (a) Write Equation 14.12. (b) Which submarine is the source of the sound? (c) Which submarine carries the observer? (d) Does the motion of the observer's submarine increase or decrease the time between the pressure maxima of the incoming sound waves? How does that affect the observed period? The observed frequency? (e) Should the sign of $v_0$ be positive or negative? (f) Does the motion of the source submarine increase or decrease the time observed between the pressure maxima? How does this motion affect the observed period? The observed frequency? (g) What sign should be chosen for $v_s$? (h) Substitute the appropriate numbers and obtain the frequency observed by the crewman on the red submarine.

## 14.7 Interference of Sound Waves

37. Two cars are stuck in a traffic jam and each sounds its horn at a frequency of 625 Hz. A bicyclist between the two cars, 4.50 m from each horn (Fig. P14.37), is disturbed to find she is at a point of constructive interference. How far backward must she move to reach the nearest point of destructive interference?

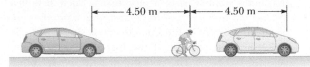

**Figure P14.37**

38. The acoustical system shown in Figure P14.38 is driven by a speaker emitting sound of frequency 756 Hz. (a) If constructive interference occurs at a particular instant, by what minimum amount should the path length in the upper U-shaped tube be increased so that destructive interference occurs instead? (b) What minimum increase in the original length of the upper tube will again result in constructive interference?

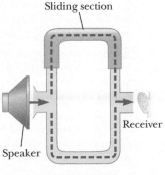

**Figure P14.38**

39. The ship in Figure P14.39 travels along a straight line parallel to the shore and a distance $d = 600$ m from it. The ship's radio receives simultaneous signals of the same frequency from antennas $A$ and $B$, separated by a distance $L = 800$ m. The signals interfere constructively at point $C$, which is equidistant from $A$ and $B$. The signal goes through the first minimum at point $D$, which is directly outward from the shore from point $B$. Determine the wavelength of the radio waves.

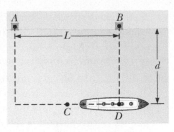

**Figure P14.39**

40. Two loudspeakers are placed above and below each other, as in Figure P14.40 and driven by the same source at a frequency of $4.50 \times 10^2$ Hz. An observer is in front of the speakers (to the right) at point $O$, at the same distance from each speaker. What minimum vertical distance upward should the top speaker be moved to create destructive interference at point $O$?

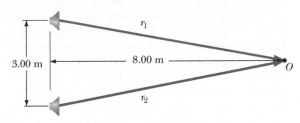

**Figure P14.40**

41. A pair of speakers separated by a distance $d = 0.700$ m are driven by the same oscillator at a frequency of 686 Hz. An observer originally positioned at one of the speakers begins to walk along a line perpendicular to the line joining the speakers as in Figure P14.41. (a) How far must the observer walk before reaching a relative maximum in intensity? (b) How far will the observer be from the speaker when the first relative minimum is detected in the intensity?

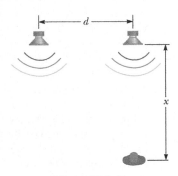

**Figure P14.41**

## 14.8 Standing Waves

42. A steel wire in a piano has a length of 0.700 0 m and a mass of $4.300 \times 10^{-3}$ kg. To what tension must this wire be stretched so that the fundamental vibration corresponds to middle C ($f_C = 261.6$ Hz on the chromatic musical scale)?

43. **V** A stretched string fixed at each end has a mass of 40.0 g and a length of 8.00 m. The tension in the string is 49.0 N. (a) Determine the positions of the nodes and antinodes for the third harmonic. (b) What is the vibration frequency for this harmonic?

44. How far, and in what direction, should a cellist move her finger to adjust a string's tone from an out-of-tune 449 Hz to an in-tune 440 Hz? The string is 68.0 cm long, and the finger is 20.0 cm from the nut for the 449-Hz tone.

45. A stretched string of length $L$ is observed to vibrate in five equal segments when driven by a 630.-Hz oscillator. What oscillator frequency will set up a standing wave so that the string vibrates in three segments?

46. A distance of 5.00 cm is measured between two adjacent nodes of a standing wave on a 20.0-cm-long string. (a) In which harmonic number $n$ is the string vibrating? (b) Find the frequency of this harmonic if the string has a mass of $1.75 \times 10^{-2}$ kg and a tension of 875 N.

47. A steel wire with mass 25.0 g and length 1.35 m is strung on a bass so that the distance from the nut to the bridge is 1.10 m. (a) Compute the linear density of the string. (b) What velocity wave on the string will produce the desired fundamental frequency of the $E_1$ string, 41.2 Hz? (c) Calculate the tension required to obtain the proper frequency. (d) Calculate the wavelength of the string's vibration. (e) What is the wavelength of the sound produced in air? (Assume the speed of sound in air is 343 m/s.)

48. S A standing wave is set up in a string of variable length and tension by a vibrator of variable frequency. Both ends of the string are fixed. When the vibrator has a frequency $f_A$, in a string of length $L_A$ and under tension $T_A$, $n_A$ antinodes are set up in the string. (a) Write an expression for the frequency $f_A$ of a standing wave in terms of the number $n_A$, length $L_A$, tension $T_A$, and linear density $\mu_A$. (b) If the length of the string is doubled to $L_B = 2L_A$, what frequency $f_B$ (written as a multiple of $f_A$) will result in the same number of antinodes? Assume the tension and linear density are unchanged. *Hint:* Make a ratio of expressions for $f_B$ and $f_A$. (c) If the frequency and length are held constant, what tension $T_B$ will produce $n_A + 1$ antinodes? (d) If the frequency is tripled and the length of the string is halved, by what factor should the tension be changed so that twice as many antinodes are produced?

49. A 12.0-kg object hangs in equilibrium from a string with total length of $L = 5.00$ m and linear mass density of $\mu = 0.001\ 00$ kg/m. The string is wrapped around two light, frictionless pulleys that are separated by the distance $d = 2.00$ m (Fig. P14.49a). (a) Determine the tension in the string. (b) At what frequency must the string between the pulleys vibrate in order to form the standing-wave pattern shown in Figure P14.49b?

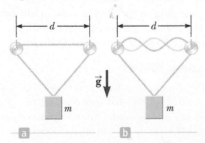

**Figure P14.49**

50. T In the arrangement shown in Figure P14.50, an object of mass $m = 5.0$ kg hangs from a cord around a light pulley. The length of the cord between point $P$ and the pulley is $L = 2.0$ m. (a) When the vibrator is set to a frequency of 150 Hz, a standing wave with six loops is formed. What must be the linear mass density of the cord? (b) How many loops (if any) will result if $m$ is changed to 45 kg? (c) How many loops (if any) will result if $m$ is changed to 10 kg?

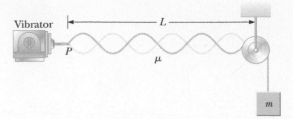

**Figure P14.50**

51. BIO A 60.00-cm guitar string under a tension of 50.000 N has a mass per unit length of 0.100 00 g/cm. What is the highest resonant frequency that can be heard by a person capable of hearing frequencies up to 20 000 Hz?

## 14.9 Forced Vibrations and Resonance

52. Standing-wave vibrations are set up in a crystal goblet with four nodes and four antinodes equally spaced around the 20.0-cm circumference of its rim. If transverse waves move around the glass at 900. m/s, an opera singer would have to produce a high harmonic with what frequency in order to shatter the glass with a resonant vibration?

53. A car's 30.0-kg front tire is suspended by a spring with spring constant $k = 1.00 \times 10^5$ N/m. At what speed is the car moving if washboard bumps on the road every 0.750 m drive the tire into a resonant oscillation?

## 14.10 Standing Waves in Air Columns

54. A pipe has a length of 0.750 m and is open at both ends. (a) Calculate the two lowest harmonics of the pipe. (b) Calculate the two lowest harmonics after one end of the pipe is closed.

55. The windpipe of a typical whooping crane is about 5.0 ft. long. What is the lowest resonant frequency of this pipe, assuming it is closed at one end? Assume a temperature of 37°C.

56. The overall length of a piccolo is 32.0 cm. The resonating air column vibrates as in a pipe that is open at both ends. (a) Find the frequency of the lowest note a piccolo can play. (b) Opening holes in the side effectively shortens the length of the resonant column. If the highest note a piccolo can sound is $4.00 \times 10^3$ Hz, find the distance between adjacent antinodes for this mode of vibration.

57. BIO V The human ear canal is about 2.8 cm long. If it is regarded as a tube that is open at one end and closed at the eardrum, what is the fundamental frequency around which we would expect hearing to be most sensitive?

58. Q|C A tunnel under a river is 2.00 km long. (a) At what frequencies can the air in the tunnel resonate? (b) Explain whether it would be good to make a rule against blowing your car horn when you are in the tunnel.

59. T A pipe open at both ends has a fundamental frequency of $3.00 \times 10^2$ Hz when the temperature is 0°C. (a) What is the length of the pipe? (b) What is the fundamental frequency at a temperature of 30.0°C?

60. Two adjacent natural frequencies of an organ pipe are found to be 550. Hz and 650. Hz. (a) Calculate the fundamental frequency of the pipe. (b) Is the pipe open at both ends or open at only one end? (c) What is the length of the pipe?

## 14.11 Beats

**61.** A guitarist sounds a tuner at 196 Hz while his guitar sounds a frequency of 199 Hz. Find the beat frequency.

**62.** Two nearby trumpets are sounded together and a beat frequency of 2 Hz is heard. If one of the trumpets sounds at a frequency of 525 Hz, what are the two possible frequencies of the other trumpet?

**63.** In certain ranges of a piano keyboard, more than one string is tuned to the same note to provide extra loudness. For example, the note at $1.10 \times 10^2$ Hz has two strings at this frequency. If one string slips from its normal tension of $6.00 \times 10^2$ N to $5.40 \times 10^2$ N, what beat frequency is heard when the hammer strikes the two strings simultaneously?

**64.** The G string on a violin has a fundamental frequency of 196 Hz. It is 30.0 cm long and has a mass of 0.500 g. While this string is sounding, a nearby violinist effectively shortens the G string on her identical violin (by sliding her finger down the string) until a beat frequency of 2.00 Hz is heard between the two strings. When that occurs, what is the effective length of her string?

**65.** Two train whistles have identical frequencies of $1.80 \times 10^2$ Hz. When one train is at rest in the station and the other is moving nearby, a commuter standing on the station platform hears beats with a frequency of 2.00 beats/s when the whistles operate together. What are the two possible speeds and directions that the moving train can have?

**66.** Two pipes of equal length are each open at one end. Each has a fundamental frequency of 480. Hz at 300. K. In one pipe the air temperature is increased to 305 K. If the two pipes are sounded together, what beat frequency results?

**67.** A student holds a tuning fork oscillating at 256 Hz. He walks toward a wall at a constant speed of 1.33 m/s. (a) What beat frequency does he observe between the tuning fork and its echo? (b) How fast must he walk away from the wall to observe a beat frequency of 5.00 Hz?

## 14.13 The Ear

**68.** BIO If a human ear canal can be thought of as resembling an organ pipe, closed at one end, that resonates at a fundamental frequency of $3.0 \times 10^3$ Hz, what is the length of the canal? Use a normal body temperature of 37.0°C for your determination of the speed of sound in the canal.

**69.** BIO Some studies suggest that the upper frequency limit of hearing is determined by the diameter of the eardrum. The wavelength of the sound wave and the diameter of the eardrum are approximately equal at this upper limit. If the relationship holds exactly, what is the diameter of the eardrum of a person capable of hearing $2.00 \times 10^4$ Hz? (Assume a body temperature of 37.0°C.)

## Additional Problems

**70.** A typical sound level for a buzzing mosquito is 40 dB, and that of a vacuum cleaner is approximately 70 dB. Approximately how many buzzing mosquitoes will produce a sound intensity equal to that of a vacuum cleaner?

**71.** Assume a 150.-W loudspeaker broadcasts sound equally in all directions and produces sound with a level of 103 dB at a distance of 1.60 m from its center. (a) Find its sound power output. If a salesperson claims the speaker is rated at 150. W,

he is referring to the maximum electrical power input to the speaker. (b) Find the efficiency of the speaker, that is, the fraction of input power that is converted into useful output power.

**72.** Two small loudspeakers emit sound waves of different frequencies equally in all directions. Speaker A has an output of 1.00 mW, and speaker B has an output of 1.50 mW. Determine the sound level (in decibels) at point C in Figure P14.72 assuming (a) only speaker A emits sound, (b) only speaker B emits sound, and (c) both speakers emit sound.

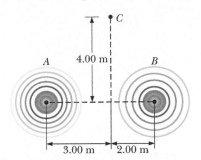

**Figure P14.72**

**73.** An interstate highway has been built through a neighborhood in a city. In the afternoon, the sound level in an apartment in the neighborhood is 80.0 dB as 100 cars pass outside the window every minute. Late at night, the traffic flow is only five cars per minute. What is the average late-night sound level?

**74.** A student uses an audio oscillator of adjustable frequency to measure the depth of a water well. He reports hearing two successive resonances at 52.0 Hz and 60.0 Hz. How deep is the well?

**75.** A stereo speaker is placed between two observers who are 36.0 m apart, along the line connecting them. If one observer records an intensity level of 60.0 dB, and the other records an intensity level of 80.0 dB, how far is the speaker from each observer?

**76.** T Two ships are moving along a line due east (Fig. P14.76). The trailing vessel has a speed relative to a land-based observation point of $v_1 = 64.0$ km/h, and the leading ship has a speed of $v_2 = 45.0$ km/h relative to that point. The two ships are in a region of the ocean where the current is moving uniformly due west at $v_{current} = 10.0$ km/h. The trailing ship transmits a sonar signal at a frequency of 1 200.0 Hz through the water. What frequency is monitored by the leading ship?

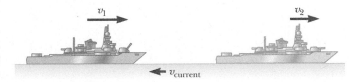

**Figure P14.76**

**77.** On a workday, the average decibel level of a busy street is 70.0 dB, with 100 cars passing a given point every minute. If the number of cars is reduced to 25 every minute on a weekend, what is the decibel level of the street?

**78.** A flute is designed so that it plays a frequency of 261.6 Hz, middle C, when all the holes are covered and the temperature is 20.0°C. (a) Consider the flute to be a pipe open at both ends and find its length, assuming the middle-C frequency is the fundamental frequency. (b) A second player, nearby in a colder room, also attempts to play middle C on an identical flute. A beat frequency of 3.00 beats/s is heard. What is the temperature of the room?

79. A block with a speaker bolted to it is connected to a spring having spring constant $k = 20.0$ N/m, as shown in Figure P14.79. The total mass of the block and speaker is 5.00 kg, and the amplitude of the unit's motion is 0.500 m. If the speaker emits sound waves of frequency 440. Hz, determine the (a) lowest and (b) highest frequencies heard by the person to the right of the speaker.

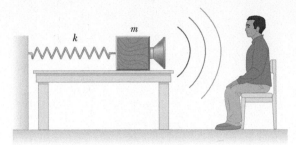

**Figure P14.79**

80. A student stands several meters in front of a smooth reflecting wall, holding a board on which a wire is fixed at each end. The wire, vibrating in its third harmonic, is 75.0 cm long, has a mass of 2.25 g, and is under a tension of 400. N. A second student, moving toward the wall, hears 8.30 beats per second. What is the speed of the student approaching the wall?

81. By proper excitation, it is possible to produce both longitudinal and transverse waves in a long metal rod. In a particular case, the rod is 1.50 m long and 0.200 cm in radius and has a mass of 50.9 g. Young's modulus for the material is $6.80 \times 10^{10}$ Pa. Determine the required tension in the rod so that the ratio of the speed of longitudinal waves to the speed of transverse waves is 8.

82. A 0.500-m-long brass pipe open at both ends has a fundamental frequency of 350. Hz. (a) Determine the temperature of the air in the pipe. (b) If the temperature is increased by 20.0°C, what is the new fundamental frequency of the pipe? Be sure to include the effects of temperature on both the speed of sound in air and the length of the pipe.

# Appendix A
## Mathematics Review

## A.1 Mathematical Notation

Many mathematical symbols are used throughout this book. These symbols are described here, with examples illustrating their use.

### Equals Sign: =

The symbol = denotes the mathematical equality of two quantities. In physics, it also makes a statement about the relationship of different physical concepts. An example is the equation $E = mc^2$. This famous equation says that a given mass $m$, *measured in kilograms*, is equivalent to a certain amount of energy, $E$, *measured in joules*. The speed of light squared, $c^2$, can be considered a constant of proportionality, neccessary because the units chosen for given quantities are rather arbitrary, based on historical circumstances.

### Proportionality: ∝

The symbol ∝ denotes a proportionality. This symbol might be used when focusing on relationships rather than an exact mathematical equality. For example, we could write $E \propto m$, which says "the energy $E$ associated with an object is proportional to the mass $m$ of the object." Another example is found in kinetic energy, which is the energy associated with an object's motion, defined by $KE = \frac{1}{2}mv^2$, where $m$ is again the mass and $v$ is the speed. Both $m$ and $v$ are variables in this expression. Hence, the kinetic energy $KE$ is proportional to $m$, $KE \propto m$, and at the same time $KE$ is proportional to the speed squared, $KE \propto v^2$. Another term used here is "directly proportional." The density $\rho$ of an object is related to its mass and volume by $\rho = m/V$. Consequently, the density is said to be directly proportional to mass and inversely proportional to volume.

### Inequalities

The symbol < means "is less than," and > means "is greater than." For example, $\rho_{Fe} > \rho_{Al}$ means that the density of iron, $\rho_{Fe}$, is greater than the density of aluminum, $\rho_{Al}$. If there is a line underneath the symbol, there is the possibility of equality: ≤ means "less than or equal to," whereas ≥ means "greater than or equal to." Any particle's speed $v$, for example, is less than or equal to the speed of light, $c$: $v \leq c$.

Sometimes the size of a given quantity greatly differs from the size of another quantity. Simple inequality doesn't convey vast differences. For such cases, the symbol << means "is much less than" and >> means "is much greater than." The mass of the Sun, $M_{Sun}$, is much greater than the mass of the Earth, $M_E$: $M_{Sun} \gg M_E$. The mass of an electron, $m_e$, is much less than the mass of a proton, $m_p$: $m_e \ll m_p$.

### Approximately Equal: ≈

The symbol ≈ indicates that two quantities are approximately equal to each other. The mass of a proton, $m_p$, is approximately the same as the mass of a neutron, $m_n$. This relationship can be written $m_p \approx m_n$.

## Equivalence: ≡

The symbol ≡ means "is defined as," which is a different statement than a simple =. It means that the quantity on the left—usually a single quantity—is another way of expressing the quantity or quantities on the right. The classical momentum of an object, $p$, is defined to be the mass of the object $m$ times its velocity $v$, hence $p \equiv mv$. Because this equivalence is by definition, there is no possibility of $p$ being equal to something else. Contrast this case with that of the expression for the velocity $v$ of an object under constant acceleration, which is $v = at + v_0$. This equation would never be written with an equivalence sign because $v$ in this context is not a defined quantity; rather it is an equality that holds true only under the assumption of constant acceleration. The expression for the classical momentum, however, is always true by definition, so it would be appropriate to write $p \equiv mv$ the first time the concept is introduced. After the introduction of a term, an ordinary equals sign generally suffices.

## Differences: Δ

The Greek letter Δ (capital delta) is the symbol used to indicate the difference in a measured physical quantity, usually at two different times. The best example is a displacement along the $x$-axis, indicated by $\Delta x$ (read as "delta x"). Note that $\Delta x$ doesn't mean "the product of Δ and $x$." Suppose a person out for a morning stroll starts measuring her distance away from home when she is 10 m from her doorway. She then continues along a straight-line path and stops strolling 50 m from the door. Her change in position during the walk is $\Delta x = 50 \text{ m} - 10 \text{ m} = 40 \text{ m}$. In symbolic form, such displacements can be written

$$\Delta x = x_f - x_i$$

In this equation, $x_f$ is the final position, and $x_i$ is the initial position. There are numerous other examples of differences in physics, such as the difference (or change) in momentum, $\Delta p = p_f - p_i$; the change in kinetic energy, $\Delta K = K_f - K_i$; and the change in temperature, $\Delta T = T_f - T_i$.

## Summation: Σ

In physics, there are often contexts in which it's necessary to add several quantities. A useful abbreviation for representing such a sum is the Greek letter Σ (capital sigma). Suppose we wish to add a set of five numbers represented by $x_1$, $x_2$, $x_3$, $x_4$, and $x_5$. In the abbreviated notation, we would write the sum as

$$x_1 + x_2 + x_3 + x_4 + x_5 = \sum_{i=1}^{5} x_i$$

where the subscript $i$ on $x$ represents any one of the numbers in the set. For example, if there are five masses in a system, $m_1$, $m_2$, $m_3$, $m_4$, and $m_5$, the total mass of the system $M = m_1 + m_2 + m_3 + m_4 + m_5$ could be expressed as

$$M = \sum_{i=1}^{5} m_i$$

The $x$-coordinate of the center of mass of the five masses, meanwhile, could be written

$$x_{CM} = \frac{\sum_{i=1}^{5} m_i x_i}{M}$$

with similar expressions for the $y$- and $z$-coordinates of the center of mass.

## Absolute Value: | |

The magnitude of a quantity $x$, written $|x|$, is simply the absolute value of that quantity. The sign of $|x|$ is always positive, regardless of the sign of $x$. For example, if $x = -5$, then $|x| = 5$; if $x = 8$, then $|x| = 8$. In physics, this sign is useful whenever the magnitude of a quantity is more important than any direction that might be implied by a sign.

# A.2 Scientific Notation

Many quantities in science have very large or very small values. The speed of light is about 300 000 000 m/s, and the ink required to make the dot over an $i$ in this textbook has a mass of about 0.000 000 001 kg. It's very cumbersome to read, write, and keep track of such numbers because the decimal places have to be counted and because a number with one significant digit may require a large number of zeros. Scientific notation is a way of representing these numbers without having to write out so many zeros, which in general are only used to establish the magnitude of the number, not its accuracy. The key is to use powers of 10. The nonnegative powers of 10 are

$$10^0 = 1$$
$$10^1 = 10$$
$$10^2 = 10 \times 10 = 100$$
$$10^3 = 10 \times 10 \times 10 = 1\ 000$$
$$10^4 = 10 \times 10 \times 10 \times 10 = 10\ 000$$
$$10^5 = 10 \times 10 \times 10 \times 10 \times 10 = 100\ 000$$

and so on. The number of decimal places following the first digit in the number and to the left of the decimal point corresponds to the power to which 10 is raised, called the **exponent** of 10. The speed of light, 300 000 000 m/s, can then be expressed as $3 \times 10^8$ m/s. Notice there are eight decimal places to the right of the leading digit, 3, and to the left of where the decimal point would be placed. Some representative numbers smaller than 1 are

$$10^{-1} = \frac{1}{10} = 0.1$$
$$10^{-2} = \frac{1}{10 \times 10} = 0.01$$
$$10^{-3} = \frac{1}{10 \times 10 \times 10} = 0.001$$
$$10^{-4} = \frac{1}{10 \times 10 \times 10 \times 10} = 0.000\ 1$$
$$10^{-5} = \frac{1}{10 \times 10 \times 10 \times 10 \times 10} = 0.000\ 01$$

In these cases, the number of decimal places to the right of the decimal point up to and including only the first nonzero digit equals the value of the (negative) exponent.

Numbers expressed as some power of 10 multiplied by another number between 1 and 10 are said to be in **scientific notation**. For example, Coulomb's constant, which is associated with electric forces, is given by 8 987 551 789 N·m²/C² and is written in scientific notation as $8.987\ 551\ 789 \times 10^9$ N·m²/C². Newton's constant of gravitation is given by 0.000 000 000 066 731 N·m²/kg², written in scientific notation as $6.673\ 1 \times 10^{-11}$ N·m²/kg².

When numbers expressed in scientific notation are being multiplied, the following general rule is very useful:

$$10^n \times 10^m = 10^{n+m} \qquad \text{[A.1]}$$

where $n$ and $m$ can be any numbers (not necessarily integers). For example, $10^2 \times 10^5 = 10^7$. The rule also applies if one of the exponents is negative: $10^3 \times 10^{-8} = 10^{-5}$.

When dividing numbers expressed in scientific notation, note that

$$\frac{10^n}{10^m} = 10^n \times 10^{-m} = 10^{n-m} \qquad\qquad [A.2]$$

### Exercises

With help from the above rules, verify the answers to the following:

1. $86\ 400 = 8.64 \times 10^4$
2. $9\ 816\ 762.5 = 9.816\ 762\ 5 \times 10^6$
3. $0.000\ 000\ 039\ 8 = 3.98 \times 10^{-8}$
4. $(4 \times 10^8)(9 \times 10^9) = 3.6 \times 10^{18}$
5. $(3 \times 10^7)(6 \times 10^{-12}) = 1.8 \times 10^{-4}$
6. $\dfrac{75 \times 10^{-11}}{5 \times 10^{-3}} = 1.5 \times 10^{-7}$
7. $\dfrac{(3 \times 10^6)(8 \times 10^{-2})}{(2 \times 10^{17})(6 \times 10^5)} = 2 \times 10^{-18}$

# A.3 Algebra

## A. Some Basic Rules

When algebraic operations are performed, the laws of arithmetic apply. Symbols such as $x$, $y$, and $z$ are usually used to represent quantities that are not specified, what are called the **unknowns**.

First, consider the equation

$$8x = 32$$

If we wish to solve for $x$, we can divide (or multiply) each side of the equation by the same factor without destroying the equality. In this case, if we divide both sides by 8, we have

$$\frac{8x}{8} = \frac{32}{8}$$

$$x = 4$$

Next consider the equation

$$x + 2 = 8$$

In this type of expression, we can add or subtract the same quantity from each side. If we subtract 2 from each side, we obtain

$$x + 2 - 2 = 8 - 2$$

$$x = 6$$

In general, if $x + a = b$, then $x = b - a$.

Now consider the equation

$$\frac{x}{5} = 9$$

If we multiply each side by 5, we are left with $x$ on the left by itself and 45 on the right:

$$\left(\frac{x}{5}\right)(5) = 9 \times 5$$

$$x = 45$$

In all cases, **whatever operation is performed on the left side of the equality must also be performed on the right side**.

The following rules for multiplying, dividing, adding, and subtracting fractions should be recalled, where $a$, $b$, and $c$ are three numbers:

| | Rule | Example |
|---|---|---|
| **Multiplying** | $\left(\dfrac{a}{b}\right)\left(\dfrac{c}{d}\right) = \dfrac{ac}{bd}$ | $\left(\dfrac{2}{3}\right)\left(\dfrac{4}{5}\right) = \dfrac{8}{15}$ |
| **Dividing** | $\dfrac{(a/b)}{(c/d)} = \dfrac{ad}{bc}$ | $\dfrac{2/3}{4/5} = \dfrac{(2)(5)}{(4)(3)} = \dfrac{10}{12} = \dfrac{5}{6}$ |
| **Adding** | $\dfrac{a}{b} \pm \dfrac{c}{d} = \dfrac{ad \pm bc}{bd}$ | $\dfrac{2}{3} - \dfrac{4}{5} = \dfrac{(2)(5) - (4)(3)}{(3)(5)} = -\dfrac{2}{15}$ |

Very often in physics we are called upon to manipulate symbolic expressions algebraically, a process most students find unfamiliar. It's very important, however, because substituting numbers into an equation too early can often obscure meaning. The following two examples illustrate how these kinds of algebraic manipulations are carried out.

## EXAMPLE

A ball is dropped from the top of a building 50.0 m tall. How long does it take the ball to fall to a height of 25.0 m?

**SOLUTION** First, write the general ballistics equation for this situation:

$$x = \tfrac{1}{2}at^2 + v_0 t + x_0$$

Here, $a = -9.80 \text{ m/s}^2$ is the acceleration of gravity that causes the ball to fall, $v_0 = 0$ is the initial velocity, and $x_0 = 50.0$ m is the initial position. Substitute only the initial velocity, $v_0 = 0$, obtaining the following equation:

$$x = \tfrac{1}{2}at^2 + x_0$$

This equation must be solved for $t$. Subtract $x_0$ from both sides:

$$x - x_0 = \tfrac{1}{2}at^2 + x_0 - x_0 = \tfrac{1}{2}at^2$$

Multiply both sides by $2/a$:

$$\left(\frac{2}{a}\right)(x - x_0) = \left(\frac{2}{a}\right)\tfrac{1}{2}at^2 = t^2$$

It's customary to have the desired value on the left, so switch the equation around and take the square root of both sides:

$$t = \pm\sqrt{\left(\frac{2}{a}\right)(x - x_0)}$$

Only the positive root makes sense. Values could now be substituted to obtain a final answer.

## EXAMPLE

A block of mass $m$ slides over a frictionless surface in the positive $x$-direction. It encounters a patch of roughness having coefficient of kinetic friction $\mu_k$. If the rough patch has length $\Delta x$, find the speed of the block after leaving the patch.

**SOLUTION** Using the work-energy theorem, we have

$$\tfrac{1}{2}mv^2 - \tfrac{1}{2}mv_0^2 = -\mu_k mg\,\Delta x$$

Add $\tfrac{1}{2}mv_0^2$ to both sides:

$$\tfrac{1}{2}mv^2 = \tfrac{1}{2}mv_0^2 - \mu_k mg\,\Delta x$$

*(Continued)*

Multiply both sides by $2/m$:

$$v^2 = v_0^2 - 2\mu_k g \, \Delta x$$

Finally, take the square root of both sides. Because the block is sliding in the positive $x$-direction, the positive square root is selected.

$$v = \sqrt{v_0^2 - 2\mu_k g \, \Delta x}$$

**Exercises**

In Exercises 1–4, solve for $x$:

**Answers**

1.  $a = \dfrac{1}{1 + x}$                          $x = \dfrac{1 - a}{a}$

2.  $3x - 5 = 13$                $x = 6$

3.  $ax - 5 = bx + 2$          $x = \dfrac{7}{a - b}$

4.  $\dfrac{5}{2x + 6} = \dfrac{3}{4x + 8}$       $x = -\dfrac{11}{7}$

5.  Solve the following equation for $v_1$:

$$P_1 + \tfrac{1}{2}\rho v_1^{\,2} = P_2 + \tfrac{1}{2}\rho v_2^{\,2}$$

Answer:   $v_1 = \pm\sqrt{\dfrac{2}{\rho}(P_2 - P_1) + v_2^{\,2}}$

## B. Powers

When powers of a given quantity $x$ are multiplied, the following rule applies:

$$x^n x^m = x^{n+m} \qquad\qquad [\text{A.3}]$$

For example, $x^2 x^4 = x^{2+4} = x^6$.

When dividing the powers of a given quantity, the rule is

$$\frac{x^n}{x^m} = x^{n-m} \qquad\qquad [\text{A.4}]$$

For example, $x^8/x^2 = x^{8-2} = x^6$.

A power that is a fraction, such as $\tfrac{1}{3}$, corresponds to a root as follows:

$$x^{1/n} = \sqrt[n]{x} \qquad\qquad [\text{A.5}]$$

For example, $4^{1/3} = \sqrt[3]{4} = 1.587\,4$. (A scientific calculator is useful for such calculations.)

Finally, any quantity $x^n$ raised to the $m$th power is

$$\left(x^n\right)^m = x^{nm} \qquad\qquad [\text{A.6}]$$

Table A.1 summarizes the rules of exponents.

**Table A.1** Rules of Exponents

| |
| --- |
| $x^0 = 1$ |
| $x^1 = x$ |
| $x^n x^m = x^{n+m}$ |
| $x^n/x^m = x^{n-m}$ |
| $x^{1/n} = \sqrt[n]{x}$ |
| $\left(x^n\right)^m = x^{nm}$ |

**Exercises**

Verify the following:

1.  $3^2 \times 3^3 = 243$
2.  $x^5 x^{-8} = x^{-3}$
3.  $x^{10}/x^{-5} = x^{15}$
4.  $5^{1/3} = 1.709\,975$ (Use your calculator.)
5.  $60^{1/4} = 2.783\,158$ (Use your calculator.)
6.  $\left(x^4\right)^3 = x^{12}$

## C. Factoring

The following are some useful formulas for factoring an equation:

$$ax + ay + az = a(x + y + z) \qquad \text{common factor}$$

$$a^2 + 2ab + b^2 = (a + b)^2 \qquad \text{perfect square}$$

$$a^2 - b^2 = (a + b)(a - b) \qquad \text{difference of squares}$$

## D. Quadratic Equations

The general form of a quadratic equation is

$$ax^2 + bx + c = 0 \qquad \text{[A.7]}$$

where $x$ is the unknown quantity and $a$, $b$, and $c$ are numerical factors referred to as coefficients of the equation. This equation has two roots, given by

$$x = \frac{-b \pm \sqrt{b^2 - 4ac}}{2a} \qquad \text{[A.8]}$$

If $b^2 - 4ac > 0$, the roots are real.

### EXAMPLE

The equation $x^2 + 5x + 4 = 0$ has the following roots corresponding to the two signs of the square-root term:

$$x = \frac{-5 \pm \sqrt{5^2 - (4)(1)(4)}}{2(1)} = \frac{-5 \pm \sqrt{9}}{2} = \frac{-5 \pm 3}{2}$$

$$x_1 = \frac{-5 + 3}{2} = \boxed{-1} \qquad x_2 = \frac{-5 - 3}{2} = \boxed{-4}$$

where $x_1$ refers to the root corresponding to the positive sign and $x_2$ refers to the root corresponding to the negative sign.

### EXAMPLE

A ball is projected upwards at 16.0 m/s. Use the quadratic formula to determine the time necessary for it to reach a height of 8.00 m above the point of release.

**SOLUTION** From the discussion of ballistics in Topic 2, we can write

$$(1) \quad x = \tfrac{1}{2}at^2 + v_0 t + x_0$$

The acceleration is due to gravity, given by $a = -9.80$ m/s$^2$; the initial velocity is $v_0 = 16.0$ m/s; and the initial position is the point of release, taken to be $x_0 = 0$. Substitute these values into Equation (1) and set $x = 8.00$ m, arriving at

$$x = -4.90t^2 + 16.00t = 8.00$$

where units have been suppressed for mathematical clarity. Rearrange this expression into the standard form of Equation A.7:

$$-4.90t^2 + 16.00t - 8.00 = 0$$

The equation is quadratic in the time, $t$. We have $a = -4.9$, $b = 16$, and $c = -8.00$. Substitute these values into Equation A.8:

$$t = \frac{-16.0 \pm \sqrt{16^2 - 4(-4.90)(-8.00)}}{2(-4.90)} = \frac{-16.0 \pm \sqrt{99.2}}{-9.80}$$

$$= 1.63 \mp \frac{\sqrt{99.2}}{9.80} = \boxed{0.614 \text{ s}, 2.65 \text{ s}}$$

Both solutions are valid in this case, one corresponding to reaching the point of interest on the way up and the other to reaching it on the way back down.

### Exercises

Solve the following quadratic equations:

**Answers**

1. $x^2 + 2x - 3 = 0$      $x_1 = 1$      $x_2 = -3$
2. $2x^2 - 5x + 2 = 0$      $x_1 = 2$      $x_2 = \frac{1}{2}$
3. $2x^2 - 4x - 9 = 0$      $x_1 = 1 + \sqrt{22}/2$      $x_2 = 1 - \sqrt{22}/2$

4. Repeat the ballistics example for a height of 10.0 m above the point of release.
   Answer: $t_1 = 0.842$ s     $t_2 = 2.42$ s

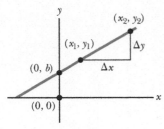

**Figure A.1**

## E. Linear Equations

A linear equation has the general form

$$y = mx + b \qquad [\text{A.9}]$$

where $m$ and $b$ are constants. This kind of equation is called linear because the graph of $y$ versus $x$ is a straight line, as shown in Figure A.1. The constant $b$, called the $y$-intercept, represents the value of $y$ at which the straight line intersects the $y$-axis. The constant $m$ is equal to the slope of the straight line. If any two points on the straight line are specified by the coordinates $(x_1, y_1)$ and $(x_2, y_2)$, as in Figure A.1, the slope of the straight line can be expressed as

$$\text{Slope} = \frac{y_2 - y_1}{x_2 - x_1} = \frac{\Delta y}{\Delta x} \qquad [\text{A.10}]$$

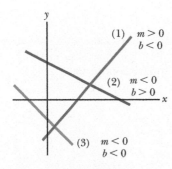

**Figure A.2**

Note that $m$ and $b$ can have either positive or negative values. If $m > 0$, the straight line has a positive slope, as in Figure A.1. If $m < 0$, the straight line has a negative slope. In Figure A.1, both $m$ and $b$ are positive. Three other possible situations are shown in Figure A.2.

---

### EXAMPLE

Suppose the electrical resistance of a metal wire is 5.00 Ω at a temperature of 20.0°C and 6.14 Ω at 80.0°C. Assuming the resistance changes linearly, what is the resistance of the wire at 60.0°C?

**SOLUTION** Find the equation of the line describing the resistance $R$ and then substitute the new temperature into it. Two points on the graph of resistance versus temperature, (20.0°C, 5.00 Ω) and (80.0°C, 6.14 Ω), allow computation of the slope:

$$(1) \quad m = \frac{\Delta R}{\Delta T} = \frac{6.14\ \Omega - 5.00\ \Omega}{80.0°C - 20.0°C} = 1.90 \times 10^{-2}\ \Omega/°C$$

Now use the point-slope formulation of a line, with this slope and (20.0°C, 5.00 Ω):

$$(2) \quad R - R_0 = m(T - T_0)$$

$$(3) \quad R - 5.00\ \Omega = (1.90 \times 10^{-2}\ \Omega/°C)(T - 20.0°C)$$

Finally, substitute $T = 60.0°$ into Equation (3) and solve for $R$, getting $R = 5.76$ Ω.

---

### Exercises

1. Draw graphs of the following straight lines:
   (a) $y = 5x + 3$     (b) $y = -2x + 4$     (c) $y = -3x - 6$
2. Find the slopes of the straight lines described in Exercise 1.
   **Answers:** (a) 5    (b) $-2$    (c) $-3$
3. Find the slopes of the straight lines that pass through the following sets of points: (a) $(0, -4)$ and $(4, 2)$   (b) $(0, 0)$ and $(2, -5)$   (c) $(-5, 2)$ and $(4, -2)$
   **Answers:** (a) $3/2$    (b) $-5/2$    (c) $-4/9$

**4.** Suppose an experiment measures the following displacements (in meters) from equilibrium of a vertical spring due to attaching weights (in Newtons): (0.025 0 m, 22.0 N), (0.075 0 m, 66.0 N). Find the spring constant, which is the slope of the line in the graph of weight versus displacement.
**Answer:** 880 N/m

## F. Solving Simultaneous Linear Equations

Consider the equation $3x + 5y = 15$, which has two unknowns, $x$ and $y$. Such an equation doesn't have a unique solution. For example, note that ($x = 0$, $y = 3$), ($x = 5$, $y = 0$), and ($x = 2$, $y = 9/5$) are all solutions to this equation.

If a problem has two unknowns, a unique solution is possible only if we have two equations. In general, if a problem has $n$ unknowns, its solution requires $n$ equations. To solve two simultaneous equations involving two unknowns, $x$ and $y$, we solve one of the equations for $x$ in terms of $y$ and substitute this expression into the other equation.

### EXAMPLE

Solve the following two simultaneous equations:

$$\textbf{(1)} \quad 5x + y = -8 \qquad \textbf{(2)} \quad 2x - 2y = 4$$

**SOLUTION** From Equation (2), we find that $x = y + 2$. Substitution of this value into Equation (1) gives

$$5(y + 2) + y = -8$$
$$6y = -18$$
$$y = \boxed{-3}$$
$$x = y + 2 = \boxed{-1}$$

**ALTERNATE SOLUTION** Multiply each term in Equation (1) by the factor 2 and add the result to Equation (2):

$$10x + 2y = -16$$
$$\underline{2x - 2y = 4}$$
$$12x = -12$$
$$x = \boxed{-1}$$
$$y = x - 2 = \boxed{-3}$$

Two linear equations containing two unknowns can also be solved by a graphical method. If the straight lines corresponding to the two equations are plotted in a conventional coordinate system, the intersection of the two lines represents the solution. For example, consider the two equations

$$x - y = 2$$
$$x - 2y = -1$$

These equations are plotted in Figure A.3. The intersection of the two lines has the coordinates $x = 5$, $y = 3$, which represents the solution to the equations. You should check this solution by the analytical technique discussed above.

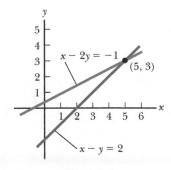

**Figure A.3**

### EXAMPLE

A block of mass $m = 2.00$ kg travels in the positive $x$-direction at $v_i = 5.00$ m/s, while a second block, of mass $M = 4.00$ kg and leading the first block, travels in the positive $x$-direction at 2.00 m/s. The surface is frictionless. What are the blocks' final velocities if the collision is perfectly elastic?

*(Continued)*

**SOLUTION** As can be seen in Topic 6, a perfectly elastic collision involves equations for the momentum and energy. With algebra, the energy equation, which is quadratic in $v$, can be recast as a linear equation. The two equations are given by

$$(1) \quad mv_i + MV_i = mv_f + MV_f$$
$$(2) \quad v_i - V_i = -(v_f - V_f)$$

Substitute the known quantities $v_i = 5.00$ m/s and $V_i = 2.00$ m/s into Equations (1) and (2):

$$(3) \quad 18 = 2v_f + 4V_f$$
$$(4) \quad 3 = -v_f + V_f$$

Multiply Equation (4) by 2 and add to Equation (3):

$$18 = 2v_f + 4V_f$$
$$6 = -2v_f + 2V_f$$
$$\overline{24 = 6V_f} \quad \rightarrow \quad V_f = \boxed{4.00 \text{ m/s}}$$

Substituting the solution for $V_f$ back into Equation (4) yields $v_f = \boxed{1.00 \text{ m/s}}$.

---

### Exercises

Solve the following pairs of simultaneous equations involving two unknowns:

**Answers**

1. $x + y = 8$       $x = 5, y = 3$
   $x - y = 2$
2. $98 - T = 10a$     $T = 65.3, a = 3.27$
   $T - 49 = 5a$
3. $6x + 2y = 6$      $x = 2, y = -3$
   $8x - 4y = 28$

## G. Logarithms and Exponentials

Suppose a quantity $x$ is expressed as a power of some quantity $a$:

$$x = a^y \qquad \text{[A.11]}$$

The number $a$ is called the **base** number. The **logarithm** of $x$ with respect to the base $a$ is equal to the exponent to which the base must be raised so as to satisfy the expression $x = a^y$:

$$y = \log_a x \qquad \text{[A.12]}$$

Conversely, the **antilogarithm** of $y$ is the number $x$:

$$x = \text{antilog}_a y \qquad \text{[A.13]}$$

The antilog expression is in fact identical to the exponential expression in Equation A.11, which is preferable for practical purposes.

In practice, the two bases most often used are base 10, called the *common* logarithm base, and base $e = 2.718 \ldots$, called the *natural* logarithm base. When common logarithms are used,

$$y = \log_{10} x \quad (\text{or } x = 10^y) \qquad \text{[A.14]}$$

When natural logarithms are used,

$$y = \ln x \quad (\text{or } x = e^y) \qquad \text{[A.15]}$$

For example, $\log_{10} 52 = 1.716$, so antilog$_{10} 1.716 = 10^{1.716} = 52$. Likewise, $\ln_e 52 = 3.951$, so antiln$_e 3.951 = e^{3.951} = 52$.

In general, note that you can convert between base 10 and base $e$ with the equality

$$\ln x = (2.302\ 585)\log_{10} x \qquad \text{[A.16]}$$

Finally, some useful properties of logarithms are

$$\log(ab) = \log a + \log b \qquad \ln e = 1$$
$$\log(a/b) = \log a - \log b \qquad \ln e^a = a$$
$$\log(a^n) = n \log a \qquad \ln\left(\frac{1}{a}\right) = -\ln a$$

Logarithms in college physics are used most notably in the definition of decibel level. Sound intensity varies across several orders of magnitude, making it awkward to compare different intensities. Decibel level converts these intensities to a more manageable logarithmic scale.

## EXAMPLE (LOGS)

Suppose a jet testing its engines produces a sound intensity of $I = 0.750$ W at a given location in an airplane hangar. What decibel level corresponds to this sound intensity?

**SOLUTION** Decibel level $\beta$ is defined by

$$\beta = 10 \log\left(\frac{I}{I_0}\right)$$

where $I_0 = 1 \times 10^{-12}$ W/m$^2$ is the standard reference intensity. Substitute the given information:

$$\beta = 10 \log\left(\frac{0.750 \text{ W/m}^2}{10^{-12} \text{ W/m}^2}\right) = 119 \text{ dB}$$

## EXAMPLE (ANTILOGS)

A collection of four identical machines creates a decibel level of $\beta = 87.0$ dB in a machine shop. What sound intensity would be created by only one such machine?

**SOLUTION** We use the equation of decibel level to find the total sound intensity of the four machines, and then we divide by 4. From Equation (1):

$$87.0 \text{ dB} = 10 \log\left(\frac{I}{10^{-12} \text{ W/m}^2}\right)$$

Divide both sides by 10 and take the antilog of both sides, which means, equivalently, to exponentiate:

$$10^{8.7} = 10^{\log(I/10^{-12})} = \frac{I}{10^{-12}}$$
$$I = 10^{-12} \cdot 10^{8.7} = 10^{-3.3} = 5.01 \times 10^{-4} \text{ W/m}^2$$

There are four machines, so this result must be divided by 4 to get the intensity of one machine:

$$I = 1.25 \times 10^{-4} \text{ W/m}^2$$

## EXAMPLE (EXPONENTIALS)

The half-life of tritium is 12.33 years. (Tritium is the heaviest isotope of hydrogen, with a nucleus consisting of a proton and two neutrons.) If a sample contains 3.0 g of tritium initially, how much remains after 20.0 years?

**SOLUTION** The equation giving the number of nuclei of a radioactive substance as a function of time is

$$N = N_0\left(\frac{1}{2}\right)^n$$

(*Continued*)

where $N$ is the number of nuclei remaining, $N_0$ is the initial number of nuclei, and $n$ is the number of half-lives. Note that this equation is an exponential expression with a base of $\frac{1}{2}$. The number of half-lives is given by $n = t/T_{1/2} = 20.0 \text{ yr}/12.33 \text{ yr} = 1.62$. The fractional amount of tritium that remains after 20.0 yr is therefore

$$\frac{N}{N_0} = \left(\frac{1}{2}\right)^{1.62} = 0.325$$

Hence, of the original 3.00 g of tritium, $0.325 \times 3.00 \text{ g} = 0.975 \text{ g}$ remains.

# A.4 Geometry

Table A.2 gives the areas and volumes for several geometric shapes used throughout this text. These areas and volumes are important in numerous physics applications. A good example is the concept of pressure $P$, which is the force per unit area. As an equation, it is written $P = F/A$. Areas must also be calculated in problems involving the volume rate of fluid flow through a pipe using the equation of continuity, the tensile stress exerted on a cable by a weight, the rate of thermal energy transfer through a barrier, and the density of current through a wire. There are numerous other applications. Volumes are important in computing the buoyant force exerted by water on a submerged object, in calculating densities, and in determining the bulk stress of fluid or gas on an object, which affects its volume. Again, there are numerous other applications.

**Table A.2** Useful Information for Geometry

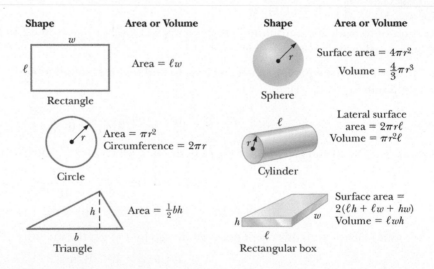

# A.5 Trigonometry

Some of the most basic facts concerning trigonometry are presented in Topic 1, and we encourage you to study the material presented there if you are having trouble with this branch of mathematics. The most important trigonometric concepts include the Pythagorean theorem:

$$\Delta s^2 = \Delta x^2 + \Delta y^2 \qquad \text{[A.17]}$$

This equation states that the square distance along the hypotenuse of a right triangle equals the sum of the squares of the legs. It can also be used to find distances between points in Cartesian coordinates and the length of a vector, where $\Delta x$ is replaced by the $x$-component of the vector and $\Delta y$ is replaced by the $y$-component of the vector. If the vector $\vec{A}$ has components $A_x$ and $A_y$, the magnitude $A$ of the vector satisfies

$$A^2 = A_x^{\,2} + A_y^{\,2} \qquad\qquad \text{[A.18]}$$

which has a form completely analogous to the form of the Pythagorean theorem. Also highly useful are the cosine and sine functions because they relate the length of a vector to its $x$- and $y$-components:

$$A_x = A \cos \theta \qquad\qquad \text{[A.19]}$$
$$A_y = A \sin \theta \qquad\qquad \text{[A.20]}$$

The direction $\theta$ of a vector in a plane can be determined by use of the tangent function:

$$\tan \theta = \frac{A_y}{A_x} \qquad\qquad \text{[A.21]}$$

A relative of the Pythagorean theorem is also frequently useful:

$$\sin^2 \theta + \cos^2 \theta = 1 \qquad\qquad \text{[A.22]}$$

Details on the above concepts can be found in the extensive discussions in Topics 1 and 3. The following are some other useful trigonometric identities:

$$\sin \theta = \cos(90° - \theta)$$

$$\cos \theta = \sin(90° - \theta)$$

$$\sin 2\theta = 2 \sin \theta \cos \theta$$

$$\cos 2\theta = \cos^2 \theta - \sin^2 \theta$$

$$\sin(\theta \pm \phi) = \sin \theta \cos \phi \pm \cos \theta \sin \phi$$

$$\cos(\theta \pm \phi) = \cos \theta \cos \phi \mp \sin \theta \sin \phi$$

The following relationships apply to any triangle, as shown in Figure A.4:

$$\alpha + \beta + \gamma = 180°$$
$$a^2 = b^2 + c^2 - 2bc \cos \alpha$$
$$b^2 = a^2 + c^2 - 2ac \cos \beta \qquad \text{law of cosines}$$
$$c^2 = a^2 + b^2 - 2ab \cos \gamma$$

$$\frac{a}{\sin \alpha} = \frac{b}{\sin \beta} = \frac{c}{\sin \gamma} \qquad \text{law of sines}$$

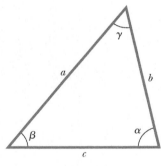

Figure A.4

# Appendix B
## An Abbreviated Table of Isotopes

| Atomic Number Z | Element | Symbol | Chemical Atomic Mass (u) | Mass Number (* Indicates Radioactive) A | Atomic Mass (u) | Percent Abundance | Half-Life (If Radioactive) $T_{1/2}$ |
|---|---|---|---|---|---|---|---|
| 0 | (Neutron) | n | | 1* | 1.008 665 | | 10.4 min |
| 1 | Hydrogen | H | 1.007 94 | 1 | 1.007 825 | 99.988 5 | |
| | Deuterium | D | | 2 | 2.014 102 | 0.011 5 | |
| | Tritium | T | | 3* | 3.016 049 | | 12.33 yr |
| 2 | Helium | He | 4.002 602 | 3 | 3.016 029 | 0.000 137 | |
| | | | | 4 | 4.002 603 | 99.999 863 | |
| 3 | Lithium | Li | 6.941 | 6 | 6.015 122 | 7.5 | |
| | | | | 7 | 7.016 004 | 92.5 | |
| 4 | Beryllium | Be | 9.012 182 | 7* | 7.016 929 | | 53.3 days |
| | | | | 9 | 9.012 182 | 100 | |
| 5 | Boron | B | 10.811 | 10 | 10.012 937 | 19.9 | |
| | | | | 11 | 11.009 306 | 80.1 | |
| 6 | Carbon | C | 12.010 7 | 10* | 10.016 853 | | 19.3 s |
| | | | | 11* | 11.011 434 | | 20.4 min |
| | | | | 12 | 12.000 000 | 98.93 | |
| | | | | 13 | 13.003 355 | 1.07 | |
| | | | | 14* | 14.003 242 | | 5 730 yr |
| 7 | Nitrogen | N | 14.006 7 | 13* | 13.005 739 | | 9.96 min |
| | | | | 14 | 14.003 074 | 99.632 | |
| | | | | 15 | 15.000 109 | 0.368 | |
| 8 | Oxygen | O | 15.999 4 | 15* | 15.003 065 | | 122 s |
| | | | | 16 | 15.994 915 | 99.757 | |
| | | | | 18 | 17.999 160 | 0.205 | |
| 9 | Fluorine | F | 18.998 403 2 | 19 | 18.998 403 | 100 | |
| 10 | Neon | Ne | 20.179 7 | 20 | 19.992 440 | 90.48 | |
| | | | | 22 | 21.991 385 | 9.25 | |
| 11 | Sodium | Na | 22.989 77 | 22* | 21.994 437 | | 2.61 yr |
| | | | | 23 | 22.989 770 | 100 | |
| | | | | 24* | 23.990 963 | | 14.96 h |
| 12 | Magnesium | Mg | 24.305 0 | 24 | 23.985 042 | 78.99 | |
| | | | | 25 | 24.985 837 | 10.00 | |
| | | | | 26 | 25.982 593 | 11.01 | |
| 13 | Aluminum | Al | 26.981 538 | 27 | 26.981 539 | 100 | |
| 14 | Silicon | Si | 28.085 5 | 28 | 27.976 926 | 92.229 7 | |
| 15 | Phosphorus | P | 30.973 761 | 31 | 30.973 762 | 100 | |
| | | | | 32* | 31.973 907 | | 14.26 days |
| 16 | Sulfur | S | 32.066 | 32 | 31.972 071 | 94.93 | |
| | | | | 35* | 34.969 032 | | 87.5 days |
| 17 | Chlorine | Cl | 35.452 7 | 35 | 34.968 853 | 75.78 | |
| | | | | 37 | 36.965 903 | 24.22 | |
| 18 | Argon | Ar | 39.948 | 40 | 39.962 383 | 99.600 3 | |
| 19 | Potassium | K | 39.098 3 | 39 | 38.963 707 | 93.258 1 | |
| | | | | 40* | 39.963 999 | 0.011 7 | $1.28 \times 10^9$ yr |
| 20 | Calcium | Ca | 40.078 | 40 | 39.962 591 | 96.941 | |
| 21 | Scandium | Sc | 44.955 910 | 45 | 44.955 910 | 100 | |
| 22 | Titanium | Ti | 47.867 | 48 | 47.947 947 | 73.72 | |

| Atomic Number Z | Element | Symbol | Chemical Atomic Mass (u) | Mass Number (* Indicates Radioactive) A | Atomic Mass (u) | Percent Abundance | Half-Life (If Radioactive) $T_{1/2}$ |
|---|---|---|---|---|---|---|---|
| 23 | Vanadium | V | 50.941 5 | 51 | 50.943 964 | 99.750 | |
| 24 | Chromium | Cr | 51.996 1 | 52 | 51.940 512 | 83.789 | |
| 25 | Manganese | Mn | 54.938 049 | 55 | 54.938 050 | 100 | |
| 26 | Iron | Fe | 55.845 | 56 | 55.934 942 | 91.754 | |
| 27 | Cobalt | Co | 58.933 200 | 59 | 58.933 200 | 100 | |
| | | | | 60* | 59.933 822 | | 5.27 yr |
| 28 | Nickel | Ni | 58.693 4 | 58 | 57.935 348 | 68.076 9 | |
| | | | | 60 | 59.930 790 | 26.223 1 | |
| 29 | Copper | Cu | 63.546 | 63 | 62.929 601 | 69.17 | |
| | | | | 65 | 64.927 794 | 30.83 | |
| 30 | Zinc | Zn | 65.39 | 64 | 63.929 147 | 48.63 | |
| | | | | 66 | 65.926 037 | 27.90 | |
| | | | | 68 | 67.924 848 | 18.75 | |
| 31 | Gallium | Ga | 69.723 | 69 | 68.925 581 | 60.108 | |
| | | | | 71 | 70.924 705 | 39.892 | |
| 32 | Germanium | Ge | 72.61 | 70 | 69.924 250 | 20.84 | |
| | | | | 72 | 71.922 076 | 27.54 | |
| | | | | 74 | 73.921 178 | 36.28 | |
| 33 | Arsenic | As | 74.921 60 | 75 | 74.921 596 | 100 | |
| 34 | Selenium | Se | 78.96 | 78 | 77.917 310 | 23.77 | |
| | | | | 80 | 79.916 522 | 49.61 | |
| 35 | Bromine | Br | 79.904 | 79 | 78.918 338 | 50.69 | |
| | | | | 81 | 80.916 291 | 49.31 | |
| 36 | Krypton | Kr | 83.80 | 82 | 81.913 485 | 11.58 | |
| | | | | 83 | 82.914 136 | 11.49 | |
| | | | | 84 | 83.911 507 | 57.00 | |
| | | | | 86 | 85.910 610 | 17.30 | |
| 37 | Rubidium | Rb | 85.467 8 | 85 | 84.911 789 | 72.17 | |
| | | | | 87* | 86.909 184 | 27.83 | $4.75 \times 10^{10}$ yr |
| 38 | Strontium | Sr | 87.62 | 86 | 85.909 262 | 9.86 | |
| | | | | 88 | 87.905 614 | 82.58 | |
| | | | | 90* | 89.907 738 | | 29.1 yr |
| 39 | Yttrium | Y | 88.905 85 | 89 | 88.905 848 | 100 | |
| 40 | Zirconium | Zr | 91.224 | 90 | 89.904 704 | 51.45 | |
| | | | | 91 | 90.905 645 | 11.22 | |
| | | | | 92 | 91.905 040 | 17.15 | |
| | | | | 94 | 93.906 316 | 17.38 | |
| 41 | Niobium | Nb | 92.906 38 | 93 | 92.906 378 | 100 | |
| 42 | Molybdenum | Mo | 95.94 | 92 | 91.906 810 | 14.84 | |
| | | | | 95 | 94.905 842 | 15.92 | |
| | | | | 96 | 95.904 679 | 16.68 | |
| | | | | 98 | 97.905 408 | 24.13 | |

(*Continued*)

| Atomic Number Z | Element | Symbol | Chemical Atomic Mass (u) | Mass Number (* Indicates Radioactive) A | Atomic Mass (u) | Percent Abundance | Half-Life (If Radioactive) $T_{1/2}$ |
|---|---|---|---|---|---|---|---|
| 43 | Technetium | Tc | | 98* | 97.907 216 | | $4.2 \times 10^6$ yr |
| | | | | 99* | 98.906 255 | | $2.1 \times 10^5$ yr |
| 44 | Ruthenium | Ru | 101.07 | 99 | 98.905 939 | 12.76 | |
| | | | | 100 | 99.904 220 | 12.60 | |
| | | | | 101 | 100.905 582 | 17.06 | |
| | | | | 102 | 101.904 350 | 31.55 | |
| | | | | 104 | 103.905 430 | 18.62 | |
| 45 | Rhodium | Rh | 102.905 50 | 103 | 102.905 504 | 100 | |
| 46 | Palladium | Pd | 106.42 | 104 | 103.904 035 | 11.14 | |
| | | | | 105 | 104.905 084 | 22.33 | |
| | | | | 106 | 105.903 483 | 27.33 | |
| | | | | 108 | 107.903 894 | 26.46 | |
| | | | | 110 | 109.905 152 | 11.72 | |
| 47 | Silver | Ag | 107.868 2 | 107 | 106.905 093 | 51.839 | |
| | | | | 109 | 108.904 756 | 48.161 | |
| 48 | Cadmium | Cd | 112.411 | 110 | 109.903 006 | 12.49 | |
| | | | | 111 | 110.904 182 | 12.80 | |
| | | | | 112 | 111.902 757 | 24.13 | |
| | | | | 113* | 112.904 401 | 12.22 | $9.3 \times 10^{15}$ yr |
| | | | | 114 | 113.903 358 | 28.73 | |
| 49 | Indium | In | 114.818 | 115* | 114.903 878 | 95.71 | $4.4 \times 10^{14}$ yr |
| 50 | Tin | Sn | 118.710 | 116 | 115.901 744 | 14.54 | |
| | | | | 118 | 117.901 606 | 24.22 | |
| | | | | 120 | 119.902 197 | 32.58 | |
| 51 | Antimony | Sb | 121.760 | 121 | 120.903 818 | 57.21 | |
| | | | | 123 | 122.904 216 | 42.79 | |
| 52 | Tellurium | Te | 127.60 | 126 | 125.903 306 | 18.84 | |
| | | | | 128* | 127.904 461 | 31.74 | $> 8 \times 10^{24}$ yr |
| | | | | 130* | 129.906 223 | 34.08 | $\leq 1.25 \times 10^{21}$ yr |
| 53 | Iodine | I | 126.904 47 | 127 | 126.904 468 | 100 | |
| | | | | 129* | 128.904 988 | | $1.6 \times 10^7$ yr |
| 54 | Xenon | Xe | 131.29 | 129 | 128.904 780 | 26.44 | |
| | | | | 131 | 130.905 082 | 21.18 | |
| | | | | 132 | 131.904 145 | 26.89 | |
| | | | | 134 | 133.905 394 | 10.44 | |
| | | | | 136* | 135.907 220 | 8.87 | $\geq 2.36 \times 10^{21}$ yr |
| 55 | Cesium | Cs | 132.905 45 | 133 | 132.905 447 | 100 | |
| 56 | Barium | Ba | 137.327 | 137 | 136.905 821 | 11.232 | |
| | | | | 138 | 137.905 241 | 71.698 | |
| 57 | Lanthanum | La | 138.905 5 | 139 | 138.906 349 | 99.910 | |
| 58 | Cerium | Ce | 140.116 | 140 | 139.905 434 | 88.450 | |
| | | | | 142* | 141.909 240 | 11.114 | $> 5 \times 10^{16}$ yr |
| 59 | Praseodymium | Pr | 140.907 65 | 141 | 140.907 648 | 100 | |
| 60 | Neodymium | Nd | 144.24 | 142 | 141.907 719 | 27.2 | |
| | | | | 144* | 143.910 083 | 23.8 | $2.3 \times 10^{15}$ yr |
| | | | | 146 | 145.913 112 | 17.2 | |

| Atomic Number Z | Element | Symbol | Chemical Atomic Mass (u) | Mass Number (* Indicates Radioactive) A | Atomic Mass (u) | Percent Abundance | Half-Life (If Radioactive) $T_{1/2}$ |
|---|---|---|---|---|---|---|---|
| 61 | Promethium | Pm | | 145* | 144.912 744 | | 17.7 yr |
| 62 | Samarium | Sm | 150.36 | 147* | 146.914 893 | 14.99 | $1.06 \times 10^{11}$ yr |
| | | | | 149* | 148.917 180 | 13.82 | $> 2 \times 10^{15}$ yr |
| | | | | 152 | 151.919 728 | 26.75 | |
| | | | | 154 | 153.922 205 | 22.75 | |
| 63 | Europium | Eu | 151.964 | 151 | 150.919 846 | 47.81 | |
| | | | | 153 | 152.921 226 | 52.19 | |
| 64 | Gadolinium | Gd | 157.25 | 156 | 155.922 120 | 20.47 | |
| | | | | 158 | 157.924 100 | 24.84 | |
| | | | | 160 | 159.927 051 | 21.86 | |
| 65 | Terbium | Tb | 158.925 34 | 159 | 158.925 343 | 100 | |
| 66 | Dysprosium | Dy | 162.50 | 162 | 161.926 796 | 25.51 | |
| | | | | 163 | 162.928 728 | 24.90 | |
| | | | | 164 | 163.929 171 | 28.18 | |
| 67 | Holmium | Ho | 164.930 32 | 165 | 164.930 320 | 100 | |
| 68 | Erbium | Er | 167.6 | 166 | 165.930 290 | 33.61 | |
| | | | | 167 | 166.932 045 | 22.93 | |
| | | | | 168 | 167.932 368 | 26.78 | |
| 69 | Thulium | Tm | 168.934 21 | 169 | 168.934 211 | 100 | |
| 70 | Ytterbium | Yb | 173.04 | 172 | 171.936 378 | 21.83 | |
| | | | | 173 | 172.938 207 | 16.13 | |
| | | | | 174 | 173.938 858 | 31.83 | |
| 71 | Lutecium | Lu | 174.967 | 175 | 174.940 768 | 97.41 | |
| 72 | Hafnium | Hf | 178.49 | 177 | 176.943 220 | 18.60 | |
| | | | | 178 | 177.943 698 | 27.28 | |
| | | | | 179 | 178.945 815 | 13.62 | |
| | | | | 180 | 179.946 549 | 35.08 | |
| 73 | Tantalum | Ta | 180.947 9 | 181 | 180.947 996 | 99.988 | |
| 74 | Tungsten (Wolfram) | W | 183.84 | 182 | 181.948 206 | 26.50 | |
| | | | | 183 | 182.950 224 | 14.31 | |
| | | | | 184* | 183.950 933 | 30.64 | $> 3 \times 10^{17}$ yr |
| | | | | 186 | 185.954 362 | 28.43 | |
| 75 | Rhenium | Re | 186.207 | 185 | 184.952 956 | 37.40 | |
| | | | | 187* | 186.955 751 | 62.60 | $4.4 \times 10^{10}$ yr |
| 76 | Osmium | Os | 190.23 | 188 | 187.955 836 | 13.24 | |
| | | | | 189 | 188.958 145 | 16.15 | |
| | | | | 190 | 189.958 445 | 26.26 | |
| | | | | 192 | 191.961 479 | 40.78 | |
| 77 | Iridium | Ir | 192.217 | 191 | 190.960 591 | 37.3 | |
| | | | | 193 | 192.962 924 | 62.7 | |
| 78 | Platinum | Pt | 195.078 | 194 | 193.962 664 | 32.967 | |
| | | | | 195 | 194.964 774 | 33.832 | |
| | | | | 196 | 195.964 935 | 25.242 | |
| 79 | Gold | Au | 196.966 55 | 197 | 196.966 552 | 100 | |
| 80 | Mercury | Hg | 200.59 | 199 | 198.968 262 | 16.87 | |
| | | | | 200 | 199.968 309 | 23.10 | |
| | | | | 201 | 200.970 285 | 13.18 | |
| | | | | 202 | 201.970 626 | 29.86 | |

*(Continued)*

| Atomic Number $Z$ | Element | Symbol | Chemical Atomic Mass (u) | Mass Number (* Indicates Radioactive) $A$ | Atomic Mass (u) | Percent Abundance | Half-Life (If Radioactive) $T_{1/2}$ |
|---|---|---|---|---|---|---|---|
| 81 | Thallium | Tl | 204.383 3 | 203 | 202.972 329 | 29.524 | |
| | | | | 205 | 204.974 412 | 70.476 | |
| | | (Th C″) | | 208* | 207.982 005 | | 3.053 min |
| | | (Ra C″) | | 210* | 209.990 066 | | 1.30 min |
| 82 | Lead | Pb | 207.2 | 204* | 203.973 029 | 1.4 | $\geq 1.4 \times 10^{17}$ yr |
| | | | | 206 | 205.974 449 | 24.1 | |
| | | | | 207 | 206.975 881 | 22.1 | |
| | | | | 208 | 207.976 636 | 52.4 | |
| | | (Ra D) | | 210* | 209.984 173 | | 22.3 yr |
| | | (Ac B) | | 211* | 210.988 732 | | 36.1 min |
| | | (Th B) | | 212* | 211.991 888 | | 10.64 h |
| | | (Ra B) | | 214* | 213.999 798 | | 26.8 min |
| 83 | Bismuth | Bi | 208.980 38 | 209 | 208.980 383 | 100 | |
| | | (Th C) | | 211* | 210.987 258 | | 2.14 min |
| 84 | Polonium | Po | | | | | |
| | | (Ra F) | | 210* | 209.982 857 | | 138.38 days |
| | | (Ra C′) | | 214* | 213.995 186 | | 164 $\mu$s |
| 85 | Astatine | At | | 218* | 218.008 682 | | 1.6 s |
| 86 | Radon | Rn | | 222* | 222.017 570 | | 3.823 days |
| 87 | Francium | Fr | | | | | |
| | | (Ac K) | | 223* | 223.019 731 | | 22 min |
| 88 | Radium | Ra | | 226* | 226.025 403 | | 1 600 yr |
| | | (Ms Th$_1$) | | 228* | 228.031 064 | | 5.75 yr |
| 89 | Actinium | Ac | | 227* | 227.027 747 | | 21.77 yr |
| 90 | Thorium | Th | 232.038 1 | | | | |
| | | (Rd Th) | | 228* | 228.028 731 | | 1.913 yr |
| | | (Th) | | 232* | 232.038 050 | 100 | $1.40 \times 10^{10}$ yr |
| 91 | Protactinium | Pa | 231.035 88 | 231* | 231.035 879 | | 32.760 yr |
| 92 | Uranium | U | 238.028 9 | 232* | 232.037 146 | | 69 yr |
| | | | | 233* | 233.039 628 | | $1.59 \times 10^5$ yr |
| | | (Ac U) | | 235* | 235.043 923 | 0.720 0 | $7.04 \times 10^8$ yr |
| | | | | 236* | 236.045 562 | | $2.34 \times 10^7$ yr |
| | | (UI) | | 238* | 238.050 783 | 99.274 5 | $4.47 \times 10^9$ yr |
| 93 | Neptunium | Np | | 237* | 237.048 167 | | $2.14 \times 10^6$ yr |
| 94 | Plutonium | Pu | | 239* | 239.052 156 | | $2.412 \times 10^4$ yr |
| | | | | 242* | 242.058 737 | | $3.73 \times 10^6$ yr |
| | | | | 244* | 244.064 198 | | $8.1 \times 10^7$ yr |

Sources: Chemical atomic masses are from T. B. Coplen, "Atomic Weights of the Elements 1999," a technical report to the International Union of Pure and Applied Chemistry, and published in *Pure and Applied Chemistry*, 73(4), 667–683, 2001. Atomic masses of the isotopes are from G. Audi and A. H. Wapstra, "The 1995 Update to the Atomic Mass Evaluation," *Nuclear Physics*, A595, vol. 4, 409–480, December 25, 1995. Percent abundance values are from K. J. R. Rosman and P. D. P. Taylor, "Isotopic Compositions of the Elements 1999," a technical report to the International Union of Pure and Applied Chemistry, and published in *Pure and Applied Chemistry*, 70(1), 217–236, 1998.

# Appendix C
## Some Useful Tables

**Table C.1** Mathematical Symbols Used in the Text and Their Meaning

| Symbol | Meaning |
|--------|---------|
| $=$ | is equal to |
| $\neq$ | is not equal to |
| $\equiv$ | is defined as |
| $\propto$ | is proportional to |
| $>$ | is greater than |
| $<$ | is less than |
| $\gg$ | is much greater than |
| $\ll$ | is much less than |
| $\approx$ | is approximately equal to |
| $\sim$ | is on the order of magnitude of |
| $\Delta x$ | change in $x$ or uncertainty in $x$ |
| $\Sigma x_i$ | sum of all quantities $x_i$ |
| $\lvert x \rvert$ | absolute value of $x$ (always a positive quantity) |

**Table C.2** Standard Symbols for Units

| Symbol | Unit | Symbol | Unit |
|--------|------|--------|------|
| A | ampere | kcal | kilocalorie |
| Å | angstrom | kg | kilogram |
| atm | atmosphere | km | kilometer |
| Bq | bequerel | kmol | kilomole |
| Btu | British thermal unit | L | liter |
| C | coulomb | lb | pound |
| °C | degree Celsius | ly | lightyear |
| cal | calorie | m | meter |
| cm | centimeter | min | minute |
| Ci | curie | mol | mole |
| d | day | N | newton |
| deg | degree (angle) | nm | nanometer |
| eV | electronvolt | Pa | pascal |
| °F | degree Fahrenheit | rad | radian |
| F | farad | rev | revolution |
| ft | foot | s | second |
| G | Gauss | T | tesla |
| g | gram | u | atomic mass unit |
| H | henry | V | volt |
| h | hour | W | watt |
| hp | horsepower | Wb | weber |
| Hz | hertz | yr | year |
| in. | inch | $\mu$m | micrometer |
| J | joule | $\Omega$ | ohm |
| K | kelvin | | |

**Table C.3** The Greek Alphabet

| | | | | | |
|--|--|--|--|--|--|
| Alpha | A | $\alpha$ | Nu | N | $\nu$ |
| Beta | B | $\beta$ | Xi | $\Xi$ | $\xi$ |
| Gamma | $\Gamma$ | $\gamma$ | Omicron | O | o |
| Delta | $\Delta$ | $\delta$ | Pi | $\Pi$ | $\pi$ |
| Epsilon | E | $\epsilon$ | Rho | P | $\rho$ |
| Zeta | Z | $\zeta$ | Sigma | $\Sigma$ | $\sigma$ |
| Eta | H | $\eta$ | Tau | T | $\tau$ |
| Theta | $\Theta$ | $\theta$ | Upsilon | $\Upsilon$ | $\upsilon$ |
| Iota | I | $\iota$ | Phi | $\Phi$ | $\phi$ |
| Kappa | K | $\kappa$ | Chi | X | $\chi$ |
| Lambda | $\Lambda$ | $\lambda$ | Psi | $\Psi$ | $\psi$ |
| Mu | M | $\mu$ | Omega | $\Omega$ | $\omega$ |

**Table C.4** Physical Data Often Used[a]

| | |
|--|--|
| Average Earth–Moon distance | $3.84 \times 10^8$ m |
| Average Earth–Sun distance | $1.496 \times 10^{11}$ m |
| Equatorial radius of Earth | $6.38 \times 10^6$ m |
| Density of air (20°C and 1 atm) | $1.20$ kg/m$^3$ |
| Density of water (20°C and 1 atm) | $1.00 \times 10^3$ kg/m$^3$ |
| Free-fall acceleration | $9.80$ m/s$^2$ |
| Mass of Earth | $5.98 \times 10^{24}$ kg |
| Mass of Moon | $7.36 \times 10^{22}$ kg |
| Mass of Sun | $1.99 \times 10^{30}$ kg |
| Standard atmospheric pressure | $1.013 \times 10^5$ Pa |

[a] These are the values of the constants as used in the text.

**Table C.5** Some Fundamental Constants

| Quantity | Symbol | Value[a] |
|---|---|---|
| Atomic mass unit | u | $1.660\ 538\ 782\ (83) \times 10^{-27}$ kg |
| | | $931.494\ 028\ (23)$ MeV/$c^2$ |
| Avogadro's number | $N_A$ | $6.022\ 141\ 79\ (30) \times 10^{23}$ particles/mol |
| Bohr magneton | $\mu_B = \dfrac{e\hbar}{2m_e}$ | $9.274\ 009\ 15\ (23) \times 10^{-24}$ J/T |
| Bohr radius | $a_0 = \dfrac{\hbar^2}{m_e e^2 k_e}$ | $5.291\ 772\ 085\ 9\ (36) \times 10^{-11}$ m |
| Boltzmann's constant | $k_B = \dfrac{R}{N_A}$ | $1.380\ 650\ 4\ (24) \times 10^{-23}$ J/K |
| Compton wavelength | $\lambda_C = \dfrac{h}{m_e c}$ | $2.426\ 310\ 217\ 5\ (33) \times 10^{-12}$ m |
| Coulomb constant | $k_e = \dfrac{1}{4\pi\epsilon_0}$ | $8.987\ 551\ 788\ldots \times 10^9$ N·m$^2$/C$^2$ (exact) |
| Deuteron mass | $m_d$ | $3.343\ 583\ 20\ (17) \times 10^{-27}$ kg |
| | | $2.013\ 553\ 212\ 724\ (78)$ u |
| Electron mass | $m_e$ | $9.109\ 382\ 15\ (45) \times 10^{-31}$ kg |
| | | $5.485\ 799\ 094\ 3\ (23) \times 10^{-4}$ u |
| | | $0.510\ 998\ 910\ (13)$ MeV/$c^2$ |
| Electron volt | eV | $1.602\ 176\ 487\ (40) \times 10^{-19}$ J |
| Elementary charge | $e$ | $1.602\ 176\ 487\ (40) \times 10^{-19}$ C |
| Gas constant | $R$ | $8.314\ 472\ (15)$ J/mol·K |
| Gravitational constant | $G$ | $6.674\ 28\ (67) \times 10^{-11}$ N·m$^2$/kg$^2$ |
| Neutron mass | $m_n$ | $1.674\ 927\ 211\ (84) \times 10^{-27}$ kg |
| | | $1.008\ 664\ 915\ 97\ (43)$ u |
| | | $939.565\ 346\ (23)$ MeV/$c^2$ |
| Nuclear magneton | $\mu_n = \dfrac{e\hbar}{2m_p}$ | $5.050\ 783\ 24\ (13) \times 10^{-27}$ J/T |
| Permeability of free space | $\mu_0$ | $4\pi \times 10^{-7}$ T·m/A (exact) |
| Permittivity of free space | $\epsilon_0 = \dfrac{1}{\mu_0 c^2}$ | $8.854\ 187\ 817\ldots \times 10^{-12}$ C$^2$/N·m$^2$ (exact) |
| Planck's constant | $h$ | $6.626\ 068\ 96\ (33) \times 10^{-34}$ J·s |
| | $\hbar = \dfrac{h}{2\pi}$ | $1.054\ 571\ 628\ (53) \times 10^{-34}$ J·s |
| Proton mass | $m_p$ | $1.672\ 621\ 637\ (83) \times 10^{-27}$ kg |
| | | $1.007\ 276\ 466\ 77\ (10)$ u |
| | | $938.272\ 013\ (23)$ MeV/$c^2$ |
| Rydberg constant | $R_H$ | $1.097\ 373\ 156\ 852\ 7\ (73) \times 10^7$ m$^{-1}$ |
| Speed of light in vacuum | $c$ | $2.997\ 924\ 58 \times 10^8$ m/s (exact) |

*Note:* These constants are the values recommended in 2006 by CODATA, based on a least-squares adjustment of data from different measurements. For a more complete list, see P. J. Mohr, B. N. Taylor, and D. B. Newell, "CODATA Recommended Values of the Fundamental Physical Constants: 2006." *Rev. Mod. Phys.* **80:**2, 633–730, 2008.

[a]The numbers in parentheses represent the uncertainties of the last two digits.

# Appendix D
## SI Units

**Table D.1** SI Base Units

| Base Quantity | SI Base Unit | |
| --- | --- | --- |
| | Name | Symbol |
| Length | meter | m |
| Mass | kilogram | kg |
| Time | second | s |
| Electric current | ampere | A |
| Temperature | kelvin | K |
| Amount of substance | mole | mol |
| Luminous intensity | candela | cd |

**Table D.2** Derived SI Units

| Quantity | Name | Symbol | Expression in Terms of Base Units | Expression in Terms of Other SI Units |
| --- | --- | --- | --- | --- |
| Plane angle | radian | rad | $m/m$ | |
| Frequency | hertz | Hz | $s^{-1}$ | |
| Force | newton | N | $kg \cdot m/s^2$ | $J/m$ |
| Pressure | pascal | Pa | $kg/m \cdot s^2$ | $N/m^2$ |
| Energy: work | joule | J | $kg \cdot m^2/s^2$ | $N \cdot m$ |
| Power | watt | W | $kg \cdot m^2/s^3$ | $J/s$ |
| Electric charge | coulomb | C | $A \cdot s$ | |
| Electric potential (emf) | volt | V | $kg \cdot m^2/A \cdot s^3$ | $W/A, J/C$ |
| Capacitance | farad | F | $A^2 \cdot s^4/kg \cdot m^2$ | $C/V$ |
| Electric resistance | ohm | $\Omega$ | $kg \cdot m^2/A^2 \cdot s^3$ | $V/A$ |
| Magnetic flux | weber | Wb | $kg \cdot m^2/A \cdot s^2$ | $V \cdot s, T \cdot m^2$ |
| Magnetic field intensity | tesla | T | $kg/A \cdot s^2$ | $Wb/m^2$ |
| Inductance | henry | H | $kg \cdot m^2/A^2 \cdot s^2$ | $Wb/A$ |

# Answers

## Quick Quizzes, Example Questions, and Odd-Numbered Conceptual Questions and Problems

### TOPIC 1

#### Quick Quizzes

1. c
2. 

| Vector | x-component | y-component |
|---|---|---|
| $\vec{A}$ | – | + |
| $\vec{B}$ | + | – |
| $\vec{A} + \vec{B}$ | – | – |

3. vector $\vec{B}$

#### Example Questions

1. False
2. True
3. $2.6 \times 10^2$ m$^2$
4. $28.0 \text{ m/s} = \left(28.0\frac{\text{m}}{\text{s}}\right)\left(\frac{2.24 \text{ mi/h}}{1.00 \text{ m/s}}\right) = 62.7 \text{ mi/h}$

   The answer is slightly different because the different conversion factors were rounded, leading to small, unpredictable differences in the final answers.
5. $\left(\frac{60.0 \text{ min}}{1.00 \text{ h}}\right)^2$
6. An answer of $10^{12}$ cells is within an order of magnitude of the given answer, corresponding to slightly different choices in the volume estimations. Consequently, $10^{12}$ cells is also a reasonable estimate. An estimate of $10^9$ cells would be suspect because (working backwards) it would imply cells that could be seen with the naked eye!
7. $\sim 4 \times 10^{11}$
8. $\sim 10^{12}$
9. Working backwards, $r = 4.998$, which further rounds to 5.00, whereas $\theta = 37.03°$, which further rounds to 37.0°. The slight differences are caused by rounding.
10. Yes. The cosine function divided into the distance to the building will give the length of the hypotenuse of the triangle in question.
11. If the vectors point in the same direction, the sum of the magnitudes of the two vectors equals the magnitude of the resultant vector.
12. Because $B_x$, $B_y$, and $B$ are all known, any of the trigonometric functions can be used to find the angle.
13. The hiker's displacement vectors are the same.
14. $R = 14.1$ km and $\theta = 71.6°$

#### Conceptual Questions

1. (a) $\sim 0.1$ m  (b) $\sim 1$ m  (c) Between 10 m and 100 m  (d) $\sim 10$ m  (e) $\sim 100$ m
3. $\sim 10^9$ s
5. (a) $\sim 10^6$ beats  (b) $\sim 10^9$ beats
7. Pulse rates vary for many reasons (e.g., activity and excitement) and differ from person to person. These differences lead to significant uncertainty in the measured time intervals.
9. b
11. (a) 1 s  (b) 3 h  (c) 3 months  (d) 2 h  (e) ⅓ s
13. (a) yes  (b) no  (c) yes  (d) no  (e) yes
15. The magnitudes add when $\vec{A}$ and $\vec{B}$ are in the same direction. The resultant will be zero when the two vectors are equal in magnitude and opposite in direction.

#### Problems

3. (b) $V_{\text{cylinder}} = \pi r^2 h$; $A_{\text{circular base}} = \pi r^2$, $V_{\text{rectangular box}}$ = length × width × height ($h$); $A_{\text{rectangular base}}$ = length × width.
5. m$^3$/(kg · s$^2$)
7. $6.8 \times 10^3$ m$^2$
9. 52 m$^2$
11. (a) three significant figures (b) four significant figures (c) three significant figures (d) two significant figures
13. (a) $(3.0 \pm 0.5)$ m$^2$  (b) $(7.0 \pm 0.6)$ m
15. $5.9 \times 10^3$ cm$^3$
17. 4.49 cubiti
19. $2 \times 10^8$ fathoms
21. 0.204 m$^3$
23. Yes. The speed of 38.0 m/s converts to a speed of 85.0 mi/h, so the driver exceeds the speed limit by 10 mi/h.
25. (a) 6.81 cm  (b) $5.83 \times 10^2$ cm$^2$  (c) $1.32 \times 10^3$ cm$^3$
27. $6.71 \times 10^8$ mi/h
29. $3.08 \times 10^4$ m$^3$
31. 9.82 cm
33. $\sim 10^8$ breaths
35. $2 \times 10^{-3}\%$
37. $\sim 10^7$ rev
39. (2.0 m, 1.4 m)
41. $r = 2.2$ m, $\theta = 27°$
43. 5.69 m
45. (a) 6.71 m  (b) 0.894  (c) 0.746
47. 3.41 m
49. (a) 3.00  (b) 3.00  (c) 0.800  (d) 0.800  (e) 1.33
51. 5.00/7.00; the angle itself is 35.5°
53. 70.0 m
55. 43 units in the negative $y$-direction
57. (a) 5 units at $-53°$  (b) 5 units at $+53°$
59. approximately 421 ft at 3° below the horizontal
61. (a) 10.3 m  (b) 119°
63. 28.7 units, 220.1 units
65. (a) 5.00 blocks at 53.1° N of E  (b) 13.0 blocks
67. 47.2 units at 122° from the positive $x$-axis
69. 157 km
71. 245 km at 21.4° W of N
73. The value of $k$, a dimensionless constant, cannot be found by dimensional analysis.
75. 152 $\mu$m
77. $1 \times 10^{10}$ gal/yr
79. (a) $1 \times 10^6$  (b) 50
81. $1 \times 10^5$

### TOPIC 2

#### Quick Quizzes

1. (a) 200 yd  (b) 0  (c) 0  (d) 8 yd/s
2. (a) False  (b) True  (c) True
3. The velocity versus time graph (a) has a constant slope, indicating a constant acceleration, which is represented by the acceleration versus time graph (e).

Graph (b) represents an object with increasing speed, but as time progresses, the lines drawn tangent to the curve have increasing slopes. Since the acceleration is equal to the slope of the tangent line, the acceleration must be increasing, and the acceleration vs. time graph that best indicates this behavior is (d).

Graph (c) depicts an object that first has a velocity that increases at a constant rate, which means that the object's acceleration is constant. The velocity then stops changing, which means that the acceleration of the object is zero. This behavior is best matched by graph (f).

4. (b)
5. (a) blue graph    (b) red graph    (c) green graph
6. (e)
7. (c)
8. (a) and (f)

## Example Questions

1. No. The object may not be traveling in a straight line. If the initial and final positions are in the same place, for example, the displacement is zero regardless of the total distance traveled during the given time.
2. No. A vertical line in a position vs. time graph would mean that an object had somehow traversed all points along the given path instantaneously, which is physically impossible.
3. No. A vertical tangent line would correspond to an infinite acceleration, which is physically impossible.
4. 34.9 m/s
5. The graphical solution is the intersection of a straight line and a parabola.
6. The coasting displacement would double to 143 m, with a total displacement of 715 m.
7. The acceleration is zero wherever the tangent to the velocity versus time graph is horizontal. Visually, that occurs from −50 s to 0 s and then again at approximately 180 s, 320 s, and 330 s.
8. The upward jump would slightly increase the ball's initial velocity, slightly increasing the maximum height.
9. 6
10. The engine should be fired again at 235 m.

## Conceptual Questions

1. Yes. If the velocity of the particle is nonzero, the particle is in motion. If the acceleration is zero, the velocity of the particle is unchanging or is constant.
3. Yes. If this occurs, the acceleration of the car is opposite to the direction of motion, and the car will be slowing down.
5. b
7. (a) Yes.   (b) Yes.
9. b
11. d

## Problems

1. ≈ 0.02 s
3. (a) 52.9 km/h   (b) 90.0 km
5. (a) Boat A wins by 60 km   (b) 0
7. (a) 180. km   (b) 63.4 km/h
9. Yes. Taking the initial speed to be zero, the final speed as 75 m/s, and $a = 1.3$ m/s$^2$, we find that the distance the plane travels before takeoff is $x = v_{to}^2/2a = 2.2$ km, which is less than the length of the runway.
11. (a) 4.4 m/s$^2$   (b) 27 m
13. (a) 2.80 h   (b) 218 km
15. 274 km/h
17. (a) 5.00 m/s   (b) −2.50 m/s   (c) 0   (d) 5.00 m/s
19. 0.18 mi west of the flagpole
21. $|\vec{a}| = 1.34 \times 10^4$ m/s$^2$
23. 3.7 s

25. (a) 70.0 mi/h · s = 31.3 m/s$^2$ = 3.19$g$   (b) 321 ft = 97.8 m
27. −16.0 cm/s$^2$
29. (a) 6.61 m/s   (b) −0.448 m/s$^2$
31. (a) 2.32 m/s$^2$   (b) 14.4 s
33. (a) 8.14 m/s$^2$   (b) 1.23 s   (c) Yes. For uniform acceleration, the velocity is a linear function of time.
35. There will be a collision only if the car and the van meet at the same place at some time. Writing expressions for the position versus time for each vehicle and equating the two gives a quadratic equation in $t$ whose solution is either 11.4 s or 13.6 s. The first solution, 11.4 s, is the time of the collision. The collision occurs when the van is 212 m from the original position of Sue's car.
37. 200 m
39. (a) 1.5 m/s   (b) 32 m
41. 958 m
43. (a) 8.2 s   (b) 1.3 × 10$^2$ m
45. (a) 31.9 m   (b) 2.55 s   (c) 2.55 s   (d) −25.0 m/s
47. (a) −12.7 m/s   (b) 38.2 m
49. Hardwood floor: $a = 2.0 \times 10^3$ m/s$^2$, $\Delta t = 1.4$ ms; carpeted floor: $a = 3.9 \times 10^2$ m/s$^2$, $\Delta t = 7.1$ ms
51. (a) It is a freely falling object, so its acceleration is −9.80 m/s$^2$ (downward) while in flight.   (b) 0   (c) 9.80 m/s   (d) 4.90 m
53. (a) It continues upward, slowing under the influence of gravity until it reaches a maximum height, and then it falls to Earth.   (b) 308 m   (c) 8.51 s   (d) 16.4 s
55. 15.0 s
57. (a) −3.50 × 10$^5$ m/s$^2$   (b) 2.86 × 10$^{-4}$ s
59. (a) 10.0 m/s upward   (b) 4.68 m/s downward
61. (a) 4.0 m/s   (b) 1.0 ms   (c) 0.82 m
63. 0.60 s
65. 6.44 m/s
67. No. In the time interval equal to David's reaction time, the $1 bill (a freely falling object) falls a distance of $gt^2/2 \cong$ 20 cm, which is about twice the distance between the top of the bill and its center.
71. (a) 7.82 m   (b) 0.782 s

# TOPIC 3

## Quick Quizzes

1. (b)
2. (a)
3. (a) 2.09 m/s   (b) 1.33 m/s
4. (c)
5. (b)

## Example Questions

1. The initial and final velocity vectors are equal in magnitude because the $x$-component doesn't change and the $y$-component changes only by a sign.
2. To the pilot, the package appears to drop straight down because the $x$-components of velocity for the plane and package are the same.
3. False
4. 45°
5. False
6. False
7. To an observer on the ground, the ball drops straight down.
8. The angle decreases with increasing speed.
9. The angle is different because relative velocity depends on both the magnitude and the direction of the velocity vectors. In Example 3.9, the boat's velocity vector forms the hypotenuse of a right triangle, whereas in Example 3.8, that vector forms a leg of a right triangle.

## Conceptual Questions

1. (a) At the top of the projectile's flight, its velocity is horizontal and its acceleration is downward. This is the only point at which the velocity and acceleration vectors are perpendicular. (b) If the projectile is thrown straight up or down, then the velocity and acceleration will be parallel throughout the motion. For any other kind of projectile motion, the velocity and acceleration vectors are never parallel.

3. (a) The acceleration is zero, because both the magnitude and direction of the velocity remain constant. (b) The particle has an acceleration because the direction of $\vec{v}$ changes.

5. (a) $v_x$, $a_x$, $a_y$   (b) $a_x$

7. For angles $\theta < 45°$, the projectile thrown at angle $\theta$ will be in the air for a shorter interval. For the smaller angle, the vertical component of the initial velocity is smaller than that for the larger angle. Thus, the projectile thrown at the smaller angle will not go as high into the air and will spend less time in the air before landing.

9. (a) Yes, the projectile is in free fall. (b) Its vertical component of acceleration is the downward acceleration of gravity. (c) Its horizontal component of acceleration is zero.

11. b

13. (i) a   (ii) b

## Problems

1. (a) $7.00 \times 10^3$ m   (b) 46.7 m/s   (c) 73.3 m/s

3. (a) 5.00 m   (b) 2.50 m

5. 7.17 m/s$^2$

7. (a) $(x, y) = (0, 50.0 \text{ m})$   (b) $v_{0x} = 18.0$ m/s, $v_{0y} = 0$
   (c) $v_x = 18.0$ m/s, $v_y = -(9.80 \text{ m/s}^2)t$
   (d) $x = (18.0 \text{ m/s})t$, $y = -(4.90 \text{ m/s}^2)t^2 + 50.0$ m
   (e) 3.19 s   (f) 36.1 m/s, $-60.1°$

9. 12 m/s

11. (a) The ball clears by 0.889 m (b) while descending

13. 25 m

15. (a) 32.5 m from the base of the cliff (b) 1.78 s

17. (a) 52.0 m/s horizontally   (b) 212 m

19. (a) 18.1 m/s   (b) 1.09 m   (c) 2.8 m

21. (a) 14.5 m/s   (b) 9.50 m/s

23. (a) $(\vec{v}_{JA})_x = 3.00 \times 10^2$ mi/h, $(\vec{v}_{JA})_y = 0$,
    (b) $(\vec{v}_{AE})_x = 86.6$ mi/h, $(\vec{v}_{AE})_y = 50.0$ mi/h
    (c) $\vec{v}_{JA} = \vec{v}_{JE} - \vec{v}_{AE}$
    (d) $\vec{v}_{JE} = 3.90 \times 10^2$ mi/h, 7.36° N of E

25. (a) 9.80 m/s$^2$ down and 2.50 m/s$^2$ south (b) 9.80 m/s$^2$ down (c) The bolt moves on a parabola with a vertical axis.

27. (a) $x = 1.81$ m/s, $y = 5.42$ m/s   (b) Yes. Because $v_0 < 6.26$ m/s, the salmon is capable of making this jump.

29. (a) $2.02 \times 10^3$ s (b) $1.67 \times 10^3$ s (c) The time savings going downstream with the current is always less than the extra time required to go the same distance against the current.

31. 18.0 s

33. (a) 0.528 s   (b) 1.62 m

35. 68.6 km/h

37. 5.37 m/s

39. (a) $1.53 \times 10^3$ m   (b) 36.2 s   (c) $4.04 \times 10^3$ m

41. (a) $R_{\text{Moon}} = 18$ m   (b) $R_{\text{Mars}} = 7.9$ m

43. (a) 42 m/s (b) 3.8 s (c) $v_x = 34$ m/s, $v_y = -13$ m/s; $v = 37$ m/s

45. (a) in the direction the ball was thrown   (b) 7.5 m/s

47. 1.56 m

49. 7.5 min

51. 10.8 m

53. (a) 4.6 m/s   (b) 0.67 s

55. (a) 35.1° or 54.9° (b) 42.2 m or 85.4 m, respectively

57. (a) 20.0° above the horizontal (b) 3.05 s

59. (a) 22.9 m/s (b) $3.60 \times 10^2$ m horizontally from the base of cliff

## TOPIC 4

### Quick Quizzes

1. (a), (c), and (d) are true.

2. (b)

3. (a) False
   (b) True
   (c) False
   (d) False

4. (c)   (d)

5. (b)

6. (b)

7. (b) By exerting an upward force component on the sled, you reduce the normal force on the ground and so reduce the force of kinetic friction.

8. (c)

9. (c)

### Example Questions

1. Other than the forces mentioned in the problem, the force of gravity pulls downwards on the boat. Because the boat doesn't sink, a force exerted by the water on the boat must oppose the gravity force. (In Topic 9 this force will be identified as the buoyancy force.)

2. False. The object's mass and the angles at which the forces are applied are also important in determining the magnitude of the acceleration vector.

3. 0.2 N

4. $3g$

5. The gravitational force of the Earth acts on the man, and an equal and opposite gravitational force of the man acts on the Earth. The normal force acts on the man, and the reaction force consists of the man pressing against the surface.

6. The static friction force is proportional to the mass, so if the mass is increased, the static friction force increases proportionately.

7. The tensions would double.

8. The magnitude of the tension force would be greater, and the magnitude of the normal force would be smaller.

9. Doubling the weight doubles the mass, which halves both the acceleration and displacement.

10. A gentler slope means a smaller angle and hence a smaller acceleration down the slope. Consequently, the car would take longer to reach the bottom of the hill.

11. The scale reading is greater than the weight of the fish during the first acceleration phase. When the velocity becomes constant, the scale reading is equal to the weight. When the elevator slows down, the scale reading is less than the weight.

12. The coefficient of kinetic friction would be larger than in the example by a factor of $1/0.35 = 2.6$.

13. Attach one end of the cable to the object to be lifted and the other end to a platform. Place lighter weights on the platform until the total mass of the weights and platform exceeds the mass of the heavy object.

14. Both the acceleration and the tension increase when $m_2$ is increased.

15. The top block would slide off the back end of the lower block.

### Conceptual Questions

1. The inertia of the suitcase would keep it moving forward as the bus stops. There would be no tendency for the suitcase to be thrown backward toward the passenger. The case should be dismissed.

3. (a) $w = mg$ and $g$ decreases with altitude. Thus, to get a good buy, purchase it in Denver. (b) If gold were sold by mass, it would not matter where you bought it.

5. (a) Two external forces act on the ball. (i) One is a downward gravitational force exerted by Earth. (ii) The second force on the ball is an upward normal force exerted by the hand. The reactions to these forces are (i) an upward gravitational force exerted by the ball on Earth and (ii) a downward force exerted by the ball on the hand. (b) After the ball leaves the hand, the only external force acting on the ball is the gravitational force exerted by Earth. The reaction is an upward gravitational force exerted by the ball on Earth.

7. (a) The force causing an automobile to move is the friction between the tires and the roadway as the automobile attempts to push the roadway backward. (b) The force driving a propeller airplane forward is the reaction force exerted by the air on the propeller as the rotating propeller pushes the air backward (the action). (c) In a rowboat, the rower pushes the water backward with the oars (the action). The water pushes forward on the oars and hence the boat (the reaction).

9. When the bus starts moving, Claudette's mass is accelerated by the force exerted by the back of the seat on her body. Clark is standing, however, and the only force acting on him is the friction between his shoes and the floor of the bus. Thus, when the bus starts moving, his feet accelerate forward, but the rest of his body experiences almost no accelerating force (only that due to his being attached to his accelerating feet!). As a consequence, his body tends to stay almost at rest, according to Newton's first law, relative to the ground. Relative to Claudette, however, he is moving toward her and falls into her lap. Both performers won Academy Awards.

11. (a) As the man takes the step, the action is the force his foot exerts on Earth; the reaction is the force exerted by Earth on his foot. (b) Here, the action is the force exerted by the snowball on the girl's back; the reaction is the force exerted by the girl's back on the snowball. (c) This action is the force exerted by the glove on the ball; the reaction is the force exerted by the ball on the glove. (d) This action is the force exerted by the air molecules on the window; the reaction is the force exerted by the window on the air molecules. In each case, we could equally well interchange the terms "action" and "reaction."

13. The tension in the rope is the maximum force that occurs in *both* directions. In this case, then, since both are pulling with a force of magnitude 200 N, the tension is 200 N. If the rope does not move, then the force on each athlete must equal zero. Therefore, each athlete exerts 200 N against the ground.

15. c

17. b

19. b

21. b

23. e

## Problems

1. $2 \times 10^4$ N

3. (a) 12 N  (b) 3.0 m/s$^2$

5. 3.71 N, 58.7 N, 2.27 kg

7. (a) 50.0 N  (b) 233°

9. (a) 13.5 m/s$^2$  (b) $1.42 \times 10^3$ N

11. (a) 0.200 m/s$^2$  (b) 10.0 m  (c) 2.00 m/s

13. 4.85 kN eastward

15. $1.11 \times 10^4$ N

17. (a) 3.75 m/s$^2$  (b) 22.5 m/s

19. (a) 147 N  (b) 127 N  (c) 192 N  (d) 84.5 N

21. (a) $-5.36$ m/s$^2$  (b) 4 690 N  (c) 84.0 m

23. (a) 0.256  (b) 0.510 m/s$^2$

25. 6 950 N

27. (a) $1.60 \times 10^3$ N  (b) 0.635  (c) 508 N

29. $\mu_s = 0.383$, $\mu_k = 0.306$

31. 32.1 N

33. 1.2 m/s$^2$ upward

35. (a) $6.00 \times 10^2$ N in vertical cable, 997 N in inclined cable, 796 N in horizontal cable  (b) If the point of attachment were moved higher up on the wall, the left cable would have a $y$-component that would help support the cat burglar, thus reducing the tension in the cable on the right. The tension in the vertical cable would stay the same.

37. (a) $T > w$  (b) $T = w$  (c) $T < w$  (d) $1.85 \times 10^4$ N; yes  (e) $1.47 \times 10^4$ N; yes  (f) $1.25 \times 10^4$ N; yes

39. 150 N in vertical cable, 75 N in right-side cable, 130 N in left-side cable

41. (a) 4.90 kN  (b) 607 N

43. (a) 14.7 m  (b) Neither mass is necessary.

45. (a) 33 m/s  (b) No. The object will speed up to 33 m/s from any lower speed and will slow down to 33 m/s from any higher speed.

47. (a) 1.11 s  (b) 0.875 s

49. (a) 0.404  (b) 45.8 lb

51. 64 N

53. (a) 78.4 N  (b) 105 N

55. (a) 3.00 m/s$^2$  (b) 48.0 N

57. (a) $T_1 = 3mg \sin \theta$  (b) $T_2 = 2mg \sin \theta$

59. (a) 7.0 m/s$^2$ horizontal and to the right  (b) 21 N  (c) 14 N horizontal and to the right

61. (a) $2.15 \times 10^3$ N forward  (b) 645 N forward  (c) 645 N rearward  (d) $1.02 \times 10^4$ N at 74.1° below horizontal and rearward

63. (a) $T_A > T_B$  (b) $a_1 = a_2$  (c) yes, by Newton's third law

65. $\mu_k = 0.287$

67. (a) $a = 0$  (b) 0.70 m/s$^2$

69. (a) 15.0 lb  (b) 5.00 lb  (c) zero

71. (a) $T_1 = 79.8$ N, $T_2 = 39.9$ N  (b) 2.34 m/s$^2$

73. (a)

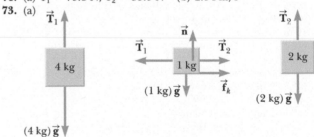

**Figure** ANS 4.73

(b) 2.31 m/s$^2$, down for the 4.00-kg object, left for the 1.00-kg object, up for the 2.00-kg object  (c) 30.0 N in the left cord, 24.2 N in the right cord  (d) Without friction, the 4-kg block falls more freely, so the tension $T_1$ in the string attached to it is reduced. The 2-kg block is accelerated upwards at a higher rate; hence, the tension force $T_2$ acting on it must be greater.

75. (a) 84.9 N upward  (b) 84.9 N downward

77. 50.0 m

79. (a) friction between box and truck  (b) 2.94 m/s$^2$

81. (a) 2.22 m  (b) 8.74 m/s down the incline

83. (a) 0.23 m/s$^2$  (b) 9.7 N

85. (a) 1.67 m/s$^2$, 16.7 N  (b) 0.687 m/s$^2$, 16.7 N

87. (a) 30.7°  (b) 0.843 N

89. 551 N

91. 72.0 N

## TOPIC 5

### Quick Quizzes

1. (c)

2. (a) true  (b) false  (c) false

3. (d)

4. (c)

5. (a) $1.13 \times 10^{-2}$ J  (b) $1.13 \times 10^{-2}$ J

6. True
7. (a) 4    (b) 9
8. (c)

## Example Questions

1. As long as the same displacement is produced by the same force, doubling the load will not change the amount of work done by the applied force.
2. Doubling the displacement doubles the amount of work done in each case.
3. The wet road would reduce the coefficient of kinetic friction, so the final velocity would be greater.
4. (c)
5. In each case, the velocity would have an additional horizontal component, meaning that the overall speed would be greater.
6. A smaller angle means that a larger initial speed would be required to allow the grasshopper to reach the indicated height.
7. In the presence of friction, a different shape slide would result in different amounts of mechanical energy lost through friction, so the final answer would depend on the slide's shape.
8. In the crouching position, there is less wind resistance. Crouching also lowers the skier's center of mass, making it easier to balance.
9. 73.5%
10. If the acrobat bends her legs and crouches immediately after contacting the springboard and then jumps as the platform pushes her back up, she can rebound to a height greater than her initial height.
11. The continuing vibration of the spring means that some energy wasn't transferred to the block. As a result, the block will go a slightly smaller distance up the ramp.
12. (a)
13. The work required would quadruple but time would double, so overall the average power would double.
14. The instantaneous power is $9.00 \times 10^4$ W, which is twice the average power.
15. False. The correct answer is one-quarter.
16. No. Using the same-size boxes is simply a matter of convenience.

## Conceptual Questions

1. Because no motion is taking place, the rope undergoes no displacement and no work is done on it. For the same reason, no work is being done on the pullers or the ground. Work is being done only within the bodies of the pullers. For example, the heart of each puller is applying forces on the blood to move blood through the body.
3. (a) When the slide is frictionless, changing the length or shape of the slide will not make any difference in the final speed of the child, as long as the difference in the heights of the upper and lower ends of the slide is kept constant. (b) If friction must be considered, the path length along which the friction force does negative work will be greater when the slide is made longer or given bumps. Thus, the child will arrive at the lower end with less kinetic energy (and hence less speed).
5. (a) Gravity does equal work on each toboggan. (b) Because gravity does equal work on each toboggan and A makes the trip in half the time of B, the average power delivered by gravity to A is twice that delivered to B.
7. (a) gravity    (b) air resistance    (c) tension
9. During the time that the toe is in contact with the ball, the work done by the toe on the ball is given by

$$W_{\text{toe}} = \tfrac{1}{2}m_{\text{ball}}v^2 - 0 = \tfrac{1}{2}m_{\text{ball}}v^2$$

where $v$ is the speed of the ball as it leaves the toe. After the ball loses contact with the toe, only the gravitational force and the retarding force due to air resistance continue to do work on the ball throughout its flight.
11. (a) Yes, the total mechanical energy of the system is conserved because the only forces acting are conservative: the force of gravity and the spring force. (b) There are two forms of potential energy in this case: gravitational potential energy and elastic potential energy stored in the spring.
13. (a) positive    (b) negative    (c) zero
15. As the satellite moves in a circular orbit about the Earth, its displacement during any small time interval is perpendicular to the gravitational force, which always acts toward the center of the Earth. Therefore, the work done by the gravitational force during any displacement is zero. (Recall that the work done by a force is defined to be $F\Delta x \cos \theta$, where $\theta$ is the angle between the force and the displacement. In this case, the angle is 90°, so the work done is zero.) Because the work–energy theorem says that the net work done on an object during any displacement is equal to the change in its kinetic energy, and the work done in this case is zero, the change in the satellite's kinetic energy is zero: hence, its speed remains constant.
17. a
19. d

## Problems

1. $7.00 \times 10^2$ J
3. (a) $2.50 \times 10^4$ J    (b) $-1.96 \times 10^4$ J
5. (a) 61.3 J    (b) $-46.3$ J    (c) 0    (d) The work done by gravity would not change, the work done by the friction force would decrease, and the work done by the normal force would not change.
7. (a) 987 N    (b) 0.495
9. (a) 2.00 m/s    (b) $2.00 \times 10^2$ N
11. (a) 879 J    (b) $2.64 \times 10^3$ J
13. (a) $-5.6 \times 10^2$ J    (b) 1.2 m
15. (a) $2.34 \times 10^4$ N    (b) $1.91 \times 10^{-4}$ s
17. (a) $4.68 \times 10^9$ J    (b) $-4.68 \times 10^9$ J    (c) $1.87 \times 10^6$ N
19. (a) 2.5 J    (b) $-9.8$ J    (c) $-12.3$ J
21. (a) 2.14 J    (b) 1.20 m/s
23. 878 kN up
25. $h = 6.94$ m
27. (a) $W_{\text{biceps}} = 120$ J    (b) $W_{\text{chin-up}} = 290$ J    (c) Additional muscles must be involved
29. (a) $4.30 \times 10^5$ J    (b) $-3.97 \times 10^4$ J    (c) 115 m/s
31. (a) The mass, spring, and Earth (including the wall) constitute the system. The mass and Earth interact through the spring force, gravity, and the normal force. (b) the point of maximum extension, $x = 0.060\ 0$ m, and the equilibrium point, $x = 0$    (c) 1.53 J at $x = 6.00$ cm; 0 J at $x = 0$    (d) $\tfrac{1}{2}mv_1^2 + \tfrac{1}{2}kx_1^2 = \tfrac{1}{2}mv_2^2 + \tfrac{1}{2}kx_2^2$

$$\rightarrow\quad v_2 = \sqrt{v_1^2 + \frac{k}{m}(x_1^2 - x_2^2)},$$

1.75 m/s (e) 1.51 m/s. This answer is not one-half the first answer because the equation is not linear.
33. 0.459 m
35. (a) 0.350 m    (b) The result would be less than 0.350 m because some of the mechanical energy is lost as a result of the force of friction between the block and track.
37. (a) 10.9 m/s    (b) 11.6 m/s
39. (a) Initially, all the energy is stored in the compressed spring. After the gun is fired and the projectile leaves the gun, the energy is transferred to the kinetic energy of the projectile, resulting in a small increase in gravitational potential energy. Once the projectile reaches its maximum height, the energy is all associated with its gravitational potential energy.    (b) 544 N/m    (c) 19.7 m/s

41. (a) Yes. There are no nonconservative forces acting on the child, so the total mechanical energy is conserved. (b) No. In the expression for conservation of mechanical energy, the mass of the child is included in every term and therefore cancels out. (c) The answer is the same in each case. (d) The expression would have to be modified to include the work done by the force of friction. (e) 15.3 m/s.

43. (a) 372 N  (b) $T_1 = 372$ N, $T_2 = T_3 = 745$ N  (c) 1.34 kJ

45. 3.8 m/s

47. 289 m

49. (a) 24.5 m/s  (b) Yes. Convert to mph to find that 24.5 m/s = 54.8 mph. A skydiver who lands at 54.8 mph will get hurt. (c) 206 m  (d) Unrealistic; the actual retarding force will vary with speed.

51. 82.6 W

53. 8.01 W

55. (a) 268 Calories  (b) 0.029 8 kg

57. (a) $2.38 \times 10^4$ W $= 32.0$ hp  (b) $4.77 \times 10^4$ W $= 63.9$ hp

59. (a) 24.0 J  (b) $-3.00$ J  (c) 21.0 J

61. (a) The graph is a straight line passing through the points (0 m, $-16$ N), (2 m, 0 N), and (3 m, 8 N).  (b) $-12.0$ J

63. (a) $PE_{\text{Ⓐ}} = 3.94 \times 10^5$ J, $PE_{\text{Ⓑ}} = 0$, $\Delta PE = -3.94 \times 10^5$ J  (b) $PE_{\text{Ⓐ}} = 5.63 \times 10^5$ J, $PE_{\text{Ⓑ}} = 1.69 \times 10^5$ J, $\Delta PE = -3.94 \times 10^5$ J

65. (a) 575 N/m  (b) 46.0 J

67. (a) 4.4 m/s  (b) $1.5 \times 10^5$ N

69. (a) 3.13 m/s  (b) 4.43 m/s  (c) 1.00 m

71. (a) 0.588 J  (b) 0.588 J  (c) 2.42 m/s  (d) $PE_C = 0.392$ J  (e) $KE_C = 0.196$ J

73. (a) 423 mi/gal  (b) 776 mi/gal

75. (a) 28.0 m/s  (b) 30.0 m  (c) 88.9 m beyond the end of the track

77. (a) 101 J  (b) 0.412 m  (c) 2.84 m/s  (d) $-9.80$ mm  (e) 2.86 m/s

79. 914 N/m

81. (a) $W_{\text{net}} = 0$  (b) $W_{\text{grav}} = -2.0 \times 10^4$ J  (c) $W_{\text{normal}} = 0$, (d) $W_{\text{friction}} = 2.0 \times 10^4$ J

83. (a) 10.2 kW  (b) 10.6 kW  (c) $5.82 \times 10^6$ J

85. 4.26 m/s

# TOPIC 6

## Quick Quizzes

1. (b)
2. (c)
3. (c)
4. (a)
5. (a) Perfectly inelastic  (b) Inelastic  (c) Inelastic
6. (a)
7. (c)

## Example Questions

1. 44 m/s
2. When one car is overtaking another, the relative velocity is small, so on impact the change in momentum is also small. In a head-on collision, however, the relative velocity is large because the cars are traveling in opposite directions. Consequently, the change in momentum of a passenger in a head-on collision is greater than when hit from behind, which implies a larger average force.
3. Assuming the kinetic energies of the two arrows are identical, the heavier arrow would have a greater momentum, because $p^2 = 2mK$. Greater arrow momentum means greater recoil speed for the archer.
4. The final velocity would be unaffected, but the change in kinetic energy would be doubled.

5. Energy can be lost due to friction during the impact, work done in deforming the bullet and block, friction in the physical mechanisms, air drag, and the creation of sound waves.
6. No. If that were the case, energy could not be conserved.
7. The blocks cannot both come to rest at the same time because then, by Equation (1), momentum would not be conserved.
8. 45°
9. $m(a + g)$

## Conceptual Questions

1. (a) No. It cannot carry more kinetic energy than it possesses. That would violate the law of energy conservation. (b) Yes. By bouncing from the object it strikes, it can deliver more momentum in a collision than it possesses in its flight.
3. (a) 0  (b) 0  (c) Answers vary, but any perfectly inelastic collision with zero initial momentum will have zero final kinetic energy.
5. Initially, the clay has momentum directed toward the wall. When it collides and sticks to the wall, neither the clay nor the wall appears to have any momentum. It is therefore tempting to (wrongfully) conclude that momentum is not conserved. The "lost" momentum, however, is actually imparted to the wall and the Earth, causing both to move. Because of the Earth's enormous mass, its recoil speed is too small to detect.
7. (a) False  (b) True
9. It will be easiest to catch the medicine ball when its speed (and kinetic energy) is lowest. The first option—throwing the medicine ball at the same velocity—will be the most difficult, because the speed will not be reduced at all. The second option, throwing the medicine ball with the same momentum, will reduce the velocity by the ratio of the masses. Since $m_t v_t = m_m v_m$, it follows that

$$v_m = v_t \left( \frac{m_t}{m_m} \right)$$

The third option, throwing the medicine ball with the same kinetic energy, will also reduce the velocity, but only by the square root of the ratio of the masses. Since

$$\tfrac{1}{2} m_t v_t^2 = \tfrac{1}{2} m_m v_m^2$$

it follows that

$$v_m = v_t \sqrt{\frac{m_t}{m_m}}$$

The slowest and easiest throw will be made when the momentum is held constant. If you wish to check this answer, try substituting in values of $v_t = 1$ m/s, $m_t = 1$ kg, and $m_m = 100$ kg. Then the same-momentum throw will be caught at 1 cm/s, while the same-energy throw will be caught at 10 cm/s.

11. b
13. Ball A
15. No. Impulse, $\vec{\mathbf{F}}\Delta t$, depends on the force and the time interval during which it is applied.
17. c

## Problems

1. (a) $8.35 \times 10^{-21}$ kg · m/s  (b) 4.50 kg · m/s  (c) 750 kg · m/s (d) $1.78 \times 10^{29}$ kg · m/s
3. (a) 31.0 m/s  (b) the bullet, $3.38 \times 10^3$ J versus 69.7 J
5. (a) 0.42 N downward (b) The hailstones would exert a larger average force on the roof because they would bounce off the roof, and the impulse acting on the hailstones would be greater than the impulse acting on the raindrops. Newton's third law then tells us that the hailstones exert a greater force on the roof.
7. (a) 22.0 m/s  (b) 1.14 kg

9. (a) $\Delta p_x = 7.83$ kg · m/s, $\Delta p_y = 5.49$ kg · m/s  (b) 191 N
11. 1.39 N · s up
13. (a) 364 kg · m/s forward  (b) 438 N forward
15. (a) 8.0 N · s  (b) 5.3 m/s  (c) 3.3 m/s
17. (a) 12 N · s  (b) 8.0 N · s  (c) 8.0 m/s  (d) 5.3 m/s
19. (a) $9.60 \times 10^{-2}$ s  (b) $3.65 \times 10^5$ N  (c) $26.6g$
21. 65.2 m/s
23. (a) 1.15 m/s  (b) 0.346 m/s directed opposite to girl's motion
25. 0.217 m/s
27. $v_{\text{thrower}} = 2.48$ m/s, $v_{\text{catcher}} = 2.25 \times 10^{-2}$ m/s
29. (a) 154 m  (b) By Newton's third law, when the astronaut exerts a force on the tank, the tank exerts a force back on the astronaut. This reaction force accelerates the astronaut towards the spacecraft.
31. (a)

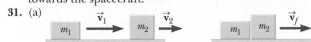

**Figure** ANS 6.31a

(b) The collision is best described as perfectly inelastic because the skaters remain in contact after the collision.
(c) $m_1 v_1 + m_2 v_2 = (m_1 + m_2) v_f$
(d) $v_f = (m_1 v_1 + m_2 v_2)/(m_1 + m_2)$  (e) 6.33 m/s
33. 15.6 m/s
35. (a) 1.80 m/s  (b) $2.16 \times 10^4$ J
37. (a) 1
39. 1.32 m
41. 57 m
43. 273 m/s
45. (a) 4.85 m/s  (b) 8.41 m
47. 17.1 cm/s (25.0-g object), 22.1 cm/s (10.0-g object)
49. (a) Over a very short time interval, external forces have no time to impart significant impulse to the players during the collision. The two players move together after the tackle, so the collision is completely inelastic. (b) 2.88 m/s at 32.3° (c) 785 J. (d) The lost kinetic energy is transformed into other forms of energy such as thermal energy and sound.
51. 5.59 m/s north
53. (a) 2.50 m/s at −60.0°  (b) elastic collision
55. $9.21 \times 10^{-2}$ N
57. 588 kg/s
59. 1.62
61. $1.78 \times 10^3$ N on truck driver, $8.89 \times 10^3$ N on car driver
63. (a) 8/3 m/s (incident particle), 32/3 m/s (target particle)
    (b) −16/3 m/s (incident particle), 8/3 m/s (target particle)
    (c) $7.1 \times 10^{-2}$ J in case (a), and $2.8 \times 10^{-3}$ J in case (b). The incident particle loses more kinetic energy in case (a), in which the target mass is 1.0 g.
65. (a) $\vec{v}_{\text{girl}} = -\left(\dfrac{m}{M-m}\right)\vec{v}_{\text{gloves}}$ (b) As she throws the gloves and exerts a force on them, the gloves exert a force of equal magnitude in the opposite direction on her (Newton's third law) that causes her to accelerate from rest to reach the velocity $\vec{v}_{\text{girl}}$.
67. 62 s
69. (a) 3  (b) 2
71. (a) −2.33 m/s, 4.67 m/s  (b) 0.277 m  (c) 2.98 m  (d) 1.49 m
73. (a) −0.667 m/s  (b) 0.952 m
75. (a) 3.54 m/s  (b) 1.77 m  (c) $3.54 \times 10^4$ N  (d) No, the normal force exerted by the rail contributes upward momentum to the system.
77. (a) No. After colliding, the cars, moving as a unit, would travel northeast, so they couldn't damage property on the southeast corner. (b) x-component: 16.3 km/h, y-component: 9.17 km/h, angle: the final velocity of the car is 18.7 km/h at 29.4° north of east, consistent with part (a).
79. (a) He moves at a velocity of $\left(\dfrac{m_g}{m_b}\right)v_g$ toward the west.

(b) $KE_g = \frac{1}{2}m_g v_g^2$; $KE_b = \left(\dfrac{m_g^2}{2m_b}\right)v_g^2$ The ratio $= \dfrac{KE_g}{KE_b} = \dfrac{m_b}{m_g}$
is greater than 1 because $m_b > m_g$. Hence, the girl has more kinetic energy. (c) Work is done by both the boy and girl as they push each other apart and the origin of this work is chemical energy in their bodies.
81. $v_0 = \left(\dfrac{M+m}{m}\right)\sqrt{2\mu gd}$
83. (a) 1.1 m/s at 30° below the positive x-axis  (b) 0.32 or 32%
85. (a) The momentum of the bullet–block system is conserved in the collision, so you can relate the speed of the block and bullet immediately after the collision to the initial speed of the bullet. Then, you can use conservation of mechanical energy for the bullet–block–Earth system to relate the speed after the collision to the maximum height.  (b) 521 m/s upward

## TOPIC 7

### Quick Quizzes

1. (c)
2. (b)
3. (b)
4. (b)
5. (a)
6. 1. (e) 2. (a) 3. (b)
7. (c)
8. (b), (c)
9. (e)
10. (d)

### Example Questions

1. Yes. The conversion factor is $180°/(\pi \text{ rad})$ or 57.3°/rad.
2. All given quantities and answers are in angular units, so altering the radius of the wheel has no effect on the answers.
3. In this case, doubling the angular acceleration doubles the angular displacement. That is true here because the initial angular speed is zero.
4. The angular acceleration of a record player during play is zero. A CD player must have nonzero acceleration because the angular velocity must change.
5. (b)
6. It would be increased.
7. The angle of the bank, the coefficient of friction, and the radius of the circle determine the minimum and maximum safe speeds.
8. The normal force is still zero.
9. Yes. The force of gravity acting on each billiard ball holds the balls against the table and assists in creating friction forces that allow the balls to roll. The gravity forces between the balls are insignificant, however.
10. First, most asteroids are irregular in shape, so Equation (1) will not apply because the acceleration may not be uniform. Second, the asteroid may be so small that there will be no significant or useful region where the acceleration is uniform. In that case, Newton's more general law of gravitation would be required.
11. (b)
12. Mechanical energy is conserved in this system. Because the potential energy at perigee is lower, the kinetic energy must be higher.
13. 5 days

### Conceptual Questions

1. (a) 1.00  (b) 2.00

3. The speedometer will be inaccurate. The speedometer measures the number of tire revolutions per second, so its readings will be too low.

5. (a) Point $C$. The total acceleration here is centripetal acceleration, straight up. (b) Point $A$. The speed at $A$ is zero where the bob is reversing direction. The total acceleration here is tangential acceleration, to the right and downward perpendicular to the cord. (c) Point $B$. The total acceleration here is to the right and pointing in a direction somewhere in between the tangential and radial directions, depending on their relative magnitudes.

7. Consider an individual standing against the inside wall of the cylinder with her head pointed toward the axis of the cylinder. As the cylinder rotates, the person tends to move in a straight-line path tangent to the circular path followed by the cylinder wall. As a result, the person presses against the wall, and the normal force exerted on her provides the radial force required to keep her moving in a circular path. If the rotational speed is adjusted such that this normal force is equal in magnitude to her weight on Earth, she will not be able to distinguish between the artificial gravity of the colony and ordinary gravity.

9. The tendency of the water is to move in a straight-line path tangent to the circular path followed by the container. As a result, at the top of the circular path, the water presses against the bottom of the pail, and the normal force exerted by the pail on the water provides the radial force required to keep the water moving in its circular path.

11. Any object that moves such that the *direction* of its velocity changes has an acceleration. A car moving in a circular path will always have a centripetal acceleration.

13. The speed changes. The component of force tangential to the path causes a tangential acceleration.

## Problems

1. (a) 0.820 rad   (b) 1.91 rev   (c) 7.85 rad/s
3. (a) $3.2 \times 10^8$ rad   (b) $5.0 \times 10^7$ rev
5. (a) 821 rad/s$^2$   (b) $4.21 \times 10^3$ rad
7. (a) 0.608 m/s$^2$   (b) 28.9 rad/s   (c) 261 rad   (d) 99.2 m
9. Main rotor: 179 m/s $= 0.522v_{sound}$
    Tail rotor: 221 m/s $= 0.644v_{sound}$
11. (a) 116 rev (b) 62.1 rad/s
13. 13.7 rad/s$^2$
15. (a) 6.53 m/s   (b) 0.285 m/s$^2$ directed toward the center of the circular arc
17. (a) 0.346 m/s$^2$   (b) 1.04 m/s   (c) 0.346 m/s$^2$   (d) 0.943 m/s$^2$
    (e) 1.00 m/s$^2$ at 20.1° forward with respect to the direction of $\vec{a}_c$
19. (a) 20.6 N (b) 3.35 m/s$^2$ downward tangent to the circle; 32.0 m/s$^2$ radially inward (c) 32.2 m/s$^2$ at 5.98° to the cord, pointing toward a location below the center of the circle. (d) No change. (e) If the object is swinging down, it is gaining speed. If it is swinging up, it is losing speed, but its acceleration has the same magnitude and its direction can be described in the same terms.
21. (a) 1.10 kN   (b) 2.04 times her weight
23. 22.6 m/s
25. (a) 18.0 m/s$^2$   (b) $9.00 \times 10^2$ N   (c) 1.84; this large coefficient is unrealistic, and she will not be able to stay on the merry-go-round.
27. (a) 9.8 N   (b) 9.8 N   (c) 6.3 m/s
29. (a) The force of static friction acting toward the road's center of curvature causes the briefcase's centripetal acceleration. When the necessary centripetal force exceeds the maximum value of the static friction force, $\mu_s n$, the briefcase begins to slide. (b) 0.370
31. (a) 1.58 m/s$^2$   (b) 455 N upward   (c) 329 N upward
    (d) 397 N directed inward and 80.8° above horizontal
33. (a) $1.72 \times 10^{20}$ N   (b) $6.38 \times 10^{-3}$ m/s$^2$   (c) $2.30 \times 10^{-5}$ m/s$^2$

35. $1.1 \times 10^{-10}$ N at 72° above the $+x$-axis
37. (a) $2.50 \times 10^{-5}$ N toward the 500-kg object (b) between the two objects and 0.245 m from the 500-kg object
39. (a) $r = \dfrac{9}{8}R_E = 7.18 \times 10^6$ m   (b) $7.98 \times 10^5$ m
41. (a) 2.43 h   (b) 6.59 km/s   (c) 4.73 m/s$^2$ toward the Earth
43. $6.3 \times 10^{23}$ kg
45. (a) 18.0 AU   (b) 35.4 AU
47. (a) $1.90 \times 10^{27}$ kg   (b) $1.89 \times 10^{27}$ kg   (c) yes
49. (a) 9.40 rev/s   (b) 44.2 rev/s$^2$; $a_r = 2\,590$ m/s$^2$; $a_t = 206$ m/s$^2$
    (c) $F_r = 515$ N; $F_t = 40.7$ N
51. (a) 2.51 m/s   (b) 7.90 m/s$^2$   (c) 4.00 m/s
53. (a) $7.76 \times 10^3$ m/s   (b) 89.3 min
55. (a) $n = m\left(g - \dfrac{v^2}{r}\right)$   (b) 17.1 m/s
57. (a) $F_{g,\text{true}} = F_{g,\text{apparent}} + mR_E\omega^2$   (b) 732 N (equator), 735 N (either pole)
59. 11.8 km/s
61. 0.835 rev/s
65. (a) 10.6 kN   (b) 14.1 m/s
67. (a) 0.71 yr (b) The departure must be timed so that the spacecraft arrives at aphelion when the target planet is there.
69. (a) 109 N   (b) 56.4 N
71. (a) 106 N   (b) 0.396
73. 0.131
75. (a) 2.1 m/s   (b) 54°   (c) 4.7 m/s

# TOPIC 8

## Quick Quizzes

1. (d)
2. (b)
3. (b)
4. (a)
5. (c)
6. (a)

## Example Questions

1. The revolving door begins to move counterclockwise instead of clockwise.
2. Placing the wedge closer to the doorknob increases its effectiveness.
3. (b)
4. The system would begin to rotate clockwise.
5. Sherry reaches the hole first, so she was moving faster. The condition for the center of mass to remain fixed is that $v_S = (m_B/m_S)v_B$. Sherry has the higher speed because she has less mass than Bob, while the forces acting on them are the same in magnitude.
6. A larger fraction of the total mass would be located in the vertical rod. With more mass having an $x$-coordinate to the left of the original system, the $x$-component of the center of mass would move to the left of 1.30 m. Similarly, the $y$-component of the center of mass would move upward.
7. For $t < 1.00$ s, the initial speed of $m_1$ must be greater than before; hence, it carries away more momentum, and $m_2$ must therefore have less momentum and a lower initial speed. It follows $m_2$ will not go as high as before, so $x_2$ will be smaller.
8. If the woman leans backwards, the torque she exerts on the seesaw increases and she begins to descend.
9. The angle made by the biceps force would still not vary much from 90°, but the length of the moment arm would be doubled, so the required biceps force would be reduced by nearly half.
10. (c)
11. The tension in the cable would increase.

12. Lengthening the rod between balls 2 and 4 would create the larger change in the moment of inertia.
13. Stepping forward transfers the momentum of the pitcher's body to the ball. Without proper timing, the transfer will not take place or will have less effect.
14. The magnitude of the acceleration would decrease; that of the tension would increase.
15. Block, ball, cylinder
16. The final answer wouldn't change.
17. His angular speed remains the same.
18. Energy conservation is not violated. The positive net change occurs because the student is doing work on the system.

## Conceptual Questions

1. In order for you to remain in equilibrium, your center of gravity must always be over your point of support, the feet. If your heels are against a wall, your center of gravity cannot remain above your feet when you bend forward, so you lose your balance.
3. No, only if its angular velocity changes.
5. The long pole has a large moment of inertia about an axis along the rope. An unbalanced torque will then produce only a small angular acceleration of the performer–pole system, to extend the time available for getting back in balance. To keep the center of mass above the rope, the performer can shift the pole left or right, instead of having to bend his body around. The pole sags down at the ends to lower the system's center of gravity, increasing the relative stability of the system.
7. (a) Clockwise   (b) 3
9. The angular momentum of the gas cloud is conserved. Thus, the product $I\omega$ remains constant. As the cloud shrinks in size, its moment of inertia decreases, so its angular speed $\omega$ must increase.
11. We can assume fairly accurately that the driving motor will run at a constant angular speed and at a constant torque. (a) As the radius of the take-up reel increases, the tension in the tape will decrease, in accordance with the equation.

$$T = \tau_{const}/R_{take\text{-}up} \qquad (1)$$

As the radius of the source reel decreases, given a decreasing tension, the torque in the source reel will decrease even faster, as the following equation shows:

$$\tau_{source} = TR_{source} = \tau_{const}R_{source}/R_{take\text{-}up} \qquad (2)$$

(b) In the case of a sudden jerk on the tape, the changing angular speed of the source reel becomes important. If the source reel is full, then the moment of inertia will be large and the tension in the tape will be large. If the source reel is nearly empty, then the angular acceleration will be large instead. Thus, the tape will be more likely to break when the source reel is nearly full. One sees the same effect in the case of paper towels: It is easier to snap a towel free when the roll is new than when it is nearly empty.
13. (a) zero   (b) zero
15. d
17. e

## Problems

1. (a) 25.0 N · m (b) 50.0 N · m
3. 168 N · m
5. (a) 30 N · m (counterclockwise)
   (b) 36 N · m (counterclockwise)
7. (a) 5.1 N · m   (b) The torque increases, because the torque is proportional to the moment arm, $L \sin \theta$, and this factor increases as $\theta$ increases.
9. 0.582 m
11. $x_{cg} = 3.33$ ft, $y_{cg} = 1.67$ ft

13. 0.100 m
15. 1.39 m
17. $F_t = 724$ N, $F_s = 716$ N
19. 312 N
21. 1.09 m
23. (a) $T = 2.71$ kN   (b) $R_x = 2.65$ kN
25. (a) 443 N   (b) 222 N (to the right), 216 N (upward)
27. $T_1 = 501$ N, $T_2 = 672$ N, $T_3 = 384$ N
29. (a) $d = \dfrac{mg}{2k\tan\theta}$   (b) $R_x = \dfrac{mg}{2\tan\theta}$; $R_y = mg$
31. $\theta = \tan^{-1}\left(\dfrac{w}{h}\right)$
33. $R = 107$ N, $T = 157$ N
35. 209 N
37. (a) 99.0 kg · m² (b) 44.0 kg · m² (c) 143 kg · m²
39. (a) 87.8 kg · m² (b) 1.61 × 10³ kg (c) 4.70 rad/s
41. (a) 0.687 kg · m² (b) 0.823 N · m
43. (a) 24.0 N · m (b) 0.035 6 rad/s² (c) 1.07 m/s²
45. 177 N
47. (a) 14.7 rad/s² (b) 7.35 m/s² (c) 14.7 m/s²
49. 276 J
51. (a) 5.47 J (b) 5.99 J
53. (a) 3.90 m/s (b) 15.6 rad/s
55. 149 rad/s
57. (a) 500. J (b) 250. J (c) 750. J
59. (a) 3.74 m/s (b) 37.4 rad/s (c) 37.4 rad/s (d) 1.21 m
61. (a) 2.50 × 10⁻² kg · m²/s (b) 10.0 rad/s
63. (a) 7.08 × 10³³ J · s (b) 2.66 × 10⁴⁰ J · s
65. 17.5 J · s counterclockwise
67. (a) 1.75 rad/s (b) up
69. 6.73 rad/s
71. 5.99 × 10⁻² J
73. (a) $\omega = \left(\dfrac{I_1}{I_1 + I_2}\right)\omega_0$  (b) $\dfrac{KE_f}{KE_i} = \dfrac{I_1}{I_1 + I_2} < 1$
75. (a) 2.6 rad/s (b) 5.1 × 10⁵ kg · m² (c) 1.7 × 10⁶ J
77. (a) As the child walks to the right end of the boat, the boat moves left (toward the pier). (b) The boat moves 1.45 m closer to the pier, so the child will be 5.55 m from the pier. (c) No. He will be short of reaching the turtle by 0.45 m.
79. 36.9°
81. (a) $Mvd$ (b) $Mv^2$ (c) $Mvd$ (d) $2v$ (e) $4Mv^2$ (f) $3Mv^2$
83. (a) 6.73 N upward   (b) $x = 0.389$ m
85. $x_{cg} = 0.459$ m, $y_{cg} = 9.55 \times 10^{-2}$ m
87. (a) $T = \dfrac{Mmg}{M + 4m}$  (b) $a_t = \dfrac{4mg}{M + 4m}$
89. 24.5 m/s
91. 9.3 kN
93. (a) 3.12 m/s² (b) $T_1 = 26.7$ N, $T_2 = 9.36$ N
95. (a) 0.73 m (b) 1.6 m/s

# TOPIC 9

## Quick Quizzes

1. (c)
2. (a)
3. (c)
4. (b)
5. (c)
6. (b)
7. (a)

## Example Questions

1. Because water is more dense than oil, the pressure exerted by a column of water is greater than the pressure exerted by a column of oil.

2. At higher altitude, the column of air above a given area is progressively shorter and less dense, so the weight of the air column is reduced. Pressure is caused by the weight of the air column, so the pressure is also reduced.

3. As fluid pours out through a single opening, the air inside the can above the fluid expands into a larger volume, reducing the pressure to below atmospheric pressure. Air must then enter the same opening going the opposite direction, resulting in disrupted fluid flow. A separate opening for air intake maintains air pressure inside the can without disrupting the flow of the fluid.

4. True

5. False

6. (a)

7. The aluminum cube would float free of the bottom.

8. The speed of the blood in the narrowed region increases.

9. A factor of 2

10. The speed decreases with time.

11. The limit is $v_1 = \sqrt{2gh}$, called Torricelli's law. (See Example 9.10).

12. The pressure difference across the wings depends linearly on the density of air. At higher altitude, the air's density decreases, so the lift force decreases as well.

13. False

14. No. There are many plants taller than 0.3 m, so there must be some additional explanation.

15. A factor of 16

16. False

17. Tungsten, steel, aluminum, rubber

18. The lineman's skull and neck would undergo compressional stress.

19. Steel, copper, mercury, water

## Conceptual Questions

1. e
3. D, A, C, B
5. (a) fluid B  (b) 4
7. a
9. At lower elevation, the water pressure is greater because pressure increases with increasing depth below the water surface in the reservoir (or water tower). The penthouse apartment is not so far below the water's surface. The pressure behind a closed faucet is weaker there and the flow weaker from an open faucet. Your fire department likely has a record of the precise elevation of every fire hydrant.
11. (a) $v_A = v_B > v_C$  (b) $P_C > P_B > P_A$
13. Opening the windows results in a smaller pressure difference between the exterior and interior of the house and, therefore, less tendency for severe damage to the structure due to the Bernoulli effect.
15. b

## Problems

1. (a) $8.27 \times 10^{-2}\,\text{m}^3$  (b) $1.77 \times 10^4\,\text{Pa}$
3. $5.17 \times 10^{19}\,\text{N}$
5. (a) $\sim 4 \times 10^{17}\,\text{kg/m}^3$  (b) The density of an atom is about $10^{14}$ times greater than the density of iron and other common solids and liquids. This shows that an atom is mostly empty space. Liquids and solids, as well as gases, are mostly empty space.
7. (a) $1.01 \times 10^6\,\text{N}$  (b) $3.88 \times 10^5\,\text{N}$  (c) $1.11 \times 10^5\,\text{Pa}$
9. (a) $3.71 \times 10^5\,\text{Pa}$  (b) $3.57 \times 10^4\,\text{N}$
11. 0.133 m
13. $1.05 \times 10^5\,\text{Pa}$
15. $1.53 \times 10^4\,\text{Pa}$
17. 0.258 N down
19. 9.41 kN
21. $1.04\,\text{kg/m}^3$

23. (a) 1.43 kN upward (b) 1.28 kN upward (c) The balloon expands because the external pressure declines with increasing altitude.
25. (a) 4.0 kN (b) 2.2 kN (c) The air pressure at this high altitude is much lower than atmospheric pressure at the surface of Earth, so the balloons expanded and eventually burst.
27. (a) 7.00 cm (b) 2.80 kg
29. (a) $1.46 \times 10^{-2}\,\text{m}^3$  (b) $2.10 \times 10^3\,\text{kg/m}^3$
31. 17.3 N (upper scale), 31.7 N (lower scale)
33. 6.57 m/s
35. (a) 80. g/s  (b) 0.27 mm/s
37. 12.6 m/s
39. (a) $9.43 \times 10^3\,\text{Pa}$ (b) 255 m/s (c) The density of air decreases with increasing height, resulting in a smaller pressure difference. Beyond the maximum operational altitude, the pressure difference can no longer support the aircraft.
41. (a) 0.553 s (b) 14.5 m/s (c) 0.145 m/s (d) $1.013 \times 10^5\,\text{Pa}$ (e) $2.06 \times 10^5\,\text{Pa}$; gravity terms can be neglected. (f) 33.0 N
43. 9.00 cm
45. 1.47 cm
47. (a) 28.0 m/s  (b) 28.0 m/s  (c) 2.11 MPa
49. $8.3 \times 10^{-2}\,\text{N/m}$
51. $5.6 \times 10^{-2}\,\text{N/m}$
53. 8.6 N
55. 2.1 MPa
57. $2.8\,\mu\text{m}$
59. 0.21 mm
61. $1.8 \times 10^{-3}\,\text{kg/m}^3$
63. $1.4 \times 10^{-5}\,\text{N} \cdot \text{s/m}^2$
65. 4.90 mm
67. $1.05 \times 10^7\,\text{Pa}$
69. $3.5 \times 10^8\,\text{Pa}$
71. 4.4 mm
73. (a) 2.5 mm  (b) 0.70 mm  (c) $6.9 \times 10^3\,\text{kg}$
75. 1.9 cm
77. (a) The buoyant forces are the same because the two blocks displace equal amounts of water. (b) The spring scale reads largest value for the iron block. (c) $B = 2.0 \times 10^3\,\text{N}$ for both blocks, $T_{\text{iron}} = 13 \times 10^3\,\text{N}$, $T_{\text{aluminum}} = 3.3 \times 10^3\,\text{N}$.
79. 0.59
81. $2.5 \times 10^7$ capillaries
83. (a) 1.57 kPa, $1.55 \times 10^{-2}$ atm, 11.8 mm Hg (b) The fluid level in the tap should rise. (c) Blockage of flow of the cerebrospinal fluid
85. 2.25 m above the level of point $B$
87. 0.605 m
89. $F = \pi R^2 (P_0 - P)$
91. 17.0 cm above the floor

# TOPIC 10

## Quick Quizzes

1. (c)
2. (b)
3. (c)
4. (c)
5. Unlike land-based ice, ocean-based ice already displaces water, so when it melts, ocean levels won't change much.
6. (b)

## Example Questions

1. A Celsius degree
2. True
3. When the temperature decreases, the tension in the wire increases.
4. The magnitude of the required temperature change would be larger because the linear expansion coefficient of steel is less than that of copper.

5. Glass, aluminum, ethyl alcohol, mercury
6. The balloon expands.
7. The pressure is slightly reduced.
8. The volume of air decreases.
9. The pressure would increase, up to double if the reflections were all elastic.
10. True

## Conceptual Questions

1. (a) An ordinary glass dish will usually break because of stresses that build up as the glass expands when heated. (b) The expansion coefficient for Pyrex glass is much lower than that of ordinary glass. Thus, the Pyrex dish will expand much less than the dish of ordinary glass and does not normally develop sufficient stress to cause breakage.
3. Mercury must have the larger coefficient of expansion. As the temperature of a thermometer rises, both the mercury and the glass expand. If they both had the same coefficient of linear expansion, the mercury and the cavity in the glass would expand by the same amount, and there would be no apparent movement of the end of the mercury column relative to the calibration scale on the glass. If the glass expanded more than the mercury, the reading would go down as the temperature went up! Now that we have argued this conceptually, we can look in a table and find that the coefficient for mercury is about 20 times as large as that for glass, so that the expansion of the glass can sometimes be ignored.
5. We can think of each bacterium as being a small bag of liquid containing bubbles of gas at a very high pressure. The ideal gas law indicates that if the bacterium is raised rapidly to the surface, then its volume must increase dramatically. In fact, the increase in volume is sufficient to rupture the bacterium.
7. Additional water vaporizes into the bubble, so that the number of moles $n$ increases.
9. (a) equal to  (b) greater than.
11. The coefficient of expansion for metal is generally greater than that of glass; hence, the metal lid loosens because it expands more than the glass.
13. As the water rises in temperature, it expands or rises in pressure or both. The excess volume would spill out of the cooling system, or else the pressure would rise very high indeed. The expansion of the radiator itself provides only a little relief, because in general solids expand far less than liquids for a given positive change in temperature.

## Problems

1. (a) $-4.60 \times 10^2 °F$  (b) $37.0°C$  (c) $-2.80 \times 10^2 °F$
3. (a) $-253°C$  (b) $-423°F$
5. $27.2$ K
7. $1.67°C$
9. (a) $107°F$ (b) Yes; the normal body temperature is $98.6°F$, so the patient has a high fever that needs immediate attention.
11. $31.3$ cm
13. $55.0°C$
15. (a) $-179°C$ (attainable)  (b) $-376°C$ (below 0 K, unattainable)
17. (a) $1.1 \times 10^4 \text{ kg/m}^3$
19. $1.02 \times 10^3$ gallons
21. (a) $0.10$ L  (b) $2.009$ L  (c) $1.0$ cm
23. $2.7 \times 10^2$ N
25. $0.548$ gal
27. (a) increase  (b) $1.603$ cm
29. (a) $0.197$ mol  (b) $1.18 \times 10^{23}$ atoms
31. (a) $627°C$  (b) $927°C$
33. (a) $2.5 \times 10^{19}$ molecules  (b) $4.1 \times 10^{-21}$ mol
35. $4.28$ atm

37. $7.1$ m
39. $16.0 \text{ cm}^3$
41. $6.21 \times 10^{-21}$ J
43. (a) $5.69 \times 10^{-21}$ J  (b) $414$ m/s  (c) $1.03 \times 10^4$ J
45. $6.64 \times 10^{-27}$ kg
47. (a) $2.01 \times 10^4$ K  (b) $901$ K
49. $16$ N
51. $0.66$ mm to the right at $78°$ below the horizontal
53. $3.55$ L
55. $35.016$ m
57. $6.57$ MPa
59. (a) $99.8$ mL (b) The change in volume of the flask is far smaller because Pyrex has a much smaller coefficient of expansion than acetone. Hence, the change in volume of the flask has a negligible effect on the answer.
61. (b) The expansion of the mercury is almost 20 times that of the flask (assuming Pyrex glass).
63. $2.7$ m
65. (a) $\theta = \dfrac{(\alpha_2 - \alpha_1)L_0(\Delta T)}{\Delta r}$

(c) The bar bends in the opposite direction.
67. (a) No torque acts on the disk so its angular momentum is constant. Yes, the angular speed increases. As the disk cools, its radius, and hence, its moment of inertia, decreases. Conservation of angular momentum then requires that its angular speed increase. (b) $25.7$ rad/s

# TOPIC 11

## Quick Quizzes

1. (a) Water, glass, iron. (b) Iron, glass, water.
2. (b) The slopes are proportional to the reciprocal of the specific heat, so a larger specific heat results in a smaller slope, meaning more energy is required to achieve a given temperature change.
3. (c)
4. (b)
5. (a) 4 (b) 16 (c) 64

## Example Questions

1. From the point of view of physics, faster repetitions don't affect the final answer; physiologically, however, the weightlifter's metabolic rate would increase.
2. (c)
3. No
4. (c)
5. No
6. The mass of ice melted would double.
7. Nickel–iron asteroids have a higher density and therefore a greater mass, which means they can deliver more energy on impact for a given speed.
8. A runner's metabolism is much higher when he is running than when he is at rest, and because muscles are only about 20% efficient, a great amount of random thermal energy is created through muscular exertion. Consequently, the runner needs to eliminate far more thermal energy when running than when resting. Once the run is over, muscular exertions cease, and the metabolism starts to return to normal, so the runner begins to feel chilled.
9. (a)
10. (a)
11. If the planet doesn't re-emit all the energy that it absorbs from its star, it will increase in temperature. As the temperature increases, the planet will radiate at a greater and greater rate until it reaches thermal equilibrium, when it emits as much as it absorbs.

## Conceptual Questions

1. When you rub the surface, you increase the temperature of the rubbed region. With the metal surface, some of this energy is transferred away from the rubbed site by conduction. Consequently, the temperature in the rubbed area is not as high for the metal as it is for the wood, and it feels relatively cooler than the wood.

3. (a) −1   (b) −2

5. A

7. (a) The operation of an immersion coil depends on the convection of water to maintain a safe temperature. As the water near a coil warms up, the warmed water floats to the top due to Archimedes' principle. The temperature of the coil cannot go higher than the boiling temperature of water, 100°C. If the coil is operated in air, convection is reduced, and the upper limit of 100°C is removed. As a result, the coil can become hot enough to be damaged. (b) If the coil is used in an attempt to warm a thick liquid like stew, convection cannot occur fast enough to carry energy away from the coil, so that it again may become hot enough to be damaged.

9. Tile is a better conductor of energy than carpet, so the tile conducts energy away from your feet more rapidly than does the carpeted floor.

11. $k_B > k_A > k_C$

13. d

15. d

## Problems

1. (a) 3.50 kcal   (b) $1.47 \times 10^4$ J

3. (a) $4.5 \times 10^3$ J (b) 910 W (c) 0.87 Cal/s (d) The excess thermal energy is transported by conduction and convection to the surface of the skin and disposed of through the evaporation of sweat.

5. (a) 81.0 W   (b) 69.7 kcal/h   (c) 8.27 m/s

7. 16.9°C

9. (a) $1.67 \times 10^{18}$ J   (b) 52.9 yr

11. 176°C

13. 88 W

15. (a) $8.90 \times 10^4$ kg   (b) $4.10 \times 10^9$ J   (c) $114

17. 0.845 kg

19. 80 g

21. (a) $1.82 \times 10^3$ J/kg · °C (b) We can't make a definite identification. It might be beryllium. (c) The material might be an unknown alloy or a material not listed in the table.

23. 0.26 kg

25. 27.1°C

27. (a) $6.66 \times 10^5$ J   (b) $2.64 \times 10^5$ J   (c) 31.5°C

29. 16°C

31. 65°C

33. 2.3 km

35. 16°C

37. (a) $t_{boil} = 2.8$ min   (b) $t_{evaporate} = 18$ min

39. (a) all ice melts, $T_f = 40$°C   (b) 8.0 g melts, $T_f = 0$°C

41. (a) The bullet loses all its kinetic energy as it is stopped by the ice. Also, thermal energy must be removed from the bullet to cool it from 30.0°C to 0°C. The sum of these two energies equals the energy it takes to melt part of the ice. The final temperature of the bullet is 0°C because not all the ice melts. (b) 0.294 g

43. $3 \times 10^3$ W

45. 402 MW

47. 709 s

49. 9.0 cm

51. $2.7 \times 10^7$ J

53. 16:1

55. 12.5 kW

57. 2.3 kg

59. $8.00 \times 10^2$ J/kg · °C. This value differs from the tabulated value by 11%, so they agree within 15%.

61. 66.7 min

63. 51.2°C

65. (a) seven times (b) As the car stops, it transforms part of its kinetic energy into internal energy due to air resistance. As soon as the brakes rise above the air temperature, they transfer energy by heat into the air. If they reach a high temperature, they transfer energy very quickly.

67. 255 K

69. 12 h

71. $3.85 \times 10^{26}$ J

73. 1.4 kg

75. (a) 75.0°C (b) 36 kJ

## TOPIC 12

### Quick Quizzes

1. (b)

2. A is isovolumetric, B is adiabatic, C is isothermal, D is isobaric.

3. (c)

4. (b)

5. The number 7 is the most probable outcome. The numbers 2 and 12 are the least probable outcomes.

### Example Questions

1. No

2. No

3. True

4. True

5. The change in temperature must always be negative because the system does work on the environment at the expense of its internal energy, and no thermal energy can be supplied to the system to compensate for the loss.

6. A diatomic gas does more work under these assumptions.

7. The carbon dioxide gas would have a final temperature lower than 380 K.

8. False

9. (a)

10. No. The efficiency improves only if the ratio $|Q_c/Q_h|$ becomes smaller. Further, too large an increase in $Q_h$ will damage the engine, so there is a limit even if $Q_c$ remains fixed.

11. If the path from B to C was a straight line, more work would be done per cycle.

12. No. The compressor does work and warms the kitchen. With the refrigerator door open, the compressor would run continuously.

13. False

14. Silver, lead, ice

15. False

16. The thermal energy created by your body during the exertion would be dissipated into the environment, increasing the entropy of the Universe.

17. Skipping meals can lower the basal metabolism, reducing the rate at which energy is used. When a large meal is eaten later, the lower metabolism means more food energy will be stored, and weight will be gained even if the same number of total calories is consumed in a day.

### Conceptual Questions

1. (a) 1   (b) 3/5   (c) 5/7

3. The energy that is leaving the body by work and heat is replaced by means of biological processes that transform chemical energy in the food that the individual ate into

internal energy. Thus, the temperature of the body can be maintained.

5. If there is no change in internal energy, then, according to the first law of thermodynamics, the heat is equal to the negative of the work done on the gas (and thus equal to the work done *by* the gas). Thus, $Q = -W = W_{\text{by gas}}$.

7. Practically speaking, it isn't possible to create a heat engine that creates no thermal pollution, because there must be both a hot heat source (energy reservoir) and a cold heat sink (low-temperature energy reservoir). The heat engine will warm the cold heat sink and will cool down the heat source. If either of those two events is undesirable, then there will be thermal pollution.

   Under some circumstances, the thermal pollution would be negligible. For example, suppose a satellite in space were to run a heat pump between its sunny side and its dark side. The satellite would intercept some of the energy that gathered on one side and would 'dump' it to the dark side. Since neither of those effects would be particularly undesirable, it could be said that such a heat pump produced no thermal pollution.

9. Although no energy is transferred into or out of the system by heat, work is done on the system as the result of the agitation. Consequently, both the temperature and the internal energy of the coffee increase.

11. The first law is a statement of conservation of energy that says that we cannot devise a cyclic process that produces more energy than we put into it. If the cyclic process takes in energy by heat and puts out work, we call the device a heat engine. In addition to the first law's limitation, the second law says that, during the operation of a heat engine, some energy must be ejected to the environment by heat. As a result, it is theoretically impossible to construct a heat engine that will work with 100% efficiency.

13. If the system is isolated, no energy enters or leaves the system by heat, work, or other transfer processes. Within the system, energy can change from one form to another, but since energy is conserved, these transformations cannot affect the total amount of energy. The total energy is constant.

15. b

17. a

## Problems

1. (a) $-465$ J  (b) The negative sign for work done on the gas indicates that the expanding gas does positive work on the surroundings.
3. (a) $-6.1 \times 10^5$ J  (b) $4.6 \times 10^5$ J
5. (a) $-810$ J  (b) $-507$ J  (c) $-203$ J
7. 96.3 mg
9. (a) $1.09 \times 10^3$ K  (b) $-6.81$ kJ
11. (a) 823 J  (b) 13.2 K
15. (a) $-88.5$ J  (b) 722 J
17. (a) 567 J  (b) 167 J
19. (a) $-180$ J  (b) $+188$ J
21. (a) 3.25 kJ  (b) 0  (c) $-3.25$ kJ  (d) The internal energy would increase, resulting in an increase in temperature of the gas.
23. (a) $-4.58 \times 10^4$ J  (b) $4.58 \times 10^4$ J  (c) 0
25. (a) $1.13 \times 10^5$ J  (b) $-1.69 \times 10^5$ J  (c) $-2.82 \times 10^5$ J  (d) $1.70 \times 10^2$ K
27. (a) 40.6 moles  (b) 506 J/K  (c) $W = 0$  (d) 16.0 kJ  (e) 31.6 K  (f) 332 K  (g) 5.53 atm
29. (a) 0.95 J  (b) $3.2 \times 10^5$ J  (c) $3.2 \times 10^5$ J
31. 405 kJ
33. 0.540 (or 54.0%)
35. (a) 0.25 (or 25%)  (b) 3/4
37. (a) 0.672 (or 67.2%)  (b) 58.8 kW
39. (a) 0.294 (or 29.4%)  (b) $5.00 \times 10^2$ J  (c) 1.67 kW

41. (a) $4.50 \times 10^6$ J  (b) $2.84 \times 10^7$ J  (c) 68.2 kg
43. 1/3
45. (a) 30.6%  (b) 985 MW
47. 143 J/K
49. (a) $-1.2$ kJ/K  (b) 1.2 kJ/K
51. 57.2 J/K
53. 3.27 J/K
55. (a)

| End Result | Possible Tosses | Total Number of Same Result |
|---|---|---|
| All H | HHHH | 1 |
| 1T, 3H | HHHT, HHTH, HTHH, THHH | 4 |
| 2T, 2H | HHTT, HTHT, THHT, HTTH, THTH, TTHH | 6 |
| 3T, 1H | TTTH, TTHT, THTT, HTTT | 4 |
| All T | TTTT | 1 |

(b) all H or all T  (c) 2H and 2T
57. $-6.5$ MJ
59. 1 300 W
61. (a) 26.2 kcal/min  (b) 629 kcal
63. 18°C
65. (a) 12.2 kJ  (b) 4.05 kJ  (c) 8.15 kJ
67. (a) 26 J  (b) $9.0 \times 10^5$ J  (c) $9.0 \times 10^5$ J
69. (a) 2.49 kJ  (b) 1.50 kJ  (c) $-990$ J
71. (a) $T_{\text{Ⓐ}} = 1.20 \times 10^2$ K; $T_{\text{Ⓑ}} = 722$ K  (b) $1.10 \times 10^5$ J  (c) $7.50 \times 10^4$ J  (d) $1.85 \times 10^5$ J
73. 0.146; 486 kcal
75. (a) $2.6 \times 10^3$ metric tons/day  (b) $\$7.6 \times 10^6$/yr  (c) $4.1 \times 10^4$ kg/s

# TOPIC 13

## Quick Quizzes

1. (d)
2. (c)
3. (b)
4. (a)
5. (c)
6. (d)
7. (c), (b)
8. (a)
9. (b)

## Example Questions

1. No. If a spring is stretched too far, it no longer satisfies Hooke's law and can become permanently deformed.
2. $k_{\text{eq}} = k_1 + k_2$
3. False
4. True
5. False
6. (b)
7. True
8. No
9. (a), (c)
10. The speed is doubled.

## Conceptual Questions

1. No. Because the total energy is $E = \frac{1}{2}kA^2$, changing the mass of the object while keeping $A$ constant has no effect on the total energy. When the object is at a displacement $x$ from equilibrium, the potential energy is $\frac{1}{2}kx^2$, independent of the mass, and the kinetic energy is $KE = E - \frac{1}{2}kx^2$, also independent of the mass.

3. (a) 2   (b) 4
5. (a) −1   (b) 1
7. We assume that the buoyant force acting on the sphere is negligible in comparison to its weight, even when the sphere is empty. We also assume that the bob is small compared with the pendulum length. Then, the frequency of the pendulum is $f = 1/T = (1/2\pi)\sqrt{g/L}$, which is independent of mass. Thus, the frequency will not change as the water leaks out.
9. (a) The bouncing ball is not an example of simple harmonic motion. The ball does not follow a sinusoidal function for its position as a function of time. (b) The daily movement of a student is also not simple harmonic motion, because the student stays at a fixed location, school, for a long time. If this motion were sinusoidal, the student would move more and more slowly as she approached her desk, and as soon as she sat down at the desk, she would start to move back toward home again.
11. The speed of a wave on a string is given by $v = \sqrt{F/\mu}$. This says the speed is independent of the frequency of the wave. Thus, doubling the frequency leaves the speed unaffected.
13. (a) I   (b) D   (c) U   (d) I

## Problems

1. (a) 17 N to the left   (b) 28 m/s² to the left
3. (a) 6.58 N   (b) 10.1 N
5. 17.8 N/m
7. (a) 0.206 m   (b) −0.042 1 m   (c) The block oscillates around the unstretched position of the spring with an amplitude of 0.248 m.
9. (a) 60.0 J   (b) 49.0 m/s
11. 0.306 m
13. 0.478 m
15. (a) 1 630 N/m   (b) 47.0 J   (c) 7.90 kg   (d) 2.57 m/s   (e) 26.1 J   (f) 20.9 J   (g) −0.201 m
17. (a) $8.00 \times 10^{-2}$ m   (b) 14.6 rad/s   (c) 2.32 Hz   (d) 0.431 s   (e) 1.17 m/s   (f) 17.0 m/s²
19. 39.2 N
21. (a) 1.53 J   (b) 1.75 m/s   (c) 1.51 m/s
23. (a) $8.27 \times 10^{-2}$ m   (b) −2.83 m/s   (c) −11.9 m/s²
25. 0.63 s
27. (a) 0.25 s   (b) 4.0 Hz   (c) 5.2 cm   (d) 21 ms
29. (a) 5.98 m/s   (b) 206 N/m   (c) 0.238 m
31. (a) 11.0 N toward the left   (b) 0.881 oscillations
33. $v = \pm\omega A \sin \omega t$, $a = -\omega^2 A \cos \omega t$
35. (a) 1.46 s   (b) 9.59 m/s²
37. (a) 9.779 m/s²   (b) 0.2477 m
39. (a) $L_{Earth}$ = 25 cm   (b) $L_{Mars}$ = 9.4 cm   (c) $m_{Earth}$ = 0.25 kg   (d) $m_{Mars}$ = 0.25 kg
41. (a) 4.13 cm   (b) 10.4 cm   (c) $5.56 \times 10^{-2}$ s   (d) 187 cm/s
43. (a) $5.45 \times 10^{14}$ Hz   (b) $1.83 \times 10^{-15}$ s
45. 31.9 cm
47. $3.53 \times 10^{-3}$ m
49. 80.0 N
51. $5.20 \times 10^{2}$ m/s
53. (a) 30.0 N   (b) 25.8 m/s
55. 28.5 m/s
57. (a) 0.051 0 kg/m   (b) 19.6 m/s
59. (a) 13.4 m/s   (b) The worker could throw an object such as a snowball at one end of the line to set up a pulse and then use a stopwatch to measure the time it takes the pulse to travel the length of the line. From this measurement, the worker would have an estimate of the wave speed, which in turn can be used to estimate the tension.
61. (a) Constructive interference gives $A$ = 0.50 m   (b) Destructive interference gives $A$ = 0.10 m
63. (a) 219 N/m   (b) 6.12 kg

65. (a) 1.68 s   (b) 16.8 N/m
67. (a) 588 N/m   (b) 0.700 m/s
69. 6.62 cm
71. (a) Using $s$ for the displacement from equilibrium along the arc, the restoring force on the balloon takes the form of Hooke's law: $F_{tangential} = -[(\rho_{air} - \rho_{He})Vg/L]s$   (b) $T$ = 1.40 s
75. (a) 15.8 rad/s   (b) 5.23 cm

## TOPIC 14

### Quick Quizzes

1. (c)
2. (c)
3. (b)
4. (b), (e)
5. (d)
6. (a)
7. (b)

### Example Questions

1. Rubber is easier to compress than solid aluminum, so aluminum must have the larger bulk modulus and by Equation 14.1, a higher sound speed.
2. 3.0 dB
3. You should increase your distance from the sound source by a factor of 5.
4. Yes. It changes because the speed of sound changes with temperature. Answer (b) is correct.
5. No
6. True
7. True
8. (b)
9. True
10. True
11. The notes are so different from each other in frequency that the beat frequency is very high and cannot be distinguished.

### Conceptual Questions

1. (a) higher   (b) lower
3. (a) 0.25   (b) 6.0
5. d, a, b, c
7. (a) The echo is Doppler shifted, and the shift is like both a moving source and a moving observer. The sound that leaves your horn in the forward direction is Doppler shifted to a higher frequency, because it is coming from a moving source. As the sound reflects back and comes toward you, you are a moving observer, so there is a second Doppler shift to an even higher frequency. (b) If the sound reflects from the spacecraft coming toward you, there is a different moving-source shift to an even higher frequency. The reflecting surface of the spacecraft acts as a moving source.
9. (i) a   (ii) c   (iii) f
11. The two engines are running at slightly different frequencies, thus producing a beat frequency between them.

### Problems

1. (a) 5.56 km (b) No. The speed of light is much greater than the speed of sound, so the time interval required for the light to reach you is negligible compared to the time interval for the sound.
3. 358 m/s
5. 515 m
7. 8.05%
9. (a) The pulse that travels through the rail   (b) 23.4 ms
11. (a) $3.16 \times 10^{-2}$ W/m²   (b) $7.91 \times 10^{-5}$ W/m²   (c) 99.0 dB
13. 150 dB

15. $3.0 \times 10^{-8} \, \text{W/m}^2$
17. (a) $1.00 \times 10^{-2} \, \text{W/m}^2$   (b) 105 dB
19. 37 dB
21. (a) $1.3 \times 10^2 \, \text{W}$   (b) 96 dB
25. $1.23 \times 10^3 \, \text{Hz}$
27. (a) 75.7 Hz drop   (b) 0.947 m
29. 596 Hz
31. 9.09 m/s
33. 19.7 m
35. (a) 56.3 s   (b) 56.6 km farther along
37. 0.137 m
39. 800 m
41. (a) 0.240 m   (b) 0.855 m
43. (a) Nodes at 0, 2.67 m, 5.33 m, and 8.00 m; antinodes at 1.33 m, 4.00 m, and 6.67 m   (b) 18.6 Hz
45. 378 Hz
47. (a) $1.85 \times 10^{-2} \, \text{kg/m}$   (b) 90.6 m/s   (c) 152 N   (d) 2.20 m   (e) 8.33 m

49. (a) 78.9 N   (b) 211 Hz
51. 19.976 kHz
53. 6.88 m/s
55. 58 Hz
57. 3.1 kHz
59. (a) 0.552 m   (b) 316 Hz
61. 3 Hz
63. 5.64 beats/s
65. 3.85 m/s away from the station or 3.77 m/s toward the station
67. (a) 1.99 beats/s   (b) 3.38 m/s
69. 1.76 cm
71. (a) 0.642 W   (b) 0.43%
73. 67.0 dB
75. $r_1 = 3.27$ m and $r_2 = 32.7$ m
77. 64 dB
79. (a) 439 Hz   (b) 441 Hz
81. $1.34 \times 10^4 \, \text{N}$

# Index

## Mechanics and Thermodynamics

Displacement and
  position vectors

Displacement and position
  component vectors

Linear ($\vec{v}$) and angular ($\vec{\omega}$)
  velocity vectors

Velocity component vectors

Force vectors ($\vec{F}$)
Force component vectors

Acceleration vectors ($\vec{a}$)
Acceleration component vectors

Energy transfer arrows

$W_{eng}$

$Q_c$

$Q_h$

Process arrow

Linear ($\vec{p}$) and
  angular ($\vec{L}$)
  momentum vectors

Linear and
  angular momentum
  component vectors

Torque vectors ($\vec{\tau}$)

Torque component
  vectors

Schematic linear or
  rotational motion
  directions

Dimensional rotational
  arrow

Enlargement arrow

Springs

Pulleys

## Electricity and Magnetism

Electric fields
Electric field vectors
Electric field component vectors

Magnetic fields
Magnetic field vectors
Magnetic field
  component vectors

Positive charges

Negative charges

Resistors

Batteries and other
  DC power supplies

Switches

Capacitors

Inductors (coils)

Voltmeters

Ammeters

AC Sources

Lightbulbs

Ground symbol

Current

## Light and Optics

Light ray

Focal light ray

Central light ray

Converging lens

Diverging lens

Mirror

Curved mirror

Objects

Images

# Periodic Table of the Elements

Transition elements

Key:

| | |
|---|---|
| Symbol → Ca ← Atomic number (20) | |
| Atomic mass† → 40.078 | |
| $4s^2$ ← Electron configuration | |

## Main Table

**Group I**

| Z | Symbol | Atomic mass | Configuration |
|---|--------|-------------|---------------|
| 1 | H | 1.0079 | $1s$ |
| 3 | Li | 6.941 | $2s^1$ |
| 11 | Na | 22.990 | $3s^1$ |
| 19 | K | 39.098 | $4s^1$ |
| 37 | Rb | 85.468 | $5s^1$ |
| 55 | Cs | 132.91 | $6s^1$ |
| 87 | Fr | (223) | $7s^1$ |

**Group II**

| Z | Symbol | Atomic mass | Configuration |
|---|--------|-------------|---------------|
| 4 | Be | 9.0122 | $2s^2$ |
| 12 | Mg | 24.305 | $3s^2$ |
| 20 | Ca | 40.078 | $4s^2$ |
| 38 | Sr | 87.62 | $5s^2$ |
| 56 | Ba | 137.33 | $6s^2$ |
| 88 | Ra | (226) | $7s^2$ |

**Transition Elements**

| Z | Symbol | Atomic mass | Configuration |
|---|--------|-------------|---------------|
| 21 | Sc | 44.956 | $3d^1 4s^2$ |
| 22 | Ti | 47.867 | $3d^2 4s^2$ |
| 23 | V | 50.942 | $3d^3 4s^2$ |
| 24 | Cr | 51.996 | $3d^5 4s^1$ |
| 25 | Mn | 54.938 | $3d^5 4s^2$ |
| 26 | Fe | 55.845 | $3d^6 4s^2$ |
| 27 | Co | 58.933 | $3d^7 4s^2$ |
| 28 | Ni | 58.693 | $3d^8 4s^2$ |
| 29 | Cu | 63.546 | $3d^{10} 4s^1$ |
| 30 | Zn | 65.39 | $3d^{10} 4s^2$ |
| 39 | Y | 88.906 | $4d^1 5s^2$ |
| 40 | Zr | 91.224 | $4d^2 5s^2$ |
| 41 | Nb | 92.906 | $4d^4 5s^1$ |
| 42 | Mo | 95.96 | $4d^5 5s^1$ |
| 43 | Tc | (98) | $4d^5 5s^2$ |
| 44 | Ru | 101.07 | $4d^7 5s^1$ |
| 45 | Rh | 102.91 | $4d^8 5s^1$ |
| 46 | Pd | 106.42 | $4d^{10}$ |
| 47 | Ag | 107.87 | $4d^{10} 5s^1$ |
| 48 | Cd | 112.41 | $4d^{10} 5s^2$ |
| 57–71* | | | |
| 72 | Hf | 178.49 | $5d^2 6s^2$ |
| 73 | Ta | 180.95 | $5d^3 6s^2$ |
| 74 | W | 183.84 | $5d^4 6s^2$ |
| 75 | Re | 186.21 | $5d^5 6s^2$ |
| 76 | Os | 190.23 | $5d^6 6s^2$ |
| 77 | Ir | 192.2 | $5d^7 6s^2$ |
| 78 | Pt | 195.08 | $5d^9 6s^1$ |
| 79 | Au | 196.97 | $5d^{10} 6s^1$ |
| 80 | Hg | 200.59 | $5d^{10} 6s^2$ |
| 89–103** | | | |
| 104 | Rf | (267) | $6d^2 7s^2$ |
| 105 | Db | (268) | $6d^3 7s^2$ |
| 106 | Sg | (271) | $6d^4 7s^2$ |
| 107 | Bh | (272) | $6d^5 7s^2$ |
| 108 | Hs | (270) | $6d^6 7s^2$ |
| 109 | Mt | (276) | $6d^7 7s^2$ |
| 110 | Ds | (281) | $6d^8 7s^2$ |
| 111 | Rg | (282) | $6d^9 7s^2$ |
| 112 | Cp | (285) | $6d^{10} 7s^2$ |

**Group III**

| Z | Symbol | Atomic mass | Configuration |
|---|--------|-------------|---------------|
| 5 | B | 10.811 | $2p^1$ |
| 13 | Al | 26.982 | $3p^1$ |
| 31 | Ga | 69.723 | $4p^1$ |
| 49 | In | 114.82 | $5p^1$ |
| 81 | Tl | 204.38 | $6p^1$ |
| 113†† | | (286) | $7p^1$ |

**Group IV**

| Z | Symbol | Atomic mass | Configuration |
|---|--------|-------------|---------------|
| 6 | C | 12.011 | $2p^2$ |
| 14 | Si | 28.086 | $3p^2$ |
| 32 | Ge | 72.64 | $4p^2$ |
| 50 | Sn | 118.71 | $5p^2$ |
| 82 | Pb | 207.2 | $6p^2$ |
| 114 | Fl | (289) | $7p^2$ |

**Group V**

| Z | Symbol | Atomic mass | Configuration |
|---|--------|-------------|---------------|
| 7 | N | 14.007 | $2p^3$ |
| 15 | P | 30.974 | $3p^3$ |
| 33 | As | 74.922 | $4p^3$ |
| 51 | Sb | 121.76 | $5p^3$ |
| 83 | Bi | 208.98 | $6p^3$ |
| 115†† | | (288) | $7p^3$ |

**Group VI**

| Z | Symbol | Atomic mass | Configuration |
|---|--------|-------------|---------------|
| 8 | O | 15.999 | $2p^4$ |
| 16 | S | 32.066 | $3p^4$ |
| 34 | Se | 78.96 | $4p^4$ |
| 52 | Te | 127.60 | $5p^4$ |
| 84 | Po | (209) | $6p^4$ |
| 116 | Lv | (293) | $7p^4$ |

**Group VII**

| Z | Symbol | Atomic mass | Configuration |
|---|--------|-------------|---------------|
| 1 | H | 1.0079 | $1s^1$ |
| 9 | F | 18.998 | $2p^5$ |
| 17 | Cl | 35.453 | $3p^5$ |
| 35 | Br | 79.904 | $4p^5$ |
| 53 | I | 126.90 | $5p^5$ |
| 85 | At | (210) | $6p^5$ |
| 117†† | | (294) | $7p^5$ |

**Group 0**

| Z | Symbol | Atomic mass | Configuration |
|---|--------|-------------|---------------|
| 2 | He | 4.0026 | $1s^2$ |
| 10 | Ne | 20.180 | $2p^6$ |
| 18 | Ar | 39.948 | $3p^6$ |
| 36 | Kr | 83.80 | $4p^6$ |
| 54 | Xe | 131.29 | $5p^6$ |
| 86 | Rn | (222) | $6p^6$ |
| 118†† | | (294) | $7p^6$ |

## *Lanthanide series

| Z | Symbol | Atomic mass | Configuration |
|---|--------|-------------|---------------|
| 57 | La | 138.91 | $5d^1 6s^2$ |
| 58 | Ce | 140.12 | $5d^1 4f^1 6s^2$ |
| 59 | Pr | 140.91 | $4f^3 6s^2$ |
| 60 | Nd | 144.24 | $4f^4 6s^2$ |
| 61 | Pm | (145) | $4f^5 6s^2$ |
| 62 | Sm | 150.36 | $4f^6 6s^2$ |
| 63 | Eu | 151.96 | $4f^7 6s^2$ |
| 64 | Gd | 157.25 | $4f^7 5d^1 6s^2$ |
| 65 | Tb | 158.93 | $4f^8 5d^1 6s^2$ |
| 66 | Dy | 162.50 | $4f^{10} 6s^2$ |
| 67 | Ho | 164.93 | $4f^{11} 6s^2$ |
| 68 | Er | 167.26 | $4f^{12} 6s^2$ |
| 69 | Tm | 168.93 | $4f^{13} 6s^2$ |
| 70 | Yb | 173.04 | $4f^{14} 6s^2$ |
| 71 | Lu | 174.97 | $4f^{14} 5d^1 6s^2$ |

## **Actinide series

| Z | Symbol | Atomic mass | Configuration |
|---|--------|-------------|---------------|
| 89 | Ac | (227) | $6d^1 7s^2$ |
| 90 | Th | 232.04 | $6d^2 7s^2$ |
| 91 | Pa | 231.04 | $5f^2 6d^1 7s^2$ |
| 92 | U | 238.03 | $5f^3 6d^1 7s^2$ |
| 93 | Np | (237) | $5f^4 6d^1 7s^2$ |
| 94 | Pu | (244) | $5f^6 7s^2$ |
| 95 | Am | (243) | $5f^7 7s^2$ |
| 96 | Cm | (247) | $5f^7 6d^1 7s^2$ |
| 97 | Bk | (247) | $5f^8 6d^1 7s^2$ |
| 98 | Cf | (251) | $5f^{10} 7s^2$ |
| 99 | Es | (252) | $5f^{11} 7s^2$ |
| 100 | Fm | (257) | $5f^{12} 7s^2$ |
| 101 | Md | (258) | $5f^{13} 7s^2$ |
| 102 | No | (259) | $5f^{14} 7s^2$ |
| 103 | Lr | (262) | $5f^{14} 6d^1 7s^2$ |

Note: Atomic mass values given are averaged over isotopes in the percentages in which they exist in nature.

† For an unstable element, mass number of the most stable known isotope is given in parentheses.

†† Elements 113, 115, 117, and 118 have not yet been officially named. Only small numbers of atoms of these elements have been observed.

††† For a description of the atomic data, visit *physics.nist.gov/PhysRefData/Elements/per_text.html*

## CONVERSION FACTORS

### Length

1 m = 39.37 in. = 3.281 ft

1 in. = 2.54 cm (exact)

1 km = 0.621 mi

1 mi = 5 280 ft = 1.609 km

1 lightyear (ly) = $9.461 \times 10^{15}$ m

1 angstrom (Å) = $10^{-10}$ m

### Mass

1 kg = $10^3$ g = $6.85 \times 10^{-2}$ slug

1 slug = 14.59 kg

1 u = $1.66 \times 10^{-27}$ kg = 931.5 MeV/$c^2$

### Time

1 min = 60 s

1 h = 3 600 s

1 day = 24 h = $1.44 \times 10^3$ min = $8.64 \times 10^4$ s

1 yr = 365.242 days = $3.156 \times 10^7$ s

### Volume

1 L = 1 000 cm$^3$ = 0.035 3 ft$^3$

1 ft$^3$ = $2.832 \times 10^{-2}$ m$^3$

1 gal = 3.786 L = 231 in.$^3$

### Angle

$180° = \pi$ rad

1 rad = 57.30°

$1° = 60$ min = $1.745 \times 10^{-2}$ rad

### Speed

1 km/h = 0.278 m/s = 0.621 mi/h

1 m/s = 2.237 mi/h = 3.281 ft/s

1 mi/h = 1.61 km/h = 0.447 m/s = 1.47 ft/s

### Force

1 N = 0.224 8 lb = $10^5$ dynes

1 lb = 4.448 N

1 dyne = $10^{-5}$ N = $2.248 \times 10^{-6}$ lb

### Work and energy

1 J = $10^7$ erg = 0.738 ft·lb = 0.239 cal

1 cal = 4.186 J

1 ft·lb = 1.356 J

1 Btu = $1.054 \times 10^3$ J = 252 cal

1 J = $6.24 \times 10^{18}$ eV

1 eV = $1.602 \times 10^{-19}$ J

1 kWh = $3.60 \times 10^6$ J

### Pressure

1 atm = $1.013 \times 10^5$ N/m$^2$ (or Pa) = 14.70 lb/in.$^2$

1 Pa = 1 N/m$^2$ = $1.45 \times 10^{-4}$ lb/in.$^2$

1 lb/in.$^2$ = $6.895 \times 10^3$ N/m$^2$

### Power

1 hp = 550 ft·lb/s = 0.746 kW

1 W = 1 J/s = 0.738 ft·lb/s

1 Btu/h = 0.293 W

## PHYSICAL CONSTANTS

| Quantity | Symbol | Value | SI unit |
|---|---|---|---|
| Avogadro's number | $N_A$ | $6.02 \times 10^{23}$ | particles/mol |
| Bohr radius | $a_0$ | $5.29 \times 10^{-11}$ | m |
| Boltzmann's constant | $k_B$ | $1.38 \times 10^{-23}$ | J/K |
| Coulomb constant, $1/4\pi\epsilon_0$ | $k_e$ | $8.99 \times 10^9$ | $N \cdot m^2/C^2$ |
| Electron Compton wavelength | $h/m_e c$ | $2.43 \times 10^{-12}$ | m |
| Electron mass | $m_e$ | $9.11 \times 10^{-31}$ | kg |
| | | $5.49 \times 10^{-4}$ | u |
| | | $0.511 \text{ MeV}/c^2$ | |
| Elementary charge | $e$ | $1.60 \times 10^{-19}$ | C |
| Gravitational constant | $G$ | $6.67 \times 10^{-11}$ | $N \cdot m^2/kg^2$ |
| Mass of Earth | $M_E$ | $5.98 \times 10^{24}$ | kg |
| Mass of Moon | $M_M$ | $7.36 \times 10^{22}$ | kg |
| Molar volume of ideal gas at STP | $V$ | 22.4 | L/mol |
| | | $2.24 \times 10^{-2}$ | $m^3$/mol |
| Neutron mass | $m_n$ | $1.674\,93 \times 10^{-27}$ | kg |
| | | $1.008\,665$ | u |
| | | $939.565 \text{ MeV}/c^2$ | |
| Permeability of free space | $\mu_0$ | $1.26 \times 10^{-6}$ | $T \cdot m/A$ |
| | | ($4\pi \times 10^{-7}$ exactly) | |
| Permittivity of free space | $\epsilon_0$ | $8.85 \times 10^{-12}$ | $C^2/N \cdot m^2$ |
| Planck's constant | $h$ | $6.63 \times 10^{-34}$ | $J \cdot s$ |
| | $\hbar = h/2\pi$ | $1.05 \times 10^{-34}$ | $J \cdot s$ |
| Proton mass | $m_p$ | $1.672\,62 \times 10^{-27}$ | kg |
| | | $1.007\,276$ | u |
| | | $938.272 \text{ MeV}/c^2$ | |
| Radius of Earth (at equator) | $R_E$ | $6.38 \times 10^6$ | m |
| Radius of Moon | $R_M$ | $1.74 \times 10^6$ | m |
| Rydberg constant | $R_H$ | $1.10 \times 10^7$ | $m^{-1}$ |
| Speed of light in vacuum | $c$ | $3.00 \times 10^8$ | m/s |
| Standard free-fall acceleration | $g$ | 9.80 | $m/s^2$ |
| Stefan-Boltzmann constant | $\sigma$ | $5.67 \times 10^{-8}$ | $W/m^2 \cdot K^4$ |
| Universal gas constant | $R$ | 8.31 | $J/mol \cdot K$ |

The values presented in this table are those used in computations in the text. Generally, the physical constants are known to much better precision.